Experientia Supplementum

Volume 107

More information about this series at http://www.springer.com/series/4822

Mario D. Cordero • Benoit Viollet
Editors

AMP-activated Protein Kinase

Springer

Editors
Mario D. Cordero
Oral Medicine Department
Universidad de Sevilla
Sevilla, Spain

Benoit Viollet
INSERM U1016 Institut Cochin
CNRS UMR 8104
Université Paris Descartes
Paris, France

ISSN 1023-294X
Experientia Supplementum
ISBN 978-3-319-43587-9 ISBN 978-3-319-43589-3 (eBook)
DOI 10.1007/978-3-319-43589-3

Library of Congress Control Number: 2016955181

Printed on acid-free paper

This Springer imprint is published by Springer Nature
The registered company is Springer International Publishing AG Switzerland

Preface

The survival of all organisms depends on the dynamic control of energy metabolism and the regulation of a plethora of biochemical and molecular processes. The availability or nonavailability of the major cellular energy resource adenosine triphosphate (ATP) determines whether cells may grow and divide or starve and die. Intracellular sensors of cellular energy and nutrient status are emerging as key player in the regulation of cell metabolism in health and disease. Among the different signaling hubs characterized in recent years, adenosine monophosphate-activated protein kinase (AMPK) signaling participates in the maintenance of intracellular ATP levels within an appropriate range. By sensing changes in AMP/ATP and ADP/ATP ratios, AMPK coordinates cellular energy balance by switching on catabolic pathways and switching off ATP-consuming processes at both the cellular and whole-body level.

According to the function of AMPK as a master regulator of cell energy levels, it is becoming evident why a dysfunction in the AMPK signaling pathway has been involved directly or indirectly in the derangement of energy metabolism in many diseases. However, caution is required when interpreting these findings showing correlation rather than formal demonstration for a role of AMPK as a driver of diseases. In addition, given the magnitude of AMPK in biomedical research, it is often difficult to decipher whether this pathway may be of significant relevance for clinicians or basic research workers. A better comprehension in the regulation of the signaling pathway but also the knowledge of available methods and models would be extremely useful. Hence, the aim of this book is to describe the state of the art of AMPK signaling and function by the authors who have been actively commited to recent developments of our understanding of how this key heterotrimeric enzyme operates to control metabolism as well as non-metabolic cellular processes at both the cellular and whole-body level.

The content of the book is distributed into five sections: beginning with basic informations about *AMPK in health*, where the principal functions of AMPK are described; then providing insights into the potential role of *AMPK in disease*, where the pathological consequences of AMPK dysfunction are highlighted; this is followed by the description of the *Pharmacology of AMPK*, a section dealing with an actualized view of the most relevant compounds used to modulate AMPK activity; then turning to *AMPK in nonmammalian systems*, where recent knowledge on AMPK function in yeast, *Drosophila melanogaster*, *Caenorhabditis elegans*, and plants is presented; and finally, in *Methods to study AMPK*, recent genetic, cellular, and molecular experimental tools in AMPK research field are depicted.

The editors and authors hope this first AMPK textbook will be a useful manual and valuable reference for a large scientific audience in biology and medicine. It is our goal to stimulate research, contribute to a better understanding of the different aspects of AMPK signaling and function, and help the generation of new ideas and scientific projects as well as a step toward translational research. Thus, this book serves as a source of information to facilitate the reading of the literature and should pave the way for improvement of pharmacological and therapeutic intervention in diseases, including aging, cancer, neurodegenerative diseases, and pathogen infection.

Seville, Spain — Mario D. Cordero
Paris, France — Benoit Viollet

Acknowledgments

The editors want to thank all the authors for the tremendous effort and dedication to the development of this book as well as the high level of chapters. This book has served to establish nice Friends.

Contents

Part I
AMPK in Health

Chapter 1
Structure and Regulation of AMPK

Ravi G. Kurumbail and Matthew F. Calabrese

Contents

Abstract AMP-activated protein kinase is a family of heterotrimeric serine/threonine protein kinases that come in twelve different flavors. They serve an essential function in all eukaryotes of conserving cellular energy levels. AMPK complexes are regulated by changes in cellular AMP:ATP or ADP:ATP ratios and by a number of neutraceuticals and some of the widely-used diabetes medications such as metformin and thiazolinonediones. Moreover, biochemical activities of AMPK are tightly regulated by phosphorylation or dephosphorylation by upstream kinases and phosphatases respectively. Efforts are underway in many pharmaceutical companies to discover direct AMPK activators for the treatment of cardiovascular and metabolic diseases such as diabetes, non-alcoholic steatohepatitis (NASH) and diabetic nephropathy. Many advances have been made in the AMPK structural biology arena over the last few years that are beginning to provide detailed molecular insights into the overall topology of these fascinating enzymes and how binding of small molecules elicit subtle conformational changes leading to their activation and protection from dephosphorylation. In the brief review below on AMPK structure and function, we have focused on the recent crystallographic results especially on specific molecular interactions of direct synthetic AMPK activators which lead to biased activation of a sub-family of AMPK isoforms.

Keywords X-ray • Crystallography • Enzyme activators • Allostery • AMPK

R.G. Kurumbail (✉) • M.F. Calabrese
Pfizer Worldwide Research and Development, Pfizer Inc, Eastern Point Road, Groton, CT 06340, USA
e-mail: ravi.g.kurumbail@pfizer.com

M.D. Cordero, B. Viollet (eds.), *AMP-activated Protein Kinase*, Experientia Supplementum 107, DOI 10.1007/978-3-319-43589-3_1

AMP-activated protein kinase (AMPK) serves an essential function in all eukaryotes as a guardian of cellular energy status (Carling and Viollet 2015; Hardie 2016; Mounier et al. 2015; Neumann et al. 2007; Oakhill et al. 2012). When the overall energy levels in cells decrease due to increased demands, AMPK gets activated through a combination of phosphorylation by upstream kinases and direct activation by AMP and ADP. Activated AMPK in turn downregulates anabolic processes that consume ATP while upregulating catabolic processes that lead to ATP synthesis. This is accomplished in part by direct inhibition of catalytic activity of a number of biosynthetic enzymes such as HMG-coA reductase or acetyl-coA carboxylase (ACC) through phosphorylation by AMPK (Carling et al. 1987). In addition, AMPK also exerts long-range effects through regulation of transcriptional co-activator PGC1-α, upregulation of which results in increased mitochondrial biogenesis (Zong et al. 2002). AMPK activation also leads to negative regulation of protein translation via phosphorylation of tuberous sclerosis complex 2 (TSC2) and regulatory associated protein of TOR (raptor) (Gwinn et al. 2008; Inoki et al. 2003). The net effect of these AMPK-mediated events is to restore the overall energy level in cells.

1.1 Structure and Topology

AMPK is a heterotrimeric serine/threonine kinase that is made up of a catalytic α subunit in complex with two regulatory subunits, β and γ. Multiple isoforms of each of these subunits exist (α1, α2, β1, β2, γ1, γ2, and γ3) that can assemble in different combinations to generate as many as 12 possible AMPK complexes (Ross et al. 2016b). Differential expression of these AMPK isoforms in tissues and species and their association with specific substrates allow for unique regulation of metabolic processes (Steinberg and Kemp 2009).

Recent years have seen tremendous progress in our structural understanding of AMPK (Table 1.1). Early structures of isolated domains and truncated complexes provided the first clues to the assembly of AMPK and revealed the interactions between the kinase and regulatory domains (Chen et al. 2009, 2013; Day et al. 2007; Handa et al. 2011; Littler et al. 2010; Mobbs et al. 2015a; Nayak et al. 2006; Polekhina et al. 2005; Rudolph et al. 2007; Walker et al. 2005). Complementary work from landmark papers revealed the structure of the AMPK regulatory core (also known as the adenylate sensor). This structural unit, comprised of the γ subunit together with the C-terminal portions of α and β, lacks catalytic activity but contains the critical nucleotide-binding domains (Amodeo et al. 2007; Chen et al. 2012; Jin et al. 2007; Townley and Shapiro 2007; Xiao et al. 2007). This work was strongly enabled by the advances made on the molecular biology front in generating a tri-cistronic expression vector that holds the coding DNA for all the three subunits on a single plasmid (Neumann et al. 2003). Detailed structural studies revealed that of the four possible nucleotide-binding sites in γ, only three appear to ever be occupied in mammalian AMPK (sites 1, 3, 4). From biophysical, biochemical, and mutagenesis experiments, we learned that these sites

Table 1.1 Structures of AMPK complexes and domains

PDB	Description	Res. (Å)	Citation
1Z0M 1Z0N	GBD (CBM) of mammalian β1	1.9 1.5	Polekhina et al. (2005)
2F15	GBD (CBM) of human β2	2.0	Walker et al. (2005)
2FH9	Kinase domain from yeast SNF1	2.8	Nayak et al. (2006)
2H6D	Kinase domain of human α2	1.8	Littler et al. (2010)
2OOX 2OOY	Yeast adenylate sensor with AMP and ATP	2.6 2.9	Townley and Shapiro (2007)
2UV4 2UV5 2UV6 2UV7	Structures of CBS domain pairs with nucleotide and ZMP	1.3 1.7 2.0 2.0	Day et al. (2007)
2QLV	Structure of heterotrimeric core of yeast SNF1	2.6	Amodeo et al. (2007)
2V8Q 2V92 2V9J	AMPKcore plus AMP AMPKcore plus ATP-AMP AMPKcore plus Mg.ATP-AMP	2.1 2.4 2.5	Xiao et al. (2007)
2QR1 2QRC 2QRD 2QRE	Yeast adenylate sensor with ADP Yeast adenylate sensor with AMP and ADP Yeast adenylate sensor with ADP and ATP Yeast adenylate sensor with ZMP	2.7 2.7 2.4 3.0	Jin et al. (2007)
2YZA 3AQV	Kinase domain T172D from human α2 Kinase domain T172D with compound C	3.0 2.1	Handa et al. (2011)
3DAE 3H4J	Phosphorylated SNF1 kinase domain Pombe AMPK KD/AID fragment	2.9 2.8	Chen et al. (2009)
3MN3	Inhibited kinase domain of yeast SNF1	2.4	Rudolph et al. (2010)
2Y8L 2Y8Q 2YA3	AMPKcore with two ADP AMPKcore with one ADP AMPKcore with coumarin ADP	2.5 2.8 2.5	Xiao et al. (2011)
3T4N 3TDH 3TE5	*S. Cerevisiae* AMPKcore plus ADP *S. Cerevisiae* AMPKcore plus AMP *S. Cerevisiae* AMPKcore plus NADH	2.3 2.3 2.5	Mayer et al. (2011)
4EAG 4EAI 4EAJ 4EAK 4EAL	Co-crystal chimeric AMPKcore plus ATP Co-crystal AMPKcore plus AMP Co-crystal AMPKcore w/AMP soaked ATP Co-crystal AMPKcore plus ATP Co-crystal AMPKcore w/ATP soaked AMP	2.7 2.3 2.6 2.5 2.5	Chen et al. (2012)
2LTU 4F2L	Solution structure of AID Structure of mammalian AID	N/A 1.5	Chen et al. (2013)

(continued)

Table 1.1 (continued)

PDB	Description	Res. (Å)	Citation
2 LU3 2 LU4	Solution structure of β2 CBM Solution structure of β2 CBM plus cyclodextrin	N/A N/A	Koay et al. (citation yet unpublished at www.rcsb.org)
4CFE 4CFF 4CFH	Full-length AMPK α2β1γ1 plus cmpd 991 Full-length AMPK α2β1γ1 plus A769662 AMPKcore plus kinase domain	3.0 3.9 3.2	Xiao et al. (2013) Xiao et al. (2011)
4QFG 4QFR 4QFS	Full-length AMPK α1β1γ1 Full-length AMPK α1β1γ1 plus Cl-A769662 Full-length AMPK α1β1γ1 plus Br-A769662core	3.5 3.3 3.5	Calabrese et al. (2014)
4RED 4RER 4REW	Human α1 KD-AID K43A Full-length AMPK α1β2γ1 Full-length non-phosphorylated AMPK α1β2γ1	2.9 4.0 4.5	Li et al. (2015)
4YEE 4Y0G 4YEF	β2 CBM with glucosyl-beta-cyclodextrin β2 CBM β1 CBM with glucosyl-beta-cyclodextrin	2.0 1.6 1.7	Mobbs et al. (2015b)
4ZHX 5EZV	Full-length AMPK α2β1γ1 in complex with compound C2	3.0	Langendorf et al. (2016)

are nonequivalent in terms of their nucleotide affinity and effect on enzyme function. These observations were significantly extended with the structure of the kinase domain plus regulatory core that revealed specific interaction of a regulatory segment from the α subunit with the nucleotide-binding sites in γ, providing a line of communication between the catalytic and regulatory machinery (Chen et al. 2013; Xiao et al. 2011).

More recently, there has been an eruption of structural data of full-length heterotrimeric AMPK complexes beginning with the structure of the α2β1γ1-complex reported by the Gamblin and Carling laboratories in late 2013 (Xiao et al. 2013). This was rapidly followed by structures of the α1β1γ1 and α1β2γ1 isoforms—representing both active and inactive enzymes in various ligand-bound states providing full-length structures for a quarter of possible heterotrimeric assemblies (Calabrese et al. 2014; Li et al. 2015). All three isoforms adopt a similarly elongated conformation (Fig. 1.1b). The catalytic protein kinase module resides on the N-terminal portion of the α subunit. This is followed by a short auto-inhibitory domain (AID) which is a three-helical bundle that negatively regulates the basal activity of the enzyme (Chen et al. 2009, 2013; Xiao et al. 2011). A long, flexible regulatory segment that contains two α-regulatory interacting motifs (α-RIM1 and α-RIM2) connects the AID to the regulatory core (Xin et al. 2013). β1 and β2, which are roughly equivalent in length, are the shortest of the three

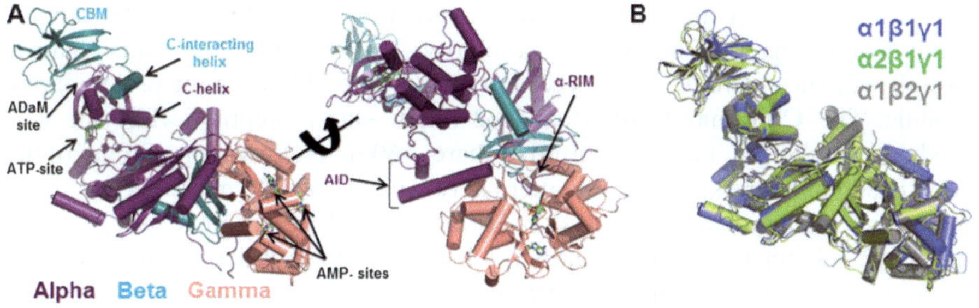

Fig. 1.1 AMPK topology and structure. (**a**) Overall topology of full-length heterotrimeric AMPK. Structure shown in *cartoon* representation for the α2β1γ1 isoform derived from 4CFE.pdb (Xiao et al. 2013). Subunits are colored as indicated and with structural modules labeled. Bound AMP in the γ subunit and staurosporine in the kinase ATP site are shown as sticks (*green*). (**b**) Global superposition of the three distinct full-length AMPK heterotrimers solved to date illustrating a conserved topology. α1β1γ1 (4QFR.pdb) (Calabrese et al. 2014), α2β1γ1 (4CFE.pdb) (Xiao et al. 2013), and α1β2γ1 (4RER.pdb) (Li et al. 2015) are shown in slate *blue*, *green*, and *gray*, respectively

AMPK subunits. They contain a flexible, glycine-rich segment with a myristoylation site at its N-terminus. The structure of this domain is as yet unclear, but it appears to have an inhibitory function on AMPK activity (Oakhill et al. 2010; Scott et al. 2008; Warden et al. 2001). The glycine-rich domain is followed by a carbohydrate-binding module (CBM) also known as glycogen-binding domain (GBD) of approximately 80 amino acids. As the name implies, CBM binds glycogen or shorter oligosaccharides which modulates the enzymatic activity of AMPK and allows it to respond to changes in sugar levels (Koay et al. 2007, 2010; McBride et al. 2009; McBride and Hardie 2009; Polekhina et al. 2003, 2005). Though able to adopt its native fold in isolation (Polekhina et al. 2005; Walker et al. 2005), the CBM of both the β1 and β2 isoforms has been shown to dock atop the N-lobe of the kinase domain in the context of full-length heterotrimers (Calabrese et al. 2014; Li et al. 2015; Xiao et al. 2013). In this environment, the CBM is followed by a novel structured segment termed the "C-interacting helix" which docks against the B- and C-helices of the α subunit. The C-terminal segment of the β subunits serves as an anchor holding together the nucleotide-binding domain of the γ subunits and the C-terminal portion of the α subunits (Xiao et al. 2007).

The γ subunits display the greatest variability among the three in terms of size. All three γ subunits contain a core nucleotide-binding module that is made up of four cystathionine-β-synthase (CBS) domains. Each of the CBS domains contains roughly ~60 amino acids and assembles as pairs called Bateman domains. Two such Bateman domains come together in a head-to-head fashion to form the core nucleotide-binding module of the γ subunits. While γ1 is comprised primarily of CBS domains alone, γ2 and γ3 contain long N-terminal extensions of unknown structure and function. As in other proteins, the CBS modules of AMPK harbor binding sites for AMP, ADP, and ATP.

The different structural and functional modules of the three AMPK subunits share a high degree of amino acid similarity. The α1 and α2 subunits share an overall sequence identity of ~77 % which increases to ~90 % for the protein kinase module. The CBM motifs of β1 and β2 retain ~80 % identity, while the CBS modules across the three γ isoforms share ~60 % conservation. Some of the AMPK subunits are known to be alternately spliced in different tissues allowing for further regulation and complexity. For example, three different splice variants of the γ2 subunit have been identified which have been described in detail in a recent review article (Steinberg and Kemp 2009).

The presence of multiple AMPK heterotrimeric complexes that are differentially expressed allows for unique regulation of metabolic processes in various tissues. The catalytic function of AMPK resides on the α1 and α2 subunits which are encoded by two separate genes. Using recombinant AMPK isoforms expressed in bacteria and quantitation of phospho SAMS peptide by careful HPLC methods, Suter et al. reported similar specific activities (~6 μmol/min/mg) for α1β1γ1- and α2β2γ1-complexes (Suter et al. 2006). Furthermore, they also reported that the α2-containing complex was much more sensitive to deactivation by the phosphatase PP2Cα compared to the α1-complex. The latter observations were recently confirmed in a study of six recombinant AMPK isoforms expressed in *Escherichia coli* which showed ~25× higher sensitivity of α2-containing heterotrimers for dephosphorylation relative to α1-complexes (Rajamohan et al. 2015). In the recent study, however, the authors observed that α1-containing isoforms in general have ~3× higher specific activities compared to α2-containing isoforms. This was true for isolated kinase domain reagents as well. Enzymatic characterization of kinetic parameters revealed that the higher specific activities for the α1 isoforms were primarily driven by reduced K_m (~3×) for the SAMS peptide. Even though the maximal stimulation by AMP of α1- and α2-complexes was approximately similar (~2×), in general α2 isoforms were much more sensitive to AMP stimulation with AC_{50} values (concentration required for half-maximal stimulation) 6–10× lower than that for α1 isoforms (Rajamohan et al. 2015). These observations were in contrast to previous reports of roughly similar sensitivity of α1- and α2-complexes to AMP stimulation (Suter et al. 2006). The precise reasons for some of these discrepancies are not clear but could be due to different experimental protocols used for generating the protein reagents and for biochemical characterization. However, both these studies unequivocally showed that bacterially expressed, non-phosphorylated AMPK has significantly less catalytic activity compared to fully phosphorylated forms, and activation by upstream kinase increases their catalytic activities 500–1000×.

As previously discussed, the nucleotide-binding regulatory domain of AMPK comes in three flavors: γ1, γ2, and γ3. Expression of recombinant γ3-containing AMPK heterotrimers has been challenging, but a robust procedure for generating human α2β2γ3 from *E. coli* has recently been described (Rajamohan et al. 2010). By immunoprecipitation of rat brain extracts using a combination of antibodies directed against α and γ subunits, it was shown that AMPK complexes that contain γ2 were the most sensitive to activation by AMP and the γ-3 complexes the least (Cheung et al. 2000). Recently, this was further probed by overexpression of

FLAG-tagged γ subunits in HEK-293 cells followed by immunoprecipitation (Ross et al. 2016a). γ1- and γ2-complexes were allosterically activated almost tenfold at physiological ATP concentration (~5 mM), whereas there was essentially no activation of γ3-complexes under these conditions. In addition, AMP also stimulated activation of γ1-complexes by the upstream kinase LKB1 while only small effects were observed with γ2- or γ3-complexes. Finally, all the three γ-complexes were protected from dephosphorylation by AMP or ADP with the γ2-complex showing the greatest sensitivity for ADP compared to γ1- or γ3-complexes. Thus, the identity of the γ subunit dictates the nucleotide dependence of different AMPK heterotrimers and this in turn is likely to influence their functional roles in vivo.

Prior to the availability of full-length AMPK crystal structures, solution-based studies had provided tantalizing glimpses into the overall structure of these proteins. Based on small-angle X-ray scattering studies (SAXS), Riek et al. showed that binding of AMP induces a 5 % decrease in the overall radius of gyration of AMPK and leads to a more compact structure (Riek et al. 2008). In addition, using scanning transmission electron microscopy, they showed that AMPK exists as a higher-order oligomer at higher concentrations. Single-particle cryo-electron microscopy of AMPK rat $\alpha1\beta1\gamma1$ revealed a trimeric arrangement and clues about the potential role of the α1 linker sequence (LS) between the kinase domain and the C-terminal module in regulation of AMPK activity (Zhu et al. 2011). Solution-based hydrogen/deuterium exchange (HDX) mass spectrometry studies showed distinct conformational changes of AMPK induced by AMP and the small molecule synthetic activator, A769662 (Landgraf et al. 2013). These suggested that AMP and A769662 activate AMPK by two radically distinct molecular mechanisms and the possible existence of multiple allosteric sites on AMPK.

1.2 AMPK Activation

Most studies indicate that AMPK has minimal basal activity in its resting state in the absence of phosphorylation on its activation loop. However, recent studies have shown that the presence of synthetic activator A769662 along with AMP can stimulate significant AMPK activity even in the absence of activation loop phosphorylation (Scott et al. 2014). Upon phosphorylation of a conserved threonine in the activation loop of the kinase module (Thr174 in α1 and Thr172 in α2), the ability of AMPK to phosphorylate downstream substrates or peptides derived from them (SAMS or AMARA peptides) increases >500-fold (Dale et al. 1995; Hawley et al. 1996; Suter et al. 2006). Two upstream kinases that promote AMPK phosphorylation are the tumor suppressor Liver Kinase B1 (LKB1) and Ca^{2+}-calmodulin-activated protein kinase kinase beta, CamKKβ (Hawley et al. 1996, 2005; Hurley et al. 2005; Shaw et al. 2004; Woods et al. 2005, 2003a, b). Of these, LKB1 is constitutively active but can be further modulated by posttranslational modifications (Hawley et al. 2003). LKB1 is activated by associations with a scaffold protein MO25 and Ste20-related adaptor protein (STRAD). It has been shown recently that binding of AMP to the γ subunit of AMPK further stimulates its activation by LKB1 (Gowans et al. 2013; Oakhill et al. 2010, 2011, 2012). The β subunit of AMPK contains an N-terminal myristoylation site which plays a role in reversible interaction with membranes and is critical for mediating the effect of AMP in phosphorylation by upstream kinases (Oakhill et al. 2010; Warden et al. 2001). It is as yet unclear how this occurs structurally as the N-terminus of the β subunit is disordered or absent in all AMPK crystal structures solved to date.

CamKKβ, in contrast, is stimulated by calcium ionophores or hormones and appears to be the primary upstream kinase of AMPK in cardiomyocytes and neurons (Hawley et al. 2005; Hurley et al. 2005; Woods et al. 2005). Activation by CamKKβ is mediated by increases in cytosolic Ca^{2+} concentrations, thus providing an alternate pathway to activate AMPK independent of changes in adenine nucleotide concentration. CamKKβ has been routinely employed for in vitro phosphorylation of bacterially expressed recombinant AMPK (Neumann et al. 2003). Conflicting data have been reported regarding the role of AMP or ADP in AMPK activation by the CamKKβ pathway. Oakhill et al. demonstrated that AMP can promote phosphorylation of AMPK on its activation loop by CamKKβ (Oakhill et al. 2010, 2011, 2012). Furthermore, they showed that the presence of a myristoyl group on the β subunit of AMPK is necessary for this effect. AMP does not promote phosphorylation by CamKKβ of *E. coli*-expressed AMPK that lacks the myristoyl group, but this can be restored by co-expression of *N*-myristoyl transferase (NMT) in *E. coli* which results in β-myristoylation (Oakhill et al. 2010). Using purified AMPK obtained from rat liver, which is myristoylated on its β subunits, Gowans et al. showed that AMP promotes phosphorylation of AMPK by LKB1 but not by CamKKβ (Gowans et al. 2013; Ross et al. 2016a). Moreover, this effect seems to be restricted to AMP as ADP did not promote phosphorylation by either LKB1 or CamKKβ (Gowans et al. 2013).

Phosphorylated AMPK is rapidly inactivated in vivo by dephosphorylation by phosphatases. The precise identity of phosphatases responsible for pThr172 dephosphorylation in different tissues remains unclear, though protein phosphatase 2-Cα (PP2C) and protein phosphatase 2-A (PP2a) can mediate this effect in vitro (Davies et al. 1995; Kudo et al. 1996). Both AMP and ADP inhibit inactivation of AMPK by inducing conformational changes, thus increasing the effective "half-life" of functionally active AMPK in cells and tissues (Oakhill et al. 2012; Ross et al. 2016a; Xiao et al. 2011). While both AMP and ADP are effective in protecting AMPK from dephosphorylation, AMP appears to be ~10× more potent (Gowans et al. 2013). ATP, in contrast, antagonizes the protective effects of AMP and ADP. Other small molecule synthetic activators such as A769662 have also been shown to inhibit AMPK dephosphorylation in vitro (Goransson et al. 2007). In addition to promoting AMPK phosphorylation by upstream kinases and inhibiting dephosphorylation by phosphatases, AMP also increases AMPK activity by allosteric activation by two- to fivefold (Carling et al. 1989). Recently, it was shown that AMPK is allosterically activated >10-fold by AMP even at physiological concentrations of ATP (Gowans et al. 2013; Ross et al. 2016a). Among the known endogenous adenine nucleotides, only AMP appears to be capable of allosteric activation, and this could be an important aspect of overall regulation of the AMPK pathway in vivo.

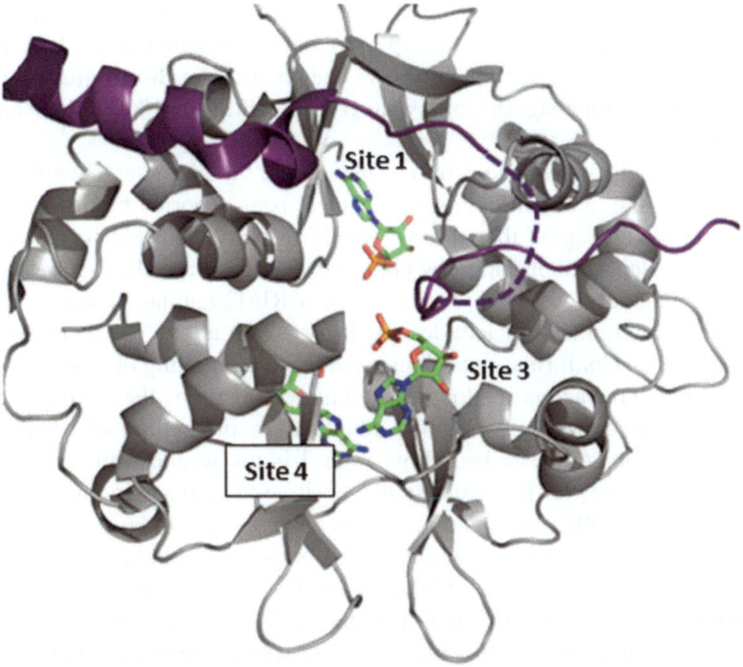

Fig. 1.2 Nucleotide-binding sites in AMPK. γ subunit (*gray*) derived from 4CFE.pdb (Xiao et al. 2013) shown in *ribbon* representation along with a portion of the regulatory α-RIM (*purple*). β subunit has been removed for clarity. AMP bound in sites 1, 3, and 4 are shown as sticks. Sites 1 and 4 lie deep within the plane of the page while site 3 lies close to the surface. The AMP in site 3 is proximal to the α-RIM

Crystallographic studies of truncated and full-length AMPK have revealed the existence of three nucleotide-binding sites on the γ subunit of mammalian AMPK (Xiao et al. 2007, 2011) (Fig. 1.2). These have been designated as sites 1, 3, and 4 based on the CBS modules that provide the acidic residue for coordination with the 2′ and 3′ hydroxyl groups of the ribose of AMP (Kemp et al. 2007). Of these, site 4 appears to be a structural site that retains a tightly held AMP molecule even during purification. This site was originally described as a non-exchangeable site although later studies have shown that under in vitro incubation at high concentrations, ATP is capable of displacing AMP and this result was reinforced by crystallographic studies (Chen et al. 2012). The physiological significance of this observation is not yet clear, and under cellular conditions, site 4 is likely to predominantly coordinate AMP.

Several studies have shed light on the significance of the remaining two nucleotide-binding sites, 1 and 3. Crystal structures have revealed that site 1 can accommodate the nucleotides AMP, ADP, or ATP (Chen et al. 2012; Xiao et al. 2007, 2011). All three nucleotides bind with an equilibrium binding constant (K_d) of 1–5 μM to site 1. However, Mg^{2+}/ATP appears to bind ~10–100× weaker than free ATP to site 1 (Rajamohan et al. 2015; Xiao et al. 2011). Site 3 is the weakest nucleotide site on the γ subunit. The equilibrium binding affinities of AMP and ADP are significantly weaker (>30×) for site 3 compared to site 1 (Rajamohan et al. 2015; Xiao et al. 2011). ATP shows even a stronger discrimination between the sites 1 and 3 with a significantly weaker K_d for site 3 (>1000 μM) (Rajamohan et al. 2015).

The functional roles of the different nucleotide-binding sites have been somewhat controversial, and conflicting results have been reported by different laboratories. Most studies agree that site 3 is likely to be responsible for protection of pThr172 from dephosphorylation by phosphatases. The crystal structures suggest a possible explanation for this observation. The α-RIM2 regulatory segment of the α subunit directly contacts nucleotides (AMP or ADP) at site 3 (Chen et al. 2013; Xiao et al. 2011). Conformational changes induced by binding of AMP or ADP can be transmitted to the activation loop of the kinase domain by the flexible regulatory segments in the α subunit, and this can lead to protection of pThr172 from dephosphorylation. As described above, site 3 is the weakest nucleotide-binding site on the γ subunit, yet it is exquisitely sensitive to changes in nucleotide levels and this allows AMPK to respond rapidly to changes in cellular energy status.

Based on crystallographic studies and competitive binding experiments of NADH or fluorescent analogs of ATP, Xiao et al. suggested that site 1 is responsible for allosteric activation of AMPK (Xiao et al. 2011). This has been questioned by Chen et al. based on site-directed mutagenesis experiments (Chen et al. 2012). The latter studies suggest that sites 3 and 4 are responsible for allosteric activation by AMP. Previous mutagenesis studies had shown that allosteric activation by AMP is modulated by all the three sites (1, 3, and 4) (Oakhill et al. 2010). These seemingly conflicting results may reflect a level of cross talk between the nucleotide-binding sites via the coordinating positively charged residues. Since AMP or ADP binds ~10× stronger than Mg^{2+}ATP at site 1, they are likely to occupy site 1 at significant proportions even though cellular ATP concentration vastly exceeds that of AMP and ADP. It is very likely that the exact binding affinity of AMP or ADP at site 3 and hence the conformational changes induced on AMPK will depend on the identity of the nucleotide that occupies site 1.

1.3 Direct and Indirect Activators of AMPK

Given the central role that AMPK plays in the maintenance of overall cellular energy status, it is not surprising that it is modulated by a variety of stimuli including plant-derived phytochemicals and “neutraceuticals” as well as reactive oxygen species (e.g., hydrogen peroxide) and mitochondrial poisons such as dinitrophenol and oligomycin. Thus, AMPK gets activated by resveratrol (from red wine), epigallocatechin (from green tea), flavonoids such as quercetin and genistein (from fruits and vegetables), capsaicin (from chili peppers), and berberine which has been used in traditional Chinese medicine for centuries (Ahn et al. 2008; Baur et al. 2006; Hwang et al. 2005; Lee et al. 2006) (Fig. 1.3). Two of the most widely

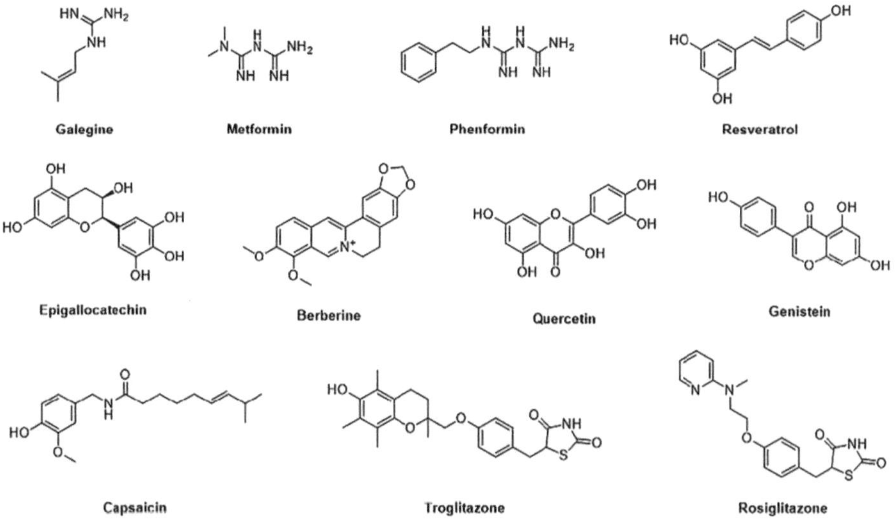

Fig. 1.3 Indirect AMPK activators. Chemical structures are shown for a representative set of reported indirect activators of AMPK (natural and synthetic)

used antidiabetic medicines, metformin and thiazolidinediones, have also been shown to activate AMPK (Fryer et al. 2002; Hawley et al. 2002; Zhou et al. 2001). The vast majority of these activate AMPK by an indirect mechanism based on an elevation in cellular AMP:ATP ratio caused by inhibition of mitochondrial function. Taking advantage of a known human mutation in AMPK γ2 subunit, Hawley et al. designed an elegant cell-based method to evaluate the mechanism of AMPK activation by a wide variety of agents (Hawley et al. 2010). They evaluated activation of AMPK in isogenic HEK293 cells that contained either wild-type AMPK γ2 or Arg513Gly γ2 variant that is insensitive to activation by AMP. Hydrogen peroxide, oligomycin, dinitrophenol, and AICAR activated AMPK in wild-type cells but not in R513G cells. AICAR is an intermediate in purine metabolism that gets phosphorylated in vivo to generate the AMP mimetic, ZMP (Corton et al. 1995). AMPK activation was observed only in wild-type cells with metformin, phenformin, troglitazone (thiazolidinedione), quercetin, resveratrol, and berberine. By measuring cellular ADP:ATP ratio and oxygen consumption, the authors showed that AMPK activation is a direct consequence of increased AMP levels in these cells. Because of the relatively low levels of AMP in cells, accurate measurement of AMP concentration is a challenge. However, they could reliably estimate AMP concentration from measured ADP levels based on the adenylate kinase reaction 2ADP ↔ ATP + AMP.

Direct activators such as A769662 (Fig. 1.4) resulted in activation of AMPK in both the wild-type and the R531G variant cells. A769662 represents the first direct synthetic activator of AMPK (Cool et al. 2006). This compound was shown to directly interact with AMPK in the nanomolar potency range and exhibit effects on levels of glucose and circulating triglycerides following dosing in *ob*/*ob* mice.

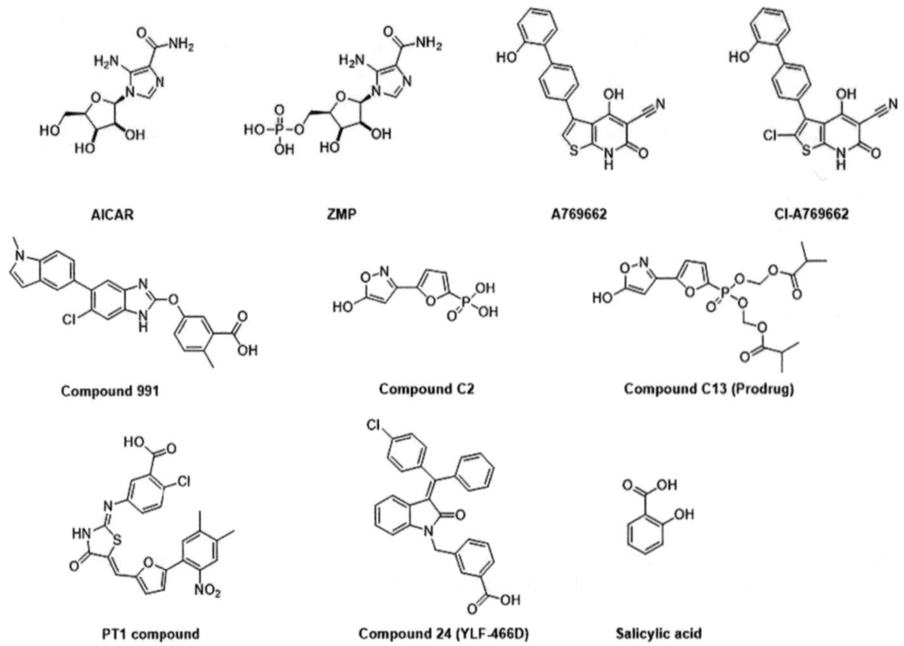

Fig. 1.4 Direct AMPK activators. Chemical structures are shown for a set of reported direct AMPK activators. Note that AICAR itself is not a direct AMPK activator but it is metabolized to a direct activator (ZMP) in vivo which then functions as a nucleotide mimetic. Compound 24 is the result of optimization of the PT1 screening hit (Yu et al. 2013)

Though the binding mode remained elusive for nearly a decade, it was quickly apparent that A769662 acted in a manner fundamentally distinct from nucleotide and appeared to show a preference for β1-containing heterotrimers with a dependence on the CBM motif (Cool et al. 2006; Goransson et al. 2007; Landgraf et al. 2013; Sanders et al. 2007; Scott et al. 2008).

Enzyme activation can be achieved through multiple kinetic mechanisms. For example, increasing the turnover number (*K*cat) or maximum enzyme velocity (V_{max}) through structural modulation can lead to an overall increase in catalytic activity of enzymes. Alternatively, lowering the Michaelis–Menten constant (K_M, the substrate concentration needed to achieve a half-maximum enzyme velocity) for one or more of the substrates will also manifest as enzyme activation. AMP exerts its allosteric effects primarily through enhancing reaction V_{max}, an observation that was made nearly three decades earlier (Carling et al. 1989). This was further confirmed recently by enzymology studies of recombinant AMPK α1β1γ1 which showed an approximate doubling of V_{max} by AMP (Calabrese et al. 2014). In contrast, A769662 appears to have very little impact on V_{max} but lowers K_M for substrate (SAMS peptide) approximately fourfold (Calabrese et al. 2014). The precise structural basis for this is not clearly understood. However, the close association of the C-interacting helix of the β subunit with the C-helix of the α

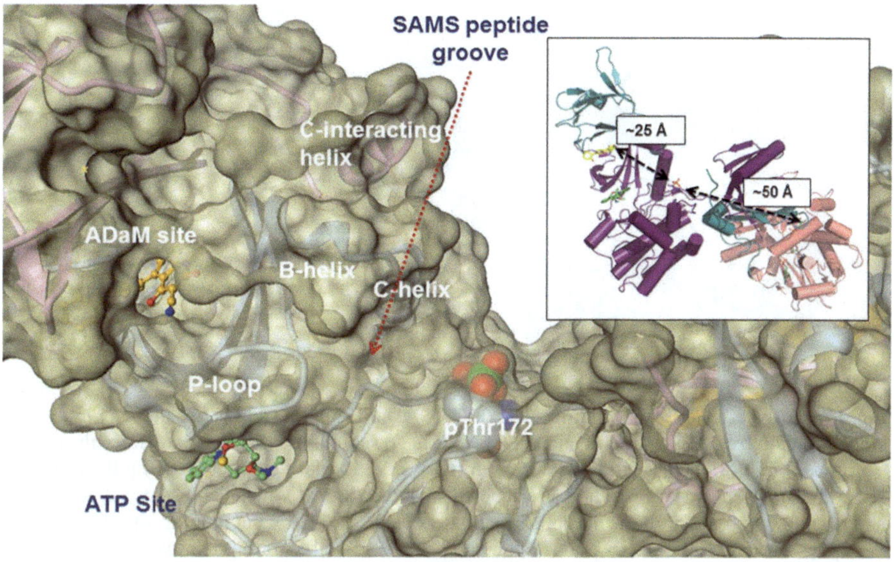

Fig. 1.5 Transparent molecular surface of AMPK α1β1γ1 (PDB accession code: 4QFR) with Cl-A769662 bound at the ADaM site. Also shown is staurosporine bound at the kinase ATP site. The C-interacting helix of the β subunit and the B- and C-helices of the kinase domain are highlighted. pThr172 is shown in space filling representation. The *dotted red arrow* points to the substrate-binding groove where SAMS peptide is likely to bind. Binding of A769662 or other allosteric activators that target the ADaM site could result in enhanced binding affinity of SAMS peptide (decrease of K_M) through structural perturbations involving the C-interacting helix and the C-helix. Inset: Overall AMPK topology colored as in (**a**) illustrating the approximate distance between the ADaM site and pThr172 (~25 Å) and the γ-nucleotide-binding sites and pThr172 (~50 Å)

subunit suggests that A769662 might promote more optimal binding of SAMS peptide in the substrate-binding groove between the N- and the C-lobes of the kinase module (Calabrese et al. 2014; Xiao et al. 2013) (Fig. 1.5).

Recent structures of AMPK bound to A769662 and related analogs as well as the benzimidazole (compound 991) revealed that these activators bind at a site that is remote from nucleotide-binding sites on the γ subunit (Calabrese et al. 2014; Xiao et al. 2013) (Fig. 1.6). The binding site, at the interface of the CMB motif of the β subunit and the N-lobe of the kinase domain of the α subunit, has been recently termed the "ADaM" (allosteric drug and metabolite) site and represents a novel binding pocket with no as-yet identified endogenous ligand (Langendorf and Kemp 2015). Close inspection of binding mode reveals that the bicyclic core of all ligands is sandwiched between Ile46 of the α subunit and a stacked Arg83 from the β subunit. Cation–π interaction formed by the positively charged guanidinium side chain of Arg83 with the delocalized π-electron cloud of the heterocyclic core of activators is an important feature of ligand binding at the ADaM site. In addition, electrostatic contributions emerge from a set of conserved outward-facing lysines from the α subunit (Lys29/Lys31/Lys51), as well as an autophosphorylation site on

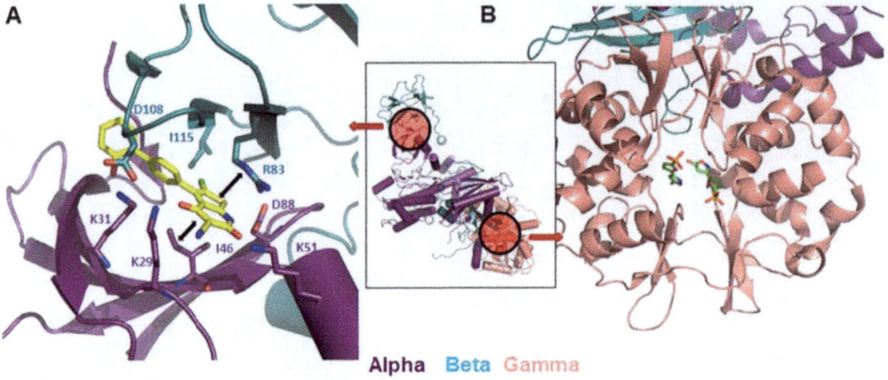

Fig. 1.6 Allosteric activator binding sites. (**a**) α/β (ADaM) binding site between the N-lobe of the kinase domain from the α subunit and the CBM of the β subunit. Structure is shown for the α1β1γ1 isoform to a chlorinated analog of A769662 (Cl-A769662) derived from 4QFR.pdb (Calabrese et al. 2014). A set of key residues are labeled including D108 which represents an engineered phosphomimetic variant of the autophosphorylated S108. (**b**) C2 ligand binding site shown bound to α2β1γ1 derived from 4ZHX.pdb (Langendorf et al. 2016) illustrating the two copies of ligand bound to γ. Inset: Overall AMPK topology illustrated for full-length AMPK α1β2γ1 derived from 4RER.pdb (Li et al. 2015). ADaM and C2 binding regions are illustrated by *red shaded circles*

the β subunit (pSer108) (Fig. 1.6a). In addition, recent work has supported that salicylate, a metabolite of aspirin, directly activates AMPK in a manner that is competitive with A769662 (Hawley et al. 2012). Low-resolution crystallographic data from iodinated-salicylate analogs lend further support to the hypothesis that this aspirin metabolite also activates AMPK via the ADaM site (Calabrese et al. 2014).

AMPK activity is modulated, in part, by the presence of an AID on the α subunit. Previous reports have shown that constructs containing the KD-AID have impaired catalytic activity relative to the isolated KD alone (Chen et al. 2009; Crute et al. 1998; Rajamohan et al. 2015). In an effort to activate AMPK through modulation of this auto-regulatory mechanism, Pang et al. carried out a high-throughput screen against a KD-AID construct and identified a hit termed PT1 compound (Pang et al. 2008) (Fig. 1.4). PT1 showed cellular potency based on phosphorylation of downstream targets and could be modeled to dock at the KD–AID interface. However, recent work has suggested that PT1's in vivo activity may derive from an indirect mechanism through inhibition of the respiratory chain affecting nucleotide levels and ratios (Jensen et al. 2015).

Recently, a new and exciting crystal structure was solved of full-length α2β1γ1 bound to the direct activator, compound C2 (Fig. 1.4) (Langendorf et al. 2016). C2, which contains a phosphonate group, and its cell-permeable diester prodrug C13 were designed as AMP mimetics that activate AMPK by directly binding to its γ subunit (Gómez-Galeno et al. 2010). In stark contrast to A769662 and compound 991, C2 binds at the center of the γ subunit and was observed to interact with a stoichiometry of two copies of ligand per AMPK heterotrimer. Though the phosphonate groups of C2 overlap the bound position previously observed for nucleotide, coordination of the remainder of the ligand is distinct and appears to include direct contacts between the two copies of compound. C2 activity appeared synergistic with A769662 raising the exciting possibility of achieving novel pharmacology through combinations of synthetic activators that function through distinct mechanisms. Furthermore, the observation that C2 is a selective activator of α1-containing isoforms suggests that tissue-specific activation of AMPK could be achieved using direct synthetic activators (Hunter et al. 2014).

1.4 Concluding Remarks and Outstanding Questions

Despite tremendous advances in knowledge of AMPK structures and activation, significant gaps in learning remain. Though crystallographic structures are available for α1 and α2, as well as β1 and β2 subunits, the only γ subunit for which we have any high-resolution structural data is γ1. Elucidation of γ2 or γ3, and additional work to dissect the role of their unique N-terminal extensions, holds the potential to further understand function and regulation. While biochemical and cellular studies have revealed the important role that β subunit myristoylation plays in AMP-mediated stimulation of AMPK activation by upstream kinases, we do not yet understand the structural basis for this. In addition, as more and more crystal structures become available, the degree of flexibility of this heterotrimer is becoming increasingly clear. As such, work to further characterize the ensemble of conformations sampled by AMPK—through additional experimental and computational studies—may help paint a clearer picture of the consequences of ligand binding at distinct sites. Finally, with the recent discovery of a novel binding site at the α/β interface, it is tempting to speculate that additional ligands, both natural and synthetic, may bind at this site to affect AMPK function. The quest to find such endogenous activators could deepen our understanding of this already complex critical metabolic sensor besides providing an exciting opportunity for the discovery of novel therapeutics for the treatment of cardiovascular and metabolic diseases.

References

Ahn J, Lee H, Kim S, Park J, Ha T (2008) The anti-obesity effect of quercetin is mediated by the AMPK and MAPK signaling pathways. Biochem Biophys Res Commun 373:545–549

Amodeo GA, Rudolph MJ, Tong L (2007) Crystal structure of the heterotrimer core of *Saccharomyces cerevisiae* AMPK homologue SNF1. Nature 449:492–495

Baur JA, Pearson KJ, Price NL, Jamieson HA, Lerin C, Kalra A, Prabhu VV, Allard JS, Lopez-Lluch G, Lewis K et al (2006) Resveratrol improves health and survival of mice on a high-calorie diet. Nature 444:337–342

Calabrese MF, Rajamohan F, Harris MS, Caspers NL, Magyar R, Withka JM, Wang H, Borzilleri KA, Sahasrabudhe PV, Hoth LR et al (2014) Structural basis for AMPK activation: natural and synthetic ligands regulate kinase activity from opposite poles by different molecular mechanisms. Structure 22:1161–1172

Carling D, Viollet B (2015) Beyond energy homeostasis: the expanding role of AMP-activated protein kinase in regulating metabolism. Cell Metab 21:799–804

Carling D, Zammit VA, Hardie DG (1987) A common bicyclic protein kinase cascade inactivates the regulatory enzymes of fatty acid and cholesterol biosynthesis. FEBS Lett 223:217–222

Carling D, Clarke PR, Zammit VA, Hardie DG (1989) Purification and characterization of the AMP-activated protein kinase. Copurification of acetyl-CoA carboxylase kinase and 3-hydroxy-3-methylglutaryl-CoA reductase kinase activities. Eur J Biochem 182:129–136

Chen L, Jiao ZH, Zheng LS, Zhang YY, Xie ST, Wang ZX, Wu JW (2009) Structural insight into the autoinhibition mechanism of AMP-activated protein kinase. Nature 459:1146–1149

Chen L, Wang J, Zhang YY, Yan SF, Neumann D, Schlattner U, Wang ZX, Wu JW (2012) AMP-activated protein kinase undergoes nucleotide-dependent conformational changes. Nat Struct Mol Biol 19:716–718

Chen L, Xin FJ, Wang J, Hu J, Zhang YY, Wan S, Cao LS, Lu C, Li P, Yan SF et al (2013) Conserved regulatory elements in AMPK. Nature 498:E8–E10

Cheung PC, Salt IP, Davies SP, Hardie DG, Carling D (2000) Characterization of AMP-activated protein kinase γ-subunit isoforms and their role in AMP binding. Biochem J 346:659–669

Cool B, Zinker B, Chiou W, Kifle L, Cao N, Perham M, Dickinson R, Adler A, Gagne G, Iyengar R et al (2006) Identification and characterization of a small molecule AMPK activator that treats key components of type 2 diabetes and the metabolic syndrome. Cell Metab 3:403–416

Corton JM, Gillespie JG, Hawley SA, Hardie DG (1995) 5-Aminoimidazole-4-Carboxamide Ribonucleoside. Eur J Biochem 229:558–565

Crute BE, Seefeld K, Gamble J, Kemp BE, Witters LA (1998) Functional domains of the alpha1 catalytic subunit of the AMP-activated protein kinase. J Biol Chem 273:35347–35354

Dale S, Wilson WA, Edelman AM, Hardie DG (1995) Similar substrate recognition motifs for mammalian AMP-activated protein kinase, higher plant HMG-CoA reductase kinase-A, yeast SNF1, and mammalian calmodulin-dependent protein kinase I. FEBS Lett 361:191–195

Davies SP, Helps NR, Cohen PT, Hardie DG (1995) 5′-AMP inhibits dephosphorylation, as well as promoting phosphorylation, of the AMP-activated protein kinase. Studies using bacterially expressed human protein phosphatase-2C alpha and native bovine protein phosphatase-2AC. FEBS Lett 377:421–425

Day P, Sharff A, Parra L, Cleasby A, Williams M, Horer S, Nar H, Redemann N, Tickle I, Yon J (2007) Structure of a CBS-domain pair from the regulatory gamma1 subunit of human AMPK in complex with AMP and ZMP. Acta Crystallogr D Biol Crystallogr 63:587–596

Fryer LG, Parbu-Patel A, Carling D (2002) The Anti-diabetic drugs rosiglitazone and metformin stimulate AMP-activated protein kinase through distinct signaling pathways. J Biol Chem 277:25226–25232

Gómez-Galeno JE, Dang Q, Nguyen TH, Boyer SH, Grote MP, Sun Z, Chen M, Craigo WA, van Poelje PD, MacKenna DA et al (2010) A potent and selective AMPK activator that inhibits de novo lipogenesis. ACS Med Chem Lett 1:478–482

Goransson O, McBride A, Hawley SA, Ross FA, Shpiro N, Foretz M, Viollet B, Hardie DG, Sakamoto K (2007) Mechanism of action of A-769662, a valuable tool for activation of AMP-activated protein kinase. J Biol Chem 282:32549–32560

Gowans GJ, Hawley SA, Ross FA, Hardie DG (2013) AMP is a true physiological regulator of AMP-activated protein kinase by both allosteric activation and enhancing net phosphorylation. Cell Metab 18:556–566

Gwinn DM, Shackelford DB, Egan DF, Mihaylova MM, Mery A, Vasquez DS, Turk BE, Shaw RJ (2008) AMPK phosphorylation of raptor mediates a metabolic checkpoint. Mol Cell 30:214–226

Handa N, Takagi T, Saijo S, Kishishita S, Takaya D, Toyama M, Terada T, Shirouzu M, Suzuki A, Lee S et al (2011) Structural basis for compound C inhibition of the human AMP-activated protein kinase alpha2 subunit kinase domain. Acta Crystallogr D Biol Crystallogr 67:480–487

Hardie DG (2016) Regulation of AMP-activated protein kinase by natural and synthetic activators. Acta Pharm Sin B 6:1–19

Hawley SA, Davison M, Davies SP, Beri RK, Carling D, Hardie DG (1996) Characterization of the AMP-activated protein kinase from rat liver and identification of threonine 172 as the major site at which it phosphorylates AMP-activated protein kinase. J Biol Chem 271:27879–27887

Hawley SA, Gadalla AE, Olsen GS, Hardie DG (2002) The antidiabetic drug metformin activates the AMP-activated protein kinase cascade via an adenine nucleotide-independent mechanism. Diabetes 51:2420–2425

Hawley SA, Boudeau J, Reid JL, Mustard KJ, Udd L, Makela TP, Alessi DR, Hardie DG (2003) Complexes between the LKB1 tumor suppressor, STRAD alpha/beta and MO25 alpha/beta are upstream kinases in the AMP-activated protein kinase cascade. J Biol 2:28

Hawley SA, Pan DA, Mustark KJ, Ross L, Bain J, Edelman AM, Frenquelli BG, Hardie DG (2005) Calmodulin-dependent protein kinase kinase-beta is an alternative upstream kinase for AMP-activated protein kinase. Cell Metab 2:9–19

Hawley SA, Ross FA, Chevtzoff C, Green KA, Evans A, Fogarty S, Towler MC, Brown LJ, Ogunbayo OA, Evans AM et al (2010) Use of cells expressing gamma subunit variants to identify diverse mechanisms of AMPK activation. Cell Metab 11:554–565

Hawley SA, Fullerton MD, Ross FA, Schertzer JD, Chevtzoff C, Walker KJ, Peggie MW, Zibrova D, Green KA, Mustard KJ et al (2012) The ancient drug salicylate directly activates AMP-activated protein kinase. Science 336:918–922

Hunter RW, Foretz M, Bultot L, Fullerton MD, Deak M, Ross FA, Hawley SA, Shpiro N, Viollet B, Barron D et al (2014) Mechanism of action of compound-13: an α1-selective small molecule activator of AMPK. Chem Biol 21:866–879

Hurley RL, Anderson KA, Franzone JM, Kemp BE, Means AR, Witters LA (2005) The Ca2+/calmodulin-dependent protein kinase kinases are AMP-activated protein kinase kinases. J Biol Chem 280:29060–29066

Hwang JT, Park IJ, Shin JI, Lee YK, Lee SK, Baik HW, Ha J, Park OJ (2005) Genistein, EGCG, and capsaicin inhibit adipocyte differentiation process via activating AMP-activated protein kinase. Biochem Biophys Res Commun 338:694–699

Inoki K, Zhu T, Guan KL (2003) TSC2 mediates cellular energy response to control cell growth and survival. Cell 115:577–590

Jensen TE, Ross FA, Kleinert M, Sylow L, Knudsen JR, Gowans GJ, Hardie DG, Richter EA (2015) PT-1 selectively activates AMPK-gamma1 complexes in mouse skeletal muscle, but activates all three gamma subunit complexes in cultured human cells by inhibiting the respiratory chain. Biochem J 467:461–472

Jin X, Townley R, Shapiro L (2007) Structural insight into AMPK regulation: ADP comes into play. Structure 15:1285–1295

Kemp BE, Oakhill JS, Scott JW (2007) AMPK structure and regulation from three angles. Structure 15:1161–1163

Koay A, Rimmer KA, Mertens HDT, Gooley PR, Stapleton D (2007) Oligosaccharide recognition and binding to the carbohydrate binding module of AMP-activated protein kinase. FEBS Lett 581:5055–5059

Koay A, Woodcroft B, Petrie EJ, Yue H, Emanuelle S, Bieri M, Bailey MF, Hargreaves M, Park J-T, Park K-H et al (2010) AMPK β subunits display isoform specific affinities for carbohydrates. FEBS Lett 584:3499–3503

Kudo N, Gillespie JG, Kung L, Witters LA, Schulz R, Clanachan AS, Lopaschuk GD (1996) Characterization of 5′AMP-activated protein kinase activity in the heart and its role in inhibiting acetyl-CoA carboxylase during reperfusion following ischemia. Biochim Biophys Acta 1301:67–75

Landgraf RR, Goswami D, Rajamohan F, Harris MS, Calabrese MF, Hoth LR, Magyar R, Pascal BD, Chalmers MJ, Busby SA et al (2013) Activation of AMP-activated protein kinase revealed by hydrogen/deuterium exchange mass spectrometry. Structure 21:1942–1953

Langendorf CG, Kemp BE (2015) Choreography of AMPK activation. Cell Res 25:5–6

Langendorf CG, Ngoei KR, Scott JW, Ling NX, Issa SM, Gorman MA, Parker MW, Sakamoto K, Oakhill JS, Kemp BE (2016) Structural basis of allosteric and synergistic activation of AMPK by furan-2-phosphonic derivative C2 binding. Nat Commun 7:10912

Lee YS, Kim WS, Kim KH, Yoon MJ, Cho HJ, Shen Y, Ye JM, Lee CH, Oh WK, Kim CT et al (2006) Berberine, a natural plant product, activates AMP-activated protein kinase with beneficial metabolic effects in diabetic and insulin-resistant states. Diabetes 55:2256–2264

Li X, Wang L, Zhou XE, Ke J, de Waal PW, Gu X, Tan MH, Wang D, Wu D, Xu HE et al (2015) Structural basis of AMPK regulation by adenine nucleotides and glycogen. Cell Res 25:50–66

Littler DR, Walker JR, Davis T, Wybenga-Groot LE, Finerty PJ Jr, Newman E, Mackenzie F, Dhe-Paganon S (2010) A conserved mechanism of autoinhibition for the AMPK kinase domain: ATP-binding site and catalytic loop refolding as a means of regulation. Acta Crystallogr Sect F Struct Biol Cryst Commun 66:143–151

Mayer FV, Heath R, Underwood E, Sanders MJ, Carmena D, McCartney RR, Leiper FC, Xiao B, Jing C, Walker PA et al (2011) ADP regulates SNF1, the Saccharomyces cerevisiae homolog of AMP-activated protein kinase. Cell Metab 14:707–714

McBride A, Hardie DG (2009) AMP-activated protein kinase—a sensor of glycogen as well as AMP and ATP? Acta Physiol (Oxf) 196:99–113

McBride A, Ghilagaber S, Nikolaev A, Hardie DG (2009) The glycogen-binding domain on the AMPK β subunit allows the kinase to act as a glycogen sensor. Cell Metab 9:23–34

Mobbs JI, Koay A, Di Paolo A, Bieri M, Petrie EJ, Gorman MA, Doughty L, Parker MW, Stapleton D, Griffin MD et al (2015a) Determinants of oligosaccharide specificity of the carbohydrate binding modules of AMP-activated protein kinase. Biochem J 468:245–257

Mobbs JI, Koay A, Di Paolo A, Bieri M, Petrie EJ, Gorman MA, Doughty L, Parker MW, Stapleton DI, Griffin MD et al (2015b) Determinants of oligosaccharide specificity of the carbohydrate-binding modules of AMP-activated protein kinase. Biochem J 468:245–257

Mounier R, Theret M, Lantier L, Foretz M, Viollet B (2015) Expanding roles for AMPK in skeletal muscle plasticity. Trends Endocrinol Metab 26:275–286

Nayak V, Zhao K, Wyce A, Schwartz MF, Lo WS, Berger SL, Marmorstein R (2006) Structure and dimerization of the kinase domain from yeast Snf1, a member of the Snf1/AMPK protein family. Structure 14:477–485

Neumann D, Woods A, Carling D, Wallimann T, Schlattner U (2003) Mammalian AMP-activated protein kinase: functional, heterotrimeric complexes by co-expression of subunits in *Escherichia coli*. Protein Expr Purif 30:230–237

Neumann D, Wallimann T, Rider MH, Tokarska-Schlattner M, Hardie DG, Schlattner U (2007) Signaling by AMP-activated protein kinase. In: Saks V (ed) Molecular system bioenergetics. Wiley, Weinheim, pp 303–338

Oakhill JS, Chen Z-P, Scott JW, Steel R, Castelli LA, Ling N, Macaulay SL, Kemp BE (2010) β-Subunit myristoylation is the gatekeeper for initiating metabolic stress sensing by AMP-activated protein kinase (AMPK). Proc Natl Acad Sci USA 107:19237–19241

Oakhill JS, Steel R, Chen Z-P, Scott JW, Ling N, Tam S, Kemp BE (2011) AMPK is a direct adenylate charge-regulated protein kinase. Science 332:1433–1435

Oakhill JS, Scott JW, Kemp BE (2012) AMPK functions as an adenylate charge-regulated protein kinase. Trends Endocrinol Metab 23:125–132

Pang T, Zhang ZS, Gu M, Qiu BY, Yu LF, Cao PR, Shao W, Su MB, Li JY, Nan FJ et al (2008) Small molecule antagonizes autoinhibition and activates AMP-activated protein kinase in cells. J Biol Chem 283:16051–16060

Polekhina G, Gupta A, Michell BJ, van Denderen B, Murthy S, Feil SC, Jennings IG, Campbell DJ, Witters LA, Parker MW et al (2003) AMPK β subunit targets metabolic stress sensing to glycogen. Curr Biol 13:867–871

Polekhina G, Gupta A, van Denderen BJ, Feil SC, Kemp BE, Stapleton D, Parker MW (2005) Structural basis for glycogen recognition by AMP-activated protein kinase. Structure 10:1453–1462

Rajamohan F, Harris MS, Frisbie RK, Hoth LR, Geoghegan KF, Valentine JJ, Reyes AR, Landro JA, Qiu X, Kurumbail RG (2010) *Escherichia coli* expression, purification and characterization of functional full-length recombinant $\alpha2\beta2\gamma3$ heterotrimeric complex of human AMP-activated protein kinase. Protein Expr Purif 73:189–197

Rajamohan F, Reyes AR, Frisbie RK, Hoth LR, Sahasrabudhe P, Magyar R, Landro JA, Withka JM, Caspers NL, Calabrese MF et al (2015) Probing the enzyme kinetics, allosteric modulation and activation of alpha-1 and alpha-2 subunit containing AMP-activated protein kinase (AMPK) heterotrimeric complexes by pharmacological and physiological activators. Biochem J 473:581–592

Riek U, Scholz R, Konarev P, Rufer A, Suter M, Nazabal A, Ringler P, Chami M, Muller SA, Neumann D et al (2008) Structural properties of AMP-activated protein kinase: dimerization, molecular shape, and changes upon ligand binding. J Biol Chem 283:18331–18343

Ross FA, Jensen TE, Hardie DG (2016a) Differential regulation by AMP and ADP of AMPK complexes containing different γ subunit isoforms. Biochem J 473:189–199

Ross FA, MacKintosh C, Hardie DG (2016b) AMP-activated protein kinase: a cellular energy sensor that comes in twelve flavours. FEBS J. doi:10.1111/febs.13698

Rudolph MJ, Amodeo GA, Iram SH, Hong SP, Pirino G, Carlson M, Tong L (2007) Structure of the Bateman2 domain of yeast Snf4: dimeric association and relevance for AMP binding. Structure 15:65–74

Rudolph MJ, Amodeo GA, Tong L (2010) An inhibited conformation for the protein kinase domain of the *Saccharomyces cerevisiae* AMPK homolog Snf1. Acta Crystallogr Sect F Struct Biol Cryst Commun 66:999–1002

Sanders MJ, Ali ZS, Hegarty BD, Heath R, Snowden MA, Carling D (2007) Defining the mechanism of activation of AMP-activated protein kinase by the small molecule A-769662, a member of the thienopyridone family. J Biol Chem 282:32539–32548

Scott JW, van Denderen BJ, Jorgensen SB, Honeyman JE, Steinberg GR, Oakhill JS, Iseli TJ, Koay A, Gooley PR, Stapleton D et al (2008) Thienopyridone drugs are selective activators of AMP-activated protein kinase β1-containing complexes. Chem Biol 15:1220–1230

Scott JW, Ling N, Issa SMA, Dite TA, O'Brien MT, Chen ZP, Galic S, Langendorf CG, Steinberg GR, Kemp BE et al (2014) Small molecule drug A-769662 and AMP synergistically activate naive AMPK independent of upstream kinase signaling. Chem Biol 21:1–9

Shaw RJ, Kosmatka M, Bardeesy N, Hurley RL, Witters LA, DePinho RA, Cantley LC (2004) The tumor suppressor LKB1 kinase directly activates AMP-activated kinase and regulates apoptosis in response to energy stress. Proc Natl Acad Sci 101:3329–3335

Steinberg GR, Kemp BE (2009) AMPK in health and disease. Physiol Rev 89:1025–1078

Suter M, Riek U, Tuerk R, Schlattner U, Wallimann T, Neumann D (2006) Dissecting the role of 5′-AMP for allosteric stimulation, activation, and deactivation of AMP-activated protein kinase. J Biol Chem 281:32207–32216

Townley R, Shapiro L (2007) Crystal structures of the adenylate sensor form fission yeast AMP-activated protein kinase. Science 315:1726–1729

Walker JR, Wybenga-Groot L, Finerty Jr PJ, Newman E, Mackenzie FM, Weigelt J, Sundstrom M, Arrowsmith C, Edwards A, Bochkarev A, Structural Genomics Consortium et al (2005) Structure of the glycogen-binding domain of the AMP-activated protein kinase beta2 subunit, PDB accession code, 2F15

Warden SM, Richardson C, O'Donnell J Jr, Stapleton D, Kemp BE, Witters LA (2001) Post-translational modifications of the β1 subunit of AMP-activated protein kinase affect enzyme activity and cellular localization. Biochem J 354:275–283

Woods A, Johnstone SR, Dickerson K, Leiper FC, Fryer LG, Neumann D, Schlattner U, Wallimann T, Carlson M, Carling D (2003a) LKB1 is the upstream kinase in the AMP-activated protein kinase cascade. Curr Biol 13:2004–2008

Woods A, Vertommen D, Neumann D, Turk R, Bayliss J, Schlattner U, Wallimann T, Carling D, Rider MH (2003b) Identification of phosphorylation sites in AMP-activated protein kinase (AMPK) for upstream AMPK kinases and study of their roles by site-directed mutagenesis. J Biol Chem 278:28434–28443

Woods A, Dickerson K, Heath R, Hong SP, Momcilovic M, Johnstone SR, Carlson M, Carling D (2005) Ca^{2+}/calmodulin-dependent protein kinase kinase-β acts upstream of AMP-activated protein kinase in mammalian cells. Cell Metab 2:21–33

Xiao B, Heath R, Saiu P, Leiper FC, Leone P, Jing C, Walker PA, Haire L, Eccleston JF, Davis CT et al (2007) Structural basis for AMP binding to mammalian AMP-activated protein kinase. Nature 449:496–500

Xiao B, Sanders MJ, Underwood E, Heath R, Mayer FV, Carmena D, Jing C, Walker PA, Eccleston JF, Haire LF et al (2011) Structure of mammalian AMPK and its regulation by ADP. Nature 472:230–233

Xiao B, Sanders MJ, Carmena D, Bright NJ, Haire LF, Underwood E, Patel BR, Heath RB, Walker PA, Hallen S et al (2013) Structural basis of AMPK regulation by small molecule activators. Nat Commun 4:2017

Xin FJ, Wang J, Zhao RQ, Wang ZX, Wu JW (2013) Coordinated regulation of AMPK activity by multiple elements in the alpha-subunit. Cell Res 10:1237–1240

Yu LF, Li YY, Su MB, Zhang M, Zhang W, Zhang LN, Pang T, Zhang RT, Liu B, Li JY et al (2013) Development of novel alkene oxindole derivatives as orally efficacious AMP-activated protein kinase activators. ACS Med Chem Lett 4:475–480

Zhou G, Myers R, Li Y, Chen Y, Shen X, Fenyk-Melody J, Wu M, Ventre J, Doebber T, Fujii N et al (2001) Role of AMP-activated protein kinase in mechanism of metformin action. J Clin Invest 108:1167–1174

Zhu L, Chen L, Zhou XM, Zhang YY, Zhang YJ, Zhao J, Ji SR, Wu JW, Wu Y (2011) Structural insights into the architecture and allostery of full-length AMP-activated protein kinase. Structure 19:515–522

Zong H, Ren JM, Young LH, Pypaert M, Mu J, Birnbaum MJ, Shulman GI (2002) AMP kinase is required for mitochondrial biogenesis in skeletal muscle in response to chronic energy deprivation. Proc Natl Acad Sci USA 99:15983–15987

Chapter 2
Regulation of Carbohydrate Metabolism, Lipid Metabolism, and Protein Metabolism by AMPK

Yeliz Angin, Christophe Beauloye, Sandrine Horman, and Luc Bertrand

Contents

Abstract This chapter summarizes AMPK function in the regulation of substrate and energy metabolism with the main emphasis on carbohydrate and lipid metabolism, protein turnover, mitochondrial biogenesis, and whole-body energy homeostasis. AMPK acts as whole-body energy sensor and integrates different signaling pathway to meet both cellular and body energy requirements while inhibiting energy-consuming processes but also activating energy-producing ones. AMPK mainly promotes glucose and fatty acid catabolism, whereas it prevents protein, glycogen, and fatty acid synthesis.

Y. Angin • S. Horman • L. Bertrand (✉)
Pole of Cardiovascular Research, Université catholique de Louvain, Institut de Recherche Expérimentale et Clinique, Brussels, Belgium
e-mail: luc.bertrand@uclouvain.be

C. Beauloye
Pole of Cardiovascular Research, Université catholique de Louvain, Institut de Recherche Expérimentale et Clinique, Brussels, Belgium

Division of Cardiology, Cliniques universitaires Saint-Luc, Brussels, Belgium

M.D. Cordero, B. Viollet (eds.), *AMP-activated Protein Kinase*, Experientia Supplementum 107, DOI 10.1007/978-3-319-43589-3_2

Keywords AMPK • Glucose metabolism • Lipid metabolism • Protein synthesis • Insulin sensitivity • Mitochondrial biogenesis

2.1 Historical Background

The AMP-activated protein kinase (AMPK) has an important place in the regulation of signaling pathways involved in energy and substrate metabolism. AMPK senses cellular energy status by monitoring ADP/ATP and AMP/ATP ratios and nutrient availability and regulates cellular events by stimulating ATP-generating cellular processes (such as glucose uptake, glycolysis, and fatty acid oxidation) while inhibiting ATP-consuming processes (such as fatty acid and protein synthesis) (Hardie et al. 2006) (Fig. 2.1). In order to adapt required energy demands, AMPK has acute effects on metabolic enzymes by direct phosphorylation as well as long-term action to change the transcriptional levels of metabolic proteins and enzymes.

The discovery of AMPK dates back to the 1970s and is clearly linked to lipid metabolism when the activity of Acetyl-CoA carboxylase (ACC) was shown to be regulated by phosphorylation and dephosphorylation reactions (Carlson and Kim 1973). Even if not yet identified, they partially purified the protein kinase responsible for ACC phosphorylation and called it ACC kinase. It was shown that ATP incorporation into liver-isolated ACC was correlated with its decreased carboxylase activity in fractions where the ACC kinase was present. ACC is a rate-limiting enzyme for fatty acid synthesis. Indeed, it carboxylates acetyl-CoA into malonyl-CoA, the first and regulatory step in fatty acid synthesis. Similar findings were simultaneously obtained with HMG-CoA reductase (HMGR), the rate-limiting enzyme of cholesterol biosynthesis pathway. HMGR was found to be negatively regulated by phosphorylation via a protein kinase present in liver cytosol (Beg et al. 1973). Inactivation of ACC and HMGR by their respective associated protein kinases and their reactivation by phosphatase treatment were confirmed later (Beg et al. 1979; Ingebritsen et al. 1981). ACC kinase and HMGR kinase were detected to be similarly positively regulated by cellular AMP levels and later on, in the late 1980s, it was find out that both enzymes were activated by the same protein kinase which has been called AMPK (Carling et al. 1987; Ferrer et al. 1985; Yeh et al. 1980; Munday et al. 1988; Carling et al. 1989).

AMPK is stimulated by AMP and inhibited by ATP, making it dependent on what is called the "adenylate energy charge," which is a monitor of the energetic level of the cell. Regulation of AMPK activity by cellular energy status makes AMPK a cellular fuel gauge and puts it into center for the coordination of cellular signaling events in different tissues. AMPK also regulates energy metabolism in the whole body via its action on hypothalamus in the brain as well as through the action of hormones and adipokines. The role of AMPK in the regulation of both cellular and whole-body energy metabolism is multifaceted and will be discussed under the

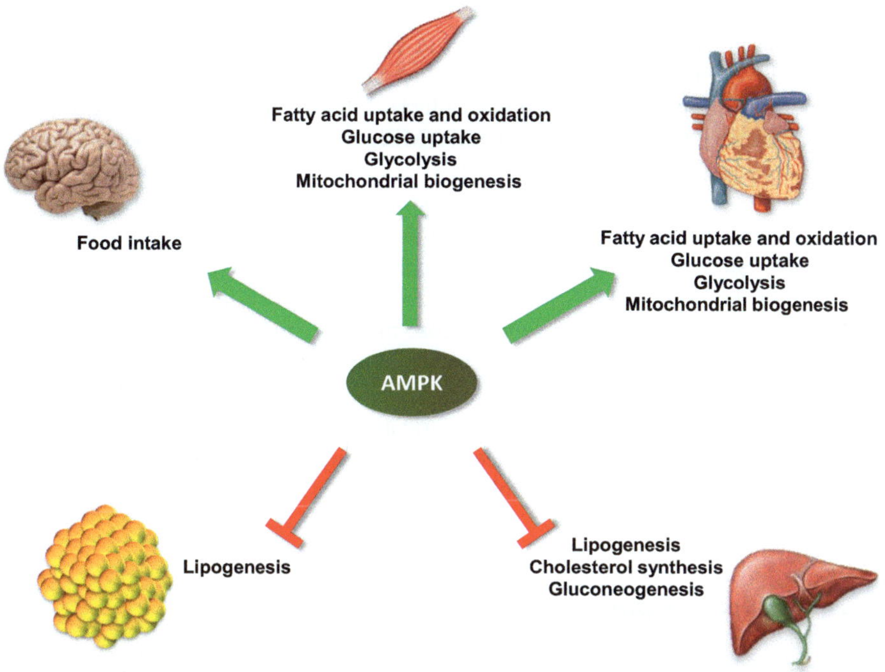

Fig. 2.1 AMPK regulates whole-body energy level. AMPK activation integrates different cellular functions to preserve energy while inhibiting energy-consuming processes and, conversely, activating energy-producing processes. Activation of AMPK by energy deprivation (fasting, exercise), by hormones, or pharmacologically leads to phosphorylation and (in)activation of numerous downstream targets in different tissues (see text for details). These actions of AMPK contribute to whole-body energy metabolism (see text for more details)

five main parts of this chapter as AMPK in carbohydrate and lipid metabolism, its role in whole-body insulin sensitivity, mitochondrial biogenesis, and protein turnover.

2.2 AMPK in Carbohydrate Metabolism

AMPK is ubiquitously expressed in the whole body and plays a vital role in substrate and energy metabolism through coordinating different cellular functions between different organs. Notably, AMPK activation is associated with whole-body insulin sensitivity. AMPK regulates carbohydrate metabolism by stimulating glucose uptake and glycolysis in insulin-sensitive tissues (heart, skeletal muscle, and adipose tissue) as well as inhibiting glucose production in liver. AMPK activation promotes cellular glucose uptake in an insulin-independent manner by translocation of glucose transporter 4 (GLUT4) to the plasma membrane as well as leading increases in the expression levels of GLUT1 and 4. Historically, it has been shown that physiological activation of AMPK via exercise and/or pharmacological activation with 5-aminoimidazole-4-carboxamide ribonucleoside (AICAR) or antidiabetic drug metformin stimulates glucose uptake into muscles. Muscle glucose uptake caused by AICAR-mediated AMPK activation and contraction was not prevented by the phosphatidylinositol (PI) 3-kinase (PI3K) inhibitor wortmannin, indicating that AMPK acts on GLUT4 translocation and subsequent glucose uptake via a different mechanism than insulin signaling (Hayashi et al. 1998; Bergeron et al. 1999). Since muscle glucose uptake constitutes the important part of glucose clearance of the body, AMPK activation via exercise and contraction and AMPK-mediated muscle glucose uptake and glucose metabolism gain more importance for whole-body glucose clearance also during some pathophysiological conditions such as disease accompanied with insulin resistance. This makes AMPK an important target for the development of therapeutic approaches to treat diabetes and other insulin resistance-associated diseases. Therapeutic approaches targeting AMPK will be discussed further in this book.

2.2.1 Glucose Uptake

GLUT4 is mainly expressed in heart, skeletal muscle, and adipose tissue, and it plays important role in controlling the whole-body glucose homeostasis. In these tissues, extracellular glucose gradient and plasma membrane, GLUT4 levels are the main rate-limiting steps for glucose uptake. Insulin, muscle contractions, and exercise stimulate translocation of GLUT4-containing vesicles to plasma membrane and thereby facilitate glucose entry. Many studies have been performed using AICAR as pharmacological AMPK activator. AICAR-mediated muscle glucose uptake requires AMPKα2 but not AMPKα1 (Jorgensen et al. 2004a). In LKB1-deficient mouse muscles, basal activity of AMPKα2 is reduced and both muscle contraction and AICAR are neither able to increase the basal activity of AMPK nor able to stimulate glucose uptake, indicating the dependence of AMPK activation to LKB1 (Sakamoto et al. 2005). However, genetic evidences linking AMPK activation observed during muscle exercise to glucose uptake are still conflicting and partially inconclusive. Contraction-mediated glucose uptake is seriously impaired in muscle of double AMPKβ1/β2-deficient mice (O'Neill et al. 2011), whereas deletion of both AMPKα1 and AMPKα2 isoforms has almost no impact (Lantier et al. 2014). Concerning the heart, both electrical stimulation and pharmacological activation of AMPK in cardiomyocytes result in increase in glucose uptake (Habets et al. 2009; Zarrinpashneh et al. 2006). Paradoxically, increased workload in a model of ex vivo perfused working heart increases glucose utilization without any modification in AMP/ATP ratio and AMPK activity but is associated with Akt signaling and increased glycolysis (Beauloye et al. 2002). However, exercise increases cardiac AMPK activity and glucose uptake (Coven et al. 2003), but this could be linked to a β-adrenergic response (An et al. 2005). In the heart, AMPK-dependent increase in glucose uptake is also linked to energetic stress situations like myocardial ischemia (Zarrinpashneh et al. 2006). In adipocytes, AMPK has been proposed to participate in the stimulation of glucose uptake mediated by adiponectin (Wu et al. 2003).

GLUT4 translocation is a complex process involving intracellular sorting of GLUT4 storing vesicles, transport of these vesicles to plasma membrane along with cytoskeletal proteins, docking, tethering, and fusion with the plasma membrane. The molecular mechanism responsible for GLUT4 translocation upon insulin stimulation has been well described (Rowland et al. 2011). Binding of insulin to its membrane receptor, which is an integral protein in cellular membranes, activates its tyrosine kinase activity, subsequently leading to tyrosine phosphorylation of insulin receptor substrate proteins (Fig. 2.2). This is followed by the recruitment and activation of PI3K, which results in the generation of second messenger PI-3,4,5-triphosphate (PIP3). PIP3 triggers the activation of Akt (also known as protein kinase B) through the action of two distinct upstream mediators, 3-phosphoinositide-dependent protein kinase-1 (PDK-1) and the mammalian target of rapamycin (mTOR) complex 2 (Destefano and Jacinto 2013; Manning and Cantley 2007). Akt substrate of 160 kDa (AS160, also known as TBC1D4) and TBC1D1 were identified as Akt target proteins. These targets are GTPase activating

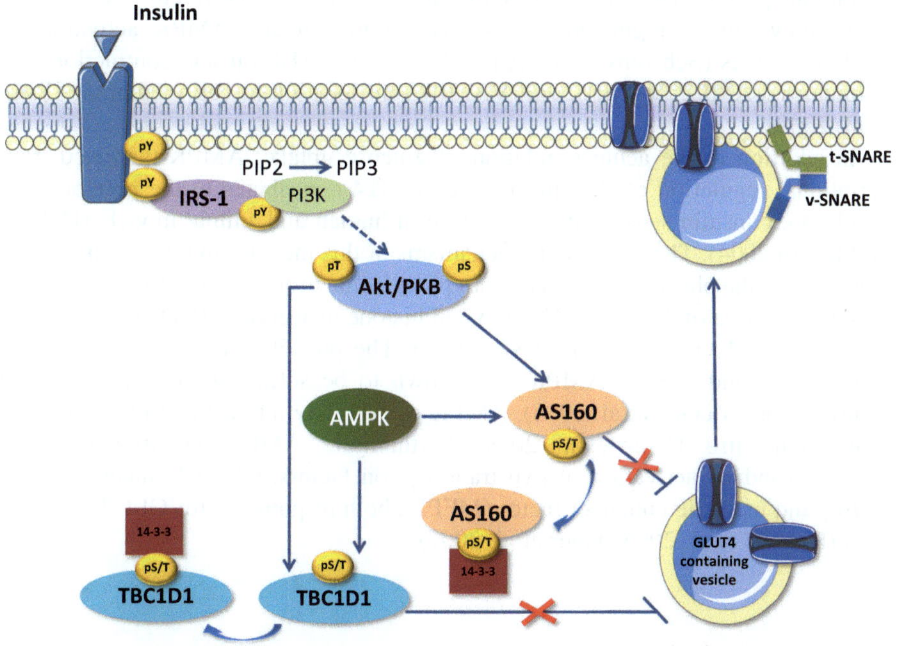

Fig. 2.2 Activation of AMPK increases glucose uptake independently of insulin signaling. AMPK mimics insulin by directly targeting AS160 and TBC1D1. Phosphorylation and inhibition of AS160 and TBC1D1 by AMPK result in translocation of GLUT4-containing vesicles to plasma membrane and concomitant glucose uptake

proteins (GAPs) for Rab, a small G protein known to be involved in vesicle formation, movement, and fusion (Zerial and McBride 2001). AS160 was first discovered in 3T3-L1 adipocytes as a downstream target of Akt under insulin stimulation (Kane et al. 2002; Sano et al. 2003). Afterward, many groups showed that activation of AMPK, either via contraction or AICAR, also promotes AS160 phosphorylation and leads to GLUT4 translocation and subsequent glucose uptake into muscle and adipose tissue (Kane et al. 2002; Sano et al. 2003; Bruss et al. 2005). Both AMPK and Akt phosphorylate several residues of AS160. A number of phosphorylation sites including Ser-588 and Thr-642 are common, others being specific to each stimulating pathway (Kramer et al. 2006). TBC1D1 is similarly phosphorylated by insulin and AMPK activators (Middelbeek et al. 2013; Pehmoller et al. 2009). Some of the phosphorylation sites in AS160 and TBC1D1 are responsible for their binding to 14-3-3 protein and their dissociation from GLUT4 storage vesicles, allowing translocation (Geraghty et al. 2007; Chen et al. 2008). It has to be mentioned that AMPK and insulin can also interact with one another to overstimulate glucose uptake independently of Akt/AS160 (Ginion et al. 2011). Fusion of GLUT4-containing vesicles with the plasma membrane requires a complex formation between target membrane-soluble *N*-ethylmaleimide-sensitive factor attachment protein receptors (t-SNAREs) and vesicle-associated SNAREs (v-SNAREs) (Sadler et al. 2015). Vesicle-associated membrane proteins (VAMPs) are subfamily of v-SNARES. Different VAMPs have been shown to be regulated in response to insulin and AMPK activators in cardiomyocytes (Schwenk et al. 2010). Besides GLUT4, muscle contraction also stimulates VAMP2, VAMP5, and VAMP7 translocation to cell membrane (Rose et al. 2009).

In addition to the acute regulation of glucose uptake, AMPK-mediated transcriptional regulation of skeletal muscle GLUT4 contributes to the regulation of whole-body insulin sensitivity by exercise in human and animal models (Holmes and Dohm 2004). The molecular mechanisms of this increase in GLUT4 transcription involve the phosphorylation of both peroxisome proliferator-activated receptor gamma coactivator-1 alpha (PGC-1α) and histone deacetylase (HDAC) 5 (McGee et al. 2008; McGee and Hargreaves 2004). The phosphorylation of HDAC5 on Serines 259 and 498 by AMPK was shown to be sufficient to induce GLUT4 transcription (McGee et al. 2008). AMPK can directly interact and phosphorylate PGC-1α in vitro (Canto et al. 2009). Furthermore, AMPK activation has been shown to induce activation of two transcription factors, GLUT4 enhancer factor (GEF) and myocyte enhancer factor (MEF2), both responsible for GLUT4 expression (Holmes et al. 2005; Jager et al. 2007).

2.2.2 *Glycolysis*

Once glucose has been taken into the cell by GLUT4, it is rapidly phosphorylated into glucose-6 phosphate (G6P) by hexokinase II. G6P is converted to fructose

6-phosphate by glucose 6-phosphate isomerase and then to fructose 1,6-bisphosphate by phosphofructokinase 1 (PFK1). PFK1 is allosterically activated by increased energy deficit (AMP levels), as well as by fructose-2,6-bisphosphate (Fru-2,6-P_2) which is synthesized by 6-phosphofructokinase-2 (PFK2). PFK-2 is activated upon phosphorylation by AMPK and promotes glycolysis in the ischemic heart (Marsin et al. 2000), in the activated monocytes (Marsin et al. 2002), and in hypoxic cancer cells (Bando et al. 2005).

2.2.3 *Glycogen Metabolism*

Glycogen levels are dynamically regulated as a function of both glycogenesis (glycogen production) and glycogenolysis (glycogen breakdown) in tissues like liver and muscle. Glycogen levels constitute an important portion of energy source for muscle energy requirements during contraction and exercise. Glycogen synthase (GS), the enzyme responsible for glycogenesis, is allosterically regulated by G6P levels but can be also regulated by phosphorylation. It has been shown that AMPK phosphorylates and inhibits GS (Jorgensen et al. 2004b; Halse et al. 2003). The relationship between glycogen and AMPK is even more complex knowing that AMPKβ subunits contain a carbohydrate-binding module (CBM) that targets AMPK to glycogen, connecting the protein kinase close to its substrate GS (Polekhina et al. 2003; Koay et al. 2010). Autophosphorylation of AMPKβ on Thr-148 located at proximity of the CBM affects its ability to bind carbohydrates (Oligschlaeger et al. 2015).

2.2.4 *Gluconeogenesis*

Blood glucose is crucial for many cells and its concentration stays in a narrow range. Lower blood glucose levels (hypoglycemia) can be detrimental for brain, whereas high blood glucose levels (hyperglycemia) can lead to insulin resistance-associated diseases. Therefore, maintaining blood glucose levels is essential and liver plays a key role by regulating glucose production either via glycogenolysis or gluconeogenesis, in response to hormones (such as insulin and glucagon) and energy status. Metformin decreases high blood glucose levels by preventing hepatic glucose production (An and He 2016). Primary findings revealed that this action partly includes AMPK activation (Shaw et al. 2005). However, controversy came later with the nice study of Foretz and colleagues showing that metformin action on gluconeogenesis similarly occurs in the absence of LKB1/AMPK (Foretz et al. 2010). Nevertheless, AMPK activation with other drugs was also shown to partly block hepatic glucose production (Guo et al. 2016).

2.3 AMPK in Lipid Metabolism

Skeletal muscle contraction and exercise lead to AMPK activation and thereby promote fatty acid uptake and oxidation into insulin-sensitive tissues mainly in muscle (skeletal and heart muscle) and adipose tissue while inhibiting lipid synthesis (lipogenesis) in liver. Adipose tissue is the main lipid storage place in human body. Accumulation of lipids other than in adipose tissue contributes to lipid-induced insulin resistance and associated diseases, which will be further discussed in this book. Lipid metabolism starts with the entry of fatty acids into cells. Depending on the cellular energy status, fatty acids taken up across the plasma membrane are converted into fatty acyl-CoAs and either directed to mitochondria for β-oxidation or stored as lipids in lipogenic tissues such as adipose tissue and liver (Bickerton et al. 2007; Glatz et al. 2010).

2.3.1 Fatty Acid Uptake

Transportation of long-chain fatty acids (LCFA) into cells requires protein type transporters. Fatty acid transport proteins (FATP1–6), plasma membrane associated fatty-acid binding protein (FABPpm), and fatty acid translocase CD36 are mostly expressed fatty acid transporters in different type of tissues (Glatz et al. 2010; Coburn et al. 2000; Luiken et al. 2003). LCFA uptake is mediated either by one or by the cooperation of more fatty acid transporters. CD36 which is present both intracellularly and also on plasma membrane is crucial for LCFA uptake in response to both insulin and muscle contraction in heart and skeletal muscle (Luiken et al. 2002; Koonen et al. 2005). In myocytes, both exercise and pharmacological activation of AMPK stimulate translocation of intracellularly stored CD36 to plasma membrane in a way similar to that of insulin signaling (Kola et al. 2008). Similarly to GLUT4 translocation, this action also requires phosphorylation and inactivation of AS160 by AMPK (Samovski et al. 2012). AMPK-mediated long-chain fatty acid uptake depends on the plasma membrane CD36 levels (Habets et al. 2007) and signals through activation of the LKB1/AMPK pathway (Habets et al. 2009).

2.3.2 Lipid Synthesis

In the historical background of this chapter, we already focused on the primordial negative regulation of AMPK on ACC and HMGR, the rate-limiting steps of fatty acid and cholesterol synthesis, respectively (Carling et al. 1987; Ferrer et al. 1985; Yeh et al. 1980; Munday et al. 1988; Carling et al. 1989). However, regulation of lipid synthesis by AMPK is multifaceted and targets several other central elements of this metabolic process. AMPK can act via the regulation of sterol regulatory element-binding protein-1 (SREBP-1) (Li et al. 2011). SREBPs are a family of transcription factors, which regulate the expression of enzymes required for endogenous cholesterol, fatty acid (FA), triacylglycerol, and phospholipid synthesis (Eberle et al. 2004). Activation of AMPK in liver and adipocytes inhibits lipogenesis by downregulating fatty acid synthase (FAS) activity through decreased SREBP-1 gene expressions (Madsen et al. 2015). It is interesting to note that adiponectin suppresses hepatic fatty acid synthesis through suppression of SREBP1c expression in an AdipoR1/LKB1/AMPK-dependent pathway (Awazawa et al. 2009).

Triacylglycerol (TAG) synthesis (esterification) and its counterpart, breakdown of lipids (lipolysis), are dynamic processes involved in the regulation of lipid levels in tissues like adipose tissue, liver, and skeletal muscle. Glycerol-3-phosphate acyltransferase (GPAT) is a rate-limiting enzyme for the synthesis of lysophosphatidic acid, the first step in TAG formation, and it is mainly expressed in adipose tissue and liver and, to a lesser extent, in skeletal muscle (Takeuchi and Reue 2009). It has been shown that GPAT can be targeted and inhibited by AMPK (Muoio et al. 1999).

2.3.3 Acetyl-CoA Carboxylase and Fatty Acid Oxidation

As mentioned previously, ACC phosphorylation/inhibition by AMPK leads to a decreased production of malonyl-CoA and subsequent inhibition of fatty acid synthesis. On the other hand, malonyl-CoA acts as an inhibitor of carnitine palmitoyl-CoA transferase-1 (CPT-1), a rate-limiting enzyme for mitochondrial fatty acid β-oxidation (Bruce et al. 2009; Nada et al. 1995). CPT-1 is located on the outer mitochondrial membrane and catalyzes the transfer of acyl moiety of LCFA-CoAs to carnitine and facilitates transport of LCFA-CoAs across the inner mitochondrial membrane. Indeed, the mitochondrial inner membrane is not permeable to acyl-CoA molecules. Following the phosphorylation and subsequent inhibition of ACC by AMPK, a fall in malonyl-CoA levels releases the inhibitory effect of malonyl-CoA on CPT-1, which results in the entry of long-chain fatty acyl-CoAs in mitochondria for β-oxidation. This release of inhibition increases the rate of β-oxidation, elevating ATP production (Fig. 2.3).

Two isoforms of ACC coexist in mammals. ACC1 and ACC2 are expressed in adipose tissue, brain, and liver, while ACC2 is predominantly expressed in heart and skeletal muscle (Kreuz et al. 2009). ACC2 has 146 residues longer in the NH_2 terminus and is located on the outer membrane of mitochondria. AMPK activation phosphorylates and inactivates ACC1 on Ser-79 and ACC2 on Ser-221 (Abu-Elheiga et al. 1995). The fate of malonyl-CoA depends on its location and production, either via ACC1 or ACC2. Studies with whole-body or tissue-specific knockout mice for ACC1 and ACC2, or using other genetic approaches, showed that malonyl-CoA produced by ACC2 is involved in the regulation of fatty acid

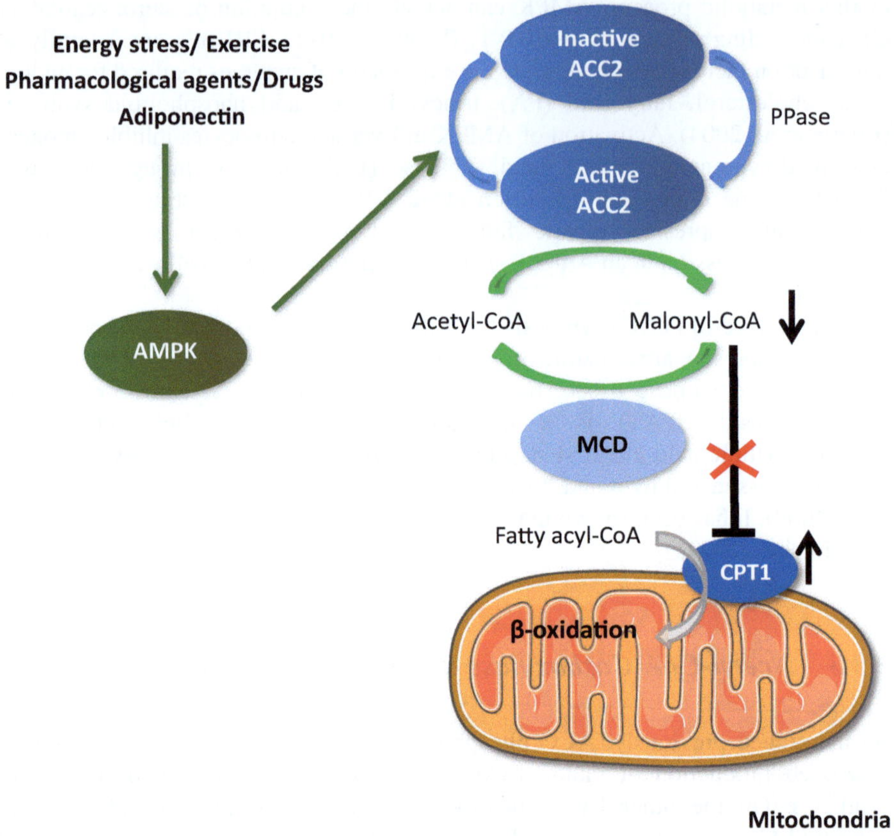

Fig. 2.3 Activation of AMPK increases fatty acid oxidation. Energy stress and pharmacological activation of AMPK increase tissue-specific ACC2 phosphorylation, which decreases malonyl-CoA production, removing its inhibitory action on CPT-1 and promoting fatty acid efflux into mitochondria for β-oxidation

oxidation, whereas malonyl-CoA produced by ACC1 is directed to fatty acid synthesis (Hardie and Pan 2002). These findings suggest that the activation of AMPK via exercise or other cellular stresses promotes fatty acid oxidation by phosphorylation and inactivation of ACC2, whereas inhibition of fatty acid synthesis is accomplished by inhibition of ACC1.

Accumulation of fatty acids outside of adipose tissues is one of the leading causes of insulin resistance and associated metabolic diseases. ACC phosphorylation by AMPK promotes lipid oxidation and thereby reduces lipid stores in muscle and liver. This action of AMPK can be counted as insulin-sensitizing effect contributing to whole-body insulin sensitivity (see next paragraph) (Awazawa et al. 2009; Yamauchi et al. 2002).

2.4 AMPK, Insulin Sensitivity, and Whole-Body Energy Balance

AMPK contributes to whole-body insulin sensitivity and energy balance either via acute phosphorylation and thereby regulation of metabolic enzymes activity or modulation of their transcriptional expression in the long term, in order to adapt gene expression to energy demands. There is a direct link between AMPK and its insulin-sensitizing effect on insulin signaling. Prolonged insulin stimulation is negatively regulated by activation of a negative feedback loop, which phosphorylates IRS-1 on Ser-636/639 and inhibits insulin signaling. By blocking this negative feedback loop, AMPK activation results in a stimulation of insulin signaling in insulin-resistant cardiomyocytes (Ginion et al. 2011). AMPK activation was also shown to increase the phosphorylation of insulin receptor substrate-1 (IRS-1) at Ser-789, which is associated with increased PI3K activity (Jakobsen et al. 2001). Furthermore, AMPK phosphorylation of IR residue Tyr-1162 is also associated with increased insulin sensitivity (Chopra et al. 2012). Finally, AMPK activation increases insulin sensitivity in muscle with a putative role of AS160 in this phenomenon (Kjobsted et al. 2015).

AMPK is also involved in the regulation of hypothalamic control of food intake and energy expenditure (Morton et al. 2006). Hypothalamus controls food intake and energy balance by modifying the synthesis of neuropeptides in response to changes in peripheral signals such as glucose levels and insulin or adipokines (Lage et al. 2008). When energy intake exceeds the energy expenditure, the expression of orexigenic (feeding promoting) neuropeptides such as AgRP and NPY decreases in contrast to the expression of anorexigenic (feeding inhibiting) neuropeptides, such as CART and POMC (Lage et al. 2008). Activity of AMPK in hypothalamus is controlled by hormonal (insulin or adipokines) and nutrient-derived anorexigenic and orexigenic signals. It has been shown that the expression of a dominant-negative form of AMPK in the hypothalamus is sufficient to reduce food intake and body weight, whereas constitutively active AMPK increases both (Minokoshi et al. 2004). Both glucose and insulin are anorexigenic signals able to inhibit AMPK activity in the hypothalamus (Minokoshi et al. 2004). Adipokines like leptin and ghrelin play an important role in hypothalamic control. Leptin regulates appetite, food intake, energy expenditure, and neuroendocrine function (Watanobe 2002). Anorexigenic effects of leptin exerted in the hypothalamus–sympathetic nervous system axis require inhibition of hypothalamic AMPK activity and thereby inhibit food intake and regulate body weight (Minokoshi et al. 2004). Similarly, orexigenic effect of ghrelin is mediated by the activation of hypothalamic AMPK (Kola et al. 2008). In summary, AMPK activation in hypothalamus in response to low leptin and high ghrelin levels leads to increased appetite.

Adiponectin is an antidiabetic adipokine. It exerts its antidiabetic properties through activation of AMPK, thereby increasing glucose utilization and fatty acid oxidation in muscle while decreasing gluconeogenesis in the liver and decreasing glucose levels in vivo (Wu et al. 2003; Yamauchi et al. 2002). Mechanistically,

AMPK activation by adiponectin is suggested to occur through the increased AMP levels sourced by the activity of FAS. FAS consumes a lot of ATP and subsequently produces AMP. FAS is suggested to act downstream of adiponectin receptors and therefore adiponectin increases the formation of fatty acyl-CoA derivatives and activates AMPK via the formation of AMP by this process in adipocytes (Liu et al. 2010).

2.5 AMPK and Mitochondrial Biogenesis

AMPK activation is associated with increased mitochondrial biogenesis, namely, growth and division of mitochondria. Factors activating AMPK such as exercise, caloric restriction, and oxidative stress have also an impact on mitochondrial biogenesis. Mitochondrial biogenesis is shown to fail in skeletal muscles of mice expressing a dominant-negative form of AMPK (Zong et al. 2002). AMPK-mediated mitochondrial gene expression involves activation of the peroxisome proliferator-activated receptors (PPARs) and PGC-1α (Jager et al. 2007; Suwa et al. 2003; Terada et al. 2002; Lee et al. 2006). PPAR-α activation is responsible for the upregulation of mitochondrial β-oxidation enzymes. PGC-1α is associated with increases in energy expenditure such as mitochondrial respiration and biogenesis as well as uptake of energy substrates (Wu et al. 1999; Lehman et al. 2000). PGC-1α exerts these effects by binding and co-activating other transcriptional factor and/or nuclear receptors to increase their expression such as ERRα, NRF1 and NRF2, and MEF2 and PPARα (Handschin and Spiegelman 2006; Vega et al. 2000). AMPK can activate PGC-1α by phosphorylation or increase its expression (Jager et al. 2007). AMPK also activates PGC-1α through its deacetylation via SIRT1, and PGC-1α deacetylation is associated with the induction of its target genes such as CPT1, pyruvate dehydrogenase kinase 4 (PDK4), or GLUT4 (Canto et al. 2009).

2.6 AMPK in Protein Turnover

Inasmuch as AMPK is a sensor promoting a global energy-saving program, it was rather logical to hypothesize that its activation should shutdown protein synthesis, a known energy-consuming pathway. This paradigm has been concomitantly established in hepatocytes (Krause et al. 2002; Horman et al. 2002; Dubbelhuis and Meijer 2002) and skeletal muscle (Bolster et al. 2002). AMPK activation leads to the inhibition of two important pathways regulating the initiation and the elongation steps of protein synthesis (Fig. 2.4). The first involves the inhibition of the mammalian target of rapamycin complex 1 (mTORC1), which is responsible for the stimulation of the p70 ribosomal S6 protein kinase (p70S6K) and of the eukaryotic initiation factor 4E (eIF4E) (Krause et al. 2002; Bolster et al. 2002).

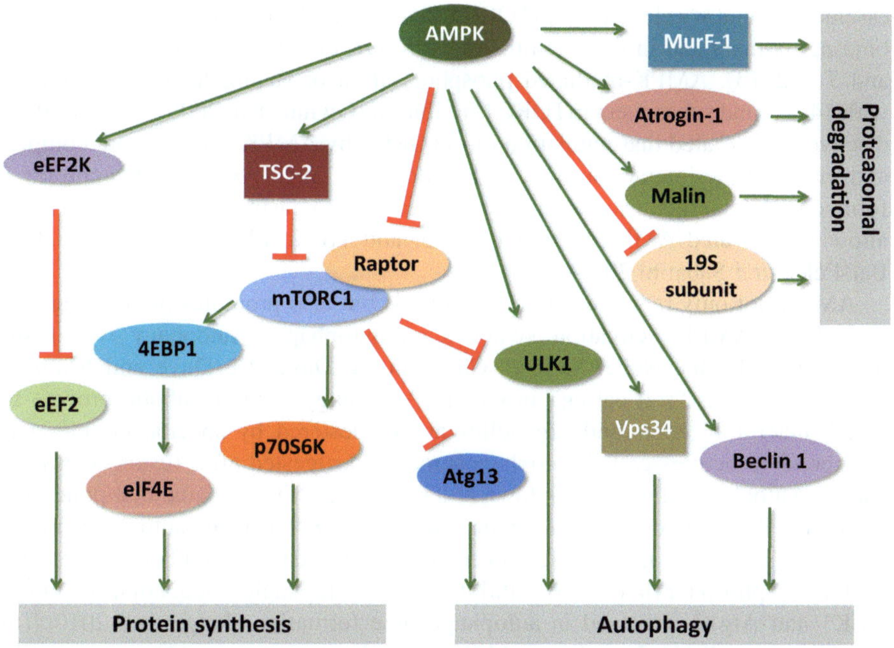

Fig. 2.4 AMPK involvement in protein turnover. This scheme represents the main downstream targets of AMPK involved in the regulation of protein synthesis, autophagy and ubiquitin-mediated proteasomal degradation (see text for more details)

Once activated by mTORC1, p70S6K phosphorylates the S6 ribosomal protein involved in the translation of 5′TOP mRNAs encoding for ribosomal proteins and translation factors. On the other side, phosphorylation of eIF4E binding protein 1 (4EBP1) by mTORC1 releases its binding partner eIF4E, which is then able to bind the mRNA cap and initiate protein synthesis. In summary, inhibition of mTORC1 by AMPK concomitantly reduces translation factor and ribosomal content as well as decreased initiation of translation. The second pathway targeted by AMPK concerns the inhibition of the eukaryotic elongation factor 2 (eEF2) implicated in the control of peptide chain elongation (Horman et al. 2002). It was rapidly established that the regulation of mTORC1 and eEF2 pathways by AMPK is indirect. Indeed, eEF2 inactivation by AMPK is mediated by the regulation of its upstream kinase called eEF2 kinase (eEF2K). AMPK phosphorylates and activates eEF2K, which, then, can phosphorylate eEF2 on an inhibitory site (Browne et al. 2004). In parallel, AMPK inhibits mTORC1 by acting on two of its partners, Raptor (Gwinn et al. 2008) and tuberous sclerosis complex 2 (TSC2) (Inoki et al. 2003). TSC2 is the GTPase-activating protein (GAP) of the small G protein Rheb involved in the activation of mTORC1. AMPK phosphorylates and activates TSC2 promoting the formation of the GDP-bound inactive state of Rheb responsible for mTORC1 inactivation (Inoki et al. 2003). Raptor makes part of the heterotrimeric central core of the mTORC1 with mTOR and mLST8. Raptor is an essential regulatory element of mTORC1 involved in the interaction with accessory binding partners and in the recruitment of mTOR downstream substrates (Dunlop and Tee 2013). AMPK-mediated phosphorylation of Raptor leads to its binding with 14-3-3 and subsequent mTORC1 inhibition (Gwinn et al. 2008). Recent salient discoveries revealed that regulation of mTORC1 by AMPK is even more complex and involved the interaction of AMPK with Axin. Such Axin/AMPK complex is formed under energetic stress and is recruited to cytoplasmic surface of lysosomes allowing activated AMPK to target Raptor and TSC2 (Zhang et al. 2013, 2014; Bar-Peled and Sabatini 2014).

AMPK not only regulates protein synthesis but also promotes protein degradation. First, AMPK activation switches on autophagy. Autophagy was firstly documented by the Nobel laureate Christian De Duve (De Duve and Wattiaux 1966) and means "self-eating" in Greek. Autophagy is the catabolic process by which macromolecules and organelles are sequestered by specific double-layer membrane structures called autophagosomes and digested after fusion with lysosomes. Autophagy allows breakdown and recycling of intracellular components to promote cell survival under energetic stress. AMPK enhances autophagy via its inhibitory action on mTORC1. Indeed, under high-energy conditions, activated mTORC1 phosphorylates and inhibits early autophagic mediators, including ULK1 and Atg13, involved in autophagosome formation (Jung et al. 2010; Tan and Miyamoto 2016). In addition, AMPK directly promotes autophagic process by phosphorylating and activating ULK1 (Kim et al. 2011; Egan et al. 2011). More recently, it has been shown that AMPK phosphorylates two other autophagic promoters, the class III phosphatidylinositol-3 kinase called Vps34 and its partner Beclin 1 (Kim et al. 2013).

Besides autophagy, AMPK also regulates protein turnover by acting on the ubiquitin-proteasome system (UPS) (Ronnebaum et al. 2014). Protein degradation by UPS is an energy-consuming mechanism, which requires approximately 150 molecules of ATP per protein molecule (Peth et al. 2013). It is then reasonable to imagine that AMPK should block such ATP-consuming process. In line with this hypothesis, it has been shown that AMPK activation inhibits proteasome activity in endothelial cells (Wang et al. 2009). The molecular mechanism proposed to explain this inhibition includes the AMPK-dependent regulation (via O-GlcNAcylation process) of the 19S subunit of the proteasome (Xu et al. 2012). AMPK has been also found to interact with PSMD11, a component of the 19S subunit (Moreno et al. 2009). Even if UPS is energy consuming, the specific and UPS-dependent degradation of particular proteins could be energetically advantageous in particular circumstances. In this context, AMPK activation has been found to promote UPS process. AMPK activation results in the increase in expression of the E3 ubiquitin ligases called MuRF1 and Atrogin-1 in muscle and cardiac tissues via the regulation of different transcription factors (FoxO and MEF2) (Krawiec et al. 2007; Tong et al. 2009; Baskin and Taegtmeyer 2011). MurF1 and Atrogin-1 mediates muscle atrophy via myofibrillar protein ubiquitination. Such atrophic program is proposed to participate in the anti-hypertrophic action of AMPK. In a similar way, inhibition of glycogen synthesis by AMPK implicates a complex called Laforin/Malin where

Malin is an E3 ubiquitin ligase, which, when activated, degrades enzymes involved in glycogen formation (Solaz-Fuster et al. 2008). Other similar AMPK-dependent activation of E3 ubiquitin ligase can be found in the literature and occur in the regulation of plasma membrane transporters and channels (Bhalla et al. 2006; Alzamora et al. 2010).

References

Abu-Elheiga L, Jayakumar A, Baldini A, Chirala SS, Wakil SJ (1995) Human acetyl-CoA carboxylase: characterization, molecular cloning, and evidence for two isoforms. Proc Natl Acad Sci USA 92(9):4011–4015, Pubmed Central PMCID: 42092

Alzamora R, Gong F, Rondanino C, Lee JK, Smolak C, Pastor-Soler NM et al (2010) AMP-activated protein kinase inhibits KCNQ1 channels through regulation of the ubiquitin ligase Nedd4-2 in renal epithelial cells. Am J Physiol Renal Physiol 299(6):F1308–F1319, Pubmed Central PMCID: 3006313

An H, He L (2016) Current understanding of metformin effect on the control of hyperglycemia in diabetes. J Endocrinol 228(3):R97–R106, Epub 2016/01/09. eng

An D, Kewalramani G, Qi D, Pulinilkunnil T, Ghosh S, Abrahani A et al (2005) beta-Agonist stimulation produces changes in cardiac AMPK and coronary lumen LPL only during increased workload. Am J Physiol Endocrinol Metab 288(6):E1120–E1127, Epub 2005/02/03. eng

Awazawa M, Ueki K, Inabe K, Yamauchi T, Kaneko K, Okazaki Y et al (2009) Adiponectin suppresses hepatic SREBP1c expression in an AdipoR1/LKB1/AMPK dependent pathway. Biochem Biophys Res Commun 382(1):51–56, Epub 2009/03/04. eng

Bando H, Atsumi T, Nishio T, Niwa H, Mishima S, Shimizu C et al (2005) Phosphorylation of the 6-phosphofructo-2-kinase/fructose 2,6-bisphosphatase/PFKFB3 family of glycolytic regulators in human cancer. Clin Cancer Res 11(16):5784–5792, Epub 2005/08/24. eng

Bar-Peled L, Sabatini DM (2014) Regulation of mTORC1 by amino acids. Trends Cell Biol 24 (7):400–406, Pubmed Central PMCID: 4074565

Baskin KK, Taegtmeyer H (2011) AMP-activated protein kinase regulates E3 ligases in rodent heart. Circ Res 109(10):1153–1161, Pubmed Central PMCID: 3254015

Beauloye C, Marsin AS, Bertrand L, Vanoverschelde JL, Rider MH, Hue L (2002) The stimulation of heart glycolysis by increased workload does not require AMP-activated protein kinase but a wortmannin-sensitive mechanism. FEBS Lett 531(2):324–328

Beg ZH, Allmann DW, Gibson DM (1973) Modulation of 3-hydroxy-3-methylglutaryl coenzyme A reductase activity with cAMP and with protein fractions of rat liver cytosol. Biochem Biophys Res Commun 54(4):1362–1369

Beg ZH, Stonik JA, Brewer HB Jr (1979) Characterization and regulation of reductase kinase, a protein kinase that modulates the enzymic activity of 3-hydroxy-3-methylglutaryl-coenzyme A reductase. Proc Natl Acad Sci USA 76(9):4375–4379, Pubmed Central PMCID: 411577

Bergeron R, Russell RR 3rd, Young LH, Ren JM, Marcucci M, Lee A et al (1999) Effect of AMPK activation on muscle glucose metabolism in conscious rats. Am J Physiol 276(5 Pt 1):E938–E944

Bhalla V, Oyster NM, Fitch AC, Wijngaarden MA, Neumann D, Schlattner U et al (2006) AMP-activated kinase inhibits the epithelial Na+ channel through functional regulation of the ubiquitin ligase Nedd4-2. J Biol Chem 281(36):26159–26169

Bickerton AS, Roberts R, Fielding BA, Hodson L, Blaak EE, Wagenmakers AJ et al (2007) Preferential uptake of dietary fatty acids in adipose tissue and muscle in the postprandial period. Diabetes 56(1):168–176, Epub 2006/12/29. eng

Bolster DR, Crozier SJ, Kimball SR, Jefferson LS (2002) AMP-activated protein kinase suppresses protein synthesis in rat skeletal muscle through down-regulated mammalian target of rapamycin (mTOR) signaling. J Biol Chem 277(27):23977–23980

Browne GJ, Finn SG, Proud CG (2004) Stimulation of the AMP-activated protein kinase leads to activation of eukaryotic elongation factor 2 kinase and to its phosphorylation at a novel site, serine 398. J Biol Chem 279(13):12220–12231

Bruce CR, Hoy AJ, Turner N, Watt MJ, Allen TL, Carpenter K et al (2009) Overexpression of carnitine palmitoyltransferase-1 in skeletal muscle is sufficient to enhance fatty acid oxidation and improve high-fat diet-induced insulin resistance. Diabetes 58(3):550–558, Pubmed Central PMCID: 2646053, Epub 2008/12/17. eng

Bruss MD, Arias EB, Lienhard GE, Cartee GD (2005) Increased phosphorylation of Akt substrate of 160 kDa (AS160) in rat skeletal muscle in response to insulin or contractile activity. Diabetes 54(1):41–50

Canto C, Gerhart-Hines Z, Feige JN, Lagouge M, Noriega L, Milne JC et al (2009) AMPK regulates energy expenditure by modulating NAD+ metabolism and SIRT1 activity. Nature 458(7241):1056–1060, Pubmed Central PMCID: 3616311

Carling D, Zammit VA, Hardie DG (1987) A common bicyclic protein kinase cascade inactivates the regulatory enzymes of fatty acid and cholesterol biosynthesis. FEBS Lett 223(2):217–222

Carling D, Clarke PR, Zammit VA, Hardie DG (1989) Purification and characterization of the AMP-activated protein kinase. Copurification of acetyl-CoA carboxylase kinase and 3-hydroxy-3-methylglutaryl-CoA reductase kinase activities. Eur J Biochem/FEBS 186 (1-2):129–136

Carlson CA, Kim KH (1973) Regulation of hepatic acetyl coenzyme A carboxylase by phosphorylation and dephosphorylation. J Biol Chem 248(1):378–380

Chen S, Murphy J, Toth R, Campbell DG, Morrice NA, Mackintosh C (2008) Complementary regulation of TBC1D1 and AS160 by growth factors, insulin and AMPK activators. Biochem J 409(2):449–459, Epub 2007/11/13. eng

Chopra I, Li HF, Wang H, Webster KA (2012) Phosphorylation of the insulin receptor by AMP-activated protein kinase (AMPK) promotes ligand-independent activation of the insulin signalling pathway in rodent muscle. Diabetologia 55(3):783–794, Pubmed Central PMCID: 4648248

Coburn CT, Knapp FF Jr, Febbraio M, Beets AL, Silverstein RL, Abumrad NA (2000) Defective uptake and utilization of long chain fatty acids in muscle and adipose tissues of CD36 knockout mice. J Biol Chem 275(42):32523–32529, Epub 2000/07/27. eng

Coven DL, Hu X, Cong L, Bergeron R, Shulman GI, Hardie DG et al (2003) Physiological role of AMP-activated protein kinase in the heart: graded activation during exercise. Am J Physiol Endocrinol Metab 285(3):E629–E636, Epub 2003/05/22. eng

De Duve C, Wattiaux R (1966) Functions of lysosomes. Annu Rev Physiol 28:435–492

Destefano MA, Jacinto E (2013) Regulation of insulin receptor substrate-1 by mTORC2 (mammalian target of rapamycin complex 2). Biochem Soc Trans 41(4):896–901, Epub 2013/07/19. eng

Dubbelhuis PF, Meijer AJ (2002) Hepatic amino acid-dependent signaling is under the control of AMP-dependent protein kinase. FEBS Lett 521(1-3):39–42

Dunlop EA, Tee AR (2013) The kinase triad, AMPK, mTORC1 and ULK1, maintains energy and nutrient homoeostasis. Biochem Soc Trans 41(4):939–943

Eberle D, Hegarty B, Bossard P, Ferre P, Foufelle F (2004) SREBP transcription factors: master regulators of lipid homeostasis. Biochimie 86(11):839–848

Egan DF, Shackelford DB, Mihaylova MM, Gelino S, Kohnz RA, Mair W et al (2011) Phosphorylation of ULK1 (hATG1) by AMP-activated protein kinase connects energy sensing to mitophagy. Science 331(6016):456–461, Pubmed Central PMCID: 3030664

Ferrer A, Caelles C, Massot N, Hegardt FG (1985) Activation of rat liver cytosolic 3-hydroxy-3-methylglutaryl coenzyme A reductase kinase by adenosine 5′-monophosphate. Biochem Biophys Res Commun 132(2):497–504

Foretz M, Hebrard S, Leclerc J, Zarrinpashneh E, Soty M, Mithieux G et al (2010) Metformin inhibits hepatic gluconeogenesis in mice independently of the LKB1/AMPK pathway via a decrease in hepatic energy state. J Clin Invest 120(7):2355–2369, Pubmed Central PMCID: 2898585

Geraghty KM, Chen S, Harthill JE, Ibrahim AF, Toth R, Morrice NA et al (2007) Regulation of multisite phosphorylation and 14-3-3 binding of AS160 in response to IGF-1, EGF, PMA and AICAR. Biochem J 407(2):231–241, Pubmed Central PMCID: 2049023

Ginion A, Auquier J, Benton CR, Mouton C, Vanoverschelde JL, Hue L et al (2011) Inhibition of the mTOR/p70S6K pathway is not involved in the insulin-sensitizing effect of AMPK on cardiac glucose uptake. Am J Physiol Heart Circ Physiol 301(2):H469–H477

Glatz JF, Luiken JJ, Bonen A (2010) Membrane fatty acid transporters as regulators of lipid metabolism: implications for metabolic disease. Physiol Rev 90(1):367–417

Guo L, Zheng X, Liu J, Yin Z (2016) Geniposide suppresses hepatic glucose production via AMPK in HepG2 cells. Biol Pharm Bull 39:484–491, Epub 2016/02/03. Eng

Gwinn DM, Shackelford DB, Egan DF, Mihaylova MM, Mery A, Vasquez DS et al (2008) AMPK phosphorylation of raptor mediates a metabolic checkpoint. Mol Cell 30(2):214–226, Pubmed Central PMCID: 2674027

Habets DD, Coumans WA, Voshol PJ, den Boer MA, Febbraio M, Bonen A et al (2007) AMPK-mediated increase in myocardial long-chain fatty acid uptake critically depends on sarcolemmal CD36. Biochem Biophys Res Commun 355(1):204–210

Habets DD, Coumans WA, El Hasnaoui M, Zarrinpashneh E, Bertrand L, Viollet B et al (2009) Crucial role for LKB1 to AMPKalpha2 axis in the regulation of CD36-mediated long-chain fatty acid uptake into cardiomyocytes. Biochim Biophys Acta 1791(3):212–219

Halse R, Fryer LG, McCormack JG, Carling D, Yeaman SJ (2003) Regulation of glycogen synthase by glucose and glycogen: a possible role for AMP-activated protein kinase. Diabetes 52(1):9–15, Epub 2002/12/28. eng

Handschin C, Spiegelman BM (2006) Peroxisome proliferator-activated receptor gamma coactivator 1 coactivators, energy homeostasis, and metabolism. Endocr Rev 27(7):728–735

Hardie DG, Pan DA (2002) Regulation of fatty acid synthesis and oxidation by the AMP-activated protein kinase. Biochem Soc Trans 30(Pt 6):1064–1070, Epub 2002/11/21. eng

Hardie DG, Hawley SA, Scott JW (2006) AMP-activated protein kinase – development of the energy sensor concept. J Physiol 574(Pt 1):7–15, Pubmed Central PMCID: 1817788

Hayashi T, Hirshman MF, Kurth EJ, Winder WW, Goodyear LJ (1998) Evidence for 5′ AMP-activated protein kinase mediation of the effect of muscle contraction on glucose transport. Diabetes 47(8):1369–1373

Holmes B, Dohm GL (2004) Regulation of GLUT4 gene expression during exercise. Med Sci Sports Exerc 36(7):1202–1206, Epub 2004/07/06. eng

Holmes BF, Sparling DP, Olson AL, Winder WW, Dohm GL (2005) Regulation of muscle GLUT4 enhancer factor and myocyte enhancer factor 2 by AMP-activated protein kinase. Am J Physiol Endocrinol Metab 289(6):E1071–E1076, Epub 2005/08/18. eng

Horman S, Browne G, Krause U, Patel J, Vertommen D, Bertrand L et al (2002) Activation of AMP-activated protein kinase leads to the phosphorylation of elongation factor 2 and an inhibition of protein synthesis. Curr Biol 12(16):1419–1423

Ingebritsen TS, Parker RA, Gibson DM (1981) Regulation of liver hydroxymethylglutaryl-CoA reductase by a bicyclic phosphorylation system. J Biol Chem 256(3):1138–1144

Inoki K, Zhu T, Guan KL (2003) TSC2 mediates cellular energy response to control cell growth and survival. Cell 115(5):577–590

Jager S, Handschin C, St-Pierre J, Spiegelman BM (2007) AMP-activated protein kinase (AMPK) action in skeletal muscle via direct phosphorylation of PGC-1alpha. Proc Natl Acad Sci USA 104(29):12017–12022, Pubmed Central PMCID: 1924552

Jakobsen SN, Hardie DG, Morrice N, Tornqvist HE (2001) 5′-AMP-activated protein kinase phosphorylates IRS-1 on Ser-789 in mouse C2C12 myotubes in response to 5-aminoimidazole-4-carboxamide riboside. J Biol Chem 276(50):46912–46916

Jorgensen SB, Viollet B, Andreelli F, Frosig C, Birk JB, Schjerling P et al (2004a) Knockout of the alpha2 but not alpha1 5′-AMP-activated protein kinase isoform abolishes 5-aminoimidazole-4-carboxamide-1-beta-4-ribofuranosidebut not contraction-induced glucose uptake in skeletal muscle. J Biol Chem 279(2):1070–1079

Jorgensen SB, Nielsen JN, Birk JB, Olsen GS, Viollet B, Andreelli F et al (2004b) The alpha2-5′AMP-activated protein kinase is a site 2 glycogen synthase kinase in skeletal muscle and is responsive to glucose loading. Diabetes 53(12):3074–3081

Jung CH, Ro SH, Cao J, Otto NM, Kim DH (2010) mTOR regulation of autophagy. FEBS Lett 584 (7):1287–1295, Pubmed Central PMCID: 2846630

Kane S, Sano H, Liu SC, Asara JM, Lane WS, Garner CC et al (2002) A method to identify serine kinase substrates. Akt phosphorylates a novel adipocyte protein with a Rab GTPase-activating protein (GAP) domain. J Biol Chem 277(25):22115–22118

Kim J, Kundu M, Viollet B, Guan KL (2011) AMPK and mTOR regulate autophagy through direct phosphorylation of Ulk1. Nat Cell Biol 13(2):132–141, Pubmed Central PMCID: 3987946

Kim J, Kim YC, Fang C, Russell RC, Kim JH, Fan W et al (2013) Differential regulation of distinct Vps34 complexes by AMPK in nutrient stress and autophagy. Cell 152(1-2):290–303, Pubmed Central PMCID: 3587159

Kjobsted R, Treebak JT, Fentz J, Lantier L, Viollet B, Birk JB et al (2015) Prior AICAR stimulation increases insulin sensitivity in mouse skeletal muscle in an AMPK-dependent manner. Diabetes 64(6):2042–2055

Koay A, Woodcroft B, Petrie EJ, Yue H, Emanuelle S, Bieri M et al (2010) AMPK beta subunits display isoform specific affinities for carbohydrates. FEBS Lett 584(15):3499–3503

Kola B, Farkas I, Christ-Crain M, Wittmann G, Lolli F, Amin F et al (2008) The orexigenic effect of ghrelin is mediated through central activation of the endogenous cannabinoid system. PLoS One 3(3):e1797, Pubmed Central PMCID: 2258435, Epub 2008/03/13. eng

Koonen DP, Glatz JF, Bonen A, Luiken JJ (2005) Long-chain fatty acid uptake and FAT/CD36 translocation in heart and skeletal muscle. Biochim Biophys Acta 1736(3):163–180, Epub 2005/10/04. eng

Kramer HF, Witczak CA, Fujii N, Jessen N, Taylor EB, Arnolds DE et al (2006) Distinct signals regulate AS160 phosphorylation in response to insulin, AICAR, and contraction in mouse skeletal muscle. Diabetes 55(7):2067–2076, Epub 2006/06/29. eng

Krause U, Bertrand L, Hue L (2002) Control of p70 ribosomal protein S6 kinase and acetyl-CoA carboxylase by AMP-activated protein kinase and protein phosphatases in isolated hepatocytes. Eur J Biochem/FEBS 269(15):3751–3759

Krawiec BJ, Nystrom GJ, Frost RA, Jefferson LS, Lang CH (2007) AMP-activated protein kinase agonists increase mRNA content of the muscle-specific ubiquitin ligases MAFbx and MuRF1 in C2C12 cells. Am J Physiol Endocrinol Metab 292(6):E1555–E1567

Kreuz S, Schoelch C, Thomas L, Rist W, Rippmann JF, Neubauer H (2009) Acetyl-CoA carboxylases 1 and 2 show distinct expression patterns in rats and humans and alterations in obesity and diabetes. Diabetes Metab Res Rev 25(6):577–586

Lage R, Dieguez C, Vidal-Puig A, Lopez M (2008) AMPK: a metabolic gauge regulating whole-body energy homeostasis. Trends Mol Med 14(12):539–549

Lantier L, Fentz J, Mounier R, Leclerc J, Treebak JT, Pehmoller C et al (2014) AMPK controls exercise endurance, mitochondrial oxidative capacity, and skeletal muscle integrity. FASEB J 28(7):3211–3224

Lee WJ, Kim M, Park HS, Kim HS, Jeon MJ, Oh KS et al (2006) AMPK activation increases fatty acid oxidation in skeletal muscle by activating PPARalpha and PGC-1. Biochem Biophys Res Commun 340(1):291–295

Lehman JJ, Barger PM, Kovacs A, Saffitz JE, Medeiros DM, Kelly DP (2000) Peroxisome proliferator-activated receptor gamma coactivator-1 promotes cardiac mitochondrial biogenesis. J Clin Invest 106(7):847–856, Pubmed Central PMCID: 517815

Li Y, Xu S, Mihaylova MM, Zheng B, Hou X, Jiang B et al (2011) AMPK phosphorylates and inhibits SREBP activity to attenuate hepatic steatosis and atherosclerosis in diet-induced insulin-resistant mice. Cell Metab 13(4):376–388, Pubmed Central PMCID: 3086578

Liu Q, Gauthier MS, Sun L, Ruderman N, Lodish H (2010) Activation of AMP-activated protein kinase signaling pathway by adiponectin and insulin in mouse adipocytes: requirement of acyl-CoA synthetases FATP1 and Acsl1 and association with an elevation in AMP/ATP ratio. FASEB J 24(11):4229–4239, Pubmed Central PMCID: 2974418

Luiken JJ, Koonen DP, Willems J, Zorzano A, Becker C, Fischer Y et al (2002) Insulin stimulates long-chain fatty acid utilization by rat cardiac myocytes through cellular redistribution of FAT/CD36. Diabetes 51(10):3113–3119, Epub 2002/09/28. eng

Luiken JJ, Koonen DP, Coumans WA, Pelsers MM, Binas B, Bonen A et al (2003) Long-chain fatty acid uptake by skeletal muscle is impaired in homozygous, but not heterozygous, heart-type-FABP null mice. Lipids 38(4):491–496, Epub 2003/07/10. eng

Madsen A, Bozickovic O, Bjune JI, Mellgren G, Sagen JV (2015) Metformin inhibits hepatocellular glucose, lipid and cholesterol biosynthetic pathways by transcriptionally suppressing steroid receptor coactivator 2 (SRC-2). Sci Rep 5:16430, Pubmed Central PMCID: 4637908

Manning BD, Cantley LC (2007) AKT/PKB signaling: navigating downstream. Cell 129 (7):1261–1274, Pubmed Central PMCID: 2756685, Epub 2007/07/03. eng

Marsin AS, Bertrand L, Rider MH, Deprez J, Beauloye C, Vincent MF et al (2000) Phosphorylation and activation of heart PFK-2 by AMPK has a role in the stimulation of glycolysis during ischaemia. Curr Biol 10(20):1247–1255

Marsin AS, Bouzin C, Bertrand L, Hue L (2002) The stimulation of glycolysis by hypoxia in activated monocytes is mediated by AMP-activated protein kinase and inducible 6-phosphofructo-2-kinase. J Biol Chem 277(34):30778–30783, Epub 2002/06/18. eng

McGee SL, Hargreaves M (2004) Exercise and myocyte enhancer factor 2 regulation in human skeletal muscle. Diabetes 53(5):1208–1214

McGee SL, van Denderen BJ, Howlett KF, Mollica J, Schertzer JD, Kemp BE et al (2008) AMP-activated protein kinase regulates GLUT4 transcription by phosphorylating histone deacetylase 5. Diabetes 57(4):860–867

Middelbeek RJ, Chambers MA, Tantiwong P, Treebak JT, An D, Hirshman MF et al (2013) Insulin stimulation regulates AS160 and TBC1D1 phosphorylation sites in human skeletal muscle. Nutr Diabetes 3:e74, Pubmed Central PMCID: 3697402, Epub 2013/06/12. eng

Minokoshi Y, Alquier T, Furukawa N, Kim YB, Lee A, Xue B et al (2004) AMP-kinase regulates food intake by responding to hormonal and nutrient signals in the hypothalamus. Nature 428 (6982):569–574

Moreno D, Viana R, Sanz P (2009) Two-hybrid analysis identifies PSMD11, a non-ATPase subunit of the proteasome, as a novel interaction partner of AMP-activated protein kinase. Int J Biochem Cell Biol 41(12):2431–2439

Morton GJ, Cummings DE, Baskin DG, Barsh GS, Schwartz MW (2006) Central nervous system control of food intake and body weight. Nature 443(7109):289–295

Munday MR, Campbell DG, Carling D, Hardie DG (1988) Identification by amino acid sequencing of three major regulatory phosphorylation sites on rat acetyl-CoA carboxylase. Eur J Biochem/FEBS 175(2):331–338

Muoio DM, Seefeld K, Witters LA, Coleman RA (1999) AMP-activated kinase reciprocally regulates triacylglycerol synthesis and fatty acid oxidation in liver and muscle: evidence that sn-glycerol-3-phosphate acyltransferase is a novel target. Biochem J 338(Pt 3):783–791, Pubmed Central PMCID: 1220117

Nada MA, Abdel-Aleem S, Schulz H (1995) On the rate-limiting step in the beta-oxidation of polyunsaturated fatty acids in the heart. Biochim Biophys Acta 1255(3):244–250, Epub 1995/04/06. eng

Oligschlaeger Y, Miglianico M, Chanda D, Scholz R, Thali RF, Tuerk R et al (2015) The recruitment of AMP-activated protein kinase to glycogen is regulated by autophosphorylation. J Biol Chem 290(18):11715–11728, Pubmed Central PMCID: 4416872

O'Neill HM, Maarbjerg SJ, Crane JD, Jeppesen J, Jorgensen SB, Schertzer JD et al (2011) AMP-activated protein kinase (AMPK) beta1beta2 muscle null mice reveal an essential role for AMPK in maintaining mitochondrial content and glucose uptake during exercise. Proc Natl Acad Sci USA 108(38):16092–16097, Pubmed Central PMCID: 3179037

Pehmoller C, Treebak JT, Birk JB, Chen S, Mackintosh C, Hardie DG et al (2009) Genetic disruption of AMPK signaling abolishes both contraction- and insulin-stimulated TBC1D1 phosphorylation and 14-3-3 binding in mouse skeletal muscle. Am J Physiol Endocrinol Metab 297(3):E665–E675, Pubmed Central PMCID: 2739697, Epub 2009/06/18. eng

Peth A, Nathan JA, Goldberg AL (2013) The ATP costs and time required to degrade ubiquitinated proteins by the 26 S proteasome. J Biol Chem 288(40):29215–29222, Pubmed Central PMCID: 3790020

Polekhina G, Gupta A, Michell BJ, van Denderen B, Murthy S, Feil SC et al (2003) AMPK β subunit targets metabolic stress sensing to glycogen. Curr Biol 13(10):867–871

Ronnebaum SM, Patterson C, Schisler JC (2014) Minireview: hey U(PS): metabolic and proteolytic homeostasis linked via AMPK and the ubiquitin proteasome system. Mol Endocrinol 28 (10):1602–1615, Pubmed Central PMCID: 4179629

Rose AJ, Jeppesen J, Kiens B, Richter EA (2009) Effects of contraction on localization of GLUT4 and v-SNARE isoforms in rat skeletal muscle. Am J Physiol Regul Integr Comp Physiol 297 (5):R1228–R1237, Epub 2009/08/14. eng

Rowland AF, Fazakerley DJ, James DE (2011) Mapping insulin/GLUT4 circuitry. Traffic 12 (6):672–681, Epub 2011/03/16. eng

Sadler JB, Bryant NJ, Gould GW (2015) Characterization of VAMP isoforms in 3T3-L1 adipocytes: implications for GLUT4 trafficking. Mol Biol Cell 26(3):530–536, Pubmed Central PMCID: 4310743, Epub 2014/12/17. eng

Sakamoto K, McCarthy A, Smith D, Green KA, Grahame Hardie D, Ashworth A et al (2005) Deficiency of LKB1 in skeletal muscle prevents AMPK activation and glucose uptake during contraction. EMBO J 24(10):1810–1820, Pubmed Central PMCID: 1142598

Samovski D, Su X, Xu Y, Abumrad NA, Stahl PD (2012) Insulin and AMPK regulate FA translocase/CD36 plasma membrane recruitment in cardiomyocytes via Rab GAP AS160 and Rab8a Rab GTPase. J Lipid Res 53(4):709–717, Pubmed Central PMCID: 3307647

Sano H, Kane S, Sano E, Miinea CP, Asara JM, Lane WS et al (2003) Insulin-stimulated phosphorylation of a Rab GTPase-activating protein regulates GLUT4 translocation. J Biol Chem 278(17):14599–14602

Schwenk RW, Dirkx E, Coumans WA, Bonen A, Klip A, Glatz JF et al (2010) Requirement for distinct vesicle-associated membrane proteins in insulin- and AMP-activated protein kinase (AMPK)-induced translocation of GLUT4 and CD36 in cultured cardiomyocytes. Diabetologia 53(10):2209–2219, Pubmed Central PMCID: 2931635

Shaw RJ, Lamia KA, Vasquez D, Koo SH, Bardeesy N, Depinho RA et al (2005) The kinase LKB1 mediates glucose homeostasis in liver and therapeutic effects of metformin. Science 310 (5754):1642–1646, Pubmed Central PMCID: 3074427, Epub 2005/11/26. eng

Solaz-Fuster MC, Gimeno-Alcaniz JV, Ros S, Fernandez-Sanchez ME, Garcia-Fojeda B, Criado Garcia O et al (2008) Regulation of glycogen synthesis by the laforin-malin complex is modulated by the AMP-activated protein kinase pathway. Hum Mol Genet 17(5):667–678

Suwa M, Nakano H, Kumagai S (2003) Effects of chronic AICAR treatment on fiber composition, enzyme activity, UCP3, and PGC-1 in rat muscles. J Appl Physiol 95(3):960–968

Takeuchi K, Reue K (2009) Biochemistry, physiology, and genetics of GPAT, AGPAT, and lipin enzymes in triglyceride synthesis. Am J Physiol Endocrinol Metab 296(6):E1195–E1209, Pubmed Central PMCID: 2692402

Tan VP, Miyamoto S (2016) Nutrient-sensing mTORC1: integration of metabolic and autophagic signals. J Mol Cell Cardiol 95:31–41

Terada S, Goto M, Kato M, Kawanaka K, Shimokawa T, Tabata I (2002) Effects of low-intensity prolonged exercise on PGC-1 mRNA expression in rat epitrochlearis muscle. Biochem Biophys Res Commun 296(2):350–354

Tong JF, Yan X, Zhu MJ, Du M (2009) AMP-activated protein kinase enhances the expression of muscle-specific ubiquitin ligases despite its activation of IGF-1/Akt signaling in C2C12 myotubes. J Cell Biochem 108(2):458–468

Vega RB, Huss JM, Kelly DP (2000) The coactivator PGC-1 cooperates with peroxisome proliferator-activated receptor alpha in transcriptional control of nuclear genes encoding mitochondrial fatty acid oxidation enzymes. Mol Cell Biol 20(5):1868–1876, Pubmed Central PMCID: 85369

Wang S, Xu J, Song P, Viollet B, Zou MH (2009) In vivo activation of AMP-activated protein kinase attenuates diabetes-enhanced degradation of GTP cyclohydrolase I. Diabetes 58 (8):1893–1901, Pubmed Central PMCID: 2712774

Watanobe H (2002) Leptin directly acts within the hypothalamus to stimulate gonadotropin-releasing hormone secretion in vivo in rats. J Physiol 545(Pt 1):255–268, Pubmed Central PMCID: 2290656

Wu Z, Puigserver P, Andersson U, Zhang C, Adelmant G, Mootha V et al (1999) Mechanisms controlling mitochondrial biogenesis and respiration through the thermogenic coactivator PGC-1. Cell 98(1):115–124

Wu X, Motoshima H, Mahadev K, Stalker TJ, Scalia R, Goldstein BJ (2003) Involvement of AMP-activated protein kinase in glucose uptake stimulated by the globular domain of adiponectin in primary rat adipocytes. Diabetes 52(6):1355–1363

Xu J, Wang S, Viollet B, Zou MH (2012) Regulation of the proteasome by AMPK in endothelial cells: the role of O-GlcNAc transferase (OGT). PLoS One 7(5):e36717, Pubmed Central PMCID: 3345026

Yamauchi T, Kamon J, Minokoshi Y, Ito Y, Waki H, Uchida S et al (2002) Adiponectin stimulates glucose utilization and fatty-acid oxidation by activating AMP-activated protein kinase. Nat Med 8(11):1288–1295

Yeh LA, Lee KH, Kim KH (1980) Regulation of rat liver acetyl-CoA carboxylase. Regulation of phosphorylation and inactivation of acetyl-CoA carboxylase by the adenylate energy charge. J Biol Chem 255(6):2308–2314

Zarrinpashneh E, Carjaval K, Beauloye C, Ginion A, Mateo P, Pouleur AC et al (2006) Role of the alpha2-isoform of AMP-activated protein kinase in the metabolic response of the heart to no-flow ischemia. Am J Physiol Heart Circ Physiol 291(6):H2875–H2883, Epub 2006/08/01. eng

Zerial M, McBride H (2001) Rab proteins as membrane organizers. Nat Rev Mol Cell Biol 2 (2):107–117

Zhang YL, Guo H, Zhang CS, Lin SY, Yin Z, Peng Y et al (2013) AMP as a low-energy charge signal autonomously initiates assembly of AXIN-AMPK-LKB1 complex for AMPK activation. Cell Metab 18(4):546–555

Zhang CS, Jiang B, Li M, Zhu M, Peng Y, Zhang YL et al (2014) The lysosomal v-ATPase-Ragulator complex is a common activator for AMPK and mTORC1, acting as a switch between catabolism and anabolism. Cell Metab 20(3):526–540

Zong H, Ren JM, Young LH, Pypaert M, Mu J, Birnbaum MJ et al (2002) AMP kinase is required for mitochondrial biogenesis in skeletal muscle in response to chronic energy deprivation. Proc Natl Acad Sci USA 99(25):15983–15987, Pubmed Central PMCID: 138551

Chapter 3
AMPK Regulation of Cell Growth, Apoptosis, Autophagy, and Bioenergetics

Marina Villanueva-Paz, David Cotán, Juan Garrido-Maraver, Manuel Oropesa-Ávila, Mario de la Mata, Ana Delgado-Pavón, Isabel de Lavera, Elizabet Alcocer-Gómez, Mónica Álvarez-Córdoba, and José A. Sánchez-Alcázar

Contents

The original version of this chapter was revised. An erratum to this chapter can be found at DOI 10.1007/978-3-319-43589-3_21.

M. Villanueva-Paz • D. Cotán • J. Garrido-Maraver • M. Oropesa-Ávila • M. de la Mata • A. Delgado-Pavón • I. de Lavera • E. Alcocer-Gómez • M. Álvarez-Córdoba • J.A. Sánchez-Alcázar (✉)
Centro Andaluz de Biología de Desarrollo (CABD), Universidad Pablo de Olavide-CSIC-Junta de Andalucía, 41013 Sevilla, Spain

Centro de Investigación Biomédica en Red de Enfermedades Raras (CIBERER), ISCIII, Madrid, Spain
e-mail: jasanalc@upo.es

M.D. Cordero, B. Viollet (eds.), *AMP-activated Protein Kinase*, Experientia Supplementum 107, DOI 10.1007/978-3-319-43589-3_3

Abstract In eukaryotic cells, AMP-activated protein kinase (AMPK) generally promotes catabolic pathways that produce ATP and at the same time inhibits anabolic pathways involved in different processes that consume ATP. As an energy sensor, AMPK is involved in the main cellular functions implicated in cell fate, such as cell growth and autophagy.

Recently, AMPK has been connected with apoptosis regulation, although the molecular mechanism by which AMPK induces and/or inhibits cell death is not clear.

This chapter reviews the essential role of AMPK in signaling pathways that respond to cellular stress and damage, highlighting the complex and reciprocal regulation between AMPK and their targets and effectors. The therapeutic implications of the role of AMPK in different pathologies such as diabetes, cancer, or mitochondrial dysfunctions are still controversial, and it is necessary to further investigate the molecular mechanisms underlying AMPK activation.

Keywords AMPK • Cell growth • Autophagy • Apoptosis • Bioenergetics

3.1 Introduction

Every cell uses ATP as the main source of energy; therefore, the ATP levels must be matched by the energy demand needed to carry on different functions. In order to maintain the equilibrium between nutrient supply and energy demand, the eukaryotic cell uses an ATP sensor coupled to different pathways that are initiated when a decrease in ATP levels is detected. AMPK exerts this function in the eukaryotic cells; therefore, it is considered a master regulator of cell metabolism (Viollet and Andreelli 2011).

Activation of AMPK occurs in response to stress circumstances such as low glucose, hypoxia, ischemia, heat shock, increased reactive oxygen species (ROS), and exercise (Luo et al. 2005), although AMPK can be activated in response to several pharmacological agents as well. Under these situations, the intracellular AMP + ADP/ATP ratio is increased, and then AMPK is phosphorylated in its kinase domain to become a catalytically competent protein. By sensing changes in adenine nucleotide ratios, AMPK is activated by different stresses that diminish cellular energy levels and/or increase cellular ATP consumption. Once activated, AMPK is able to stimulate catabolic processes, ATP-generating pathways, and inhibits anabolic processes such as synthesis of lipids, carbohydrates, and proteins in order to assure cell survival.

The involvement of AMPK in many regulatory pathways makes it considered as a master metabolic switch. In fact, in mammals, AMPK has an essential role in coordinating survival, growth, and metabolism under conditions of low energy. Therefore, AMPK regulates in a coordinate way catabolism, cell growth, autophagy, and apoptosis, since all these cell functions cannot be balanced independently.

The AMPK signaling network covers numerous genes involved in survival and cell growth. Among them, mammalian Target Of Rapamycin (mTOR) regulation is the axis of numerous cell processes since it plays an essential role in metabolism regulation, autophagy, cell growth, and apoptosis (Xu et al. 2012). This relation is complex but essential to understand the dialogue and the interplay between autophagy and apoptosis, which has been gaining importance in the last years and is closely related with cell growth and survival.

3.2 AMPK Regulation of Cell Growth

AMPK is situated in the center of an energy-sensing cascade that is activated by ATP depletion. AMPK promotes catabolic pathways that produce ATP, and at the same time AMPK inhibits anabolic pathways involved in cell growth and other processes that consume ATP (Grahame Hardie 2014).

Moreover, AMPK is part of a tumor suppressor network that regulates cell growth and proliferation in stress situations; in fact, it is assumed that AMPK is responsible for the tumor suppressor effects of LKB1 (Liver Kinase B1) (Hardie 2011). Consistent with these functions, AMPK acts as a cell growth inhibitor in different stress situations, such as glucose deprivation, energy depletion, tumor proliferation, or DNA damage.

How does activation of AMPK result in cell growth suppression? Cell growth requires balanced nutrient availability and energy production. Therefore, protein, rRNA, and lipid synthesis are needed, in addition to replication of DNA (Vander Heiden et al. 2009). In this sense, AMPK prevents cell growth by inhibiting protein, rRNA, and lipid synthesis, and it also induces cell cycle arrest in G1, preventing DNA replication (Table 3.1).

3.2.1 AMPK Inhibits Protein and rRNA Synthesis

AMPK is considered an antigrowth molecule, among other reasons, due to its relationship with the tumor suppressor gene Tuberous Sclerosis Complex 2 (TSC2), also known as Tuberin. AMPK directly phosphorylates TSC2 on T1227 and S1345 to enhance its Rheb-GAP activity, which causes an inhibition

Table 3.1 Inhibitory effects of AMPK on cell growth

Pathway	Targets	Consequence
Protein synthesis	TSC2/Raptor → mTORC1	↓ Translation initiation
	eEF2 kinase	↓ Translation elongation
rRNA synthesis	TSC2/Raptor → mTORC1	↓ RNA Pol I-mediated transcription
	TIF-1A	
Regulation of cell cycle	MDMX/p53	Cell cycle arrest
		↑ P53 stabilization and activation
Lipid synthesis	ACC1/FAS/SREBP-1c/HMG-CoA reductase/GPAT	↓ Substrate for fatty acid synthesis ↓ Lipogenesis and lipid accumulation ↓ Cholesterol synthesis ↓ Glycerolipid metabolism

of mammalian Target Of Rapamycin Complex 1 (mTORC1) kinase activity (Inoki et al. 2003). Moreover, AMPK is able to directly inhibit mTORC1 by phosphorylating raptor, the mTORC1-binding partner (Gwinn et al. 2008). The phosphorylation of two conserved serines (Ser722 and Ser792) of raptor by AMPK blocks the ability of the mTORC1 kinase complex to phosphorylate its substrates.

By inhibiting mTORC1, AMPK prevents two major biosynthetic pathways required for cell growth, such as protein and rRNA synthesis. The objective of the relationship between AMPK and mTORC1 is adjusting energy required by anabolic process to energy availability (Alexander and Walker 2011).

mTORC1 regulates two key proteins which are enhancers of protein synthesis: p70S6K (p70S6 Kinase) and 4EBP1 (eIF-4E-binding protein 1) (Zoncu et al. 2011). p70S6K is a positive regulator of translation initiation and elongation. When it is phosphorylated by mTORC1, p70S6K phosphorylates multiple substrates including eukaryotic translation initiation factor 4B (eIF4B) (Holz et al. 2005). In contrast, 4EBP1 inhibits mRNA translation. When phosphorylated by mTORC1, it dissociates from eIF4E, allowing the latest to recruit eukaryotic translation initiation factor 4 gamma (eIF4G) to the 5′ end of most mRNAs (Ma and Blenis 2009).

AMPK inhibits protein synthesis in an mTOR-independent manner by promoting phosphorylation of eukaryotic elongation factor 2 (eEF2). eEF2 phosphorylation regulates the peptide chain elongation (Browne and Proud 2002). Phosphorylation of eEF2, which inhibits its activity, is catalyzed by a specific calcium/calmodulin-dependent protein kinase termed eEF2 kinase. AMPK inactivates eEF2 by directly phosphorylating eEF2 kinase (Browne et al. 2004).

By this way, AMPK and mTOR provide the link between cellular energy status and protein synthesis, a major consumer of metabolic energy.

Related to rRNA synthesis, mTORC1 regulates RNA Pol I-mediated transcription by enhancing the activity and localization of TIF-IA, a transcription factor for RNA polymerase-1 that senses nutrient availability (Mayer et al. 2004).

AMPK also inhibits rRNA synthesis in an mTOR-independent manner: it directly phosphorylates TIF-1A at Ser635, and by this way the assembly of functional transcription initiation complexes becomes impaired (Hoppe et al. 2009).

By the mechanisms described above, AMPK allows the adaptation of protein and rRNA synthesis (and finally cell growth) to changes in cellular energy supply.

3.2.2 AMPK Induces Cell Cycle Arrest

Besides protein and rRNA synthesis, cell growth needs DNA replication. AMPK uses different mechanisms to induce cell cycle arrest. First, AMPK is able to phosphorylate MDMX, which regulates p53 proteasomal turnover since it is a component of the E3 ubiquitin ligase complex (He et al. 2014). AMPK-mediated phosphorylation of MDMX on Ser342 promotes the association between MDMX and 14-3-3. This inhibits p53 ubiquitylation, preventing its turnover by the proteasome. Moreover, inhibition of p53 ubiquitylation induces cell cycle arrest by promoting stabilization and activation of p53.

Second, AMPK directly phosphorylates p53 on Ser15 to stabilize it and initiate AMPK-dependent cell cycle arrest (Jones et al. 2005). In stress conditions, such as glucose deprivation, cell cycle arrest promoted by p53 allows cell to survive. Upon glucose restoration, cells reenter the cell cycle. Interestingly, p53 activation also regulates AMPK. It has been demonstrated that the products of Sestrin1 and Sestrin2, which are two p53 target genes, activate AMPK (Budanov and Karin 2008). Furthermore, p53 inhibits mTORC1 by increasing expression of genes that negatively regulate its function, such as insulin-like growth factor-binding protein 3 (igf-bp3), PTEN (phosphatase and tensin homolog), and TSC2 (Buckbinder et al. 1995; Feng et al. 2007). Thus, the stabilization of p53 by AMPK not only induces cell cycle arrest but also indirectly promotes protein and rRNA synthesis inhibition.

At this point, we can consider that the connection between AMPK, mTORC1, and p53 balances the growth-inhibiting response to cellular stress.

3.2.3 AMPK Inhibits Lipid Synthesis

During cell division, cells must double their lipid content. Among different lipid classes, fatty acids are particularly important for cell growth, since cells need them for membrane generation, protein modification, and bioenergetic requirements. Therefore, fatty acid synthesis plays an essential role in cell growth.

It has been described that AMPK inhibits fatty acid synthesis by different mechanisms. First, AMPK inhibits lipogenesis by phosphorylating and inactivating the acetyl-CoA carboxylase ACC1 on Ser79 (Davies et al. 1992; Hardie and Pan 2002). ACC1 catalyzes the synthesis of malonyl-CoA, the substrate for fatty acid synthesis. Moreover, AMPK also downregulates the expression of fatty acid synthase (FAS), which is also important in fatty acid synthesis (Foretz et al. 1998).

Second, AMPK downregulates the expression of enzymes involved in fatty acid synthesis at the transcriptional level by phosphorylating the transcription factor Sterol regulatory element-binding protein-1c (SREBP-1c). SBREP activity controls fatty acid and sterol synthesis; therefore, activation of SREBP and expression of SREBP target genes are required for efficient cell growth in mammalian cells (Porstmann et al. 2008). AMPK phosphorylates SREBP-1c on Ser372 to prevent its translocation to the nucleus, leading to inhibition of lipogenesis and lipid accumulation (Li et al. 2011).

Besides, AMPK phosphorylates and inactivates other enzymes involved in lipid synthesis, such as HMG-CoA reductase, essential for cholesterol synthesis (Henin et al. 1995), or glycerol-3-phosphate acyltransferase (GPAT), which participates in glycerolipid and glycerophospholipid metabolism (Muoio et al. 1999).

3.3 AMPK Regulation of Autophagy

As we have exposed above, AMPK mainly stimulates catabolic pathways and prevents anabolic pathways, and it exerts its function via direct phosphorylation of different downstream enzymes or indirect regulation of transcription factors activities (Hardie 2007). Autophagy is one of the most important processes in cell catabolism, since cytosol and organelles are sequestered within double-membrane vesicles (pre-autophagosomes) that deliver the contents to the lysosome/vacuole (autolysosome) for hydrolytic degradation and recycling of the resulting macromolecules. Moreover, autophagy is considered a survival mechanism of the cell in stress situations (Goldman et al. 2010); hence, an appropriate regulation of autophagy is essential for cellular homeostasis, since excessive self-digestion can be harmful (Mizushima 2007).

Autophagy can be a nonselective or selective process because autophagy proteins and receptors of the autophagosome can interact specifically with the cytoplasmic component that needs to be eliminated (Johansen and Lamark 2011). Whereas autophagy usually refers to the nonselective elimination of any component of the cytoplasm, different terms have been coined to define the selective degradation of different organelles such as mitochondria (mitophagy) (Lemasters 2005), peroxisomes (pexophagy) (Till et al. 2012), or even lipids (lipophagy) (Liu and Czaja 2013).

Autophagy is generally activated by starvation and nutrient deprivation to generate metabolic intermediates to maintain ATP production. In fact, the autophagy checkpoint is a major mechanism for the maintenance of intracellular homeostasis that can be upregulated by nutrient deprivation and/or organelle damage. Therefore, autophagy is controlled by several metabolites, including the ATP/ADP ratio and the availability of acetyl-CoA, which affect the activity of various acetyltransferases (Green et al. 2014).

Since AMPK was primarily characterized as a kinase allosterically activated by AMP (Yeh et al. 1980), it was logical to hypothesize that AMPK would have a critical role in autophagy regulation.

Energy levels manifested in the form of ADP/ATP ratio trigger autophagy by activating AMPK. AMP enhances kinase activity of AMPK by binding to its γ subunits. On the contrary, ATP inhibits the allosteric activation of AMPK by AMP (Hardie 2007). Moreover, oxidative stress, which manifests in the form of high levels of ROS or oxidizing agents, activates AMPK by different pathways, some of them in an AMP-independent manner (Hwang et al. 2004; Emerling et al. 2009; Mungai et al. 2011).

In addition to energy levels, it has been hypothesized that other stimuli trigger autophagy by AMPK activation, under normal conditions indeed. Intracellular calcium (Ca^{2+}) levels seem to be crucial in AMPK activity, since Calmodulin-Dependent Kinase Kinase β (CaMKKβ) is suggested as an upstream kinase that activates AMPK (Hawley et al. 2005; Woods et al. 2005). It has been observed that different Ca^{2+} mobilizing agents, such as vitamin D compounds, induce autophagy by CaMKKβ and AMPK phosphorylation and activation. Moreover, inhibition of CaMKKβ by siRNA causes the inactivation of AMPK and an attenuation of autophagy (Høyer-Hansen et al. 2007).

Therefore, although AMPK activation is necessary to trigger autophagy, it is not known if it is sufficient or there are other molecular pathways which induce the change between basal autophagy and a huge autophagy in response to a cellular stress.

3.3.1 AMPK Regulation of Nonselective Autophagy

Independently of the stimuli, AMPK activation can induce the autophagic process through two different mechanisms: inhibition of mTOR and direct phosphorylation of ULK1 (Unc-51 Like Kinase 1, a mammalian orthologue of Atg1).

mTOR complex is a main regulator of diverse intracellular pathways that controls growth, proliferation, and survival. It is an effector in the PI3K/AKT pathway, and it carries out its action by two different complexes called mTORC1 and mTORC2. mTORC1 promotes protein synthesis, lipid biogenesis, cell growth, and anabolism and inhibits cellular catabolism by preventing autophagy. Instead, mTORC2 regulates cell survival, cell proliferation, and metabolism.

Although mTOR name comes from its sensitivity to rapamycin, rapamycin inhibits autophagy by affecting solely mTORC1, which is formed by the serine–threonine kinase mTOR, mLST8, PRAS40, and raptor, a regulatory protein which recruits downstream substrates such as 4EBP1 and ribosomal S6 kinase. mTORC1 inhibits autophagy by blocking the autophagy initiator complex activity, which is formed by ULK1, Atg13, Atg101, and FIP200 (Hosokawa et al. 2009; Jung et al. 2009). ULK1 is the essential protein for autophagy initiation in mammalian cells (Kuroyanagi et al. 1998; Chan et al. 2007; Wong et al. 2013).

The autophagy initiator complex is assembled independently of mTOR activity or nutrient conditions. However, under fed conditions, mTORC1 phosphorylates ULK1 in Ser638 and Ser758 and inhibits its kinase activity, preventing autophagy initiation (Shang and Wang 2011). Atg13 is also a substrate of mTOR kinase, and its phosphorylation has negative effects on ULK1 activity (Hosokawa et al. 2009). Moreover, ULK1 phosphorylation by mTOR disrupts its interaction with AMPK, preventing autophagy. Under starvation conditions, mTORC1-dependent phosphorylation of ULK1 is removed and ULK1 autophosphorylates and phosphorylates Atg13 and FIP200, initiating autophagy.

Since activation of mTORC1 is the principal mechanism to inhibit autophagy in mammalian cells, one of the mechanisms of AMPK-dependent induction of autophagy is inhibition of mTORC1 activity. AMPK is able to inhibit mTORC1 in a direct way by phosphorylating raptor (Gwinn et al. 2008) and in an indirect way by phosphorylating Tuberous Sclerosis Complex 2 (TSC2 or Tuberin) (Inoki et al. 2003).

When energy levels are low or there is a starvation situation, AMPK is phosphorylated and activated AMPK is able to bind and phosphorylate raptor, the mTORC1-binding partner (Gwinn et al. 2008). Gwinn et al. firstly revealed by bioinformatics techniques that raptor could be a direct substrate of AMPK since it presents two serine residues (Ser722 and Ser792) which match with the AMPK consensus motif. Secondly, the bioinformatics data were corroborated by biological evidences in vivo. Under energy stress conditions (low ATP levels), activated AMPK phosphorylates raptor in the predicted serine residues and this phosphorylation causes mTORC1 inhibition, so the autophagy initiator ULK1 can start the autophagic cascade. The mechanism by which raptor phosphorylation inhibits mTORC1 kinase function is a common mechanism for phosphorylation-based inactivation of target proteins. Phosphorylation of Ser722 and Ser792 induces 14-3-3 binding to raptor. 14-3-3 binding do not cause a disruption of mTORC1 complex but an inhibition of mTORC1 kinase activity.

Moreover, activated AMPK phosphorylates TSC2, an upstream inhibitor of mTORC1, since it inactivates GTPase Rheb, an activator of mTORC1 (Inoki et al. 2002). TSC2 forms a functional complex with TSC1 to inhibit two key regulators of translation: S6K and 4EBP1, which are phosphorylated by mTOR. It has been observed that under energy stress or starvation conditions, AMPK directly phosphorylates TSC2 on T1227 and S1345 to enhance its Rheb-GAP activity, which causes an inhibition of mTOR kinase activity (Inoki et al. 2003).

Therefore, energy stress results in AMPK phosphorylation and subsequently AMPK activation, which phosphorylates the mTORC1 inhibitor TSC2 and the mTORC1 regulatory subunit Raptor to inhibit mTORC1 and regulate catabolism, principally by initiating autophagy (Hardie 2008; Shaw 2009).

More recently, it has been observed that AMPK can also activate autophagy by direct phosphorylation of ULK1 (Guan et al. 2011; Egan et al. 2011b). However, there are some important discrepancies between the two investigations. The principal one is that the key phosphorylation sites identified by the different groups are not the same. Moreover, recently it was published that ULK1 forms a stable complex with AMPK (Behrends et al. 2010). Guan et al. described that this complex is disrupted when mTOR is activated and phosphorylates ULK1, but Egan et al. did not observe this mechanism. Therefore, more investigation about regulation of autophagy by AMPK is needed to resolve these differences.

Furthermore, in recent times it has been suggested that phosphorylation of ULK1 by AMPK stabilizes a negative regulation of autophagy initiation (Löffler et al. 2011). ULK1 is able to phosphorylate the three subunits of AMPK, and this effect reduces the phosphorylation of AMPKα Thr172, preventing AMPK kinase activity. By this way, ULK1 participates in the elimination of the initial autophagy signal, in addition to initiate the autophagic cascade.

Thus, we can conclude that AMPK uses a dual mechanism to activate autophagy by inactivating mTOR and also directly activating ULK1 by direct phosphorylation. Interestingly, the patterns of activation of AMPK and mTOR are opposite, regulating by this way the activation of autophagy depending on the nutrient conditions (Egan et al. 2011a; Alers et al. 2012) (Fig. 3.1).

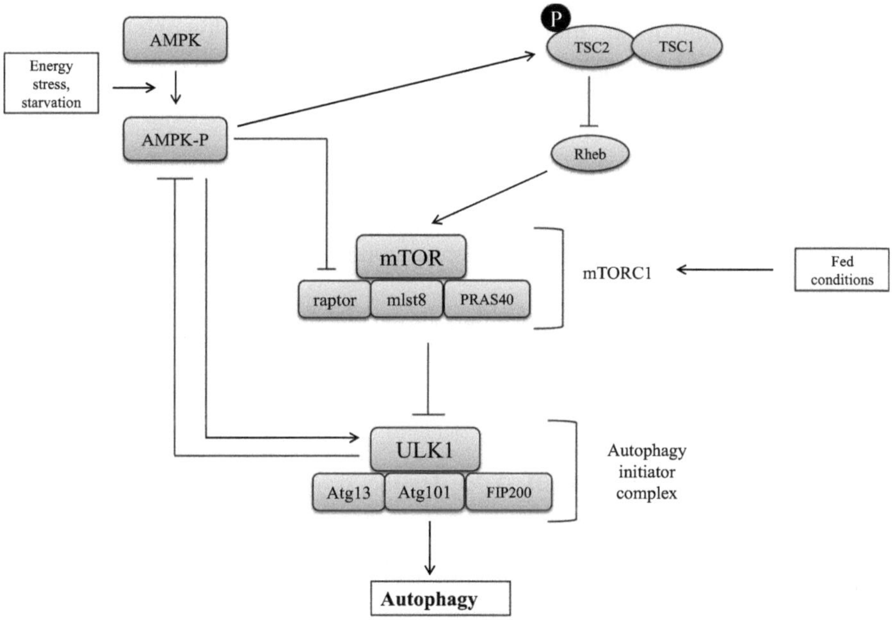

Fig. 3.1 Dual regulation of autophagy by mTOR and AMPK

3.3.2 AMPK Activation of Selective Autophagy of Mitochondria

Besides activating macroautophagy, AMPK promotes a selective form of autophagy called mitophagy, by which dysfunctional mitochondria are engulfed in autophagosomes and transported to lysosomes where they are degraded to recycle their components. In mammals, loss of AMPK resulted in aberrant accumulation of the autophagy adaptor p62 and defective mitophagy (Egan et al. 2011b). p62 binds to specific cargo targeted for autophagy-mediated degradation. Especially, p62 is bound to dysfunctional mitochondria targeted for mitophagy and is involved in mitochondrial aggregation and clearance (Geisler et al. 2010). Moreover, number of mitochondria per cell is significantly increased in AMPK-deficient cells, suggesting that AMPK plays an important role in mitochondria turnover.

Moreover, although the best-known function of the autophagy-initiating factor ULK1 is the autophagy induction (Mizushima 2010; Wirth et al. 2013), recent studies suggest that ULK1 has a more selective function in mitophagy (Kundu et al. 2008; Itakura et al. 2012). Recently, it has been described that ULK1 translocates to dysfunctional mitochondria in response to hypoxia or FCCP. Translocated ULK1 phosphorylates FUNDC1 (a mitophagy receptor located in external mitochondrial membrane) allowing its interaction with LC3 and inducing mitochondria degradation by autophagy (Wu et al. 2014).

In relation with this novel role of ULK1, it has been demonstrated that specific phosphorylation of ULK1 at Ser555 by activated AMPK is crucial for ULK translocation to the mitochondria (Tian et al. 2015). Therefore, AMPK-mediated phosphorylation of ULK1 is essential for mitophagy to initiate under stress conditions such as hypoxia.

For these reasons, AMPK is considered essential for the regulation of selective autophagy, besides its role as a macroautophagy inductor.

3.4 AMPK Regulation of Apoptosis

Although the best-known functions of AMPK are maintaining ATP homeostasis and regulating metabolism, AMPK has recently been proposed as a regulator of cell apoptosis or survival under stress conditions.

However, it is not clear whether AMPK is a proapoptotic or pro-survival molecule. In fact, AMPK participates in cell death or survival depending on the kind of stress, the cell type, and the duration of the activation of the signaling cascade. All these features make the role of AMPK in apoptosis controversial and complex.

3.4.1 AMPK as a Proapoptotic Molecule

Several metabolic checkpoints convert metabolic changes (signals), which are detected by specific systems (sensors), into stimuli which regulate the function of components of the cell death-regulatory machinery (effectors).

AMPK activation has been shown to mediate its proapoptosis effects primarily through modulation of several downstream signaling events by regulating JNK (c-Jun N-terminal protein kinase) and p53, inhibiting mTORC1 or directly enhancing the activation of proapoptotic proteins, among other different pathways (Fig. 3.2).

3.4.1.1 AMPK Induces Apoptosis by Activating JNK Pathway

The first demonstration of a proapoptotic effect of AMPK was observed in liver rat cells. It was observed that AMPK activation by high concentrations of AICAR triggers apoptosis in liver cells by activating JNK and Caspase 3 (Meisse et al. 2002). JNK pathway regulates proteins involved in apoptosis such as p53, c-myc, and members of the Bcl-2 family.

AICAR activation of apoptosis mediated by JNK and caspase 3 has been observed in different cell types. Kefas et al. observed that prolonged stimulation of AMPK induced apoptosis of insulin-producing cells (MIN6) of mouse. It is interesting that apoptosis induction was proportional to concentration of AICAR and time of exposure. Prolonged exposure to low glucose concentration induced apoptosis by AMPK activation too (Kefas et al. 2003a). The same response was observed in pancreatic beta cells (Kefas et al. 2003b).

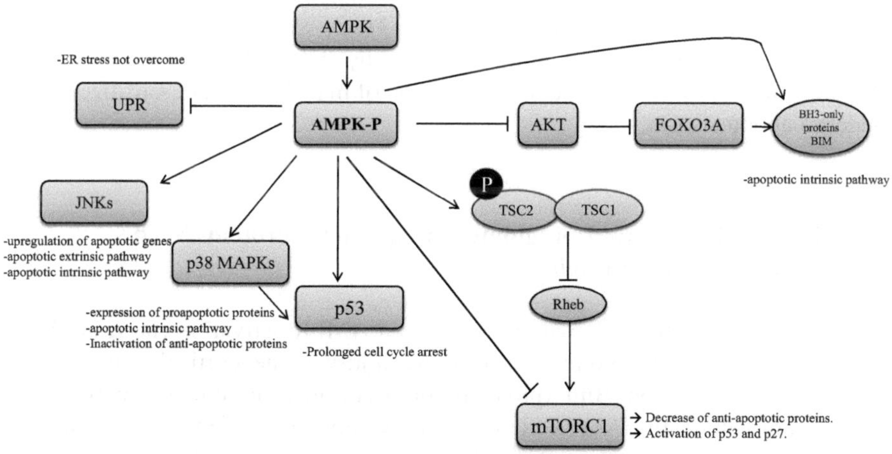

Fig. 3.2 Different pathways by which AMPK can induce apoptosis

3.4.1.2 AMPK Induces Apoptosis by Regulating and Stabilizing p53

Apart from its involvement in JNK activation, it has been described how AICAR induced p53 activation by AMPK stimulation. One of the first evidences that AMPK could be involved in regulating p53 function was observed in human hepatocellular carcinoma cell lines (HepG2) (Imamura et al. 2001). Later, it has been described how AMPK relates glucose availability to the p53 pathway, a master regulator of survival and proliferation (Jones et al. 2005). Prolonged cell cycle arrest induced by p53 may be followed by apoptosis activation. Glucose deprivation also induces phosphorylation of AMPK and subsequent activation of p53 and bax, a proapoptotic protein, promoting thymocytes and osteosarcoma cells to undergo apoptosis (Okoshi et al. 2008). Moreover, AICAR also induces apoptosis in other cancer cells such as glioblastomas due to AMPK-dependent inhibition of mTORC1 (Guo et al. 2009). Therefore, the same compound can activate AMPK and induce apoptosis in different types of cells by various mechanisms.

AMPK also contributes to UV- and H_2O_2-induced apoptosis in human skin keratinocytes (Cao et al. 2008). Chiefly, on UV- or H_2O_2-treated cells, AMPK induces apoptosis by inhibiting mTORC1 and positively regulating p53 and p38 MAPK, which mediates apoptosis elicited by both energy imbalance and pro-oxidant conditions.

3.4.1.3 AMPK Induces Apoptosis by Inhibiting mTORC1

One of the most important metabolic checkpoints that control cell death is the AMPK–mTORC1 pathway, which is based on the short half-life of antiapoptotic proteins such as FLIPL and MCL-1. In this sense, mTORC1 inhibition by AMPK activation causes the inhibition of protein translation; therefore, the abundance of short-lived antiapoptotic proteins such as MCL1 decreases and the cell is sensitized to mitochondrial apoptosis (Meynet et al. 2013).

Moreover, since mTOR inhibits apoptosis by the mediation of p53 and p27 (Faivre et al. 2006), AMPK-induced inhibition of mTORC1 could sensitize cells to undergo apoptosis too.

3.4.1.4 AMPK Directly Upregulates Proapoptotic Proteins Such as BH3-Only Proteins

For instance, it has been shown that prolonged AMPK activation can result in the direct induction of apoptotic excitotoxicity injury in neocortical neurons. This effect is dependent on Bim (proapoptotic BH3-only protein essential for the initiation of cell death). AMPK couples energy depletion to Bim mRNA induction and the subsequent activation of the Bcl-2-regulated apoptotic pathway in neurons (Concannon et al. 2010).

Moreover, AMPK activates the proapoptotic BH3-only proteins either through an indirect mechanism, by the AKT–FOXO3A pathway, because it has been suggested that FOXO3A is able to induce the translocation of Bim to mitochondria (Weisová et al. 2011).

3.4.1.5 AMPK Can Induce Apoptosis by Inhibiting the Unfolded Protein Response Pathway

Apart from inhibition of mTORC1, p53 stabilization, and JNK pathway activation, recently it has been proposed a new mechanism to explain AMPK-induced apoptosis. Metformin-induced activation of AMPK triggers apoptosis in acute lymphoblastic leukemia (ALL) cells by inhibiting the Unfolded Protein Response (UPR) pathway (Leclerc et al. 2013), which is triggered in response to the accumulation of unfolded/misfolded proteins in the ER lumen. First, metformin induces ER stress by accumulation of unfolded/misfolded proteins in the ER as a consequence of ATP depletion. At the same time, metformin activates AMPK, which acts as a negative regulator of the UPR, preventing cells from effectively engaging the UPR to overcome ER stress, leading to apoptotic death by inositol-requiring enzyme 1α (IRE1α), C/EBP homologous protein (CHOP), caspases, and JNK activation.

3.4.1.6 AMPK Involvement in Apoptosis Is Controversial

Recently, AICAR-induced apoptosis has been controversial since it has been shown that AICAR can induce an apoptosis in an AMPK- and p53-independent mechanism (Santidrián et al. 2010; Gonzalez-Girones et al. 2013). It has been shown that AICAR induced apoptosis in chronic lymphoid leukemia (CLL) through the mitochondrial apoptotic pathway, by inducing the modulation of RNA levels of Bcl-2, Bim, and Noxa, independently of AMPK and p53.

In the same way, AMPK sustained activation causes apoptosis induction in beta cells due to enhanced production of mitochondria-derived oxygen radicals which induce the activation of the intrinsic mitochondrial apoptosis pathway (Cai et al. 2007). Besides this mechanism, it is suggested an induction of the intrinsic apoptosis pathway by sustained AMPK activation, since dysregulation of BH3 members of the Bcl-2 family can be observed.

Other AMPK activators such as metformin have been demonstrated to induce apoptosis by different mechanisms that could be dependent or independent on AMPK. Metformin induces apoptosis in different cancer cell types (Leclerc et al. 2013; Queiroz et al. 2014), but it is important to notice that metformin has other targets apart from AMPK, for example, it functions as an inhibitor of complex I of the electron transport chain (Batandier et al. 2006). Moreover, the AMPK inhibitor, compound C, has been shown to compromise cell survival by AMPK-dependent and AMPK-independent mechanisms, since inhibition of AMPK is not sufficient to induce apoptosis in some cell types (Vucicevic et al. 2009).

Therefore, role of AMPK in AICAR/metformin-mediated apoptosis remains unclear and more investigations are needed to understand the real function and importance of AMPK in apoptosis induction (Bonini and Gantner 2013). What is clear is that AMPK is necessary (although in some cases is not sufficient) to promote apoptosis in different cell types.

3.4.1.7 Involvement of AMPK in Anticancer Drug Effects

Besides metformin, other anticancer drugs have been demonstrated to activate AMPK-dependent cell apoptosis pathway, although the molecular mechanisms involved in apoptosis initiation are different between the diverse compounds.

Vincristine, a drug that binds to tubulin and causes microtubule depolymerization, induces apoptosis in cells undergoing mitosis by AMPK activation. In B16 melanoma cells, high levels of ROS and LKB1 and AMPK activation have been observed, which induce both p53 activation and stabilization and mTORC1 inhibition, which are necessary mechanisms to mediate melanoma cell apoptosis (Chen et al. 2011).

Taxol (paclitaxel) is another drug that also affects microtubules but by a different mechanism. Taxol stabilizes the microtubule polymer and protects it from disassembly. Due to its action, chromosomes are unable to achieve a metaphase spindle configuration. It has been observed that taxol activates AMPK and downstream ACC in ovarian cancer cells and AMPK activation is implicated in the apoptosis induction mechanism (Sun et al. 2011). Apoptosis induction could be caused by ROS augmentation, mTORC1 inhibition, and activation of JNK pathway.

Temozolomide (TMZ) is another anticancer drug that exerts its action by a different mechanism, due to its ability to alkylate/methylate DNA. This methylation damages the DNA and has been demonstrated to induce apoptosis in primary cultured human glioblastoma cell lines. TMZ-mediated apoptosis is induced by AMPK activation, among other mechanisms. AMPK phosphorylation and activation in turn stabilize and activate p53 and induce the upregulation of p21, Noxa, and Bax. Activation of AMPK by TMZ also inhibits mTORC1 signaling and promotes antiapoptosis protein Bcl-2 downregulation (Zhang et al. 2010).

However, other drugs that induce DNA damage responses, such as doxorubicin (intercalating agent), induce AMPK activation in a ROS-dependent manner, which in turn contributes to apoptotic cell death by mTORC1 inhibition (Ji et al. 2010). It is interesting that inhibition of ROS production and AMPK activation by antioxidants such as *N*-acetylcysteine (NAC) or manganese (III) tetrakis (4-benzoic acid) porphyrins (MnTBAP) inhibits doxorubicin-induced AMPK activation and cell death. Therefore, the high level of ROS is the activating signal for AMPK in this situation.

Therefore, there are so many signals that are able to activate AMPK under stress situations for cells to undergo apoptosis.

3.4.2 *AMPK as a Pro-survival Molecule*

AMPK-mediated apoptosis has been reported in numerous cell types as we have exposed above. Conversely, it is recognized that activation of this kinase leads to metabolic alterations that can prevent ATP depletion in certain cell types, resulting in improved cell survival under different stress situations and protection from apoptosis. In fact, in conditions like starvation, energy deprivation, and oxidative stress, AMPK activation is required for cell survival. The molecular targets and effectors of the AMPK-mediated cell survival are diverse and different depending on the stress which induces the ATP depletion situation (Fig. 3.3).

3.4.2.1 AMPK Induces Cell Survival in Oxidative Stress Conditions

In oxidative stress conditions, for example, cell exposure to H_2O_2, AMPK is required for cell survival and inhibits apoptosis in osteoblasts (She et al. 2014). AMPK activation inhibits H_2O_2-induced oxidative stress through inhibiting NADPH depletion. Moreover, H_2O_2 is demonstrated to induce autophagy in MG63 cells. AMPK-dependent ULK1 activation and mTORC1 inactivation involve autophagy activation, which exerts a pro-survival effect instead of apoptosis induction. Therefore, activation of AMPK by H_2O_2 is pro-survival in osteoblasts through two mechanisms: activation of autophagy and inhibition of NADPH depletion.

It is important to observe that in the presence of the same condition (e.g., oxidative stress), AMPK can act as a pro-survival or proapoptotic molecule. It depends on the type of cell, the time of exposure, and the duration of the AMPK

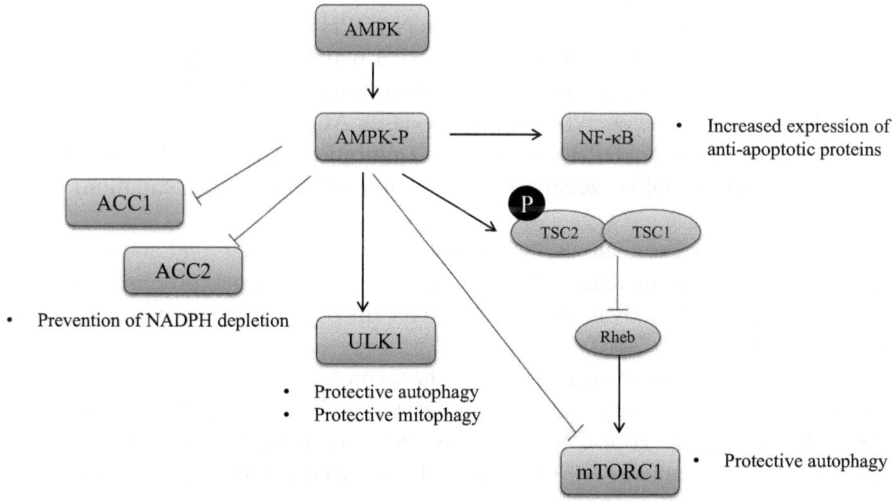

Fig. 3.3 Different pathways by which AMPK can prevent apoptosis

activation, among other features. Therefore, AMPK regulation of cell survival and death is very complex, and it is important to study every case separately.

3.4.2.2 AMPK Induces Cell Survival in Starvation Conditions

TSC2 and its phosphorylation by AMPK protect cells from glucose deprivation-induced apoptosis. As we have mentioned above, one of the major cellular functions of TSC1/TSC2 is to prevent protein translation by inhibiting the phosphorylation of S6K and 4EBP1. AMPK directly phosphorylates TSC2 on T1227 and S1345. Although TSC2 phosphorylation by AMPK plays an essential role in inhibiting cell growth, it is important for regulating cell survival under glucose starvation conditions too. Energy starvation activates TSC2, which then prevents cell growth and promotes cell survival. Activation of TSC2 by AMPK-dependent phosphorylation prepares cells for an unfavorable growth environment and results in protection from cell death. In fact, a mutant version of TSC2, TSC2-3A, which cannot be phosphorylated by AMPK, has demonstrated to be incapable to protect cells from glucose deprivation-induced apoptosis (Inoki et al. 2003).

3.4.2.3 AMPK Protects from Energy Stress-Induced Apoptosis

AMPK activation under an energy stress situation extends cell survival by redox regulation. In glucose limitation conditions, NADPH generation by the pentose phosphate pathway (PPP) is reduced, and this situation generates oxidative stress, since NADPH is needed for the regeneration of reduced glutathione (GSH), which is used by glutathione peroxidase (GPX) to eliminate H_2O_2. However, AMPK is able to induce alternative routes to maintain NADPH and inhibit cell death. By this way, AMPK plays an essential role in NADPH maintenance, which is critical for cancer cell survival under energy stress conditions. AMPK maintains NADPH levels by inhibiting the acetyl-CoA carboxylases ACC1 and ACC2. These effects induce a decrease in NADPH consumption (inhibition of fatty acid synthesis) and an increase in NADPH generation (induction of fatty acid oxidation) (Jeon et al. 2012).

Recently, it has been shown that AMPK promotes cell survival and suppresses apoptosis by directly increasing the expression of antiapoptotic proteins (Blc2 and Survivin) via NF-κB activation in endothelial cells exposed to hypoxia and glucose deprivation (Liu et al. 2010).

NF-κB has been proposed to induce the expression of different genes whose products can enhance cell survival and protect cells from apoptosis (Mitsiades 2002; Angileri et al. 2008). Nevertheless, the underlying mechanisms by which AMPK activation induces NF-κB-mediated cell survival remain poorly defined.

3.5 AMPK Regulation of Bioenergetics

Bioenergetics is the study of the transformation of energy in living organisms and the different cellular processes that can lead to production and usage of energy in different forms such as ATP.

AMPK has an essential role in bioenergetics regulation because it is a cellular energy sensor activated by conditions of metabolic stress characterized by an increase in the AMP/ATP ratio. AMPK activation restores cellular bioenergetics by inhibiting anabolic pathways that consume ATP and by activating specific catabolic reactions that generate ATP (Hardie 2011; Hindupur et al. 2015). Under stress conditions, the objective of AMPK activation is the conservation of cellular energy in order to avoid bioenergetic catastrophe (Faubert et al. 2015).

AMPK activation changes the way used by the cells to obtain energy in the form of ATP molecules. Predominantly, AMPK upregulates the production of ATP by glycolysis and fatty acid oxidation, for this reason, AMPK is crucial for the metabolic adaptation of cells which present some dysfunctions of the OXPHOS (Oxidative Phosphorylation) system (Wu and Wei 2012; Garrido-Maraver et al. 2015).

Activated AMPK switches off anabolic pathways such as fatty acid, glycogen, RNA, and protein synthesis to avoid ATP waste. However, in this section, we are focusing on the different catabolic pathways that are upregulated upon AMPK activation, in order to maintain energetic homeostasis (Fig. 3.4) (Hardie 2014), such as glucose uptake, fatty acid uptake, glycolysis, and fatty acid oxidation.

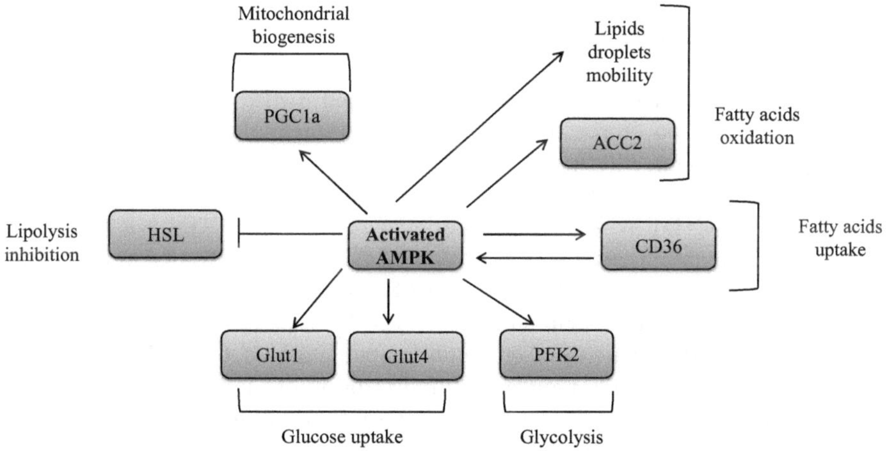

Fig. 3.4 Different catabolic pathways activated by AMPK

3.5.1 Activation of Glucose Uptake by AMPK

AMPK activation induces the upregulation of two glucose transporter: GLUT1 and GLUT4. GLUT1 is responsible for the low level of basal glucose uptake required to maintain respiration in cells, so AMPK is involved in basal glucose uptake maintenance (GLUT1 upregulation). Moreover, AMPK is responsible for insulin-dependent glucose uptake in adipose tissues and striated muscle, due to its role in GLUT4 upregulation.

3.5.1.1 GLUT1 Upregulation by AMPK

First studies in glucose uptake regulation by AMPK demonstrated that AMPK activation increased the Vmax of GLUT1-dependent glucose uptake (Barnes et al. 2002). More recently, a possible explanation for these results has been found. It has been suggested that AMPK activation increases the levels of GLUT1 mRNA due to the AMPK-dependent phosphorylation of Thioredoxin-interacting protein (TXNIP) (Wu et al. 2013).

TXNIP is induced in response to glucose elevation and suppresses glucose uptake directly, by binding to GLUT1 and inducing GLUT1 internalization, as well as indirectly, by reducing the level of GLUT1 mRNA. AMPK phosphorylation of TXNIP induces its rapid degradation by proteasome system. Therefore, TXNIP is released from GLUT1, and there is an increase in GLUT1 mRNA levels and function.

3.5.1.2 GLUT4 Upregulation by AMPK

First studies in glucose uptake regulation showed that AMPK activation induced the GLUT4 translocation from an intracellular location to the plasma membranes (Kurth-Kraczek et al. 1999). Recently, it has been demonstrated that AMPK-mediated translocation of GLUT4 is due in part to phosphorylation of the Rab GAP protein TBC1D1 by AMPK (Pehmøller et al. 2009). TBC1D1 is a component of the insulin-signaling cascade downstream of Akt that regulates GLUT4 translocation. Phosphorylation of TBC1D1 by AMPK promotes its 14-3-3 binding, which could increase basal and insulin-stimulated GLUT4 translocation.

Moreover, AMPK activation promotes upregulation of GLUT4 by a second mechanism because AMPK induces the transcription of the GLUT4 gene, due to phosphorylation of histone deacetylase-5 (HDAC5), a transcriptional repressor of GLUT4 (McGee et al. 2008). AMPK-mediated phosphorylation of HDAC5 reduces HDAC5 association with the GLUT4 promoter, increasing the GLUT4 mRNA levels.

3.5.2 AMPK Activation of Fatty Acids Uptake

Long chain fatty acid (LCFA) uptake is mostly dependent on the translocase CD36 (Koonen et al. 2005; Habets et al. 2007). AMPK activation stimulates LCFA uptake inducing the translocation of CD36 to the sarcolemma, which is essential for AMPK-induced fatty acid (FA) metabolism (Luiken et al. 2003; Bonen et al. 2007; Habets et al. 2009). Recently, it has been proposed that CD36 also interacts with AMPK to coordinate FA uptake. When exogenous FA levels are low, CD36 maintains AMPK inactive by indirectly promoting LKB1 transport to the nucleus. When exogenous FA levels increase, CD36-induced LKB1 transport to the nucleus is disrupted, and the cytosolic enrichment in LKB1 promotes AMPK activation and FA uptake (Samovski et al. 2015).

3.5.3 AMPK Activation of Glycolysis

Under cellular stress conditions (glucose deprivation, hypoxia, or oxidative stress), AMPK signaling drives glycolytic flux to maintain energy homeostasis and cellular ATP levels through glucose catabolism. AMPK promotes glycolysis via phosphorylation and activation of 6-phosphofructo-2-kinase/fructose-2,6-biphosphatase (PFKFB). PFKFB is a bifunctional enzyme with different tissue-dependent isoforms. The phosphofructo-2-kinase (PFK-2) domain functions to promote glycolysis by increasing levels of fructose-2,6-bisphosphate through phosphorylation of fructose-6-phosphate, and the fructose-2,6-biphosphatase (FB) domain catalyzes the reverse reaction producing fructose-6-phosphate.

In fact, AMPK is responsible for the increased glycolytic flux present after an ischemia in the heart, since it phosphorylates PFK-2 at Ser466 (Marsin et al. 2000). AMPK also induces glycolysis under hypoxia conditions (Marsin et al. 2002). It is important to remark that oxidative stress also promotes AMPK to increase glycolytic flux (Wu and Wei 2012).

3.5.4 AMPK Activation of Fatty Acid Oxidation

AMPK plays an important role in the regulation of FA oxidation because AMPK activation induces a change in different steps involved in FA oxidation pathway. First, as we commented above, AMPK controls fatty acid transport across the cell membrane by inducing CD-36 translocation.

Second, activated AMPK phosphorylates and inactivates the ACC2 isoform of acetyl-CoA carboxylase, which catalyzes the carboxylation of acetyl-CoA to malonyl-CoA, the limiting step in fatty acid synthesis. Malonyl-CoA is also an inhibitor of fatty acid absorption into mitochondria (Merrill et al. 1997). Therefore, AMPK activation increases the rate of fatty acid oxidation.

Third, very recently, it has been shown that AMPK phosphorylation promotes lipid droplets (LDs) mobility, dispersion, and consumption by affecting the microtubules network, increasing the rate of FA oxidation (Herms et al. 2015). By this way, AMPK promotes FAs from LDs to be channeled preferentially to different metabolic fates depending on the cellular energetic status: in the absence of glucose, FAs are directed to beta oxidation in mitochondria.

3.5.5 Induction of Mitochondrial Biogenesis by AMPK

Besides promoting different catabolic pathways, AMPK activates mitochondrial biogenesis to maintain the overall ATP-generating capacity of cells and avoid cell death. The most important regulator of mitochondrial biogenesis is PGC1α (Hardie et al. 2012). It has been observed that PGC1α is directly phosphorylated and transactivated by AMPK (Jäger et al. 2007). PGC1α can also be activated by Sirt-1-mediated deacetylation, which is induced by AMPK activation too (Cantó et al. 2009). PGC1α upregulates the activity of transcription factors involved in mitochondrial biogenesis, such as NRF-1, which in turn modulates the expression of other factors such as mtTFA, important for the regulation of mtDNA replication and transcription.

3.5.6 Inhibition of Lipolysis by AMPK

In response to glucose deprivation, AMPK inhibits lipolysis in adipocytes. One of the most important enzymes involved in lipolysis, hormone-sensitive lipase (HSL) is phosphorylated by AMPK on a key serine residue that antagonizes phosphorylation of HSL by cyclic AMP-dependent protein kinase (PKA) (Garton and Yeaman 1990), causing suppression of lipolysis. Although it is a catabolic pathway, its inhibition by AMPK is considered as a way to maintain cell energy homeostasis because an excessive accumulation of fatty acids promotes the synthesis of triglycerides, an ATP-consuming process (Garton 1989; Daval 2005).

3.6 Conclusion

AMPK is a critical regulator of energy homeostasis at both the cellular and whole-body levels. However, as a result of the wide range of pathways and signaling in which AMPK is involved, diverse contradictions arise about the role of AMPK in cell death and survival. For example, AMPK has different roles in apoptosis regulation depending on the duration of stimulation, dosage of induction, the type of cell, and the type of stress that induce its activation. Thus, AMPK phosphorylation induces apoptosis in mouse astrocytoma but may protect normal brain cells from apoptosis under similar energy stress condition (Mukherjee et al. 2008). Therefore, for now, it is necessary to study every case separately because we do not have all the knowledge about AMPK pathways, signaling, and targets.

Understanding the complex role of AMPK in regulation of ATP balance is important because disturbances in energy homoeostasis underlie a number of diseases in humans, e.g., cancer, mitochondrial dysfunctions, diabetes, etc. In this sense, discovering new targets of AMPK is a research field which is gaining importance in the last years. For example, the involvement of AMPK in the insulin/IGF1 and MAPK–ERK pathways to control cell growth is being extensively studied.

As more targets of AMPK are known, we can discover which effectors are responsible for the beneficial effects of AMPK activation seen in diverse diseases. It is accepted that AMPK is emerged as a potential therapeutic target in different human diseases, but it is necessary to investigate about the molecular mechanisms underlying its therapeutic function.

Moreover, it is logical to think that AMPK regulation of cell growth, bioenergetics, and cell fate (death or survival) is closely related with the aging process regulation. Some investigations suggest that aging is linked to a loss of sensibility of AMPK—it loses its capacity to sense changes in AMP/ATP ratio—which induce a generalized energy imbalance situation. Then, the complete understanding of AMPK-mediated regulation of cell metabolism might lead to improve the quality of life in the elderly and to fight different diseases which have not curative treatments yet.

References

Alers S, Loffler AS, Wesselborg S, Stork B (2012) Role of AMPK-mTOR-Ulk1/2 in the regulation of autophagy: cross talk, shortcuts, and feedbacks. Mol Cell Biol 32:2–11. doi:10.1128/MCB.06159-11

Alexander A, Walker CL (2011) The role of LKB1 and AMPK in cellular responses to stress and damage. FEBS Lett 585:952–957. doi:10.1016/j.febslet.2011.03.010

Angileri FF, Aguennouz M, Conti A et al (2008) Nuclear factor-κB activation and differential expression of survivin and Bcl-2 in human grade 2-4 astrocytomas. Cancer 112:2258–2266. doi:10.1002/cncr.23407

Barnes K, Ingram JC, Porras OH et al (2002) Activation of GLUT1 by metabolic and osmotic stress: potential involvement of AMP-activated protein kinase (AMPK). J Cell Sci 115:2433–2442

Batandier C, Guigas B, Detaille D et al (2006) The ROS production induced by a reverse-electron flux at respiratory-chain complex 1 is hampered by metformin. J Bioenerg Biomembr 38:33–42. doi:10.1007/s10863-006-9003-8

Behrends C, Sowa ME, Gygi SP, Harper JW (2010) Network organization of the human autophagy system. Nature 466:68–76. doi:10.1038/nature09204

Bonen A, Han X-X, Habets DDJ et al (2007) A null mutation in skeletal muscle FAT/CD36 reveals its essential role in insulin- and AICAR-stimulated fatty acid metabolism. Am J Physiol Endocrinol Metab 292:E1740–E1749. doi:10.1152/ajpendo.00579.2006
Bonini MG, Gantner BN (2013) The multifaceted activities of AMPK in tumor progression – why the "one size fits all" definition does not fit at all? IUBMB Life 65:889–896. doi:10.1002/iub.1213
Browne GJ, Proud CG (2002) Regulation of peptide-chain elongation in mammalian cells. Eur J Biochem 269:5360–5368. doi:10.1046/j.1432-1033.2002.03290.x
Browne GJ, Finn SG, Proud CG (2004) Stimulation of the AMP-activated protein kinase leads to activation of eukaryotic elongation factor 2 kinase and to its phosphorylation at a novel site, serine 398. J Biol Chem 279:12220–12231. doi:10.1074/jbc.M309773200
Buckbinder L, Talbott R, Velasco-Miguel S et al (1995) Induction of the growth inhibitor IGF-binding protein 3 by p53. Nature 377:646–649. doi:10.1038/377646a0
Budanov AV, Karin M (2008) p53 target genes sestrin1 and sestrin2 connect genotoxic stress and mTOR signaling. Cell 134:451–460. doi:10.1016/j.cell.2008.06.028
Cai Y, Martens GA, Hinke SA et al (2007) Increased oxygen radical formation and mitochondrial dysfunction mediate beta cell apoptosis under conditions of AMP-activated protein kinase stimulation. Free Radic Biol Med 42:64–78. doi:10.1016/j.freeradbiomed.2006.09.018
Cantó C, Gerhart-Hines Z, Feige JN et al (2009) AMPK regulates energy expenditure by modulating NAD+ metabolism and SIRT1 activity. Nature 458:1056–1060. doi:10.1038/nature07813
Cao C, Lu S, Kivlin R et al (2008) AMP-activated protein kinase contributes to UV- and H_2O_2-induced apoptosis in human skin keratinocytes. J Biol Chem 283:28897–28908. doi:10.1074/jbc.M804144200
Chan EYW, Kir S, Tooze SA (2007) siRNA screening of the kinome identifies ULK1 as a multidomain modulator of autophagy. J Biol Chem 282:25464–25474. doi:10.1074/jbc.M703663200
Chen M-B, Shen W-X, Yang Y et al (2011) Activation of AMP-activated protein kinase is involved in vincristine-induced cell apoptosis in B16 melanoma cell. J Cell Physiol 226:1915–1925. doi:10.1002/jcp.22522
Concannon CG, Tuffy LP, Weisová P et al (2010) AMP kinase-mediated activation of the BH3-only protein Bim couples energy depletion to stress induced apoptosis. J Cell Biol 189:83–94. doi:10.1083/jcb.200909166
Daval M (2005) Anti-lipolytic action of AMP-activated protein kinase in rodent adipocytes. J Biol Chem 280:25250–25257. doi:10.1074/jbc.M414222200
Davies SP, Carling D, Munday MR, Hardie DG (1992) Diurnal rhythm of phosphorylation of rat liver acetyl-CoA carboxylase by the AMP-activated protein kinase, demonstrated using freeze-clamping. Effects of high fat diets. Eur J Biochem 203:615–623
Egan D, Kim J, Shaw RJ, Guan KL (2011a) The autophagy initiating kinase ULK1 is regulated via opposing phosphorylation by AMPK and mTOR. Autophagy 7:643–644
Egan DF, Shackelford DB, Mihaylova MM et al (2011b) Phosphorylation of ULK1 (hATG1) by AMP-activated protein kinase connects energy sensing to mitophagy. Science 331:456–461. doi:10.1126/science.1196371
Emerling BM, Weinberg F, Snyder C et al (2009) Hypoxic activation of AMPK is dependent on mitochondrial ROS but independent of an increase in AMP/ATP ratio. Free Radic Biol Med 46:1386–1391. doi:10.1016/j.freeradbiomed.2009.02.019
Faivre S, Kroemer G, Raymond E (2006) Current development of mTOR inhibitors as anticancer agents. Nat Rev Drug Discov 5:671–688. doi:10.1038/nrd2062
Faubert B, Vincent EE, Poffenberger MC, Jones RG (2015) The AMP-activated protein kinase (AMPK) and cancer: many faces of a metabolic regulator. Cancer Lett 356:165–170. doi:10.1016/j.canlet.2014.01.018
Feng Z, Hu W, de Stanchina E et al (2007) The regulation of AMPK 1, TSC2, and PTEN expression by p53: stress, cell and tissue specificity, and the role of these gene products in

modulating the IGF-1-AKT-mTOR pathways. Cancer Res 67:3043–3053. doi:10.1158/0008-5472.CAN-06-4149

Foretz M, Carling D, Guichard C et al (1998) AMP-activated protein kinase inhibits the glucose-activated expression of fatty acid synthase gene in rat hepatocytes. J Biol Chem 273:14767–14771

Garrido-Maraver J, Paz MV, Cordero MD et al (2015) Critical role of AMP-activated protein kinase in the balance between mitophagy and mitochondrial biogenesis in MELAS disease. Biochim Biophys Acta 1852:2535–2553. doi:10.1016/j.bbadis.2015.08.027

Garton A (1989) Phosphorylation of bovine hormone-sensitive lipase by the AMP-activated protein kinase. Eur J Biochem 254:249–254

Garton AJ, Yeaman SJ (1990) Identification and role of the basal phosphorylation site on hormone-sensitive lipase. Eur J Biochem 191:245–250

Geisler S, Holmström KM, Skujat D et al (2010) PINK1/Parkin-mediated mitophagy is dependent on VDAC1 and p62/SQSTM1. Nat Cell Biol 12:119–131. doi:10.1038/ncb2012

Goldman SJ, Taylor R, Zhang Y, Jin S (2010) Autophagy and the degradation of mitochondria. Mitochondrion 10:309–315. doi:10.1016/j.mito.2010.01.005

Gonzalez-Girones DM, Moncunill-Massaguer C, Iglesias-Serret D et al (2013) AICAR induces Bax/Bak-dependent apoptosis through upregulation of the BH3-only proteins Bim and Noxa in mouse embryonic fibroblasts. Apoptosis 18:1008–1016. doi:10.1007/s10495-013-0850-610.1007/s10495-013-0850-6

Grahame Hardie D (2014) AMP-activated protein kinase: a key regulator of energy balance with many roles in human disease. J Intern Med 276:543–559. doi:10.1111/joim.12268

Green DR, Galluzzi L, Kroemer G (2014) Cell biology. Metabolic control of cell death. Science 345(80):1250256. doi:10.1126/science.1250256

Guan JK, Mondira K, Benoit V, Kun L (2011) AMPK and mTOR regulate autophagy through direct phosphorylation of Ulk1. Nat Cell Biol 13:132–141. doi:10.1038/ncb2152

Guo D, Hildebrandt IJ, Prins RM et al (2009) The AMPK agonist AICAR inhibits the growth of EGFRvIII-expressing glioblastomas by inhibiting lipogenesis. Proc Natl Acad Sci 106:12932–12937. doi:10.1073/pnas.0906606106

Gwinn DM, Shackelford DB, Egan DF et al (2008) AMPK phosphorylation of raptor mediates a metabolic checkpoint. Mol Cell 30:214–226. doi:10.1016/j.molcel.2008.03.003

Habets DDJ, Coumans WA, Voshol PJ et al (2007) AMPK-mediated increase in myocardial long-chain fatty acid uptake critically depends on sarcolemmal CD36. Biochem Biophys Res Commun 355:204–210. doi:10.1016/j.bbrc.2007.01.141

Habets DDJ, Coumans WA, El Hasnaoui M et al (2009) Crucial role for LKB1 to AMPKalpha2 axis in the regulation of CD36-mediated long-chain fatty acid uptake into cardiomyocytes. Biochim Biophys Acta 1791:212–219. doi:10.1016/j.bbalip.2008.12.009

Hardie DG (2007) AMP-activated/SNF1 protein kinases: conserved guardians of cellular energy. Nat Rev Mol Cell Biol 8:774–785. doi:10.1038/nrm2249

Hardie DG (2008) AMPK and raptor: matching cell growth to energy supply. Mol Cell 30:263–265. doi:10.1016/j.molcel.2008.04.012

Hardie DG (2011) AMP-activated protein kinase: an energy sensor that regulates all aspects of cell function. Genes Dev 25:1895–1908. doi:10.1101/gad.17420111

Hardie DG (2014) AMP-activated protein kinase: maintaining energy homeostasis at the cellular and whole-body levels. Annu Rev Nutr 34:31–55. doi:10.1146/annurev-nutr-071812-161148

Hardie DG, Pan DA (2002) Regulation of fatty acid synthesis and oxidation by the AMP-activated protein kinase. Biochem Soc Trans 30:1064–1070. doi:10.1042/

Hardie DG, Ross FA, Hawley SA (2012) AMPK: a nutrient and energy sensor that maintains energy homeostasis. Nat Rev Mol Cell Biol 13:251–262. doi:10.1038/nrm3311

Hawley SA, Pan DA, Mustard KJ et al (2005) Calmodulin-dependent protein kinase kinase-beta is an alternative upstream kinase for AMP-activated protein kinase. Cell Metab 2:9–19. doi:10.1016/j.cmet.2005.05.009

He G, Zhang Y-W, Lee J-H et al (2014) AMP-activated protein kinase induces p53 by phosphorylating MDMX and inhibiting its activity. Mol Cell Biol 34:148–157. doi:10.1128/MCB.00670-13

Henin N, Vincent MF, Gruber HE, Van den Berghe G (1995) Inhibition of fatty acid and cholesterol synthesis by stimulation of AMP-activated protein kinase. FASEB J 9:541–546

Herms A, Bosch M, Reddy BJN et al (2015) AMPK activation promotes lipid droplet dispersion on detyrosinated microtubules to increase mitochondrial fatty acid oxidation. Nat Commun 6:7176. doi:10.1038/ncomms8176

Hindupur SK, González A, Hall MN (2015) The opposing actions of target of rapamycin and AMP-activated protein kinase in cell growth control. Cold Spring Harb Perspect Biol. doi:10.1101/cshperspect.a019141

Holz MK, Ballif BA, Gygi SP, Blenis J (2005) mTOR and S6K1 mediate assembly of the translation preinitiation complex through dynamic protein interchange and ordered phosphorylation events. Cell 123:569–580. doi:10.1016/j.cell.2005.10.024

Hoppe S, Bierhoff H, Cado I et al (2009) AMP-activated protein kinase adapts rRNA synthesis to cellular energy supply. Proc Natl Acad Sci USA 106:17781–17786. doi:10.1073/pnas.0909873106

Hosokawa N, Hara T, Kaizuka T et al (2009) Nutrient-dependent mTORC1 association with the ULK1-Atg13-FIP200 complex required for autophagy. Mol Biol Cell 20:1981–1991. doi:10.1091/mbc.E08-12-1248

Høyer-Hansen M, Bastholm L, Szyniarowski P et al (2007) Control of macroautophagy by calcium, calmodulin-dependent kinase kinase-β, and Bcl-2. Mol Cell 25:193–205. doi:10.1016/j.molcel.2006.12.009

Hwang J-T, Lee M, Jung S-N et al (2004) AMP-activated protein kinase activity is required for vanadate-induced hypoxia-inducible factor 1alpha expression in DU145 cells. Carcinogenesis 25:2497–2507. doi:10.1093/carcin/bgh253

Imamura K, Ogura T, Kishimoto A et al (2001) Cell cycle regulation via p53 phosphorylation by a 5′-AMP activated protein kinase activator, 5-aminoimidazole- 4-carboxamide-1-beta-D-ribofuranoside, in a human hepatocellular carcinoma cell line. Biochem Biophys Res Commun 287:562–567. doi:10.1006/bbrc.2001.5627

Inoki K, Li Y, Zhu T et al (2002) TSC2 is phosphorylated and inhibited by Akt and suppresses mTOR signalling. Nat Cell Biol 4:648–657. doi:10.1038/ncb839

Inoki K, Zhu T, Guan K-L (2003) TSC2 mediates cellular energy response to control cell growth and survival. Cell 115:577–590

Itakura E, Kishi-Itakura C, Koyama-Honda I, Mizushima N (2012) Structures containing Atg9A and the ULK1 complex independently target depolarized mitochondria at initial stages of Parkin-mediated mitophagy. J Cell Sci 125:1488–1499. doi:10.1242/jcs.094110

Jäger S, Handschin C, St-Pierre J, Spiegelman BM (2007) AMP-activated protein kinase (AMPK) action in skeletal muscle via direct phosphorylation of PGC-1alpha. Proc Natl Acad Sci USA 104:12017–12022. doi:10.1073/pnas.0705070104

Jeon S-M, Chandel NS, Hay N (2012) AMPK regulates NADPH homeostasis to promote tumour cell survival during energy stress. Nature 485:661–665. doi:10.1038/nature11066

Ji C, Yang B, Yang Y-L et al (2010) Exogenous cell-permeable C6 ceramide sensitizes multiple cancer cell lines to Doxorubicin-induced apoptosis by promoting AMPK activation and mTORC1 inhibition. Oncogene 29:6557–6568. doi:10.1038/onc.2010.379

Johansen T, Lamark T (2011) Selective autophagy mediated by autophagic adapter proteins. Autophagy 7:279–296

Jones RG, Plas DR, Kubek S et al (2005) AMP-activated protein kinase induces a p53-dependent metabolic checkpoint. Mol Cell 18:283–293. doi:10.1016/j.molcel.2005.03.027

Jung CH, Jun CB, Ro S-H et al (2009) ULK-Atg13-FIP200 complexes mediate mTOR signaling to the autophagy machinery. Mol Biol Cell 20:1992–2003. doi:10.1091/mbc.E08-12-1249

Kefas BA, Cai Y, Ling Z et al (2003a) AMP-activated protein kinase can induce apoptosis of insulin-producing MIN6 cells through stimulation of c-Jun-N-terminal kinase. J Mol Endocrinol 30:151–161. doi:10.1677/jme.0.0300151

Kefas BA, Heimberg H, Vaulont S et al (2003b) AICA-riboside induces apoptosis of pancreatic beta cells through stimulation of AMP-activated protein kinase. Diabetologia 46:250–254. doi:10.1007/s00125-002-1030-3

Koonen DPY, Glatz JFC, Bonen A, Luiken JJFP (2005) Long-chain fatty acid uptake and FAT/CD36 translocation in heart and skeletal muscle. Biochim Biophys Acta 1736:163–180. doi:10.1016/j.bbalip.2005.08.018

Kundu M, Lindsten T, Yang C-Y et al (2008) Ulk1 plays a critical role in the autophagic clearance of mitochondria and ribosomes during reticulocyte maturation. Blood 112:1493–1502. doi:10.1182/blood-2008-02-137398

Kuroyanagi H, Yan J, Seki N et al (1998) Human ULK1, a novel serine/threonine kinase related to UNC-51 kinase of *Caenorhabditis elegans*: cDNA cloning, expression, and chromosomal assignment. Genomics 51:76–85

Kurth-Kraczek EJ, Hirshman MF, Goodyear LJ, Winder WW (1999) 5′ AMP-activated protein kinase activation causes GLUT4 translocation in skeletal muscle. Diabetes 48:1667–1671

Leclerc GM, Leclerc GJ, Kuznetsov JN et al (2013) Metformin induces apoptosis through AMPK-dependent inhibition of UPR signaling in ALL lymphoblasts. PLoS One 8:e74420. doi:10.1371/journal.pone.0074420

Lemasters JJ (2005) Selective mitochondrial autophagy, or mitophagy, as a targeted defense against oxidative stress, mitochondrial dysfunction, and aging. Rejuvenation Res 8:3–5. doi:10.1089/rej.2005.8.3

Li Y, Xu S, Mihaylova MM et al (2011) AMPK phosphorylates and inhibits SREBP activity to attenuate hepatic steatosis and atherosclerosis in diet-induced insulin-resistant mice. Cell Metab 13:376–388. doi:10.1016/j.cmet.2011.03.009

Liu K, Czaja MJ (2013) Regulation of lipid stores and metabolism by lipophagy. Cell Death Differ 20:3–11. doi:10.1038/cdd.2012.63

Liu C, Liang B, Wang Q et al (2010) Activation of AMP-activated protein kinase α1 alleviates endothelial cell apoptosis by increasing the expression of anti-apoptotic proteins Bcl-2 and survivin. J Biol Chem 285:15346–15355. doi:10.1074/jbc.M110.102491

Löffler AS, Alers S, Dieterle AM et al (2011) Ulk1-mediated phosphorylation of AMPK constitutes a negative regulatory feedback loop. Autophagy 7:696–706. doi:10.4161/auto.7.7.15451

Luiken JJFP, Coort SLM, Willems J et al (2003) Contraction-induced fatty acid translocase/CD36 translocation in rat cardiac myocytes is mediated through AMP-activated protein kinase signaling. Diabetes 52:1627–1634. doi:10.2337/diabetes.52.7.1627

Luo Z, Saha AK, Xiang X, Ruderman NB (2005) AMPK, the metabolic syndrome and cancer. Trends Pharmacol Sci 26:69–76. doi:10.1016/j.tips.2004.12.011

Ma XM, Blenis J (2009) Molecular mechanisms of mTOR-mediated translational control. Nat Rev Mol Cell Biol 10:307–318. doi:10.1038/nrm2672

Marsin AS, Bertrand L, Rider MH et al (2000) Phosphorylation and activation of heart PFK-2 by AMPK has a role in the stimulation of glycolysis during ischaemia. Curr Biol 10:1247–1255

Marsin A-S, Bouzin C, Bertrand L, Hue L (2002) The stimulation of glycolysis by hypoxia in activated monocytes is mediated by AMP-activated protein kinase and inducible 6-phosphofructo-2-kinase. J Biol Chem 277:30778–30783. doi:10.1074/jbc.M205213200

Mayer C, Zhao J, Yuan X, Grummt I (2004) mTOR-dependent activation of the transcription factor TIF-IA links rRNA synthesis to nutrient availability. Genes Dev 18:423–434. doi:10.1101/gad.285504

McGee SL, van Denderen BJW, Howlett KF et al (2008) AMP-activated protein kinase regulates GLUT4 transcription by phosphorylating histone deacetylase 5. Diabetes 57:860–867. doi:10.2337/db07-0843

Meisse D, Van de Casteele M, Beauloye C et al (2002) Sustained activation of AMP-activated protein kinase induces c-Jun N-terminal kinase activation and apoptosis in liver cells. FEBS Lett 526:38–42. doi:10.1016/S0014-5793(02)03110-1

Merrill GF, Kurth EJ, Hardie DG, Winder WW (1997) AICA riboside increases AMP-activated protein kinase, fatty acid oxidation, and glucose uptake in rat muscle. Am J Physiol 273: E1107–E1112

Meynet O, Zunino B, Happo L et al (2013) Caloric restriction modulates Mcl-1 expression and sensitizes lymphomas to BH3 mimetic in mice. Blood 122:2402–2411. doi:10.1182/blood-2013-01-478651

Mitsiades N (2002) Biologic sequelae of nuclear factor-kappa B blockade in multiple myeloma: therapeutic applications. Blood 99:4079–4086. doi:10.1182/blood.V99.11.4079

Mizushima N (2007) Autophagy: process and function. Genes Dev 21:2861–2873. doi:10.1101/gad.1599207

Mizushima N (2010) The role of the Atg1/ULK1 complex in autophagy regulation. Curr Opin Cell Biol 22:132–139. doi:10.1016/j.ceb.2009.12.004

Mukherjee P, Mulrooney TJ, Marsh J et al (2008) Differential effects of energy stress on AMPK phosphorylation and apoptosis in experimental brain tumor and normal brain. Mol Cancer 7:37. doi:10.1186/1476-4598-7-37

Mungai PT, Waypa GB, Jairaman A et al (2011) Hypoxia triggers AMPK activation through reactive oxygen species-mediated activation of calcium release-activated calcium channels. Mol Cell Biol 31:3531–3545. doi:10.1128/MCB.05124-11

Muoio DM, Seefeld K, Witters LA, Coleman RA (1999) AMP-activated kinase reciprocally regulates triacylglycerol synthesis and fatty acid oxidation in liver and muscle: evidence that sn-glycerol-3-phosphate acyltransferase is a novel target. Biochem J 338(Pt 3):783–791

Okoshi R, Ozaki T, Yamamoto H et al (2008) Activation of AMP-activated protein kinase induces p53-dependent apoptotic cell death in response to energetic stress. J Biol Chem 283:3979–3987. doi:10.1074/jbc.M705232200

Pehmøller C, Treebak JT, Birk JB et al (2009) Genetic disruption of AMPK signaling abolishes both contraction- and insulin-stimulated TBC1D1 phosphorylation and 14-3-3 binding in mouse skeletal muscle. Am J Physiol Endocrinol Metab 297:E665–E675. doi:10.1152/ajpendo.00115.2009

Porstmann T, Santos CR, Griffiths B et al (2008) SREBP activity is regulated by mTORC1 and contributes to Akt-dependent cell growth. Cell Metab 8:224–236. doi:10.1016/j.cmet.2008.07.007

Queiroz EAIF, Puukila S, Eichler R et al (2014) Metformin induces apoptosis and cell cycle arrest mediated by oxidative stress, AMPK and FOXO3a in MCF-7 breast cancer cells. PLoS One 9: e98207. doi:10.1371/journal.pone.0098207

Samovski D, Sun J, Pietka T et al (2015) Regulation of AMPK activation by CD36 links fatty acid uptake to β-oxidation. Diabetes 64:353–359. doi:10.2337/db14-0582

Santidrián AF, González-gironès DM, Iglesias-serret D et al (2010) AICAR induces apoptosis independently ofAMPK and p53 through up-regulation of the BH3-only proteins BIM and NOXAin chronic lymphocytic leukemia cells. Blood 116:3023–3032. doi:10.1182/blood-2010-05-283960

Shang L, Wang X (2011) AMPK and mTOR coordinate the regulation of Ulk1 and mammalian autophagy initiation. Autophagy 7:924–926

Shaw RJ (2009) LKB1 and AMP-activated protein kinase control of mTOR signalling and growth. Acta Physiol 196:65–80. doi:10.1111/j.1748-1716.2009.01972.x

She C, Zhu L, Zhen Y et al (2014) Activation of AMPK protects against hydrogen peroxide-induced osteoblast apoptosis through autophagy induction and NADPH maintenance: new implications for osteonecrosis treatment? Cell Signal 26:1–8. doi:10.1016/j.cellsig.2013.08.046

Sun H, Yu T, Li J (2011) Co-administration of perifosine with paclitaxel synergistically induces apoptosis in ovarian cancer cells: more than just AKT inhibition. Cancer Lett 310:118–128. doi:10.1016/j.canlet.2011.06.010

Tian W, Li W, Chen Y et al (2015) Phosphorylation of ULK1 by AMPK regulates translocation of ULK1 to mitochondria and mitophagy. FEBS Lett 589:1847–1854. doi:10.1016/j.febslet.2015.05.020

Till A, Lakhani R, Burnett SF, Subramani S (2012) Pexophagy: the selective degradation of peroxisomes. Int J Cell Biol 2012:512721. doi:10.1155/2012/512721

Vander Heiden MG, Cantley LC, Thompson CB (2009) Understanding the Warburg effect: the metabolic requirements of cell proliferation. Science 324:1029–1033. doi:10.1126/science.1160809

Viollet B, Andreelli F (2011) AMP-activated protein kinase and metabolic control. Handb Exp Pharmacol 303–330. doi:10.1007/978-3-642-17214-4_13

Vucicevic L, Misirkic M, Janjetovic K et al (2009) AMP-activated protein kinase-dependent and -independent mechanisms underlying in vitro antiglioma action of compound C. Biochem Pharmacol 77:1684–1693. doi:10.1016/j.bcp.2009.03.005

Weisová P, Dávila D, Tuffy LP et al (2011) Role of 5′-adenosine monophosphate-activated protein kinase in cell survival and death responses in neurons. Antioxid Redox Signal 14:1863–1876. doi:10.1089/ars.2010.3544

Wirth M, Joachim J, Tooze SA (2013) Autophagosome formation – the role of ULK1 and Beclin1-PI3KC3 complexes in setting the stage. Semin Cancer Biol 23:301–309. doi:10.1016/j.semcancer.2013.05.007

Wong PM, Puente C, Ganley IG, Jiang X (2013) The ULK1 complex: sensing nutrient signals for autophagy activation. Autophagy 9:124–137

Woods A, Dickerson K, Heath R et al (2005) Ca2+/calmodulin-dependent protein kinase kinase-beta acts upstream of AMP-activated protein kinase in mammalian cells. Cell Metab 2:21–33. doi:10.1016/j.cmet.2005.06.005

Wu S-B, Wei Y-H (2012) AMPK-mediated increase of glycolysis as an adaptive response to oxidative stress in human cells: Implication of the cell survival in mitochondrial diseases. Biochim Biophys Acta 1822:233–247. doi:10.1016/j.bbadis.2011.09.014

Wu N, Zheng B, Shaywitz A et al (2013) AMPK-dependent degradation of TXNIP upon energy stress leads to enhanced glucose uptake via GLUT1. Mol Cell 49:1167–1175. doi:10.1016/j.molcel.2013.01.035

Wu W, Tian W, Hu Z et al (2014) ULK1 translocates to mitochondria and phosphorylates FUNDC1 to regulate mitophagy. EMBO Rep 15:566–575. doi:10.1002/embr.201438501

Xu J, Ji J, Yan X-H (2012) Cross-talk between AMPK and mTOR in regulating energy balance. Crit Rev Food Sci Nutr 52:373–381. doi:10.1080/10408398.2010.500245

Yeh L, Lee K, Kim K (1980) Regulation of rat liver acetyl-coA carboxylase. J Biol Chem 255:2308–2315

Zhang W-B, Wang Z, Shu F et al (2010) Activation of AMP-activated protein kinase by temozolomide contributes to apoptosis in glioblastoma cells via p53 activation and mTORC1 inhibition. J Biol Chem 285:40461–40471. doi:10.1074/jbc.M110.164046

Zoncu R, Efeyan A, Sabatini DM (2011) mTOR: from growth signal integration to cancer, diabetes and ageing. Nat Rev Mol Cell Biol 12:21–35. doi:10.1038/nrm3025

Chapter 4
AMPK and Placental Progenitor Cells

Melissa R. Kaufman and Thomas L. Brown

Contents

Abstract AMPK is important in numerous physiological systems but plays a vital role in embryonic and placental development. The placenta is a unique organ that is the essential lifeline between the mother and baby during pregnancy and gestation. During placental development, oxygen concentrations are very low until cells differentiate to establish the appropriate lineages that take on new functions required for placental and embryonic survival. Balancing the oxygen regulatory environment with the demands for energy and need to maintain metabolism during this process places AMPK at the center of maintaining placental cellular homeostasis as it integrates and responds to numerous complex stimuli. AMPK plays a critical role in sensing metabolic and energy changes. Once activated, it turns on pathways that produce energy and shuts down catabolic processes. AMPK coordinates cell growth, differentiation, and nutrient transport to maintain cell survival. Appropriate regulation of AMPK is essential for normal placental and embryonic development, and its dysregulation may lead to pregnancy-associated disorders such as intrauterine growth restriction, placental insufficiency, or preeclampsia.

Keywords Placenta • Trophoblast • AMPK • Pregnancy • Stem cells • Differentiation

M.R. Kaufman • T.L. Brown, Ph.D. (✉)
Department of Neuroscience, Cell Biology and Physiology, Wright State University Boonshoft School of Medicine, 3640 Colonel Glenn Highway, 457 Neuroscience Engineering Collaboration Building, Dayton, OH 45435, USA
e-mail: Thomas.L.brown@wright.edu

M.D. Cordero, B. Viollet (eds.), *AMP-activated Protein Kinase*, Experientia Supplementum 107, DOI 10.1007/978-3-319-43589-3_4

4.1 AMPK

AMP-activated protein kinase (AMPK) is a heterotrimeric, serine/threonine protein kinase that is present in almost every cell in the body and is composed of a catalytic alpha 1 (AMPKα1, PrkAA1) or alpha 2 (AMPKα2, PrkAA2) subunit, in combination with beta/gamma regulatory subunits (Viollet et al. 2003, 2009; Steinberg and Kemp 2009). AMPK is evolutionarily conserved and homologues have been identified in all eukaryotes, including protozoa (Steinberg and Kemp 2009; Viollet et al. 2009). Such a highly conserved gene and its subsequent signaling mechanisms highlight the essential role it plays in physiological homeostasis and cell survival.

AMPK acts as a critical sensor of cellular energy, metabolism, and stress. It functions to coordinate cell growth, differentiation, and apoptosis as well as regulates autophagy and mitochondrial biogenesis to maintain cell survival (Mihaylova and Shaw 2011; Hardie 2011; Rafalski et al. 2012). It is activated by energy depletion and mitochondrial stress that can lead to mitophagy (Toyama et al. 2016; Kim et al. 2011). AMPK also directly activates the first step in the autophagy pathway and responds to low nutrient and metabolic status as a means to maintain cell survival (Egan et al. 2011). Under chronic stress, AMPK is capable of stimulating the production of new mitochondria to replace those lost or damaged (Toyama et al. 2016). AMPK is activated by hypoxia or when a decrease in energy levels occur, i.e., when there is an increased ratio of cellular AMP:ATP, and in response triggers signaling pathways to increase the cells' production of its energy source, ATP (Evans et al 2006). Thus, AMPK activates energy-producing signaling pathways while inhibiting energy-consuming processes to maintain homeostasis (Fig. 4.1).

4.2 AMPK in Development

AMPK has been shown to be important in numerous physiological systems; however, its vital role in placental and embryonic development has only recently begun to emerge (Carey et al. 2014). The AMPK catalytic subunits mediate energy utilization during development and gene knockout of AMPKα1 or AMPKα2 alone have shown distinct physiological phenotypes in mice, but are viable (Viollet et al. 2003, 2009). In contrast, AMPKα1/AMPKα2 double knockout mice are lethal at embryonic day 10.5 in utero and this signifies the importance of AMPK in development (Viollet et al. 2009). Recent reports indicate that AMPK is an important regulator and determinant of cell lineage during the differentiation of embryonic stem cells (Young et al. 2016; Louden et al. 2014; Vazquez-Martin et al. 2012); however, the role of AMPK in placental development has only recently come under study.

The placenta is a unique organ that is generated within the mother's uterus only during pregnancy and is essential for embryonic development and survival.

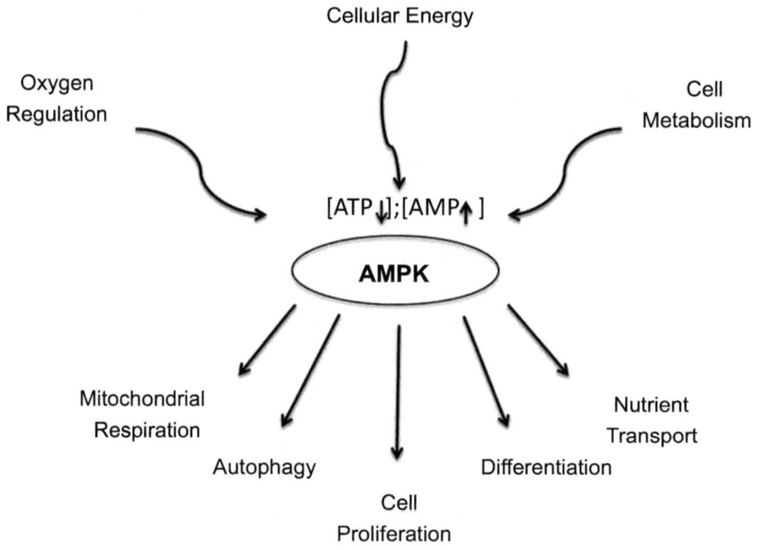

Fig. 4.1 AMPK acts as a critical sensor of oxygen regulation, cellular energy, and metabolism. AMPK recognizes the decrease in energy as a reduction in ATP or elevation in AMP/ADP. In response to low-energy conditions, AMPK plays a major role in regulating mitochondrial respiration, autophagy, differentiation, nutrient transport, and cell proliferation to promote cell survival

The placenta originates from specialized cells of the developing embryo (trophoblasts) and serves as the lifeline between the mother and developing baby to carry out functions that the embryo/fetus cannot. The placenta functions to facilitate the uptake and delivery of nutrients and aids in the elimination of wastes. The placenta is also important in mediating the exchange of critical gases, such as oxygen and carbon dioxide, via the mother's blood supply and produces hormones necessary to sustain pregnancy. In addition, the placenta provides an immunoprivileged site that not only provides a barrier against viral infection but also prevents the rejection of the fetus as a foreign body.

Placental development is regulated by a critical balance between stem cell proliferation and the differentiation of trophoblasts into distinct placental cell lineages. The mouse placenta, which shares many attributes and functions with the human placenta, is made up of three lineages and numerous cellular subtypes that are important for trophoblast invasion, stem cell/progenitor growth, and nutrient transport (Watson and Cross 2005). Labyrinthine cells are responsible for nutrient transport and the exchange of gases and waste at the maternal fetal interface (Natale et al. 2006; Jansson and Powell 2013; Lager and Powell 2012). As proper placental transport is necessary for a healthy pregnancy and to allow for normal fetal development, alterations in AMPK signaling and trophoblast differentiation could impair fetal viability and result in pregnancy-associated disorders.

4.3 Pathophysiology of Pregnancy-Associated Disorders

Abnormal placental development is known to be associated with a number of pregnancy-associated disorders including intrauterine growth restriction (IUGR), placental insufficiency, and preeclampsia (Caniggia et al. 2000; Chaddha et al. 2004; Red-Horse et al. 2004). In humans, alterations in the AMPKα1 genotype are associated with birth weight, oxygen regulation, and metabolic homeostasis and may be implicated in the pathophysiology of IUGR and high altitude adaptation (Bigham et al. 2014). Furthermore, AMPK is present in placental tissue in humans and mice, was increased under low oxygen or hypoxic conditions, and facilitated maternal uterine artery blood flow (Skeffington et al. 2016). These results suggest that AMPK provides a protective role by sensing and rescuing cellular energy depletion to maintain maternal/fetal blood flow and prevent placental insufficiency. The role of AMPK in the pathogenesis of preeclampsia has been reported using a reduced uteroplacental perfusion pressure (RUPP) rat model (Banek et al. 2013). RUPP mice have an angiogenic imbalance and develop high blood pressure. Administration of an AMPK activator, AICAR, was able to prevent the development of hypertension and normalize angiogenesis (Banek et al. 2013). These results suggest that reduced levels of AMPK can lead to significant pathological conditions during pregnancy and indicate that decreasing oxidative stress and maintaining proper metabolic and energy levels are critical to normal placental development and fetal outcome, as well as maternal well-being.

4.4 AMPK in Placental Progenitor Cells and Differentiation

Trophoblast cells experience environmental changes as they divide and migrate, such as increasing oxygen concentrations and increased metabolic demands, which leads to corresponding changes in gene regulation. These changes in gene expression determine placental progenitor differentiation into the appropriate cell lineages at the proper time during development. Recent reports have suggested that placental stem cell proliferation and differentiation can be regulated by AMPK (Carey et al. 2014). To examine the effect of AMPK inhibition, we created a small interfering RNA (siRNA) and short hairpin RNA (shRNA) capable of targeting both catalytic isoforms of AMPKα1/2. Causing significant knockdown of both AMPKα isoforms in mouse and human cells resulted in a significant reduction in AMPK activity (Tangeman et al. 2012). The major metabolic and energy demands on the placenta would occur in labyrinthine nutrient-transporting cells (Jansson and Powell 2013). Therefore, we examined the effects of AMPK knockdown in the mouse labyrinthine trophoblast progenitor cell line, SM10 (Selesniemi et al. 2005a, b). Inhibition of AMPK results in the loss of appropriate placental differentiation with significant alterations in cell morphology, inhibition of cell growth, and reduced nutrient transport (Carey et al. 2014). Our results are supported by two studies in blastocysts and early trophoblast stem cells that show that AMPK is important in controlling trophoblast stem cell (TS) differentiation (Xie et al. 2013; Zhong et al. 2010). While these studies considered AMPK signaling in the context of a stress response and the manner in which trophoblast cells may respond to changes in their microenvironment, these findings could also be considered in the context of normal development and differentiation. As trophoblasts differentiate, their glycolytic need and mitochondrial respiratory capacity change in concert with their energy requirements and ability to produce ATP. AMPK could play a key role in controlling metabolic homeostasis in placental trophoblasts and its dysregulation could therefore negatively impact differentiation of these placental progenitor cells. Interestingly, mitochondrial DNA has been shown to be increased in human IUGR placentas and was inversely correlated with oxygen levels (Cetin and Alvino 2009). This could be due to a compensatory mechanism for a hypoxic environment or a metabolic response to reduced nutrient transport.

4.5 Future Directions

While the importance of AMPK in placental differentiation and pregnancy-associated disorders has only recently become evident, many questions remain. The role of mitochondria and their response to the energy demands of trophoblast differentiation and maintenance of homeostasis have not been reported. Furthermore, examining the interactions AMPK encounters and the signaling in response to a low oxygen environment are of great interest as they represent the in vivo scenario that occurs during pregnancy as the placenta develops. The comparative response of AMPK to chronic and high-dose stress compared to low stress levels that may allow for developmental adaptation may provide valuable information as to how placental plasticity may be involved and at what point plasticity may be lost or beyond adaptation (Mansouri et al. 2012).

Analysis of genes capable of interacting with or in response to AMPK levels to mediate differentiation may provide new information on the regulatory controls mediating placental development (Zhong et al. 2010; Selesniemi et al. 2016). Determining the signal transduction pathways that govern the hypoxic response as they relate to the control of cellular metabolism and nutrient transport during development should provide new findings about the integrated network necessary to keep cells alive and maintain metabolism in the very low oxygen state in which the placenta develops and eventually undergoes differentiation.

While gene knockout studies can be highly valuable in determining the function of genes, they suffer from the ability to distinguish whether the critical function is in the embryo or placenta, as both are targeted by this technique. Studies directed at examining placental-specific gene expression of AMPK using Cre-lox technology, Crisper/Cas, and/or lentiviral blastocyst transduction would provide novel insights into understanding AMPK function (Kaufman et al. 2014; Okada et al. 2007).

Placental transgenesis studies would accelerate our understanding of the role of placental AMPK and pregnancy-associated disorders and provide new possibilities for therapeutic advances.

Acknowledgements We would like to thank Dr. David Natale (University of California San Diego) for critical reading of the manuscript and helpful input. This work was supported in part by a grant from the National Institutes of Health NICHD-R01 HD059969 (TLB) and The Wright State University Endowment for Research on Pregnancy Associated Disorders (www.wright.edu/give/pregnancyassociateddisorders) (TLB).

References

Banek CT, Bauer AJ, Needham KM, Dreyer HC, Gilbert JS (2013) AICAR administration ameliorates hypertension and angiogenic imbalance in a model of preeclampsia in the rat. Am J Physiol Heart Circ Physiol 304:H1159–H1165

Bigham AW, Julian CG, Wilson MJ, Vargas E, Browne VA, Shriver MD, Moore LG (2014) Maternal PRKAA1 and EDNRA genotypes are associated with birth weight, and PRKAA1 with uterine artery diameter and metabolic homeostasis at high altitude. Physiol Genomics 46:687–697

Caniggia I, Winter J, Lye SJ, Post M (2000) Oxygen and placental development during the first trimester: implications for the pathophysiology of pre-eclampsia. Placenta 21(Suppl A):S25–S30

Carey EAK, Albers RE, Doliboa SR, Hughes M, Wyatt CN, Natale DRC, Brown TL (2014) AMPK knockdown in placental trophoblast cells results in altered morphology and function. Stem Cells Dev 23:2921–2930

Cetin I, Alvino G (2009) Intrauterine growth restriction: implications for placental metabolism and transport. A review. Placenta 30(Suppl A):S77–S82

Chaddha V, Viero S, Huppertz B, Kingdom J (2004) Developmental biology of the placenta and the origins of placental insufficiency. Semin Fetal Neonatal Med 9:357–369

Egan DF, Shackelford DB, Mihaylova MM, Gelino S, Kohnz RA, Mair W, Vasquez DS, Joshi A, Gwinn DM, Taylor R, Asara JM, Fitzpatrick J, Dillin A, Viollet B, Kundu M, Hansen M, Shaw RJ (2011) Phosphorylation of ULK1 (hATG1) by AMP-activated protein kinase connects energy sensing to mitophagy. Science 331:456–461

Evans AM, Hardie DG, Galione A, Peers C, Kumar P, Wyatt CN (2006) AMP-activated protein kinase couples mitochondrial inhibition by hypoxia to cell-specific Ca2+ signalling mechanisms in oxygen-sensing cells. Novartis Found Symp 272:234–252

Hardie DG (2011) AMP-activated protein kinase: an energy sensor that regulates all aspects of cell function. Genes Dev 25:1895–1908

Jansson T, Powell TL (2013) Role of placental nutrient sensing in developmental programming. Clin Obstet Gynecol 56:591–601

Kaufman MR, Albers RE, Keoni C, Kulkarni-Datar K, Natale DR, Brown TL (2014) Important aspects of placental-specific gene transfer. Theriogenology 82:1043–1048

Kim J, Kundu M, Viollet B, Guan KL (2011) AMPK and mTOR regulate autophagy through direct phosphorylation of Ulk1. Nat Cell Biol 13:132–141

Lager S, Powell TL (2012) Regulation of nutrient transport across the placenta. J Pregnancy 2012:179827

Louden ED, Luzzo KM, Jimenez PT, Chi T, Chi M, Moley KH (2014) TallyHO obese female mice experience poor reproductive outcomes and abnormal blastocyst metabolism that is reversed by metformin. Reprod Fertil Dev 27:31–39

Mansouri L, Xie Y, Rappolee DA (2012) Adaptive and pathogenic responses to stress by stem cells during development. Cells 1:1197–1224

Mihaylova MM, Shaw RJ (2011) The AMPK signaling pathway coordinates cell growth, autophagy and metabolism. Nat Cell Biol 13:1016–1023

Natale DR, Starovic M, Cross JC (2006) Phenotypic analysis of the mouse placenta. Methods Mol Med 121:275–293

Okada Y, Ueshin Y, Isotani A, Saito-Fujita T, Nakashima H, Kimura K, Mizoguchi A, Oh-Hora M, Mori Y, Ogata M, Oshima RG, Okabe M, Ikawa M (2007) Complementation of placental defects and embryonic lethality by trophoblast-specific lentiviral gene transfer. Nat Biotechnol 25:233–237

Rafalski VA, Mancini E, Brunet A (2012) Energy metabolism and energy-sensing pathways in mammalian embryonic and adult stem cell fate. J Cell Sci 125:5597–5608

Red-Horse K, Zhou Y, Genbacev O, Prakobphol A, Foulk R, McMaster M, Fisher SJ (2004) Trophoblast differentiation during embryo implantation and formation of the maternal-fetal interface. J Clin Invest 114:744–745

Selesniemi KL, Reedy MA, Gultice AD, Brown TL (2005a) Identification of committed placental stem cell lines for studies of differentiation. Stem Cells Dev 14:535–547

Selesniemi K, Reedy M, Gultice A, Guilbert LJ, Brown TL (2005b) Transforming growth factor-beta induces differentiation of the labyrinthine trophoblast stem cell line SM10. Stem Cells Dev 14:697–711

Selesniemi KL, Albers RE, Brown TL (2016) Id2 mediates differentiation of labyrinthine placental progenitor cell line, SM10. Stem Cells Dev. ePub

Skeffington KL, Higgins JS, Mahmoud AD, Evans AM, Sferruzzi-Perri AN, Fowden AL, Yung HW, Burton GJ, Giussani DA, Moore LG (2016) Hypoxia, AMPK activation and uterine artery vasoreactivity. J Physiol 594:1357–1369

Steinberg GR, Kemp BE (2009) AMPK in health and disease. Physiol Rev 89:1025–1078

Tangeman L, Wyatt CN, Brown TL (2012) Knockdown of AMP-activated protein kinase alpha 1 and alpha 2 catalytic subunits. J RNAi Gene Silencing 8:470–478

Toyama EQ, Herzig S, Courchet J, Lewis TL Jr, Losón OC, Hellberg K, Young NP, Chen H, Polleux F, Chan DC, Shaw RJ (2016) Metabolism. AMP-activated protein kinase mediates mitochondrial fission in response to energy stress. Science 351:275–281

Vazquez-Martin A, Vellon L, Quirós PM, Cufí S, Ruiz de Galarreta E, Oliveras-Ferraros C, Martin AG, Martin-Castillo B, López-Otín C, Menendez JA (2012) Activation of AMP-activated protein kinase (AMPK) provides a metabolic barrier to reprogramming somatic cells into stem cells. Cell Cycle 11:974–989

Viollet B, Andreelli F, Jørgensen SB, Perrin C, Flamez D, Mu J, Wojtaszewski JF, Schuit FC, Birnbaum M, Richter E, Burcelin R, Vaulont S (2003) Physiological role of AMP-activated protein kinase (AMPK): insights from knockout mouse models. Biochem Soc Trans Part 1:216–219

Viollet B, Athea Y, Mounier R, Guigas B, Zarrinpashneh E, Horman S, Lantier L, Hebrard S, Devin-Leclerc J, Beauloye C, Foretz M, Andreelli F, Ventura-Clapier R, Bertrand L (2009) AMPK: lessons from transgenic and knockout animals. Front Biosci 14:19–44

Watson ED, Cross JC (2005) Development of structures and transport functions in the mouse placenta. Physiology (Bethesda) 20:180–193

Xie Y, Awonuga A, Liu J, Rings E, Puscheck EE, Rappolee DA (2013) Stress induces AMPK-dependent loss of potency factors Id2 and Cdx2 in early embryos and stem cells. Stem Cells Dev 22:1564–1575

Young NP, Kamireddy A, Van Nostrand JL, Eichner LJ, Shokhirev MN, Dayn Y, Shaw RJ (2016) AMPK governs lineage specification through Tfeb-dependent regulation of lysosomes. Genes Dev 30:535–552

Zhong W, Xie Y, Abdallah M, Awonuga AO, Slater JA, Sipahi L, Puscheck EE, Rappolee DA (2010) Cellular stress causes reversible, PRKAA1/2-, and proteasome-dependent ID2 protein loss in trophoblast stem cells. Reproduction 140:921–930

Chapter 5
Role of AMP-Activated Protein Kinase for Regulating Post-exercise Insulin Sensitivity

Rasmus Kjøbsted, Jørgen F.P. Wojtaszewski, and Jonas T. Treebak

Contents

R. Kjøbsted
Novo Nordisk Foundation Center for Basic Metabolic Research, Section of Integrative Physiology, Faculty of Health and Medical Sciences, University of Copenhagen, Blegdamsvej 3b, 2200 Copenhagen, Denmark

Section of Molecular Physiology, Department of Nutrition, Exercise and Sports, Faculty of Science, University of Copenhagen, 2100 Copenhagen, Denmark

J.F.P. Wojtaszewski
Section of Molecular Physiology, Department of Nutrition, Exercise and Sports, Faculty of Science, University of Copenhagen, 2100 Copenhagen, Denmark

J.T. Treebak (✉)
Novo Nordisk Foundation Center for Basic Metabolic Research, Section of Integrative Physiology, Faculty of Health and Medical Sciences, University of Copenhagen, Blegdamsvej 3b, 2200 Copenhagen, Denmark
e-mail: jttreebak@sund.ku.dk

M.D. Cordero, B. Viollet (eds.), *AMP-activated Protein Kinase*, Experientia Supplementum 107, DOI 10.1007/978-3-319-43589-3_5

Abstract Skeletal muscle insulin resistance precedes development of type 2 diabetes (T2D). As skeletal muscle is a major sink for glucose disposal, understanding the molecular mechanisms involved in maintaining insulin sensitivity of this tissue could potentially benefit millions of people that are diagnosed with insulin resistance. Regular physical activity in both healthy and insulin-resistant individuals is recognized as the single most effective intervention to increase whole-body insulin sensitivity and thereby positively affect glucose homeostasis. A single bout of exercise has long been known to increase glucose disposal in skeletal muscle in response to physiological insulin concentrations. While this effect is identified to be restricted to the previously exercised muscle, the molecular basis for an apparent convergence between exercise- and insulin-induced signaling pathways is incompletely known. In recent years, we and others have identified the Rab GTPase-activating protein, TBC1 domain family member 4 (TBC1D4) as a target of key protein kinases in the insulin- and exercise-activated signaling pathways. Our working hypothesis is that the AMP-activated protein kinase (AMPK) is important for the ability of exercise to insulin sensitize skeletal muscle through TBC1D4. Here, we aim to provide an overview of the current available evidence linking AMPK to post-exercise insulin sensitivity.

Keywords AMP-activated protein kinase • Exercise • Skeletal muscle • Glucose uptake • TBC1D4 • AS160 • Insulin sensitivity

5.1 Introduction

The prevalence of type 2 diabetes (T2D) and its comorbidities has reached epidemic proportions worldwide. T2D is a heterogeneous disorder that is often characterized by the coexistence of insulin deficiency and insulin resistance in multiple tissues (DeFronzo 1988). Insulin resistance in major metabolic organs such as liver, skeletal muscle, and adipose tissue has been recognized as being one of the first abnormalities in the development of T2D (Abdul-Ghani and DeFronzo 2010). Insulin resistance can be defined as the inability of normal levels of insulin to induce a normal biological response, and traditionally this is demonstrated by an impairment of insulin to regulate whole-body glucose homeostasis (Lebovitz 2001).

Insulin is an anabolic hormone that plays a key role in controlling glucose homeostasis. Insulin represses glucose production from the liver while stimulating glucose uptake in peripheral tissues such as skeletal muscle and adipose tissue (Saltiel and Kahn 2001). Since skeletal muscle constitutes around 40–45 % of whole-body mass in humans and is responsible for up to ~80 % of total body glucose uptake during insulin stimulation (DeFronzo et al. 1983; Baron et al. 1988), it is considered an important organ for maintaining glucose homeostasis. It has been argued that skeletal muscle insulin resistance plays a pivotal role in the pathogenesis of T2D (DeFronzo and Tripathy 2009). Thus, to combat T2D, an understanding of the molecular mechanisms that regulate insulin sensitivity in skeletal muscle is evidently important.

Depending on both time and intensity, muscle contractile activity possesses the ability to increase muscle glucose uptake by ~20-fold (Rose and Richter 2005). Similarly to insulin, this is in part mediated by an increased translocation of glucose transporter (GLUT) 4 proteins to the cell surface membrane (Hirshman et al. 1988; Douen et al. 1989). Several studies suggest that contraction increases muscle glucose uptake via intracellular mechanisms distinct from those of insulin (Lee et al. 1995; Lund et al. 1995; Yeh et al. 1995; Wojtaszewski et al. 1996, 1999). This is also supported by fully additive effects of insulin and contraction on muscle glucose uptake (Nesher et al. 1985; Ploug et al. 1987), indicating recruitment of different intracellular pools of GLUT4 by insulin and contraction, respectively (Douen et al. 1990; Coderre et al. 1995; Ploug et al. 1998). Nevertheless, contraction and insulin signaling pathways seem to converge at some point(s), since skeletal muscle displays a marked increase in insulin sensitivity to stimulate glucose uptake in a prolonged period after a single bout of exercise (Richter et al. 1982, 1989). Interestingly, this is observed both in healthy individuals and in insulin-resistant T2D patients (Devlin et al. 1987). However, we still do not completely understand the molecular mechanisms governing this phenomenon. Elucidating such mechanisms may serve as a valuable source of knowledge for future pharmacological intervention in the treatment of muscle insulin resistance in various conditions. Our hypothesis is that activation of AMP-activated protein kinase (AMPK) is important for the ability of exercise to insulin sensitize skeletal muscle, and we propose that pharmacological targeting of AMPK could be a viable approach to prevent muscle insulin resistance.

AMPK is considered a key sensor of cellular energy status and found in essentially all eukaryotic genomes (Hardie 2007). AMPK is activated in response to metabolic stresses that cause an imbalance in the cellular energy homeostasis, which is recognized by an increase in the ratios of AMP:ATP and ADP:ATP (Gowans and Hardie 2014). In skeletal muscle, AMPK is commonly activated in response to exercise, muscle contraction, hypoxia, and ischemia (Kahn et al. 2005), but also myokines and adipokines (e.g., IL-6, leptin, and adiponectin) have been observed to regulate AMPK activity in skeletal muscle (Minokoshi et al. 2002; Yamauchi et al. 2002; Kelly et al. 2004). In general, AMPK switches on catabolic processes that increase ATP production while switching off anabolic processes that consume ATP in an attempt to restore cellular energy homeostasis (Hardie et al. 2012). Thus, in skeletal muscle AMPK promotes glucose uptake and lipid oxidation (Jørgensen et al. 2004b; Fentz et al. 2015) while suppressing glycogen synthase activity and lipid and protein synthesis (Muoio et al. 1999; Inoki et al. 2003; Jørgensen et al. 2004a). AMPK is a heterotrimeric complex consisting of a catalytic α subunit and regulatory β and γ subunits of which several isoforms exist (α1, α2, β1, β2, γ1, γ2, γ3) (Steinberg and Kemp 2009). While all isoforms have been found in sample preparations of human and mouse skeletal muscle (Wojtaszewski et al. 2005; Treebak et al. 2009a), only a subgroup of heterotrimeric complexes are present in human vastus lateralis (i.e., α2β2γ1, α2β2γ3, and α1β2γ1) (Wojtaszewski et al. 2005; Birk and Wojtaszewski 2006) and in mouse skeletal muscle (i.e., α2β2γ3, α2β2γ1, α2β1γ1, α1β2γ1, and α1β1γ1) (Treebak et al. 2009a).

In the following, we provide an overview of how AMPK activity is modulated in skeletal muscle. Subsequently, we will discuss the role of AMPK in mediating glucose uptake, and finally we will review the evidence for a role of AMPK as a central player regulating increased skeletal muscle insulin sensitivity observed after an acute bout of exercise.

5.2 Regulation of AMPK Activity

5.2.1 Adenosine Nucleotides

AMP and ADP are the primary intracellular activators of AMPK, whereas ATP antagonizes this effect. Thus, the intracellular AMP:ATP and ADP:ATP ratios may be regarded as predictors of AMPK activation (Hardie et al. 2011). During metabolic stress, when ATP consumption is accelerated (e.g., muscle contraction) or ATP synthesis is inhibited (e.g., inhibition of mitochondrial respiration), the intracellular ADP concentration increases. This drives the reversible adenylate kinase reaction (2ADP ↔ ATP + AMP) towards ATP (and AMP) in an attempt to maintain intracellular ATP concentrations. By virtue of the adenylate kinase reaction during metabolic stress, the AMP:ATP ratio will rise and consequently increase allosteric activation of AMPK by up to tenfold (Suter et al. 2006). Besides allosteric activation, binding of AMP to AMPK promotes phosphorylation of Thr172 by upstream kinases and hinders Thr172 dephosphorylation by protein phosphatases (Xiao et al. 2011). This has been shown to increase AMPK activity by more than 100-fold (Suter et al. 2006; Sanders et al. 2007), indicating that phosphorylation of Thr172 is a major regulator of AMPK activity. Although allosteric activation of AMPK is triggered by AMP only, binding of ADP mimics the effects of AMP on regulating Thr172 phosphorylation (Oakhill et al. 2011).

5.2.2 Glycogen

In addition to the regulatory role of adenine nucleotides, muscle glycogen levels seem to be an important factor in regulating AMPK activity. Thus, AMPK may not only be viewed as an energy sensor but also as a fuel sensor in skeletal muscle. The AMPK β subunit contains a glycogen-binding domain, which seems to be responsible for associating AMPK to glycogen particles (Hudson et al. 2003; Polekhina et al. 2003). Studies in both humans and rodents have shown that glycogen acts as a negative regulator of AMPK since glycogen loading of skeletal muscle suppresses basal AMPK activity and reduces activation of AMPK in response to activating stimuli such as exercise, contraction, and AICAR (Derave et al. 2000a; Wojtaszewski et al. 2002a, 2003b). However, other observations argue against a direct role of glycogen in regulating muscle AMPK activity in response to exercise (Nielsen et al. 2002; McConell et al. 2005). Hence, in muscle from McArdle patients, which exhibits chronic high glycogen levels, exercise-mediated AMPK activation is not suppressed compared to control subjects (Nielsen et al. 2002). Despite these opposing observations, the inverse relationship between contraction-induced AMPK activation and muscle glycogen content is not found in the AMPKγ3 KO mouse model indicating that only AMPKγ3-associated activity is subjected to inhibitory regulation by glycogen (Barnes et al. 2005). Interestingly, after a period of exercise training, muscle glycogen content increases, which may play a role in the reduced activation of AMPK observed in response to a single bout of exercise in the trained muscle (Durante et al. 2002; Nielsen et al. 2003; Mortensen et al. 2013). Moreover, AMPKγ3 protein abundance is reduced in skeletal muscle following exercise training, which may further explain the reduced effect of exercise on AMPK activation (Durante et al. 2002; Nielsen et al. 2003; Frøsig et al. 2004; Wojtaszewski et al. 2005; Mortensen et al. 2013). Nevertheless, the highly significant correlations between exercise-induced AMPKγ3 activity and muscle glycogen content observed before and after exercise training (Mortensen et al. 2013) support the idea that increasing levels of AMPKγ3 activity during glycogen-depleting exercise is related to a relief from glycogen-associated inhibition.

5.2.3 Upstream Kinases

AMPK activity is regulated by phosphorylation of Thr172 of the α subunit by upstream kinases. Three AMPK kinases have been identified to phosphorylate Thr172, including Serine/threonine-protein kinase 11 (also known as LKB1), Ca $^{2+}$/calmodulin-dependent kinase kinase (CaMKK), and transforming growth factor β-activating kinase 1 (TAK1). The exact physiological significance in skeletal muscle of the latter remains unsure and will not be dealt with further (Momcilovic et al. 2006; Xie et al. 2006; Herrero-Martín et al. 2009). LKB1 is a heterotrimeric kinase that is active only when forming a complex with two regulatory subunits, mouse protein 25 (MO25) and STE20-related adaptor protein (STRAD) (Hawley et al. 2003; Woods et al. 2003a). Besides regulating activity, binding of STRAD and MO25 translocate the complex from the nucleus to the cytosol where it targets downstream effectors (Baas et al. 2003; Boudeau et al. 2003). LKB1 appears to be constitutively active in skeletal muscle, since LKB1 activity does not change in response to AMPK activating stimuli (e.g., exercise, contraction, and AICAR) (Sakamoto et al. 2004; Jeppesen et al. 2013). Both in vitro and animal studies suggest that LKB1 functions as an AMPKK. Thus, in LKB1-deficient HeLa cells, AMPK is only activated by AICAR if the cells are allowed to express stabile LKB1 (Hawley et al. 2003), and in muscle-specific LKB1 knockout (KO) mice, activation of AMPKα2 by contraction or AICAR is abolished with only minor effects on AMPKα1 activity (Sakamoto et al. 2004, 2005). Similarly, activation of AMPKα2 by exercise is blunted in muscle-specific LKB1 KO mice while AMPKα1 activity seems unaffected (Jeppesen et al. 2013). Thus, LKB1 is likely the primary regulator of AMPKα2 activity in skeletal muscle, whereas AMPKα1 activity may be regulated by a kinase different from LKB1.

Three independent groups have reported that CaMKK increases activation and phosphorylation of AMPK in the absence of a nucleotide imbalance, indicating that AMPK activity is regulated by Ca^{2+} signaling (Hawley et al. 2005; Hurley et al. 2005; Woods et al. 2005). The studies showed that AMPK activity increased in HeLa cells stimulated with Ca^{2+} ionophores while failing to appear in cells pretreated with CaMKK inhibitor STO-609 or CaMKK siRNA. In support of this, using CaMKK inhibitors STO-609 and KN-93, it has been reported that CaMKK acts as a regulator of AMPKα1 activity in mouse skeletal muscle at the onset of mild tetanic contraction in a time- and/or intensity-dependent manner (Jensen et al. 2007a, b). Also, overexpression of constitutively active CaMKK(α) in mouse skeletal muscle by in vivo gene electro-transfer increases AMPKα1 and AMPKα2 activity by 2.5-fold (Witczak et al. 2007).

Taken together, LKB1 appears to mainly activate AMPKα2-containing trimers, and since these complexes have been shown to play important roles for skeletal muscle metabolism and mitochondrial biogenesis (Jørgensen et al. 2004a, b; Garcia-Roves et al. 2008), LKB1 appears pivotal for regulating AMPK activity in skeletal muscle. Since LKB1 is presumed constitutively active in skeletal muscle this indicates that AMPK is regulated by allosteric activation and/or by changes in the rate by which AMPK is dephosphorylated at Thr172 by protein phosphatase (PP) 2A and 2C (Davies et al. 1995; Woods et al. 2003b; Suter et al. 2006; Sanders et al. 2007). In addition, a highly conserved nuclear export sequence has been found in the catalytic subunits of AMPK, which regulates activity and subcellular localization of the heterotrimer complex (Kazgan et al. 2010). This may also affect the ability of upstream kinases to regulate AMPK.

5.2.4 Exercise/Muscle Contraction

During contractile activity (e.g., exercise) muscle energy consumption may increase up to 100-fold (Sahlin et al. 1998). This highlights the huge dynamics of muscle energy metabolism, which represents a major metabolic challenge for the muscle cell. Yet, the intracellular ATP concentration varies surprisingly little even during high-intensity exercise (Hellsten et al. 1998). Thus, maintaining a constant level of ATP in the working muscle seems essential to sustain muscle function. The finding that mice lacking AMPK in skeletal muscle have a reduced exercise capacity (Maarbjerg et al. 2009; O'Neill et al. 2011; Jeppesen et al. 2013; Fentz et al. 2015) highlights the importance of AMPK for maintaining energy homeostasis.

In 1996, Winder and Hardie were the first to report that exercise increases AMPK activity in rat skeletal muscle (Winder and Hardie 1996). Later, it was shown that exercise also increases AMPK activity in human skeletal muscle (Chen et al. 2000; Fujii et al. 2000; Wojtaszewski et al. 2000b) and that specific AMPK heterotrimer complexes were regulated in an intensity- and time-dependent manner. As such, activation of AMPK in human skeletal muscle has been observed to occur at exercise intensities of at least 60 % of $VO2_{peak}$ (Wojtaszewski et al. 2000b, 2003b; Stephens et al. 2002; Chen et al. 2003), although activation at lower intensity (~45 % of $VO2_{peak}$) has been reported during prolonged exercise (~3.5 h) (Wojtaszewski et al. 2002b). Furthermore, most studies (Fujii et al. 2000; Stephens et al. 2002; Wojtaszewski et al. 2002b) but not all (Musi et al. 2001a) report a time-dependent increase in exercise-induced AMPK activation. When dividing AMPK activity into either AMPKα1- or α2-associated activity, it becomes evident that these complexes display differential regulation in response to exercise. Thus, activation of AMPK in human skeletal muscle during exercise is mainly associated with increased α2 activity (Chen et al. 2000; Fujii et al. 2000; Wojtaszewski et al. 2000b, 2002b, 2003b; Musi et al. 2001a; Stephens et al. 2002; Chen et al. 2003; Birk and Wojtaszewski 2006; Treebak et al. 2007; Mortensen et al. 2013; Kjøbsted et al. 2016) while α1 activity has been reported to decrease (Birk and Wojtaszewski 2006; Mortensen et al. 2013; Kjøbsted et al. 2016), increase (Chen et al. 2000, 2003; Treebak et al. 2007), or even remain unchanged (Wojtaszewski et al. 2000b, 2002b, 2003b; Fujii et al. 2000; Musi et al. 2001a; Stephens et al. 2002; Birk and Wojtaszewski 2006) depending on exercise mode, time, and intensity. In addition, several reports from our group have shown that the initial increase in muscle AMPK activity during exercise ($\leq$30 min) is restricted to the $\alpha2\beta2\gamma3$ complex (Birk and Wojtaszewski 2006; Treebak et al. 2007), whereas the $\alpha2\beta2\gamma1$ and $\alpha1\beta2\gamma1$ complexes seem to increase at later time points ($\geq$60 min) (Treebak et al. 2007) or even in exercise recovery (Kjøbsted et al. 2016). There is consensus that activation of AMPK is a local phenomenon that occurs in the working muscle. In support of this, a human study applying one-legged knee extensor exercise with and without simultaneous arm cranking exercise did not find that humoral factors released during exercise increased AMPK activity in resting muscle or potentiated AMPK activity in exercising muscle (Kristensen et al. 2007). Studies in isolated intact rodent muscle induced to contract are in line with this observation (Vavvas et al. 1997; Derave et al. 2000a; Sakamoto et al. 2005).

In contrast to observations in humans, activation of AMPK in rodent skeletal muscle during wheel running, treadmill exercise, in situ, and ex vivo contraction seems to involve changes in both AMPKα1- and α2-associated activity (Jørgensen et al. 2004b, 2005; Jensen et al. 2007b; Koh et al. 2008; Maarbjerg et al. 2009; O'Neill et al. 2011; Jeppesen et al. 2013), although two studies reported unchanged α1-associated activity in response to in situ contraction (10–20 min) of rat gastrocnemius muscle (Sakamoto et al. 2004, 2005). In mice, ex vivo muscle contraction appears to increase AMPK-associated activity of all five AMPK trimer complexes, but the absolute level of activity is fiber type-dependent (Treebak et al. 2014). Future studies should determine the pattern of AMPK trimer activation in both human and rodent skeletal muscle, in order to assign AMPK complex-specific roles to specific cellular functions.

5.2.5 AICAR

5-Aminoimidazole-4-carboxamide-1-β-D-ribofuranoside (AICAR) is an adenosine analogue that is taken up into cells by the adenosine transport system and subsequently phosphorylated by adenosine kinase to yield 5-aminoimidazole-4-carboxamide-1-β-D-ribonucleotide (ZMP), a close analogue of AMP. ZMP is part of the de novo synthesis of purine metabolites and is predominantly metabolized into inosine monophosphate (IMP) within the cell (Sabina et al. 1985). In response to AICAR treatment, the intracellular ZMP levels can increase to mM concentrations depending on stimulation time (Sabina et al. 1985; Corton et al. 1995). Although ZMP is thought to mimic most effects of AMP, several studies provide evidence to support that accumulation of ZMP and its metabolites do not perturb the intracellular content of AMP, ADP, and ATP (Corton et al. 1995; Merrill et al. 1997). Merrill and colleagues demonstrated that perfusion of the rat hindlimb with increasing concentrations of AICAR (0.5–2 mM) or increasing time intervals (7½, 15, and 45 min) did not cause a significant change in muscle adenosine nucleotides (Merrill et al. 1997). Furthermore, one study reported that the adenosine nucleotide pool in dog skeletal muscle remained unchanged 60 min after infusion of AICAR (8 mM) (Sabina et al. 1982). Interestingly, at this time point the authors also reported that the accumulation of AICAR and ZMP in skeletal muscle had returned to near pre-infusion levels. Taken together, the intracellular accumulation of ZMP in skeletal muscle is likely determined by AICAR concentration and stimulation time, which does not seem to affect the adenosine nucleotide pools. At least in the perfused animal the acute effect of AICAR on ZMP formation in skeletal muscle seems to be gone within ~1 h.

The effect of ZMP on AMPK activation was originally described in a patent from 1993 (Van den Berghe and Gruber 1993). The following year, it was demonstrated that AICAR increased AMPK activity in a time- and dose-dependent manner in rat adipocytes (Sullivan et al. 1994). Because ZMP binds to the γ subunit similarly to AMP, it likely mimics all the effects of AMP in regulating AMPK complex activity (Day et al. 2007). However, at physiological ATP concentrations (4 mM) the ZMP Km value is approximately 50-fold higher compared to AMP

(Corton et al. 1995). Furthermore, in rat hepatocytes high concentrations of ZMP (>5–10 mM) have been shown to decrease AMPK activity likely due to competing binding between ZMP and ATP at the AMPK catalytic site (Henin et al. 1996). High concentrations of AMP (>1 mM) has similar effects on AMPK activity, but since AMP is a more potent activator, inhibition of AMPK activity is not observed in skeletal muscle at concentrations normally seen during exercise (AMP_{free}: ~6–10 μM, assuming that intracellular water comprises 80 % of skeletal muscle) (Birk and Wojtaszewski 2006). In light of these findings, it seems apparent that ZMP acts similar to AMP in regulating AMPK activity albeit at much lower potency. This is presumably due to decreased binding affinity of ZMP for AMPK (AMP > ADP > ATP > ZMP) (Ramanathan et al. 2010).

AICAR has been shown to increase muscle AMPK activity in several rodent models including ex vivo and in vivo studies in mice (Mu et al. 2001; Jørgensen et al. 2004b; Steinberg et al. 2010; Kjøbsted et al. 2015) and rats (Hayashi et al. 1998; Bergeron et al. 1999; Musi et al. 2001b; Wojtaszewski et al. 2002a). Surprisingly, several studies in humans have consistently found no increase in muscle AMPK activity in response to continuous intravenous AICAR infusion (10–45 mg kg^{-1} h^{-1}) (Cuthbertson et al. 2007; Boon et al. 2008; Bosselaar et al. 2009; Babraj et al. 2009), although one study reported increased ACC Ser79 phosphorylation—a known marker of endogenous AMPK activity (Boon et al. 2008). This may indicate that AICAR delivery to resting human muscle was inadequate as AICAR has been shown to increase AMPK activity (Thr172 phosphorylation) in isolated human muscle strips (Koistinen et al. 2003). Similar to exercise, the AMPK heterotrimeric complexes display differential regulation in response to AICAR. Thus, in rat skeletal muscle AICAR increases AMPKα2-associated activity while AMPKα1-associated activity seems to increase in some (Wojtaszewski et al. 2002a; Miyamoto et al. 2007) but not all (Aschenbach et al. 2002; Jessen et al. 2003) studies. Interestingly, AICAR appears to increase α1-associated activity only in rat muscle high in glycolytic fibers, and this seems to depend on muscle glycogen content (Wojtaszewski et al. 2002a). In mouse skeletal muscle, AICAR increases both AMPKα2- and α1-associated activity regardless of muscle type (Jørgensen et al. 2004b; Fujii et al. 2005; Sakamoto et al. 2005; Lefort et al. 2008; Maarbjerg et al. 2009; Jensen et al. 2015). Currently, it is unknown why AICAR acts more potently on AMPKα1-associated activity in mouse muscle compared to rat muscle. However, it may be related to muscle glycogen levels as mouse skeletal muscle contains roughly 25–50 % of glycogen content compared to rat muscle (Jensen et al. 2009). Observations from our group indicate that AICAR increases the activity of all AMPK trimer complexes in mouse skeletal muscle (Fig. 5.1; Treebak, Birk and Wojtaszewski, unpublished observations). Thus, phosphorylation of AMPK Thr172 as well as total AMPK activity increases similarly in extensor digitorum longus (EDL) and soleus (SOL) muscle after 2 h of AICAR stimulation (Fig. 5.1a + b). These observations are mirrored by α2β2γ1 activity (Fig. 5.1c) but are highly different for measurements of α2β2γ3, α1β2γ1, and α1β1γ1 + α2β1γ1 activities (Fig. 5.1d–f). Based on these findings, it is evident

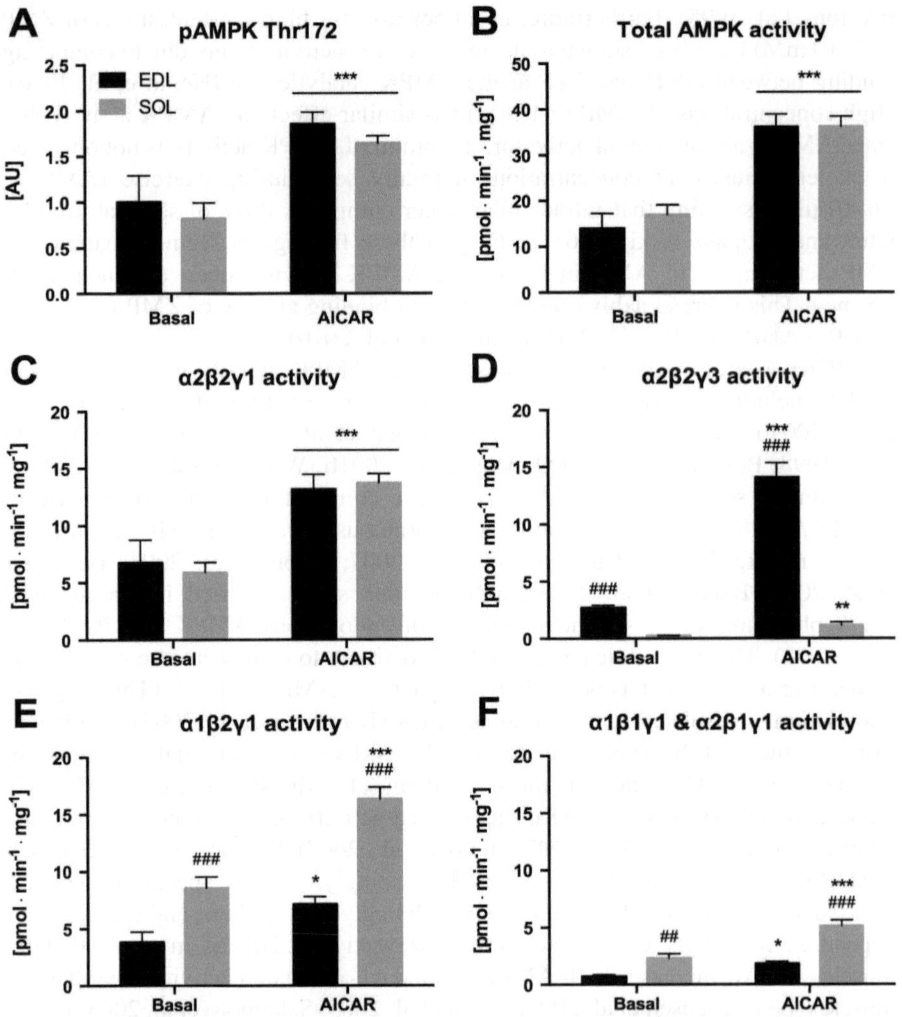

Fig. 5.1 AMPK Thr172 phosphorylation and activity of mouse EDL and SOL muscle in response to acute AICAR stimulation. EDL (*black bars*) and SOL (*gray bars*) muscles from C57BL/6 mice were incubated ± AICAR (2 mM) for 2 h. Following muscle homogenization and preparation, muscle lysates were subjected to Western blot analyses for detection of phosphorylated AMPK Thr172 (**a**). Furthermore, muscle lysates were used for immunoprecipitation experiments to estimate AMPK total (**b**) and associated α2β2γ1 (**c**), α2β2γ3 (**d**), α1β2γ1 (**e**), and α1β1γ1 & α2β1γ1 (**f**) activity. Data are presented as means ± SEM and analyzed by two-way ANOVAs ($n = 6$). (**a–c**): $^{***}p < 0.001$ main effect of AICAR. (**d–f**): Interaction ($p < 0.05$ muscle type × treatment). $^{*}p < 0.05$, $^{**}p < 0.01$, $^{***}p < 0.001$ effect of AICAR within muscle type. $^{\#\#}p < 0.01$, $^{\#\#\#}p < 0.001$ effect of muscle type within treatment. *AU* arbitrary units

that measures of total AMPK activity or phosphorylation mask differential regulation among the AMPK heterotrimer complexes. By comparison, these differences likely reflect the relative distribution of the different heterotrimer complexes in soleus and EDL muscle (Treebak et al. 2009a). However, in other circumstances it may reflect changes in the stoichiometry of heterotrimeric activity, as elevated total AMPK activity and Thr172 phosphorylation following acute exercise can be a result of increased, decreased, or unaltered activity of specific heterotrimeric complexes as seen in human skeletal muscle (Birk and Wojtaszewski 2006; Kjøbsted et al. 2016).

5.3 Role of AMPK in Regulating Glucose Uptake in Skeletal Muscle

Merrill and colleagues provided the first indication of a link between AMPK and its ability to regulate muscle glucose uptake. Not only did AICAR increase AMPK activity but it also resulted in an increase in muscle glucose uptake (Merrill et al. 1997). This was later confirmed by others (Hayashi et al. 1998; Bergeron et al. 1999) and in 2001 observations in mice lacking functional AMPKα2 in skeletal muscle provided genetic evidence to support a role of AMPK in regulating muscle glucose uptake in response to AICAR (Mu et al. 2001). Studies in several other AMPK-deficient mouse models as well as LKB1-deficient mice have generated similar results (Jørgensen et al. 2004b; Barnes et al. 2004; Fujii et al. 2005; Sakamoto et al. 2005; Steinberg et al. 2010). The current thinking is that the $\alpha2\beta2\gamma3$ heterotrimer complex is responsible for AICAR-induced glucose uptake in skeletal muscle. In support of this, AICAR increases glucose uptake more potently in EDL compared to SOL muscle (Jørgensen et al. 2004b; Jensen et al. 2015; Kjøbsted et al. 2015), which is in line with the relative distribution of the $\alpha2\beta2\gamma3$ trimer complex in the two muscles (Treebak et al. 2009a). Furthermore, in most studies (Balon and Jasman 2001; Kaushik et al. 2001; Ai et al. 2002; Wright et al. 2005) but not all (Wojtaszewski et al. 2002a), AICAR fails to increase glucose uptake in rat SOL muscle which expresses very low levels of AMPKγ3 (Mahlapuu et al. 2004).

Since exercise/contraction increases both muscle glucose uptake and AMPK activity in an intensity- and time-dependent manner (Rose and Richter 2005; Birk and Wojtaszewski 2006), it has long been reasoned that AMPK regulates muscle glucose uptake during exercise/contraction. However, despite this association several ex vivo, in situ, and in vivo studies in AMPK-deficient mice have demonstrated that exercise-/contraction-induced muscle glucose uptake is not (Barnes et al. 2004; Fujii et al. 2005; Maarbjerg et al. 2009; Merry et al. 2010; Steinberg et al. 2010; Lantier et al. 2014; Fentz et al. 2015) or only partially (Mu et al. 2001; Jørgensen et al. 2004b; Jensen et al. 2008; Lefort et al. 2008; Lee-Young et al. 2009; O'Neill et al. 2011) mediated by AMPK. The reason for this discrepancy may be related to differences in the experimental design including background strain, gender, mutagenesis methodology, exercise/contraction protocol, protocol for measuring glucose uptake, and ex vivo settings (temperature, buffer composition, and oxygenation). Supporting a negligible role of AMPK in regulating muscle glucose uptake during acute exercise/contraction, we recently demonstrated in conventional AMPKα1/α2 skeletal muscle-specific double knockout mice (AMPK mdKO) that muscle glucose uptake and muscle glucose clearance are not impaired during ex vivo contraction and in vivo treadmill exercise, respectively (Lantier et al. 2014; Fentz et al. 2015). These observations suggest a dispensable role of AMPK or that AMPK may be involved but not necessary for regulating muscle glucose uptake during exercise/contraction. In fact, AMPK-independent mechanisms (e.g., actin remodeling by Rac1 during mechanical stretch) may also be important for the ability of contraction/exercise to increase muscle glucose uptake (Sylow et al. 2015).

In order to assess a possible role of AMPK in regulating contraction/exercise-induced glucose uptake, studies investigating the impact of contraction and AMPK activation in GLUT4 trafficking may provide clues to this challenge. A study in cardiomyocytes demonstrated that contraction increases GLUT4 exocytosis while activation of AMPK in response to oligomycin decreases the rate of GLUT4 endocytosis (Yang and Holman 2005). Similarly, studies have demonstrated that activation of AMPK by AICAR increases muscle glucose uptake by inhibiting GLUT4 endocytosis (Karlsson et al. 2009; Fazakerley et al. 2010). Considering the effect of contraction on muscle AMPK activity, it could be argued that contraction-induced glucose uptake in skeletal muscle involves increased exocytosis in combination with decreased endocytosis of GLUT4. Thus, discrepancies in exercise/contraction-induced glucose uptake between different AMPK transgenic mouse models may be related to the potential effect of AMPK deficiency on GLUT4 exocytosis (either positive or negative). Alternatively and somewhat speculative, one may find an explanation to the contradicting reports on AMPK and muscle glucose uptake by evaluating the different protocols used to assess glucose uptake. Thus, most (Mu et al. 2001; Jørgensen et al. 2004b; Jensen et al. 2008; Lefort et al. 2008; Lee-Young et al. 2009) but not all (O'Neill et al. 2011) studies reporting impaired exercise-/contraction-induced glucose uptake in AMPK and LKB1-deficient mice measure glucose uptake in the post-exercise/contraction period. This is in contrast to observations in the AMPK mdKO mouse model, in which muscle glucose uptake was measured during exercise/contraction (Fentz et al. 2015; Lantier et al. 2014).

5.3.1 Downstream Signaling to Glucose Uptake

Studies in both skeletal and cardiac muscle have shown that AMPK (AICAR) induces glucose uptake by increasing the amount of GLUT4 in the cell surface membrane (Kurth-Kraczek et al. 1999; Russell et al. 1999; Mu et al. 2001;

Koistinen et al. 2003). However, the signaling events mediating the effect of AMPK on GLUT4 translocation are not fully described but may involve phosphorylation of the downstream targets TBC1D4 and/or TBC1D1 (for review see Cartee 2015b). Thus, in cell-free assays AMPK strongly phosphorylates TBC1D4 Ser588 (Thr642 weakly), which has also been observed in HEK-293 cells after AICAR treatment (Geraghty et al. 2007). Also, recombinant AMPK phosphorylates Ser704 on TBC1D4 purified from mouse tibialis anterior (TA) muscle (Treebak et al. 2010). In addition to these findings, several studies have demonstrated that phosphorylation of TBC1D4 increases in human and rodent skeletal muscle during exercise/contraction as well as in rodent skeletal muscle upon AICAR stimulation (Bruss et al. 2005; Kramer et al. 2006a; Funai et al. 2009, 2010; Treebak et al. 2010, 2014; Ducommun et al. 2012; Schweitzer et al. 2012; Castorena et al. 2014). This suggests that AMPK may act as an upstream regulator of TBC1D4.

Although TBC1D4 is phosphorylated on several sites in response to exercise/contraction, observations in different AMPK-deficient mouse models indicate that AMPK only regulates phosphorylation of TBC1D4 Thr642 (Kramer et al. 2006a) and Ser704 (Treebak et al. 2010) in mature skeletal muscle (at least in mice). In healthy human skeletal muscle, we recently reported strong correlations between TBC1D4 phosphorylation (Ser341 and Ser704) and AMPK $\alpha2\beta2\gamma1$ activity ($R^2 = 0.84$, $p < 0.01$) after one-legged knee extensor exercise (Treebak et al. 2014), indicating that this particular AMPK complex may regulate TBC1D4 phosphorylation during exercise. On the other hand, 1 h of cycle exercise only increased phosphorylation of TBC1D4 Ser704 and $\alpha2\beta2\gamma3$ activity in skeletal muscle of type 2 diabetic and weight-matched control subjects (Kjøbsted et al. 2016). Thus, the exercise-induced phosphorylation signature on TBC1D4 may depend on exercise mode and the metabolic characteristics of the muscle in a specific population.

Using in vivo gene electro-transfer to overexpress a mutant construct of TBC1D4 containing four different Ser/Thr to Ala mutations in skeletal muscle, contraction-stimulated glucose uptake was reduced by ~20 % compared to muscle overexpressing wild type (WT) TBC1D4 (Kramer et al. 2006b). Since one of the sites mutated in the construct was Thr642, impairments in contraction-induced glucose uptake could have been due to the absence of AMPK-mediated phosphorylation of Thr642. However, in skeletal muscle of mice carrying a single TBC1D4 Thr642Ala knock-in mutation, neither AICAR- nor contraction-stimulated glucose uptake is affected (Ducommun et al. 2012). Moreover, a TBC1D4 Ser704Ala mutant overexpressed in mouse TA muscle by in vivo gene electro-transfer did not exert impairments in either AICAR- or contraction-stimulated glucose uptake (Treebak et al. 2010). Taken together, the current available data from different transgenic mouse models question the role of TBC1D4 in facilitating AMPK-mediated glucose uptake in skeletal muscle.

In cell systems and in cell-free assays, AICAR and recombinant AMPK increase phosphorylation of TBC1D1 Ser237 and Thr596, respectively (Chen et al. 2008). Furthermore, several studies employing isolated skeletal muscle from different AMPK-deficient mouse models have provided genetic evidence to support that phosphorylation of TBC1D1 (PAS, Ser237, Thr596, Ser660, and Ser700) is regulated by AMPK during contraction and possibly the AMPKα2 heterotrimer complex (Pehmøller et al. 2009; Frøsig et al. 2010; Vichaiwong et al. 2010; O'Neill et al. 2011; Treebak et al. 2014). In addition, several human studies have demonstrated that phosphorylation of TBC1D1 increases in skeletal muscle in response to exercise (Frøsig et al. 2010; Jessen et al. 2011; Treebak et al. 2014; Vendelbo et al. 2014; Kjøbsted et al. 2016), and correlative evidence strongly supports that the α2β2γ3 complex induces phosphorylation of TBC1D1 Ser237 (Treebak et al. 2014; Kristensen et al. 2015; Kjøbsted et al. 2016). These results indicate a functional role of TBC1D1 in regulating AMPK-mediated glucose uptake, although this needs to be verified. Nevertheless, this idea is supported by two studies using in vivo gene electro-transfer to overexpress TBC1D1 mutated at four predicted AMPK phosphorylation sites. Both studies observed that the contraction-stimulated glucose uptake was impaired by ~20–35 % in mouse TA muscle overexpressing the TBC1D1 mutant (An et al. 2010; Vichaiwong et al. 2010). Although impairments in glucose uptake may seem low, one have to consider that only a fraction of the muscle is transfected following electroporation (~50–85 %) (McMahon et al. 2001; Fujii et al. 2004). Thus, a relative small decrease in glucose uptake of electroporated muscle could in fact become quite significant if transfection efficiency is improved. Interestingly, single point mutations of TBC1D1 did not affect glucose uptake in response to contraction (Vichaiwong et al. 2010).

Within recent years, several transgenic mouse models, which lack functional TBC1D1 or have reduced levels of TBC1D1 in skeletal muscle, have been described. Common for these models is an impaired effect of insulin, exercise, contraction, and AICAR to stimulate glucose uptake in glycolytic skeletal muscle (with high expression of TBC1D1) (Szekeres et al. 2012; Dokas et al. 2013; Stöckli et al. 2015; Chadt et al. 2015). However, all the TBC1D1-deficient mouse models have reduced expression of GLUT4 protein in skeletal muscle, which may be responsible for the observed defects in glucose uptake. Taken together, the abovementioned results indicate that AMPK-mediated glucose uptake involves multisite phosphorylation of TBC1D1. Investigations of TBC1D1 in the regulation of exocytosis and endocytosis of GLUT4 in skeletal muscle have not been performed, and further studies are warranted to assess the exact mechanism by which AMPK, if at all, induces glucose uptake in skeletal muscle during/after exercise and contraction.

5.4 Effect of Prior Contractile Activity on Muscle Insulin Sensitivity: A Role for AMPK?

A biological response generated by insulin may be defined by two separate terms: (1) *insulin responsiveness*, which is defined as the maximal response that can be achieved at any insulin concentration and (2) *insulin sensitivity*, which is defined as the insulin concentration that produces a half-maximal response. An increase in insulin responsiveness results in a greater biological response to a maximal effective insulin concentration, which translates into an upward shift of the insulin dose–response curve. On the other hand, an increase in insulin sensitivity decreases the insulin concentration needed to produce a half-maximal biological response. This is observed as a leftward shift of the insulin dose–response curve. It is important to differentiate between the two terms since the mechanisms responsible for improving insulin sensitivity and responsiveness may be quite distinct. It is also important to specify the biological response to which insulin sensitivity and insulin responsiveness refer, since insulin produces numerous effects in skeletal muscle. Thus, in this context, the terms insulin sensitivity and insulin responsiveness refer to the effect of insulin on muscle glucose transport/uptake.

In the landmark study from Neil Ruderman's laboratory, Richter et al. showed, using the perfused rat hindlimb model, that a single bout of treadmill exercise increased insulin sensitivity to stimulate glucose uptake in skeletal muscle (Richter et al. 1982). Improvements in muscle insulin sensitivity were found 1, 2, and 4 h after exercise and were associated with enhanced insulin-stimulated glycogen accumulation. Based on these findings, the authors suggested that an increase in muscle insulin sensitivity following acute glycogen-depleting exercise was part of a mechanism ensuring resynthesis of muscle glycogen for future use. The molecular mechanisms governing this phenomenon remain, however, incompletely understood.

It is generally believed that increased GLUT4 translocation to the plasma membrane is responsible for the enhanced muscle insulin sensitivity after exercise. Evidence to support this has been generated by the use of the ATB-[2-^{3}H]BMPA exofacial photolabeling technique. Hansen and colleagues demonstrated in rat skeletal muscle that the increase in GLUT4 translocation was similar in magnitude to the increase in muscle insulin sensitivity 3 h after acute exercise (Hansen et al. 1998). Although these results indicate that the ability of insulin to stimulate glucose uptake following exercise associates with the amount of GLUT4 in the cell surface membrane, it has been argued that changes in muscle insulin sensitivity may also be explained by increased intrinsic activity of GLUT4 (Thong et al. 2003). As for now this has not been confirmed, but measurements of maximal insulin-stimulated glucose uptake may indirectly provide an indication. Thus, if GLUT4 intrinsic activity is elevated in exercise recovery, insulin responsiveness to stimulate glucose transport would be elevated as well. Some studies do not observe an increase in insulin responsiveness after exercise/contraction (Gulve et al. 1990; Iwabe et al. 2014), but it is not a universal finding (Richter et al. 1982, 1989; Garetto et al. 1984; Wallberg-Henriksson et al. 1988). Improvements in muscle insulin sensitivity may therefore be explained by increased GLUT4 translocation and/or changes in intrinsic GLUT4 activity.

Several observations from cell culture studies implicate AMPK in the regulation of muscle insulin sensitivity. In C_2C_{12} myotubes, greater insulin-stimulated glucose uptake is observed 2 h after AICAR stimulation (Smith et al. 2005). Furthermore, incubation of C_2C_{12} myotubes in hyperosmotic medium (AMPK-activating treatment) acts synergistically with insulin to increase glucose uptake, an effect that is prevented by inclusion of compound C (AMPK inhibitor) (Smith et al. 2005). Another study found increased insulin sensitivity in C_2C_{12} myotubes when these were transfected with constitutively active AMPK (Ju et al. 2007). In contrast, prior AICAR stimulation failed to increase insulin sensitivity if myotubes were transfected with dominant negative AMPK (Ju et al. 2007). Intriguingly, prior serum- and AICAR-induced AMPK activation did not improve insulin sensitivity in primary human myocytes (Al-Khalili et al. 2004). The authors suggested that differences in species and/or experimental models could explain the absence of enhanced insulin sensitivity, with a low abundance of GLUT4 protein in human myocytes as an obvious candidate (Al-Khalili et al. 2004). Taken together, results obtained from cell culture systems suggest that AMPK plays an important role in regulating muscle insulin sensitivity. In the following sections, we will focus our discussion on studies performed to assess the insulin-sensitizing effects of prior exercise in intact skeletal muscle, and we will specifically highlight the studies linking AMPK to this phenomenon.

5.4.1 Muscle Insulin Sensitivity Is Increased Up To 48 h After Acute Exercise

It has long been known that exercise/contraction increases glucose uptake in skeletal muscle by an insulin-independent mechanism (Ploug et al. 1984; Richter et al. 1985). After cessation of exercise/contraction, muscle glucose uptake gradually decreases back to resting levels depending on the muscle type and intensity of the preceding exercise bout. Thus, glucose uptake has been observed to reverse within ~½, ~1, and ~2 h following exercise/contraction in rat soleus, red gastrocnemius, and epitrochlearis muscle, respectively (Ploug et al. 1987; Young et al. 1987). In human thigh, glucose uptake returns to resting levels ~1 h after 60 min of cycling exercise (Wojtaszewski et al. 2003b) and ~3–4 h after 60 min of one-legged knee extensor exercise (Richter et al. 1989; Thong et al. 2002; Wojtaszewski et al. 1997, 2000a). On the other hand, an increase in muscle insulin sensitivity has been observed to persist for up to 48 h after prior exercise in rats depending on substrate availability (Cartee et al. 1989). Similarly, improvements in whole-body insulin sensitivity have been observed in untrained subjects 48 h after 60 min of cycling exercise (64 % of VO_2 max) (Mikines et al. 1988). The post-exercise period can generally be characterized by two phases (Garetto et al. 1984): *phase I*, a remaining effect of prior exercise on muscle glucose uptake which occur independently of insulin, and *phase II*, increased insulin sensitivity to stimulate muscle glucose uptake. Interestingly, many studies (Bogardus et al. 1983; Bonen et al. 1984; Davis et al. 1986; Devlin and Horton 1985; Mikines et al. 1988; Richter et al. 1982, 1985, 1989; Zorzano et al. 1986; Wallberg-Henriksson et al. 1988) but not all (Devlin et al. 1987; Cartee et al. 1989; Mikines et al. 1989; Gulve et al. 1990; Iwabe et al. 2014) also observe an increase in insulin responsiveness in the period after acute exercise/contraction. Although this topic is not within the scope of the present review, changes in insulin responsiveness early in exercise recovery (~3 h) could be explained by an increase in GLUT4 intrinsic activity and/or enhanced glucose phosphorylation that may become limiting for glucose disposal under specific conditions (Wasserman et al. 2011).

A large number of studies have demonstrated that AMPK Thr172 phosphorylation and activity increase in skeletal muscle in response to exercise and contraction. Following exercise/contraction, AMPK activity decreases reaching levels observed in resting muscle 3–7 h into recovery. A process that likely depends on exercise duration and intensity (Sriwijitkamol et al. 2007; Lee-Young et al. 2008; Frøsig et al. 2009; Mortensen et al. 2013; Kjøbsted et al. 2016). Assuming that phosphorylation of AMPK Thr172 is a reliable marker of AMPK activity, several studies indicate that the exercise-induced activation of AMPK has reversed to resting levels at times of increased insulin sensitivity (Arias et al. 2007; Frøsig et al. 2009; Funai et al. 2010; Pehmøller et al. 2012; Schweitzer et al. 2012). These observations are supported by the finding that AMPK activity of all three heterotrimeric complexes in human skeletal muscle had reversed to resting levels 4 h after acute exercise concomitant with enhanced muscle insulin sensitivity (Frøsig et al. 2009). The same study also noticed a small increase in ACC Ser221 phosphorylation 4 h after exercise suggesting a persistent effect of previously activated AMPK. Thus, although AMPK activity in skeletal muscle during increased insulin sensitivity may have returned to levels seen in resting muscle, a persistent effect of prior AMPK activation on downstream targets could still mediate the enhanced insulin sensitivity.

5.5 Improvements in Muscle Insulin Sensitivity Following Exercise Is Modulated by Local Contraction-Induced Factors

By in situ contraction of a single hindlimb via electric stimulation of the sciatic nerve followed by measurements of muscle insulin sensitivity in both hindlimbs by hindquarter perfusion, it was shown that increased muscle insulin sensitivity occurs in the prior contracted hindlimb only (Richter et al. 1984). These observations indicated that a local contraction-induced mechanism was sufficient to enhance insulin sensitivity in rat skeletal muscle. This was later verified in human skeletal muscle by several studies employing one-legged knee extensor exercise and a hyperinsulinemic euglycemic clamp in combination with arterial and bilateral femoral vein catheterization (Richter et al. 1989; Wojtaszewski et al. 1997, 2000a). Similarly, others have found enhanced insulin sensitivity in leg but not in forearm muscle following exercise (combined cycling and treadmill) (Annuzzi et al. 1991). Improved insulin sensitivity is also observed in incubated muscle from prior exercised rats (Davis et al. 1986; Wallberg-Henriksson et al. 1988; Cartee et al. 1989; Cartee and Holloszy 1990; Gulve et al. 1990) and in incubated rat muscle previously subjected to in situ contraction (Kim et al. 2004), further supporting the idea of an intrinsic skeletal muscle phenomenon. Interestingly, an intact systemic circulation does not seem to be necessary, since contraction of skeletal muscle ex vivo induces a subsequent increase in insulin sensitivity as long as the muscle has been immersed in serum during contraction (Gao et al. 1994). Although this indicates that exercise/contraction increases insulin sensitivity locally in skeletal muscle, it also demonstrates the presence of a factor in serum indispensable for contraction (and AICAR) to subsequently improve insulin sensitivity (Cartee and Holloszy 1990; Dumke et al. 2002; Fisher et al. 2002). It has been suggested that the kallikrein–kinin system may contribute to the insulin-sensitizing effect of exercise on muscle glucose uptake (Dumke et al. 2002). The system is activated during muscle contractile activity, and findings suggest that it may enhance insulin sensitivity (Rett et al. 1989; Kishi et al. 1998). However, more recent studies do not support this notion (Schweitzer et al. 2011; Schweitzer and Cartee 2011).

It should be recognized that local improvements in muscle insulin sensitivity within 3–4 h following acute exercise most likely does not implicate changes in abundance of proteins involved in insulin-stimulated glucose uptake. In support of this, inhibition of protein synthesis by adding cycloheximide to the incubation medium containing acutely exercised muscle does not prevent a subsequent increase in insulin sensitivity 3.5 h later (Fisher et al. 2002). However, at later time points (48 h) changes in protein expression may play a role. In fact, studies in both rat and human have shown that skeletal muscle GLUT4 protein levels increase 1.5–24 h after a single bout of exercise (Ren et al. 1994; Kuo et al. 1999a, b; Chibalin et al. 2000; Greiwe et al. 2000; Kraniou et al. 2006), although these findings are not universal (Frøsig et al. 2009; Leick et al. 2010; Steensberg et al. 2002; Wojtaszewski et al. 2000a). Thus, mechanisms regulating muscle insulin sensitivity may be different depending on the duration of recovery.

If prior exercise increases insulin sensitivity by modulating local factors in skeletal muscle, any possible involvement of AMPK should reflect such notion. We have found, using the one-legged knee extensor exercise model, that acute exercise only increases AMPK activity in active working skeletal muscle (Kristensen et al. 2007; Treebak et al. 2014). To determine the relationship between acute changes in AMPK Thr172 phosphorylation and post-contraction-enhanced insulin sensitivity during various in situ contraction protocols, it was reported that phosphorylation of AMPK Thr172 is consistently increased by contraction protocols that induce an increase in muscle insulin sensitivity (Kim et al. 2004). However, contraction protocols that do not trigger significant improvements in muscle insulin sensitivity may still induce increases in AMPK Thr172 phosphorylation (Kim et al. 2004). Similarly, after high- and low-intensity swimming exercise in rats a dissociation between the level of exercise-induced AMPK Thr172 phosphorylation and increase in muscle insulin sensitivity was found (Koshinaka et al. 2008, 2009). Since phosphorylation of AMPK Thr172 may conceal differential regulation among the AMPK heterotrimer complexes (e.g., Fig. 5.1), it cannot be ruled out that only a subset of AMPK trimer complexes were activated by the exercise/contraction protocols. This may explain discrepancies in post-exercise/contraction-enhanced muscle insulin sensitivity. Nevertheless, these results could also indicate that activation of AMPK may be necessary but not sufficient for improving muscle insulin sensitivity after contractile activity.

5.5.1 Regulation of Muscle Insulin Sensitivity After Exercise by Carbohydrate Availability and Glycogen

There is evidence to support that high-carbohydrate feeding accelerates the reversal of not only the exercise-induced increase in cell permeability to glucose (Young et al. 1983) but also post-exercise improvements in muscle insulin sensitivity. Accordingly, muscle insulin sensitivity was reversed 18 h after an acute bout of exercise if animals were given free access to chow or a high-carbohydrate diet in the post-exercise period (Cartee et al. 1989). In contrast, when the animals were provided a high-fat diet, enhanced muscle insulin sensitivity was observed as long as 48 h after exercise (Cartee et al. 1989). In keeping with this, incubating muscle from acutely exercised rats in 20 or 40 mM glucose for 3 h alleviates the insulin-sensitizing effect of prior exercise on glucose transport (Gulve et al. 1990). Interestingly, muscle exposed to the non-metabolized glucose analog 2-deoxyglucose (36 mM) retained the potentiating effect of prior exercise on insulin sensitivity. Comparing muscles incubated in 20 mM glucose versus 36 mM 2-deoxyglucose following acute exercise, 2-deoxyglucose was transported across the sarcolemma to a similar extent as glucose (Gulve et al. 1990). Assuming that 2-deoxyglucose and glucose have comparable effects on the plasma membrane and the glucose transport system, this indicates that the impact of glucose availability on muscle insulin sensitivity following exercise/contraction is founded in glucose metabolism at or beyond the hexokinase reaction. Since the majority of glucose (~80 %) taken up by human skeletal muscle during a hyperinsulinemic euglycemic clamp is converted into muscle glycogen (Shulman et al. 1990), it is reasonable to assume that glycogen may regulate muscle insulin sensitivity in the post-exercise period.

Studies manipulating muscle glycogen concentrations by means of fasting (Jensen et al. 2006), adrenaline (Jensen et al. 1997; Kolnes et al. 2015), or exercise-diet protocols (Derave et al. 2000b) show that insulin-stimulated glucose uptake is increased in muscle with reduced glycogen levels, which may be related to increased translocation of GLUT4 to the plasma membrane (Derave et al. 2000b). By correlating the exercise-induced increase in insulin sensitivity with the amount of glycogen broken down during the preceding exercise bout, Wojtaszewski and colleagues proposed that glycogen breakdown may be part of the regulatory mechanism responsible for enhancing post-exercise muscle insulin sensitivity in man (Wojtaszewski et al. 2003a). However, a contraction-induced decrease in muscle glycogen content does not necessarily elicit a subsequent increase in insulin sensitivity (Kim et al. 2004), which is also supported by the finding that prior AICAR stimulation induces a subsequent increase in muscle insulin sensitivity (Fisher et al. 2002; Kjøbsted et al. 2015) despite comparable muscle glycogen content between control and prior AICAR-stimulated muscles (Fisher et al. 2002). Moreover, rats continue to exhibit enhanced muscle insulin sensitivity following exercise, although glycogen levels return to resting values (Richter et al. 1982; Cartee et al. 1989). Collectively, decreased muscle glycogen content per se may not be required for post-exercise improvements in muscle insulin sensitivity, but it does not exclude that reductions in glycogen content may initiate intracellular events that enhance muscle insulin sensitivity.

In contrast to the potentiating effect of low glycogen content, high concentrations of muscle glycogen do not seem to compromise the ability of insulin to stimulate glucose uptake compared to muscle harboring glycogen levels found in the normal state. Accordingly, the relationship between glycogen concentration and glucose uptake seems to track a hyperbolic function (Jensen et al. 1997). Mice with glycogen supercompensation in skeletal muscle, due to overexpression of glycogen synthase, have normal insulin-stimulated glucose uptake (2-deoxyglucose) during hindlimb perfusion (Fogt et al. 2004). Furthermore, inducing muscle glycogen supercompensation by acute fasting–refeeding does not significantly compromise insulin-stimulated glucose uptake in sedentary animals (Jensen et al. 1997, 2006). On the other hand, a study in muscle from rats fed carbohydrates for various time intervals following acute exercise showed that only during glycogen supercompensation the potentiating effect of prior exercise on muscle insulin sensitivity was lost. Conversely, when the rats were fed a high-fat diet for 48 h, muscle glycogen content returned to resting levels concomitant with a persistent increase in muscle insulin sensitivity (Cartee et al. 1989). This indicates that glycogen supercompensation may play a role in reversing improvements in muscle insulin sensitivity following exercise.

Considering that high concentrations of glycogen in skeletal muscle diminish both AMPK activity and enhanced muscle insulin sensitivity following exercise, this may suggest that AMPK function as a link between muscle glycogen and its ability to regulate insulin sensitivity (Richter et al. 2001). Supporting such notion, Kraegen and colleagues observed that intravenous glucose infusion, inducing hyperglycemia (~18 mM) for 3 h in fed rats, increased muscle glycogen content far above resting levels without altering insulin sensitivity or AMPK Thr172 phosphorylation (Kraegen et al. 2006). However, 5 h of glucose infusion decreased AMPKα2 (but not α1) activity concomitant with a decrease in insulin sensitivity (Hoy et al. 2007). A possible link between AMPK, glycogen, and muscle insulin sensitivity may also be derived from studies in the trained state. Accordingly, in response to exercise training ($\geq$3 weeks), glycogen levels and AMPK-α1 and -α2 activity increase in skeletal muscle (Frøsig et al. 2004; Mortensen et al. 2013). Interestingly, activation of the AMPK $\alpha 2\beta 2\gamma 3$ complex in response to acute exercise is severely blunted in trained human skeletal muscle concomitant with a marked decrease in AMPKγ3 protein levels (Frøsig et al. 2004; Mortensen et al. 2013; Wojtaszewski et al. 2005). Furthermore, although muscle insulin sensitivity in general is increased in trained human muscle, the potentiating effect of acute exercise on muscle insulin sensitivity seems to be lost (Mikines et al. 1989). Taken together, this may indicate that high glycogen levels reverse the enhancing effect of prior acute exercise on muscle insulin sensitivity by regulating AMPKγ3-associated activity through a negative feedback mechanism.

5.5.2 *Involvement of the AMPK α2β2γ3 Complex for Regulating Muscle Insulin Sensitivity After Exercise*

We have previously found that in EDL muscle AMPKγ3-associated activity was increased 6 h after AICAR treatment concomitant with an enhanced AMPKγ3-dependent increase in muscle insulin sensitivity (Kjøbsted et al. 2015). As AICAR may initiate cellular mechanisms different from those turned on by exercise/contraction, we cannot infer that the AMPK α2β2γ3 complex also regulates muscle insulin sensitivity after exercise. Arguing for such a role, however, prior AICAR stimulation fails to increase insulin sensitivity in soleus muscle in which the α2β2γ3 complex represent less than 2 % of all AMPK heterotrimer complexes (Treebak et al. 2009a; Kjøbsted et al. 2015). These later observations differ somewhat from other findings in which prior exercise increased insulin sensitivity in soleus muscle from rats and mice (Hamada et al. 2006; Iwabe et al. 2014; Xiao et al. 2013), suggesting that exercise may regulate insulin sensitivity in soleus muscle by mechanisms different from those employed by AMPK. The potential involvement of the AMPK α2β2γ3 complex is further strengthened by additional discoveries in the transgenic AMPKγ3 mouse models. Thus, expression of an activating AMPKγ3 mutant (R225Q) enhances, and knockout of AMPKγ3 impairs glycogen resynthesis in gastrocnemius muscle following glucose injection after a single bout of exercise (Barnes et al. 2004). Since the majority of glucose taken up by the muscle during insulin stimulation is stored as glycogen in the post-exercise period, these results support the notion that the AMPK α2β2γ3 complex acts to regulate muscle insulin sensitivity after exercise.

5.5.3 *Regulation of Muscle Insulin Sensitivity After Exercise: Impact of Glycogen Synthase Activity*

Studies in muscle-specific glycogen synthase (GS) knockout mice indicate that GS is important for regulating insulin-stimulated glucose uptake in muscle. Thus, during a hyperinsulinemic euglycemic clamp skeletal muscle glucose uptake is reduced demonstrating that the ability to store glucose as glycogen is central for the continuous uptake of glucose in response to insulin (Pederson et al. 2005). Observations in type 2 diabetic patients may also indicate that GS is a regulator of glucose uptake. Thus, several studies have shown that insulin-stimulated muscle glucose uptake is reduced by T2D. This can largely be accounted for by reductions in muscle glycogen synthesis in part due to a defective action of insulin on GS activity (Shulman et al. 1990; Højlund et al. 2003, 2009; DeFronzo and Tripathy 2009). Decreased insulin signaling to GLUT4 translocation as well as muscle protein content of HKII may, however, also play a significant role (Vind et al. 2011). Interestingly, acute exercise increases muscle GS activity and substrate affinity most likely by decreasing muscle glycogen levels and inducing dephosphorylation of GS site 2 + 2a and 3a + 3b (for review see Jensen and Lai 2009). The exercise-induced effects on GS are sustained into exercise recovery in which several studies in both humans and rodents report higher GS activity during insulin stimulation possibly due to a further reduction in site 3a + 3b phosphorylation (Richter et al. 1982, 1984, 1989; Bogardus et al. 1983; Devlin and Horton 1985; Wojtaszewski et al. 2000a; Pedersen et al. 2015). Taken together, this may suggest that greater GS activity contributes to improvements in muscle insulin sensitivity following acute exercise. However, observations from a transgenic mouse model that overexpresses a constitutively active form of GS do not support such relationship. Following in situ contraction of muscle from GS transgenic and WT mice, insulin-stimulated glucose uptake was similar between genotypes (Fogt et al. 2004). Although lack of GS activity may affect glucose uptake negatively, these results indicate that elevated GS activity per se does not further increase insulin-stimulated glucose uptake during exercise recovery. However, it seems reasonable to assume that an increase in GS activity is important for repartitioning intracellular glucose intermediates towards glucose storage during improved insulin-stimulated glucose uptake following acute exercise/contraction.

5.5.4 Improved Muscle Insulin Sensitivity After Exercise Despite Normal Proximal Insulin Signaling

The mechanism for increased GLUT4 translocation in response to insulin in the post-exercise period could be explained by a potentiation of proximal insulin signaling. Based on several studies in both human and rodents, it has been reasoned that this is probably not the case. It was shown that binding of insulin to the insulin receptor on the muscle membrane was not enhanced after exercise (Bonen et al. 1984). Furthermore, in a series of human studies evaluating muscle insulin sensitivity in response to one-legged knee extensor exercise, it was demonstrated that although muscle glucose uptake was enhanced, prior exercise did not improve insulin-stimulated insulin receptor tyrosine kinase activity, IRS1 tyrosine phosphorylation, IRS1-associated PI 3-Kinase activity, Akt Thr308 and Ser473 phosphorylation, and GSK-3 activity (Wojtaszewski et al. 1997, 2000a; Thong et al. 2002; Frøsig et al. 2007). Similarly, in studies of incubated rodent skeletal muscle prior exercise/contraction or AICAR stimulation does not potentiate a subsequent effect of insulin on insulin receptor and IRS1 phosphorylation (Hansen et al. 1998; Castorena et al. 2014), PI 3-Kinase activity (Fisher et al. 2002), and Akt Ser473 phosphorylation (Hamada et al. 2006; Arias et al. 2007; Koshinaka et al. 2008; Schweitzer et al. 2012; Iwabe et al. 2014; Castorena et al. 2014; Kjøbsted et al. 2015). Some studies report increased insulin-stimulated phosphorylation of Akt Thr308 (Arias et al. 2007; Funai et al. 2009; Koshinaka et al. 2009; Kjøbsted et al. 2015); however, this is not reflected in measurements of Akt activity (Funai et al. 2009), indicating that it may not be of any functional relevance. Although these in vitro analyses indicate changes in protein activity, they do not take into account possible changes in protein subcellular localization, which may play a role for the insulin-sensitizing effect of exercise. In spite of this, Wilson and colleagues did not find evidence to support that the ratio of insulin signaling proteins (insulin receptor, IRS1, IRS2, p85, Akt, and GSK-3) was altered between the plasma membrane, cytosol, and nuclear muscle cell fraction immediately after or 3 h after 60 min of cycling exercise (~67 % of VO2 peak) (Wilson et al. 2006). Taken together, these findings suggest that improvements in muscle insulin sensitivity following exercise are mediated by mechanisms downstream of the canonical PI 3-Kinase cascade.

5.5.5 Potential Role of TBC1D4 and TBC1D1 for Regulating Muscle Insulin Sensitivity After Exercise

With the identification of TBC1D4 and TBC1D1 as convergence points for exercise/AMPK- and insulin-associated signaling, these two proteins have emerged as possible candidates for regulating muscle insulin sensitivity after exercise. In skeletal muscle of healthy subjects, phosphorylation of TBC1D4 increases at Ser341, Ser588, Thr642, Ser704, and Ser751 in response to acute exercise (Treebak et al. 2014). Although TBC1D4 has several known kinases, these effects may likely have been mediated by increased Akt activity since we detected an increased phosphorylation of Akt Thr308 and Ser473 immediately after exercise. Supporting this notion, following 60 min of cycling exercise only TBC1D4 Ser704 phosphorylation increased in skeletal muscle of type 2 diabetic and weight-matched control subjects concomitant with unchanged or slightly decreased phosphorylation of Akt Thr308 and Ser473, respectively (Kjøbsted et al. 2016). This suggests that phosphorylation of TBC1D4 Ser704 is regulated by "typical" exercise signaling in human skeletal muscle as also indicated by ex vivo and in situ contraction of mouse skeletal muscle (Treebak et al. 2010, 2014). In rat skeletal muscle, contraction increases phosphorylation of TBC1D4 but analogous to observations in humans these changes may be mediated by increased PI 3-K/Akt signaling (Funai and Cartee 2009). Nevertheless, it has been hypothesized that the exercise-induced phosphorylation of skeletal muscle TBC1D4 is sustained for many hours into exercise recovery, which would affect a subsequent effect of insulin leading to enhanced glucose uptake. Accordingly, two studies demonstrated that acute exercise increased TBC1D4 PAS and Thr642 phosphorylation in rat skeletal muscle, which were maintained 3–27 h into exercise recovery. Following insulin stimulation, phosphorylation of TBC1D4 reached even higher levels in prior exercised muscle compared to rested muscle concomitant with enhanced muscle insulin sensitivity (Arias et al. 2007; Funai et al. 2009). Similar observations have been reported for skeletal muscle of lean healthy humans. Studies from our laboratory have found increased phosphorylation of TBC1D4 on Ser318, Ser341, and Ser751 (tendency on Ser588) 4 h after acute exercise (Treebak et al. 2009b) and increased phosphorylation of TBC1D4 on Ser318, Ser341, Ser588, Thr642, Ser704, and Ser751 5 h into exercise recovery (Pehmøller et al. 2012). Following hyperinsulinemia, phosphorylation of TBC1D4 on all abovementioned sites increased to higher levels in previously exercised muscle compared to rested muscle in concert with enhanced muscle insulin sensitivity (Treebak et al. 2009b; Pehmøller et al. 2012). It should be emphasized that some studies of rat skeletal muscle do not detect a sustained effect of exercise on phosphorylation of TBC1D4 Ser588 and Thr642 ~3–4 h into exercise recovery (Xiao et al. 2013; Castorena et al. 2014; Iwabe et al. 2014; Sharma et al. 2015). However, several of these studies report a significant insulin-mediated potentiation of phosphorylated TBC1D4 Ser588 and Thr642 concomitant with enhanced muscle insulin sensitivity (Xiao et al. 2013; Castorena et al. 2014; Iwabe et al. 2014). Taken together, these observations indicate that maintaining phosphorylation of TBC1D4 in the recovery period after exercise is not a prerequisite for prior exercise to increase muscle insulin sensitivity. However, enhanced insulin-stimulated phosphorylation of TBC1D4 in prior exercised muscle seems essential for mediating improvements in muscle insulin sensitivity.

Since AICAR- and contraction-induced phosphorylation of TBC1D4 Ser704 is dependent on AMPK (Treebak et al. 2010), we previously proposed that prior AICAR stimulation may enhance muscle insulin sensitivity through AMPK-dependent changes in TBC1D4 Ser704 phosphorylation during acute AICAR stimulation (Kjøbsted et al. 2015). However, phosphorylation of TBC1D4 Ser704 does not seem important for insulin-stimulated glucose uptake (Treebak et al. 2010), and we therefore hypothesized that the potentiated effect of insulin on TBC1D4 Ser704 phosphorylation by prior AICAR stimulation relays insulin-signaling to enhance phosphorylation of Thr642, which subsequently may facilitate improvements in muscle insulin sensitivity. This hypothesis is supported by the correlative relationship between the insulin-stimulated increase in TBC1D4 Thr642 and Ser704 phosphorylation as well as the correlative relationships between the insulin-stimulated increase in glucose uptake and phosphorylation of TBC1D4 Thr642/Ser704 after AICAR stimulation (Kjøbsted et al. 2015) Moreover, insulin-stimulated phosphorylation of TBC1D4 Thr642 is reduced in skeletal muscle overexpressing a phospho-mutant of Ser704 (Ser704Ala) and insulin-stimulated phosphorylation of TBC1D4 Ser704 is attenuated in skeletal muscle overexpressing a phospho-mutant of Thr642 (T642A) (Kjøbsted et al. 2015). This indicates mutual dependency between the two sites and strengthens the notion that AMPK facilitates improvements in muscle insulin sensitivity by conveying phospho-signaling at TBC1D4 Ser704. Based on the assumption that prior AICAR stimulation regulates muscle insulin sensitivity through TBC1D4 Ser704, it seems plausible that the AMPK $\alpha 2\beta 2\gamma 3$ complex may be involved (Kjøbsted et al. 2015). If correct, improvements in insulin-stimulated phosphorylation of TBC1D4 Thr642 and Ser704 following AICAR stimulation would likely be attenuated in the AMPKγ3 KO mouse model. Indeed, we found increased insulin-stimulated phosphorylation of TBC1D4 Thr642 and Ser704 in WT muscle but not in muscle of AMPKγ3 KO mice following AICAR stimulation (Kjøbsted, Treebak, Wojtaszewski, unpublished observations).

Studies finding no effect of acute exercise on muscle insulin sensitivity may also indirectly provide indications as to whether or not TBC1D4 is involved in regulating muscle insulin sensitivity following exercise/contraction. Funai and colleagues reported that muscle insulin sensitivity as well as phosphorylation of TBC1D4 PAS and Thr642 reversed to resting levels 3 h after exercise if rats were given free access to chow in the intervening period (Funai et al. 2009). Moreover, neither insulin sensitivity nor insulin-stimulated phosphorylation of TBC1D4 Ser588 and Thr642 were found to be potentiated in soleus muscle from 30-months-old rats following acute exercise (Wang et al. 2015). Although these observations may indicate that insulin sensitivity is not regulated by contractile activity in soleus muscle, others have previously reported increased insulin sensitivity in soleus muscle of rats and mice during recovery from acute exercise (Hamada et al. 2006; Iwabe et al. 2014). Based on current results, it may be reasoned that TBC1D4 is primed by AMPK, which enhances a subsequent effect of insulin leading to improvements in muscle insulin sensitivity. So far, no studies have investigated post-exercise insulin sensitivity in transgenic models of TBC1D4, and causality remains to be established between enhanced muscle insulin sensitivity and greater phosphorylation of TBC1D4.

Studies in rat, mouse, and human skeletal muscle do not find greater phosphorylation of TBC1D1 PAS, Ser237, and Thr596, 3–27 h after acute exercise/contraction, and enhanced phosphorylation of TBC1D1 is not observed in response to a subsequent effect of insulin (Funai et al. 2009, 2010; Pehmøller et al. 2012; Iwabe et al. 2014). Conversely, we observed increased phosphorylation of TBC1D1 Ser237 6 h after AICAR stimulation in muscle of WT and AMPK-deficient mice but similar insulin-induced phosphorylation of TBC1D1 Thr596 regardless of prior treatment (Kjøbsted et al. 2015). Given that prior AICAR- and insulin-induced phosphorylation of TBC1D1 was similar between WT and AMPK-deficient mice, but enhanced muscle insulin sensitivity was evident in WT mice only; this suggests that phosphorylation of TBC1D1 may not be involved in regulating improvements in muscle insulin sensitivity after exercise.

5.5.6 *Effect of Insulin Resistance and T2D in Regulating Muscle Insulin Sensitivity After Exercise: Impact of AMPK and TBC1D4*

Similar to healthy subjects, improved whole-body insulin sensitivity following a single bout of exercise has been demonstrated for insulin-resistant individuals as well as patients with T2D (Devlin and Horton 1985; Devlin et al. 1987; Perseghin et al. 1996; Sharoff et al. 2010; Pehmøller et al. 2012). The insulin-sensitizing effect of an acute bout of exercise has also been shown to occur in insulin-resistant rats (Ropelle et al. 2006; Tanaka et al. 2007; Castorena et al. 2014), while no studies have examined this in insulin-resistant mice (for review see Cartee 2015a). Following exercise in insulin-resistant humans and rats, muscle insulin sensitivity improves to levels seen in the healthy rested controls. Thus, when comparing previously exercised controls with equivalent insulin-resistant models, post-exercise enhancement of muscle insulin sensitivity in the insulin-resistant state fails to reach levels seen in the controls (Cartee 2015a). This may suggest that potential triggers mediating improvements in muscle insulin sensitivity after exercise differ from those inducing insulin resistance.

As observed in healthy subjects, improvements in muscle insulin sensitivity have been observed for up to ~48 h following a single bout of exercise in insulin-resistant individuals (Perseghin et al. 1996). Furthermore, glycogen levels and AMPK activity have been found to be similar between type 2 diabetic and control subjects immediately after exercise in most (Musi et al. 2001a; Pedersen et al. 2015; Kjøbsted et al. 2016) but not all studies (Sriwijitkamol et al. 2007), suggesting that insulin resistance/T2D does not compromise glycogen breakdown or activation of AMPK during exercise. We observed that activity of all AMPK heterotrimeric complexes in skeletal muscle of type 2 diabetic patients was increased 3 h into exercise recovery (Kjøbsted et al. 2016), a time point when post-exercise improvements in muscle insulin sensitivity prevail in healthy lean subjects (Wojtaszewski et al. 1997; Thong et al. 2002). Interestingly, whole-body insulin sensitivity has been shown to be attenuated in short-term metformin-treated insulin-resistant subjects following acute exercise (Sharoff et al. 2010). This was explained by a parallel attenuation of AMPKα2 activation in skeletal muscle during exercise.

If we assume that exercise improves muscle insulin sensitivity by intracellular mechanisms as previously suggested (e.g., phosphorylation of TBC1D4), it is reasonable to believe that acute exercise alleviates muscle insulin resistance by enhancing processes (e.g., AMPK activation) that work in parallel (e.g., Akt) or are located distal to the effect triggering insulin resistance (e.g., serine phosphorylation of insulin receptor). This is also supported by the fact that acute exercise never completely rescues the primary defect inducing muscle insulin resistance (Cartee 2015a). No available data argues against that improvements in post-exercise muscle insulin sensitivity are regulated differently between insulin-resistant/type 2 diabetics and controls. To illustrate this, prior exercise did not enhance proximal insulin signaling at the level of insulin receptor tyrosine phosphorylation, IRS1-

associated PI 3-Kinase activity, Akt Thr308 and Ser473 phosphorylation, as well as Akt activity in insulin-resistant/type 2 diabetic subjects and animal models (Tanaka et al. 2007; Pehmøller et al. 2012; Castorena et al. 2014; Kjøbsted et al. 2016). This indicates that improvements in muscle insulin sensitivity following an acute bout of exercise may be found downstream of Akt as suggested for insulin-sensitive controls.

Only a few studies have evaluated post-exercise insulin signaling to muscle glucose transport downstream of Akt in insulin-resistant animals and humans. A study in high-fat diet fed rats reported greater phosphorylation of TBC1D4 Ser588 in skeletal muscle after acute exercise, which increased above resting levels during insulin stimulation in parallel with enhanced muscle insulin sensitivity (Castorena et al. 2014). Also, in a study from our laboratory in which insulin resistance was induced by lipid infusion in otherwise healthy, lean individuals reveals greater phosphorylation of TBC1D4 Ser318, Ser341, Ser588, Ser704, and Ser751 5 h after acute one-legged knee extensor exercise (Pehmøller et al. 2012). Moreover, in overweight/obese type 2 diabetic and weight-matched control subjects, phosphorylation of TBC1D4 Ser318, Ser341, and Ser704 was increased 3 h after acute cycling exercise (Kjøbsted et al. 2016). However, in contrast to results obtained in lipid-induced insulin-resistant subjects (Pehmøller et al. 2012), 2 h of insulin infusion did not further enhance phosphorylation of TBC1D4 (and whole-body insulin sensitivity) in prior exercised muscle vs. rested muscle of overweight/obese type 2 diabetic and weight-matched control subjects (Kjøbsted et al. 2016). Although differences in insulin infusion rates, exercise intensity, and mode as well as methods for determining insulin sensitivity (muscle vs. whole-body insulin sensitivity) make it difficult to compare the two studies (Pehmøller et al. 2012; Kjøbsted et al. 2016), the results indicate that overweight/obesity may per se impair the ability of insulin to further phosphorylate TBC1D4 in exercise recovery. In line with this, 2 weeks on a high-fat diet increases body mass in rats and blunts the potentiating effect of prior exercise on insulin-stimulated phosphorylation of TBC1D4 Thr642 (Castorena et al. 2014).

5.5.7 Exercise-Independent Modulators of Muscle Insulin Sensitivity

It has been proposed that improvements in muscle insulin sensitivity is a general phenomenon that is not regulated by exercise only but occurs during internalization of GLUT4 following any stimulus that induces GLUT4 translocation to the cell surface membrane (Holloszy 2005). In keeping with this, it is thought that after termination of a stimulus that increases GLUT4 translocation, subsequent endocytosis of GLUT4 causes a transient increase in GLUT4 protein content in specific intracellular compartments where GLUT4 is more prone to recruitment when exposed to new stimuli that facilitate translocation. Accordingly, improvement in muscle insulin sensitivity is observed after various stimuli regulating GLUT4 translocation including exercise, contraction, hypoxia, AICAR, and maximal insulin concentrations with or without previous exercise (Richter et al. 1982, 1984; Fisher et al. 2002; Geiger et al. 2006; Kjøbsted et al. 2015). In addition, hypoxia-, vanadate-, and hydrogen peroxide-induced glucose uptake is also improved after acute exercise (Cartee and Holloszy 1990) supporting the proposed idea that any stimulus facilitating GLUT4 translocation sensitizes the muscle for a subsequent stimulus. Intriguing, all these stimuli either activate Akt (insulin) or AMPK (exercise, hypoxia, AICAR, vanadate, and hydrogen peroxide) in skeletal muscle (Fisher et al. 2002; Hwang et al. 2004; Jensen et al. 2008). Thus, it can be hypothesized that the sustained action of any Akt- and AMPK-regulated protein involved in GLUT4 translocation are contributing to retaining GLUT4 in specific intracellular compartments that are highly susceptible to recruitment by a subsequent signal. It is interesting that TBC1D4 regulates GLUT4 translocation and is phosphorylated on Ser704 in response to elevated AMPK and insulin signaling (Treebak et al. 2010). In this context, an increase in muscle insulin sensitivity following exercise could be mediated by increased phosphorylation of TBC1D4 Ser704, which facilitates movement of specific GLUT4 pools into certain regions. At this location insulin signaling may increase overall phosphorylation of TBC1D4 allowing a greater proportion of the intracellular GLUT4 content to fuse with the cell surface membrane facilitating improved glucose uptake.

5.6 Key Aspects in the Regulation of Muscle Insulin Sensitivity Following Exercise

On the basis of current literature, potential mechanisms regulating the effect of acute exercise on muscle insulin sensitivity were assessed in health and various aspects of muscle insulin resistance. Although much has been learned since the first discoveries in 1982, the exact regulatory mechanisms are still not fully understood but may involve activation of AMPK and downstream phosphorylation of TBC1D4. Presented below is a summary of current evidence that characterizes improvements in muscle insulin sensitivity following acute exercise.

1. Insulin sensitivity is increased up to 48 h after a single bout of exercise in both human and rodent skeletal muscle.
2. Improved muscle insulin sensitivity is limited to the prior working muscles.
3. Improvements in muscle insulin sensitivity are not dependent on an intact circulation.
4. Improvements in muscle insulin sensitivity require the presence of a serum factor (protein) during exercise/contraction.
5. Elevated carbohydrate availability enhances reversal of improvements in muscle insulin sensitivity.
6. Muscle glycogen supercompensation alleviates the exercised-induced increase in muscle insulin sensitivity.

7. Improvements in muscle insulin sensitivity are not dependent on protein synthesis in the post-exercise period.
8. Improved muscle insulin sensitivity can be explained by increased GLUT4 protein content in the muscle surface membrane.
9. Improvements in muscle insulin sensitivity do not require enhanced proximal insulin signaling.
10. Glycogen synthase activity is likely not limiting for the enhanced effect of insulin on muscle glucose uptake in the post-exercise period.
11. Increased AMPK activity and/or signaling in the post-exercise period coincide with enhanced muscle insulin sensitivity.
12. Enhanced site-specific phosphorylation of TBC1D4 associates with improved muscle insulin sensitivity.
13. Acute exercise likely improves insulin sensitivity in insulin-resistant skeletal muscle by mechanisms similar to those induced by exercise in healthy skeletal muscle, although overweight/obesity may reduce this effect of exercise.

5.7 Maybe It Is not All About AMPK and TBC1D4

Ex vivo and in situ contraction as well as in vivo exercise all lead to improvements in muscle insulin sensitivity (Richter et al. 1982, 1984; Gao et al. 1994). Furthermore, all these stimuli increase AMPK activity and phosphorylation of TBC1D4 (Funai et al. 2010; Treebak et al. 2010, 2014). Although it may seem reasonable to believe that muscle insulin sensitivity may be regulated by the same mechanisms following each of the three different contraction stimulus, results from both our and other research groups may indicate that this is not the case. In order to determine the potential effect of AMPK in regulating muscle insulin sensitivity following exercise, we developed a treadmill exercise protocol followed by ex vivo incubation of mouse skeletal muscle in many ways similar to that described by Hamada and colleagues (Hamada et al. 2006). Soleus and EDL muscle from exercised (30 min) and rested WT and AMPKα2 kinase-dead (KD) mice were incubated for determination of submaximal insulin-stimulated glucose uptake ~75 min after completion of exercise. Similar to previous findings (Hamada et al. 2006), acute treadmill exercise increased insulin sensitivity in soleus but not in EDL muscle from WT mice. Surprisingly, acute exercise also increased insulin sensitivity in soleus muscle from the AMPKα2 KD mouse model (Fig. 5.2; Treebak and Wojtaszewski, unpublished observations). These results imply that activation of AMPK in skeletal muscle is not necessary for the ability of acute exercise to improve muscle insulin sensitivity in mouse soleus muscle. Unfortunately, muscle AMPK activity was not determined in these experiments, and since soleus muscle from AMPKα2 KD mice still retain α1-containing trimer complexes that are activated upon ex vivo contraction (Treebak et al. 2014), it cannot be excluded that increased AMPKα1-associated activity in prior exercised soleus muscle facilitates improvements in insulin sensitivity.

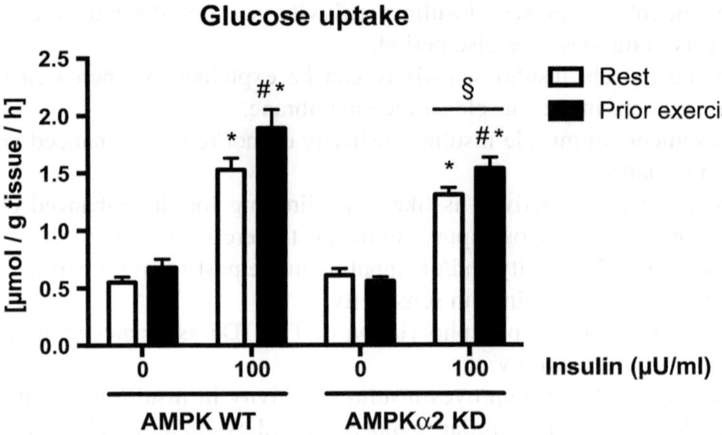

Fig. 5.2 Ex vivo insulin-stimulated glucose uptake in soleus muscle from female AMPKα2 KD mice and WT littermates ~75 min after 30 min of treadmill exercise. Mice were fasted 4 h before they were assigned to a sedentary or an exercise group. The exercise protocol consisted of 30 min of running (WT, 21 m/min; KD, 15 m/min). After the exercise, mice were anesthetized by an intraperitoneal injection of sodium pentobarbital. Soleus muscles were excised and suspended in incubation chambers containing Krebs–Ringer buffer. Approximately 75 min after exercise cessation, submaximal (100 µU/ml) insulin-stimulated glucose uptake was assessed by the accumulation of ^{3}H-labeled 2-deoxyglucose. Data are presented as means ± SEM. Exercise × insulin interaction ($p < 0.05$), $^{*}p < 0.001$ effect of insulin within group, and $^{\#}p < 0.05$ effect of exercise within insulin. $^{\S}p < 0.05$ Genotype × insulin interaction

Phosphorylation of TBC1D4 in the post-exercise period has not been determined in mouse treadmill experiments, but based on a recent study in rats, the ability of treadmill exercise to improve insulin sensitivity in soleus muscle seems to involve enhanced insulin-stimulated phosphorylation of TBC1D4 Thr642 and Ser588 (Iwabe et al. 2014). However, in contrast to observations in prior exercised skeletal muscle (Funai et al. 2009; Schweitzer et al. 2012), insulin does not induce enhanced phosphorylation of TBC1D4 PAS and Thr642 in rat epitrochlearis muscle 3 h after ex vivo contraction, although improvement in insulin sensitivity is observed (Funai et al. 2010). This suggests that exercise may regulate muscle insulin sensitivity by a mechanism different from that used by ex vivo contraction. Supporting this, combining exercise with ex vivo contraction has an additive effect on muscle insulin sensitivity compared to either stimulus alone (Funai et al. 2010). Assuming that the intense ex vivo contraction protocol recruits all muscle fibers, greater muscle insulin sensitivity induced by combining exercise with ex vivo contraction is likely not mediated by a difference in muscle fiber recruitment. This rationale is also strengthened by the fact that increasing exercise duration from 1 to 2 h does not further increase muscle insulin sensitivity, although one would expect additional muscle fibers to have been recruited during the 2 h exercise (Funai et al. 2010). While the study by Funai and colleagues does not provide an obvious explanation for the additive effect of exercise and ex vivo contraction on muscle insulin sensitivity (Funai et al. 2010), their results on TBC1D4 phosphorylation indicate that prior muscle contractile activity may enhance insulin sensitivity by TBC1D4-dependent and -independent mechanisms.

5.8 Concluding Remarks and Future Directions

Here we sought to discuss the underlying mechanisms involved in regulating muscle insulin sensitivity after acute exercise. Results obtained from animal studies have provided genetic evidence to support that AMPK plays an important role in regulating muscle insulin sensitivity, which has not previously been directly assessed (Kjøbsted et al. 2015). We believe that this is mediated by enhanced insulin-stimulated site-specific phosphorylation of TBC1D4 that act as a convergence point for insulin (Akt) and exercise (AMPK) signaling, relaying cooperative imposed phosphorylation to increase GLUT4 protein content in the cell surface membrane. As exercise-induced AMPK activation and phosphorylation of TBC1D4 are not compromised by T2D per se (Kjøbsted et al. 2016), this suggests that AMPK may also improve insulin sensitivity in insulin-resistant skeletal muscle.

There are still unanswered questions that need addressing in order to unravel the exact mechanisms responsible for enhancing the effect of insulin on muscle glucose uptake following acute exercise. Studies in EDL muscle of the TBC1D4 KO mouse model, which exhibits normal insulin- and AICAR-mediated glucose uptake (Chadt et al. 2015), could potentially provide important answers as to whether or not TBC1D4 is necessary for improving muscle insulin sensitivity after prior AICAR stimulation and exercise. Moreover, generation of a TBC1D4 Ser704Ala knock-in mouse model could potentially elucidate whether the enhanced phosphorylation of Ser704 observed several hours after prior AICAR stimulation (Kjøbsted et al. 2015) and exercise (Pehmøller et al. 2012; Treebak et al. 2009a, b) plays a pivotal role in regulating muscle insulin sensitivity. Interestingly, a common Greenlandic p. Arg684Ter variant in the gene encoding TBC1D4 was recently described. This truncated variant lacks phosphorylation of Ser704 and homozygous carriers have impaired glucose tolerance and a tenfold risk of developing diabetes (Moltke et al. 2014). Subjecting these carriers to an experimental protocol involving acute exercise in combination with a hyperinsulinemic euclycemic clamp could shed light on whether or not TBC1D4 is also involved in regulating insulin sensitivity in human skeletal muscle.

Although much evidence favors a functional role of AMPK in regulating muscle insulin sensitivity, results presented here suggest that mechanisms distinct from those regulated by AMPK may also be involved in the potentiating effect of exercise on muscle insulin sensitivity. Application of global and unbiased mass spectrometry analyses of control and insulin-sensitized skeletal muscle could provide valuable information on differentially regulated candidate proteins that could be further investigated in transgenic animal models.

It is a common belief that improved muscle insulin sensitivity following acute exercise is mediated by increased translocation of GLUT4 to the cell surface membrane. However, this has only been reported once for rat skeletal muscle (Hansen et al. 1998), and currently it is not known whether such observations also apply to human and mouse skeletal muscle. Clearly, this issue needs further attention in order to substantiate its existence in other species. Furthermore, the effect and extent of intracellular GLUT4 compartmentalization following acute exercise still await thorough investigation; however, it may be hypothesized that increased phosphorylation of TBC1D4 in the post-exercise period may locate GLUT4 to "high-susceptibility regions" within the cell (Holloszy 2005). This may be examined by pre-embedding immunogold electron microscopy of single muscle fibers, which makes it possible to visualize and quantify possible dynamics in cellular localization of GLUT4 (Ploug et al. 1998).

Based on the early studies of incubated muscle from prior exercised rats there is no doubt that improvement in muscle insulin sensitivity is dependent on a local intracellular mechanism (Davis et al. 1986; Wallberg-Henriksson et al. 1988; Cartee et al. 1989). However, in an in vivo setting, enhanced insulin sensitivity post-exercise may not only depend on the ability to increase glucose transport but may also depend on an increased supply of glucose to the "sensitive" muscle. In this context, changes in microvascular perfusion seem highly important for glucose delivery to the muscle bed. Insulin and contraction likely increase microvascular perfusion by different mechanisms, which are supported by findings of an additive effect of insulin and contraction on microvascular perfusion similar to that observed for glucose uptake (Inyard et al. 2009). Interestingly, it has been shown that exercise training increases insulin-stimulated microvascular perfusion in skeletal muscle, which may in part be responsible for improvements in muscle glucose uptake (Rattigan et al. 2001). Considering the fact that the insulin-stimulated increase in microvascular perfusion precedes both insulin signaling and glucose uptake in rat skeletal muscle (Vincent et al. 2004), enhanced microvascular perfusion following acute exercise may help explain improvements in muscle insulin sensitivity, although this needs to be investigated further in detail.

The potential ability of AMPK to enhance muscle insulin sensitivity would expectedly be beneficial for individuals with muscle insulin resistance and T2D since well-maintained whole-body glucose homeostasis is a prerequisite for preservation of health. Because the prevalence of diabetes and comorbidities continues to grow worldwide, there is a need for safe and efficacious antidiabetic medication. In light of this, AMPK may be considered an attractive target for antidiabetic drug development. However, since activation of hypothalamic AMPK has been shown to increase feeding (Minokoshi et al. 2004), the challenge lies in developing drugs that are tissue and/or trimer complex-specific given the possible undesirable consequences of global AMPK activation. Targeting the AMPKγ3 isoform may provide ways for tissue-specific regulation of insulin sensitivity in skeletal muscle. Recently, it was demonstrated that AMPK complexes containing different γ isoforms respond differently to changes in intracellular adenosine nucleotide levels (Ross et al. 2016). Although further research needs to clarify the physiological role of these findings, it highlights the potential for developing selective drugs targeting the AMPKγ3 isoform in skeletal muscle, which may be effective in the treatment of muscle insulin resistance in various conditions.

Acknowledgements Jesper B. Birk and Rasmus Quistgaard are acknowledged for skilled technical assistance at experiments presented in Figs. 5.1 and 5.2, respectively. RK and JTT were supported by the Novo Nordisk Foundation Center for Basic Metabolic Research. The Novo Nordisk Foundation Center for Basic Metabolic Research is an independent research center at the University of Copenhagen partially funded by an unrestricted donation from the Novo Nordisk Foundation (http://metabol.ku.dk). RK was also supported by Department of Nutrition, Exercise and Sports, University of Copenhagen. JFPW and the studies presented in this chapter were supported by the Danish Council for Independent Research | Medical Sciences, the Novo Nordisk Foundation, The Lundbeck Foundation, and The research program "Physical activity and nutrition for improvement of health" funded by the University of Copenhagen Excellence Program for Interdisciplinary Research. Some paragraphs from this text are in part based on a Ph.D. thesis written by RK.

References

Abdul-Ghani MA, DeFronzo RA (2010) Pathogenesis of insulin resistance in skeletal muscle. J Biomed Biotechnol 2010:476279. doi:10.1155/2010/476279

Ai H, Ihlemann J, Hellsten Y et al (2002) Effect of fiber type and nutritional state on AICAR- and contraction-stimulated glucose transport in rat muscle. Am J Physiol Endocrinol Metab 282: E1291–E1300

Al-Khalili L, Krook A, Zierath JR, Cartee GD (2004) Prior serum- and AICAR-induced AMPK activation in primary human myocytes does not lead to subsequent increase in insulin-stimulated glucose uptake. Am J Physiol Endocrinol Metab 287:E553–E557. doi:10.1152/ajpendo.00161.2004

An D, Toyoda T, Taylor EB et al (2010) TBC1D1 regulates insulin- and contraction-induced glucose transport in mouse skeletal muscle. Diabetes 59:1358–1365. doi:10.2337/db09-1266

Annuzzi G, Riccardi G, Capaldo B, Kaijser L (1991) Increased insulin-stimulated glucose uptake by exercised human muscles one day after prolonged physical exercise. Eur J Clin Invest 21:6–12

Arias EB, Kim J, Funai K, Cartee GD (2007) Prior exercise increases phosphorylation of Akt substrate of 160 kDa (AS160) in rat skeletal muscle. Am J Physiol Endocrinol Metab 292: E1191–E1200. doi:10.1152/ajpendo.00602.2006

Aschenbach WG, Hirshman MF, Fujii N et al (2002) Effect of AICAR treatment on glycogen metabolism in skeletal muscle. Diabetes 51:567–573

Baas AF, Boudeau J, Sapkota GP et al (2003) Activation of the tumour suppressor kinase LKB1 by the STE20-like pseudokinase STRAD. EMBO J 22:3062–3072. doi:10.1093/emboj/cdg292

Babraj JA, Mustard K, Sutherland C et al (2009) Blunting of AICAR-induced human skeletal muscle glucose uptake in type 2 diabetes is dependent on age rather than diabetic status. Am J Physiol Endocrinol Metab 296:E1042–E1048. doi:10.1152/ajpendo.90811.2008

Balon TW, Jasman AP (2001) Acute exposure to AICAR increases glucose transport in mouse EDL and soleus muscle. Biochem Biophys Res Commun 282:1008–1011. doi:10.1006/bbrc.2001.4677

Barnes BR, Marklund S, Steiler TL et al (2004) The 5′-AMP-activated protein kinase gamma3 isoform has a key role in carbohydrate and lipid metabolism in glycolytic skeletal muscle. J Biol Chem 279:38441–38447. doi:10.1074/jbc.M405533200

Barnes BR, Glund S, Long YC et al (2005) 5′-AMP-activated protein kinase regulates skeletal muscle glycogen content and ergogenics. FASEB J 19:773–779. doi:10.1096/fj.04-3221com

Baron AD, Brechtel G, Wallace P, Edelman SV (1988) Rates and tissue sites of non-insulin- and insulin-mediated glucose uptake in humans. Am J Physiol 255:E769–E774

Bergeron R, Russell RR, Young LH et al (1999) Effect of AMPK activation on muscle glucose metabolism in conscious rats. Am J Physiol 276:E938–E944

Birk JB, Wojtaszewski JFP (2006) Predominant alpha2/beta2/gamma3 AMPK activation during exercise in human skeletal muscle. J Physiol 577:1021–1032. doi:10.1113/jphysiol.2006.120972

Bogardus C, Thuillez P, Ravussin E et al (1983) Effect of muscle glycogen depletion on in vivo insulin action in man. J Clin Invest 72:1605–1610. doi:10.1172/JCI111119

Bonen A, Tan MH, Watson-Wright WM (1984) Effects of exercise on insulin binding and glucose metabolism in muscle. Can J Physiol Pharmacol 62:1500–1504

Boon H, Bosselaar M, Praet SFE et al (2008) Intravenous AICAR administration reduces hepatic glucose output and inhibits whole body lipolysis in type 2 diabetic patients. Diabetologia 51:1893–1900. doi:10.1007/s00125-008-1108-7

Bosselaar M, Boon H, van Loon LJC et al (2009) Intra-arterial AICA-riboside administration induces NO-dependent vasodilation in vivo in human skeletal muscle. Am J Physiol Endocrinol Metab 297:E759–E766. doi:10.1152/ajpendo.00141.2009

Boudeau J, Baas AF, Deak M et al (2003) MO25α/β interact with STRADα/β enhancing their ability to bind, activate and localize LKB1 in the cytoplasm. EMBO J 22:5102–5114. doi:10.1093/emboj/cdg490

Bruss MD, Arias EB, Lienhard GE, Cartee GD (2005) Increased phosphorylation of Akt substrate of 160 kDa (AS160) in rat skeletal muscle in response to insulin or contractile activity. Diabetes 54:41–50. doi:10.2337/diabetes.54.1.41

Cartee GD (2015a) Roles of TBC1D1 and TBC1D4 in insulin- and exercise-stimulated glucose transport of skeletal muscle. Diabetologia 58:19–30. doi:10.1007/s00125-014-3395-5

Cartee GD (2015b) Mechanisms for greater insulin-stimulated glucose uptake in normal and insulin resistant skeletal muscle after acute exercise. Am J Physiol Endocrinol Metab. doi:10.1152/ajpendo.00416.2015

Cartee GD, Holloszy JO (1990) Exercise increases susceptibility of muscle glucose transport to activation by various stimuli. Am J Physiol 258:E390–E393

Cartee GD, Young DA, Sleeper MD et al (1989) Prolonged increase in insulin-stimulated glucose transport in muscle after exercise. Am J Physiol 256:E494–E499

Castorena CM, Arias EB, Sharma N, Cartee GD (2014) Post-exercise improvement in insulin-stimulated glucose uptake occurs concomitant with greater AS160 phosphorylation in muscle from normal and insulin resistant rats. Diabetes 63:1–37. doi:10.2337/db13-1686

Chadt A, Immisch A, de Wendt C et al (2015) Deletion of both Rab-GTPase–activating proteins TBC1D1 and TBC1D4 in mice eliminates insulin- and AICAR-stimulated glucose transport [corrected]. Diabetes 64:746–759. doi:10.2337/db14-0368

Chen ZP, McConell GK, Michell BJ et al (2000) AMPK signaling in contracting human skeletal muscle: acetyl-CoA carboxylase and NO synthase phosphorylation. Am J Physiol Endocrinol Metab 279:E1202–E1206. doi:10.1007/s003390201284

Chen ZP, Stephens TJ, Murthy S et al (2003) Effect of exercise intensity on skeletal muscle AMPK signaling in humans. Diabetes 52:2205–2212. doi:10.2337/diabetes.52.9.2205

Chen S, Murphy J, Toth R et al (2008) Complementary regulation of TBC1D1 and AS160 by growth factors, insulin and AMPK activators. Biochem J 409:449–459. doi:10.1042/BJ20071114

Chibalin AV, Yu M, Ryder JW et al (2000) Exercise-induced changes in expression and activity of proteins involved in insulin signal transduction in skeletal muscle: differential effects on insulin-receptor substrates 1 and 2. Proc Natl Acad Sci USA 97:38–43. doi:10.1073/pnas.97.1.38

Coderre L, Kandror KV, Vallega G, Pilch PF (1995) Identification and characterization of an exercise-sensitive pool of glucose transporters in skeletal muscle. J Biol Chem 270:27584–27588. doi:10.1074/jbc.270.46.27584

Corton JM, Gillespie JG, Hawley SA, Hardie DG (1995) 5-aminoimidazole-4-carboxamide ribonucleoside. A specific method for activating AMP-activated protein kinase in intact cells? Eur J Biochem 229:558–565. doi:10.1111/j.1432-1033.1995.0558k.x

Cuthbertson DJ, Babraj JA, Mustard KJW et al (2007) 5-aminoimidazole-4-carboxamide 1-beta-D-ribofuranoside acutely stimulates skeletal muscle 2-deoxyglucose uptake in healthy men. Diabetes 56:2078–2084. doi:10.2337/db06-1716

Davies SP, Helps NR, Cohen PTW, Hardie DG (1995) 5′-AMP inhibits dephosphorylation, as well as promoting phosphorylation, of the AMP-activated protein kinase. Studies using bacterially expressed human protein phosphatase-2Cα and native bovine protein phosphatase-2AC. FEBS Lett 377:421–425. doi:10.1016/0014-5793(95)01368-7

Davis TA, Klahr S, Tegtmeyer ED et al (1986) Glucose metabolism in epitrochlearis muscle of acutely exercised and trained rats. Am J Physiol 250:E137–E143

Day P, Sharff A, Parra L et al (2007) Structure of a CBS-domain pair from the regulatory γ1 subunit of human AMPK in complex with AMP and ZMP. Acta Crystallogr Sect D Biol Crystallogr 63:587–596. doi:10.1107/S0907444907009110

DeFronzo RA (1988) Lilly lecture 1987. The triumvirate: beta-cell, muscle, liver. A collusion responsible for NIDDM. Diabetes 37:667–687

DeFronzo R, Tripathy D (2009) Skeletal muscle insulin resistance is the primary defect in type 2 diabetes. Diabetes Care 32(Suppl 2):S157–S163. doi:10.2337/dc09-S302

DeFronzo RA, Ferrannini E, Hendler R et al (1983) Regulation of splanchnic and peripheral glucose uptake by insulin and hyperglycemia in man. Diabetes 32:35–45

Derave W, Ai H, Ihlemann J et al (2000a) Dissociation of AMP-activated protein kinase activation and glucose transport in contracting slow-twitch muscle. Diabetes 49:1281–1287. doi:10.2337/diabetes.49.8.1281

Derave W, Hansen BF, Lund S et al (2000b) Muscle glycogen content affects insulin-stimulated glucose transport and protein kinase B activity. Am J Physiol Endocrinol Metab 279:E947–E955

Devlin JT, Horton ES (1985) Effects of prior high-intensity exercise on glucose metabolism in normal and insulin-resistant men. Diabetes 34:973–979. doi:10.2337/diabetes.34.10.973

Devlin JT, Hirshman M, Horton ED, Horton ES (1987) Enhanced peripheral and splanchnic insulin sensitivity in NIDDM men after single bout of exercise. Diabetes 36:434–439. doi:10.2337/diabetes.36.4.434

Dokas J, Chadt A, Nolden T et al (2013) Conventional knockout of Tbc1d1 in mice impairs insulin- and AICAR-stimulated glucose uptake in skeletal muscle. Endocrinology 154:3502–3514. doi:10.1210/en.2012-2147

Douen AG, Ramlal T, Klip A et al (1989) Exercise-induced increase in glucose transporters in plasma membranes of rat skeletal muscle. Endocrinology 124(6):449–454

Douen AG, Ramlal T, Cartee GD, Klip A (1990) Exercise modulates the insulin-induced translocation of glucose transporters in rat skeletal muscle. FEBS Lett 261:256–260. doi:10.1016/0014-5793(90)80566-2

Ducommun S, Wang HY, Sakamoto K et al (2012) Thr649Ala-AS160 knock-in mutation does not impair contraction/AICAR-induced glucose transport in mouse muscle. Am J Physiol Endocrinol Metab 302:E1036–E1043. doi:10.1152/ajpendo.00379.2011

Dumke CL, Kim J, Arias EB, Cartee GD (2002) Role of kallikrein-kininogen system in insulin-stimulated glucose transport after muscle contractions. J Appl Physiol 92:657–664. doi:10.1152/japplphysiol.00854.2001

Durante PE, Mustard KJ, Park S-H et al (2002) Effects of endurance training on activity and expression of AMP-activated protein kinase isoforms in rat muscles. Am J Physiol Endocrinol Metab 283:E178–E186. doi:10.1152/ajpendo.00404.2001

Fazakerley DJ, Holman GD, Marley A et al (2010) Kinetic evidence for unique regulation of GLUT4 trafficking by insulin and AMP-activated protein kinase activators in L6 myotubes. J Biol Chem 285:1653–1660. doi:10.1074/jbc.M109.051185

Fentz J, Kjøbsted R, Birk JB et al (2015) AMPKα is critical for enhancing skeletal muscle fatty acid utilization during in vivo exercise in mice. FASEB J 29:1725–1738. doi:10.1096/fj.14-266650

Fisher JS, Gao J, Han D-H et al (2002) Activation of AMP kinase enhances sensitivity of muscle glucose transport to insulin. Am J Physiol Endocrinol Metab 282:E18–E23. doi:10.1152/ajpendo.00477.2004

Fogt DL, Pan S, Lee S et al (2004) Effect of glycogen synthase overexpression on insulin-stimulated muscle glucose uptake and storage. Am J Physiol Endocrinol Metab 286:E363–E369. doi:10.1152/ajpendo.00115.2003\r00115.2003 [pii]

Frøsig C, Jørgensen SB, Hardie DG et al (2004) 5′-AMP-activated protein kinase activity and protein expression are regulated by endurance training in human skeletal muscle. Am J Physiol Endocrinol Metab 286:E411–E417. doi:10.1152/ajpendo.00317.2003

Frøsig C, Sajan MP, Maarbjerg SJ et al (2007) Exercise improves phosphatidylinositol-3,4,5-trisphosphate responsiveness of atypical protein kinase C and interacts with insulin signalling to peptide elongation in human skeletal muscle. J Physiol 582:1289–1301. doi:10.1113/jphysiol.2007.136614

Frøsig C, Roepstorff C, Brandt N et al (2009) Reduced malonyl-CoA content in recovery from exercise correlates with improved insulin-stimulated glucose uptake in human skeletal muscle. Am J Physiol Endocrinol Metab 296:E787–E795. doi:10.1152/ajpendo.90556.2008

Frøsig C, Pehmøller C, Birk JB et al (2010) Exercise-induced TBC1D1 Ser237 phosphorylation and 14-3-3 protein binding capacity in human skeletal muscle. J Physiol 588:4539–4548. doi:10.1113/jphysiol.2010.194811

Fujii N, Hayashi T, Hirshman MF et al (2000) Exercise induces isoform-specific increase in 5′AMP-activated protein kinase activity in human skeletal muscle. Biochem Biophys Res Commun 273:1150–1155. doi:10.1006/bbrc.2000.3073

Fujii N, Boppart MD, Dufresne SD et al (2004) Overexpression or ablation of JNK in skeletal muscle has no effect on glycogen synthase activity. Am J Physiol Cell Physiol 287:C200–C208. doi:10.1152/ajpcell.00415.2003

Fujii N, Hirshman MF, Kane EM et al (2005) AMP-activated protein kinase alpha2 activity is not essential for contraction- and hyperosmolarity-induced glucose transport in skeletal muscle. J Biol Chem 280:39033–39041. doi:10.1074/jbc.M504208200

Funai K, Cartee GD (2009) Inhibition of contraction-stimulated AMP-activated protein kinase inhibits contraction-stimulated increases in PAS-TBC1D1 and glucose transport without altering PAS-AS160 in rat skeletal muscle. Diabetes 58:1096–1104. doi:10.2337/db08-1477

Funai K, Schweitzer GG, Sharma N et al (2009) Increased AS160 phosphorylation, but not TBC1D1 phosphorylation, with increased postexercise insulin sensitivity in rat skeletal muscle. Am J Physiol Endocrinol Metab 297:E242–E251. doi:10.1152/ajpendo.00194.2009

Funai K, Schweitzer GG, Castorena CM et al (2010) In vivo exercise followed by in vitro contraction additively elevates subsequent insulin-stimulated glucose transport by rat skeletal muscle. Am J Physiol Endocrinol Metab 298:E999–E1010. doi:10.1152/ajpendo.00758.2009

Gao J, Gulve EA, Holloszy JO (1994) Contraction-induced increase in muscle insulin sensitivity: requirement for a serum factor. Am J Physiol 266:E186–E192

Garcia-Roves PM, Osler ME, Holmström MH, Zierath JR (2008) Gain-of-function R225Q mutation in AMP-activated protein kinase γ3 subunit increases mitochondrial biogenesis in glycolytic skeletal muscle. J Biol Chem 283:35724–35734. doi:10.1074/jbc.M805078200

Garetto LP, Richter EA, Goodman MN, Ruderman NB (1984) Enhanced muscle glucose metabolism after exercise in the rat: the two phases. Am J Physiol 246:E471–E475. doi:10.1172/JCI110517

Geiger PC, Han DH, Wright DC, Holloszy JO (2006) How muscle insulin sensitivity is regulated: testing of a hypothesis. Am J Physiol Endocrinol Metab 291:E1258–E1263. doi:10.1152/ajpendo.00273.2006

Geraghty KM, Chen S, Harthill JE et al (2007) Regulation of multisite phosphorylation and 14-3-3 binding of AS160 in response to IGF-1, EGF, PMA and AICAR. Biochem J 407:231–241. doi:10.1042/BJ20070649

Gowans GJ, Hardie DG (2014) AMPK: a cellular energy sensor primarily regulated by AMP. Biochem Soc Trans 42:71–75. doi:10.1042/BST20130244

Greiwe JS, Holloszy JO, Semenkovich CF (2000) Exercise induces lipoprotein lipase and GLUT-4 protein in muscle independent of adrenergic-receptor signaling. J Appl Physiol 89:176–181

Gulve EA, Cartee GD, Zierath JR et al (1990) Reversal of enhanced muscle glucose transport after exercise: roles of insulin and glucose. Am J Physiol 259:E685–E691

Hamada T, Arias EB, Cartee GD (2006) Increased submaximal insulin-stimulated glucose uptake in mouse skeletal muscle after treadmill exercise. J Appl Physiol 101:1368–1376. doi:10.1152/japplphysiol.00416.2006

Hansen PA, Nolte LA, Chen MM, Holloszy JO (1998) Increased GLUT-4 translocation mediates enhanced insulin sensitivity of muscle glucose transport after exercise. J Appl Physiol 85:1218–1222

Hardie DG (2007) AMP-activated/SNF1 protein kinases: conserved guardians of cellular energy. Nat Rev Mol Cell Biol 8:774–785. doi:10.1038/nrm2249

Hardie DG, Carling D, Gamblin SJ (2011) AMP-activated protein kinase: also regulated by ADP? Trends Biochem Sci 36:470–477. doi:10.1016/j.tibs.2011.06.004

Hardie DG, Ross FA, Hawley SA (2012) AMPK: a nutrient and energy sensor that maintains energy homeostasis. Nat Rev Mol Cell Biol 13:251–262. doi:10.1038/nrm3311

Hawley SA, Boudeau J, Reid JL et al (2003) Complexes between the LKB1 tumor suppressor, STRAD alpha/beta and MO25 alpha/beta are upstream kinases in the AMP-activated protein kinase cascade. J Biol 2:28. doi:10.1186/1475-4924-2-28

Hawley SA, Pan DA, Mustard KJ et al (2005) Calmodulin-dependent protein kinase kinase-beta is an alternative upstream kinase for AMP-activated protein kinase. Cell Metab 2:9–19. doi:10.1016/j.cmet.2005.05.009

Hayashi T, Hirshman MF, Kurth EJ et al (1998) Evidence for 5′ AMP-activated protein kinase mediation of the effect of muscle contraction on glucose transport. Diabetes 47:1369–1373

Hellsten Y, Sjödin B, Richter EA, Bangsbo J (1998) Urate uptake and lowered ATP levels in human muscle after high-intensity intermittent exercise. Am J Physiol 274:E600–E606

Henin N, Vincent MF, Van den Berghe G (1996) Stimulation of rat liver AMP-activated protein kinase by AMP analogues. Biochim Biophys Acta 1290:197–203

Herrero-Martín G, Høyer-Hansen M, García-García C et al (2009) TAK1 activates AMPK-dependent cytoprotective autophagy in TRAIL-treated epithelial cells. EMBO J 28:677–685. doi:10.1038/emboj.2009.8

Hirshman MF, Wallberg-Henriksson H, Wardzala LJ et al (1988) Acute exercise increases the number of plasma membrane glucose transporters in rat skeletal muscle. FEBS Lett 238:235–239

Højlund K, Stæhr P, Hansen BF et al (2003) Increased phosphorylation of skeletal muscle glycogen synthase at NH2-terminal sites during physiological hyperinsulinemia in type 2 diabetes. Diabetes 52:1393–1402. doi:10.2337/diabetes.52.6.1393

Højlund K, Birk JB, Klein DK et al (2009) Dysregulation of glycogen synthase COOH- and NH2-terminal phosphorylation by insulin in obesity and type 2 diabetes mellitus. J Clin Endocrinol Metab 94:4547–4556. doi:10.1210/jc.2009-0897

Holloszy JO (2005) Exercise-induced increase in muscle insulin sensitivity. J Appl Physiol 99:338–343. doi:10.1152/japplphysiol.00123.2005

Hoy AJ, Bruce CR, Cederberg A et al (2007) Glucose infusion causes insulin resistance in skeletal muscle of rats without changes in Akt and AS160 phosphorylation. Am J Physiol Endocrinol Metab 293:E1358–E1364. doi:10.1152/ajpendo.00133.2007

Hudson ER, Pan DA, James J et al (2003) A novel domain in AMP-activated protein kinase causes glycogen storage bodies similar to those seen in hereditary cardiac arrhythmias. Curr Biol 13:861–866. doi:10.1016/S0960-9822(03)00249-5

Hurley RL, Anderson KA, Franzone JM et al (2005) The Ca2+/calmodulin-dependent protein kinase kinases are AMP-activated protein kinase kinases. J Biol Chem 280:29060–29066. doi:10.1074/jbc.M503824200

Hwang J-T, Lee M, Jung S-N et al (2004) AMP-activated protein kinase activity is required for vanadate-induced hypoxia-inducible factor 1alpha expression in DU145 cells. Carcinogenesis 25:2497–2507. doi:10.1093/carcin/bgh253

Inoki K, Zhu T, Guan K-L (2003) TSC2 mediates cellular energy response to control cell growth and survival. Cell 115:577–590. doi:10.1016/S0092-8674(03)00929-2

Inyard AC, Chong DG, Klibanov AL, Barrett EJ (2009) Muscle contraction, but not insulin, increases microvascular blood volume in the presence of free fatty acid-induced insulin resistance. Diabetes 58:2457–2463. doi:10.2337/db08-1077

Iwabe M, Kawamoto E, Koshinaka K, Kawanaka K (2014) Increased postexercise insulin sensitivity is accompanied by increased AS160 phosphorylation in slow-twitch soleus muscle. Physiol Rep 2:1–9. doi:10.14814/phy2.12162

Jensen J, Lai Y-C (2009) Regulation of muscle glycogen synthase phosphorylation and kinetic properties by insulin, exercise, adrenaline and role in insulin resistance. Arch Physiol Biochem 115:13–21. doi:10.1080/13813450902778171

Jensen J, Aslesen R, Ivy JL, Brørs O (1997) Role of glycogen concentration and epinephrine on glucose uptake in rat epitrochlearis muscle. Am J Physiol 272:E649–E655

Jensen J, Jebens E, Brennesvik EO et al (2006) Muscle glycogen inharmoniously regulates glycogen synthase activity, glucose uptake, and proximal insulin signaling. Am J Physiol Endocrinol Metab 290:E154–E162. doi:10.1152/ajpendo.00330.2005

Jensen TE, Rose AJ, Hellsten Y et al (2007a) Caffeine-induced Ca(2+) release increases AMPK-dependent glucose uptake in rodent soleus muscle. Am J Physiol Endocrinol Metab 293:E286–E292. doi:10.1152/ajpendo.00693.2006

Jensen TE, Rose AJ, Jørgensen SB et al (2007b) Possible CaMKK-dependent regulation of AMPK phosphorylation and glucose uptake at the onset of mild tetanic skeletal muscle contraction. Am J Physiol Endocrinol Metab 292:E1308–E1317. doi:10.1152/ajpendo.00456.2006

Jensen TE, Schjerling P, Viollet B et al (2008) AMPK alpha1 activation is required for stimulation of glucose uptake by twitch contraction, but not by H2O2, in mouse skeletal muscle. PLoS One 3:e2102. doi:10.1371/journal.pone.0002102

Jensen TE, Wojtaszewski JFP, Richter EA (2009) AMP-activated protein kinase in contraction regulation of skeletal muscle metabolism: necessary and/or sufficient? Acta Physiol (Oxf) 196:155–174. doi:10.1111/j.1748-1716.2009.01979.x

Jensen TE, Ross FA, Kleinert M et al (2015) PT-1 selectively activates AMPK-γ1 complexes in mouse skeletal muscle, but activates all three γ subunit complexes in cultured human cells by inhibiting the respiratory chain. Biochem J 467:461–472. doi:10.1042/BJ20141142

Jeppesen J, Maarbjerg SJ, Jordy AB et al (2013) LKB1 regulates lipid oxidation during exercise independently of AMPK. Diabetes 62:1490–1499. doi:10.2337/db12-1160

Jessen N, Pold R, Buhl ES et al (2003) Effects of AICAR and exercise on insulin-stimulated glucose uptake, signaling, and GLUT-4 content in rat muscles. J Appl Physiol 94:1373–1379. doi:10.1152/japplphysiol.00250.2002

Jessen N, An D, Lihn AS et al (2011) Exercise increases TBC1D1 phosphorylation in human skeletal muscle. Am J Physiol Endocrinol Metab 301:E164–E171. doi:10.1152/ajpendo.00042.2011

Jørgensen SB, Nielsen JN, Birk JB et al (2004a) The α2-5′ AMP-activated protein kinase is a site 2 glycogen synthase kinase in skeletal muscle and is responsive to glucose loading. Diabetes 53:3074–3081. doi:10.2337/diabetes.53.12.3074

Jørgensen SB, Viollet B, Andreelli F et al (2004b) Knockout of the alpha2 but not alpha1 5′-AMP-activated protein kinase isoform abolishes 5-aminoimidazole-4-carboxamide-1-beta-4-ribofuranoside but not contraction-induced glucose uptake in skeletal muscle. J Biol Chem 279:1070–1079. doi:10.1074/jbc.M306205200

Jørgensen SB, Wojtaszewski JFP, Viollet B et al (2005) Effects of alpha-AMPK knockout on exercise-induced gene activation in mouse skeletal muscle. FASEB J 19:1146–1148. doi:10.1096/fj.04-3144fje

Ju J-S, Gitcho MA, Gitcho CA et al (2007) Potentiation of insulin-stimulated glucose transport by the AMP-activated protein kinase. Am J Physiol Cell Physiol 292:C564–C572. doi:10.1152/ajpcell.00269.2006

Kahn BB, Alquier T, Carling D, Hardie DG (2005) AMP-activated protein kinase: ancient energy gauge provides clues to modern understanding of metabolism. Cell Metab 1:15–25. doi:10.1016/j.cmet.2004.12.003

Karlsson HKR, Chibalin AV, Koistinen HA et al (2009) Kinetics of GLUT4 trafficking in rat and human skeletal muscle. Diabetes 58:847–854. doi:10.2337/db08-1539

Kaushik VK, Young ME, Dean DJ et al (2001) Regulation of fatty acid oxidation and glucose metabolism in rat soleus muscle: effects of AICAR. Am J Physiol Endocrinol Metab 281: E335–E340

Kazgan N, Williams T, Forsberg LJ, Brenman JE (2010) Identification of a nuclear export signal in the catalytic subunit of AMP-activated protein kinase. Mol Biol Cell 21:3433–3442. doi:10.1091/mbc.E10-04-0347

Kelly M, Keller C, Avilucea PR et al (2004) AMPK activity is diminished in tissues of IL-6 knockout mice: the effect of exercise. Biochem Biophys Res Commun 320:449–454. doi:10.1016/j.bbrc.2004.05.188

Kim J, Solis RS, Arias EB, Cartee GD (2004) Postcontraction insulin sensitivity: relationship with contraction protocol, glycogen concentration, and 5′ AMP-activated protein kinase phosphorylation. J Appl Physiol 96:575–583. doi:10.1152/japplphysiol.00909.2003

Kishi K, Muromoto N, Nakaya Y et al (1998) Bradykinin directly triggers GLUT4 translocation via an insulin-independent pathway. Diabetes 47:550–558

Kjøbsted R, Treebak JT, Fentz J et al (2015) Prior AICAR stimulation increases insulin sensitivity in mouse skeletal muscle in an AMPK-dependent manner. Diabetes 64:2042–2055. doi:10.2337/db14-1402

Kjøbsted R, Pedersen AJT, Hingst JR et al (2016) Intact regulation of the AMPK signaling network in response to exercise and insulin in skeletal muscle of male patients with type 2 diabetes – illumination of AMPK activation in recovery from exercise. Diabetes 65:1219–1230. doi:10.2337/db15-1034

Koh H-J, Brandauer J, Goodyear LJ (2008) LKB1 and AMPK and the regulation of skeletal muscle metabolism. Curr Opin Clin Nutr Metab Care 11:227–232. doi:10.1097/MCO.0b013e3282fb7b76

Koistinen HA, Galuska D, Chibalin AV et al (2003) 5-amino-imidazole carboxamide riboside increases glucose transport and cell-surface GLUT4 content in skeletal muscle from subjects with type 2 diabetes. Diabetes 52:1066–1072. doi:10.2337/diabetes.52.5.1066

Kolnes AJ, Birk JB, Eilertsen E et al (2015) Epinephrine-stimulated glycogen breakdown activates glycogen synthase and increases insulin-stimulated glucose uptake in epitrochlearis muscles. Am J Physiol Endocrinol Metab 308:E231–E240. doi:10.1152/ajpendo.00282.2014

Koshinaka K, Sano A, Howlett KF et al (2008) Effect of high-intensity intermittent swimming on postexercise insulin sensitivity in rat epitrochlearis muscle. Metabolism 57:749–756. doi:10.1016/j.metabol.2008.01.014

Koshinaka K, Kawasaki E, Hokari F, Kawanaka K (2009) Effect of acute high-intensity intermittent swimming on post-exercise insulin responsiveness in epitrochlearis muscle of fed rats. Metabolism 58:246–253. doi:10.1016/j.metabol.2008.09.021

Kraegen EW, Saha AK, Preston E et al (2006) Increased malonyl-CoA and diacylglycerol content and reduced AMPK activity accompany insulin resistance induced by glucose infusion in muscle and liver of rats. Am J Physiol Endocrinol Metab 290:E471–E479. doi:10.1152/ajpendo.00316.2005

Kramer HF, Witczak CA, Taylor EB et al (2006a) AS160 regulates insulin- and contraction-stimulated glucose uptake in mouse skeletal muscle. J Biol Chem 281:31478–31485. doi:10.1074/jbc.M605461200

Kramer HF, Witczak CA, Fujii N et al (2006b) Distinct signals regulate AS160 phosphorylation in response to insulin, AICAR, and contraction in mouse skeletal muscle. Diabetes 55:2067–2076. doi:10.2337/db06-0150

Kraniou GN, Cameron-Smith D, Hargreaves M (2006) Acute exercise and GLUT4 expression in human skeletal muscle: influence of exercise intensity. J Appl Physiol 101:934–937. doi:10.1152/japplphysiol.01489.2005

Kristensen JM, Johnsen AB, Birk JB et al (2007) Absence of humoral mediated 5′AMP-activated protein kinase activation in human skeletal muscle and adipose tissue during exercise. J Physiol 585:897–909. doi:10.1113/jphysiol.2007.143123

Kristensen DE, Albers PH, Prats C et al (2015) Human muscle fibre type-specific regulation of AMPK and downstream targets by exercise. J Physiol 593:2053–2069. doi:10.1113/jphysiol.2014.283267

Kuo CH, Browning KS, Ivy JL (1999a) Regulation of GLUT4 protein expression and glycogen storage after prolonged exercise. Acta Physiol Scand 165:193–201. doi:10.1046/j.1365-201x.1999.00489.x

Kuo CH, Hunt DG, Ding Z, Ivy JL (1999b) Effect of carbohydrate supplementation on postexercise GLUT-4 protein expression in skeletal muscle. J Appl Physiol 87:2290–2295

Kurth-Kraczek EJ, Hirshman MF, Goodyear LJ, Winder WW (1999) 5′ AMP-activated protein kinase activation causes GLUT4 translocation in skeletal muscle. Diabetes 48:1667–1671

Lantier L, Fentz J, Mounier R et al (2014) AMPK controls exercise endurance, mitochondrial oxidative capacity, and skeletal muscle integrity. FASEB J 28:3211–3224. doi:10.1096/fj.14-250449

Lebovitz HE (2001) Insulin resistance: definition and consequences. Exp Clin Endocrinol Diabetes 109:S135–S148. doi:10.1055/s-2001-18576

Lee AD, Hansen PA, Holloszy JO (1995) Wortmannin inhibits insulin-stimulated but not contraction-stimulated glucose transport activity in skeletal muscle. FEBS Lett 361:51–54. doi:10.1016/0014-5793(95)00147-2

Lee-Young RS, Koufogiannis G, Canny BJ, McConell GK (2008) Acute exercise does not cause sustained elevations in AMPK signaling or expression. Med Sci Sports Exerc 40:1490–1494. doi:10.1249/MSS.0b013e318173a037

Lee-Young RS, Griffee SR, Lynes SE et al (2009) Skeletal muscle AMP-activated protein kinase is essential for the metabolic response to exercise in vivo. J Biol Chem 284:23925–23934. doi:10.1074/jbc.M109.021048

Lefort N, St-Amand E, Morasse S et al (2008) The alpha-subunit of AMPK is essential for submaximal contraction-mediated glucose transport in skeletal muscle in vitro. Am J Physiol Endocrinol Metab 295:E1447–E1454. doi:10.1152/ajpendo.90362.2008

Leick L, Plomgaard P, Grønløkke L et al (2010) Endurance exercise induces mRNA expression of oxidative enzymes in human skeletal muscle late in recovery. Scand J Med Sci Sports 20:593–599. doi:10.1111/j.1600-0838.2009.00988.x

Lund S, Holman GD, Schmitz O, Pedersen O (1995) Contraction stimulates translocation of glucose transporter GLUT4 in skeletal muscle through a mechanism distinct from that of insulin. Proc Natl Acad Sci USA 92:5817–5821. doi:10.1073/pnas.92.13.5817

Maarbjerg SJ, Jørgensen SB, Rose AJ et al (2009) Genetic impairment of AMPKalpha2 signaling does not reduce muscle glucose uptake during treadmill exercise in mice. Am J Physiol Endocrinol Metab 297:E924–E934. doi:10.1152/ajpendo.90653.2008

Mahlapuu M, Johansson C, Lindgren K et al (2004) Expression profiling of the gamma-subunit isoforms of AMP-activated protein kinase suggests a major role for gamma3 in white skeletal muscle. Am J Physiol Endocrinol Metab 286:E194–E200. doi:10.1152/ajpendo.00147.2003

McConell GK, Lee-Young RS, Chen Z-P et al (2005) Short-term exercise training in humans reduces AMPK signalling during prolonged exercise independent of muscle glycogen. J Physiol 568:665–676. doi:10.1113/jphysiol.2005.089839

McMahon JM, Signori E, Wells KE et al (2001) Optimisation of electrotransfer of plasmid into skeletal muscle by pretreatment with hyaluronidase – increased expression with reduced muscle damage. Gene Ther 8:1264–1270. doi:10.1038/sj.gt.3301522

Merrill GF, Kurth EJ, Hardie DG, Winder WW (1997) AICA riboside increases AMP-activated protein kinase, fatty acid oxidation, and glucose uptake in rat muscle. Am J Physiol 273: E1107–E1112

Merry TL, Steinberg GR, Lynch GS, McConell GK (2010) Skeletal muscle glucose uptake during contraction is regulated by nitric oxide and ROS independently of AMPK. Am J Physiol Endocrinol Metab 298:E577–E585. doi:10.1152/ajpendo.00239.2009

Mikines KJ, Sonne B, Farrell PA et al (1988) Effect of physical exercise on sensitivity and responsiveness to insulin in humans. Am J Physiol 254:E248–E259

Mikines KJ, Sonne B, Tronier B, Galbo H (1989) Effects of acute exercise and detraining on insulin action in trained men. J Appl Physiol 66:704–711

Minokoshi Y, Kim Y-B, Peroni OD et al (2002) Leptin stimulates fatty-acid oxidation by activating AMP-activated protein kinase. Nature 415:339–343. doi:10.1038/415339a

Minokoshi Y, Alquier T, Furukawa N et al (2004) AMP-kinase regulates food intake by responding to hormonal and nutrient signals in the hypothalamus. Nature 428:569–574. doi:10.1038/nature02440

Miyamoto L, Toyoda T, Hayashi T et al (2007) Effect of acute activation of 5′-AMP-activated protein kinase on glycogen regulation in isolated rat skeletal muscle. J Appl Physiol 102:1007–1013. doi:10.1152/japplphysiol.01034.2006

Moltke I, Grarup N, Jørgensen ME et al (2014) A common Greenlandic TBC1D4 variant confers muscle insulin resistance and type 2 diabetes. Nature 512:190–193. doi:10.1038/nature13425

Momcilovic M, Hong SP, Carlson M (2006) Mammalian TAK1 activates Snf1 protein kinase in yeast and phosphorylates AMP-activated protein kinase in vitro. J Biol Chem 281:25336–25343. doi:10.1074/jbc.M604399200

Mortensen B, Hingst JR, Frederiksen N et al (2013) Effect of birth weight and 12 weeks of exercise training on exercise-induced AMPK signaling in human skeletal muscle. Am J Physiol Endocrinol Metab 304:E1379–E1390. doi:10.1152/ajpendo.00295.2012

Mu J, Brozinick JT, Valladares O et al (2001) A role for AMP-activated protein kinase in contraction- and hypoxia-regulated glucose transport in skeletal muscle. Mol Cell 7:1085–1094

Muoio DM, Seefeld K, Witters LA, Coleman RA (1999) AMP-activated kinase reciprocally regulates triacylglycerol synthesis and fatty acid oxidation in liver and muscle: evidence that sn-glycerol-3-phosphate acyltransferase is a novel target. Biochem J 338(Pt 3):783–791. doi:10.1042/0264-6021:3380783

Musi N, Fujii N, Hirshman MF et al (2001a) AMP-activated protein kinase (AMPK) is activated in muscle of subjects with type 2 diabetes during exercise. Diabetes 50:921–927. doi:10.2337/diabetes.50.5.921

Musi N, Hayashi T, Fujii N et al (2001b) AMP-activated protein kinase activity and glucose uptake in rat skeletal muscle. Am J Physiol Endocrinol Metab 280:E677–E684

Nesher R, Karl IE, Kipnis DM (1985) Dissociation of effects of insulin and contraction on glucose transport in rat epitrochlearis muscle. Am J Physiol 249:C226–C232

Nielsen JN, Vissing J, Wojtaszewski JFP et al (2002) Decreased insulin action in skeletal muscle from patients with McArdle's disease. Am J Physiol Endocrinol Metab 282:E1267–E1275. doi:10.1152/ajpendo.00526.2001

Nielsen JN, Mustard KJW, Graham DA et al (2003) 5′-AMP-activated protein kinase activity and subunit expression in exercise-trained human skeletal muscle. J Appl Physiol 94:631–641. doi:10.1152/japplphysiol.00642.2002

O'Neill HM, Maarbjerg SJ, Crane JD et al (2011) AMP-activated protein kinase (AMPK) beta1beta2 muscle null mice reveal an essential role for AMPK in maintaining mitochondrial content and glucose uptake during exercise. Proc Natl Acad Sci USA 108:16092–16097. doi:10.1073/pnas.1105062108

Oakhill JS, Steel R, Chen Z-P et al (2011) AMPK is a direct adenylate charge-regulated protein kinase. Science 332(80):1433–1435. doi:10.1126/science.1200094

Pedersen AJT, Hingst JR, Friedrichsen M et al (2015) Dysregulation of muscle glycogen synthase in recovery from exercise in type 2 diabetes. Diabetologia 58:1569–1578. doi:10.1007/s00125-015-3582-z

Pederson BA, Schroeder JM, Parker GE et al (2005) Glucose metabolism in mice lacking muscle glycogen synthase. Diabetes 54:3466–3473. doi:10.1007/s10346-005-0027-7

Pehmøller C, Treebak JT, Birk JB et al (2009) Genetic disruption of AMPK signaling abolishes both contraction- and insulin-stimulated TBC1D1 phosphorylation and 14-3-3 binding in mouse skeletal muscle. Am J Physiol Endocrinol Metab 297:E665–E675. doi:10.1152/ajpendo.00115.2009

Pehmøller C, Brandt N, Birk J (2012) Exercise alleviates lipid-induced insulin resistance in human skeletal muscle–signaling interaction at the level of TBC1 domain family member 4. Diabetes 61:2743–2752. doi:10.2337/db11-1572

Perseghin G, Price TB, Petersen KF et al (1996) Increased glucose transport-phosphorylation and muscle glycogen synthesis after exercise training in insulin-resistant subjects. N Engl J Med 335:1357–1362. doi:10.1056/NEJM199610313351804

Ploug T, Galbo H, Richter EA (1984) Increased muscle glucose uptake during contractions: no need for insulin. Am J Physiol 247:E726–E731. doi:10.2337/diabetes.34.10.1041

Ploug T, Galbo H, Vinten J et al (1987) Kinetics of glucose transport in rat muscle: effects of insulin and contractions. Am J Physiol 253:E12–E20

Ploug T, Van Deurs B, Ai H et al (1998) Analysis of GLUT4 distribution in whole skeletal muscle fibers: identification of distinct storage compartments that are recruited by insulin and muscle contractions. J Cell Biol 142:1429–1446. doi:10.1083/jcb.142.6.1429

Polekhina G, Gupta A, Michell BJ et al (2003) AMPKß subunit targets metabolic stress sensing to glycogen. Curr Biol 13:867–871. doi:10.1016/S

Ramanathan L, Sheth PR, Ogas P et al (2010) Purification and characterization of truncated human AMPK alpha 2 beta 2 gamma 3 heterotrimer from baculovirus-infected insect cells. Protein Expr Purif 70:13–22. doi:10.1016/j.pep.2009.10.007

Rattigan S, Wallis MG, Youd JM, Clark MG (2001) Exercise training improves insulin-mediated capillary recruitment in association with glucose uptake in rat hindlimb. Diabetes 50:2659–2665. doi:10.2337/diabetes.50.12.2659

Ren JM, Semenkovich CF, Gulve EA et al (1994) Exercise induces rapid increases in GLUT4 expression, glucose transport capacity, and insulin-stimulated glycogen storage in muscle. J Biol Chem 269:14396–14401. doi:10.1210/me.2009-0437

Rett K, Wicklmayr M, Fink E et al (1989) Local generation of kinins in working skeletal muscle tissue in man. Biol Chem Hoppe Seyler 370:445–449

Richter EA, Garetto LP, Goodman MN, Ruderman NB (1982) Muscle glucose metabolism following exercise in the rat: increased sensitivity to insulin. J Clin Invest 69:785–793

Richter EA, Garetto LP, Goodman MN, Ruderman NB (1984) Enhanced muscle glucose metabolism after exercise: modulation by local factors. Am J Physiol 246:E476–E482

Richter EA, Ploug T, Galbo H (1985) Increased muscle glucose uptake after exercise. No need for insulin during exercise. Diabetes 34:1041–1048. doi:10.2337/diabetes.34.10.1041

Richter EA, Mikines KJ, Galbo H, Kiens B (1989) Effect of exercise on insulin action in human skeletal muscle. J Appl Physiol 66:876–885

Richter EA, Derave W, Wojtaszewski JFP (2001) Glucose, exercise and insulin: emerging concepts. J Physiol 535:313–322. doi:10.1111/j.1469-7793.2001.t01-2-00313.x

Ropelle ER, Pauli JR, Prada PO et al (2006) Reversal of diet-induced insulin resistance with a single bout of exercise in the rat: the role of PTP1B and IRS-1 serine phosphorylation. J Physiol 577:997–1007. doi:10.1113/jphysiol.2006.120006

Rose AJ, Richter EA (2005) Skeletal muscle glucose uptake during exercise: how is it regulated? Physiology (Bethesda) 20:260–270. doi:10.1152/physiol.00012.2005

Ross FA, Jensen TE, Hardie DG (2016) Differential regulation by AMP and ADP of AMPK complexes containing different γ subunit isoforms. Biochem J 473:189–199. doi:10.1042/BJ20150910

Russell RR, Bergeron R, Shulman GI, Young LH (1999) Translocation of myocardial GLUT-4 and increased glucose uptake through activation of AMPK by AICAR. Am J Physiol 277: H643–H649

Sabina RL, Kernstine KH, Boyd RL et al (1982) Metabolism of 5-amino-4-imidazolecarboxamide riboside in cardiac and skeletal muscle. Effects on purine nucleotide synthesis. J Biol Chem 257:10178–10183

Sabina RL, Patterson D, Holmes EW (1985) 5-Amino-4-imidazolecarboxamide riboside (Z-riboside) metabolism in eukaryotic cells. J Biol Chem 260:6107–6114

Sahlin K, Tonkonogi M, Söderlund K (1998) Energy supply and muscle fatigue in humans. Acta Physiol Scand 162:261–266. doi:10.1046/j.1365-201X.1998.0298f.x

Sakamoto K, Göransson O, Hardie DG, Alessi DR (2004) Activity of LKB1 and AMPK-related kinases in skeletal muscle: effects of contraction, phenformin, and AICAR. Am J Physiol Endocrinol Metab 287:E310–E317. doi:10.1152/ajpendo.00074.2004

Sakamoto K, McCarthy A, Smith D et al (2005) Deficiency of LKB1 in skeletal muscle prevents AMPK activation and glucose uptake during contraction. EMBO J 24:1810–1820. doi:10.1038/sj.emboj.7600667

Saltiel AR, Kahn CR (2001) Insulin signalling and the regulation of glucose and lipid metabolism. Nature 414:799–806. doi:10.1038/414799a

Sanders MJ, Grondin PO, Hegarty BD et al (2007) Investigating the mechanism for AMP activation of the AMP-activated protein kinase cascade. Biochem J 403:139–148. doi:10.1042/BJ20061520

Schweitzer GG, Cartee GD (2011) Postexercise skeletal muscle glucose transport is normal in kininogen-deficient rats. Med Sci Sports Exerc 43:1148–1153. doi:10.1249/MSS.0b013e31820a7f65

Schweitzer GG, Castorena CM, Hamada T et al (2011) The B2 receptor of bradykinin is not essential for the post-exercise increase in glucose uptake by insulin-stimulated mouse skeletal muscle. Physiol Res 60:511–519, doi: 932085 [pii]

Schweitzer GG, Arias EB, Cartee GD (2012) Sustained postexercise increases in AS160 Thr642 and Ser588 phosphorylation in skeletal muscle without sustained increases in kinase phosphorylation. J Appl Physiol 113:1852–1861. doi:10.1152/japplphysiol.00619.2012

Sharma N, Wang H, Arias EB et al (2015) Mechanisms for independent and combined effects of calorie restriction and acute exercise on insulin-stimulated glucose uptake by skeletal muscle of old rats. Am J Physiol Endocrinol Metab. doi:10.1152/ajpendo.00618.2014

Sharoff CG, Hagobian TA, Malin SK et al (2010) Combining short-term metformin treatment and one bout of exercise does not increase insulin action in insulin-resistant individuals. Am J Physiol Endocrinol Metab 298:E815–E823. doi:10.1152/ajpendo.00517.2009

Shulman GI, Rothman DL, Jue T et al (1990) Quantitation of muscle glycogen synthesis in normal subjects and subjects with non-insulin-dependent diabetes by 13C nuclear magnetic resonance spectroscopy. N Engl J Med 322:223–228. doi:10.1056/NEJM199001253220403

Smith JL, Patil PB, Fisher JS (2005) AICAR and hyperosmotic stress increase insulin-stimulated glucose transport. J Appl Physiol 99:877–883. doi:10.1152/japplphysiol.01297.2004

Sriwijitkamol A, Coletta DK, Wajcberg E et al (2007) Effect of acute exercise on AMPK signaling in skeletal muscle of subjects with type 2 diabetes: a time-course and dose-response study. Diabetes 56:836–848. doi:10.2337/db06-1119

Steensberg A, van Hall G, Keller C et al (2002) Muscle glycogen content and glucose uptake during exercise in humans: influence of prior exercise and dietary manipulation. J Physiol 541:273–281. doi:10.1113/jphysiol.2001.015594
Steinberg GR, Kemp BE (2009) AMPK in health and disease. Physiol Rev 89:1025–1078. doi:10.1152/physrev.00011.2008
Steinberg GR, O'Neill HM, Dzamko NL et al (2010) Whole body deletion of AMP-activated protein kinase β2 reduces muscle AMPK activity and exercise capacity. J Biol Chem 285:37198–37209. doi:10.1074/jbc.M110.102434
Stephens TJ, Chen Z-P, Canny BJ et al (2002) Progressive increase in human skeletal muscle AMPKalpha2 activity and ACC phosphorylation during exercise. Am J Physiol Endocrinol Metab 282:E688–E694. doi:10.1152/ajpendo.00101.2001
Stöckli J, Meoli CC, Hoffman NJ et al (2015) The RabGAP TBC1D1 plays a central role in exercise-regulated glucose metabolism in skeletal muscle. Diabetes 64:1914–1922. doi:10.2337/db13-1489
Sullivan JE, Brocklehurst KJ, Marley AE et al (1994) Inhibition of lipolysis and lipogenesis in isolated rat adipocytes with AICAR, a cell-permeable activator of AMP-activated protein kinase. FEBS Lett 353:33–36. doi:10.1016/0014-5793(94)01006-4
Suter M, Riek U, Tuerk R et al (2006) Dissecting the role of 5′-AMP for allosteric stimulation, activation, and deactivation of AMP-activated protein kinase. J Biol Chem 281:32207–32216. doi:10.1074/jbc.M606357200
Sylow L, Moller LL, Kleinert M et al (2015) Stretch-stimulated glucose transport in skeletal muscle is regulated by Rac1. J Physiol 593:645–656. doi:10.1113/jphysiol.2014.284281
Szekeres F, Chadt A, Tom RZ et al (2012) The Rab-GTPase-activating protein TBC1D1 regulates skeletal muscle glucose metabolism. AJP Endocrinol Metab 303:E524–E533. doi:10.1152/ajpendo.00605.2011
Tanaka S, Hayashi T, Toyoda T et al (2007) High-fat diet impairs the effects of a single bout of endurance exercise on glucose transport and insulin sensitivity in rat skeletal muscle. Metabolism 56:1719–1728. doi:10.1016/j.metabol.2007.07.017
Thong FSL, Derave W, Kiens B et al (2002) Caffeine-induced impairment of insulin action but not insulin signaling in human skeletal muscle is reduced by exercise. Diabetes 51:583–590. doi:10.2337/diabetes.51.3.583
Thong FSL, Derave W, Urso B et al (2003) Prior exercise increases basal and insulin-induced p38 mitogen-activated protein kinase phosphorylation in human skeletal muscle. J Appl Physiol 94:2337–2341. doi:10.1152/japplphysiol.00036.2003
Treebak JT, Birk JB, Rose AJ et al (2007) AS160 phosphorylation is associated with activation of alpha2beta2gamma1- but not alpha2beta2gamma3-AMPK trimeric complex in skeletal muscle during exercise in humans. Am J Physiol Endocrinol Metab 292:E715–E722. doi:10.1152/ajpendo.00380.2006
Treebak JT, Birk JB, Hansen BF et al (2009a) A-769662 activates AMPK beta1-containing complexes but induces glucose uptake through a PI3-kinase-dependent pathway in mouse skeletal muscle. Am J Physiol Cell Physiol 297:C1041–C1052. doi:10.1152/ajpcell.00051.2009
Treebak JT, Frøsig C, Pehmøller C et al (2009b) Potential role of TBC1D4 in enhanced post-exercise insulin action in human skeletal muscle. Diabetologia 52:891–900. doi:10.1007/s00125-009-1294-y
Treebak JT, Taylor EB, Witczak CA et al (2010) Identification of a novel phosphorylation site on TBC1D4 regulated by AMP-activated protein kinase in skeletal muscle. Am J Physiol Cell Physiol 298:C377–C385. doi:10.1152/ajpcell.00297.2009
Treebak JT, Pehmøller C, Kristensen JM et al (2014) Acute exercise and physiological insulin induce distinct phosphorylation signatures on TBC1D1 and TBC1D4 proteins in human skeletal muscle. J Physiol 592:351–375. doi:10.1113/jphysiol.2013.266338
Van den Berghe G, Gruber H (1993) International patent WO 93/03734

Vavvas D, Apazidis A, Saha AK et al (1997) Contraction-induced changes in acetyl-CoA carboxylase and 5′-AMP-activated kinase in skeletal muscle. J Biol Chem 272:13255–13261. doi:10.1074/jbc.272.20.13255

Vendelbo MH, Møller AB, Treebak JT et al (2014) Sustained AS160 and TBC1D1 phosphorylations in human skeletal muscle 30 min after a single bout of exercise. J Appl Physiol 117:289–296. doi:10.1152/japplphysiol.00044.2014

Vichaiwong K, Purohit S, An D et al (2010) Contraction regulates site-specific phosphorylation of TBC1D1 in skeletal muscle. Biochem J 431:311–320. doi:10.1042/BJ20101100

Vincent MA, Clerk LH, Lindner JR et al (2004) Microvascular recruitment is an early insulin effect that regulates skeletal muscle glucose uptake in vivo. Diabetes 53:1418–1423. doi:10.2337/diabetes.53.6.1418

Vind BF, Pehmøller C, Treebak JT et al (2011) Impaired insulin-induced site-specific phosphorylation of TBC1 domain family, member 4 (TBC1D4) in skeletal muscle of type 2 diabetes patients is restored by endurance exercise-training. Diabetologia 54:157–167. doi:10.1007/s00125-010-1924-4

Wallberg-Henriksson H, Constable SH, Young DA, Holloszy JO (1988) Glucose transport into rat skeletal muscle: interaction between exercise and insulin. J Appl Physiol 65:909–913

Wang H, Sharma N, Arias EB, Cartee GD (2015) Insulin signaling and glucose uptake in the soleus muscle of 30-month-old rats after calorie restriction with or without acute exercise. J Gerontol A Biol Sci Med Sci 1–10. doi:10.1093/gerona/glv142

Wasserman DH, Kang L, Ayala JE et al (2011) The physiological regulation of glucose flux into muscle in vivo. J Exp Biol 214:254–262. doi:10.1242/jeb.048041

Wilson C, Hargreaves M, Howlett KF (2006) Exercise does not alter subcellular localization, but increases phosphorylation of insulin-signaling proteins in human skeletal muscle. Am J Physiol Endocrinol Metab 290:E341–E346. doi:10.1152/ajpendo.00314.2005

Winder WW, Hardie DG (1996) Inactivation of acetyl-CoA carboxylase and activation of AMP-activated protein kinase in muscle during exercise. Am J Physiol 270:E299–E304

Witczak CA, Fujii N, Hirshman MF, Goodyear LJ (2007) Ca2+/calmodulin-dependent protein kinase kinase-alpha regulates skeletal muscle glucose uptake independent of AMP-activated protein kinase and Akt activation. Diabetes 56:1403–1409. doi:10.2337/db06-1230

Wojtaszewski JF, Hansen BF, Ursø B, Richter EA (1996) Wortmannin inhibits both insulin- and contraction-stimulated glucose uptake and transport in rat skeletal muscle. J Appl Physiol 81:1501–1509

Wojtaszewski JFP, Hansen BF, Kiens B, Richter EA (1997) Insulin signaling in human skeletal muscle: time course and effect of exercise. Diabetes 46:1775–1781. doi:10.2337/diab.46.11.1775

Wojtaszewski JF, Higaki Y, Hirshman MF et al (1999) Exercise modulates postreceptor insulin signaling and glucose transport in muscle-specific insulin receptor knockout mice. J Clin Invest 104:1257–1264. doi:10.1172/JCI7961

Wojtaszewski JF, Hansen BF, Gade et al (2000a) Insulin signaling and insulin sensitivity after exercise in human skeletal muscle. Diabetes 49:325–331

Wojtaszewski JF, Nielsen P, Hansen BF et al (2000b) Isoform-specific and exercise intensity-dependent activation of 5′-AMP-activated protein kinase in human skeletal muscle. J Physiol 528(Pt 1):221–226

Wojtaszewski JFP, Jørgensen SB, Hellsten Y et al (2002a) Glycogen-dependent effects of 5-aminoimidazole-4-carboxamide (AICA)-riboside on AMP-activated protein kinase and glycogen synthase activities in rat skeletal muscle. Diabetes 51:284–292. doi:10.1186/1471-213X-4-9

Wojtaszewski JFP, Mourtzakis M, Hillig T et al (2002b) Dissociation of AMPK activity and ACCbeta phosphorylation in human muscle during prolonged exercise. Biochem Biophys Res Commun 298:309–316

Wojtaszewski JFP, Jørgensen SB, Frøsig C et al (2003a) Insulin signalling: effects of prior exercise. Acta Physiol Scand 178:321–328. doi:10.1046/j.1365-201X.2003.01151.x

Wojtaszewski JFP, MacDonald C, Nielsen JN et al (2003b) Regulation of 5′AMP-activated protein kinase activity and substrate utilization in exercising human skeletal muscle. Am J Physiol Endocrinol Metab 284:E813–E822. doi:10.1152/ajpendo.00436.2002

Wojtaszewski JFP, Birk JB, Frøsig C et al (2005) 5′AMP activated protein kinase expression in human skeletal muscle: effects of strength training and type 2 diabetes. J Physiol 564:563–573. doi:10.1113/jphysiol.2005.082669

Woods A, Johnstone SR, Dickerson K et al (2003a) LKB1 is the upstream kinase in the AMP-activated protein kinase cascade. Curr Biol 13:2004–2008. doi:10.1016/j.cub.2003.10.031

Woods A, Vertommen D, Neumann D et al (2003b) Identification of phosphorylation sites in AMP-activated protein kinase (AMPK) for upstream AMPK kinases and study of their roles by site-directed mutagenesis. J Biol Chem 278:28434–28442. doi:10.1074/jbc.M303946200

Woods A, Dickerson K, Heath R et al (2005) Ca2+/calmodulin-dependent protein kinase kinase-beta acts upstream of AMP-activated protein kinase in mammalian cells. Cell Metab 2:21–33. doi:10.1016/j.cmet.2005.06.005

Wright DC, Geiger PC, Holloszy JO, Han D-H (2005) Contraction- and hypoxia-stimulated glucose transport is mediated by a Ca2+-dependent mechanism in slow-twitch rat soleus muscle. Am J Physiol Endocrinol Metab 288:E1062–E1066. doi:10.1152/ajpendo.00561.2004

Xiao B, Sanders MJ, Underwood E et al (2011) Structure of mammalian AMPK and its regulation by ADP. Nature 472:230–233. doi:10.1038/nature09932

Xiao Y, Sharma N, Arias EB et al (2013) A persistent increase in insulin-stimulated glucose uptake by both fast-twitch and slow-twitch skeletal muscles after a single exercise session by old rats. Age (Dordr) 35:573–582. doi:10.1007/s11357-012-9383-0

Xie M, Zhang D, Dyck JRB et al (2006) A pivotal role for endogenous TGF-beta-activated kinase-1 in the LKB1/AMP-activated protein kinase energy-sensor pathway. Proc Natl Acad Sci USA 103:17378–17383. doi:10.1073/pnas.0604708103

Yamauchi T, Kamon J, Minokoshi Y et al (2002) Adiponectin stimulates glucose utilization and fatty-acid oxidation by activating AMP-activated protein kinase. Nat Med 8:1288–1295. doi:10.1038/nm788

Yang J, Holman GD (2005) Insulin and contraction stimulate exocytosis, but increased AMP-activated protein kinase activity resulting from oxidative metabolism stress slows endocytosis of GLUT4 in cardiomyocytes. J Biol Chem 280:4070–4078. doi:10.1074/jbc.M410213200

Yeh JI, Gulve EA, Rameh L, Birnbaum MJ (1995) The effects of wortmannin on rat skeletal muscle. Dissociation of signaling pathways for insulin- and contraction-activated hexose transport. J Biol Chem 270:2107–2111

Young JC, Garthwaite SM, Bryan JE et al (1983) Carbohydrate feeding speeds reversal of enhanced glucose uptake in muscle after exercise. Am J Physiol 245:R684–R688

Young DA, Wallberg-Henriksson H, Sleeper MD, Holloszy JO (1987) Reversal of the exercise-induced increase in muscle permeability to glucose. Am J Physiol 253:E331–E335

Zorzano A, Balon TW, Goodman MN, Ruderman NB (1986) Additive effects of prior exercise and insulin on glucose and AIB uptake by rat muscle. Am J Physiol 251:E21–E26

Part II
AMPK in the Diseases

Chapter 6
AMPK/Mitochondria in Metabolic Diseases

Pedro Bullon, Fabiola Marin-Aguilar, and Lourdes Roman-Malo

Contents

Abstract The obtaining of nutrients is the most important task in our lives. Energy is central to life's evolutions; this was one of the aspect that induced the selection of the more adaptable and more energetically profitable species. Nowadays things

P. Bullon (✉)
Department of Oral Medicine and Periodontology, Dental School, University of Seville, C/Avicena s/n, Seville 41009, Spain
e-mail: pbullon@us.es

F. Marin-Aguilar
Departament of Oral Medicine and Periodontology, Ministry of Education, Culture and Sport, c/Avicena s/n, Seville 41009, Spain

L. Roman-Malo
Departament of Oral Medicine and Periodontology, University of Seville, c/Avicena s/n, Seville 41009, Spain

M.D. Cordero, B. Viollet (eds.), *AMP-activated Protein Kinase*, Experientia Supplementum 107, DOI 10.1007/978-3-319-43589-3_6

have changed in our modern society. A high proportion of people has access to plenty amount of food and the obesity appear as one of the pathological characteristics of our society. Energy is obtained essentially in the mitochondria with the transfer of protons across the inner membrane that produce ATP. The exactly regulation of the synthesis and degradation of ATP (ATP $\leftrightarrow$ ADP + phosphate) is essential to all form of life. This task is performed by the 5' adenosine monophosphate-activated protein kinase (AMPK). mtDNA is highly exposed to oxidative damage and could play a central role in human health and disease. This high potential rate of abnormalities is controlled by one of the most complex mechanism: the autophagy. AMPK appears to be the key cellular energy sensor involved in multiple cellular mechanisms and is essential to have a good metabolic homeostasis to face all the aggression and start the inflammatory reaction. Therefore its disturbances have been related with multiple diseases. Recent findings support the role of AMPK in inflammation and immunity such as Metabolic Syndrome, Obesity and Diabetes. All these Metabolic Disorders are considered pandemics and they need an adequate control and prevention. One important way to achieve it is deepen in the pathogenic mechanisms. Mitochondria and AMPK are the key elements through which it happen, their knowledge and research allow us to a better management. The discovery and use of drugs that can modulate them is imperative to improve our way of manage the metabolic disorders.

Keywords AMPK • Mitochondria • mtDNA • Metabolic Diseases • Metabolic Syndrome • Obesity • Diabetes • Inflammation • Inflammation treatment

6.1 Introduction

All forms of life need energy to survive. Energy is central to life's evolutions; we can only understand the properties of life if we bring energy to equation. The way the energy is produced gives advantage to one organism against the other one. During the evolution of all the form of lives in the earth, this was one of the aspects that induced the selection of the more adaptable and more energetically profitable species. The Homo sapiens as an animal species has suffered famine period like other animal species, but we could survive. The obtaining of nutrients is the most important task in our lives. This was essential in the past because large periods of nutrient shortage were very common. But nowadays things have changed in our modern society. A high proportion of people have access to plenty amount of food. The way to deal with this time of plenty is to consume the calories with exercise or accumulate the nutrients in adipose tissue. In a sedentary attitude of the population, the obesity appears as one of the pathological characteristics of our society. Our entire metabolism, controlled by genes, needs to be adapted to this abundance of nutrient but usually it cannot be achieved. The breaks of the homeostasis and equilibrium in the metabolism involve most of the times to the development of metabolic disorder (Ezzati et al. 2002).

Essentially, all living cells power themselves through the flow of protons (positively charged hydrogen atoms), in what amounts to a kind of electricity with protons in place of electrons. The energy we gain from burning food in respiration is used to pump protons across a membrane, forming a reservoir on one side of the membrane. The flow of protons back from this reservoir can be used to power work in the same way as a turbine in a hydroelectric dam. This transfer of protons across a membrane is known as chemiosmotic hypothesis (Mitchell 1961). This use of proton gradients is universal across life on earth; proton power is as much an integral part of all life as the universal genetic code. This mechanism produces energy, but the requirements of energy to live is extremely high. The energy currency used by all living cells is a molecule called ATP (adenosine triphosphate). A single cell consumes around ten million molecules of ATP every second. If we calculate 40 trillion cells contained in the human body, we used about 60–100 kg per day, roughly our own body weight. ATP is usually split into two unequal pieces, ADP (adenosine diphosphate) and inorganic phosphate (PO_4^{3-}). The energy of respiration (the energy released from the reaction of food with oxygen) is used to make ATP from ADP and P_i. This energy comes from just one particular type of chemical reaction known as a redox reaction in which electrons are transferred from one molecule to another. This takes place in the inner membrane of the mitochondria in the electron transport chain that comprise five protein membrane-bound complexes. The electrons pass through this respiratory complexes, protons are ferried across the membrane, the enzyme ATP synthase is activated, and ATP is produced (Rich 2003).

An appreciation of the rising global burden of chronic, noncommunicable diseases and its influence in mortality and global health cost is a main concern in our society. It has been calculated that 35 million people died in 2005 from heart disease, stroke, cancer, and other chronic diseases (Strong et al. 2005). Physicians and health managers have applied effective measures, including behavioral interventions and pharmaceutical treatment, in the prevention and management of chronic diseases, but these are neither widely used nor equitably distributed. Most of them are related to modern life style, mainly with sedentary way of living and obesity. Therefore, the most effective treatment is the prevention of changing these parameters. But the results are not adequate, for instance, overweight and obesity continue to grow its prevalence, even in children and adolescent (Brown et al. 2015). The only way to improve our knowledge of the mechanism of metabolic diseases is to deepen the pathogenic mechanism involved in the energy production and its control. Energy production inside a cell is a very complex and accurate process that needs the proper function of the electron transport chain placed in the internal mitochondrial membrane. It is essential to attend all the environmental requirements as soon as possible, with the exact amount that does not permit to waste energy.

6.2 Mitochondria

Mitochondria are ubiquitous in eukaryotes and are essential for survival. Their primary function is to support aerobic respiration and to provide energy substrates (such as ATP) for intracellular metabolic pathways. Mitochondria have also been shown to play an important role in iron and calcium homeostasis, amino acid, fatty acid and steroid hormone metabolism, and in cell signaling, particularly in signaling for apoptotic cell death. Mitochondria host several metabolic pathways, including the Krebs cycle, β-oxidation, and lipid and cholesterol synthesis (Schapira 2006). Mitochondria are intracellular double membrane-bound structures. Although traditionally considered as small isolated organelles within the cell, it is more likely that mitochondria form a complex branching network. They are derived from prokaryotic cells (bacteria) that were assimilated by another cell (archaea) through a process called endosymbiosis. Although these organelles usually retain some DNA, most of their original genome has moved to the host nucleus through endosymbiotic gene transfer (Archibald 2015). Human mtDNA is a circular double-stranded molecule about 16 · 6 Kb long. It is much smaller than most nuclear genes. MtDNA codes for 22 transfer and 2 ribosomal RNAs and for 13 proteins. Human mtDNA is extremely compact and contains virtually non-intronic (noncoding) regions. By encoding proteins of the respiratory chain, mtDNA allows individual mitochondria to respond, by gene expression, to changes in membrane potential. Although human mtDNA encodes the basic machinery for protein synthesis, it remains entirely dependent upon the nucleus for the provision of enzymes for replication, repair, transcription, and translation. This dependency lies at the heart of several newly recognized human diseases that are characterized by secondary abnormalities of mtDNA.

There are good reasons to believe that genes affecting the mitochondria could play a central role in human health and disease. Most of the genes that have remained in the mitochondrion have been linked to a series of devastating diseases, indicating the importance of fully functional mitochondria to human health. Genes residing in the mitochondria pose a particular problem, in part because they are unusually prone to damage. Unlike nuclear genes, which are wrapped in protective proteins and stored safely away in the nucleus, mitochondrial genes are vulnerable to attack from highly reactive molecules called free radicals; these are generated during energy production. In mammals, the mutation rate of mitochondrial genes is 10–20 times higher than that of the nuclear genes. The idea that mutations in mitochondrial DNA could cause metabolic diseases, or even aging, is widely accepted (Lane 2006). These alterations range from changes to single DNA bases to deletions of large sections of the genome. Mitochondria that lose their genome (hydrogenosomes and mitosomes) lose the ability to synthesize ATP by chemiosmotic coupling (Lane and Martin 2010). Respiration rates do correlate with the amount of mtDNA in the cell (Williams 1986), and mutations that deplete mtDNA usually cause mitochondrial diseases (Schapira 2006). Oxidative phosphorylation is under tight control by the amount of mtDNA in the cell, and the full complement of mtDNA is necessary to maintain a normal energy production level (Rocher et al. 2008). Mitochondria, along with their tiny genomes, are normally inherited only from the mother—they are present in huge numbers in the egg, whereas the handful in sperm is marked up for destruction in the fertilized egg. This gives at least some mitochondrial diseases a maternal-inheritance pattern. Even so, trying to spot mitochondrial diseases by looking to the mother can be grossly misleading and has own played the importance of these organelles in disease. More than 80 % of diseases known to be linked to faulty mitochondria do not follow a maternal-inheritance pattern at all. Why not? At least partly because some mitochondrial diseases may be caused by mutations in the nuclear genes encoding mitochondrial proteins. So far, mutations in more than 30 nuclear genes have been shown to give rise to mitochondrial disease (Lynch and Conery 2003).

This high potential rate of abnormalities is controlled by one of the most complex mechanism: the autophagy. Autophagy is the most efficient mitochondrial turnover mechanism, providing for the complete removal of irreversibly damaged mitochondria (mitophagy). It is believed that mitochondria are normally replaced every 2–4 weeks in rat brain, heart, liver, and kidneys, although recent studies have shown that the turnover rate might be considerably higher. Mitophagy is particularly important for long-lived postmitotic cells, whose mitochondria have pronounced oxidative damage (Terman et al. 2010), and also prevents an excessive accumulation of mitochondria. It should be mentioned that the changes in mitochondriogenesis are tissue specific (Johnson et al. 2007), being more dramatic in the central nervous system, which is consistent with the generally higher susceptibility of neurons and other postmitotic cells to the aging process (Lopez-Lluch et al. 2008). Mitochondria are dynamic structures that show different morphologies: small spheres, short rods, or long tubules that depend on cell type and also cell status. In most eukaryotic cells, mitochondria move along cytoskeletal tracks, and their overall morphology depends on the balance between fusion and fission events. For example, an extent of fusion activities leads to interconnected mitochondrial networks, and on the contrary, an extent of fission events generates numerous different small spherical organelles (Westermann 2008). Mitochondrial membrane fusion/fission processes are clearly involved in mitochondria dynamics and that these events are undoubtedly critical for several cell functions and the balance cell life/cell death. Perturbations of mitochondrial dynamics can have tremendous consequences on cell metabolism and therefore on cell life/cell death (Benard and Rossignol 2008). The frequent fusion/fission events undergone by the mitochondrial network appear clearly linked to the bioenergetic state of mitochondria (Twig et al. 2008) and involve a protein machinery and a group of lipids (Furt and Moreau 2009). Disturbed mitochondrial dynamics is involved in the most important chronic disease such as neurodegenerative disorders (Burté et al. 2015), aging process (Gonzalez-Freire et al. 2015), and cardiovascular disease (Biala et al. 2015).

6.3 AMPK

The electron transport chain placed in the inner mitochondrial membrane has a main function of producing ATP, which is the battery of our cells. Therefore, the exact regulation of the synthesis and degradation of ATP (ATP ↔ ADP + phosphate) is essential to all forms of life. This reaction is maintained by catabolism many orders of magnitude away from equilibrium, yielding a high ratio of ATP to ADP that is used to drive energy-requiring processes. ATP generation needs to remain in balance with ATP consumption, and regulatory proteins that sense ATP and ADP levels would be a logical way to achieve this. This task is performed by the 5′ adenosine monophosphate-activated protein kinase (AMPK). It is a highly conserved sensor of cellular energy status, expressed in essentially all eukaryotic cells. AMPK is switched on by metabolic stresses and xenobiotic compounds that cause a cellular energy imbalance, which is detected as increases in the ratios of ADP–ATP and AMP–ATP (Gowans et al. 2013). Because the energy status of the cell is a crucial factor in all aspects of cell function, it is not surprising that AMPK has many downstream targets whose phosphorylation mediates dramatic changes in cell metabolism, cell growth, and other functions. In general, AMPK switches on catabolic processes that provide alternative pathways to generate ATP, while switching off anabolic pathways and other processes consuming ATP, thus acting to restore cellular energy homeostasis. The kinase evolved in single-celled eukaryotes and is still involved in multicellular organisms in regulating energy balance in a cell-autonomous manner (Hardie et al. 2012).

AMPK and its orthologues seem to exist universally as heterotrimeric complexes comprising a catalytic α-subunit and regulatory β- and γ-subunits. Each of these three subunits takes on a specific role in both the stability and activity of AMPK. Specifically, the γ-subunit includes four particular cystathionine beta synthase (CBS) domains giving AMPK its ability to sensitively detect shifts in the AMP–ATP ratio (Stapleton et al. 1996). In mammalian cells, AMPK is activated by various types of metabolic stresses (starvation for glucose or oxygen or addition of a metabolic poison, muscle contraction), drugs (metformin, phenformin, resveratrol, epigallocatechin, capsaicin, curcumin), and xenobiotics through the mechanisms described above, which involve increases in cellular AMP, ADP, or Ca^{2+}. These can now be regarded as the classical or "canonical" AMPK activation mechanisms. However, recent work suggests that other stimuli activate AMPK via mechanisms that do not involve changes in the levels of AMP, ADP, and Ca^{2+}, which can therefore be termed "noncanonical" mechanisms such as those triggered by ROS and DNA-damaging agents. AMPK acts as a metabolic master switch regulating several catabolic intracellular systems, including the cellular uptake of glucose and fatty acids, the β-oxidation of fatty acids, the biogenesis of glucose transporter 4 (GLUT4), mitochondrial biogenesis, and mitophagy. Also conserves ATP by switching off almost all anabolic pathways, including the biosynthesis of lipids, carbohydrates, proteins, and ribosomal RNA. It achieves this in part by phosphorylating and/or regulating enzymes or regulatory proteins that are directly involved in these pathways (Shirwany and Zou 2014). AMPK appears to have evolved early in the evolution of unicellular eukaryotes as a signaling pathway that orchestrated responses to glucose starvation in a cell-autonomous manner. But also it is intriguing that hormones that modulate energy balance at the whole-body level, which clearly arose later during the evolution of multicellular organisms, appear to have adapted to interact with the AMPK system, especially in the hypothalamus. Adiponectin, leptin, ghrelin, insulin, and triiodothyronine can influence the AMPK production in the hypothalamus (Hardie and Ashford 2014). In mammals, it also regulates metabolism and helps to maintain energy balance at the whole-body level. It does this by mediating effects of hormones and other agents acting on neurons in different hypothalamic regions, which regulate intake of food (and hence energy) and energy expenditure. AMPK also regulates diurnal rhythms of feeding and metabolism. By switching off biosynthetic pathways required for cell growth, AMPK activation exerts a cytostatic effect, helping to explain why its upstream activator, LKB1, is a tumor suppressor. Commensurate with its role in preserving cellular energy homeostasis, AMPK also downregulates ATP-requiring processes outside metabolism, including progress through the cell cycle (another potential tumor suppressor effect) and firing of action potentials in neurons (Hardie 2014). Some works demonstrate that the AMPK signaling pathway also plays a role in bone physiology. Activation of AMPK promotes bone formation in vitro and the deletion of α- or β-subunit of AMPK decreases bone mass in mice (Jeyabalan et al. 2012).

AMPK appears to be the key cellular energy sensor involved in multiple cellular mechanisms and is essential to have a good metabolic homeostasis to face all the aggression and start the inflammatory reaction. Therefore, its disturbances have been related with multiple diseases. Age-related decrease of mitochondrial biogenesis is related, at least in part, to diminished AMPK activity (Reznick et al. 2007). AMPK appears to be the key cellular energy sensor, linking decreased mitochondriogenesis to several aging-associated changes, including insulin resistance and deficient lipid metabolism (Qiang et al. 2007). AMPK activity and autophagy of monocytes were significantly decreased in Acute Coronary Syndrome patients due to a decrease in plaque vulnerability and subsequent plaque rupture (Cheng et al. 2015). Recent findings support the role of AMPK in inflammation and immunity, providing the enticing prospect of new therapeutic approaches for inflammatory diseases. All of the AMPK activators have been reported to inhibit inflammatory responses in various model systems, and AMPK-activating drugs do have anti-inflammatory actions in animal models (O'Neill and Hardie 2013). Tumor cells appear to be under selection pressure to downregulate AMPK, thus limiting its restraining influence on cell growth and proliferation. Paradoxically, however, a complete loss of AMPK function, which appears to be rare in human cancers, may be deleterious to survival of tumor cells. AMPK can therefore be either a friend or a foe in cancer, depending on the context (Hardie 2015).

6.4 Metabolic Syndrome and Obesity

The metabolic syndrome (MetS) is a major and escalating public health and clinical challenge worldwide in the wake of urbanization, surplus energy intake, increasing obesity, and sedentary life habits. MetS confers a fivefold increase in the risk of type 2 diabetes mellitus (T2DM) and twofold the risk of developing cardiovascular disease (CVD) over the next 5–10 years (Alberti et al. 2009). Further, patients with the MetS are at two to fourfold increased risk of stroke, a three to fourfold increased risk of myocardial infarction (MI), and twofold the risk of dying from such an event compared with those without the syndrome (Alberti and Zimmet 2005) regardless of a previous history of cardiovascular events (Olijhoek et al. 2004).

MetS is a state of chronic low-grade inflammation as a consequence of complex interplay between genetic and environmental factors. Insulin resistance, visceral adiposity, atherogenic dyslipidemia, endothelial dysfunction, genetic susceptibility, elevated blood pressure, hypercoagulable state, and chronic stress are the several factors that constitute the syndrome (Table 6.1).

There have been several definitions of MetS, but the most commonly used criteria for definition at present are from the World Health Organization (WHO) (Table 6.2).

Hypercaloric diets initially result in obesity due to the storage of extra energy in the adipose tissue. However, the continuous caloric overload eventually results in the aberrant accumulation of lipids in non-adipose tissues (Virtue et al. 2012). The direct pathological consequence of chronic hypercaloric diets is actually a multisystemic deterioration known as metabolic syndrome (Kaur 2014). Of note, the comorbidities associated with metabolic syndrome also overlap with some of the most important aging-associated diseases, namely, diabetes, cardiovascular and cerebrovascular diseases, and cancer (Gurevich-Panigrahi et al. 2009).

6.4.1 *Inflammation in Obesity and MetS*

Considering the "obesity epidemic" as mainly responsible for the rising prevalence of MetS due to the strong connection between this syndrome and, especially,

Table 6.1 Clinical identification of the metabolic syndrome

Risk factor	Defining level
Abdominal obesity, given as waist circumference	Men >102 cm Women >88 cm
Triglycerides	≥150 mg/dL
HDL cholesterol	Men <40 mg/dL Women <50 mg/dL
Blood pressure	≥130/≥85 mmHg
Fasting glucose	≥110 mg/dL

Table 6.2 WHO clinical criteria for metabolic syndrome

Clinical measures	WHO (1998)
Insulin resistance	IGT, IFG, T2DM, or lowered insulin Sensitivity[a] Plus any two of the following
Body weight	Men: waist-to-hip ratio >90 cm Women: waist-to-hip ratio >85 cm and/or BMI > 30 kg/m^2
Lipids	TGs ≥150 mg/dL and/or HDL-C Men: <35 mg/dL Women: <39 mg/dL
Blood pressure	≥140/90 mmHg
Glucose	IGT, IFG, or T2DM
Other	Microalbuminuria Urinary excretion rate of >20 mg/min or Albumin: creatinine ratio of >30 mg/g

[a]Insulin sensitivity measured under hyperinsulinemic euglycemic conditions, glucose uptake below lowest quartile for background population under investigation

BMI body mass index, *HDL-C* high-density lipoprotein cholesterol, *IFG* impaired fasting glucose, *IGT* impaired glucose tolerance, *TGs* triglycerides, *T2DM* type 2 diabetes mellitus, *WC* waist circumference

abdominal obesity, it is vital to bear in mind that obesity is not only a risk factor but also a disease in itself. Obesity contributes to hypertension, high serum cholesterol, low HDL cholesterol, and hyperglycemia, and it otherwise associates with higher CDV risk. Excess adipose tissue releases several products that apparently exacerbate these risk factors. They include free fatty acids (FFA), cytokines: interleukin-6 and Tumor Necrosis Factor-α (IL-6 and TNF-α), plasminogen activator inhibitor 1 (PAI-1), and adiponectin. A high plasma FFA level overloads muscle and liver with lipids, which is an event commonly found in abdominal obesity (Miles and Jensen 2005), enhancing insulin resistance (Boden et al. 2001). Evidence suggests that TNF-α induces adipocytes apoptosis (Xydakis et al. 2004) and promotes insulin resistance by the inhibition of the insulin receptor substrate 1 signaling pathway (Lau et al. 2005). Plasma TNF-α is positively associated with the body weight and triglycerides (TGs), while a negative association exists between plasma TNF-α levels and high-density lipoprotein-cholesterol (HDL-C) (Xydakis et al. 2004). In addition, IL-6 has been shown to be positively associated with body mass index (BMI), fasting insulin, and the development of T2DM (Pradhan et al. 2001) and negatively associated with HDL-C (Zuliani et al. 2007).

Plasminogen activator inhibitor-1 (PAI-1) levels are increased in abdominally obese subjects (Cigolini et al. 1996), and an elevated PAI-1 contributes to a prothrombotic state, whereas low adiponectin levels that accompany obesity correlate with worsening of metabolic risk factors, since it regulates the lipid and glucose metabolism, increases insulin sensitivity, regulates food intake and body weight, and protects against a chronic inflammation (Liu and Liu 2010). Adiponectin is seen to be "protective," not only in its inverse relationship with the features of MetS (Engeli et al. 2003) but also through its antagonism of TNF-α action (Ouchi et al. 2000). Considering all these aspects, new therapies for metabolic diseases have been investigated.

6.4.2 *Mitochondrial DNA Alterations in Obesity and MetS*

Obesity results from an energy surplus, greater energy intake than expenditure. Excess energy is stored as fat in white adipose tissue (WAT), dysfunction of which lies at the core of obesity and associated metabolic disorders. By contrast, brown adipose tissue (BAT) burns fat and dissipates chemical energy as heat. The development and activation of "brown-like" adipocytes, also known as beige cells, result in WAT browning and thermogenesis (heat production). Both "white" and "brown" adipose cells (color is due to the quantity of mitochondria) play a crucial role in thermogenesis. "Brown-like" adipocytes, which have a rich sympathetic innervation, are present in rodents and infants, whereas in adults are usually interspersed with WAT. Due to the numerous mitochondria existing in BAT, these cells are an important heat source, generating more heat and less ATP than "white-like" adipocytes. This event occurs because of the presence of mitochondrial uncoupling proteins (UCP), as it has already been demonstrated in murine models. Three isoforms have been identified, UCP1, 2, and 3, which are differently distributed in BAT.

These proteins "uncouple" oxidative phosphorylation, and those way mitochondria continues oxidizing substrates while producing minor quantities of ATP levels, which favors the net loss of energy in the form of heat. As one would expect, cold exposure or leptin administration increases UCP activity and quantity in BAT.

Noradrenalin is a hormone which acts on β_3-adrenergic receptors in BAT and boosts peroxisome proliferator-activated receptor gamma (PPAR-γ) activity, which activates, at the same time, the gene that encodes UCP-1. Curiously, β_3-adrenergic receptors expression is lower in mice that are genetically obese.

Until recently, studies on mtDNA biology in adipose tissue have been limited to the analysis of mitochondrial biogenesis in BAT. However, recent studies have highlighted the importance of mitochondrial biogenesis in WAT and the potential for mitochondrial alterations to disturb white adipocyte development and function. Studies by Corvera and collaborators (Wilson-Fritch et al. 2003) have shown that mitochondrial biogenesis is directly associated with white adipocyte differentiation; genetically obese mice (*ob*/*ob*) displayed impaired mitochondrial mass and function in white fat and thiazolidinediones, PPAR-γ activators that favor adipocyte differentiation, ameliorated these alterations (Wilson-Fritch et al. 2004). It has been also shown that white adipocyte differentiation is associated with increases in the relative abundance of mtDNA, and upregulation of components of the mtDNA replication and transcription machinery, such as TFAM (Mitochondrial Transcription Factor A) (Shi et al. 2008), and components of deoxynucleotide metabolism are required for mtDNA replication (Rylova et al. 2005). Agents that promote white adipocyte differentiation in vitro, such as glitazones, also increase mtDNA levels in human adipocytes in vitro (Bogacka et al. 2005a).

Studies of potential alterations in WAT mtDNA as they relate to obesity have focused on two aspects: changes in mtDNA levels that underlie obese phenotypes and the occurrence of mutated, polymorphic, forms of mtDNA that are specifically associated with obesity. In experimental models of obesity, such as *ob/ob* or *db/db* mice, abnormally low levels of mtDNA have been reported (Choo et al. 2006; Rong et al. 2007). As noted above for in vitro studies, treatment of obese mice with glitazones increases mtDNA levels in white fat (Rong et al. 2007). A study in which mice were treated with a diet enriched in polyunsaturated fatty acids identified mtDNA encoded transcripts and proteins among the most upregulated genes in WAT; this upregulation was associated with an enhancement of fatty acid oxidation in WAT (Flachs et al. 2006).

In humans, the scenario appears to be more complex. It has been reported that mtDNA levels in adipose tissue are lowered in type 2 diabetic patients (Bogacka et al. 2005b), and studies by Arner and collaborators (Dahlman et al. 2006) have confirmed that mtDNA levels are not associated with obesity per se, but rather with type 2 diabetes phenotypes. Moreover, mtDNA levels were found to be strongly related to lipogenesis in WAT, rather than to BMI.

The mechanism by which mtDNA copy number in white adipose tissue could affect lipogenesis rate remains to be established, but it stands in contrast to the expected relationship between mtDNA level variations and energy expenditure and fat oxidation. Moreover, in humans, as in rodents, pioglitazone treatment causes an increase in mtDNA levels in WAT of type 2 diabetes patients (Bogacka et al. 2005b), but not in nondiabetic obese individuals (Bogacka et al. 2007). These findings highlight the potential role of mtDNA levels in WAT mass.

6.4.3 Homeostatic Control of Energy Balance in Obesity

The arcuate nucleus (ARC) is a key hypothalamic nucleus that regulates appetite, eating behavior, and energy state (Schwartz et al. 2000). It receives afferent pathways from digestive tract and contains leptin, an adipokine involved in the regulation of satiety and energy intake (Lau et al. 2005), and other significant hormone receptors. A reduction in leptin levels activates orexigenic neurons, leading to an increase in food intake, fat synthesis, and storage (anabolism) and to a decrease in energy expenditure. In reverse, when increases in leptin levels take place, another group of neurons are activated, triggering an anorexigenic effect and catabolism.

Energy balance depends directly on food intake, energy storage in adipose tissue, and energy expenditure (Spiegelman and Flier 2001). For the majority of people, this process is closely connected by an homeostatic system which integrates different hormones, such as leptin (Ahima and Flier 2000), considered as an indicator of lipid reserves, since an augmentation in lipid deposits enhances leptin releasing in plasma by adipocytes. Ghrelin, which is released to the intestine during food intake, provides us the feeling of hunger, and cholecystokinin (CCK), secreted into the duodenum as a response to the process of food intake and digestion, providing satiety, as well.

Since homeostatic control in energy balance is extremely complex, it is not easy to precisely determine what does not work in obesity. In fact, when leptin was discovered (Zhang et al. 1994), it was thought that an alteration in its kinetics would provide a simple explanation to this disease, but there is a notable variability in leptin sensibility among individuals, and some people seem to synthesize insufficient amounts of this hormone. However, plasmatic leptin usually reaches higher levels in obese than in normal weight people. What prevail in obese is a greater resistance to leptin and not an insufficient production of it (Hutley and Prins 2005). This resistance may obey to alterations in blood circulation leptin transport, in its transport to Central Nervous System (CNS), or in hypothalamic leptin receptors (like *db*/*db* mice).

Obesity, could be involved in other alteration in mediators apart from leptin, for example, TNF-α, cytokine that transmits information from adipose tissue to brain, is increased in insulin-resistant obese adipose tissue. Another common pathophysiological alteration in obesity is a decreasing insulin sensibility in skeletal muscle and adipose tissue.

Thus, events causing obesity depends on diet, exercise, social, economic and cultural factors, and genetic predisposition (Barsh et al. 2000). Although other causes have been related to alterations in leptin action or synthesis, such as a thermogenesis decreasing in adipocytes or a reduction the energy metabolic consumption. A key player at regulating energy balance is AMP-activated protein kinase (AMPK), at both cellular and whole-body levels, placing it at the center stage in studies of obesity, diabetes, and the metabolic syndrome (Hardie 2008).

6.4.4 AMPK: A Master Metabolic Regulator

6.4.4.1 Lipid Metabolism Controlled by AMPK

AMPK plays a key role in lipid metabolism, being involved in Acetyl-CoA carboxylase (ACC) phosphorylation and inactivation. ACC catalyzes the transformation from Acetyl-CoA into malonyl-CoA, which is the very first reaction in fatty acid biosynthesis in liver and adipose tissue. By inactivating ACC, AMPK is then responsible for the inhibition of fatty acid synthesis in lipogenic tissues. AMPK also has long-term effects on transcriptional genes involved in lipogenesis, ACC and fatty acid synthesis, interfering with the expression, and the activity of transcriptional factors, such as sterol regulatory element-binding protein 1c (SREBP1c) (Hardie 2008; Foretz et al. 2005) and carbohydrate response element-binding protein (ChREBP) (Zhou et al. 2001).

In addition, in both liver and striated muscle (skeletal and myocardial), malonyl-CoA produced by ACC plays a regulatory role. It blocks, in fact, fatty acids transport from cytosol to mitochondria by inhibiting carnitine palmityl transferase (CPT-1). AMPK activation in those tissues triggers a decrease in cytosolic concentration of malonyl-CoA, enabling that way the fatty acids penetration into the mitochondria and its consequent oxidation. By this mechanism, recent data reveal that the adipokines, leptin, and adiponectin stimulate fatty acid oxidation in both liver and skeletal muscle secondarily to the AMPK activation in these tissues (Minokoshi et al. 2002; Tomas et al. 2002). So, the effect of lipid depletion on these tissues improves metabolic parameters in different insulin-resistant rodent models. In fact, the accumulation of triglycerides in liver and skeletal muscle is linked to the pathophysiology of insulin resistance in human and animals, as well as lipid depletion in these tissues ameliorates insulin sensitivity (lipotoxicity concept). So here lies the metabolic interest of the AMPK activation on lipotoxicity reduction.

AMPK may be also involved in the hypothalamus satiety central control; this brain region plays an essential role in energy homeostasis by controlling food intake and energy expenditure. Hypothalamic AMPK activity varies according to the nutritional status; AMPK is activated in fasting and inhibited in satiety period (Minokoshi et al. 2004). At the same time, it is also interesting that the existing relationship between the activity of hypothalamic AMPK and food intake in response to different hormones or metabolites is known to be modulated by nutritional state. For example, ghrelin and endocannabinoids activate AMPK and induce food intake while insulin, glucose, or leptin acts in the opposite manner (Andersson et al. 2004; Kola et al. 2005). Several studies have demonstrated that the variations in hypothalamic energetic state are able to directly modulate AMPK activity, which suggests that hypothalamic AMPK may be a therapeutic target in the numerous factors that affect eating behavior. But, it is important to bear in mind that AMPK is differently regulated in the hypothalamus and in peripheral tissues. For example, leptin activates AMPK in skeletal muscle and inhibits it in the hypothalamus, which suggests that regulation mechanisms are not the same. Therefore, drawbacks in peripheral tissues must be taken into account when considering AMPK as a pharmacological inhibitor for the treatment of metabolic disorders (Fig. 6.1).

The obesity epidemic and its complications continue rising as a global health challenge, despite the increasing public awareness and the use of lifestyle and medical interventions; the main treatment for obesity consists of lifestyle modifications based on a suitable diet and physical exercise. Since the current antiobesity drugs, such as Orlistat (lipase inhibitor), suffer from numerous disadvantages, gastrointestinal symptoms, elevated blood pressure, abdominal pain, dyspepsia, diarrhea, flatulence, etc (Bray and Tartaglia 2000), the biomedical community is urged to develop new treatments for the metabolic diseases. AMP-activated protein kinase (AMPK), as a master regulator of energy homeostasis, is a potential target for therapeutic agents that may meet this challenge. For these reasons, novel drugs

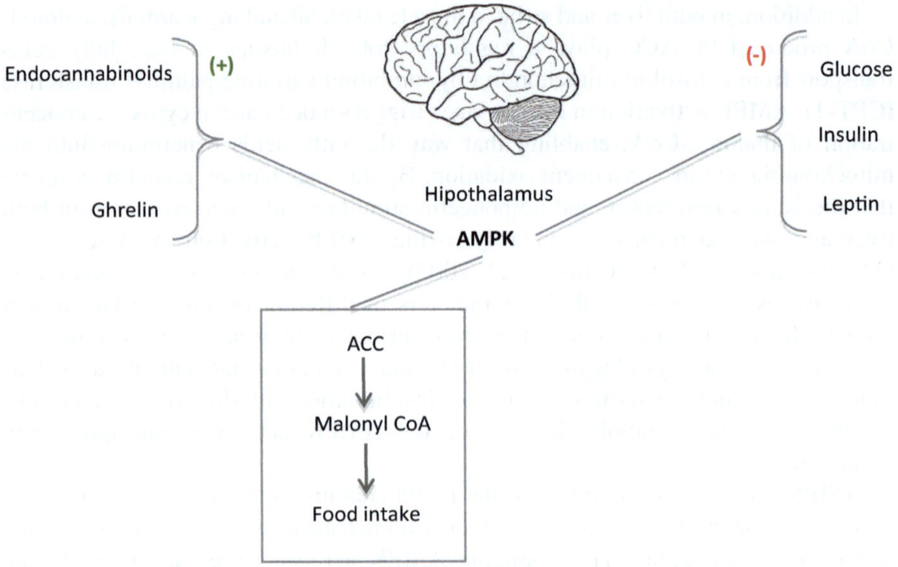

Fig. 6.1 Hypothalamic AMPK regulation by hormonal and nutritional signals

activating AMPK may have potential for the treatment of obesity, T2DM, and MetS.

Some compounds have already shown to have an effect on AMPK. 5-Aminoimidazole-4-carboxamide-1-β-D-ribofuranoside (AICAR) is a known activator of AMPK that induces allosteric changes in AMPK conformation and thereby leading to kinase activation. In 3T3L1 cells, AICAR inhibits adipocyte differentiation by downregulating key transcriptional factors such as sterol regulatory element-binding protein 1 (SREBP1), CCAAT-enhancer binding proteins (C/EBPα), and peroxisome proliferator-activated receptor gamma (PPARγ), which strictly regulates adipocyte differentiation. AICAR proved remarkably effective in maintaining body weight and epididymal fat content, improving insulin sensitivity and glucose tolerance in diet-induced obese mice models. However, in rats, the chronic administration of AICAR resulted in significant changes, in skeletal muscle that included an increase in GLUT4 and glycogen stores, and increased activity of hexokinase and mitochondrial oxidative enzymes (Giri et al. 2004, 2006).

The potential to reduce hypertriglyceridemia and elevated storage of triglycerides by inhibiting triglyceride and fatty acid synthesis and stimulating fatty acid oxidation and also the ability to lower blood glucose by activation of AMPK suggest that modulators of AMPK kinase activity might prove effective remedy for treating obesity and related metabolic disorders. Reports have also shown that certain strains of mouse that are resistant to diet-induced obesity (mice overexpressing uncoupling protein-1 in white adipocytes, stearoyl-CoA desaturase-1 knockouts, and mice overexpressing uncoupling protein-3 in skeletal muscle) exhibit increased basal level of AMPK activity (Matejkova et al. 2004). These findings have led to an intense interest in designing AMPK activators as potential therapies for type II diabetes and obesity.

Thus, AMPK plays a key role in regulating a wide range of activities in lipid and glucose metabolism; it appears as a promising strategy for the treatment of obesity and related metabolic disorders.

Keeping in view the epidemic of obesity, the role of AMPK pharmacotherapy seems to be the most obvious tool to combat the disease and attempts to develop novel therapies via AMPK-mediated mechanisms are worthy of pursuit.

6.5 Diabetes

The incidence of diabetes is increasing worldwide approaching epidemic proportions. According to National Diabetes Statistics, the number of people diagnosed and undiagnosed with diabetes in the USA reached 29.1 million people, which is 9.3 % of the general population in 2012.

Diabetes is considered a metabolic disease that is characterized by high blood sugar levels over a prolonged period (hyperglycemia). We can differentiate three main types: type 1 which results from the pancreas failure to produce enough insulin, type 2 when cells fail to respond to insulin properly (insulin resistance), and gestational diabetes occurs when pregnant women without a previous history of diabetes develop a high blood sugar level.

Type 2 diabetes (T2DM) is the most common form of this pathology and is directly related to obesity and metabolic syndrome. When obesity is established and body weight increases with age, a parallel state of chronic inflammation, characterized by an elevation of proinflammatory cytokines, can induce changes and switch the metabolic homeostatic set points, leading to T2DM (Medzhitov 2010). In T2DM, the major insulin-resistant organs include liver, muscle, and adipose tissue. In a state of insulin resistance, glucose uptake and utilization are dramatically decreased, and skeletal muscle becomes metabolically inflexible, unable to switch between glucose and fatty acid use. The main complications of diabetes are the microvascular and macrovascular complications such as retinopathy, nephropathy, and cardiovascular diseases, which are mediated by inflammatory processes.

6.5.1 Inflammation in Diabetes

Nowadays, it is well known that inflammation plays a key role in the natural history of diabetes (Agrawal and Kant 2014). Oxidative stress and inflammation are key players in insulin resistance progression and the establishment of T2DM. We know that inflammation is able to increase insulin resistance. The major cell involved in inflammation and insulin resistance is the adipocyte. Insulin regulates glucose uptake and triglyceride storage by adipocytes. The various adipocytokines, especially leptin, adiponectin, omentin, resistin, and visfatin, contribute to beta cell dysfunction in the pancreas. Adipose tissue also secretes dipeptidyl peptidase-4 (DDP-4) which enhances the degradation of glucagon like peptide-1 (GLP-1) and has an insulinotropic effect on beta cells (insulin secreting cells) (Lamers et al. 2011). Cytokines including tumor necrosis factor-alpha (TNF-α), interleukin beta (IL-1β), and interferon-gamma (IFN-ɣ) disrupt the regulation of intracellular calcium in the beta cells and hence insulin release. TNF-α acts on beta cells leading to their accelerated death (Cai et al. 2011).

Oxidative stress is another pathway that leads to inflammation through activation of cytokines (Lamb and Goldstein 2008). Pancreatic islets have low antioxidant defense and are hence vulnerable to oxidative stress. Marselli found differential regulation of oxidative stress genes in cells of T2DM subject compared with healthy controls (Marselli et al. 2010).

An emerging body of evidence also suggests that insulin suppresses the inflammatory process not only through preventing hyperglycemia but also by modulating key inflammatory molecules (Hyun et al. 2011). With all this in mind, the search for anti-inflammatory therapies for diabetes was started.

6.5.2 Mitochondrion and Insulin Resistance

The concept of insulin resistance was introduced like the previous step of T2DM. Although, it became evident that insulin resistance was not confined to T2DM, has been regarded as the centerpiece of the pathophysiologic mechanism of T2DM.

The relation between mitochondrial dysfunction and insulin resistance is known since the 1960s. To demonstrate the importance of this relationship the assays of Jucker and his colleagues (Jucker et al. 2001), who using a sophisticated nuclear magnetic resonance spectroscopy, measure the mitochondrial function in the human liver and muscle. They showed that the insulin resistance in elderly people can be explained about 40 % reduction in the mitochondrial oxidative phosphorylation capability compared to young people and demonstrated that T2DM patients showed approximately 30 % reduction in mitochondrial phosphorylation activity compared to the insulin-sensitive control subjects (Petersen et al. 2003, 2004). A recent review established a direct relation between defective mitochondrial dysfunction, reduction on fatty acid oxidation, and inhibition of glucose transport. This is the hallmark of insulin resistance and T2DM. The chronic production of excess ROS and inflammation result in mitochondrial dysfunction potentially inducing lipid accumulation and the endless vicious cycle of insulin resistance (Hernández-Aguilera et al. 2013). Altered mitochondrial function is the major factor that leads to increased muscular lipid accumulation and decreased insulin sensitivity. Mitochondrial dysfunction could provide important implications for the diagnosis and treatment of T2DM and other related disease like obesity and metabolic syndrome (Lee et al. 2010).

6.5.3 Role of AMPK in Diabetes

A number of physiological processes have been shown to stimulate AMPK, including conditions that lead to alterations of the intracellular AMP/ATP ratio (e.g., hypoxia, glucose deprivation) and calcium concentration, as well as the action of various hormones, cytokines, and adipokines. The activated form of the enzyme is responsible for metabolic changes via phosphorylation of various downstream substrates. The net effect is a change in local and whole-body energy utilization from an energy-consuming state to an energy-producing state in order to restore energy balance.

We know AMPK plays a key role in the interconversion of glucose (the primary cellular energy substrate) and its storage forms by affecting the transcription and translocation of the GLUT4 glucose transporter, glycogen synthesis, glycolysis, and gluconeogenesis.

Beginning with mice clinical trials, Hardie was the first to relate a possible role of disturbed AMPK signaling in diabetes (Winder and Hardie 1999). After that, a growing body of evidence has begun to validate this idea. For example, studies in AMPKα2 knockout mice observed hyperglycemia, glucose intolerance, and increased hepatic glucose production (Andreelli et al. 2006). A recent review suggests that AMPK mediates glucose uptake and is complementary to insulin as well as possibly independent of this hormone, thereby implicating the kinase in diabetes pathophysiology (Shirwany and Zou 2014).

AMPK is a key regulator of energy balance and plays many roles in human health and disease. Activation of AMPK by pharmacological agents holds a considerable potential to reverse the metabolic abnormalities associated with T2DM. So, AMPK could be a potential target for novel agents that may meet this global epidemic (Zhang et al. 2009).

6.5.4 AMPK and Inflammation: Targets of Treatment

A growing body of evidence is emerging to show that metabolic diseases are intimately related to chronic inflammation. The new pharmacological strategies are focused to reduce silent inflammation (Scheen et al. 2015). The considered insulin sensitizers or glucose-lowering agents appear to have greater anti-inflammatory activity than insulin-secreting agents (Pfützner et al. 2005).

Several glucose-lowering agents currently used as antidiabetic medications exert anti-inflammatory actions that may contribute to improved T2DM patients' outcomes. This effect may result from correction of hyperglycemia, but may also be due to direct effects of the drug, independent of improvement of glucose control (Scheen et al. 2015). This is demonstrated for metformin and thiazolidinediones (TZDs).

6.5.4.1 Metformin

This is the first-choice drug for the management of T2DM. This biguanide acts as an AMP-activated protein kinase (AMPK) activator (Foretz et al. 2014). Activation of AMPK has a high number of potentially antiatherosclerotic effects, including reducing inflammatory cell adhesion to blood vessel endothelium, reducing lipid accumulation, proliferation of inflammatory cells caused by oxidized lipids, stimulation of gene expression responsible for cellular antioxidant defenses, and stimulation of enzymes responsible for nitric oxide formation (Ewart and Kennedy 2011).

Metformin can inhibit proinflammatory responses and cytokine-induced nuclear factor kappab (NF-κB) activation via AMPK activation in vascular endothelial cells (Hattori et al. 2006; Isoda et al. 2006) and also inhibit inflammatory responses via the AMPK-phosphatase and tensin homologue (PTEN) pathway in vascular smooth muscle cells (Kim and Choi 2012). AMPK activity can also inhibit monocyte-to-macrophage differentiation (Vasamsetti et al. 2015), while other anti-inflammatory mechanisms have been proposed with lysosomes as a target of metformin (Lockwood 2010).

Although, surveys with metformin in human showed less pronounced efficacy than TZDs in reducing various inflammatory markers (Erem et al. 2014; Hanefeld et al. 2011), it seems clear that metformin may confer benefits in chronic inflammatory diseases independent of its ability to normalize blood glucose. There is now growing interest in identifying and exploiting AMPK anti-inflammatory effects with the development of new compounds that are currently under investigation (Scheen et al. 2015).

6.5.4.2 Thiazolidinediones

TZDs or glitazones are a group of agonists of peroxisome proliferator-activated receptor gamma (PPARγ). PPARs have been implicated as a molecular pathway in insulin resistance, T2DM, and atherosclerosis (Szanto and Nagy 2008). When the rat muscle cells were isolated and incubated in culture medium containing TDZ for 15 min, they significantly increased phosphorylation of AMPK (LeBrasseur et al. 2006), as well as the AMP/ATP ratio. It is suggested that TDZs can activate AMPK by a mechanism that is likely independent of PPARγ-regulated gene transcription. However, the major effect of TDZs is likely to be on the release of adiponectin by adipocytes, leading to activation of AMPK in liver to reduce glucose production (Kubota et al. 2007). Several studies have compared the anti-inflammatory effects of glitazones with other glucose-lowering agents and found glitazones to be superior (Marfella et al. 2006).

Many other specific anti-inflammatory approaches have been investigated in recent years to treat T2DM and its associated vascular complications, but none has yet emerged for use in clinical practice (Esser et al. 2015).

6.6 Conclusions

Energy production and its control are main tasks in every form of lives and are essential to have a proper homeostasis to adapt to the environment and to face every aggression. The Metabolic Disorders are considered pandemics and they need an adequate control and prevention. One important way to achieve it is to deepen the pathogenic mechanisms. Mitochondria and AMPK are the key elements through which it happens; their knowledge and research allow us to a better management. The discovery and use of drugs that can modulate them are imperative to improve our way of managing the metabolic disorders.

References

Agrawal NK, Kant S (2014) Targeting inflammation in diabetes: newer therapeutic options. World J Diabetes 5:697–710

Ahima RS, Flier JS (2000) Leptin. Annu Rev Physiol 62:413–437

Alberti KGMM, Zimmet P (2005) The metabolic syndrome—a new worldwide definition. Lancet 366:1059–1062

Alberti KGMM, Eckel RH, Grundy SM et al (2009) Harmonizing the metabolic syndrome: a joint interim statement of the international diabetes federation task force on epidemiology and prevention; National heart, lung, and blood institute; American heart association; World heart federation; International atherosclerosis society; And international association for the study of obesity. Circulation 120:1640–1645

Andersson U, Filipsson K, Abbott CR et al (2004) AMP-activated protein kinase plays a role in the control of food intake. J Biol Chem 279:12005–12008

Andreelli F, Foretz M, Knauf C, Cani PD, Perrin C, Iglesias MA, Pillot B, Bado A, Tronche F, Mithieux G, Vaulont S, Burcelin R, Viollet B (2006) Liver adenosine monophosphate-activated kinasealpha2 catalytic subunit is a key target for the control of hepatic glucose production by adiponectin and leptin but not insulin. Endocrinology 147:2432–2441

Archibald JM (2015) Evolution: gene transfer in complex cells. Nature 524:423–424

Barsh GS, Farooqi IS, O'Rahilly S (2000) Genetics of body-weight regulation. Nature 404:644–651

Benard G, Rossignol R (2008) Ultrastructure of the mitochondrion and its bearing on function and bioenergetics. Antioxid Redox Signal 10:1313–13142

Biala AK, Dhingra R, Kirshenbaum LA (2015) Mitochondrial dynamics: orchestrating the journey to advanced age. J Mol Cell Cardiol 83:37–43

Boden G, Lebed B, Schatz M, Homko C, Lemieux S (2001) Effects of acute changes of plasma free fatty acids on intramyocellular fat content and insulin resistance in healthy subjects. Diabetes 50:1612–1617

Bogacka I, Ukropcova B, McNeil M, Gimble JM, Smith SR (2005a) Structural and functional consequences of mitochondrial biogenesis in human adipocytes in vitro. J Clin Endocrinol Metab 90:6650–6656

Bogacka I, Xie H, Bray GA, Smith SR (2005b) Pioglitazone induces mitochondrial biogenesis in human subcutaneous adipose tissue in vivo. Diabetes 54:1392–1399

Bogacka I, Gettys TW, de Jonge L et al (2007) The effect of β-adrenergic and peroxisome proliferator-activated receptor-γ stimulation on target genes related to lipid metabolism in human subcutaneous adipose tissue. Diabetes Care 30:1179–1186

Bray GA, Tartaglia LA (2000) Medicinal strategies in the treatment of obesity. Nature 404:672–677
Brown CL, Halvorson EE, Cohen GM, Lazorick S, Skelton JA (2015) Addressing childhood obesity: opportunities for prevention. Pediatr Clin North Am 62:1241–1261
Burté F, Carelli V, Chinnery PF, Yu-Wai-Man P (2015) Disturbed mitochondrial dynamics and neurodegenerative disorders. Nat Rev Neurol 11:11–24
Cai K, Qi D, Wang O, Chen J, Liu X, Deng B, Qian L, Liu X, Le Y (2011) TNF-α acutely upregulates amylin expression in murine pancreatic beta cells. Diabetologia 54:617–626
Cheng J, Qiao L, Xu X, Zhai C, Zhao K, Ji X, Chen W (2015) Lower AMP-activated protein kinase level is associated with the vulnerability of coronary atherosclerotic plaques by attenuating the expression of monocyte autophagy. Coron Artery Dis 26:322–327
Choo HJ, Kim JH, Kwon OB et al (2006) Mitochondria are impaired in the adipocytes of type 2 diabetic mice. Diabetologia 49:784–791
Cigolini M, Targher G, Andreis IAB, Tonoli M, Agostino G, De Sandre G (1996) Visceral fat accumulation and its relation to plasma hemostatic factors in healthy men. Arterioscler Thromb Vasc Biol 16:368–374
Dahlman I, Forsgren M, Sjögren A et al (2006) Downregulation of electron transport chain genes in visceral adipose tissue in type 2 diabetes independent of obesity and possibly involving tumor necrosis factor-α. Diabetes 55:1792–1799
Engeli S, Feldpausch M, Gorzelniak K et al (2003) Association between adiponectin and mediators of inflammation in obese women. Diabetes 52:942–947
Erem C, Ozbas HM, Nuhoglu I, Deger O, Civan N, Ersoz HO (2014) Comparison of effects of gliclazide, metformin and pioglitazone monotherapies on glycemic control and cardiovascular risk factors in patients with newly diagnosed uncontrolled type 2 diabetes mellitus. Exp Clin Endocrinol Diabetes 122:295–302
Esser N, Paquot N, Scheen AJ (2015) Anti-inflammatory agents to treat or prevent type 2 diabetes, metabolic syndrome and cardiovascular disease. Expert Opin Investig Drugs 24:283–307
Ewart MA, Kennedy S (2011) AMPK and vasculoprotection. Pharmacol Ther 131:242–253
Ezzati M, Lopez AD, Rodgers A, Vander Hoorn S, Murray CJ, Comparative Risk Assessment Collaborating Group (2002) Selected major risk factors and global and regional burden of disease. Lancet 360:1347–1360
Flachs P, Mohamed-Ali V, Horakova O et al (2006) Polyunsaturated fatty acids of marine origin induce adiponectin in mice fed a high-fat diet. Diabetologia 49:394–397
Foretz M, Ancellin N, Andreelli F et al (2005) Short-term over expression of a constitutively active form of AMP-activated protein kinase in the liver leads to mild hypoglycemia and fatty liver. Diabetes 54:1331–1339
Foretz M, Guigas B, Bertrand L, Pollak M, Viollet B (2014) Metformin: from mechanisms of action to therapies. Cell Metab 20:953–966
Furt F, Moreau P (2009) Importance of lipid metabolism for intracellular and mitochondrial membrane fusion/fission processes. Int J Biochem Cell Biol 41:1828–1836
Giri S, Nath N, Smith B, Viollet B, Singh AK, Singh I (2004) 5-aminoimidazole-4-carboxamide-1-beta-4-ribofuranoside inhibits proinflammatory response in glial cells: a possible role of AMP-activated protein kinase. J Neurosci 24:479–487
Giri S, Rattan R, Haq E et al (2006) AICAR inhibits adipocyte differentiation in 3T3L1 and restores metabolic alterations in diet-induced obesity mice model. Nutr Metab 3:1743–7075
Gonzalez-Freire M, de Cabo R, Bernier M, Sollott SJ, Fabbri E, Navas P, Ferrucci L (2015) Reconsidering the role of mitochondria in aging. J Gerontol A Biol Sci Med Sci 70:1334–1342
Gowans GJ, Hawley SA, Ross FA, Hardie DG (2013) AMP is a true physiological regulator of AMP-activated protein kinase, both by allosteric activation and by enhancing net phosphorylation. Cell Metab 18:556–566
Gurevich-Panigrahi T, Panigrahi S, Wiechec E, Los M (2009) Obesity: pathophysiology and clinical management. Curr Med Chem 16:506–521

Hanefeld M, Pfutzner A, Forst T, Kleine I, Fuchs W (2011) Double-blind, randomized, multicentre, and active comparator controlled investigation of the effect of pioglitazone, metformin, and the combination of both on cardiovascular risk in patients with type 2 diabetes receiving stable basal insulin therapy: the Piocomb study. Cardiovasc Diabetol 10:65

Hardie DG (2008) AMPK: a key regulator of energy balance in the single cell and the whole organism. Int J Obes 32:S7–S12

Hardie DG (2014) AMP-activated protein kinase: maintaining energy homeostasis at the cellular and whole-body levels. Annu Rev Nutr 34:31–55

Hardie DG (2015) Molecular pathways: is AMPK a friend or a foe in cancer? Clin Cancer Res 21:3836–3840

Hardie DG, Ashford ML (2014) AMPK: regulating energy balance at the cellular and whole body levels. Physiology (Bethesda) 29:99–107

Hardie DG, Ross FA, Hawley SA (2012) AMPK: a nutrient and energy sensor that maintains energy homeostasis. Nat Rev Mol Cell Bio 13:251–262

Hattori Y, Suzuki K, Hattori S, Kasai K (2006) Metformin inhibits cytokine-induced nuclear factor kappaB activation via AMP-activated protein kinase activation in vascular endothelial cells. Hypertension 47:1183–1188

Hernández-Aguilera A, Rull A, Rodríguez-Gallego E, Riera-Borrull M, Luciano-Mateo F, Camps J, Menéndez JA, Joven J (2013) Mitochondrial dysfunction: a basic mechanism in inflammation-related non-communicable diseases and therapeutic opportunities. Mediators Inflamm 2013:135698

Hutley L, Prins JB (2005) Fat as an endocrine organ: relationship to the metabolic syndrome. Am J Med Sci 330:280–289

Hyun E, Ramachandran R, Hollenberg MD, Vergnolle N (2011) Mechanisms behind the anti-inflammatory actions of insulin. Crit Rev Immunol 31:307–340

Isoda K, Young JL, Zirlik A, MacFarlane LA, Tsuboi N, Gerdes N, Schönbeck U, Libby P (2006) Metformin inhibits proinflammatory responses and nuclear factor-kappaB in human vascular wall cells. Arterioscler Thromb Vasc Biol 26:611–617

Jeyabalan J, Shah M, Viollet B, Chenu C (2012) AMP-activated protein kinase pathway and bone metabolism. J Endocrinol 212:277–290

Johnson DT, Harris RA, French S, Blair PV, You J, Bemis KG, Wang M, Balaban R (2007) Tissue heterogeneity of the mammalian mitochondrial proteome. Am J Physiol Cell Physiol 292: C689–C697

Jucker BM, Dufour S, Ren J, Cao X, Previs SF, Underhill B, Cadman KS, Shulman GI (2001) Assessment of mitochondrial energy coupling in vivo by 13C/31P NMR. Proc Natl Acad Sci USA 97:6880–6884

Kaur J (2014) A comprehensive review on metabolic syndrome. Cardiol Res Pract 2014:943162

Kim SA, Choi HC (2012) Metformin inhibits inflammatory response via AMPK-PTEN pathway in vascular smooth muscle cells. Biochem Biophys ResCommun 425:866–872

Kola B, Hubina E, Tucci SA et al (2005) Cannabinoids and ghrelin have both central and peripheral metabolic and cardiac effects via AMP-activated protein kinase. J Biol Chem 280:25196–25201

Kubota N, Yano W, Kubota T, Yamauchi T, Itoh S, Kumagai H, Kozono H, Takamoto I, Okamoto S, Shiuchi T, Suzuki R, Satoh H, Tsuchida A, Moroi M, Sugi K, Noda T, Ebinuma H, Ueta Y, Kondo T, Araki E, Ezaki O, Nagai R, Tobe K, Terauchi Y, Ueki K, Minokoshi Y, Kadowaki T (2007) Adiponectin stimulates AMP activated protein kinase in the hypothalamus and increases food intake. Cell Metab 6:55–68

Lamb RE, Goldstein BJ (2008) Modulating an oxidative-inflammatory cascade: potential new treatment strategy for improving glucose metabolism, insulin resistance, and vascular function. Int J Clin Pract 62:1087–1095

Lamers D, Famulla S, Wronkowitz N, Hartwig S, Lehr S, Ouwens DM, Eckardt K, Kaufman JM, Ryden M, Müller S, Hanisch FG, Ruige J, Arner P, Sell H, Eckel J (2011) Dipeptidyl peptidase

4 is a novel adipokine potentially linking obesity to the metabolic syndrome. Diabetes 60:1917–1925

Lane N (2006) Powerhouse of disease. Nature 440:600–602

Lane N, Martin W (2010) The energetics of genome complexity. Nature 21:929–934

Lau DCW, Dhillon B, Yan H, Szmitko PE, Verma S (2005) Adipokines: molecular links between obesity and atheroslcerosis. Am J Physiol Heart Circ Physiol 288:H2031–H2041

LeBrasseur N, Kelly M, Tsao TS, Farmer SR, Saha AK, Ruderman NB, Tomas E (2006) Thiazolidinediones can rapidly activate AMP-activated protein kinase in mammalian tissues. Am J Physiol Endocrinol Metab 291:E175–E181

Lee HK, Cho YM, Kwak SH, Lim S, Park KS, Shim EB (2010) Mitochondrial dysfunction and metabolic syndrome—looking for environmental factors. Biochim Biophys Acta 1800:282–289

Liu M, Liu F (2010) Transcriptional and post-translational regulation of adiponectin. Biochem J 425:41–52

Lockwood TD (2010) The lysosome among targets of metformin: new anti-inflammatory uses for an old drug? Expert Opin Ther Targets 14:467–478

Lopez-Lluch G, Irusta PM, Navas P, de Cabo R (2008) Mitochondrial biogenesis and healthy aging. Exp Gerontol 43:813–819

Lynch M, Conery JS (2003) The origins of genome complexity. Science 302:1401–1404

Marfella R, D'Amico M, Esposito K, Baldi A, Di Filippo C, Siniscalchi M, Sasso FC, Portoghese M, Cirillo F, Cacciapuoti F, Carbonara O, Crescenzi B, Baldi F, Ceriello A, Nicoletti GF, D'Andrea F, Verza M, Coppola L, Rossi F, Giugliano D (2006) The ubiquitin-proteasome system and inflammatory activity in dia-betic atherosclerotic plaques: effects of rosiglitazone treatment. Diabetes 55:622–632

Marselli L, Thorne J, Dahiya S, Sgroi DC, Sharma A, Bonner-Weir S, Marchetti P, Weir GC (2010) Gene expression profiles of Beta-cell enriched tissue obtained by laser capture micro-dissection from subjects with type 2 diabetes. PLoS One 5:e11499

Matejkova O, Mustard KJ, Sponarova J et al (2004) Possible involvement of AMP-activated protein kinase in obesity resistance induced by respiratory uncoupling in white fat. FEBS Lett 569:245–248

Medzhitov R (2010) Inflammation 2010: new adventures of an old flame. Cell 140:771–776

Miles JM, Jensen MD (2005) Counterpoint: visceral adiposity is not causally related to insulin resistance. Diabetes Care 28:2326–2328

Minokoshi Y, Kim YB, Peroni OD et al (2002) Leptin stimulates fatty-acid oxidation by activating AMP-activated protein kinase. Nature 415:339–343

Minokoshi Y, Alquier T, Furukawa N et al (2004) AMP-kinase regulates food intake by responding to hormonal and nutrient signals in the hypothalamus. Nature 428:569–574

Mitchell P (1961) Coupling of phosphorylation to electron and hydrogen transfer by a chemi-osmotic type of mechanism. Nature 191:144–148

O'Neill LA, Hardie DG (2013) Metabolism of inflammation limited by AMPK and pseudo-starvation. Nature 493:346–355

Olijhoek JK, Van Der Graaf Y, Banga J-D, Algra A, Rabelink TJ, Visseren FLJ (2004) The metabolic syndrome is associated with advanced vascular damage in patients with coronary heart disease, stroke, peripheral arterial disease or abdominal aortic aneurysm. Eur Heart J 25:342–348

Ouchi N, Kihara S, Arita Y et al (2000) Adiponectin, an adipocytederived plasma protein, inhibits endothelial NF-kB signaling through a cAMP-dependent pathway. Circulation 102:1296–1301

Petersen KF, Befroy D, Dufour S, Dziura J, Ariyan C, Rothman DL, DiPietro L, Cline GW, Shulman GI (2003) Mitochondrial dysfunction in the elderly: possible role in insulin resistance. Science 300:1140–1142

Petersen KF, Dufour S, Befroy D, Garcia R, Shulman GI (2004) Impaired mitochondrial activity in the insulin-resistant offspring of patients with type 2 diabetes. N Engl J Med 350:664–671

Pfützner A, Marx N, Lübben G, Langenfeld M, Walcher D, Konrad T, Forst T (2005) Improvement of cardiovascular risk markers by pioglitazone is independent from glycemic control: results from the pioneer study. J Am Coll Cardiol 45:1925–1931

Pradhan AD, Manson JE, Rifai N, Buring JE, Ridker PM (2001) C-reactive protein, interleukin 6, and risk of developing type 2 diabetes mellitus. JAMA 286:327–334

Qiang W, Weiqiang K, Qing Z, Pengju Z, Yi L (2007) Aging impairs insulin-stimulated glucose uptake in rat skeletal muscle via suppressing AMPKalpha. Exp Mol Med 39:535–543

Reznick RM, Zong H, Li J, Morino K, Moore IK, Yu HJ, Liu ZX, Dong J, Mustard KJ, Hawley SA, Befroy D, Pypaert M, Hardie DG, Young LH, Shulman GI (2007) Aging-associated reductions in AMP-activated protein kinase activity and mitochondrial biogenesis. Cell Metab 5:151–156

Rich PR (2003) The cost of living. Nature 421:583

Rocher C, Taanman JW, Pierron D, Faustin B, Benard G, Rossignol R, Malgat M, Pedespan L, Letellier T (2008) Influence of mitochondrial DNA level on cellular energy metabolism: implications for mitochondrial diseases. J Bioenerg Biomembr 40:59–67

Rong JX, Qiu Y, Hansen MK et al (2007) Adipose mitochondrial biogenesis is suppressed in db/db and high-fat diet-fed mice and improved by rosiglitazone. Diabetes 56:1751–1760

Rylova SN, Albertioni F, Flygh G, Eriksson S (2005) Activity profiles of deoxynucleoside kinases and 5′-nucleotidases in cultured adipocytes and myoblastic cells: insights into mitochondrial toxicity of nucleoside analogs. Biochem Pharmacol 69:951–960

Schapira AH (2006) Mitochondrial disease. Lancet 368:70–82

Scheen AJ, Esser N, Paquot N (2015) Antidiabetic agents: potential anti-inflammatory activity beyond glucose control. Diabetes Metab 41:183–194

Schwartz MW, Woods SC, Porte DJ et al (2000) Central nervous control of food intake. Nature 404:661–671

Shi X, Burkart A, Nicoloro SM et al (2008) Paradoxical effect of mitochondrial respiratory chain impairment on insulin signaling and glucose transport in adipose cells. J Biol Chem 283:30658–30667

Shirwany NA, Zou MH (2014) AMPK: a cellular metabolic and redox sensor. A minireview. Front Biosci (Landmark Ed) 19:447–474

Spiegelman BM, Flier JS (2001) Obesity regulation and energy balance. Cell 104:531–543

Stapleton D, Mitchelhill KI, Gao G, Widmer J, Michell BJ, Teh T, House CM, Fernandez CS, Cox T, Witters LA, Kemp BE (1996) Mammalian AMP-activated protein kinase subfamily. J Biol Chem 271:611–614

Strong K, Mathers C, Leeder S, Beaglehole R (2005) Preventing chronic diseases: how many lives can we save? Lancet 366:1578–1582

Szanto A, Nagy L (2008) The many faces of PPARgamma: anti-inflammatory by any means? Immunobiology 213:789–803

Terman A, Kurz T, Navratil M, Arriaga EA, Brunk UT (2010) Mitochondrial turnover and aging of long-lived postmitotic cells: the mitochondrial-lysosomal axis theory of aging. Antioxid Redox Signal 12:503–535

Tomas E, Tsao TS, Saha AK et al (2002) Enhanced muscle fat oxidation and glucose transport by ACRP30 globular domain: acetyl-CoA carboxylase inhibition and AMP-activated protein kinase activation. Proc Natl Acad Sci 99:16309–16313

Twig G, Hyde B, Shirihai OS (2008) Mitochondrial fusion, fission and autophagy as a quality control axis: the bioenergetic view. Biochim Biophys Acta 1777:1092–1097

Vasamsetti SB, Karnewar S, Kanugula AK, Thatipalli AR, Kumar JM, Kotamraju S (2015) Metformin inhibits monocyte-to-macrophage differentiation via AMPK mediated inhibition of STAT3 activation: potential role in atherosclerosis. Diabetes 64:2028–2041

Virtue S, Even P, Vidal-Puig A (2012) Below thermoneutrality, changes in activity do not drive changes in total daily energy expenditure between groups of mice. Cell Metab 16:665–671

Westermann B (2008) Molecular machinery of mitochondrial fusion and fission. J Biol Chem 283:13501–13505

Williams RS (1986) Mitochondrial gene expression in mammalian striated muscle: evidence that variation in gene dosage is the major regulatory event. J Biol Chem 261:12390–12394

Wilson-Fritch L, Burkart A, Bell G et al (2003) Mitochondrial biogenesis and remodeling during adipogenesis and in response to the insulin sensitizer rosiglitazone. Mol Cell Biol 23:1085–1094

Wilson-Fritch L, Nicoloro S, Chouinard M et al (2004) Mitochondrial remodeling in adipose tissue associated with obesity and treatment with rosiglitazone. J Clin Invest 114:1281–1289

Winder WW, Hardie DG (1999) AMP-activated protein kinase, a metabolic master switch: possible roles in type 2 diabetes. Am J Physiol 277:E1–E10

Xydakis AM, Case CC, Jones PH et al (2004) Adiponectin, inflammation, and the expression of the metabolic syndrome in obese individuals: the impact of rapid weight lose through caloric restriction. J Clin Endocrinol Metab 89:2697–2703

Zhang Y, Proenca R, Maffei M et al (1994) Positional cloning of the mouse obese gene and its human homologue. Nature 372:425–432

Zhang BB, Zhou G, Li C (2009) AMPK: an emerging drug target for diabetes and the metabolic syndrome. Cell Metab 5:407–416

Zhou G, Myers R, Li Y et al (2001) Role of AMP-activated protein kinase in mechanism of metformin action. J Clin Invest 108:1167–1174

Zuliani G, Volpato S, Blé A, Bandinelli S, Corsi AM, Lauretani F, Paolisso G, Fellin R, Ferrucci L (2007) High interleukin-6 plasma levels are associated with low HDL-C levels in communitydwelling older adults: the InChianti study. Atherosclerosis 192:384–390

Chapter 7
AMPK in Neurodegenerative Diseases

Manon Domise and Valérie Vingtdeux

Contents

Abstract Alzheimer's disease, Parkinson's disease, Huntington's disease, and amyotrophic lateral sclerosis are neurodegenerative disorders that are characterized by a progressive degeneration of nerve cells eventually leading to dementia. While these diseases affect different neuronal populations and present distinct clinical features, they share in common several features and signaling pathways. In particular, energy metabolism defects, oxidative stress, and excitotoxicity are commonly described and might be correlated with AMP-activated protein kinase (AMPK) deregulation. AMPK is a master energy sensor which was reported to be overactivated in the brain of patients affected by these neurodegenerative disorders. While the exact role played by AMPK in these diseases remains to be clearly established, several studies reported the implication of AMPK in various signaling pathways that are involved in these diseases' progression. In this chapter, we review the current literature regarding the involvement of AMPK in the development of these diseases and discuss the common pathways involved.

Keywords AMPK • Neurodegeneration • Alzheimer's disease • Parkinson's disease • Huntington's disease • Amyotrophic lateral sclerosis

M. Domise • V. Vingtdeux (✉)
Univ. Lille, Inserm, CHU Lille, UMR-S 1172 – JPArc – Centre de Recherche Jean-Pierre AUBERT, F-59000 Lille, France
e-mail: valerie.vingtdeux@inserm.fr

M.D. Cordero, B. Viollet (eds.), *AMP-activated Protein Kinase*, Experientia Supplementum 107, DOI 10.1007/978-3-319-43589-3_7

7.1 Introduction

Neurodegenerative diseases including Alzheimer's (AD), Parkinson's (PD), Huntington's (HD), and amyotrophic lateral sclerosis (ALS) are characterized by the progressive degeneration of nerve cells eventually leading to dementia. While these disorders affect different neuronal populations, they share in common several features. For instance, they are characterized by the presence of protein aggregates in degenerating neurons that likely result from defective clearance mechanisms including proteasomal dysfunction and lysosomal clearance. In addition, metabolic alterations, excitotoxicity, and oxidative stress are often described. All of the latter could participate in the deregulation of AMP-activated protein kinase (AMPK) that was reported to occur in these diseases (Fig. 7.1). AMPK is a heterotrimer composed of one α, β, and γ subunits, often referred to as a master energy sensor. Indeed, AMPK possesses on its regulatory γ subunit four CBS (cystathionine-beta-synthase) domains which are binding sites for adenine nucleotides. Three of these sites can bind AMP, ADP, and ATP (Sanders et al. 2007; Gowans et al. 2013; Xiao et al. 2011). Metabolic stresses that increase the AMP:ATP ratio will allow the preferential binding of AMP to the γ subunit, thereby inducing a conformational change and favoring the phosphorylation of the residue Thr^{172} located on the catalytic α subunit by upstream AMPKs (Sanders et al. 2007; Gowans et al. 2013). The liver kinase B1 (LKB1) seems to be mostly responsible for AMPK phosphorylation in these conditions (Hawley et al. 2003; Woods et al. 2003; Shaw et al. 2004). At least two other kinases were reported to phosphorylate AMPK on Thr^{172}, the calcium/calmodulin-dependent protein kinase kinase II (CamKKII) that is regulated by an increase in intracellular calcium levels (Woods et al. 2005; Hawley et al. 2005; Hurley et al. 2005; Connolly et al. 2014) and the transforming growth factor β-activated kinase 1 (TAK1) that was reported to phosphorylate AMPK under oxidative stress conditions (Momcilovic et al. 2006; Chen et al. 2013). While not much is known about AMPK function in neuronal cells, studies realized in other cell types demonstrated that AMPK is a very important hub involved in the regulation of many intracellular pathways. In order to preserve energy levels, AMPK was described to downregulate many energy-consuming pathways. These include protein synthesis in particular through the regulation of mTORC1-mediated translational control (Inoki et al. 2003; Gwinn et al. 2008) and eukaryotic elongation factor 2 (eEF2)-mediated translation (Browne et al. 2004; Horman et al. 2002) and fatty acid synthesis through the direct phosphorylation of acetyl CoA carboxylase 1 (ACC1) and the expression of enzymes involved in fatty acid synthesis by inhibition of the lipogenic transcription factor sterol regulatory element-binding protein C1 [SREBP1C; Li et al. (2011)]. On the opposite, AMPK upregulates energy-producing pathways such as mitochondrial biogenesis through the activation of the PGC-1α (peroxisome proliferator-activated receptor-γ coactivator 1α) pathway (Jager et al. 2007); glucose uptake

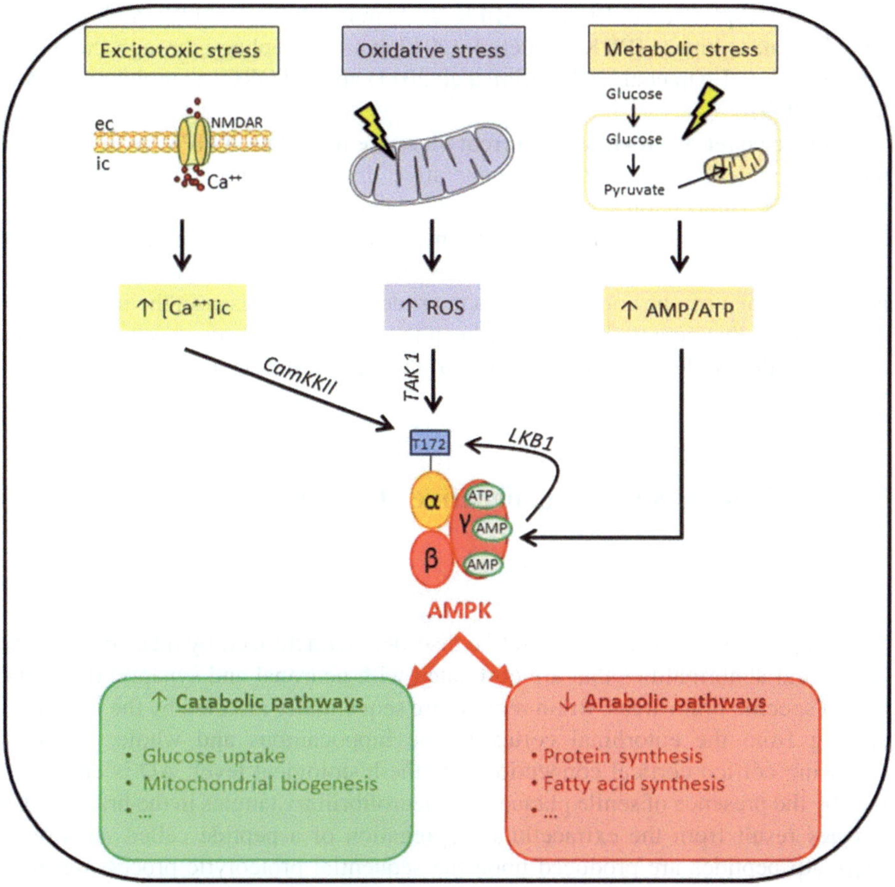

Fig. 7.1 Regulation of AMPK. AMPK is a metabolic sensor which is activated by different stresses. Excitotoxic and oxidative stresses promote, respectively, the activation of CamKKII and TAK1 that phosphorylate AMPK on its residue Thr^{172} which is necessary for its activation. Metabolic stress induces an increase of the AMP/ATP ratio that promotes AMP binding to the γ subunit of AMPK. This induces a conformational change that allows the phosphorylation of AMPK by LKB1. Once activated, AMPK triggers catabolic pathways and represses anabolic pathways in order to maintain energetic homeostasis. *ROS* reactive oxygen species, *CamKKII* Calcium/calmodulin kinase kinase II, *LKB1* liver kinase B1, *TAK1* transforming growth factor β-activated kinase 1, *NMDAR N*-methyl D-aspartate receptor. Figure was produced in part using Servier MedicalArt

through the regulation of glucose transporters expression (Zheng et al. 2001) and cell surface localization (Russell et al. 1999; Abbud et al. 2000; Weisova et al. 2009); glucose utilization through the direct phosphorylation of enzymes involved in the glycolytic pathway including hexokinase (Abnous and Storey

2008), 6-phosphofructo-2-kinase [PFK-2, Marsin et al. (2000)], and pyruvate dehydrogenase kinase [PDK, Wu et al. (2013)]; and autophagy through the inhibition of ULK1 (Egan et al. 2011; Kim et al. 2011) and mTORC1 complex [review in Shaw (2009)].

In this chapter, we review the current literature regarding AMPK involvement in the development of main neurodegenerative diseases that include Alzheimer's, Parkinson's, Huntington's, and amyotrophic lateral sclerosis and discuss the possible common pathological mechanisms involved. It is also important to note that AMPK is also studied in the context of ischemic stroke in animal models. While ischemic stroke can be considered as a neurodegenerative disease, the involvement of AMPK in this context has already been the subject of many reviews (Manwani and McCullough 2013; Weisova et al. 2011) and will not be discussed here.

7.2 AMPK in Neurodegenerative Diseases

7.2.1 Alzheimer's Disease

AD is a progressive neurodegenerative disorder characterized by memory loss and behavioral abnormalities that are correlated with neuronal and synaptic degeneration in specific brain areas. Brain regions are sequentially affected by the pathology starting from the entorhinal cortex to the hippocampus and whole neocortex following cortico-cortical connections. At the histological level, AD is characterized by the presence of senile plaques and neurofibrillary tangles in the brain. Senile plaques result from the extracellular aggregation of a peptide called Amyloid-β (Aβ). Aβ peptides are produced upon the sequential proteolytic processing of its precursor protein (APP) by β- and γ-secretases. Neurofibrillary tangles are composed of paired helical filaments that result from the intracellular aggregation of hyper- and abnormally phosphorylated tau proteins. Tau is a microtubule-associated protein whose main function which is regulated by phosphorylation consists of the regulation of microtubule dynamics. While the exact role of APP remains to be clearly established, there are some rare familial forms of AD which present mutations in APP, Presenilin-1 or Presenilin-2 genes; the latter two being the core components of the γ-secretase complex. However, the vast majorities of AD cases are of sporadic origin and are likely driven by a combination of genetic and environmental factors. The main genetic risk factor is the allele ε4 of *APOE* (coding for Apolipoprotein E). In addition, other risks factors have been identified following genome-wide association studies and include *CLU* (coding for clusterin), *CR1* (coding for the complement component receptor 1), *PICALM*, and *BIN1* (Lambert et al. 2009; Harold et al. 2009; Seshadri et al. 2010). Environmental factors include age, arterial hypertension, obesity, diabetes, and metabolic syndrome [review in Barberger-Gateau et al. (2013)].

Besides senile plaques and neurofibrillary tangles, perturbations in calcium homeostasis, oxidative stress, and energy metabolism defects are observed in the brain of AD patients (Bezprozvanny and Mattson 2008; Green and LaFerla 2008; Mattson 2007; Sayre et al. 2008). For instance, positron emission tomography (PET) imaging with the 2-[18F]-fluorodeoxyglucose (FDG) tracer is used as a diagnostic marker in AD where reduced glucose energy metabolism can be observed even at early stages of the disease (Mosconi 2005; Ferreira et al. 2010). Additionally, mitochondrial dysfunctions are also commonly described to be associated with AD [for a review, see Cabezas-Opazo et al. (2015)]. These include mitochondrial morphology, dynamics, and bioenergetics defects (DuBoff et al. 2013; Bubber et al. 2005; Garcia-Escudero et al. 2013). Interestingly, these mitochondrial abnormalities were found to be restricted to vulnerable neurons and to occur in neurons lacking neurofibrillary tangles, thus suggesting that they could represent an early event in AD (Hirai et al. 2001). Additionally, mitochondrial axonal transport is also impaired (Wang et al. 2015; Sheng 2014). Both amyloid and tau proteins have been shown to induce mitochondrial dysfunctions (Grimm et al. 2016). Conversely, studies also report that mitochondrial complexes I and III dysfunctions associated with reactive oxygen species (ROS) generation enhance Aβ production both in vitro and in vivo (Leuner et al. 2012).

AMPK was found to be deregulated in AD brains where immunohistochemistry studies revealed that activated AMPK co-localized with hyper-phosphorylated tau in pre-tangle and tangle-bearing neurons (Vingtdeux et al. 2011b). In addition, AMPK activation in AD was also demonstrated by Western blotting where phosphorylated AMPK was significantly upregulated in AD brains as well as in $APP^{SWE,IND}$(J20) and $APP^{SWE}/PS1^{dE9}$ mice models of the disease (Ma et al. 2014; Mairet-Coello et al. 2013; Son et al. 2012). AMPK deregulation was also observed in Tauopathies, a subset of neurodegenerative disorders characterized by the presence of abnormally and hyper-phosphorylated tau proteins, including tangle-predominant dementia, Guam Parkinson dementia complex, Pick's disease, frontotemporal dementia with Parkinsonism linked to chromosome 17, corticobasal degeneration, progressive supranuclear palsy, and argyrophilic grain disease (Vingtdeux et al. 2011b).

The exact role played by AMPK in AD remains controversial. The fact that AMPK co-localizes with hyper-phosphorylated tau in AD led to the hypothesis that AMPK could represent a new tau kinase. Indeed, in vitro studies using recombinant proteins showed that AMPK could phosphorylate tau at several epitopes including Thr^{231}, Ser^{262}, Ser^{356}, and $Ser^{396/404}$ (Thornton et al. 2011; Vingtdeux et al. 2011b). In cellular models, AMPK was also found to phosphorylate tau under stress conditions (Domise et al. 2016; Thornton et al. 2011). More particularly, Aβ oligomers were found to induce specifically AMPK α1 subunit activation by increasing intracellular calcium concentration and subsequent CamKKII activation. This Aβ oligomer-mediated AMPK activation was suggested to induce tau phosphorylation at epitopes Ser^{262} and $Ser^{396/404}$ in primary neuronal cultures (Thornton et al. 2011). In addition, it was postulated that this pathway was responsible for the toxic effects induced by Aβ oligomers on translational block (Yoon et al. 2012), dendritic spines (Mairet-Coello et al. 2013), and synaptic plasticity (Ma et al. 2014).

Indeed, AMPK activation following 2-deoxy-D-glucose (2-DG) or Aβ oligomers treatment was found to impair long-term potentiation (LTP) in ex vivo hippocampal slices (Potter et al. 2010; Ma et al. 2014). These results were corroborated in $APP^{SWE}/PS1^{dE9}$ transgenic animals where AMPK inhibition was found to rescue the LTP impairments mediated by Aβ (Ma et al. 2014). In these studies, AMPK negative effects on synaptic plasticity were found to be the result of decreased protein synthesis through mTORC1 and eEF2 pathways, respectively (Potter et al. 2010; Ma et al. 2014). In addition, AMPK was recently found to modulate tau pathology in vivo (Domise et al. 2016). On the contrary, other studies reported that AMPK activation induced by leptin or metformin reduced tau phosphorylation (Greco et al. 2009, 2011; Kickstein et al. 2010). The effect of metformin might, however, be AMPK independent. Indeed, metformin was suggested to induce protein phosphatase 2A (PP2A) activation, thereby leading to tau dephosphorylation (Kickstein et al. 2010). In a recent study, AMPK modulation was also related to tau dephosphorylation and rather correlated to AMPK phosphorylation at Ser^{485}, which is thought to be an inhibitory AMPK phosphorylation site prohibiting further phosphorylation at epitope Thr^{172} (Horman et al. 2006). In conditions of metabolic syndrome, insulin resistance or glucose depletion, tau phosphorylation might be differently regulated either because AMPK activation status could differ or because other tau kinases and phosphatases might be involved (Kim et al. 2015). While these findings are somehow controversial, it is clear that tau is an AMPK target either direct or indirect depending on the environmental conditions. Tau epitopes regulated by AMPK include Ser^{262} and Ser^{356} which are KXGS domains located in tau microtubules binding repeat regions. Phosphorylation of these particular epitopes regulates tau affinity for microtubules (Fischer et al. 2009). As a consequence, AMPK-mediated tau phosphorylation might control tau binding with microtubules and thereby axonal transport of cargos including mitochondria (Sato-Harada et al. 1996; Reddy 2011). Tau Thr^{231} is another central epitope since it was reported to serve as a priming site for GSK3β, a very important tau kinase participating to tau hyper-phosphorylation and aggregation (Lin et al. 2007).

AMPK was also found to be involved in APP metabolism. A decrease of Aβ production was reported in primary neurons after AICAR (5-aminoimidazole-4-carboxamide ribonucleotide)-dependent AMPK stimulation; conversely, Aβ levels were increased in primary neurons lacking the AMPK α2 subunit (Won et al. 2010). Opposite results have also been obtained, and for instance, AMPK activation following metformin treatment was reported to increase the transcription of BACE1, one of the enzymes involved in Aβ production and hence to be associated with increased Aβ levels (Chen et al. 2009). The effect of AMPK on Aβ production and/or degradation is likely to be controlled by energy status given that depending on the extracellular glucose concentrations opposite results are obtained (Yang et al. 2015). As a master regulator of autophagy, AMPK activation following resveratrol or AICAR treatment was found to reduce Aβ secretion by increasing its degradation through the autophagic/lysosomal pathway (Vingtdeux et al. 2010, 2011a). In general, AMPK activation might be beneficial by helping clearing protein aggregates through autophagy induction. However, in latter stages of the disease, lysosomal-mediated degradation is impaired (Nixon and Yang 2011), consequently, increasing autophagosomes production without increasing autophagic flux might have deleterious consequences. Indeed, inhibition of autophagic flux will decrease the degradation of misfolded proteins including Aβ and tau (Pickford et al. 2008; Wang et al. 2010) as well as dysfunctional mitochondria. In addition, autophagosomes accumulation might be a source for Aβ production (Yu et al. 2005), thereby inducing a vicious circle.

In conclusion, these data support a role for AMPK in AD as an upstream player in the pathology development. Overall, AMPK could play a role in AD by participating in Aβ production and/or clearance as well as on tau phosphorylation, the two hallmarks of AD. Additionally, AMPK was found to mediate the toxic effects of Aβ on synapses number and synaptic plasticity. These detrimental effects of AMPK in the latter stages of AD are summarized in Fig. 7.2.

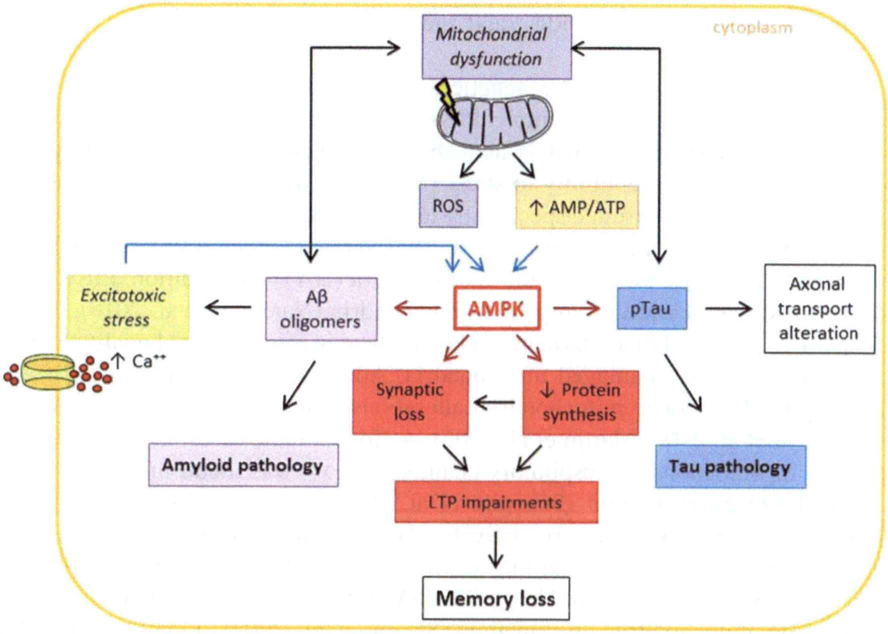

Fig. 7.2 Harmful roles of AMPK in the late stages of Alzheimer's disease. Alzheimer's disease is characterized by excitotoxicity as well as metabolic and oxidative stresses. Mitochondrial dysfunction eventually leads to the production of ROS and to the increase of the AMP/ATP ratio that correspond, respectively, to oxidative and metabolic stresses. These two events activate AMPK which in turn decreases protein synthesis ultimately leading to synaptic loss and LTP impairments that contribute to memory loss. AMPK is also involved in tau and amyloid pathologies. On one side, AMPK phosphorylates tau protein thereby altering microtubules assembly and as a result axonal transport of vesicles and mitochondria. On the other side, AMPK plays a part in the production and degradation of Aβ peptides. Finally, Aβ and tau might contribute to the chronic activation of AMPK by inducing mitochondrial impairments and excitotoxicity. *LTP* long-term potentiation, *ROS* reactive oxygen species. Figure was produced in part using Servier Medical Art

7.2.2 *Parkinson's Disease*

PD is characterized by resting tremor, rigidity, bradykinesia, gait disturbance, and postural instability. Pathological features include loss of dopaminergic neurons in the substantia nigra associated with Lewy bodies inclusions (Beitz 2014). These Lewy bodies are mainly composed of aggregated α-synuclein. PD etiology involves many genetic and environmental factors (Olanow and Tatton 1999; Verstraeten et al. 2015). While the majority of cases are sporadic, mutations in a number of genes were identified to be responsible for rare familial forms of the disease. These genes include *SNCA* (coding for α-synuclein), *Park2* (coding for the cytosolic E3 ubiquitin ligase Parkin), and *PINK1* (coding for PTEN-induced kinase 1). In addition, genetic variants have been identified as PD risk alleles in *LRRK2* (leucine-rich repeat kinase 2), *SNCA*, H1 haplotype of microtubule-associated protein tau, and *GBA* (coding for beta acid glucosidase) [for a review, see Verstraeten et al. (2015)]. Environmental factors include exposure to environmental toxins (pesticides, herbicides, and industrial chemicals) and drugs of abuse (Olanow and Tatton 1999).

Interestingly, many of these genetic and environmental factors are linked to mitochondrial function. For example, PINK1 is localized to the mitochondria where it exerts a protective role that is abolished by mutations, overall resulting in a cellular increased susceptibility to stress (Valente et al. 2004). Parkin is a protein that was found to be recruited specifically to dysfunctional mitochondria to promote their degradation by the autophagic pathway (Narendra et al. 2008), referred to as mitophagy [for a review, see Youle and Narendra (2011)]. In addition, PINK1 was found to activate Parkin on impaired mitochondria (Narendra et al. 2010). Therefore, it was proposed that Parkin might be involved in mitochondrial quality control as a way to remove damaged mitochondria. Additionally, α-synuclein itself was also reported to induce mitochondrial alterations in neuronal cells and transgenic mice (Hsu et al. 2000; Martin et al. 2006). As for sporadic cases, a decrease in the activity of mitochondrial respiratory chain complex I was found in the substantia nigra of PD patients brain (Schapira et al. 1990). Complex I was found to be functionally impaired, i.e., oxidatively damaged and misassembled (Keeney et al. 2006). In addition, regarding environmental risk factors, many pesticides and 1-methyl-4-1,2,3,6-tetrahydropyridine (MPTP) share the common mechanism of causing mitochondrial dysfunction (Sherer et al. 2002). Finally, FDG–PET studies also demonstrated marked reductions in glucose metabolism in the brain of PD patients (Eckert et al. 2005).

AMPK deregulation was observed in the brain of PD patients where activated AMPK was found near the rim of Lewy bodies in the cytoplasm as opposed to control individuals where AMPK was mainly nuclear (Jiang et al. 2013). AMPK activation was also reported in animal models of PD induced by intra-striatal injection of 6-hydroxydopamine (6-OHDA) or MPTP (Kim et al. 2013; Choi et al. 2010). On the contrary, α-synuclein expression in cell models was reported to downregulate AMPK activation (Dulovic et al. 2014). Whether AMPK activation is beneficial or detrimental in PD remains controversial. AMPK activation was reported in some instance to be detrimental given that further activation of AMPK, for example, following metformin administration significantly enhanced dopaminergic neuron degeneration induced by 6-OHDA, whereas overexpression of a dominant-negative AMPK in the striatum reduced dopaminergic neuron degeneration following 6-OHDA (Kim et al. 2013). In cellular models, PD toxins (6-OHDA, MPP+, or rotenone) induced AMPK activation and Akt inactivation that cooperatively contributed to the downregulation of mTOR-mediated S6K1 (ribosomal p70 S6 kinase) and 4E-BP1 (eukaryotic initiation factor 4E binding protein 1), thereby leading to neuronal cell death (Xu et al. 2014). AMPK might also participate in Lewy bodies' accumulation through direct phosphorylation of α-synuclein (Jiang et al. 2013) that could impair the clearance of its aggregates (Tenreiro et al. 2014). On the opposite, AMPK activation using AICAR or metformin was reported to reduce the toxicity mediated by α-synuclein (Dulovic et al. 2014). AMPK also protected cells against rotenone toxicity by enhancing autophagy (Hou et al. 2015). This AMPK-induced autophagic pathway also regulates α-synuclein degradation following resveratrol treatment (Wu et al. 2011). AMPK might also participate in mitochondrial function regulation in PD. Results obtained in *Drosophila melanogaster* models suggest that AMPK activation could be beneficial for familial forms of PD that present mutations in Parkin or LRRK2. Indeed, genetic inactivation of AMPK was reported to reduce the beneficial effects of epigallocatechin gallate (EGCG), an antioxidant found in green tea, in mutant LRRK2 and Parkin-null flies (Ng et al. 2012). In addition, results obtained from patient's primary fibroblasts presenting Park2 mutations also suggest that the beneficial effects on mitochondrial function and autophagy induced by resveratrol were due to AMPK activation (Ferretta et al. 2014).

Altogether, these studies highlight the potential double role that can be played by AMPK in PD (Fig. 7.3). On one side, AMPK could be neuroprotective by participating, for example, in mitochondrial quality control; yet under other circumstances, AMPK could participate in neurodegeneration.

7.2.3 Huntington's Disease

Clinical manifestations of HD include motor disturbances comprising chorea and dystonia and cognitive and behavioral dysfunctions. HD is characterized by the loss of medium spiny neurons in the striatum and eventually more widespread loss of cortical, thalamic, hippocampal, and hypothalamic neurons. Another characteristic of the disease is the appearance of nuclear and cytoplasmic inclusions that contain mutant huntingtin and polyglutamine (Walker 2007). HD is an autosomal dominant genetic disease that is induced by the repetition of a polyglutamine CAG triplet repeat in the exon 1 of the huntingtin (Htt) gene with 41 or more polyQ repeats being fully penetrant. (The Huntington's Disease Collaborative Research Group 1993.) These repeats might confer a toxic gain of function for mutant Htt (mHtt) or

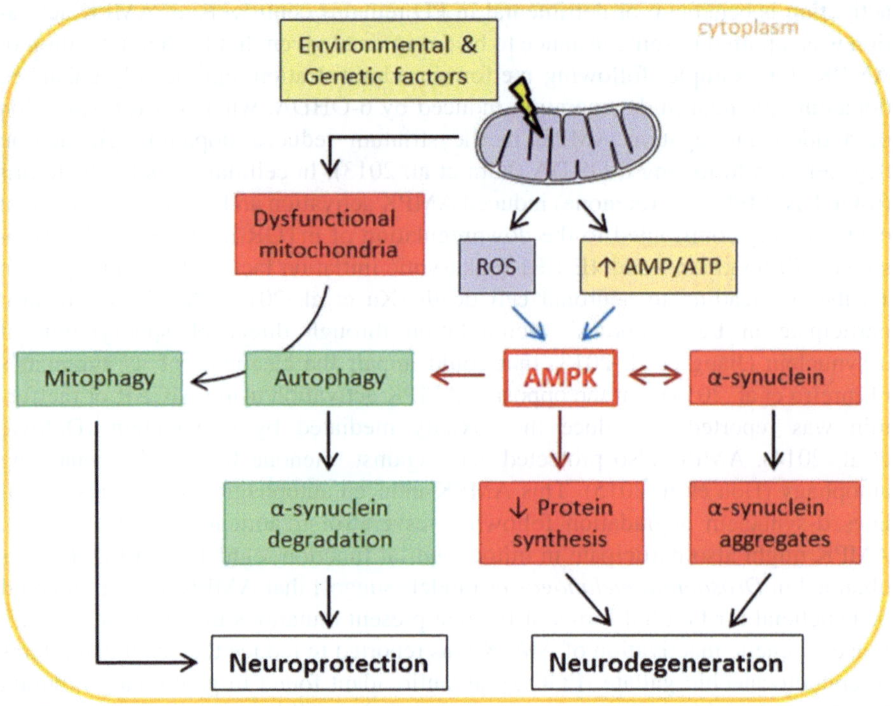

Fig. 7.3 Dual role of AMPK in Parkinson's disease. Environmental and genetic risk factors are involved in the buildup of mitochondrial alterations. These alterations eventually lead to oxidative stress through the production of ROS and metabolic stress via an increase of the AMP/ATP ratio. These stresses induce the activation of AMPK which phosphorylates α-synuclein, the latter promoting its aggregation and ultimately neurodegeneration. Neurodegeneration might also result from decreased protein synthesis triggered by AMPK activation. On the contrary, AMPK could also exert a neuroprotective effect in particular by inducing the degradation of damaged mitochondria and α-synuclein aggregates via autophagy. *ROS* reactive oxygen species. Figure was produced in part using Servier Medical Art

a loss of normal Htt function (Zuccato et al. 2010). The physiological role of Htt remains poorly understood; however, it was suggested to be involved in axonal, vesicular, and mitochondrial transport (Smith et al. 2009; Tian et al. 2014).

In HD, mitochondrial dynamics, fusion and fission mechanisms as well as the activity of enzymes involved in oxidative phosphorylation are disturbed (Shirendeb et al. 2011; Song et al. 2011; Browne et al. 1997; Gu et al. 1996). These perturbations have for consequence to increase the accumulation of fragmented and damaged mitochondria eventually leading to oxidative stress. Additionally, mitophagy defects were also proposed to participate in the disease progression (Wong and Holzbaur 2014). A selective impairment of glycolytic metabolism in the striatum of HD patient early in the course of their disease was observed by in vivo PET measurements (Powers et al. 2007). This glucose hypometabolism in the early stages of the disease was also reported in the cerebral cortex and in the brain caudate (Shin et al. 2013; Ciarmiello et al. 2012). Deficits in glycolysis have also been reported in striatal neurons in a rat model of the disease (Gouarne et al. 2013). Huntingtin itself might play a role in glycolysis. Indeed, Htt was found to interact with the glycolytic enzyme GAPDH (glyceraldehyde-3-phosphate dehydrogenase) (Burke et al. 1996). However, studies of HD patients' brains did not conclusively find an alteration of GAPDH activity (Browne et al. 1997; Tabrizi et al. 1999; Kish et al. 1998; Olah et al. 2008). GAPDH was described to bear additional functions unrelated to its energetic role. GAPDH might act in concert with the ubiquitin-E3-ligase Siah1 to induce mHtt neurotoxicity by assisting its nuclear translocation (Bae et al. 2006). Huntingtin could also be involved in fast axonal transport by scaffolding GAPDH on vesicles, thereby providing onboard energy (Zala et al. 2013). Finally, a recent study demonstrated that mHtt interfered with mitophagy. Indeed, mHtt was found to affect GAPDH-driven mitophagy, thereby leading to the accumulation of damaged mitochondria (Hwang et al. 2015).

The $\alpha 1$ subunit of AMPK seems to be particularly involved in HD pathogenesis. Indeed, it was found to be activated in the nucleus of striatal neurons where it was suggested to downregulate the antiapoptotic protein Bcl2, thus inducing cell death (Ju et al. 2011) (Fig. 7.4). Accumulation of activated AMPK was also reported in the striatum of transgenic mouse models of HD, R6/2 mice harboring exon 1 of the human Htt gene with 144 CAG repeats (Chou et al. 2005; Mochel et al. 2012; Ju et al. 2014). This overactivation of AMPK could be reversed by activating A_{2A} receptors using an agonist, additionally diminishing the HD-like pathology in these animals (Chou et al. 2005; Ju et al. 2011). A_{2A} receptors signaling pathway involves PKA activation. PKA was reported to phosphorylate AMPK $\alpha 1$ at residue Ser^{173}, thereby preventing the activating phosphorylation at Thr^{172} (Djouder et al. 2010). Additionally, AMPK activation in this mouse model might also result from increased oxidative stress (Ju et al. 2014). On the contrary, in cellular models, AMPK activation through viniferin treatment was reported to provide neuroprotection against mHtt (Fu et al. 2012). Finally, metformin, which can activate AMPK was reported to be beneficial in male R6/2 mice (Ma et al. 2007). However, the exact mechanism behind metformin's beneficial effects remains to be determined.

Overall, these studies also highlight AMPK signaling pathway as a potential player in the pathology of HD.

7.2.4 Amyotrophic Lateral Sclerosis

Amyotrophic Lateral Sclerosis (ALS) is characterized by the progressive loss of upper and lower motor neurons at the spinal or bulbar level (Rowland and Shneider 2001). The most common symptoms of ALS are muscle weakness, muscular atrophy, spasticity, and eventually paralysis. While the exact cause of the disease

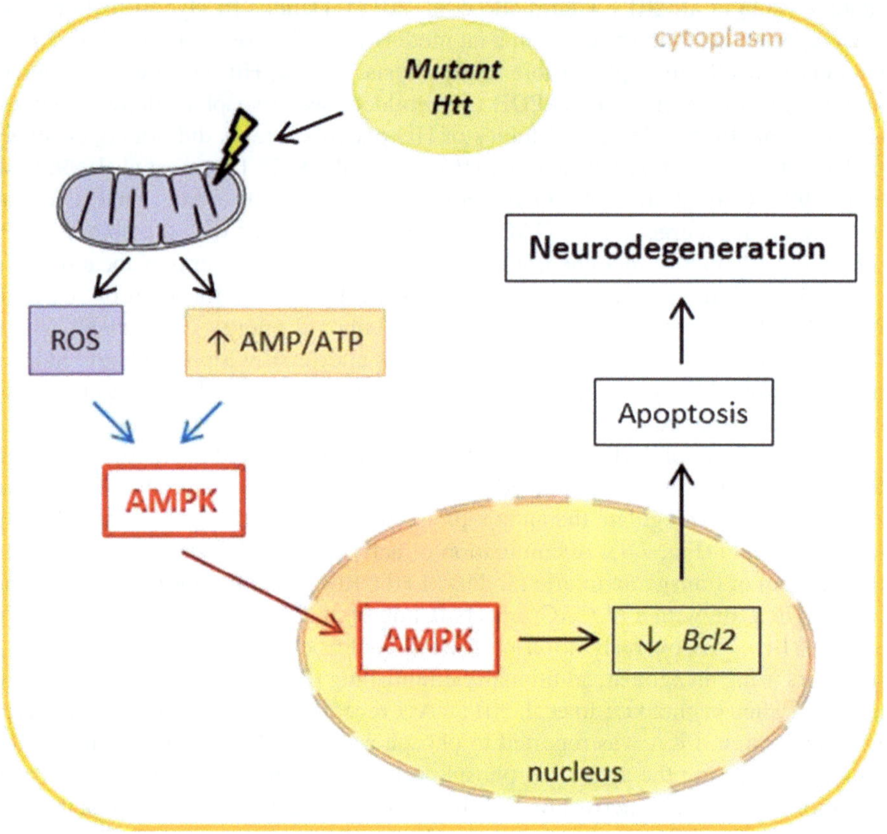

Fig. 7.4 Model of AMPK-mediated apoptosis in Huntington's disease. Mutant Huntingtin induces mitochondrial alterations that lead to oxidative stress and hypometabolism. These participate in the activation of AMPK and its translocation from the cytoplasm to the nucleus where AMPK downregulates the antiapoptotic protein Bcl2. This pathway promotes apoptosis and thereby neurodegeneration. *ROS* reactive oxygen species, *Htt* Huntingtin. Schematic is adapted from Ju et al. (2011). Figure was produced in part using Servier Medical Art

is unknown, around 10 % of familial forms exist involving, for example, the *SOD1* gene (superoxide dismutase 1), *TARDBP* (encoding TAR DNA-binding protein 43), *FUS* (fused in sarcoma), and hexanucleotide repeat expansion in *C9ORF72* (Zarei et al. 2015; Renton et al. 2014). The sporadic forms of the disease might be driven by genetic and lifestyle risk factors (Ingre et al. 2015). At the histological level, ALS is characterized by the aggregation of ubiquitinated proteins that can include TDP43, p62, and FUS in affected neurons (Blokhuis et al. 2013). ALS is associated with defects in energy metabolism comprising weight loss, increased resting energy expenditure (hypermetabolism), and hyperlipidemia (Dupuis et al. 2011). The precise origin of these metabolic dysfunctions remains unclear.

AMPK activation was found to be deregulated in motor neurons of ALS patients (Liu et al. 2015b). In cells and mouse models of the disease, AMPK regulation differs according to the model used. In the mSOD1^{G93A} mouse model, AMPK activity is increased in spinal cords from symptom onset (Lim et al. 2012; Perera et al. 2014; Zhao et al. 2015). Similar results were obtained in vitro, in spinal cord cultures, in motor neuron cell lines expressing mutant SOD1, and in embryonic neural stem cells derived from SOD1^{G93A} mice (Lim et al. 2012; Perera et al. 2014; Sui et al. 2014). On the opposite, AMPK activation was reported to be downregulated in mutant TDP43^{A315T} mouse models of spinal cord and brain (Perera et al. 2014). Similar results were also obtained in motor neuronal cell lines expressing mutant TDP-43, probably as a consequence of increased PP2A activity (Perera et al. 2014). On the contrary, AMPK activity was reported to be increased in the spinal cord of a mouse model overexpressing wild-type TDP43 (Liu et al. 2015b). AMPK was also suggested to be involved in TDP-43 mislocalization from the nucleus to the cytoplasm (Liu et al. 2015a). Similarly, AMPK activation was described to induce the human antigen R [HuR, a major mRNA stabilizer recently shown to regulate TDP-43 and FUS (Lu et al. 2014)] delocalization by directly phosphorylating importin-α1 (Liu et al. 2015b). The impact of AMPK activation in this disease remains a matter of debate. Indeed, modulation of AMPK activity in these various models has given conflicting data. Metformin administration in SOD1^{G93A} mice accelerated disease onset and progression in females only (Davis and Lin 2011), while resveratrol was found to provide beneficial effects (Mancuso et al. 2014; Song et al. 2014). The beneficial effect of resveratrol could act in part through an increase of Sirtuin 1 expression, normalization of autophagic flux, and reduced oxidative stress (Mancuso et al. 2014; Song et al. 2014). Similarly, preconditioning with latrepirdine, a small molecule shown to activate AMPK (Weisova et al. 2013), was reported to delay symptoms onset and increase the lifespan of SOD^{G93A} mice (Coughlan et al. 2015). Decreasing AMPK activity in cell cultures or in *Caenorhabditis elegans* expressing mutant SOD1 or TDP43 was reported to be beneficial (Mancuso et al. 2014) and to rescue TDP43 mislocalization in motor neuronal cells and to delay disease progression in TDP43 wild-type mice (Liu et al. 2015b). Finally, AMPK α2-deficient mice were recently described to present gait abnormalities resembling early stages of ALS supporting a key role for AMPK in the development of this disease (Vergouts et al. 2015).

In conclusion, the role played by AMPK in ALS might vary according to the nature of the disease as mutations in SOD1 and TDP43 were reported to affect differently the kinase.

7.3 General Considerations

The specific vulnerability of the neuronal populations affected in each of these diseases is likely to be driven by both genetic and environmental factors. Interestingly, many of these factors converge to an impairment of cellular energy metabolism. This is the case, for instance, of mutations in genes that are directly involved in mitochondrial function or clearance (such as SOD1, PINK1, and Parkin). These mitochondrial dysfunctions might contribute to the increase in neuronal excitotoxicity and AMPK deregulation. In addition to impairing energy metabolism, mitochondrial insults can cause an imbalance between ROS production and removal, thereby participating in oxidative stress, another common factor of these diseases (Sayre et al. 2008). This oxidative stress through the activation of TAK1 might also contribute to the chronic activation of AMPK. Conversely, given its role on mitochondria function, biogenesis, and degradation, it is also possible that AMPK participates in the establishment of mitochondrial dysfunctions that are observed in these diseases. Whether AMPK deregulation is triggered by these metabolic perturbations or could be involved in their development will be an important issue to investigate.

While AMPK is highly expressed in neurons, its physiological function remains poorly studied. Nonetheless, AMPK is vital for neuronal survival. Indeed, results obtained in *Drosophila* demonstrated that genetic ablation of AMPK subunits γ [*lochrig* mutant, Tschape et al. (2002)] or β [*alicorn* mutant, Spasic et al. (2008)] induces progressive neurodegeneration. Although it is becoming increasingly evident that AMPK might participate in these neurodegenerative diseases development, whether this activation is beneficial or detrimental remains matter of debate. In general, the contradictory results that have been obtained in vivo regarding the beneficial or detrimental role of AMPK could also be due to peripheral AMPK activity. Several papers reported beneficial effects of peripheral AMPK activation on cognition. For instance, it was shown that AMPK activation following AICAR administration in mice enhanced endurance and spatial memory in the Morris water maze (Kobilo et al. 2014). AICAR blood–brain barrier permeability is very low (Marangos et al. 1990); therefore, its effects on cognition or on the brain in general are likely to be indirect. The beneficial effects of AICAR reported in the Kobilo et al.'s study were demonstrated to be mediated by muscle AMPK activation since mice overexpressing a muscle-specific dominant negative of AMPK α2 did not show any improvements following AICAR administration. These behavioral improvements were suggested to result from enhanced dentate gyrus neurogenesis in AICAR-treated animals (Kobilo et al. 2011). On the contrary, direct administration of AICAR in the brain by means of intracerebral infusions was found to impair memory functions (Dash et al. 2006) and lead to excitotoxicity in an HD mouse model (Ju et al. 2011). As a consequence, it is very important to take into account the drug used to activate or inhibit AMPK and its administration route to determine the impact of peripheral AMPK activation in addition to its central regulation before drawing conclusions.

It is very likely that AMPK could act both as a friend and as a foe during the course of these neurodegenerative diseases' progression. Indeed, AMPK might be activated in the early stages of these diseases to help maintain or restore neuronal energy metabolism. However, chronic AMPK activation would eventually become detrimental to brain functions by repressing pathways that consume energy. Overall, several common mechanisms regulated by AMPK can be identified and are summarized in Fig. 7.5. For instance, the beneficial effects of AMPK often involve an increase of the autophagy pathway that might be involved in the clearance of misfolded proteins, protein aggregates, or defective mitochondria. It was also reported that AMPK might activate PP2A, thereby reducing the phosphorylation status of tau and α-synuclein. On the opposite, the deleterious impact of AMPK implies the phosphorylation of proteins which aggregates represent the common hallmarks of these diseases, including tau, Aβ, and α-synuclein. Additionally, AMPK chronic activation by repressing protein synthesis could, on the long term, impair synaptic integrity and plasticity and eventually lead to cell death.

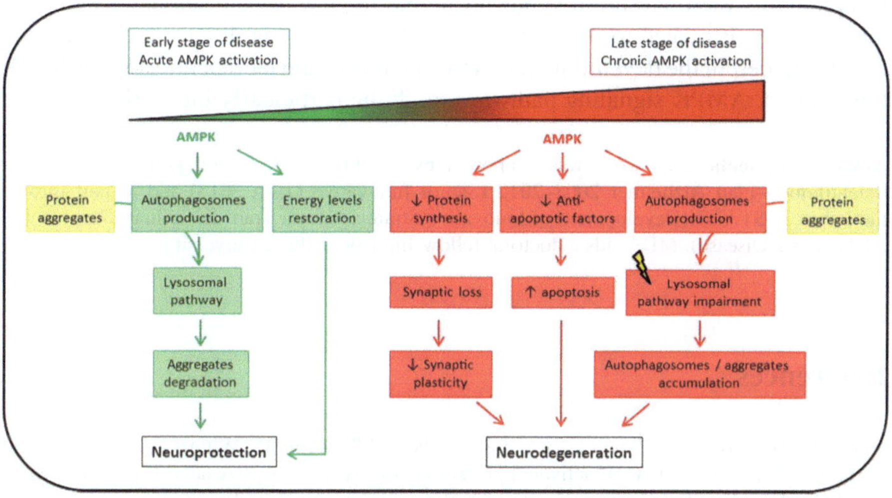

Fig. 7.5 AMPK in neurodegenerative diseases, friend or foe? At the onset of neurodegenerative diseases, activation of AMPK might be beneficial since it allows the restoration of energetic homeostasis and the elimination of protein aggregates which are often reported to be toxic for neurons. Indeed, AMPK promotes the formation of autophagosomes in order to induce protein aggregates and impaired mitochondria degradation through the autophagy/lysosomal pathway. On the other hand, in the late stages of these diseases, chronic AMPK activation becomes disadvantageous for neurons. This overactivation of AMPK could lead to neurodegeneration through several signaling pathways. Decreasing protein synthesis could drive synaptic loss and impair synaptic plasticity subsequently leading to neurodegeneration. Decreasing antiapoptotic factors could lead to induction of apoptosis and neurodegeneration. Finally, the production of autophagosomes combined with an alteration of lysosomal clearance (which is often reported to occur in these disorders), in the end, leads to the accumulation of autophagosomes and contributes to upsurge the levels of toxic protein aggregates and defective mitochondria

7.4 Conclusion

While the clinical manifestations, neuronal populations affected and proteins involved differ widely between these diseases, energy metabolism perturbations are often reported early in the course of these diseases' progression. These metabolic perturbations might result from the various environmental and genetic risk factors that drive these pathologies as it is already well acknowledged for mutations that affect directly mitochondrial functions. As a consequence, one can expect AMPK overactivation to be an additional early feature of these disorders. Hence, AMPK was suggested to participate in these diseases' progression by contributing in the establishment of the observed lesions mainly by regulating the clearance and posttranslational modifications of the proteins forming the respective aggregates. Additionally, AMPK chronic overactivation might participate in neurodegeneration by repressing energy-consuming pathways.

Given the demographic trend towards an aging population, the prevalence of these neurodegenerative diseases and thus their socioeconomic burden will continue to increase dramatically in the next decades. The current treatments are only symptomatic; there are no therapies available to cure these diseases. As a consequence, there is a need to better understand the underlying disease mechanisms in order to underpin the development of new diagnostic and therapeutic approaches. In this context, AMPK signaling pathways might be particularly interesting.

Acknowledgments This work was supported by the French Fondation pour la cooperation Scientifique—Plan Alzheimer 2008–2012 (Senior Innovative Grant 2013) and in part through the Labex DISTALZ (Development of Innovative Strategies for a Transdisciplinary Approach to Alzheimer's Disease). MD holds a doctoral fellowship from Lille 2 University.

References

Abbud W, Habinowski S, Zhang JZ, Kendrew J, Elkairi FS, Kemp BE, Witters LA, Ismail-Beigi F (2000) Stimulation of AMP-activated protein kinase (AMPK) is associated with enhancement of Glut1-mediated glucose transport. Arch Biochem Biophys 380(2):347–352

Abnous K, Storey KB (2008) Skeletal muscle hexokinase: regulation in mammalian hibernation. Mol Cell Biochem 319(1–2):41–50

Bae BI, Hara MR, Cascio MB, Wellington CL, Hayden MR, Ross CA, Ha HC, Li XJ, Snyder SH, Sawa A (2006) Mutant huntingtin: nuclear translocation and cytotoxicity mediated by GAPDH. Proc Natl Acad Sci USA 103(9):3405–3409

Barberger-Gateau P, Lambert JC, Feart C, Peres K, Ritchie K, Dartigues JF, Alperovitch A (2013) From genetics to dietetics: the contribution of epidemiology to understanding Alzheimer's disease. J Alzheimers Dis 33(Suppl 1):S457–S463

Beitz JM (2014) Parkinson's disease: a review. Front Biosci (Schol Ed) 6:65–74

Bezprozvanny I, Mattson MP (2008) Neuronal calcium mishandling and the pathogenesis of Alzheimer's disease. Trends Neurosci 31(9):454–463

Blokhuis AM, Groen EJ, Koppers M, van den Berg LH, Pasterkamp RJ (2013) Protein aggregation in amyotrophic lateral sclerosis. Acta Neuropathol 125(6):777–794

Browne SE, Bowling AC, MacGarvey U, Baik MJ, Berger SC, Muqit MM, Bird ED, Beal MF (1997) Oxidative damage and metabolic dysfunction in Huntington's disease: selective vulnerability of the basal ganglia. Ann Neurol 41(5):646–653

Browne GJ, Finn SG, Proud CG (2004) Stimulation of the AMP-activated protein kinase leads to activation of eukaryotic elongation factor 2 kinase and to its phosphorylation at a novel site, serine 398. J Biol Chem 279(13):12220–12231

Bubber P, Haroutunian V, Fisch G, Blass JP, Gibson GE (2005) Mitochondrial abnormalities in Alzheimer brain: mechanistic implications. Ann Neurol 57(5):695–703

Burke JR, Enghild JJ, Martin ME, Jou YS, Myers RM, Roses AD, Vance JM, Strittmatter WJ (1996) Huntingtin and DRPLA proteins selectively interact with the enzyme GAPDH. Nat Med 2(3):347–350

Cabezas-Opazo FA, Vergara-Pulgar K, Perez MJ, Jara C, Osorio-Fuentealba C, Quintanilla RA (2015) Mitochondrial dysfunction contributes to the pathogenesis of Alzheimer's disease. Oxid Med Cell Longev 2015:509654

Chen Y, Zhou K, Wang R, Liu Y, Kwak YD, Ma T, Thompson RC, Zhao Y, Smith L, Gasparini L, Luo Z, Xu H, Liao FF (2009) Antidiabetic drug metformin (GlucophageR) increases biogenesis of Alzheimer's amyloid peptides via up-regulating BACE1 transcription. Proc Natl Acad Sci USA 106(10):3907–3912

Chen Z, Shen X, Shen F, Zhong W, Wu H, Liu S, Lai J (2013) TAK1 activates AMPK-dependent cell death pathway in hydrogen peroxide-treated cardiomyocytes, inhibited by heat shock protein-70. Mol Cell Biochem 377(1–2):35–44

Choi JS, Park C, Jeong JW (2010) AMP-activated protein kinase is activated in Parkinson's disease models mediated by 1-methyl-4-phenyl-1,2,3,6-tetrahydropyridine. Biochem Biophys Res Commun 391(1):147–151

Chou SY, Lee YC, Chen HM, Chiang MC, Lai HL, Chang HH, Wu YC, Sun CN, Chien CL, Lin YS, Wang SC, Tung YY, Chang C, Chern Y (2005) CGS21680 attenuates symptoms of Huntington's disease in a transgenic mouse model. J Neurochem 93(2):310–320

Ciarmiello A, Giovacchini G, Orobello S, Bruselli L, Elifani F, Squitieri F (2012) 18F-FDG PET uptake in the pre-Huntington disease caudate affects the time-to-onset independently of CAG expansion size. Eur J Nucl Med Mol Imaging 39(6):1030–1036

Connolly NM, Dussmann H, Anilkumar U, Huber HJ, Prehn JH (2014) Single-cell imaging of bioenergetic responses to neuronal excitotoxicity and oxygen and glucose deprivation. J Neurosci 34(31):10192–10205

Coughlan KS, Mitchem MR, Hogg MC, Prehn JH (2015) "Preconditioning" with latrepirdine, an adenosine 5′-monophosphate-activated protein kinase activator, delays amyotrophic lateral sclerosis progression in SOD1(G93A) mice. Neurobiol Aging 36(2):1140–1150

Dash PK, Orsi SA, Moore AN (2006) Spatial memory formation and memory-enhancing effect of glucose involves activation of the tuberous sclerosis complex-Mammalian target of rapamycin pathway. J Neurosci 26(31):8048–8056

Davis JD, Lin SY (2011) DNA damage and breast cancer. World J Clin Oncol 2(9):329–338

Djouder N, Tuerk RD, Suter M, Salvioni P, Thali RF, Scholz R, Vaahtomeri K, Auchli Y, Rechsteiner H, Brunisholz RA, Viollet B, Makela TP, Wallimann T, Neumann D, Krek W (2010) PKA phosphorylates and inactivates AMPKalpha to promote efficient lipolysis. EMBO J 29(2):469–481

Domise M, Didier S, Marinangeli C, Zhao H, Chandakkar P, Buée L, Viollet B, Davies P, Marambaud P, Vingtdeux V (2016) AMP-activated protein kinase modulates tau phosphorylation and tau pathology in vivo. Sci Rep 6:26758

DuBoff B, Feany M, Gotz J (2013) Why size matters—balancing mitochondrial dynamics in Alzheimer's disease. Trends Neurosci 36(6):325–335

Dulovic M, Jovanovic M, Xilouri M, Stefanis L, Harhaji-Trajkovic L, Kravic-Stevovic T, Paunovic V, Ardah MT, El-Agnaf OM, Kostic V, Markovic I, Trajkovic V (2014) The protective role of AMP-activated protein kinase in alpha-synuclein neurotoxicity in vitro. Neurobiol Dis 63:1–11

Dupuis L, Pradat PF, Ludolph AC, Loeffler JP (2011) Energy metabolism in amyotrophic lateral sclerosis. Lancet Neurol 10(1):75–82

Eckert T, Barnes A, Dhawan V, Frucht S, Gordon MF, Feigin AS, Eidelberg D (2005) FDG PET in the differential diagnosis of parkinsonian disorders. Neuroimage 26(3):912–21

Egan DF, Shackelford DB, Mihaylova MM, Gelino S, Kohnz RA, Mair W, Vasquez DS, Joshi A, Gwinn DM, Taylor R, Asara JM, Fitzpatrick J, Dillin A, Viollet B, Kundu M, Hansen M, Shaw RJ (2011) Phosphorylation of ULK1 (hATG1) by AMP-activated protein kinase connects energy sensing to mitophagy. Science 331(6016):456–461

Ferreira IL, Resende R, Ferreiro E, Rego AC, Pereira CF (2010) Multiple defects in energy metabolism in Alzheimer's disease. Curr Drug Targets 11(10):1193–1206

Ferretta A, Gaballo A, Tanzarella P, Piccoli C, Capitanio N, Nico B, Annese T, Di Paola M, Dell'aquila C, De Mari M, Ferranini E, Bonifati V, Pacelli C, Cocco T (2014) Effect of resveratrol on mitochondrial function: implications in parkin-associated familiar Parkinson's disease. Biochim Biophys Acta 1842(7):902–915

Fischer D, Mukrasch MD, Biernat J, Bibow S, Blackledge M, Griesinger C, Mandelkow E, Zweckstetter M (2009) Conformational changes specific for pseudophosphorylation at serine 262 selectively impair binding of tau to microtubules. Biochemistry 48(42):10047–10055

Fu J, Jin J, Cichewicz RH, Hageman SA, Ellis TK, Xiang L, Peng Q, Jiang M, Arbez N, Hotaling K, Ross CA, Duan W (2012) trans-(-)-epsilon-Viniferin increases mitochondrial sirtuin 3 (SIRT3), activates AMP-activated protein kinase (AMPK), and protects cells in models of Huntington Disease. J Biol Chem 287(29):24460–24472

Garcia-Escudero V, Martin-Maestro P, Perry G, Avila J (2013) Deconstructing mitochondrial dysfunction in Alzheimer disease. Oxid Med Cell Longev 2013:162152

Gouarne C, Tardif G, Tracz J, Latyszenok V, Michaud M, Clemens LE, Yu-Taeger L, Nguyen HP, Bordet T, Pruss RM (2013) Early deficits in glycolysis are specific to striatal neurons from a rat model of huntington disease. PLoS One 8(11):e81528

Gowans GJ, Hawley SA, Ross FA, Hardie DG (2013) AMP is a true physiological regulator of AMP-activated protein kinase by both allosteric activation and enhancing net phosphorylation. Cell Metab 18(4):556–566

Greco SJ, Sarkar S, Johnston JM, Tezapsidis N (2009) Leptin regulates tau phosphorylation and amyloid through AMPK in neuronal cells. Biochem Biophys Res Commun 380(1):98–104

Greco SJ, Hamzelou A, Johnston JM, Smith MA, Ashford JW, Tezapsidis N (2011) Leptin boosts cellular metabolism by activating AMPK and the sirtuins to reduce tau phosphorylation and beta-amyloid in neurons. Biochem Biophys Res Commun 414(1):170–174

Green KN, LaFerla FM (2008) Linking calcium to Abeta and Alzheimer's disease. Neuron 59 (2):190–194

Grimm A, Schmitt K, Eckert A (2016) Advanced mitochondrial respiration assay for evaluation of mitochondrial dysfunction in Alzheimer's disease. Methods Mol Biol 1303:171–183

Gu M, Gash MT, Mann VM, Javoy-Agid F, Cooper JM, Schapira AH (1996) Mitochondrial defect in Huntington's disease caudate nucleus. Ann Neurol 39(3):385–389

Gwinn DM, Shackelford DB, Egan DF, Mihaylova MM, Mery A, Vasquez DS, Turk BE, Shaw RJ (2008) AMPK phosphorylation of raptor mediates a metabolic checkpoint. Mol Cell 30 (2):214–226

Harold D, Abraham R, Hollingworth P, Sims R, Gerrish A, Hamshere ML, Pahwa JS, Moskvina V, Dowzell K, Williams A, Jones N, Thomas C, Stretton A, Morgan AR, Lovestone S, Powell J, Proitsi P, Lupton MK, Brayne C, Rubinsztein DC, Gill M, Lawlor B, Lynch A, Morgan K, Brown KS, Passmore PA, Craig D, McGuinness B, Todd S, Holmes C, Mann D, Smith AD, Love S, Kehoe PG, Hardy J, Mead S, Fox N, Rossor M, Collinge J, Maier W, Jessen F, Schurmann B, Heun R, van den Bussche H, Heuser I, Kornhuber J, Wiltfang J, Dichgans M, Frolich L, Hampel H, Hull M, Rujescu D, Goate AM, Kauwe JS, Cruchaga C, Nowotny P, Morris JC, Mayo K, Sleegers K, Bettens K, Engelborghs S, De Deyn PP, Van Broeckhoven C, Livingston G, Bass NJ, Gurling H, McQuillin A, Gwilliam R, Deloukas P, Al-Chalabi A, Shaw CE, Tsolaki M, Singleton AB, Guerreiro R, Muhleisen TW, Nothen MM, Moebus S, Jockel

KH, Klopp N, Wichmann HE, Carrasquillo MM, Pankratz VS, Younkin SG, Holmans PA, O'Donovan M, Owen MJ, Williams J (2009) Genome-wide association study identifies variants at CLU and PICALM associated with Alzheimer's disease. Nat Genet 41 (10):1088–1093

Hawley SA, Boudeau J, Reid JL, Mustard KJ, Udd L, Makela TP, Alessi DR, Hardie DG (2003) Complexes between the LKB1 tumor suppressor, STRAD alpha/beta and MO25 alpha/beta are upstream kinases in the AMP-activated protein kinase cascade. J Biol 2(4):28

Hawley SA, Pan DA, Mustard KJ, Ross L, Bain J, Edelman AM, Frenguelli BG, Hardie DG (2005) Calmodulin-dependent protein kinase kinase-beta is an alternative upstream kinase for AMP-activated protein kinase. Cell Metab 2(1):9–19

Hirai K, Aliev G, Nunomura A, Fujioka H, Russell RL, Atwood CS, Johnson AB, Kress Y, Vinters HV, Tabaton M, Shimohama S, Cash AD, Siedlak SL, Harris PL, Jones PK, Petersen RB, Perry G, Smith MA (2001) Mitochondrial abnormalities in Alzheimer's disease. J Neurosci 21 (9):3017–3023

Horman S, Browne G, Krause U, Patel J, Vertommen D, Bertrand L, Lavoinne A, Hue L, Proud C, Rider M (2002) Activation of AMP-activated protein kinase leads to the phosphorylation of elongation factor 2 and an inhibition of protein synthesis. Curr Biol 12(16):1419–1423

Horman S, Vertommen D, Heath R, Neumann D, Mouton V, Woods A, Schlattner U, Wallimann T, Carling D, Hue L, Rider MH (2006) Insulin antagonizes ischemia-induced Thr172 phosphorylation of AMP-activated protein kinase alpha-subunits in heart via hierarchical phosphorylation of Ser485/491. J Biol Chem 281(9):5335–5340

Hou YS, Guan JJ, Xu HD, Wu F, Sheng R, Qin ZH (2015) Sestrin2 protects dopaminergic cells against rotenone toxicity through AMPK-dependent autophagy activation. Mol Cell Biol 35 (16):2740–2751

Hsu LJ, Sagara Y, Arroyo A, Rockenstein E, Sisk A, Mallory M, Wong J, Takenouchi T, Hashimoto M, Masliah E (2000) alpha-synuclein promotes mitochondrial deficit and oxidative stress. Am J Pathol 157(2):401–410

Hurley RL, Anderson KA, Franzone JM, Kemp BE, Means AR, Witters LA (2005) The Ca2+/calmodulin-dependent protein kinase kinases are AMP-activated protein kinase kinases. J Biol Chem 280(32):29060–29066

Hwang S, Disatnik MH, Mochly-Rosen D (2015) Impaired GAPDH-induced mitophagy contributes to the pathology of Huntington's disease. EMBO Mol Med 7(10):1307–1326

Ingre C, Roos PM, Piehl F, Kamel F, Fang F (2015) Risk factors for amyotrophic lateral sclerosis. Clin Epidemiol 7:181–193

Inoki K, Zhu T, Guan KL (2003) TSC2 mediates cellular energy response to control cell growth and survival. Cell 115(5):577–590

Jager S, Handschin C, St-Pierre J, Spiegelman BM (2007) AMP-activated protein kinase (AMPK) action in skeletal muscle via direct phosphorylation of PGC-1alpha. Proc Natl Acad Sci USA 104(29):12017–12022

Jiang P, Gan M, Ebrahim AS, Castanedes-Casey M, Dickson DW, Yen SH (2013) Adenosine monophosphate-activated protein kinase overactivation leads to accumulation of alpha-synuclein oligomers and decrease of neurites. Neurobiol Aging 34(5):1504–1515

Ju TC, Chen HM, Lin JT, Chang CP, Chang WC, Kang JJ, Sun CP, Tao MH, Tu PH, Chang C, Dickson DW, Chern Y (2011) Nuclear translocation of AMPK-alpha1 potentiates striatal neurodegeneration in Huntington's disease. J Cell Biol 194(2):209–227

Ju TC, Chen HM, Chen YC, Chang CP, Chang C, Chern Y (2014) AMPK-alpha1 functions downstream of oxidative stress to mediate neuronal atrophy in Huntington's disease. Biochim Biophys Acta 1842(9):1668–1680

Keeney PM, Xie J, Capaldi RA, Bennett JP Jr (2006) Parkinson's disease brain mitochondrial complex I has oxidatively damaged subunits and is functionally impaired and misassembled. J Neurosci 26(19):5256–5264

Kickstein E, Krauss S, Thornhill P, Rutschow D, Zeller R, Sharkey J, Williamson R, Fuchs M, Kohler A, Glossmann H, Schneider R, Sutherland C, Schweiger S (2010) Biguanide metformin acts on tau phosphorylation via mTOR/protein phosphatase 2A (PP2A) signaling. Proc Natl Acad Sci USA 107(50):21830–21835

Kim J, Kundu M, Viollet B, Guan KL (2011) AMPK and mTOR regulate autophagy through direct phosphorylation of Ulk1. Nat Cell Biol 13(2):132–141

Kim TW, Cho HM, Choi SY, Suguira Y, Hayasaka T, Setou M, Koh HC, Hwang EM, Park JY, Kang SJ, Kim HS, Kim H, Sun W (2013) (ADP-ribose) polymerase 1 and AMP-activated protein kinase mediate progressive dopaminergic neuronal degeneration in a mouse model of Parkinson's disease. Cell Death Dis 4:e919

Kim B, Figueroa-Romero C, Pacut C, Backus C, Feldman EL (2015) Insulin resistance prevents AMPK-induced tau dephosphorylation through Akt-mediated increase in AMPKSer-485 phosphorylation. J Biol Chem 290(31):19146–19157

Kish SJ, Lopes-Cendes I, Guttman M, Furukawa Y, Pandolfo M, Rouleau GA, Ross BM, Nance M, Schut L, Ang L, DiStefano L (1998) Brain glyceraldehyde-3-phosphate dehydrogenase activity in human trinucleotide repeat disorders. Arch Neurol 55(10):1299–1304

Kobilo T, Yuan C, van Praag H (2011) Endurance factors improve hippocampal neurogenesis and spatial memory in mice. Learn Mem 18(2):103–107

Kobilo T, Guerrieri D, Zhang Y, Collica SC, Becker KG, van Praag H (2014) AMPK agonist AICAR improves cognition and motor coordination in young and aged mice. Learn Mem 21 (2):119–126

Lambert JC, Heath S, Even G, Campion D, Sleegers K, Hiltunen M, Combarros O, Zelenika D, Bullido MJ, Tavernier B, Letenneur L, Bettens K, Berr C, Pasquier F, Fievet N, Barberger-Gateau P, Engelborghs S, De Deyn P, Mateo I, Franck A, Helisalmi S, Porcellini E, Hanon O, European Alzheimer's Disease Initiative Investigators, de Pancorbo MM, Lendon C, Dufouil C, Jaillard C, Leveillard T, Alvarez V, Bosco P, Mancuso M, Panza F, Nacmias B, Bossu P, Piccardi P, Annoni G, Seripa D, Galimberti D, Hannequin D, Licastro F, Soininen H, Ritchie K, Blanche H, Dartigues JF, Tzourio C, Gut I, Van Broeckhoven C, Alperovitch A, Lathrop M, Amouyel P (2009) Genome-wide association study identifies variants at CLU and CR1 associated with Alzheimer's disease. Nat Genet 41(10):1094–1099

Leuner K, Schutt T, Kurz C, Eckert SH, Schiller C, Occhipinti A, Mai S, Jendrach M, Eckert GP, Kruse SE, Palmiter RD, Brandt U, Drose S, Wittig I, Willem M, Haass C, Reichert AS, Muller WE (2012) Mitochondrion-derived reactive oxygen species lead to enhanced amyloid beta formation. Antioxid Redox Signal 16(12):1421–1433

Li Y, Xu S, Mihaylova MM, Zheng B, Hou X, Jiang B, Park O, Luo Z, Lefai E, Shyy JY, Gao B, Wierzbicki M, Verbeuren TJ, Shaw RJ, Cohen RA, Zang M (2011) AMPK phosphorylates and inhibits SREBP activity to attenuate hepatic steatosis and atherosclerosis in diet-induced insulin-resistant mice. Cell Metab 13(4):376–388

Lim MA, Selak MA, Xiang Z, Krainc D, Neve RL, Kraemer BC, Watts JL, Kalb RG (2012) Reduced activity of AMP-activated protein kinase protects against genetic models of motor neuron disease. J Neurosci 32(3):1123–1141

Lin YT, Cheng JT, Liang LC, Ko CY, Lo YK, Lu PJ (2007) The binding and phosphorylation of Thr231 is critical for Tau's hyperphosphorylation and functional regulation by glycogen synthase kinase 3beta. J Neurochem 103(2):802–813

Liu YJ, Ju TC, Chen HM, Jang YS, Lee LM, Lai HL, Tai HC, Fang JM, Lin YL, Tu PH, Chern Y (2015a) Activation of AMP-activated protein kinase alpha1 mediates mislocalization of TDP-43 in amyotrophic lateral sclerosis. Hum Mol Genet 24(3):787–801

Liu YJ, Lee LM, Lai HL, Chern Y (2015b) Aberrant activation of AMP-activated protein kinase contributes to the abnormal distribution of HuR in amyotrophic lateral sclerosis. FEBS Lett 589(4):432–439

Lu L, Zheng L, Si Y, Luo W, Dujardin G, Kwan T, Potochick NR, Thompson SR, Schneider DA, King PH (2014) Hu antigen R (HuR) is a positive regulator of the RNA-binding proteins TDP-43 and FUS/TLS: implications for amyotrophic lateral sclerosis. J Biol Chem 289 (46):31792–31804

Ma TC, Buescher JL, Oatis B, Funk JA, Nash AJ, Carrier RL, Hoyt KR (2007) Metformin therapy in a transgenic mouse model of Huntington's disease. Neurosci Lett 411(2):98–103

Ma T, Chen Y, Vingtdeux V, Zhao H, Viollet B, Marambaud P, Klann E (2014) Inhibition of AMP-activated protein kinase signaling alleviates impairments in hippocampal synaptic plasticity induced by amyloid beta. J Neurosci 34(36):12230–12238

Mairet-Coello G, Courchet J, Pieraut S, Courchet V, Maximov A, Polleux F (2013) The CAMKK2-AMPK kinase pathway mediates the synaptotoxic effects of Abeta oligomers through Tau phosphorylation. Neuron 78(1):94–108

Mancuso R, del Valle J, Modol L, Martinez A, Granado-Serrano AB, Ramirez-Nunez O, Pallas M, Portero-Otin M, Osta R, Navarro X (2014) Resveratrol improves motoneuron function and extends survival in SOD1(G93A) ALS mice. Neurotherapeutics 11(2):419–432

Manwani B, McCullough LD (2013) Function of the master energy regulator adenosine monophosphate-activated protein kinase in stroke. J Neurosci Res 91(8):1018–1029

Marangos PJ, Loftus T, Wiesner J, Lowe T, Rossi E, Browne CE, Gruber HE (1990) Adenosinergic modulation of homocysteine-induced seizures in mice. Epilepsia 31 (3):239–246

Marsin AS, Bertrand L, Rider MH, Deprez J, Beauloye C, Vincent MF, Van den Berghe G, Carling D, Hue L (2000) Phosphorylation and activation of heart PFK-2 by AMPK has a role in the stimulation of glycolysis during ischaemia. Curr Biol 10(20):1247–1255

Martin LJ, Pan Y, Price AC, Sterling W, Copeland NG, Jenkins NA, Price DL, Lee MK (2006) Parkinson's disease alpha-synuclein transgenic mice develop neuronal mitochondrial degeneration and cell death. J Neurosci 26(1):41–50

Mattson MP (2007) Calcium and neurodegeneration. Aging Cell 6(3):337–350

Mochel F, Durant B, Meng X, O'Callaghan J, Yu H, Brouillet E, Wheeler VC, Humbert S, Schiffmann R, Durr A (2012) Early alterations of brain cellular energy homeostasis in Huntington disease models. J Biol Chem 287(2):1361–1370

Momcilovic M, Hong SP, Carlson M (2006) Mammalian TAK1 activates Snf1 protein kinase in yeast and phosphorylates AMP-activated protein kinase in vitro. J Biol Chem 281 (35):25336–25343

Mosconi L (2005) Brain glucose metabolism in the early and specific diagnosis of Alzheimer's disease. FDG-PET studies in MCI and AD. Eur J Nucl Med Mol Imaging 32(4):486–510

Narendra D, Tanaka A, Suen DF, Youle RJ (2008) Parkin is recruited selectively to impaired mitochondria and promotes their autophagy. J Cell Biol 183(5):795–803

Narendra DP, Jin SM, Tanaka A, Suen DF, Gautier CA, Shen J, Cookson MR, Youle RJ (2010) PINK1 is selectively stabilized on impaired mitochondria to activate Parkin. PLoS Biol 8(1): e1000298

Ng CH, Guan MS, Koh C, Ouyang X, Yu F, Tan EK, O'Neill SP, Zhang X, Chung J, Lim KL (2012) AMP kinase activation mitigates dopaminergic dysfunction and mitochondrial abnormalities in Drosophila models of Parkinson's disease. J Neurosci 32(41):14311–14317

Nixon RA, Yang DS (2011) Autophagy failure in Alzheimer's disease—locating the primary defect. Neurobiol Dis 43(1):38–45

Olah J, Klivenyi P, Gardian G, Vecsei L, Orosz F, Kovacs GG, Westerhoff HV, Ovadi J (2008) Increased glucose metabolism and ATP level in brain tissue of Huntington's disease transgenic mice. FEBS J 275(19):4740–4755

Olanow CW, Tatton WG (1999) Etiology and pathogenesis of Parkinson's disease. Annu Rev Neurosci 22:123–144

Perera ND, Sheean RK, Scott JW, Kemp BE, Horne MK, Turner BJ (2014) Mutant TDP-43 deregulates AMPK activation by PP2A in ALS models. PLoS One 9(4):e95549

Pickford F, Masliah E, Britschgi M, Lucin K, Narasimhan R, Jaeger PA, Small S, Spencer B, Rockenstein E, Levine B, Wyss-Coray T (2008) The autophagy-related protein beclin 1 shows reduced expression in early Alzheimer disease and regulates amyloid beta accumulation in mice. J Clin Invest 118(6):2190–2199

Potter WB, O'Riordan KJ, Barnett D, Osting SM, Wagoner M, Burger C, Roopra A (2010) Metabolic regulation of neuronal plasticity by the energy sensor AMPK. PLoS One 5(2):e8996

Powers WJ, Videen TO, Markham J, McGee-Minnich L, Antenor-Dorsey JV, Hershey T, Perlmutter JS (2007) Selective defect of in vivo glycolysis in early Huntington's disease striatum. Proc Natl Acad Sci USA 104(8):2945–2949

Reddy PH (2011) Abnormal tau, mitochondrial dysfunction, impaired axonal transport of mitochondria, and synaptic deprivation in Alzheimer's disease. Brain Res 1415:136–148

Renton AE, Chio A, Traynor BJ (2014) State of play in amyotrophic lateral sclerosis genetics. Nat Neurosci 17(1):17–23

Rowland LP, Shneider NA (2001) Amyotrophic lateral sclerosis. N Engl J Med 344 (22):1688–1700

Russell RR 3rd, Bergeron R, Shulman GI, Young LH (1999) Translocation of myocardial GLUT-4 and increased glucose uptake through activation of AMPK by AICAR. Am J Physiol 277(2 Pt 2):H643–H649

Sanders MJ, Grondin PO, Hegarty BD, Snowden MA, Carling D (2007) Investigating the mechanism for AMP activation of the AMP-activated protein kinase cascade. Biochem J 403 (1):139–148

Sato-Harada R, Okabe S, Umeyama T, Kanai Y, Hirokawa N (1996) Microtubule-associated proteins regulate microtubule function as the track for intracellular membrane organelle transports. Cell Struct Funct 21(5):283–295

Sayre LM, Perry G, Smith MA (2008) Oxidative stress and neurotoxicity. Chem Res Toxicol 21 (1):172–188

Schapira AH, Cooper JM, Dexter D, Clark JB, Jenner P, Marsden CD (1990) Mitochondrial complex I deficiency in Parkinson's disease. J Neurochem 54(3):823–827

Seshadri S, Fitzpatrick AL, Ikram MA, DeStefano AL, Gudnason V, Boada M, Bis JC, Smith AV, Carassquillo MM, Lambert JC, Harold D, Schrijvers EM, Ramirez-Lorca R, Debette S, Longstreth WT Jr, Janssens AC, Pankratz VS, Dartigues JF, Hollingworth P, Aspelund T, Hernandez I, Beiser A, Kuller LH, Koudstaal PJ, Dickson DW, Tzourio C, Abraham R, Antunez C, Du Y, Rotter JI, Aulchenko YS, Harris TB, Petersen RC, Berr C, Owen MJ, Lopez-Arrieta J, Varadarajan BN, Becker JT, Rivadeneira F, Nalls MA, Graff-Radford NR, Campion D, Auerbach S, Rice K, Hofman A, Jonsson PV, Schmidt H, Lathrop M, Mosley TH, Au R, Psaty BM, Uitterlinden AG, Farrer LA, Lumley T, Ruiz A, Williams J, Amouyel P, Younkin SG, Wolf PA, Launer LJ, Lopez OL, van Duijn CM, Breteler MM, CHARGE Consortium, GERAD1 Consortium, EADI1 Consortium (2010) Genome-wide analysis of genetic loci associated with Alzheimer disease. JAMA 303(18):1832–1840

Shaw RJ (2009) LKB1 and AMP-activated protein kinase control of mTOR signalling and growth. Acta Physiol (Oxf) 196(1):65–80

Shaw RJ, Kosmatka M, Bardeesy N, Hurley RL, Witters LA, DePinho RA, Cantley LC (2004) The tumor suppressor LKB1 kinase directly activates AMP-activated kinase and regulates apoptosis in response to energy stress. Proc Natl Acad Sci USA 101(10):3329–3335

Sheng ZH (2014) Mitochondrial trafficking and anchoring in neurons: new insight and implications. J Cell Biol 204(7):1087–1098

Sherer TB, Betarbet R, Greenamyre JT (2002) Environment, mitochondria, and Parkinson's disease. Neuroscientist 8(3):192–197

Shin H, Kim MH, Lee SJ, Lee KH, Kim MJ, Kim JS, Cho JW (2013) Decreased metabolism in the cerebral cortex in early-stage Huntington's disease: a possible biomarker of disease progression? J Clin Neurol 9(1):21–25

Shirendeb U, Reddy AP, Manczak M, Calkins MJ, Mao P, Tagle DA, Reddy PH (2011) Abnormal mitochondrial dynamics, mitochondrial loss and mutant huntingtin oligomers in Huntington's disease: implications for selective neuronal damage. Hum Mol Genet 20(7):1438–1455

Smith R, Bacos K, Fedele V, Soulet D, Walz HA, Obermuller S, Lindqvist A, Bjorkqvist M, Klein P, Onnerfjord P, Brundin P, Mulder H, Li JY (2009) Mutant huntingtin interacts with {beta}-tubulin and disrupts vesicular transport and insulin secretion. Hum Mol Genet 18 (20):3942–3954

Son SM, Jung ES, Shin HJ, Byun J, Mook-Jung I (2012) Abeta-induced formation of autophagosomes is mediated by RAGE-CaMKKbeta-AMPK signaling. Neurobiol Aging 33 (5):1006e11–1006e23

Song W, Chen J, Petrilli A, Liot G, Klinglmayr E, Zhou Y, Poquiz P, Tjong J, Pouladi MA, Hayden MR, Masliah E, Ellisman M, Rouiller I, Schwarzenbacher R, Bossy B, Perkins G, Bossy-Wetzel E (2011) Mutant huntingtin binds the mitochondrial fission GTPase dynamin-related protein-1 and increases its enzymatic activity. Nat Med 17(3):377–382

Song L, Chen L, Zhang X, Li J, Le W (2014) Resveratrol ameliorates motor neuron degeneration and improves survival in SOD1(G93A) mouse model of amyotrophic lateral sclerosis. Biomed Res Int 2014:483501

Spasic MR, Callaerts P, Norga KK (2008) Drosophila alicorn is a neuronal maintenance factor protecting against activity-induced retinal degeneration. J Neurosci 28(25):6419–6429

Sui Y, Zhao Z, Liu R, Cai B, Fan D (2014) Adenosine monophosphate-activated protein kinase activation enhances embryonic neural stem cell apoptosis in a mouse model of amyotrophic lateral sclerosis. Neural Regen Res 9(19):1770–1778

Tabrizi SJ, Cleeter MW, Xuereb J, Taanman JW, Cooper JM, Schapira AH (1999) Biochemical abnormalities and excitotoxicity in Huntington's disease brain. Ann Neurol 45(1):25–32

Tenreiro S, Reimao-Pinto MM, Antas P, Rino J, Wawrzycka D, Macedo D, Rosado-Ramos R, Amen T, Waiss M, Magalhaes F, Gomes A, Santos CN, Kaganovich D, Outeiro TF (2014) Phosphorylation modulates clearance of alpha-synuclein inclusions in a yeast model of Parkinson's disease. PLoS Genet 10(5):e1004302

The Huntington's Disease Collaborative Research Group (1993) A novel gene containing a trinucleotide repeat that is expanded and unstable on Huntington's disease chromosomes. Cell 72(6):971–983

Thornton C, Bright NJ, Sastre M, Muckett PJ, Carling D (2011) AMP-activated protein kinase (AMPK) is a tau kinase, activated in response to amyloid beta-peptide exposure. Biochem J 434(3):503–512

Tian J, Yan YP, Zhou R, Lou HF, Rong Y, Zhang BR (2014) Soluble N-terminal fragment of mutant Huntingtin protein impairs mitochondrial axonal transport in cultured hippocampal neurons. Neurosci Bull 30(1):74–80

Tschape JA, Hammerschmied C, Muhlig-Versen M, Athenstaedt K, Daum G, Kretzschmar D (2002) The neurodegeneration mutant lochrig interferes with cholesterol homeostasis and Appl processing. EMBO J 21(23):6367–6376

Valente EM, Abou-Sleiman PM, Caputo V, Muqit MM, Harvey K, Gispert S, Ali Z, Del Turco D, Bentivoglio AR, Healy DG, Albanese A, Nussbaum R, Gonzalez-Maldonado R, Deller T, Salvi S, Cortelli P, Gilks WP, Latchman DS, Harvey RJ, Dallapiccola B, Auburger G, Wood NW (2004) Hereditary early-onset Parkinson's disease caused by mutations in PINK1. Science 304(5674):1158–1160

Vergouts M, Marinangeli C, Ingelbrecht C, Genard G, Schakman O, Sternotte A, Calas AG, Hermans E (2015) Early ALS-type gait abnormalities in AMP-dependent protein kinase-deficient mice suggest a role for this metabolic sensor in early stages of the disease. Metab Brain Dis 30(6):1369–1377

Verstraeten A, Theuns J, Van Broeckhoven C (2015) Progress in unraveling the genetic etiology of Parkinson disease in a genomic era. Trends Genet 31(3):140–149

Vingtdeux V, Giliberto L, Zhao H, Chandakkar P, Wu Q, Simon JE, Janle EM, Lobo J, Ferruzzi MG, Davies P, Marambaud P (2010) AMP-activated protein kinase signaling activation by resveratrol modulates amyloid-beta peptide metabolism. J Biol Chem 285(12):9100–9113

Vingtdeux V, Chandakkar P, Zhao H, d'Abramo C, Davies P, Marambaud P (2011a) Novel synthetic small-molecule activators of AMPK as enhancers of autophagy and amyloid-beta peptide degradation. FASEB J 25(1):219–231

Vingtdeux V, Davies P, Dickson DW, Marambaud P (2011b) AMPK is abnormally activated in tangle- and pre-tangle-bearing neurons in Alzheimer's disease and other tauopathies. Acta Neuropathol 121(3):337–349

Walker FO (2007) Huntington's disease. Lancet 369(9557):218–228

Wang Y, Martinez-Vicente M, Kruger U, Kaushik S, Wong E, Mandelkow EM, Cuervo AM, Mandelkow E (2010) Synergy and antagonism of macroautophagy and chaperone-mediated autophagy in a cell model of pathological tau aggregation. Autophagy 6(1):182–183

Wang ZX, Tan L, Yu JT (2015) Axonal transport defects in Alzheimer's disease. Mol Neurobiol 51(3):1309–1321

Weisova P, Concannon CG, Devocelle M, Prehn JH, Ward MW (2009) Regulation of glucose transporter 3 surface expression by the AMP-activated protein kinase mediates tolerance to glutamate excitation in neurons. J Neurosci 29(9):2997–3008

Weisova P, Davila D, Tuffy LP, Ward MW, Concannon CG, Prehn JH (2011) Role of 5′-adenosine monophosphate-activated protein kinase in cell survival and death responses in neurons. Antioxid Redox Signal 14(10):1863–1876

Weisova P, Alvarez SP, Kilbride SM, Anilkumar U, Baumann B, Jordan J, Bernas T, Huber HJ, Dussmann H, Prehn JH (2013) Latrepirdine is a potent activator of AMP-activated protein kinase and reduces neuronal excitability. Transl Psychiatry 3:e317

Won JS, Im YB, Kim J, Singh AK, Singh I (2010) Involvement of AMP-activated-protein-kinase (AMPK) in neuronal amyloidogenesis. Biochem Biophys Res Commun 399(4):487–491

Wong YC, Holzbaur EL (2014) The regulation of autophagosome dynamics by huntingtin and HAP1 is disrupted by expression of mutant huntingtin, leading to defective cargo degradation. J Neurosci 34(4):1293–1305

Woods A, Johnstone SR, Dickerson K, Leiper FC, Fryer LG, Neumann D, Schlattner U, Wallimann T, Carlson M, Carling D (2003) LKB1 is the upstream kinase in the AMP-activated protein kinase cascade. Curr Biol 13(22):2004–2008

Woods A, Dickerson K, Heath R, Hong SP, Momcilovic M, Johnstone SR, Carlson M, Carling D (2005) Ca2+/calmodulin-dependent protein kinase kinase-beta acts upstream of AMP-activated protein kinase in mammalian cells. Cell Metab 2(1):21–33

Wu Y, Li X, Zhu JX, Xie W, Le W, Fan Z, Jankovic J, Pan T (2011) Resveratrol-activated AMPK/SIRT1/autophagy in cellular models of Parkinson's disease. Neurosignals 19(3):163–174

Wu CA, Chao Y, Shiah SG, Lin WW (2013) Nutrient deprivation induces the Warburg effect through ROS/AMPK-dependent activation of pyruvate dehydrogenase kinase. Biochim Biophys Acta 1833(5):1147–1156

Xiao B, Sanders MJ, Underwood E, Heath R, Mayer FV, Carmena D, Jing C, Walker PA, Eccleston JF, Haire LF, Saiu P, Howell SA, Aasland R, Martin SR, Carling D, Gamblin SJ (2011) Structure of mammalian AMPK and its regulation by ADP. Nature 472(7342):230–233

Xu Y, Liu C, Chen S, Ye Y, Guo M, Ren Q, Liu L, Zhang H, Xu C, Zhou Q, Huang S, Chen L (2014) Activation of AMPK and inactivation of Akt result in suppression of mTOR-mediated S6K1 and 4E-BP1 pathways leading to neuronal cell death in in vitro models of Parkinson's disease. Cell Signal 26(8):1680–1689

Yang TT, Shih YS, Chen YW, Kuo YM, Lee CW (2015) Glucose regulates amyloid beta production via AMPK. J Neural Transm 122(10):1381–1390

Yoon SO, Park DJ, Ryu JC, Ozer HG, Tep C, Shin YJ, Lim TH, Pastorino L, Kunwar AJ, Walton JC, Nagahara AH, Lu KP, Nelson RJ, Tuszynski MH, Huang K (2012) JNK3 perpetuates metabolic stress induced by Abeta peptides. Neuron 75(5):824–837

Youle RJ, Narendra DP (2011) Mechanisms of mitophagy. Nat Rev Mol Cell Biol 12(1):9–14

Yu WH, Cuervo AM, Kumar A, Peterhoff CM, Schmidt SD, Lee JH, Mohan PS, Mercken M, Farmery MR, Tjernberg LO, Jiang Y, Duff K, Uchiyama Y, Naslund J, Mathews PM, Cataldo AM, Nixon RA (2005) Macroautophagy—a novel Beta-amyloid peptide-generating pathway activated in Alzheimer's disease. J Cell Biol 171(1):87–98

Zala D, Hinckelmann MV, Yu H, Lyra da Cunha MM, Liot G, Cordelieres FP, Marco S, Saudou F (2013) Vesicular glycolysis provides on-board energy for fast axonal transport. Cell 152 (3):479–491

Zarei S, Carr K, Reiley L, Diaz K, Guerra O, Altamirano PF, Pagani W, Lodin D, Orozco G, Chinea A (2015) A comprehensive review of amyotrophic lateral sclerosis. Surg Neurol Int 6:171

Zhao Z, Sui Y, Gao W, Cai B, Fan D (2015) Effects of diet on adenosine monophosphate-activated protein kinase activity and disease progression in an amyotrophic lateral sclerosis model. J Int Med Res 43(1):67–79

Zheng D, MacLean PS, Pohnert SC, Knight JB, Olson AL, Winder WW, Dohm GL (2001) Regulation of muscle GLUT-4 transcription by AMP-activated protein kinase. J Appl Physiol (1985) 91(3):1073–1083

Zuccato C, Valenza M, Cattaneo E (2010) Molecular mechanisms and potential therapeutical targets in Huntington's disease. Physiol Rev 90(3):905–981

Chapter 8
AMPK in Cardiovascular Diseases

Evangelos P. Daskalopoulos, Cécile Dufeys, Christophe Beauloye, Luc Bertrand, and Sandrine Horman

Contents

E.P. Daskalopoulos
Université catholique de Louvain, Institut de Recherche Expérimentale et Clinique, Pole of Cardiovascular Research, Brussels, Belgium

Cardiovascular Research (Care) Institute, Athens, Ioannina, Greece

C. Dufeys • L. Bertrand (✉) • S. Horman
Université catholique de Louvain, Institut de Recherche Expérimentale et Clinique, Pole of Cardiovascular Research, Brussels, Belgium
e-mail: luc.bertrand@uclouvain.be

C. Beauloye
Université catholique de Louvain, Institut de Recherche Expérimentale et Clinique, Pole of Cardiovascular Research, Brussels, Belgium

Division of Cardiology, Cliniques universitaires Saint-Luc, Brussels, Belgium

M.D. Cordero, B. Viollet (eds.), *AMP-activated Protein Kinase*, Experientia Supplementum 107, DOI 10.1007/978-3-319-43589-3_8

Abstract This chapter summarizes the implication of AMP-activated protein kinase (AMPK) in the regulation of various physiological and pathological cellular events of great importance for the maintenance of cardiac function. These include the control of both metabolic and non-metabolic elements targeting the different cellular components of the cardiac tissue, i.e., cardiomyocytes, fibroblasts, and vascular cells. The description of the multifaceted action of the two AMPK catalytic isoforms, α1 and α2, emphasizes the general protective action of this protein kinase against the development of critical pathologies like myocardial ischemia, cardiac hypertrophy, diabetic cardiomyopathy, and heart failure.

Keywords AMPK • Heart • Ischemia • Ischemia-reperfusion • Metabolism • Myocardial infarction • Hypertrophy • Fibrosis • Left ventricular remodeling • Heart failure

8.1 Historical View

AMP-activated protein kinase (AMPK) has been firstly investigated in the heart by Lopaschuck's lab in the mid-1990s (Kudo et al. 1995; Kudo et al. 1996). At that time, AMPK was known to be activated under various stress conditions elevating cellular AMP concentration. They nicely and logically showed that cardiac ischemia, a notable metabolic stress condition, induced potent AMPK activation, being responsible for the decrease in malonyl-CoA, the inhibition of acetyl-CoA carboxylase, and the concomitant increase in fatty acid oxidation rate that occur during reperfusion phase. Several other putative metabolic downstream targets of AMPK were identified in the ischemic heart thereafter. Amongst them, we can cite 6-phosphofructo-2-kinase (PFK2), which promotes anaerobic glycolysis, the sole way to generate ATP during an ischemic episode (Marsin et al. 2000). Glucose uptake was also similarly demonstrated to be stimulated by AMPK activators in the heart (Russell et al. 2004). AMPKα2 is the main catalytic subunit expressed in the heart, and its activation during ischemia is fully dependent on the upstream kinase Liver Kinase B1 (LKB1) (Cheung et al. 2000; Sakamoto et al. 2006). However, both AMPKα1 and α2 catalytic subunits are activated by cardiac ischemia (Marsin et al. 2000). Construction of varied transgenic mouse models has permitted us to partially decipher the role of AMPKα2 catalytic subunit in metabolic control and more generally to reveal the protective action of AMPK during myocardial ischemia (Russell et al. 2004; Xing et al. 2003; Zarrinpashneh et al. 2006). But cardiac AMPK operates beyond metabolism. Indeed, it has been firstly demonstrated that AMPK can phosphorylate and activate the endothelial nitric oxide synthase (eNOS) (Chen et al. 1999). Subsequently, numerous studies illustrated non-metabolic functions of AMPK, particularly AMPKα1, in fibroblasts, endothelial, and smooth muscle cells. Currently, it is well established that AMPK has a broader scope than the simple metabolic protective action during myocardial ischemia (Daskalopoulos et al. 2016).

8.2 Actions and Targets of AMPK in the Heart

8.2.1 Energy Metabolism

Once activated by energetic stress, AMPK modulates a variety of physiological processes in order to maintain sufficient ATP levels for sustaining the contractile function and membrane ionic gradients that are both important for the preservation of cell function and viability (Weiss and Hiltbrand 1985; Horman et al. 2012). This involves the stimulation of glucose utilization, fatty acid oxidation, and mitochondrial biogenesis through the phosphorylation of key metabolic enzymes and transcription factors (see Fig. 8.1 that outlines the different AMPK targets in the heart). AMPK promotes the uptake and anaerobic metabolism of glucose by phosphorylating Akt substrate of 160 kDa (AS160) and 6-phosphofructo-2-kinase (PFK2). AS160 is a Rab GTPase-activating protein (RabGAP) regulating trafficking of storage vesicles containing the glucose transporter type 4 (GLUT4) to the cell surface (Fukuda 2011). AMPK-induced phosphorylation of AS160 inhibits its GAP activity, elevates GTP-bound Rab protein, and thereby promotes GLUT4 translocation to the cell surface and glucose import into the cardiomyocytes

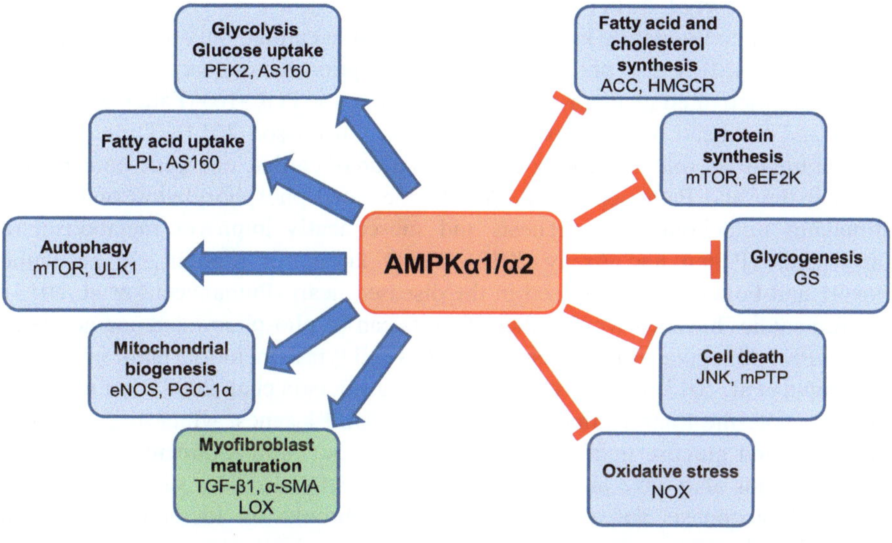

Fig. 8.1 *Main molecular targets of AMPK in the heart.* See text for more details. Abbreviations: *α-SMA* alpha-smooth muscle actin, *ACC* acetyl-coenzyme A carboxylase, *AMPK* AMP-activated protein kinase, *AS160* Akt substrate of 160 kDa, *eEF2K* eukaryotic elongation factor 2 kinase, *eNOS* endothelial nitric oxide synthase, *GS* glycogen synthase, *HMGCR* 3-hydroxy-3-methyl-glutaryl-CoA reductase, *JNK* c-Jun terminal kinase, *LOX* lysyl oxidase, *LPL* lipoprotein lipase, *mPTP* mitochondrial permeability transition pore, *mTOR* mammalian target of rapamycin, *NOX* NADPH oxidase, *PFK2* 6-phosphofructo-2-kinase, *PGC-1α* peroxisome proliferator-activated receptor-gamma coactivator 1α, *TGF-β1* transforming growth factor-β1, *ULK1* Unc-51 like autophagy activating kinase 1

(Treebak et al. 2006; Kramer et al. 2006). In addition, AMPK inhibits the endocytotic cycling of GLUT4, which contributes to increase its membrane content (Yang and Holman 2005). AMPK not only provides substrate for glycolysis but also promotes the production of ATP through the direct phosphorylation and activation of PFK2 (Marsin et al. 2000). The phosphorylation of PFK2 leads to an enhancement of fructose 2,6-bisphosphate, the most potent stimulator of 6-phosphofructo-1-kinase which is the key regulating enzyme of the glycolytic flux. In addition, AMPK is critical for facilitating both fatty acid uptake and oxidation. It promotes the delivery of fatty acids to cardiomyocytes through the translocation of lipoprotein lipase (LPL) on the endothelial surface in the heart (An et al. 2005). Moreover, AS160 phosphorylation also recruits the fatty acid transporter CD36 to the cell surface (Samovski et al. 2012). Lastly, acetyl-coenzyme A (CoA) carboxylase (ACC) phosphorylation and inhibition lead to a decrease in the intracellular concentration of malonyl-CoA, an inhibitor of fatty acid import into mitochondria, accelerating fatty acid oxidation (Kudo et al. 1995). The rise in fatty acid oxidation takes place shortly after the end of ischemia, during the first moment of reperfusion, a period during which oxygen is available and AMPK still activated. AMPK may also inhibit ATP-consuming pathways by phosphorylating glycogen synthase (GS), ACC, or 3-hydroxy-3-methyl-glutaryl-CoA reductase (HMGCR), involved in glycogen, lipid, and cholesterol synthesis, respectively (Hardie 2007; Hue and Rider 2007; Beauloye et al. 2011).

In addition to regulating key enzymes involved in cardiac fatty acid and glucose metabolism, AMPK also phosphorylates transcription factors such as peroxisome proliferator-activated receptor-gamma coactivator 1α (PGC-1α) (Patten and Arany 2012) and forkhead box O (FoxO) (Kubli and Gustafsson 2014). These proteins play an important role in the long-term regulation of cardiac energetic homeostasis (Ronnebaum and Patterson 2010). AMPK-mediated phosphorylation of PGC-1α stimulates mitochondrial biogenesis and subsequently improves metabolism of fatty acids (Patten and Arany 2012). FoxO family of proteins, in particular FoxO1 and FoxO3, are expressed in the diseased heart (Puthanveetil et al. 2013). In vitro studies have shown that these factors can be phosphorylated and activated by AMPK and regulate metabolism in various cell types, including cardiomyocytes (Sengupta et al. 2013; Dixit et al. 2008; Evans-Anderson et al. 2008). For example, they promote the transcription of CD36 and GLUT4 genes, which may improve fatty acid and glucose uptake into cardiomyocytes (Philip-Couderc et al. 2008). FoxO proteins also have an indirect impact on metabolism by stimulating mitochondrial biogenesis and fatty acid oxidation through an increase in PGC-1α (Armoni et al. 2006).

8.2.2 Protein Synthesis

Mammalian target of rapamycin (mTOR) is a central mediator of protein synthesis and cell growth (Tan and Miyamoto 2016; Sciarretta et al. 2014). Its inhibition has a critical relevance for cardiac hypertrophy and results from the AMPK-mediated phosphorylation of upstream factors, such as tuberous sclerosis complex 2 (TSC2) and Raptor (Inoki et al. 2003; Gwinn et al. 2008). mTOR inhibition leads to the suppression of p70 ribosomal S6 protein kinase (p70S6K) and 4E-binding protein-1 (4E-BP1), both involved in the initiation of protein synthesis and cell growth (Bertrand et al. 2008). In addition, AMPK may directly phosphorylate and activate eukaryotic elongation factor 2 kinase (eEF2K), thereby inhibiting the elongation phase of protein synthesis through eEF2 phosphorylation and inhibition (Horman et al. 2012). In vitro experiments have further shown that activation of AMPK by adenoviral infection blocks cardiac hypertrophy and the calcineurin-nuclear factor of activated T cells (NFAT), nuclear factor kappa B (NF-kB), and mitogen-activated protein kinase (MAPK) signal pathways (Chan et al. 2008; Li et al. 2007).

8.2.3 Cell Survival and Oxidative Stress

In addition to regulating metabolism, FoxO1 induction protects cultured cardiomyocytes against the oxidative stress response by mediating autophagy (Ronnebaum and Patterson 2010; Ning et al. 2015). Autophagy promotes cell survival by removing malfunctioning mitochondria and recycling substrates for cellular metabolism. AMPK also promotes autophagy by inhibiting mTOR, known to be a master regulator of this pathway (Meley et al. 2006; Kim et al. 2011a). Finally, AMPK can induce cardiomyocyte autophagy via direct phosphorylation of the Unc-51 like autophagy activating kinase 1 (ULK1), an inducer of autophagy (Kim et al. 2011a; Herrero-Martin et al. 2009).

Endoplasmic reticulum (ER) stress resulting in the accumulation of unfolded or misfolded proteins is another feature of cardiac injury (Kaufman 2002). ER stress induces apoptosis associated with transcriptional induction of C/EBP homologous protein (CHOP), activation of caspase 12, and c-Jun terminal kinase (JNK) pathways (Oyadomari and Mori 2004; Nakagawa et al. 2000; Urano et al. 2000). By inhibiting protein synthesis, intrinsic AMPK activation may indirectly protect the heart through an attenuation of ER stress and its subsequent cell death pathways.

AMPK has also an important action in regulating mitochondrial reactive oxygen species (ROS) production (Zaha et al. 2015). It has been suggested that AMPK is involved in the transcriptional regulation of anti-oxidative enzymes, including thioredoxin reductase 2, superoxide dismutase, glutathione peroxidase, and NAD (P)H dehydrogenase (Howden 2013). However, analysis of mitochondrial proteome indicates that AMPK does not alter the expression of anti-oxidative components in the heart and is more likely involved in the preservation of mitochondrial integrity during ischemia (Zaha et al. 2015). By preventing an excessive mitochondrial ROS production, AMPK reduces the subsequent JNK activation and thereby protects against mitochondrial permeability transition pore (mPTP) opening (Zaha et al. 2015; Paiva et al. 2011). The latter can disrupt mitochondrial function and is therefore a crucial event in triggering cell death during an ischemic episode (Lemasters 1999).

Finally, AMPK can regulate cytosolic ROS production in isolated cardiomyocytes by indirectly inhibiting NADPH oxidase (NOX) activity (Balteau et al. 2014). It can also be a regulator of NOX2/4 expression in the heart, limiting ROS production (Balteau et al. 2014).

8.2.4 Non-myocyte Targets

AMPK is expressed in cardiomyocytes but also in non-myocyte cell types, namely, fibroblasts (Hermida et al. 2013; Noppe et al. 2014), endothelial cells (Castanares-Zapatero et al. 2013), smooth muscle cells (Horman et al. 2008; Schneider et al. 2015), and mesenchymal stem cells (de Meester et al. 2014). In cardiac fibroblasts (CF), the transforming growth factor-β (TGF-β) signaling is in close association with AMPK (Noppe et al. 2014). TGF-β1, the predominant isoform in the myocardium, is the major determinant of the CF activation/differentiation following injury (Desmouliere et al. 1993). AMPK deletion leads to an inhibition of the TGF-β pathway and subsequent impaired maturation of myofibroblasts (Noppe et al. 2014). Conversely, a role for AMPK activation in the inhibition of the TGF-β-mediated differentiation of myofibroblasts has been described in several studies (Mishra et al. 2008; Wu et al. 2015). In endothelial cells, AMPK phosphorylates and activates the eNOS (Chen et al. 1999; Reihill et al. 2007). A mechanism involving the AMPK-dependent inhibition of the 26S proteasome through its O-GlcNAcylation has also been proposed to protect against endothelial dysfunction (Xu et al. 2012). In addition, activated AMPK preserves the endothelial barrier integrity by regulating the expression/phosphorylation of proteins involved in intercellular junctions (Castanares-Zapatero et al. 2013). Finally, in vascular smooth muscle, AMPK participates in the control of vascular tone through phosphorylation of myosin light chain kinase (MLCK) (Horman et al. 2012) and phospholamban (Schneider et al. 2015), a regulator of the sarcoplasmic/endoplasmic Ca(2+)-ATPase (SERCA).

8.3 AMPK in Ischemia–Reperfusion Injury

AMPK is activated during low-energy cellular states, such as myocardial ischemia (see Fig. 8.2 that summarizes cardioprotective actions of AMPK). It preserves the energetic equilibrium by promoting glucose uptake and glycolysis, while decreasing energy-consuming processes, through the various mechanisms described above. The beneficial role of AMPK activation in the ischemic heart has been clearly demonstrated in various mouse models where AMPK activation has been genetically impaired (Russell et al. 2004; Zarrinpashneh et al. 2006; Carvajal et al. 2007).

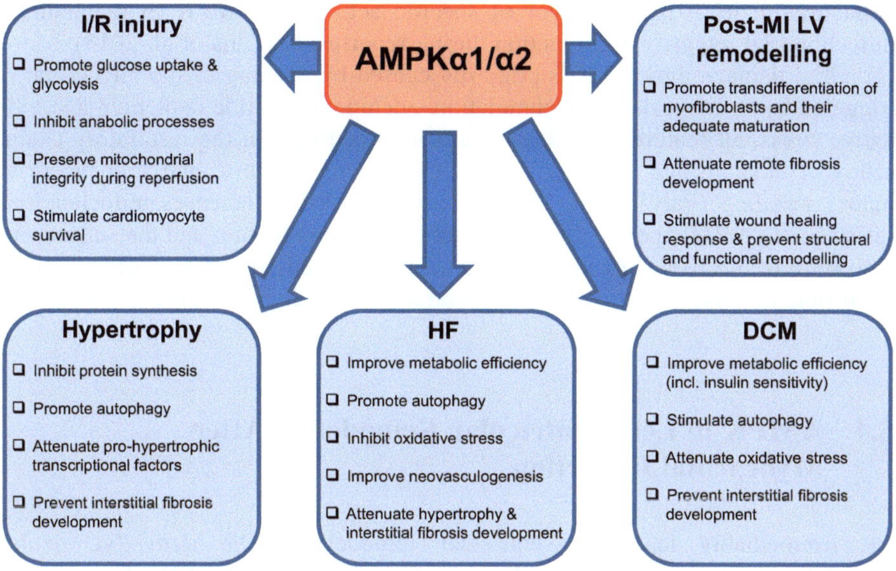

Fig. 8.2 *Main cardioprotective actions of AMPK*. See text for more details. *Abbreviations*: *AMPK* AMP-activated protein kinase, *DCM* diabetic cardiomyopathy, *HF* heart failure, *I/R* ischemia/reperfusion, *LV* left ventricle, *MI* myocardial infarction

However, when oxygen delivery is restored during reperfusion, the residual AMPK activation increases fatty acid oxidation (accounting for the majority of ATP production in the normal aerobic heart) by phosphorylating and inactivating ACC, the enzyme that catalyzes the formation of malonyl-CoA, an inhibitor of fatty acid import into mitochondria (Kudo et al. 1995). The effect of AMPK on fatty acid oxidation has raised concerns regarding its protective role during early reperfusion (Dyck and Lopaschuk 2006). Indeed, fatty acid oxidation may attenuate glucose oxidation via the Randle cycle, with a potential for increased oxygen demand (Lopaschuk 2008). In addition, it should promote a detrimental uncoupling of glycolysis and glucose oxidation during reperfusion and lead to acidosis (due to increased lactate and proton production) and detrimental calcium overload (due to the combination of sodium-proton exchange and sodium-calcium exchange) in the reperfused heart (Dyck and Lopaschuk 2006). Nevertheless, in mice lacking AMPKα2 or expressing a dominant-negative AMPK isoform, the myocardial recovery is blunted after ischemia and during ischemia–reperfusion (I/R) (Russell et al. 2004; Carvajal et al. 2007; Kim et al. 2011b). These studies support the notion that a beneficial effect of AMPK prevails over its putative deleterious action during reperfusion. In this context, it is not surprising that pharmacological agents targeting AMPK, such as metformin and the A-769662 compound, have cardioprotective effects when administered in various animal models of I/R (Kim et al. 2011b; Calvert et al. 2008; Foretz et al. 2014). A critical mechanism for explaining the protective action of AMPK has been proposed by Zaha and collaborators in 2016 and involves specific actions of AMPK in maintaining mitochondrial integrity and function during reperfusion (Zaha et al. 2015). Mitochondrial damage during I/R is primarily caused by opening of the mPTP that is triggered by matrix calcium accumulation, adenine nucleotide depletion, and oxidative stress, all features of I/R. It results in uncoupling of the respiratory chain, efflux of calcium and small proteins such as cytochrome c, and mitochondrial matrix swelling (Paiva et al. 2011). Activated AMPK increases mitochondrial resistance to mPTP opening by counteracting ROS production and the subsequent activation of the JNK signaling pathway (Zaha et al. 2015). These specific actions contribute to maintain mitochondrial respiratory capacity in hearts exposed to I/R.

8.4 AMPK in Left Ventricular Remodeling After Myocardial Infarction

I/R irremediably leads to ventricular remodeling. The term "ventricular remodeling" refers to structural changes of the left ventricle (LV) that occur in response to cardiac injury. Following myocardial infarction (MI), an inflammatory response is initiated and drives a fibrogenetic process resulting in the replacement of dead cardiomyocytes within the necrotic area by a stable fibrotic scar (reparative fibrosis) (Daskalopoulos et al. 2014). This initial response is protective and critical for wound healing and survival (adaptive remodeling). However, after several weeks, fibrosis expands to the non-infarcted myocardium (interstitial and perivascular reactive fibrosis) and can induce myocardial stiffness, resulting in diastolic dysfunction. It may even impact the entire ventricle, causing its dilatation, systolic dysfunction, and eventually leading to heart failure (HF) and death (Masci et al. 2014; Weber et al. 1989).

AMPK is a central player in cardiac metabolic adaptation during I/R by promoting, amongst other, cardiomyocytes survival (see Sect. 8.3). The protective role of the AMPKα2 catalytic isoform has been demonstrated in vivo using specific AMPKα2-inactivated mouse models of coronary occlusion (Carvajal et al. 2007; Wang et al. 2009). However, it should be highlighted that AMPKα1 in cardiomyocytes does not play a major role in limiting cell death during ischemia (Noppe et al. 2014). This concept is supported by the work of Noppe and colleagues showing that in AMPKα1-deficient mice subjected to a permanent coronary artery ligation, infarct size and cardiomyocyte apoptosis are not affected compared to wild-type mice. However, AMPKα1 plays a critical role during post-MI remodeling through the regulation of CF fibrotic properties (Noppe et al. 2014). CFs are pivotal for the maintenance of the extracellular matrix. They are also crucial in the reparative response to MI and are involved in the pathogenesis of cardiac remodeling (Daskalopoulos et al. 2012; Weber et al. 2013). They are activated into myofibroblasts (normally absent from the healthy myocardium) that repair necrotic areas, forming stable scars to preserve myocardial integrity and prevent cardiac rupture development. They express alpha-smooth muscle actin (α-SMA) microfilaments and rapidly mount healing responses that involve contracting granulation tissue deposition and collagen fibrillogenesis (Daskalopoulos et al. 2012). This transdifferentiation is induced by TGF-β, mechanical stress, and various other growth factors (Porter and Turner 2009). AMPKα1 deficiency has been associated with increased number of proliferative CFs in infarcted areas (Noppe et al. 2014). However, these fibroblasts exhibit a myodifferentiation defect reflected by a robust decline in the expression of α-SMA. Impaired myodifferentiation in AMPKα1-deficient mouse heart infarcts is associated with a downregulation of the TGF-β/p38 MAPK signaling. In addition, lysyl oxidase (LOX) expression, a critical factor for collagen cross-linking, is dramatically reduced in the infarct zone of AMPKα1-deficient mouse hearts. These changes result in compromised scar contractility, defective scar collagen maturation, and exacerbated LV dilatation 30 days after MI (Noppe et al. 2014). In agreement, Cieslik and collaborators have demonstrated that, in old infarcted mice, AMPK activation can increase LOX and α-SMA expression and enhance collagen content in the necrotic area, contributing to a better scar maturation (Cieslik et al. 2013). In addition, it decreases interstitial collagen deposition in the remote myocardium. Other groups have demonstrated this CF-suppressive effect of AMPK activation in various animal models (Hermida et al. 2013; Wu et al. 2015; Chen et al. 2014). Importantly, AMPK activation might also confer beneficial anti-fibrotic effects through its action on cardiomyocyte survival or hypertrophy. Additionally, an indirect effect of myocytes to fibrosis cannot be excluded, through paracrine signals or via intercellular junction proteins. This makes assigning AMPK-mediated effects solely on cardiomyocytes or CFs quite challenging. Further research using CF- or cardiomyocyte-specific AMPK-deficient mice will help understand more about the fibrotic-mediating actions of AMPK in the heart.

8.5 AMPK in Cardiac Hypertrophy

8.5.1 Cardiac Hypertrophy

Sustained mechanical stress of the myocardium, of variable etiology, ranging from myocardial ischemia to volume and/or pressure overload, is the key stimulus for the development of cardiac hypertrophy. Thus, cardiomyocyte hypertrophy is nothing else than a cellular response to increase load on the heart muscle. It has to be noted that hypertrophy initially serves an important adaptive role in order to preserve an adequate cardiac output. However, if the presence of the injurious stimulus is extended (e.g., chronic untreated hypertension), then hypertrophy becomes maladaptive and leads to remodeling and dysfunction that may even progress to HF (Frey and Olson 2003). There is a myriad of signaling pathways driving the cardiac hypertrophic response (Heineke and Molkentin 2006). Furthermore, cardiac hypertrophy is associated with drastic changes in the metabolic pathways of the myocyte, which can cause defects in cardiac energetics, also leading to cardiac dysfunction (Nascimben et al. 2004). AMPK is a principal regulator of the metabolic pathways of the cardiomyocyte (Horman et al. 2012) and so it has attracted great interest as a target in anti-hypertrophic strategies.

8.5.2 Role of AMPK in Cardiac Hypertrophy

The most characteristic attribute of hypertrophy is an augmentation of cardiomyocyte size, which eventually leads to a thickening of the LV walls (concentric hypertrophy). Cardiac metabolism dysregulation is a pivotal component of the hypertrophic response, and actually the metabolic dysfunction appears to be a crucial inhibitory factor regulating the structural and functional dysfunction of the heart muscle (Kundu et al. 2015). AMPK, being the "fuel gauge" of the cardiomyocyte, can thus play a paramount role in this complex and progressive process. At first, the finding that AMPK expression is induced in the hypertrophic heart was regarded as a sign of a noxious role of AMPK in cardiac hypertrophy (Tian et al. 2001), coming in complete disagreement with the theory for AMPK's cardioprotective action. It was later realized that the increase in AMPK activity in hypertrophy occurs lately and is due to the disturbances in energy status of the heart and is practically an adaptive response to the increased energy needs of the anabolic processes involved in hypertrophy (Horman et al. 2012). The most important upstream kinase controlling cardiac AMPK is LKB1. It has been shown that deletion (Ikeda et al. 2009) or inhibition (Dolinsky et al. 2009) of LKB1 in cardiomyocytes can have detrimental pro-hypertrophic effects and this is associated with a suppression of AMPK activation (Ikeda et al. 2009), while on the other hand stimulation of AMPK can reverse them. The observed effects were also found to be related to a modulation of protein translation process under the control of both mTOR/p70S6K (Ikeda et al. 2009; Dolinsky et al. 2009) and eEF2 (Ikeda et al. 2009) signaling pathways (discussed below). Quite recently, it was confirmed that any disturbance in the LKB1/AMPK activation can produce robust pro-hypertrophic effects. This includes the deletion of the C terminus of HSC70-interacting protein—a chaperone protein linked to LKB1—which leads to inhibited AMPKα activation and a dramatic hypertrophic response (Schisler et al. 2013). Metformin and losartan can activate AMPK and produce robust suppression of hypertrophy in isolated cardiomyocytes with Hernandez and colleagues also providing evidence for a role of mitochondria in this process (Hernandez et al. 2014). Similarly, chronic activation of AMPK after aortic banding confers cardioprotective effects by preventing the deleterious effects of hypertrophy (Li et al. 2007).

AMPK appears to hold an essential role in hypertrophy. Nevertheless, two burning questions emerge. How does it attain these cardioprotective effects and which processes and cross talking mechanisms are involved in the anti-hypertrophic effects of AMPK? The major prerequisites for the initiation and propagation of cardiac hypertrophy are cell growth, protein synthesis, transcription of hypertrophy-related genes, cytoskeleton amplification, and expanded organization of the sarcomere (Frey and Olson 2003), but also adequate energy levels to fuel these anabolic processes. All these elements have been linked to AMPK (Horman et al. 2012). First, signaling pathways modulating protein synthesis and involved in the cardiac hypertrophic response are influenced by AMPK. The eEF2 plays a prominent role in protein synthesis by assisting in the translocation of the ribosome during peptide-chain elongation. eEF2 is phosphorylated and then inactivated by AMPK, leading to a suppression of protein synthesis (Horman et al. 2002). There is strong evidence (Chan et al. 2008; Chan et al. 2004) proving that the anti-hypertrophic effect of pharmacological activation of AMPK is, at least partially, eEF2 mediated. The Akt/mTOR/p70S6K axis is another major regulator of protein synthesis (Zanchi and Lancha 2008), with p70S6K acting as the promoter of mRNA translation (Proud 1996). The role of mTOR/p70S6K pathway in the anti-hypertrophic action of AMPK has been established in vitro (Chan et al. 2004) but also in vivo. AMPKα2-deficient mice were dramatically affected by LV hypertrophy after pharmacologically induced pressure overload (using isoproterenol) (Zarrinpashneh et al. 2008) or following aortic banding (Zhang et al. 2008). In both cases, hypertrophy was enhanced by AMPKα2 deletion, and elevated levels of p70S6K were found to be the culprit for the induction of protein synthesis and cell growth. In addition, AMPK activation by metformin was able to suppress the Akt/mTOR axis and prevent hypertrophy following aortic banding, but only in mice with intact AMPKα2, signifying the decisive importance of this signaling pathway in the regulation of LV hypertrophy (Fu et al. 2011).

An essential role of autophagy has been established during compensatory hypertrophy, as well as in later stages, when hypertrophy becomes maladaptive and progresses to HF. Autophagy is generally considered to be a protective mechanism, promoting cell survival and cellular homeostasis. Autophagy is shown to be a key player for the protein quality control and the clearance of malfunctioning mitochondria, both crucial factors in the hypertrophic heart (Li et al. 2015). AMPK activation by 5-aminoimidazole-4-carboxamide-1-β-D-ribofuranoside (AICAR) or metformin confers mTOR-dependent pro-autophagic actions and suppresses cardiac hypertrophy following transverse aortic banding (Li et al. 2014). Furthermore, high-fat diet in adiponectin-deficient mice leads to an inhibition of AMPK activity and suppressed autophagy which possibly mediate hypertrophy and LV dysfunction development (Guo et al. 2013).

Transcriptional remodeling is another major point where AMPK can operate in order to confer anti-hypertrophic effects. The NFAT axis is a typical example (Chan et al. 2008). Calcineurin dephosphorylates NFAT; this leads to the translocation of the latter into the myocyte's nucleus, where it activates the transcription of a wide range of target genes that are associated with hypertrophy (Wilkins and Molkentin 2004). AICAR (Li et al. 2007) and resveratrol (Chan et al. 2008) have shown very promising results activating AMPK and thus preventing cardiomyocyte hypertrophy, most probably by indirectly preventing NFAT translocation into the nucleus. Furthermore, AMPK cross talks with other (but less extensively studied) factors, which play important roles in the transcriptional control of hypertrophy, namely, the extracellular signal-regulated kinases (ERK), the FoxO1/muscle RING-finger protein-1 (FoxO1/MuRF1) axis, and c-myc. ERK is known to be induced following pressure overload. Transgenic animals that possess attenuated Gq-signaling (including an inactivated ERK signaling pathway) show suppressed ERK activity and an inhibition of hypertrophy after transverse aortic constriction (Esposito et al. 2001). Adiponectin KO mice (showing depressed AMPK activity) demonstrate exacerbated hypertrophy—associated with increased ERK signaling—after pressure overload, but adenoviral adiponectin treatment reverses these effects (Shibata et al. 2004). MuRF1 has been suggested as a crucial player in regulating pro- and anti-hypertrophic signals in the cardiomyocytes. It is localized in various important areas of the myocyte (microtubules, sarcomere, around and inside the nucleus, etc.) and can antagonize hypertrophy in vitro (Arya et al. 2004). FoxO1, MuRF1, and AMPK are upregulated after cardiomyocytes are treated with AICAR, leading to an attenuation of hypertrophy, while silencing MuRF1 prevents the AICAR/AMPK-mediated effects (Chen et al. 2010). Lastly, c-myc and AMPK have been suggested to cooperate tightly in the control of metabolic functions and actions on mitochondria (Edmunds et al. 2015). Li and collaborators confirmed the importance of adiponectin and its receptors (AdipoR1/R2) in cardiac hypertrophy and provided surprising evidence for a complex axis involving Angiotensin II type 1 (AT1) receptors, oxidative stress, ERK, c-myc, and AMPK in the hypertrophic response (Li et al. 2013).

Various groups have focused on AMPK as a promising target for novel anti-hypertrophic strategies that could attenuate the development and further progression of hypertrophy. Treatment with the antidiabetic drug metformin attenuates hypertrophy following TAC (Fu et al. 2011; Zhang et al. 2011). It should be noted that the mode of action of metformin is still not completely understood. AMPKα2 is shown to be essential for the anti-hypertrophic actions of metformin (Fu et al. 2011); however two groups have challenged this theory (Xu et al. 2014; Yin et al. 2011). Thus, more research is urgently needed to delineate the roles and mode(s) of action of metformin in hypertrophy. The AMPK specific activator, AICAR, can attenuate hypertrophy and improve cardiac function in a rat transverse aortic constriction model (Li et al. 2007). Furthermore, adenoviral overexpression of adiponectin leads to induced AMPK activity and an attenuation of hypertrophy already 1 week after transverse aortic constriction. The adiponectin-mediated increase in AMPK activity demonstrates beneficial anti-hypertrophic effects in adiponectin-deficient animal model as well as in wild-type and *db/db* diabetic mice, manifesting the dexterity of AMPK's action (Shibata et al. 2004). Lastly, Dolinsky et al. provided robust evidence on the LKB1/AMPK-mediated cardioprotective effects of resveratrol in hypertensive rodents (Dolinsky et al. 2013), confirming earlier in vitro-generated results of the same group (Chan et al. 2008). This evidence provides a basis for the investigation of AMPK as a potential pharmacological target in order to halt/prevent cardiac hypertrophy and its pernicious repercussion, HF.

8.6 AMPK in the Transition From Cardiac Injury to Heart Failure

Similarly to MI, hypertrophy frequently leads irremediably to adverse cardiac remodeling and HF. The transition from hypertrophy (compensated dysfunction) to organ failure (decompensated dysfunction) is associated with several molecular signals and cascades that play essential roles (Lips 2003). Hypertrophy and HF are associated with energy depletion (Neubauer 2007), and this has sparked extensive interest in the actions of AMPK and its potential as a novel target to prevent HF development (Beauloye et al. 2011). The first report on the crucial role of AMPKα2 following pressure overload was given by Zhang et al. They showed that the deletion of AMPKα2 can have detrimental effects on fibrosis and cardiac function, aggravate hypertrophy, and promote HF development following aortic banding in mice (Zhang et al. 2008). Concerning MI, metformin decreases mortality in a mouse I/R model. These effects were associated with induction of the AMPK/eNOS/PGC-1α axis (Gundewar et al. 2009) and were lost in AMPKα2 KO mice, confirming the findings of Zhang et al. Lastly, Sasaki and colleagues demonstrated the cardioprotective effects of AMPK activation in the canine rapid-pacing model, which is a well-established model of HF. It has to be noted here that the dog-pacing model does not lead to hypertrophy; however, the HF-halting effects of treatment are undisputable. Metformin or AICAR treatment activated AMPK; suppressed fibrosis, apoptosis, and oxidative stress; and led to a halting of the progression of HF (Sasaki et al. 2009).

The pleiotropic cardioprotective actions of AMPK extend well beyond its effects on energy deficiency, protein synthesis, and cell growth. The transition of LV compensated hypertrophy to decompensated hypertrophy and HF is a multifactorial and complex process. It involves several factors that run parallel with hypertrophy, such as fibrosis, oxidative stress, autophagy, and angiogenesis.

AMPK has been established as an essential player in the regulation of fibrosis in various cardiovascular pathologies (see Sect. 8.5). Adiponectin deficiency demonstrates suppressed AMPK activity, which is associated with hypertrophy, interstitial fibrosis, and adverse LV remodeling after transverse aortic constriction (Shimano et al. 2010). Lastly, pioglitazone (a thiazolidinedione antidiabetic) treatment can affect AMPK levels and attenuate Angiotensin-II-mediated fibrosis and adverse cardiac remodeling (Li et al. 2010), underscoring the role of AMPK as a truly pleiotropic regulator (Daskalopoulos et al. 2016).

Oxidative stress is a well-established factor playing a noxious role in the development of hypertrophy and HF. ROS produced by NOX isoforms are implicated in hypertrophy development and its progression to failure (Murdoch et al. 2006). In vitro work with cardiomyocytes has shown that Angiotensin-II-mediated hypertrophy is driven by Nox2 (Nakagami et al. 2003). Furthermore, there is strong evidence that deletion of Nox2 in mice leads to dramatically suppressed hypertrophy following Angiotensin-II infusions (Bendall et al. 2002). AMPK activation has been known to confer important oxidative stress-mediator roles in the heart, and thus it can protect the myocardium from hypertrophy. The AMPK activator A-769662 suppresses ROS production and prevents cardiomyocyte cell death in an in vitro model of ischemia (Timmermans et al. 2014). AMPK activation can attenuate NOX2 in cardiomyocytes (AMPKα2 isoform) (Balteau et al. 2014), while metformin treatment has anti-hypertrophic effects on myocytes via an inductive effect on eNOS (Hernandez et al. 2014). Finally, exercise training can reduce NOX4 levels, limit ROS production, and suppress isoproterenol-induced hypertrophy but only in mice with intact AMPKα2 (Ma et al. 2015), signifying its importance for the hypertrophic process leading to HF.

A basal level of autophagy is paramount for the normal functioning of the cardiomyocyte, since it can regulate oxidative stress, cell death, energy sources, and protein quality control. Nevertheless, dysregulation can lead to hypertrophy and has been linked to HF development (Li et al. 2015), while restabilization of aberrant autophagy was shown to halt HF (Sun et al. 2013). AMPK has been suggested as a regulator of postischemic autophagy and cardiomyocyte survival by eradicating damaged (ROS producing) mitochondria (mitophagy) (Matsui et al. 2007; Takagi et al. 2007). ULK1 is an inducer of autophagy and can be suppressed by mTOR. By inhibiting mTOR and by phosphorylating and activating ULK1, AMPK induces autophagy, suppresses hypertrophy, and prevents cardiac dysfunction following pressure overload (Li et al. 2014; Lee et al. 2010).

Strong evidence suggests that angiogenesis plays important roles in the transition of hypertrophy to HF (Oka et al. 2014). As mentioned earlier, hypertrophy leads to increased cardiomyocyte mass and contractile function. This increases the demands in oxygen and nutrients, highlighting neovasculogenesis as a key player in the hypertrophic response. It is characteristic that reduced or disturbed capillary density (in contrast to normal blood vessel patterns) is associated with end-stage HF (Karch et al. 2005; Shiojima et al. 2005). It appears that AMPK and the vascular endothelial growth factor (VEGF), a key factor in the angiogenetic process (Isner and Losordo 1999), are closely linked, as AMPK can stimulate VEGF (Ouchi et al. 2005) and vice versa (Stahmann et al. 2010). Shimano and collaborators have provided proof that AMPK suppression leads to dysfunctional angiogenesis (Shimano et al. 2010). This in turn causes dramatically increased hypertrophy and speeds up the transition towards HF after pressure overload. The authors highlighted the effects on neovasculogenesis and VEGF expression and showed that activation of AMPK stimulates VEGF secretion in cardiomyocytes.

As a conclusion, it appears that AMPK plays a decisive role by halting the progression of cardiac hypertrophy to heart failure. The cardioprotective effects of AMPK might be conferred via actions that extend beyond its metabolic control, such as effects on neovasculogenesis, autophagy, oxidative stress, fibrosis, and others. Nevertheless, we have to shed more light into the exact effects of AMPK activation during the various stages of hypertrophy development and get an insight into which cardioprotective effects of AMPK play the most prominent role and what the optimal timing for intervention is. More research is urgently required in order to shed light into the mysteries of AMPK, its anti-hypertrophic effects, and its potential as a novel target in strategies against HF.

8.7 AMPK in Diabetic Cardiomyopathy

Diabetic cardiomyopathy (DCM) is a serious LV pathology that establishes in the setting of diabetes mellitus (DM), independently of other comorbidities such as coronary heart disease and hypertension. It is characterized by diastolic (at first) and systolic (which usually follows) LV dysfunction (Falcao-Pires and Leite-Moreira 2012), and without proper treatment it can progress to HF (Kannel et al. 1974). There is a wide range of factors involved in the progression of DCM including insulin resistance (Hintz and Ren 2002), hyperglycemia (which leads to further disturbances in free fatty acid utilization and calcium handling) (Falcao-Pires and Leite-Moreira 2012), hypertrophy (Levy et al. 1990), and fibrosis (Martinez et al. 2003).

Daniels and collaborators established an association between low AMPK activity, cardiac metabolic inefficiency, and reduced contractile ability in *db/db* mice (a model of type 2 DM) (Daniels et al. 2010). They postulated that DM affects AMPK activation, which in turn negatively affects the energetic capacity of the myocardium, leading to depressed contractile function. In OVE26 mice, a type 1 DM model, authors provided evidence for an autophagy-related role of AMPK (Xie et al. 2011). These mice demonstrate LV dysfunction, suppressed AMPK activity, and reduced autophagy compared to their wild-type littermates. They also showed that when DM is induced in mice with deficient AMPKα2, this has detrimental effects on autophagy levels, LV function, and mortality. Treatment with metformin attenuated the cardiac dysfunction only in wild-type but not in AMPKα2-deficient mice, indicating the crucial role of the latter in DCM development. Fibrosis is another important contributor in LV dysfunction and AMPK demonstrates anti-fibrotic effects in both diabetic mice (He et al. 2013) and rats (Lee et al. 2012). It has been also shown that AMPK activation improves insulin sensitivity of insulin-resistant cardiomyocytes (Bertrand et al. 2006; Ginion et al. 2011). Lastly, ROS production is a well-established etiology for the development of DCM (Seddon et al. 2007). Balteau et al. showed that AMPK activation in cardiomyocytes can suppress NOX2-mediated ROS and inhibit glucotoxicity (Balteau et al. 2014). Similar anti-oxidative stress actions conferred by AMPK activation have been shown in animal studies (Zhang et al. 2014).

References

An D, Pulinilkunnil T, Qi D, Ghosh S, Abrahani A, Rodrigues B (2005) The metabolic "switch" AMPK regulates cardiac heparin-releasable lipoprotein lipase. Am J Physiol Endocrinol Metab 288(1):E246–E253

Armoni M, Harel C, Karni S, Chen H, Bar-Yoseph F, Ver MR, Quon MJ, Karnieli E (2006) FOXO1 represses peroxisome proliferator-activated receptor-gamma1 and -gamma2 gene promoters in primary adipocytes. A novel paradigm to increase insulin sensitivity. J Biol Chem 281(29):19881–19891

Arya R, Kedar V, Hwang JR, McDonough H, Li HH, Taylor J, Patterson C (2004) Muscle ring finger protein-1 inhibits PKC{epsilon} activation and prevents cardiomyocyte hypertrophy. J Cell Biol 167(6):1147–1159

Balteau M, Van Steenbergen A, Timmermans AD, Dessy C, Behets-Wydemans G, Tajeddine N, Castanares-Zapatero D, Gilon P, Vanoverschelde JL, Horman S, Hue L, Bertrand L, Beauloye C (2014) AMPK activation by glucagon-like peptide-1 prevents NADPH oxidase activation induced by hyperglycemia in adult cardiomyocytes. Am J Physiol Heart Circ Physiol 307(8): H1120–H1133

Beauloye C, Bertrand L, Horman S, Hue L (2011) AMPK activation, a preventive therapeutic target in the transition from cardiac injury to heart failure. Cardiovasc Res 90(2):224–233

Bendall JK, Cave AC, Heymes C, Gall N, Shah AM (2002) Pivotal role of a gp91(phox)-containing NADPH oxidase in angiotensin II-induced cardiac hypertrophy in mice. Circulation 105(3):293–296

Bertrand L, Ginion A, Beauloye C, Hebert AD, Guigas B, Hue L, Vanoverschelde JL (2006) AMPK activation restores the stimulation of glucose uptake in an in vitro model of insulin-resistant cardiomyocytes via the activation of protein kinase B. Am J Physiol Heart Circ Physiol 291(1):H239–H250

Bertrand L, Horman S, Beauloye C, Vanoverschelde JL (2008) Insulin signalling in the heart. Cardiovasc Res 79(2):238–248

Calvert JW, Gundewar S, Jha S, Greer JJ, Bestermann WH, Tian R, Lefer DJ (2008) Acute metformin therapy confers cardioprotection against myocardial infarction via AMPK-eNOS-mediated signaling. Diabetes 57(3):696–705

Carvajal K, Zarrinpashneh E, Szarszoi O, Joubert F, Athea Y, Mateo P, Gillet B, Vaulont S, Viollet B, Bigard X, Bertrand L, Ventura-Clapier R, Hoerter JA (2007) Dual cardiac contractile effects of the alpha2-AMPK deletion in low-flow ischemia and reperfusion. Am J Physiol Heart Circ Physiol 292(6):H3136–H3147

Castanares-Zapatero D, Bouleti C, Sommereyns C, Gerber B, Lecut C, Mathivet T, Horckmans M, Communi D, Foretz M, Vanoverschelde JL, Germain S, Bertrand L, Laterre PF, Oury C, Viollet B, Horman S, Beauloye C (2013) Connection between cardiac vascular permeability, myocardial edema, and inflammation during sepsis: role of the alpha1AMP-activated protein kinase isoform. Crit Care Med 41(12):e411–e422

Chan AY, Soltys CL, Young ME, Proud CG, Dyck JR (2004) Activation of AMP-activated protein kinase inhibits protein synthesis associated with hypertrophy in the cardiac myocyte. J Biol Chem 279(31):32771–32779

Chan AY, Dolinsky VW, Soltys CL, Viollet B, Baksh S, Light PE, Dyck JR (2008) Resveratrol inhibits cardiac hypertrophy via AMP-activated protein kinase and Akt. J Biol Chem 283 (35):24194–24201

Chen ZP, Mitchelhill KI, Michell BJ, Stapleton D, Rodriguez-Crespo I, Witters LA, Power DA (1999) Ortiz de Montellano PR, Kemp BE. AMP-activated protein kinase phosphorylation of endothelial NO synthase. FEBS Lett 443(3):285–289

Chen BL, Ma YD, Meng RS, Xiong ZJ, Wang HN, Zeng JY, Liu C, Dong YG (2010) Activation of AMPK inhibits cardiomyocyte hypertrophy by modulating of the FOXO1/MuRF1 signaling pathway in vitro. Acta Pharmacol Sin 31(7):798–804

Chen KH, Hsu HH, Lee CC, Yen TH, Ko YC, Yang CW, Hung CC (2014) The AMPK agonist AICAR inhibits TGF-beta1 induced activation of kidney myofibroblasts. PLoS One 9(9), e106554

Cheung PC, Salt IP, Davies SP, Hardie DG, Carling D (2000) Characterization of AMP-activated protein kinase gamma-subunit isoforms and their role in AMP binding. Biochem J 346 (Pt 3):659–669

Cieslik KA, Taffet GE, Crawford JR, Trial J, Mejia Osuna P, Entman ML (2013) AICAR-dependent AMPK activation improves scar formation in the aged heart in a murine model of reperfused myocardial infarction. J Mol Cell Cardiol 63:26–36

Daniels A, van Bilsen M, Janssen BJ, Brouns AE, Cleutjens JP, Roemen TH, Schaart G, van der Velden J, van der Vusse GJ, van Nieuwenhoven FA (2010) Impaired cardiac functional reserve in type 2 diabetic db/db mice is associated with metabolic, but not structural, remodelling. Acta Physiol (Oxf) 200(1):11–22

Daskalopoulos EP, Janssen BJ, Blankesteijn WM (2012) Myofibroblasts in the infarct area: concepts and challenges. Microsc Microanal 18(1):35–49

Daskalopoulos EP, Hermans KCM, van Delft L, Altara R, Blankesteijn WM (2014) Inflammation in heart failure, 1st edn. Academic, New York

Daskalopoulos EP, Dufeys C, Bertrand L, Beauloye C, Horman S (2016) AMPK in cardiac fibrosis and repair: Actions beyond metabolic regulation. J Mol Cell Cardiol 91:188–200

de Meester C, Timmermans AD, Balteau M, Ginion A, Roelants V, Noppe G, Porporato PE, Sonveaux P, Viollet B, Sakamoto K, Feron O, Horman S, Vanoverschelde JL, Beauloye C, Bertrand L (2014) Role of AMP-activated protein kinase in regulating hypoxic survival and proliferation of mesenchymal stem cells. Cardiovasc Res 101(1):20–29

Desmouliere A, Geinoz A, Gabbiani F, Gabbiani G (1993) Transforming growth factor-beta 1 induces alpha-smooth muscle actin expression in granulation tissue myofibroblasts and in quiescent and growing cultured fibroblasts. J Cell Biol 122(1):103–111

Dixit M, Bess E, Fisslthaler B, Hartel FV, Noll T, Busse R, Fleming I (2008) Shear stress-induced activation of the AMP-activated protein kinase regulates FoxO1a and angiopoietin-2 in endothelial cells. Cardiovasc Res 77(1):160–168

Dolinsky VW, Chan AY, Robillard Frayne I, Light PE, Des Rosiers C, Dyck JR (2009) Resveratrol prevents the prohypertrophic effects of oxidative stress on LKB1. Circulation 119 (12):1643–1652

Dolinsky VW, Chakrabarti S, Pereira TJ, Oka T, Levasseur J, Beker D, Zordoky BN, Morton JS, Nagendran J, Lopaschuk GD, Davidge ST, Dyck JR (2013) Resveratrol prevents hypertension and cardiac hypertrophy in hypertensive rats and mice. Biochim Biophys Acta 1832 (10):1723–1733

Dyck JR, Lopaschuk GD (2006) AMPK alterations in cardiac physiology and pathology: enemy or ally? J Physiol 574(Pt 1):95–112

Edmunds LR, Sharma L, Wang H, Kang A, d'Souza S, Lu J, McLaughlin M, Dolezal JM, Gao X, Weintraub ST, Ding Y, Zeng X, Yates N, Prochownik EV (2015) c-Myc and AMPK Control Cellular Energy Levels by Cooperatively Regulating Mitochondrial Structure and Function. PLoS One 10(7), e0134049

Esposito G, Prasad SV, Rapacciuolo A, Mao L, Koch WJ, Rockman HA (2001) Cardiac overexpression of a G(q) inhibitor blocks induction of extracellular signal-regulated kinase and c-Jun NH(2)-terminal kinase activity in in vivo pressure overload. Circulation 103 (10):1453–1458

Evans-Anderson HJ, Alfieri CM, Yutzey KE (2008) Regulation of cardiomyocyte proliferation and myocardial growth during development by FOXO transcription factors. Circ Res 102 (6):686–694

Falcao-Pires I, Leite-Moreira AF (2012) Diabetic cardiomyopathy: understanding the molecular and cellular basis to progress in diagnosis and treatment. Heart Fail Rev 17(3):325–344

Foretz M, Guigas B, Bertrand L, Pollak M, Viollet B (2014) Metformin: from mechanisms of action to therapies. Cell Metab 20(6):953–966

Frey N, Olson EN (2003) Cardiac hypertrophy: the good, the bad, and the ugly. Annu Rev Physiol 65:45–79

Fu YN, Xiao H, Ma XW, Jiang SY, Xu M, Zhang YY (2011) Metformin attenuates pressure overload-induced cardiac hypertrophy via AMPK activation. Acta Pharmacol Sin 32 (7):879–887

Fukuda M (2011) TBC proteins: GAPs for mammalian small GTPase Rab? Biosci Rep 31 (3):159–168

Ginion A, Auquier J, Benton CR, Mouton C, Vanoverschelde JL, Hue L, Horman S, Beauloye C, Bertrand L (2011) Inhibition of the mTOR/p70S6K pathway is not involved in the insulin-sensitizing effect of AMPK on cardiac glucose uptake. Am J Physiol Heart Circ Physiol 301 (2):H469–H477

Gundewar S, Calvert JW, Jha S, Toedt-Pingel I, Ji SY, Nunez D, Ramachandran A, Anaya-Cisneros M, Tian R, Lefer DJ (2009) Activation of AMP-activated protein kinase by metformin improves left ventricular function and survival in heart failure. Circ Res 104(3):403–411

Guo R, Zhang Y, Turdi S, Ren J (2013) Adiponectin knockout accentuates high fat diet-induced obesity and cardiac dysfunction: role of autophagy. Biochim Biophys Acta 1832(8):1136–1148

Gwinn DM, Shackelford DB, Egan DF, Mihaylova MM, Mery A, Vasquez DS, Turk BE, Shaw RJ (2008) AMPK phosphorylation of raptor mediates a metabolic checkpoint. Mol Cell 30 (2):214–226

Hardie DG (2007) AMP-activated/SNF1 protein kinases: conserved guardians of cellular energy. Nat Rev Mol Cell Biol 8(10):774–785

He C, Zhu H, Li H, Zou MH, Xie Z (2013) Dissociation of Bcl-2-Beclin1 complex by activated AMPK enhances cardiac autophagy and protects against cardiomyocyte apoptosis in diabetes. Diabetes 62(4):1270–1281

Heineke J, Molkentin JD (2006) Regulation of cardiac hypertrophy by intracellular signalling pathways. Nat Rev Mol Cell Biol 7(8):589–600

Hermida N, Markl A, Hamelet J, Van Assche T, Vanderper A, Herijgers P, van Bilsen M, Hilfiker-Kleiner D, Noppe G, Beauloye C, Horman S, Balligand JL (2013) HMGCoA reductase inhibition reverses myocardial fibrosis and diastolic dysfunction through AMP-activated protein kinase activation in a mouse model of metabolic syndrome. Cardiovasc Res 99 (1):44–54

Hernandez JS, Barreto-Torres G, Kuznetsov AV, Khuchua Z, Javadov S (2014) Crosstalk between AMPK activation and angiotensin II-induced hypertrophy in cardiomyocytes: the role of mitochondria. J Cell Mol Med 18(4):709–720

Herrero-Martin G, Hoyer-Hansen M, Garcia-Garcia C, Fumarola C, Farkas T, Lopez-Rivas A, Jaattela M (2009) TAK1 activates AMPK-dependent cytoprotective autophagy in TRAIL-treated epithelial cells. EMBO J 28(6):677–685

Hintz KK, Ren J (2002) Prediabetic insulin resistance is not permissive to the development of cardiac resistance to insulin-like growth factor I in ventricular myocytes. Diabetes Res Clin Pract 55(2):89–98

Horman S, Browne G, Krause U, Patel J, Vertommen D, Bertrand L, Lavoinne A, Hue L, Proud C, Rider M (2002) Activation of AMP-activated protein kinase leads to the phosphorylation of elongation factor 2 and an inhibition of protein synthesis. Curr Biol 12(16):1419–1423

Horman S, Morel N, Vertommen D, Hussain N, Neumann D, Beauloye C, El Najjar N, Forcet C, Viollet B, Walsh MP, Hue L, Rider MH (2008) AMP-activated protein kinase phosphorylates and desensitizes smooth muscle myosin light chain kinase. J Biol Chem 283(27):18505–18512

Horman S, Beauloye C, Vanoverschelde JL, Bertrand L (2012) AMP-activated protein kinase in the control of cardiac metabolism and remodeling. Curr Heart Fail Rep 9(3):164–173

Howden R (2013) Nrf2 and cardiovascular defense. Oxid Med Cell Longev 2013:104308

Hue L, Rider MH (2007) The AMP-activated protein kinase: more than an energy sensor. Essays Biochem 43:121–137

Ikeda Y, Sato K, Pimentel DR, Sam F, Shaw RJ, Dyck JR, Walsh K (2009) Cardiac-specific deletion of LKB1 leads to hypertrophy and dysfunction. J Biol Chem 284(51):35839–35849

Inoki K, Zhu T, Guan KL (2003) TSC2 mediates cellular energy response to control cell growth and survival. Cell 115(5):577–590

Isner JM, Losordo DW (1999) Therapeutic angiogenesis for heart failure. Nat Med 5(5):491–492

Kannel WB, Hjortland M, Castelli WP (1974) Role of diabetes in congestive heart failure: the Framingham study. Am J Cardiol 34(1):29–34

Karch R, Neumann F, Ullrich R, Neumuller J, Podesser BK, Neumann M, Schreiner W (2005) The spatial pattern of coronary capillaries in patients with dilated, ischemic, or inflammatory cardiomyopathy. Cardiovasc Pathol 14(3):135–144

Kaufman RJ (2002) Orchestrating the unfolded protein response in health and disease. J Clin Invest 110(10):1389–1398

Kim J, Kundu M, Viollet B, Guan KL (2011a) AMPK and mTOR regulate autophagy through direct phosphorylation of Ulk1. Nat Cell Biol 13(2):132–141

Kim AS, Miller EJ, Wright TM, Li J, Qi D, Atsina K, Zaha V, Sakamoto K, Young LH (2011b) A small molecule AMPK activator protects the heart against ischemia-reperfusion injury. J Mol Cell Cardiol 51(1):24–32

Kramer HF, Witczak CA, Fujii N, Jessen N, Taylor EB, Arnolds DE, Sakamoto K, Hirshman MF, Goodyear LJ (2006) Distinct signals regulate AS160 phosphorylation in response to insulin, AICAR, and contraction in mouse skeletal muscle. Diabetes 55(7):2067–2076

Kubli DA, Gustafsson AB (2014) Cardiomyocyte health: adapting to metabolic changes through autophagy. Trends Endocrinol Metab 25(3):156–164

Kudo N, Barr AJ, Barr RL, Desai S, Lopaschuk GD (1995) High rates of fatty acid oxidation during reperfusion of ischemic hearts are associated with a decrease in malonyl-CoA levels due to an increase in 5’-AMP-activated protein kinase inhibition of acetyl-CoA carboxylase. J Biol Chem 270(29):17513–17520

Kudo N, Gillespie JG, Kung L, Witters LA, Schulz R, Clanachan AS, Lopaschuk GD (1996) Characterization of 5’AMP-activated protein kinase activity in the heart and its role in inhibiting acetyl-CoA carboxylase during reperfusion following ischemia. Biochim Biophys Acta 1301(1–2):67–75

Kundu BK, Zhong M, Sen S, Davogustto G, Keller SR, Taegtmeyer H (2015) Remodeling of glucose metabolism precedes pressure overload-induced left ventricular hypertrophy: review of a hypothesis. Cardiology 130(4):211–220

Lee JW, Park S, Takahashi Y, Wang HG (2010) The association of AMPK with ULK1 regulates autophagy. PLoS One 5(11), e15394

Lee JE, Yi CO, Jeon BT, Shin HJ, Kim SK, Jung TS, Choi JY, Roh GS (2012) Alpha-Lipoic acid attenuates cardiac fibrosis in Otsuka Long-Evans Tokushima Fatty rats. Cardiovasc Diabetol 11:111

Lemasters JJ (1999) The mitochondrial permeability transition and the calcium, oxygen and pH paradoxes: one paradox after another. Cardiovasc Res 44(3):470–473

Levy D, Garrison RJ, Savage DD, Kannel WB, Castelli WP (1990) Prognostic implications of echocardiographically determined left ventricular mass in the Framingham Heart Study. N Engl J Med 322(22):1561–1566

Li HL, Yin R, Chen D, Liu D, Wang D, Yang Q, Dong YG (2007) Long-term activation of adenosine monophosphate-activated protein kinase attenuates pressure-overload-induced cardiac hypertrophy. J Cell Biochem 100(5):1086–1099

Li P, Shibata R, Unno K, Shimano M, Furukawa M, Ohashi T, Cheng X, Nagata K, Ouchi N, Murohara T (2010) Evidence for the importance of adiponectin in the cardioprotective effects of pioglitazone. Hypertension 55(1):69–75

Li L, Zhang ZG, Lei H, Wang C, Wu LP, Wang JY, Fu FY, Zhu WG, Wu LL (2013) Angiotensin II reduces cardiac AdipoR1 expression through AT1 receptor/ROS/ERK1/2/c-Myc pathway. PLoS One 8(1), e49915

Li Y, Chen C, Yao F, Su Q, Liu D, Xue R, Dai G, Fang R, Zeng J, Chen Y, Huang H, Ma Y, Li W, Zhang L, Liu C, Dong Y (2014) AMPK inhibits cardiac hypertrophy by promoting autophagy via mTORC1. Arch Biochem Biophys 558:79–86

Li Z, Wang J, Yang X (2015) Functions of autophagy in pathological cardiac hypertrophy. Int J Biol Sci 11(6):672–678

Lips DJ (2003) deWindt LJ, van Kraaij DJ, Doevendans PA. Molecular determinants of myocardial hypertrophy and failure: alternative pathways for beneficial and maladaptive hypertrophy. Eur Heart J 24(10):883–896

Lopaschuk GD (2008) AMP-activated protein kinase control of energy metabolism in the ischemic heart. Int J Obes (Lond) 32(Suppl 4):S29–S35

Ma X, Fu Y, Xiao H, Song Y, Chen R, Shen J, An X, Shen Q, Li Z, Zhang Y (2015) Cardiac fibrosis alleviated by exercise training is AMPK-dependent. PLoS One 10(6), e0129971

Marsin AS, Bertrand L, Rider MH, Deprez J, Beauloye C, Vincent MF, Van den Berghe G, Carling D, Hue L (2000) Phosphorylation and activation of heart PFK-2 by AMPK has a role in the stimulation of glycolysis during ischaemia. Curr Biol 10(20):1247–1255

Martinez DA, Guhl DJ, Stanley WC, Vailas AC (2003) Extracellular matrix maturation in the left ventricle of normal and diabetic swine. Diabetes Res Clin Pract 59(1):1–9

Masci PG, Doulaptsis C, Bertella E, Del Torto A, Symons R, Pontone G, Barison A, Droogne W, Andreini D, Lorenzoni V, Gripari P, Mushtaq S, Emdin M, Bogaert J, Lombardi M (2014) Incremental prognostic value of myocardial fibrosis in patients with non-ischemic cardiomyopathy without congestive heart failure. Circ Heart Fail 7(3):448–456

Matsui Y, Takagi H, Qu X, Abdellatif M, Sakoda H, Asano T, Levine B, Sadoshima J (2007) Distinct roles of autophagy in the heart during ischemia and reperfusion: roles of AMP-activated protein kinase and Beclin 1 in mediating autophagy. Circ Res 100(6):914–922

Meley D, Bauvy C, Houben-Weerts JH, Dubbelhuis PF, Helmond MT, Codogno P, Meijer AJ (2006) AMP-activated protein kinase and the regulation of autophagic proteolysis. J Biol Chem 281(46):34870–34879

Mishra R, Cool BL, Laderoute KR, Foretz M, Viollet B, Simonson MS (2008) AMP-activated protein kinase inhibits transforming growth factor-beta-induced Smad3-dependent transcription and myofibroblast transdifferentiation. J Biol Chem 283(16):10461–10469

Murdoch CE, Zhang M, Cave AC, Shah AM (2006) NADPH oxidase-dependent redox signalling in cardiac hypertrophy, remodelling and failure. Cardiovasc Res 71(2):208–215

Nakagami H, Takemoto M, Liao JK (2003) NADPH oxidase-derived superoxide anion mediates angiotensin II-induced cardiac hypertrophy. J Mol Cell Cardiol 35(7):851–859

Nakagawa T, Zhu H, Morishima N, Li E, Xu J, Yankner BA, Yuan J (2000) Caspase-12 mediates endoplasmic-reticulum-specific apoptosis and cytotoxicity by amyloid-beta. Nature 403 (6765):98–103

Nascimben L, Ingwall JS, Lorell BH, Pinz I, Schultz V, Tornheim K, Tian R (2004) Mechanisms for increased glycolysis in the hypertrophied rat heart. Hypertension 44(5):662–667

Neubauer S (2007) The failing heart--an engine out of fuel. N Engl J Med 356(11):1140–1151

Ning Y, Li Z, Qiu Z (2015) FOXO1 silence aggravates oxidative stress-promoted apoptosis in cardiomyocytes by reducing autophagy. J Toxicol Sci 40(5):637–645

Noppe G, Dufeys C, Buchlin P, Marquet N, Castanares-Zapatero D, Balteau M, Hermida N, Bouzin C, Esfahani H, Viollet B, Bertrand L, Balligand JL, Vanoverschelde JL, Beauloye C, Horman S (2014) Reduced scar maturation and contractility lead to exaggerated left ventricular dilation after myocardial infarction in mice lacking AMPKalpha1. J Mol Cell Cardiol 74:32–43

Oka T, Akazawa H, Naito AT, Komuro I (2014) Angiogenesis and cardiac hypertrophy: maintenance of cardiac function and causative roles in heart failure. Circ Res 114(3):565–571

Ouchi N, Shibata R, Walsh K (2005) AMP-activated protein kinase signaling stimulates VEGF expression and angiogenesis in skeletal muscle. Circ Res 96(8):838–846

Oyadomari S, Mori M (2004) Roles of CHOP/GADD153 in endoplasmic reticulum stress. Cell Death Differ 11(4):381–389

Paiva MA, Rutter-Locher Z, Goncalves LM, Providencia LA, Davidson SM, Yellon DM, Mocanu MM (2011) Enhancing AMPK activation during ischemia protects the diabetic heart against reperfusion injury. Am J Physiol Heart Circ Physiol 300(6):H2123–H2134

Patten IS, Arany Z (2012) PGC-1 coactivators in the cardiovascular system. Trends Endocrinol Metab 23(2):90–97

Philip-Couderc P, Tavares NI, Roatti A, Lerch R, Montessuit C, Baertschi AJ (2008) Forkhead transcription factors coordinate expression of myocardial KATP channel subunits and energy metabolism. Circ Res 102(2):e20–e35

Porter KE, Turner NA (2009) Cardiac fibroblasts: at the heart of myocardial remodeling. Pharmacol Ther 123(2):255–278

Proud CG (1996) p70 S6 kinase: an enigma with variations. Trends Biochem Sci 21(5):181–185

Puthanveetil P, Wan A, Rodrigues B (2013) FoxO1 is crucial for sustaining cardiomyocyte metabolism and cell survival. Cardiovasc Res 97(3):393–403

Reihill JA, Ewart MA, Hardie DG, Salt IP (2007) AMP-activated protein kinase mediates VEGF-stimulated endothelial NO production. Biochem Biophys Res Commun 354(4):1084–1088

Ronnebaum SM, Patterson C (2010) The FoxO family in cardiac function and dysfunction. Annu Rev Physiol 72:81–94

Russell RR 3rd, Li J, Coven DL, Pypaert M, Zechner C, Palmeri M, Giordano FJ, Mu J, Birnbaum MJ, Young LH (2004) AMP-activated protein kinase mediates ischemic glucose uptake and prevents postischemic cardiac dysfunction, apoptosis, and injury. J Clin Invest 114(4):495–503

Sakamoto K, Zarrinpashneh E, Budas GR, Pouleur AC, Dutta A, Prescott AR, Vanoverschelde JL, Ashworth A, Jovanovic A, Alessi DR, Bertrand L (2006) Deficiency of LKB1 in heart prevents ischemia-mediated activation of AMPKalpha2 but not AMPKalpha1. Am J Physiol Endocrinol Metab 290(5):E780–E788

Samovski D, Su X, Xu Y, Abumrad NA, Stahl PD (2012) Insulin and AMPK regulate FA translocase/CD36 plasma membrane recruitment in cardiomyocytes via Rab GAP AS160 and Rab8a Rab GTPase. J Lipid Res 53(4):709–717

Sasaki H, Asanuma H, Fujita M, Takahama H, Wakeno M, Ito S, Ogai A, Asakura M, Kim J, Minamino T, Takashima S, Sanada S, Sugimachi M, Komamura K, Mochizuki N, Kitakaze M (2009) Metformin prevents progression of heart failure in dogs: role of AMP-activated protein kinase. Circulation 119(19):2568–2577

Schisler JC, Rubel CE, Zhang C, Lockyer P, Cyr DM, Patterson C (2013) CHIP protects against cardiac pressure overload through regulation of AMPK. J Clin Invest 123(8):3588–3599

Schneider H, Schubert KM, Blodow S, Kreutz CP, Erdogmus S, Wiedenmann M, Qiu J, Fey T, Ruth P, Lubomirov LT, Pfitzer G, Mederos YSM, Hardie DG, Gudermann T, Pohl U (2015) AMPK Dilates Resistance Arteries via Activation of SERCA and BKCa Channels in Smooth Muscle. Hypertension 66(1):108–116

Sciarretta S, Volpe M, Sadoshima J (2014) Mammalian target of rapamycin signaling in cardiac physiology and disease. Circ Res 114(3):549–564

Seddon M, Looi YH, Shah AM (2007) Oxidative stress and redox signalling in cardiac hypertrophy and heart failure. Heart 93(8):903–907

Sengupta A, Kalinichenko VV, Yutzey KE (2013) FoxO1 and FoxM1 transcription factors have antagonistic functions in neonatal cardiomyocyte cell-cycle withdrawal and IGF1 gene regulation. Circ Res 112(2):267–277

Shibata R, Ouchi N, Ito M, Kihara S, Shiojima I, Pimentel DR, Kumada M, Sato K, Schiekofer S, Ohashi K, Funahashi T, Colucci WS, Walsh K (2004) Adiponectin-mediated modulation of hypertrophic signals in the heart. Nat Med 10(12):1384–1389

Shimano M, Ouchi N, Shibata R, Ohashi K, Pimentel DR, Murohara T, Walsh K (2010) Adiponectin deficiency exacerbates cardiac dysfunction following pressure overload through disruption of an AMPK-dependent angiogenic response. J Mol Cell Cardiol 49(2):210–220

Shiojima I, Sato K, Izumiya Y, Schiekofer S, Ito M, Liao R, Colucci WS, Walsh K (2005) Disruption of coordinated cardiac hypertrophy and angiogenesis contributes to the transition to heart failure. J Clin Invest 115(8):2108–2118

Stahmann N, Woods A, Spengler K, Heslegrave A, Bauer R, Krause S, Viollet B, Carling D, Heller R (2010) Activation of AMP-activated protein kinase by vascular endothelial growth factor mediates endothelial angiogenesis independently of nitric-oxide synthase. J Biol Chem 285 (14):10638–10652

Sun M, Ouzounian M, de Couto G, Chen M, Yan R, Fukuoka M, Li G, Moon M, Liu Y, Gramolini A, Wells GJ, Liu PP (2013) Cathepsin-L ameliorates cardiac hypertrophy through activation of the autophagy-lysosomal dependent protein processing pathways. J Am Heart Assoc 2(2), e000191

Takagi H, Matsui Y, Hirotani S, Sakoda H, Asano T, Sadoshima J (2007) AMPK mediates autophagy during myocardial ischemia in vivo. Autophagy 3(4):405–407

Tan VP, Miyamoto S (2016) Nutrient-sensing mTORC1: Integration of metabolic and autophagic signals. J Mol Cell Cardiol

Tian R, Musi N, D'Agostino J, Hirshman MF, Goodyear LJ (2001) Increased adenosine monophosphate-activated protein kinase activity in rat hearts with pressure-overload hypertrophy. Circulation 104(14):1664–1669

Timmermans AD, Balteau M, Gelinas R, Renguet E, Ginion A, de Meester C, Sakamoto K, Balligand JL, Bontemps F, Vanoverschelde JL, Horman S, Beauloye C, Bertrand L (2014) A-769662 potentiates the effect of other AMP-activated protein kinase activators on cardiac glucose uptake. Am J Physiol Heart Circ Physiol 306(12):H1619–H1630

Treebak JT, Glund S, Deshmukh A, Klein DK, Long YC, Jensen TE, Jorgensen SB, Viollet B, Andersson L, Neumann D, Wallimann T, Richter EA, Chibalin AV, Zierath JR, Wojtaszewski JF (2006) AMPK-mediated AS160 phosphorylation in skeletal muscle is dependent on AMPK catalytic and regulatory subunits. Diabetes 55(7):2051–2058

Urano F, Wang X, Bertolotti A, Zhang Y, Chung P, Harding HP, Ron D (2000) Coupling of stress in the ER to activation of JNK protein kinases by transmembrane protein kinase IRE1. Science 287(5453):664–666

Wang Y, Gao E, Tao L, Lau WB, Yuan Y, Goldstein BJ, Lopez BL, Christopher TA, Tian R, Koch W, Ma XL (2009) AMP-activated protein kinase deficiency enhances myocardial ischemia/reperfusion injury but has minimal effect on the antioxidant/antinitrative protection of adiponectin. Circulation 119(6):835–844

Weber KT, Pick R, Jalil JE, Janicki JS, Carroll EP (1989) Patterns of myocardial fibrosis. J Mol Cell Cardiol 21(Suppl 5):121–131

Weber KT, Sun Y, Bhattacharya SK, Ahokas RA, Gerling IC (2013) Myofibroblast-mediated mechanisms of pathological remodelling of the heart. Nat Rev Cardiol 10(1):15–26

Weiss J, Hiltbrand B (1985) Functional compartmentation of glycolytic versus oxidative metabolism in isolated rabbit heart. J Clin Invest 75(2):436–447

Wilkins BJ, Molkentin JD (2004) Calcium-calcineurin signaling in the regulation of cardiac hypertrophy. Biochem Biophys Res Commun 322(4):1178–1191

Wu D, Lei H, Wang JY, Zhang CL, Feng H, Fu FY, Li L, Wu LL (2015) CTRP3 attenuates post-infarct cardiac fibrosis by targeting Smad3 activation and inhibiting myofibroblast differentiation. J Mol Med (Berl) 93(12):1311–1325

Xie Z, Lau K, Eby B, Lozano P, He C, Pennington B, Li H, Rathi S, Dong Y, Tian R, Kem D, Zou MH (2011) Improvement of cardiac functions by chronic metformin treatment is associated with enhanced cardiac autophagy in diabetic OVE26 mice. Diabetes 60(6):1770–1778

Xing Y, Musi N, Fujii N, Zou L, Luptak I, Hirshman MF, Goodyear LJ, Tian R (2003) Glucose metabolism and energy homeostasis in mouse hearts overexpressing dominant negative alpha2 subunit of AMP-activated protein kinase. J Biol Chem 278(31):28372–28377

Xu J, Wang S, Viollet B, Zou MH (2012) Regulation of the proteasome by AMPK in endothelial cells: the role of O-GlcNAc transferase (OGT). PLoS One 7(5), e36717

Xu X, Lu Z, Fassett J, Zhang P, Hu X, Liu X, Kwak D, Li J, Zhu G, Tao Y, Hou M, Wang H, Guo H, Viollet B, McFalls EO, Bache RJ, Chen Y (2014) Metformin protects against systolic overload-induced heart failure independent of AMP-activated protein kinase alpha2. Hypertension 63(4):723–728

Yang J, Holman GD (2005) Insulin and contraction stimulate exocytosis, but increased AMP-activated protein kinase activity resulting from oxidative metabolism stress slows endocytosis of GLUT4 in cardiomyocytes. J Biol Chem 280(6):4070–4078
Yin M, van der Horst IC, van Melle JP, Qian C, van Gilst WH, Sillje HH, de Boer RA (2011) Metformin improves cardiac function in a nondiabetic rat model of post-MI heart failure. Am J Physiol Heart Circ Physiol 301(2):H459–H468
Zaha VG, Qi D, Su KN, Palmeri M, Lee HY, Hu X, Wu X, Shulman GI, Rabinovitch PS, Russell RR 3rd, Young LH (2015) AMPK is critical for mitochondrial function during reperfusion after myocardial ischemia. J Mol Cell Cardiol 91:104–113
Zanchi NE, Lancha AH Jr (2008) Mechanical stimuli of skeletal muscle: implications on mTOR/p70s6k and protein synthesis. Eur J Appl Physiol 102(3):253–263
Zarrinpashneh E, Carjaval K, Beauloye C, Ginion A, Mateo P, Pouleur AC, Horman S, Vaulont S, Hoerter J, Viollet B, Hue L, Vanoverschelde JL, Bertrand L (2006) Role of the alpha2-isoform of AMP-activated protein kinase in the metabolic response of the heart to no-flow ischemia. Am J Physiol Heart Circ Physiol 291(6):H2875–H2883
Zarrinpashneh E, Beauloye C, Ginion A, Pouleur AC, Havaux X, Hue L, Viollet B, Vanoverschelde JL, Bertrand L (2008) AMPKalpha2 counteracts the development of cardiac hypertrophy induced by isoproterenol. Biochem Biophys Res Commun 376(4):677–681
Zhang P, Hu X, Xu X, Fassett J, Zhu G, Viollet B, Xu W, Wiczer B, Bernlohr DA, Bache RJ, Chen Y (2008) AMP activated protein kinase-alpha2 deficiency exacerbates pressure-overload-induced left ventricular hypertrophy and dysfunction in mice. Hypertension 52(5):918–924
Zhang CX, Pan SN, Meng RS, Peng CQ, Xiong ZJ, Chen BL, Chen GQ, Yao FJ, Chen YL, Ma YD, Dong YG (2011) Metformin attenuates ventricular hypertrophy by activating the AMP-activated protein kinase-endothelial nitric oxide synthase pathway in rats. Clin Exp Pharmacol Physiol 38(1):55–62
Zhang Z, Wang S, Zhou S, Yan X, Wang Y, Chen J, Mellen N, Kong M, Gu J, Tan Y, Zheng Y, Cai L (2014) Sulforaphane prevents the development of cardiomyopathy in type 2 diabetic mice probably by reversing oxidative stress-induced inhibition of LKB1/AMPK pathway. J Mol Cell Cardiol 77:42–52

Chapter 9
AMPK and Cancer

Zhiyu Wang, Neng Wang, Pengxi Liu, and Xiaoming Xie

Contents

Abstract This chapter focuses on the role of AMPK as a stress-response molecule with an emphasis on its duplex implication in carcinogenesis and cancer drug resistance. AMPK is closely correlated to the tumor-suppressive functions of LKB1 and P53, consequently modulating the activity of cellular survival signaling such as mTOR and Akt, leading to cell growth inhibition and cell cycle arrest. On

Z. Wang (✉) • P. Liu
Department of Mammary Disease, Guangdong Provincial Hospital of Chinese Medicine, The Second Clinical Collage of Guangzhou University of Chinese Medicine, Guangzhou, China
e-mail: wangzhiyu976@126.com

N. Wang • X. Xie
Department of Breast Oncology, Sun Yat-sen University Cancer Center, State Key Laboratory of Oncology in South China, Collaborative Innovation Center for Cancer Medicine, Guangzhou, Guangdong, China

M.D. Cordero, B. Viollet (eds.), *AMP-activated Protein Kinase*, Experientia Supplementum 107, DOI 10.1007/978-3-319-43589-3_9

the contrary, AMPK is tightly involved in cancer drug resistance via interacting with multiple known mechanisms of chemoresistance such as ABCG2 expression, autophagy induction, and cancer stem cells enrichment. Targeting AMPK has become a novel strategy for cancer prevention and treatment.

Keywords AMPK • Carcinogenesis • Cancer drug resistance • Stress-response • Cancer prevetion

9.1 Introduction

According to GLOBOCAN 2012, an estimated 14.1 million new cancer cases and 8.2 million cancer-related deaths occurred in 2012. Meanwhile, it is expected that the global cancer burden would grow to 21.4 million new cancer cases and 13.2 million cancer deaths by 2030 (Ferlay et al. 2012). Although great success have been made in the treatment of some cancer, including chronic myeloid leukemia, acute promyelocytic leukemia, breast cancer, Hodgkin's lymphomas, and testicular cancers, prevalence estimates for 2012 showed that only 32.6 million cancer patients survive in the previous five years (Ferlay et al. 2012). Therefore, a deeper investigation into molecular network underlying cancer initiation and development is urgently needed in the future.

Since the 1920s, it has been known that cancer cells possess the unique metabolic characteristics, which are distinguishable from benign cells. However, it was only in the past decade that the reprogramming of energy metabolism has been acknowledged as one of cancer hallmarks (Cantor and Sabatini 2012). Besides, cancer is a stress-related disease. During evolution from normal cells to malignant cells, cells must face various types of stress stimulus, such as oncogene activation, hypoxia condition, nutrient deprivation, and chemo- and radiotherapy. Under stress stimulation, cells have to maintain energy homeostasis to provide enough nutrients to support cancer survival and growth (Pazarentzos and Bivona 2015). One of the representative molecules linking cell metabolism and cancer is 5′-AMP-activated protein kinase (AMPK). AMPK governs overall cellular homeostasis by modulating the metabolism of sugars, fats, and proteins. Once activated, AMPK can promote catabolic processes to generate ATP, thereby assisting cells evading from death and resulting in drug resistance and metastasis. On the other hand, AMPK is also reported to be positively associated with tumor suppressor genes such as p53 and LKB1. Therefore, AMPK activation will lead to cell cycle arrest and tumor growth inhibition, thus playing a critical role in cancer prevention (Li et al. 2015). AMPK activator metformin is also reported to be effective in decreasing cancer risk. Meanwhile, emerging natural chemoprevention compounds are also found active in upregulating AMPK. Based on the current findings, AMPK may play a duplex role in cancer development. In the early phase, AMPK may function as a tumor suppressive molecule and its loss would lead to tumor initiation and growth, whereas in the later phase, AMPK activation would protect cancer cells from death-inducing events by maintaining intracellular homeostasis and finally lead to cancer drug resistance and metastasis.

Herein, we summarized AMPK signaling in cancer development and discussed its duplex role in carcinogenesis and drug resistance. Finally, we introduced drug discovery targeting AMPK. An improved understanding of AMPK molecular network will shed novel light in cancer prevention and treatment.

9.2 AMPK Signaling

AMPK has been widely found in plants, yeast, and mammalian cells and was the first kinase that was shown to be regulated by cellular energy state. In humans, AMPK is a heterotrimer consisting of a catalytic α (α1, α2) subunit and regulatory β (β1, β2) and γ (γ1, γ2, γ3) subunits and therefore could form at least 12 possible heterotrimeric combinations (Fig. 9.1) (Ahmed and Li 2007). AMPK is activated during situations in which the cellular level of ATP is decreased and the relative level of ADP/AMP is increased, such as stressful conditions induced by glucose deprivation, hypoxia, tissue ischemia, or muscle excise (Shirwany and Zou 2014a). During activation, AMP or ADP can directly bind to the γ subunit of AMPK and causes a conformational change that promotes phosphorylation of Thr-172 by upstream kinases while inhibiting dephosphorylation by phosphatases. Once activated, AMPK exerts stimulatory effects on ATP-generation events including glucose uptake, glycolysis, and fatty acid oxidation, while inhibits ATP-consuming cellular events, such as protein synthesis or fatty acid synthesis. (Luo et al. 2005; Hardie 2013; Xiao et al. 2011). Therefore, AMPK response to acute metabolic stress is important for maintaining cell viability and escaping from death-related crisis. In addition, chronic activation of AMPK not only brings metabolic adaptation but also can contribute to genetic changes that assist the occurrence of metabolic disorders such as diabetes, obesity, and cancer. (Hardie 2014).

AMPK activation by Thr172 phosphorylation is largely dependent on its upstream kinase LKB1, which is a tumor suppressor gene lost in the inherited cancer order Peutz–Jeghers syndrome (Shirwany and Zou 2014b). It has been demonstrated that in LKB1-deficient mice, the AMPK activity is greatly reduced (Sanli et al. 2014). Meanwhile, metformin or excise-induced AMPK activation was also blocked by LKB1 inhibition (Mihaylova and Shaw 2011). In addition to regulating AMPK phosphorylation, LKB1 was also demonstrated to be effective in phosphorylating another 12 kinases relating to AMPK, including MARKs (microtubule affinity regulating kinases; 1–4), SIKs (salt-inducible kinases; 1–3), BRSKs (brain-specific kinases; 1 and 2), and NUAKs (1 and 2) (Wan et al. 2014). Besides LKB1, Thr-172 can be also phosphorylated by the Ca2+/calmodulin-

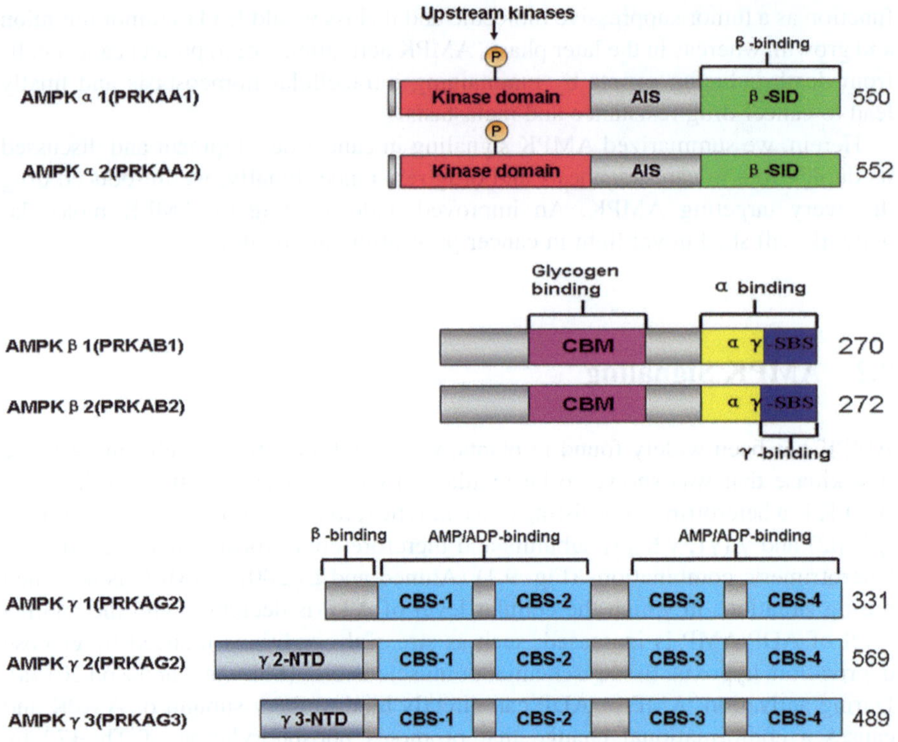

Fig. 9.1 Domain organization of AMPK subunits. AMPK is composed of a catalytic subunit (α) and two regulatory subunits (β and γ). The α subunit consists of an N-terminal kinase domain, an autoinhibitory sequence (AIS), and a β-subunit interacting domain (β-SID). The β subunit is N-terminally myristoylated and contains a carbohydrate-binding module (CBM) and C-terminal αγ subunit-binding sequence (SBS). The γ subunit contains four cystathione β-synthase (CBS) domains

dependent kinase kinases (CaMKKs); this mechanism might be responsible for AMPK activation upon agonists that increase intracellular Ca2+ such as alpha-lipoic acid or nicotine treatment (Hardie 2013; Lizcano et al. 2004). In addition, some studies also suggested that the MAPKKK (mitogen-activated protein kinase kinase kinase) family member TAK1 may be also involved in Thr-172 phosphorylation, but further in vivo validation is still needed (Sakamoto et al. 2005; Shaw et al. 2005).

The downstream enzymes inhibited by AMPK include mammalian homolog target of rapamycin (mTOR), acetyl-CoA carboxylase (ACC), fatty acid synthase (FAS), and glycerol phosphate acyltransferase (GPAT), which are key regulators of protein, fatty acid, and glycerolipid synthesis, respectively (Hawley et al. 2005). By inhibiting fat synthesis and promoting fat oxidation, AMPK activation would generate enough ATP and NADH for maintaining cell survival. In addition,

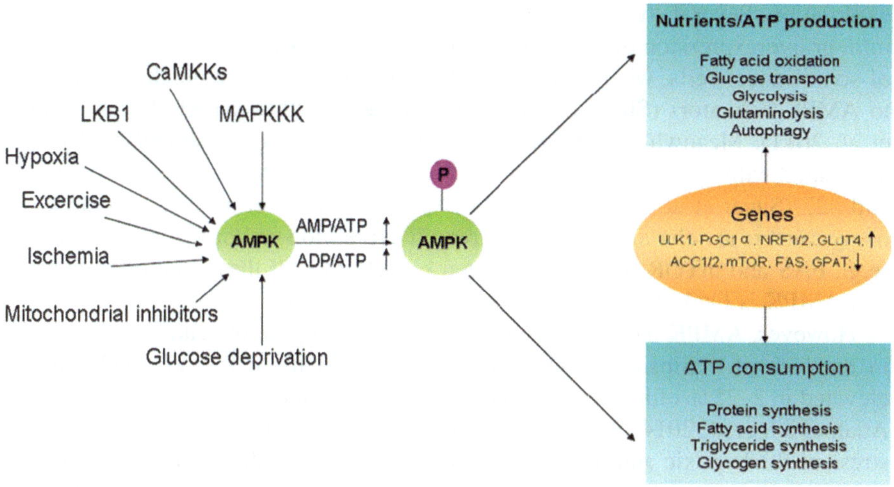

Fig. 9.2 AMPK activation and downstream cellular energy state. AMPK is activated when AMP and ADP levels in the cells rise due to a variety of upstream stresses such as glucose deprivation, hypoxia, or physical exercise. LKB1 is the upstream kinase activating AMPK by phosphorylating its Thr172 site, whereas CAMKK2 activates AMPK in response to calcium increase. AMPK activation leads to an increase in cellular process that generate ATP and a decrease in metabolism process that consume ATP. AMPK mediates these effects by either direct actions on enzymes or indirect regulation of gene expression listed in the *middle yellow box*

AMPK activation was found to promote mitochondrial biogenesis through directly interacting with PGC1α (PPAR gamma coactivator 1alpha) and NRF1/2 (nuclear respiratory factor 1 and 2), which are important transcriptors controlling genes encoding respiratory chain complexes (Hurley et al. 2005; Xie et al. 2006). Furthermore, it was found that AMPK activation could promote glucose uptake via accurately translocating GLUT4 from intracellular storage vesicles to the plasma membrane and chronically upregulating GLUT4 expression (Herrero-Martín et al. 2009; Hardie and Alessi 2013). Finally, AMPK activation could inhibit mTORC1 signaling via phosphorylating TSC2 or directly disrupting the combination between mTORC1 and Raptor. Since mTOR is critical in promoting protein synthesis and autophagy inhibition, mTOR suppression would be favorable for nutrients recycle to support cell survival under stressful conditions (Wan et al. 2014; Yu and Yang 2010). We summarized the AMPK signaling pathway in Fig. 9.2.

9.3 Duplex Role of AMPK in Cancer Development

Considering the tumor suppressor gene LKB1 served as the major upstream kinase controlling AMPK activation, a strong link between AMPK and cancer was introduced, and emerging studies indicated that this might be an intriguing area for exploration. Several studies suggested that AMPK activators, AICAR and metformin, or overexpression of either α or β subunits of AMPK could inhibit the growth of some cancer cells, whereas LKB1-deficient cancer cells are much less sensitive to AMPK activators (Shi et al. 2011; Nishino et al. 2004; Xu et al. 2012; Green et al. 2011). Meanwhile, AMPK activation was found to cause cell cycle arrest associated with stabilization of p53 and the cyclin-dependent kinase inhibitors p21 and p27 (Sauer et al. 2012; Kasznicki et al. 2014; Xiao et al. 2012). In addition, many cancers are characterized by increased expression and/or activity of enzymes (e.g., FAS and mTOR) that are inhibited by AMPK. All these findings suggested that AMPK is likely to exert tumor suppressor function.

However, AMPK has also recently been shown to promote cancer cell survival in the face of extrinsic and intrinsic stressors, and its α, β, and γ subunits are elevated in 2–25 % of human cancers and cancer cell lines (Shackelford et al. 2013; Adamovich et al. 2014; Short et al. 2010; Nam et al. 2008). Due to the oncogenic stress or the hypoxic and nutrient-deficient conditions in the tumor microenvironment, cancer cells may exhibit increased reliance on AMPK function. Loss of bioenergetics homeostasis leads to programmed cell death in cancer cells lacking either LKB1 or AMPK but not normal cells identically stressed (Buckendahl et al. 2011; Li et al. 2012). Meanwhile, numerous studies demonstrated that AMPK activation could trigger autophagy induction and protect cancer cells resistant to growth factor deprivation and anticancer therapy (Park et al. 2009). On the other hand, AMPK activation can inhibit fatty acid biosynthesis via suppressing acetyl-CoA carboxylase (ACC) that produces malonyl-CoA, an inhibitor of carnitine palmitoyltransferase 1 (CPT1) that governs the transfer of long-chain fatty acylcoA molecules from the cytosol into mitochondrion where they are oxidized (Kim et al. 2012). Besides, AMPK also induces upregulation of CPT1C in response to glucose starvation and hypoxia to facilitate fatty acid catabolism and ATP production (Inoki et al. 2003). In addition, accumulating NADPH is generated during the accelerated fatty acid oxidation and therefore providing cellular protection from oxidative stress damage. Furthermore, AMPK is required for the expression of mitochondrial respiratory chain complexes that support use of glutamine as energy source during metabolic reprogramming in some cancer cells. Thus, AMPK activation in cancer cells results in increased mitochondrial biogenesis, ATP production, and NADPH accumulation, supplying the energy source for rapid cancer growth and maintain cellular survival under stress condition.

Based on the above findings, AMPK may act as a duplex molecule in cancer development and progression depending on the degree of AMPK activation, the particular AMPK isoform, and other compensatory processes activated in cells. It is possible that modest activation of AMPK by slight stress may exert protective effects and result in oncogene-like activities, while excessive stress stimulus could induce AMPK exhibiting tumor suppressor function and lead to cancer cell death. Considering the chronic development of cancer drug resistance, AMPK might be one of the critical molecules making cancer cells adaptation to the long course treatment of chemo- or radiotherapy.

9.4 AMPK and Carcinogenesis

Carcinogenesis is a complex process originated from long-term interactions between host genome and harmful stresses including chemical, physical, or biological carcinogens. AMPK, as a stress-response molecule, is demonstrated to be a critical factor contributing to tumor formation. AMPK activation will promote catabolic pathways generating ATP while inhibiting ATP-consuming processes. The latter include most anabolic bioactivities such as synthesis of lipids, RNAs, and proteins; therefore, AMPK trigger will result in cell growth inhibition and provide a period for DNA damage repair. Besides, AMPK is closely correlated to the tumor suppressive functions of LKB1 and P53, consequently modulating the activity of cellular survival signaling such as mTOR, Akt, and Cox-2, leading to cell growth inhibition and cell cycle arrest (Li et al. 2015). A number of studies also demonstrated that AMPK was often downregulated in tumors by mechanisms other than somatic mutations. For example, the expression of AMPK-α subunit was significantly reduced in human breast cancer biopsies compared with surrounding normal tissue, indicating the tumor suppressive role of AMPK (Hadad et al. 2009). Meanwhile, AMPK activator metformin was revealed closely correlating with a decreased cancer incidence. Libby et al. demonstrated that cancer incidence ratio was only 7.3 % among 4085 metformin users, while the ratio reached as high as 11.6 % of 4085 controls ($P < 0.001$, HR 0.46), indicating a potential protective effect of metformin from cancer occurrence (Libby et al. 2009). More recently, a randomized pilot study also proved that low-dose metformin (250 mg/day) application suppressed formation of aberrant crypt foci, a representative marker for colon cancer, after only 1 month of administration (Hosono et al. 2010). What's more important, metformin application is associated with better outcomes after chemotherapy in various types of cancer (Li et al. 2009). All these findings indicated that AMPK may play tumor suppressive function in cancer development.

9.4.1 AMPK and Tumor Suppressor LKB1

The *LKB1* gene was first identified in 1996 and is broadly expressed in all fetal and adult tissues at different levels (Hemminki et al. 1998). Greater interest toward *LKB1* was initiated through identification of germ line mutation of *LKB1* in Peutz–Jeghers syndrome (PJS) where patients develop multiple hamartomatous polyps in the gastrointestinal tract and distinctive mucocutaneous pigmentation (Hemminki et al. 1997). What's more important, somatic *LKB1* mutations are also found in various types of cancer patients without PJS, such as non-small cell lung cancer, ovarian and breast cancer, cervical cancer, and pancreatic cancer (Giardiello et al. 2000). Meanwhile, PJS patients were also recognized to have an increased risk of cancer. The frequency of *LKB1* heterozygosity loss was higher in carcinomas than that in polyp lesions. In addition, several studies reported the biallelic loss of *LKB1* in liver, colorectal, breast, pancreas, and sex cord tumors (Kim et al. 2004; Connolly et al. 2000; Nakanishi et al. 2004; Sato et al. 2001). Therefore, *LKB1* is considered as a tumor suppressor gene. Although it had been clear for many years that phosphorylation of Thr172 in the activation loop of AMPK is required for AMPK activation, LKB1 was found to be its crucial upstream kinase in 2003, and LKB1 appears to be critically involved in the activation of AMPKα2 but not AMPKα1 (Hardie 2004). The discovery was further validated by failure of AMPK activation by AICAR (5-aminoimidazole 4-carboxamide riboside, a metformin analogue) in mammalian cells that lacked LKB1 expression (Hardie and Alessi 2013). Mice lacking LKB1 expression also exhibit greatly reduced AMPK activity, supporting the notion that AMPK may play a tumor-suppressive function following LKB1 activation (Carretero et al. 2007).

AMPK mediates the tumor-suppressive function of LKB1 via the mTOR signaling, which is a central regulator of cell survival, protein synthesis, and transcription (Guertin and Sabatini 2007). mTOR forms two functionally distinct complexes, mTORC1 and mTORC2. mTORC1 is nutrient sensitive and acutely inhibited by rapamycin. Additionally, it controls the translation of various types of cell growth regulators such as cyclin D1, hypoxia-inducible factor 1α, and c-Myc, which in turn promote cell cycle progression, cell growth, and angiogenesis (Shaw and Cantley 2006). In contrast, mTORC2 is neither sensitive to nutrients nor acutely inhibited by rapamycin. One mechanism of AMPK inhibiting mTORC1 is by direct phosphorylation of the tumor suppressor TSC2 on Ser 1387. However, in $TSC2^{-/-}$ mouse embryonic fibroblasts, AMPK activation still suppresses mTORC1, leading to the discovery that AMPK also directly phosphorylates the mTORC1 subunit Raptor on two conserved serines and subsequently blocks the signal transduction (Inoki et al. 2003; Gwinn et al. 2008). As well as cell growth, AMPK-mTORC1 axis can also regulate autophagy, a stress-response process within the cell. mTORC1 inhibition is sufficient to induce autophagy, which degrades intracellular organelles to provide nutrients for cell survival (Mao and Klionsky 2011). Therefore, AMPK activation could trigger autophagy efflux via inhibiting mTORC1. Interestingly, autophagy inhibition was frequently observed during cancer initiation, which might be closely correlated to a downregulated level of AMPK.

9.4.2 AMPK and p53

p53 is a tumor suppressor star that prevents cancer development by responding to a variety of cellular stresses, including DNA damage, oncogene activation, and stimulus controlling cell cycle and apoptosis. AMPK has been also shown to directly phosphorylate p53 on serine 15, which is the common phosphorylation site by DNA damage response kinases such as ATM and ATR (Jones et al. 2005). AMPK activation by AICAR administration or glucose deprivation could also result in p53 upregulation as well as its phosphorylation. Meanwhile, mouse embryonic fibroblasts (MEFs) bearing wild-type p53 are arrested at G1/S phase following AMPK activation, whereas ineffective in p53-deficient cell (Jones et al. 2005). It is also reported that overexpression of the mutated AMPKα1 subunit in prostate cancer cells could accelerate tumor growth, accompanying with decreased level of p53 (Zhou et al. 2009). On the other hand, several studies also indicated that AMPK is also a downstream signaling of p53. AMPKβ1 subunit is upregulated by p53 and two downstream targets of p53, sestrin 1 and sestrin 2, are implicated in the activation of AMPK and concordant inhibition of mTOR (Budanov and Karin 2008). These findings suggest the existence of a feedback loop between p53 and AMPK that enhance their suppressive functions.

Besides p53, AMPK has been also shown to phosphorylate several other molecules that regulate cell metabolism, growth, and transformation, such as FOXO3, p300 histone acetyltransferase, and the cell cycle inhibitor p27 (Yang et al. 2001; Greer et al. 2007; Liang et al. 2007). Of note is that AMPK phosphorylates FOXO3 and p27 at the same sites as Akt. It is still remain unclear that why these pro-growth and antigrowth signals would both target the same phosphorylation site; future studies are needed to explore the complex molecular networks underlying AMPK using more rigorously validated phosphor-specific antibodies and genetic knockout or knock-in mouse and cell models.

9.4.3 AMPK and Akt

The protein kinase Akt is hyperactivated in many tumors by gain-of-function mutations in PI3K or loss-of-function mutations in PTEN. Akt has been shown to be a negative regulator of AMPK and upstream positive regulator of mTOR. Both pathways involve direct phosphorylation of TSC2 at different sites. AMPK activates TSC2 and Akt inhibits TSC2, both leading to mTOR activation and subsequent increase in protein synthesis and other cellular processes. On the other hand, Akt is able to phosphorylate AMPKα1 at Ser 485 (Ser 487 in humans) which subsequently inhibits Thr 172 phosphorylation, thereby blocking AMPK activation (Antico Arciuch et al. 2013). In PTEN knockout mouse model, the Ser 485 phosphorylation of AMPK was increased and Thr 172 phosphorylation level decreased, accompanying with reduced level of total AMPK expression in the hyperplasia lesions of thyroid gland (Antico Arciuch et al. 2013). Consistent with this, the phosphorylation level of Ser 487 site of AMPK was also increased in several PTEN-deficient glioblastoma and breast cancer cell lines, and therefore it is more difficult to activate AMPK (Tao et al. 2010). However, several studies also reported that AMPK could induce Akt phosphorylation and activation, indicating that the cross talk between AMPK and Akt is bidirectional, yet the feedback regulation mechanisms still remain unknown.

9.5 AMPK and Cancer Drug Resistance

Upon drug treatment, cancer cells must activate its self-protection system to defend the hazardous stress and to survive through the battle. Therefore, stress-related proteins should be the first-line signaling to induce drug resistance. Actually, numerous stress-related proteins have already been demonstrated to be closely correlated to drug resistance, such as GRP78, GRP94, and the protein kinase ataxia-telangiectasia mutated (ATM) (Ahmed and Li 2007; Dong et al. 2005; Zhang and Wang 2009). As one of the key physiological energy sensors, AMPK is a major regulator of cellular and organismal energy homeostasis that coordinate multiple metabolic pathways to balance energy supply and ultimately modulate cellular and organ growth (Shirwany and Zou 2014a). Recent numerous studies also proved that AMPK is tightly involved in cancer drug resistance via interacting with multiple known mechanisms of chemoresistance; these include regulating ABCG2 expression, triggering autophagy induction, altering drug metabolism, and modulating CSCs (Luo et al. 2005; Hardie 2013). Targeting AMPK has become a novel approach to overcome cancer drug resistance.

9.5.1 AMPK and Cancer Metabolism

Cancer cells have a fundamentally different bioenergetic metabolism from that of nonneoplastic cells. In normal cells, energetic metabolism mostly relies upon the process of mitochondrial oxidative phosphorylation which consumes glucose and oxygen to produce energy. However, due to the mitochondrial gene mutations and the hypoxia microenvironment, cancer cells depend predominantly on glycolysis to produce energy regardless of the availability of oxygen, and this unique metabolism is known as the "Warburg Effect" (Kim and Dang 2006; Ferreira 2010). The accelerated aerobic glycolysis not only provides ATP and lactate for supporting rapid cancer cells growth but also activates the pentose phosphate pathway (PPP) to facilitate nucleic acid synthesis and support cell division (Jeon et al. 2012). More recently, mounting evidence supports the idea that Warburg effect is linked to drug resistance in cancer therapy. WZB117, an inhibitor of GLUT1 (glucose transporter 1), decreases glucose uptake and displays synergistic effects with cisplatin or paclitaxel to inhibit breast cancer growth (Liu et al. 2012). Meanwhile, the GLUT1 inhibitor phloretin significantly overcomes hypoxia-conferred drug resistance and sensitized P-glycoprotein overexpressed doxorubicin-resistant cells to daunorubicin (Cao et al. 2007). In addition, the hexokinase inhibitor 3-BrPA causes ATP depletion, decreases ABC transporter activity, and drug efflux, therefore increasing drug concentration in cells and resulting in preferential cell death in malignant cells (Hulleman et al. 2009). Pyruvate kinase (PK) is the last rate-limiting enzyme in the glycolytic pathway and catalyzes the conversion of phosphoenolpyruvate and ADP into pyruvate and ATP. There are four isoforms of PK in mammals, among which PKM2 is overexpressed in cancer cells. Increased PKM2 expression was observed in sera and tissues from colorectal cancer patients with poor response to 5-FU, whereas PKM2 silencing improves the therapeutic efficacy of cisplatin by increasing cancer cells apoptosis and inhibiting proliferation (Shi et al. 2010). Lactate dehydrogenase A (LDHA) is responsible for transforming pyruvate and NADH to lactate and NAD^+ and has a critical role in tumor maintenance. LDHA silencing in cancer cells leads to increased mitochondrial respiration, downregulated cellular viability, and suppressed tumorigenicity (Wang et al. 2012). It was found that LDH-A expression and activity is higher in taxol-resistant breast cancer cells than in taxol-sensitive cells, and LDH-A inhibition by siRNA or FX11 induces significant cell death via oxidative stress burst (Zhou et al. 2010; Le et al. 2010). Furthermore, LDH-A expression was closely linked to trastuzumab resistance; combination of trastuzumab and LDH-A inhibitor oxamate was found to synergistically inhibit both trastuzumab-sensitive and trastuzumab-resistant cancer growth in vitro and in vivo (Zhao et al. 2011). Therefore, targeting aerobic glycolysis has become a novel strategy to reverse cancer drug resistance.

Intriguingly, AMPK activation was recently shown to mediate the Warburg effect in cancer cells. Starvation of HeLa cells induced apoptosis, accompanying with ROS production and AMPK phosphorylation. Meanwhile, the intracellular level of lactate and pyruvate increased, while the mitochondrial oxygen consumption decreased, indicating that the Warburg effect was induced (Wu 1833). In addition, AMPK was found to regulate glycolysis by phosphorylating PFKFB3 (6-phosphofructo-2-kinase/fructose-2,6-biphosphatase 3), which increases the levels of fructose-2,6-bisphosphate and thereafter activates the rate-limiting glycolysis enzyme PFK1 (Mendoza et al. 2012). Another study further reported that AMPK knockdown in orthotopic tumors prepared from MDA-MB-231 cells exhibited suppressed cancer growth and decreased lactate production, indicating that AMPK might contribute to the core glucose metabolism of aggressive tumors (Mouradian et al. 2014). What's more important, it was found that loss of tumor suppressor gene FLCN (folliculin) constitutively activates AMPK, resulting in PGC-1α-mediated mitochondrial biogenesis and increased ROS production, which thereafter activates HIF transcription and drives Warburg metabolic reprogramming (Yan et al. 2014). All these findings suggested that targeting AMPK might inhibit cancer Warburg effect and therefore increase tumor chemosensitivity.

9.5.2 AMPK and Cancer Stem Cells

In recent years, CSCs are emerging as the ultimate reason determining cancer chemosensitivity. CSCs theory holds that cancer is initiated from or driven by a small population of stem-like cells that are capable of continuous self-renewal and differentiation. The hypothesis was supported by increasing evidence of CSCs existence in various types of malignancies. Meanwhile, more and more evidences indicated that CSCs are relatively resistant to both radiation and chemotherapy (Dean 2009). Only 0.5 ± 0.3 % of tumor cells before chemotherapy, but 5.9 ± 1.7 % of cells obtained after chemotherapy, were found to form mammospheres (Yu et al. 2007). Similarly, the proportion of $CD44^{+}/CD24^{-/low}$ cells was 9.5-fold higher in tumor samples after chemotherapy (Zhang and Wang 2009). In addition, a higher percentage of stem-like cancer cells at baseline were also revealed with a lower pathologic complete response to chemotherapy. Neumeister et al. demonstrated that breast cancer patients with high CSC populations showed worse outcome ($P = 0.0003$) independent of tumor size, histological grade, nodal status, hormone receptors, and HER2 status (Neumeister et al. 2010). Some studies indicated that both circulating tumor cells and disseminated tumor cells expressed CSCs phenotype (Książkiewicz et al. 2012). Mechanism exploration revealed that CSCs express high levels of ABC transporters. In fact, the ability of CSCs to actively exclude the Hochest 33342 dye has been exploited to facilitate their isolation and purification (Tirino et al. 2013). Meanwhile, drug-resistant cancer cells were also found to possess stem-like properties and a molecular portrait dominated by epithelial-to-mesenchymal (EMT)-related and self-renewal-related genes (Xin et al. 2013). In addition, due to the relatively quiescent characteristics of CSCs, they have strong ability to repair the damaged genes and restart the cell cycle to reestablish the damaged tissue after chemo- or radiotherapy. Combination of chemotherapy with Chk1 inhibitors significantly enhances the chemosensitivity of lung CSCs and limits their tumorigenicity (Venkatesha et al. 2012). Moreover, the consistent activation of Wnt/β-catenin, Notch and Hedgehog pathways were also demonstrated to be correlated with the drug-resistant capacity of CSCs, and inhibitors targeting these aberrant signaling pathways greatly induced cell death of CSCs (Maugeri-Saccà et al. 2011). What's more, recent findings also suggested that the hypoxic and low pH condition in the tumor microenvironment also contribute to the drug-resistant property of CSCs (Geng et al. 2014). Overall, several mechanisms underlying the therapeutic resistance of CSCs have been suggested, including (i) stem cells remain quiescent, making them resistant to cell-cycle active chemodrugs; (ii) high DNA repair ability; (iii) high expression of drug efflux proteins such as ABCG2; and (iv) most current therapies do not target the signaling pathways that regulate CSCs self-renewal. Therefore, further elucidation of the aberrant signaling pathways underlying the chemoresistant property of CSCs may provide an efficient way to increase cancer chemosensitivity and finally eliminate cancer cells.

More recently, emerging findings also suggested a potential link between AMPK signaling and the chemoresistance of cancer stem cells. Ye et al. reported that radioresistant glioblastoma CSCs exhibited increased fatty acid oxidation-dependent ATP maintenance, elevated ROS, and increased AMPK activity (Sato et al. 2012). Meanwhile, another study also found that AMPK is critical in mediating mammosphere formation. AMPK could directly phosphorylate PEA15 at ser ^{116} residue, which promotes its antiapoptotic functions, whereas inhibition or knockdown of AMPK impairs the mammosphere formation (Hindupur et al. 2014). In addition, the two catalytic subunits of AMPK, *Prkaa1* and *Prkaa2*, are more highly expressed in hematopoietic stem cells populations compared with unfractionated whole bone marrow, and AMPK inhibition significantly reduces the population of hematopoietic stem cells (Corominas-Faja et al. 2013). All these findings suggested that AMPK might be critical in modulating the self-renewal and chemoresistance of CSCs. However, several studies also claimed that AMPK activator metformin could increase the chemosensitivity and limit the population of CSCs in breast and prostate cancers (Kim et al. 2014). The phenomenon might be attributed to the overactivation of AMPK and may result in the excessive catabolic metabolism and unavoidable cell death. In addition, it was possible that the chronic stressful conditions in the CSCs niche, such as the low oxygen concentration, inflammation infiltration and stromal cells, were not taken into consideration in the in vitro and in vivo experiments. Therefore, further investigation is still needed to elucidate the precise effects of AMPK on modulating CSCs functions.

9.5.3 AMPK and Autophagy

Autophagy is now emerging as a crucial player in response to metabolic and therapeutic stresses, which attempts to maintain cellular homeostasis through eliminating excessive or unnecessary proteins and damaged organelles via lysosome degradation. Although autophagy was implicated in suppressing carcinogenesis, increasing findings indicated that autophagy may protect cancer cells from anticancer therapies. Emerging studies observed that clinical cancer therapeutics, as well as experimental anticancer strategies, including cytotoxic chemotherapy, radiation therapy, and endocrine or monoclonal antibody therapies, may induce autophagy in cell culture and animal models (Sui et al. 2013). Meanwhile, it was found that autophagy activity was significantly elevated in drug-resistant cancer cells. Autophagy induction by inhibiting aurora kinase A was revealed to trigger drug resistance in breast cancer cells, whereas genetic or pharmacological inhibition of autophagy could enhance the therapeutic efficacy of chemo- or radiotherapy and resensitize chemoresistant cancer cells (Zou et al. 2012; Levy et al. 2014; Chang et al. 2014). Several mechanisms have been put forward to explain the modulating effects of autophagy on cancer drug resistance. First, autophagy-mediated degradation and recycling could provide key nutrients to meet the metabolic requirements for cancer cells' survival and progression. In addition, some study reported that autophagy inhibition could induce a significant downregulation of ABC transporters (Wang et al. 2014). Finally, autophagy induction seems to be essential for cancer cells to maintain genomic stability and organelle homeostasis. A previous study has shown that autophagy induction contributed to sustained survival of breast cancer cells through increased DNA repair regulated by ATM-mediated activation of DNA-PKcs and PARP-1 (Yoon et al. 2012). Therefore, elucidating the molecular mechanisms involved in autophagy regulation and development of targeting agents has become a novel strategy to overcome cancer drug resistance.

Autophagy is a complicated regulatory process. Genetic screens in yeast have identified a large family of core autophagy-related genes (ATG), such as Atg1, Atg4, LC3/Atg8, and BECN1. Nevertheless, no matter in normal or cancer cells, the mTOR serves as the main regulator of autophagy. Under nutrient deprivation, hypoxia, genomic instability, or other stressful conditions, mTOR suppression triggers the autophagic cascade and inhibits the cell proliferation. Mechanisms study revealed that mTORC1 activation can negatively regulate autophagy by phosphorylating a complex of autophagy proteins such as the Unc51-like kinases (ULK1/2), which interfere with the formation of autophagosomes (Chan 2009). As a central sensor of cellular nutrient status or energy levels, AMPK is also involved in autophagy regulation and served as one of the upstream regulators of mTORC1. In particular, AMPK activation can subsequently trigger tuberous sclerosis protein 2 (TSC2) to repress mTORC1 and upregulate autophagy (Tripathi et al. 2013). Meanwhile, recent several study demonstrated that AMPK could directly phosphorylate multiple sites in ULK1 (S317, S467, S555, T575, S637, and S777) and promote ULK1 function in autophagy (Zhao and Klionsky 2011). ULK1 cannot be activated when AMPK-knockout MEFs (mouse embryonic fibroblasts) are subjected to glucose deprivation, and ULK1 mutant cells are also defective in autophagy induction following AMPK activation (Löffler et al. 2011). These observations demonstrated the functional importance of ULK1 phosphorylation by AMPK in autophagy induction. In addition, a recent report revealed that mTORC1 can directly phosphorylate ULK1 on S757, which is located in the AMPK binding motif (711-828) on ULK1, indicating that mTORC1 may directly inhibit the interaction between ULK1 and AMPK (Kim et al. 2011a). Therefore, targeting AMPK-mediated autophagy has become a novel direction to reverse cancer drug resistance. AMPK activity was found significantly elevated after cisplatin administration in human gastric carcinoma cell line AGS and colon carcinoma cells HCT116. AMPK inhibition resulted in a remarkable increase in cisplatin-induced apoptosis (Kim et al. 2008). Meanwhile, another study also reported that AMPK inhibition in MLL-rearranged cell lines synergistically enhanced the antiproliferative effects of vincristine, daunorubicin, cytarabine, dexamethasone and L-asparaginase in most of the evaluated conditions (Accordi et al. 2013). Although molecular elucidation of AMPK in mediating cancer drug resistance is only at its initial stage, AMPK has become an important drug discovery target to increase cancer chemosensitivity.

9.6 Drug Discovery Targeting AMPK

9.6.1 AMPK Inhibitors

Compound C is a cell-permeable pyrazolopyrimidine compound that can act as a reversible and ATP competitive inhibitor of AMPK. Structural elucidation revealed that compound C could dramatically bind with the phosphorylated-state mimic T172D mutant kinase domain from the human AMPKα2 subunit (Handa et al. 2011). Increasing studies also demonstrated that compound C could induce apoptotic cell death in myeloma, breast cancer cells, and glioma cells. However, several studies indicated that the cell death-inducing effects of Compound C seemed to be independent of AMPK activity.

For example, Compound C-mediated apoptosis in cancer cells was found associating with the decreased expression of Bcl-2 and Bcl-xl, accompanying with overexpression of P53. Notably, AMPK inhibition by siRNA had no effects on compound C-induced apoptosis, indicating that the apoptosis-inducing effects of compound C was AMPK independent (Dai et al. 2013). Meanwhile, another study even demonstrated that compound C could induce autophagic response in cancer cells via AMPK inhibition-independent downregulation of Akt/mTOR pathway (Vucicevic et al. 2011). Furthermore, Compound C does not inhibit AMPK activation in response to all stimuli. Several studies revealed that this pharmacological inhibitor blunted the AICAR-induced but not the dinitrophenol-induced or the LPS-induced activation of AMPK (Vucicevic et al. 2009; Kim et al. 2011b). Lastly, Compound C appears to inhibit a number of other protein kinases with lower IC50 values than AMPK, indicating that this compound could certainly have "off-target" effects. Therefore, these findings warrant caution when using compound C to inhibit AMPK-dependent cellular response, and further exploration of compound C in AMPK inhibition and anticancer effects is still needed.

NUAK1 and NUAK2 are members of the AMPK family of protein kinases that are activated by LKB1. Recent work suggested that they play critical roles in regulating key biological processes, including Myc-driven tumorigenesis, senescence, cell adhesion, and neuronal polarity. Through kinase profiling assay, it was found that WZ4003 inhibits NUAK1 with an IC50 of 20 nM and NUAK2 with an IC50 of 100 nM, while it has little effects on other 139 kinds of kinases, indicating that WZ4003 might be a highly specific NUAK1 inhibitor (Banerjee et al. 2014). Biofunction assays also demonstrated that WZ4003 could significantly inhibit the proliferation of MEFs and invasion ability of U2OS cells to the same extent as NUAK1 knockout or shRNA knockdown (Owen et al. 2000). Further study is still needed to elucidate the effects of WZ4003 on autophagy initiating process and chemosensitivity modulation.

9.6.2 AMPK Activators

Metformin, the most widely prescribed Type 2 diabetes drug, has been shown to activate AMPK since 2001. The first clue to the underlying molecular mechanisms came in 2000, when it was shown that metformin and other biguanides could inhibit Complex I of the mitochondrial respiratory chain and therefore leading to an increase in the ADP:ATP and AMP:ATP ratios and thus activate AMPK directly (Ouyang et al. 2011). Meanwhile, it was also proposed that metformin might act by inhibiting AMP deaminase, an enzyme that degrades cellular AMP (Ouyang et al. 2011). Interestingly, a series of studies also presented that metformin administration could significantly inhibit cancer growth and increase cancer chemosensitivity in vitro and in vivo (Snima et al. 2014). Anisimov et al. was the first to discover that metformin was able to inhibit the incidence and size and increase mean latency to development of mammary adenocarcinomas in transgenic HER-2/neu-positive mice (Anisimov et al. 2005). Thereafter, metformin was found to inhibit translation initiation in breast cancer cell lines by decreasing phosphorylation of S6 kinase, ribosomal protein S6, and eIF4E-binding protein 1, which was greatly associated with mTOR inhibition following AMPK activation (Lee et al. 2014). Some studies also proposed that metformin could attenuate the acquired resistance to lapatinib for HER-2 positive cancer through suppressing mTOR activity (Shell et al. 2008). Since mTOR plays a central role in regulating cell growth by controlling mRNA translation, ribosome biogenesis, and metabolism, it is understandable that mTOR inhibition may suppress cancer growth and induce cellular death of drug-resistant tumor. Clinical studies also revealed a close association between metformin use and decreased cancer incidence. In a large retrospective study by Libby et al., cancer was diagnosed among 7.3 % of 4085 metformin users compared with 11.6 % of 4085 controls, with median times of cancer of 3.5 and 2.6 years, respectively ($P < 0.001$, HR 0.46 (95 % CI 0.40–0.53)) (Libby et al. 2009). Meanwhile, diabetic patients who had taken metformin had a significantly lower risk of pancreatic cancer compared with those who had not taken metformin (odds ratio, 0.38; 95 % CI, 0.22–0.96; $P = 0.001$) (Li et al. 2009). More recently, Jiralesping et al. showed that diabetic patients receiving metformin and neoadjuvant taxane chemotherapy had a higher pathologic complete remission (pCR) than those diabetic patients not receiving metformin (Jiralerspong et al. 2009). Based on these findings, it is worthwhile to further explore the potential application of metformin in cancer prevention and therapy. It might be better to study the anticancer effects of metformin under stressful conditions and to identify whether the tumor-suppressing effects of metformin is AMPK dependent and the possibility to involve other molecular targets, such as miRNA or DNA methylation.

Besides metformin, a number of other natural compounds have been reported to activate AMPK. These include resveratrol from red grapes, ginsenoside from ginseng, curcumin from curcuma longa, berberine from coptis chinensis, epigallocatechin gallate from green tea, theaflavin from black tea, and quercetin existing in many fruits and vegetables (Hardie 2013). We summarized the chemical structures of these AMPK activators and inhibitors in Fig. 9.3. Many of these compounds were well recorded with anticancer activities, and some of them presented excellent effects in reversing cancer drug resistance, such as berberine or quercetin. Although multiple molecular mechanisms have been provided to explain their bioactivities against cancer, AMPK seems to be involved in the molecular network. Based on the findings that berberine inhibited the mitochondrial respiratory chain and resveratrol inhibited the F1 ATP synthase, it was possible that many of the above compounds may indirectly inhibit the mitochondrial ATP production and thus increase the AMP:ATP and/or ADP:ATP ratios, resulting in

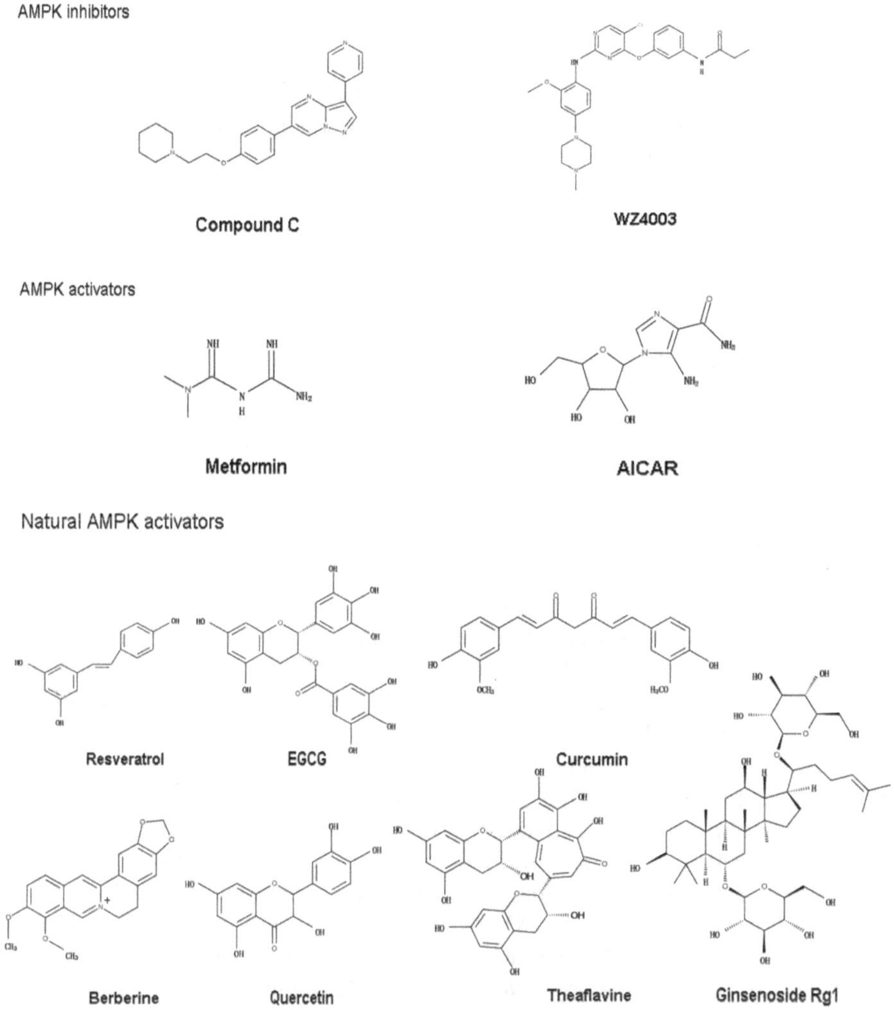

Fig. 9.3 Chemical structures of AMPK inhibitors and activators

AMPK activation (Diogo et al. 2011; Gledhill et al. 2007). Notably, AMPK activation by resveratrol, berberine, and quercetin was reduced or eliminated in cells expressing the AMP/ADP-insensitive AMPK mutants, indicating that the dysfunction of mitochondrial respiratory chain might be the upstream reason responsible for the AMPK activation effects of these phytochemicals. However, it still remained unclear whether these natural compounds act in an AMPK-dependent manner and sufficiently activate AMPK in vivo at a clinically effective dose.

9.7 Concluding Marks and Perspective

Extensive biological studies confirm that AMPK plays a central role in stress response and that its activation promotes energy-producing catabolism and inhibits energy-consuming anabolism. Although tumor suppressor gene *LKB1* serves as the upstream molecule controlling AMPK activation and emerging evidences continue to show conflicting roles of AMPK in regulating tumorigenesis, most studies tend to support that AMPK promotes cancer cell survival under stress even when a tumor suppressor role has been proposed. Therefore, elucidating the underlying mechanisms of AMPK in protecting cancer cells escaping from stressful damage caused by chemo- or radiotherapy will advance our current understanding on drug resistance. Emerging evidence has also suggested that AMPK activation plays a central role in reprogramming metabolism in drug-resistant cancer cells, such as promoting Warburg effect, fatty acid oxidation, and glutaminolysis. In addition, AMPK was also found effective in maintaining the self-renewal ability of CSCs and critical in autophagy induction, which were demonstrated as rooting factors influencing cancer chemosensitivity. Although some studies suggested opposite effects of AMPK on drug resistant modulation, these may be attributed to the differences in AMPK isoenzymes expression, the degree and duration of AMPK action, and experimental condition setting. Therefore, a much more comprehensive analysis of AMPK using conditional knockout models and deletion of individual AMPK subunits in mice is needed to decode the precise effects of AMPK on chemosensitivity regulation.

Considering the significant role of AMPK in adapting cancer cells to intrinsic or extrinsic harmful stress, AMPK inhibition may be more promising in reversing drug resistance than AMPK activation. However, current AMPK inhibitors seem to not act in an AMPK-dependent molecular mechanism and thus cannot fully reflect the actual role of AMPK in regulating bioenergetic homeostasis. Therefore, development of the highly selective AMPK inhibitors will not only facilitate further understanding on the molecular function of AMPK but also provide a potential approach for cancer therapy. Notably, AMPK inhibition may lead to the overactivation of its downstream signal mTOR cascade, which could mitigate the effects of AMPK inhibitors. It is expected to be solved by combination therapy with mTOR inhibitors. Interestingly, AMPK agonists are also currently being considered for cancer prevention and treatment. Metformin was demonstrated to cause growth arrest in a variety of cancers. However, most of the AMPK activators seem to activate AMPK via disrupting the mitochondrial respiratory chain, indicating that AMPK activation might be only a subsequent response event following mitochondrial dysfunction, and the cellular death-inducing effects might not be AMPK dependent. Therefore, it is necessary to clearly distinguish functional outcomes of AMPK activation from the secondary consequences induced by therapeutic intervention of other targets.

Taken together, an improved understanding of AMPK regulation as well as its role in chemosensitivity modulation is still needed to facilitate better elucidation of

AMPK in cancer prevention and development. Meanwhile, new studies are expected to develop novel drugs directly targeting AMPK to provide powerful tools for laboratory validation of its molecular function. In addition, clinical studies are also required to demonstrate the pathological significance of AMPK in cancer development and response to conventional therapy.

Conflict of Interests No potential conflicts of interest exist.

Acknowledgements This work was supported by the National Natural Science Foundation of China (81402173 and 81573651) and Pearl River S&T Nova Program of Guangzhou (201506010098).

References

Accordi B, Galla L, Milani G et al (2013) AMPK inhibition enhances apoptosis in MLL-rearranged pediatric B-acute lymphoblastic leukemia cells. Leukemia 27(5):1019–1027

Adamovich Y, Adler J, Meltser V, Reuven N, Shaul Y (2014) AMPK couples p73 with p53 in cell fate decision. Cell Death Differ 21:1451–1459

Ahmed KM, Li JJ (2007) ATM-NF-kappaB connection as a target for tumor radiosensitization. Curr Cancer Drug Targets 7:335–342

Anisimov VN, Egormin PA, Bershtein LM et al (2005) Metformin decelerates aging and development of mammary tumors in HER-2/neu transgenic mice. Bull Exp Biol Med 139 (6):721–723

Antico Arciuch VG, Russo MA, Kang KS, Di Cristofano A (2013) Inhibition of AMPK and Krebs cycle gene expression drives metabolic remodeling of Pten-deficient preneoplastic thyroid cells. Cancer Res 73:5459–5472

Banerjee S, Buhrlage SJ, Huang HT, Deng X, Zhou W, Wang J, Traynor R, Prescott AR, Alessi DR, Gray NS (2014) Characterization of WZ4003 and HTH-01-015 as selective inhibitors of the LKB1-tumour-suppressor-activated NUAK kinases. Biochem J 457(1):215–225

Buckendahl AC, Budczies J, Fiehn O et al (2011) Prognostic impact of AMP-activated protein kinase expression in ovarian carcinoma: correlation of protein expression and GC/TOF-MS-based metabolomics. Oncol Rep 25:1005–1012

Budanov AV, Karin M (2008) p53 target genes sestrin1 and sestrin2 connect genotoxic stress and mTOR signaling. Cell 134(3):451–460

Cantor JR, Sabatini DM (2012) Cancer cell metabolism: one hallmark, many faces. Cancer Discov 2(10):881–898

Cao X, Fang L, Gibbs S et al (2007) Glucose uptake inhibitor sensitizes cancer cells to daunorubicin and overcomes drug resistance in hypoxia. Cancer Chemother Pharmacol 59:495–505

Carretero J, Medina PP, Blanco R et al (2007) Dysfunctional AMPK activity, signaling through mTOR and survival in response to energetic stress in LKB1-deficient lung cancer. Oncogene 26(11):1616–1625

Chan EY (2009) mTORC1 phosphorylates the ULK1-mAtg13-FIP200 autophagy regulatory complex. Sci Signal 2(84), e51

Chang Z, Huo L, Li K, Wu Y, Hu Z (2014) Blocked autophagy by miR-101 enhances osteosarcoma cell chemosensitivity in vitro. Sci World J 2014:794756

Connolly DC, Katabuchi H, Cliby WA, Cho KR (2000) Somatic mutations in the STK11/LKB1 gene are uncommon in rare gynecological tumor types associated with Peutz-Jegher's syndrome. Am J Pathol 156(1):339–345

Corominas-Faja B, Cufí S, Oliveras-Ferraros C et al (2013) Nuclear reprogramming of luminal-like breast cancer cells generates Sox2-overexpressing cancer stem-like cellular states harboring transcriptional activation of the mTOR pathway. Cell Cycle 12:3109–3124

Dai RY, Zhao XF, Li JJ et al (2013) Implication of transcriptional repression in compound C-induced apoptosis in cancer cells. Cell Death Dis 4, e883

Dean M (2009) ABC transporters, drug resistance, and cancer stem cells. J Mammary Gland Biol Neoplasia 14(1):3–9

Diogo CV, Machado NG, Barbosa IA et al (2011) Berberine as a promising safe anti-cancer agent - is there a role for mitochondria? Curr Drug Targets 12(6):850–859

Dong D, Ko B, Baumeister P et al (2005) Vascular targeting and antiangiogenesis agents induce drug resistance effector GRP78 within the tumor microenvironment. Cancer Res 65:5785–5791

Ferlay J, Soerjomataram I, Ervik M, et al (2012) GLOBOCAN 2012 v1.0, Cancer Incidence and Mortality Worldwide: IARC CancerBase No. 11 [Internet]. Lyon, France: International Agency for Research on Cancer. Available from http://globocan.iarc.fr

Ferreira LM (2010) Cancer metabolism: the Warburg effect today. Exp Mol Pathol 89:372–380

Geng SQ, Alexandrou AT, Li JJ (2014) Breast cancer stem cells: Multiple capacities in tumor metastasis. Cancer Lett 349:1–7

Giardiello FM, Brensinger JD, Tersmette AC et al (2000) Very high risk of cancer in familial Peutz-Jeghers syndrome. Gastroenterology 119(6):1447–1453

Gledhill JR, Montgomery MG, Leslie AG, Walker JE (2007) Mechanism of inhibition of bovine F1-ATPase by resveratrol and related polyphenols. Proc Natl Acad Sci U S A 104:13632–13637

Green AS, Chapuis N, Lacombe C et al (2011) LKB1/AMPK/mTOR signaling pathway in hematological malignancies: from metabolism to cancer cell biology. Cell Cycle 10:2115–2120

Greer EL, Oskoui PR, Banko MR et al (2007) The energy sensor AMP-activated protein kinase directly regulates the mammalian FOXO3 transcription factor. J Biol Chem 282 (41):30107–30119

Guertin DA, Sabatini DM (2007) Defining the role of mTOR in cancer. Cancer Cell 12:9–22

Gwinn DM, Shackelford DB, Egan DF et al (2008) AMPK phosphorylation of raptor mediates a metabolic checkpoint. Mol Cell 30(2):214–226

Hadad SM, Baker L, Quinlan PR et al (2009) Histological evaluation of AMPK signalling in primary breast cancer. BMC Cancer 1(9):307

Handa N, Takagi T, Saijo S et al (2011) Structural basis for compound C inhibition of the human AMP-activated protein kinase α2 subunit kinase domain. Acta Crystallogr D Biol Crystallogr 67(Pt 5):480–487

Hardie DG (2004) The AMP-activated protein kinase pathway--new players upstream and downstream. J Cell Sci 117(Pt 23):5479–5487

Hardie DG (2013) AMPK: a target for drugs and natural products with effects on both diabetes and cancer. Diabetes 62:2164–2172

Hardie DG (2014) AMPK: positive and negative regulation, and its role in whole-body energy homeostasis. Curr Opin Cell Biol 33C:1–7

Hardie DG, Alessi DR (2013) LKB1 and AMPK and the cancer-metabolism link - ten years after. BMC Biol 11:36

Hawley SA, Pan DA, Mustard KJ et al (2005) Calmodulin-dependent protein kinase kinase-beta is an alternative upstream kinase for AMP-activated protein kinase. Cell Metab 2:9–19

Hemminki A, Tomlinson I, Markie D et al (1997) Localization of a susceptibility locus for Peutz-Jeghers syndrome to 19p using comparative genomic hybridization and targeted linkage analysis. Nat Genet 15(1):87–90

Hemminki A, Avizienyte E, Roth S et al (1998) A serine/threonine kinase gene defective in Peutz-Jeghers syndrome. Duodecim 114(7):667–668

Herrero-Martín G, Høyer-Hansen M, García-García C et al (2009) TAK1 activates AMPK-dependent cytoprotective autophagy in TRAIL-treated epithelial cells. EMBO J 28(6):677–685

Hindupur SK, Balaji SA, Saxena M et al (2014) Identification of a novel AMPK-PEA15 axis in the anoikis-resistant growth of mammary cells. Breast Cancer Res 16:420

Hosono K, Endo H, Takahashi H et al (2010) Metformin suppresses colorectal aberrant crypt foci in a short-term clinical trial. Cancer Prev Res (Phila) 3(9):1077–1083

Hulleman E, Kazemier KM, Holleman A et al (2009) Inhibition of glycolysis modulates prednisolone resistance in acute lymphoblastic leukemia cells. Blood 113:2014–2021

Hurley RL, Anderson KA, Franzone JM et al (2005) The Ca2+/calmodulin-dependent protein kinase kinases are AMP-activated protein kinase kinases. J Biol Chem 280:29060–29066

Inoki K, Zhu T, Guan KL (2003) TSC2 mediates cellular energy response to control cell growth and survival. Cell 115:577–590

Jeon SM, Chandel NS, Hay N (2012) AMPK regulates NADPH homeostasis to promote tumour cell survival during energy stress. Nature 485:661–665

Jiralerspong S, Palla SL, Giordano SH et al (2009) Metformin and pathologic complete responses to neoadjuvant chemotherapy in diabetic patients with breast cancer. J Clin Oncol 27:3297–3302

Jones RG, Plas DR, Kubek S et al (2005) AMP-activated protein kinase induces a p53-dependent metabolic checkpoint. Mol Cell 18(3):283–293

Kasznicki J, Sliwinska A, Drzewoski J (2014) Metformin in cancer prevention and therapy. Ann Transl Med 2(6):57

Kim JW, Dang CV (2006) Cancer's molecular sweet tooth and the Warburg effect. Cancer Res 66:8927–8930

Kim CJ, Cho YG, Park JY et al (2004) Genetic analysis of the LKB1/STK11 gene in hepatocellular carcinomas. Eur J Cancer 40(1):136–141

Kim HS, Hwang JT, Yun H et al (2008) Inhibition of AMP-activated protein kinase sensitizes cancer cells to cisplatin-induced apoptosis via hyper-induction of p53. J Biol Chem 283 (7):3731–3742

Kim J, Kundu M, Viollet B, Guan KL (2011a) AMPK and mTOR regulate autophagy through direct phosphorylation of Ulk1. Nat Cell Biol 13:132–141

Kim YM, Kim MY, Kim HJ et al (2011b) Compound C independent of AMPK inhibits ICAM-1 and VCAM-1 expression in inflammatory stimulants-activated endothelial cells in vitro and in vivo. Atherosclerosis 219(1):57–64

Kim YH, Liang H, Liu X et al (2012) AMPKα modulation in cancer progression: multilayer integrative analysis of the whole transcriptome in Asian gastric cancer. Cancer Res 72:2512–2521

Kim TH, Suh DH, Kim MK, Song YS (2014) Metformin against cancer stem cells through the modulation of energy metabolism: special considerations on ovarian cancer. Biomed Res Int 2014:132702

Książkiewicz M, Markiewicz A, Zaczek AJ (2012) Epithelial-mesenchymal transition: a hallmark in metastasis formation linking circulating tumor cells and cancer stem cells. Pathobiology 79:195–208

Le A, Cooper CR, Gouw AM et al (2010) Inhibition of lactate dehydrogenase A induces oxidative stress and inhibits tumor progression. Proc Natl Acad Sci U S A 107(5):2037–2042

Lee H, Park HJ, Park CS et al (2014) Response of breast cancer cells and cancer stem cells to metformin and hyperthermia alone or combined. PLoS One 9(2), e87979

Levy JM, Thompson JC, Griesinger AM et al (2014) Autophagy inhibition improves chemosensitivity in BRAF(V600E) brain tumors. Cancer Discov 4:773–780

Li D, Yeung SC, Hassan MM, Konopleva M, Abbruzzese JL (2009) Antidiabetic therapies affect risk of pancreatic cancer. Gastroenterology 137(2):482–488

Li C, Liu VW, Chiu PM, Chan DW, Ngan HY (2012) Over-expressions of AMPK subunits in ovarian carcinomas with significant clinical implications. BMC Cancer 12:357

Li W, Saud SM, Young MR, Chen G, Hua B (2015) Targeting AMPK for cancer prevention and treatment. Oncotarget 6(10):7365–7378

Liang J, Shao SH, Xu ZX et al (2007) The energy sensing LKB1–AMPK pathway regulates p27 (kip1) phosphorylation mediating the decision to enter autophagy or apoptosis. Nat Cell Biol 9 (2):218–224

Libby G, Donnelly LA, Donnan PT et al (2009) New users of metformin are at low risk of incident cancer: a cohort study among people with type 2 diabetes. Diabetes Care 32(9):1620–1625

Liu Y, Cao Y, Zhang W et al (2012) A small-molecule inhibitor of glucose transporter 1 downregulates glycolysis, induces cell-cycle arrest, and inhibits cancer cell growth in vitro and in vivo. Mol Cancer Ther 11:1672–1682

Lizcano JM, Göransson O, Toth R et al (2004) LKB1 is a master kinase that activates 13 kinases of the AMPK subfamily, including MARK/PAR-1. EMBO J 23:833–843

Löffler AS, Alers S, Dieterle AM et al (2011) Ulk1-mediated phosphorylation of AMPK constitutes a negative regulatory feedback loop. Autophagy 7:696–706

Luo Z, Saha AK, Xiang X, Ruderman NB (2005) AMPK, the metabolic syndrome and cancer. Trends Pharmacol Sci 26(2):69–76

Mao K, Klionsky DJ (2011) AMPK activates autophagy by phosphorylating ULK1. Circ Res 108:787–788

Maugeri-Saccà M, Vigneri P, De Maria R (2011) Cancer stem cells and chemosensitivity. Clin Cancer Res 17:4942–4947

Mendoza EE, Pocceschi MG, Kong X et al (2012) Control of Glycolytic Flux by AMP-Activated Protein Kinase in Tumor Cells Adapted to Low pH. Transl Oncol 5:208–216

Mihaylova MM, Shaw RJ (2011) The AMPK signalling pathway coordinates cell growth, autophagy and metabolism. Nat Cell Biol 13:1016–1023

Mouradian M, Kikawa KD, Dranka BP et al (2014) Docosahexaenoic acid attenuates breast cancer cell metabolism and the Warburg phenotype by targeting bioenergetic function. Mol Carcinog. doi:10.1002/mc.22151

Nakanishi C, Yamaguchi T, Iijima T et al (2004) Germline mutation of the LKB1/STK11 gene with loss of the normal allele in an aggressive breast cancer of Peutz-Jeghers syndrome. Oncology 67(5-6):476–479

Nam M, Lee WH, Bae EJ, Kim SG (2008) Compound C inhibits clonal expansion of preadipocytes by increasing p21 level irrespectively of AMPK inhibition. Arch Biochem Biophys 479:74–81

Neumeister V, Agarwal S, Bordeaux J, Camp RL, Rimm DL (2010) In situ identification of putative cancer stem cells by multiplexing ALDH1, CD44, and cytokeratin identifies breast cancer patients with poor prognosis. Am J Pathol 176:2131–2138

Nishino Y, Miura T, Miki T et al (2004) Ischemic preconditioning activates AMPK in a PKC-dependent manner and induces GLUT4 up-regulation in the late phase of cardioprotection. Cardiovasc Res 61(3):610–619

Ouyang J, Parakhia RA, Ochs RS (2011) Metformin activates AMP kinase through inhibition of AMP deaminase. J Biol Chem 286(1):1–11

Owen MR, Doran E, Halestrap AP (2000) Evidence that metformin exerts its anti-diabetic effects through inhibition of complex 1 of the mitochondrial respiratory chain. Biochem J 348 (Pt 3):607–614

Park HU, Suy S, Danner M et al (2009) AMP-activated protein kinase promotes human prostate cancer cell growth and survival. Mol Cancer Ther 8:733–741

Pazarentzos E, Bivona TG (2015) Adaptive stress signaling in targeted cancer therapy resistance. Oncogene 34(45):5599–5606

Sakamoto K, McCarthy A, Smith D et al (2005) Deficiency of LKB1 in skeletal muscle prevents AMPK activation and glucose uptake during contraction. EMBO J 24:1810–1820

Sanli T, Steinberg GR, Singh G, Tsakiridis T (2014) AMP-activated protein kinase (AMPK) beyond metabolism: a novel genomic stress sensor participating in the DNA damage response pathway. Cancer Biol Ther 15:156–169

Sato N, Rosty C, Jansen M et al (2001) STK11/LKB1 Peutz-Jeghers gene inactivation in intraductal papillary-mucinous neoplasms of the pancreas. Am J Pathol 159 (6):2017–2022

Sato A, Sunayama J, Okada M et al (2012) Glioma-initiating cell elimination by metformin activation of FOXO3 via AMPK. Stem Cells Transl Med 1:811–824

Sauer H, Engel S, Milosevic N, Sharifpanah F, Wartenberg M (2012) Activation of AMP-kinase by AICAR induces apoptosis of DU-145 prostate cancer cells through generation of reactive oxygen species and activation of c-Jun N-terminal kinase. Int J Oncol 40:501–508

Shackelford DB, Abt E, Gerken L et al (2013) LKB1 inactivation dictates therapeutic response of non-small cell lung cancer to the metabolism drug phenformin. Cancer Cell 23:143–158

Shaw RJ, Cantley LC (2006) Ras, PI(3)K and mTOR signalling controls tumour cell growth. Nature 441:424–430

Shaw RJ, Lamia KA, Vasquez D et al (2005) The kinase LKB1 mediates glucose homeostasis in liver and therapeutic effects of metformin. Science 310:1642–1646

Shell SA, Lyass L, Trusk PB et al (2008) Activation of AMPK is necessary for killing cancer cells and sparing cardiac cells. Cell Cycle 7(12):1769–1775

Shi HS, Li D, Zhang J et al (2010) Silencing of pkm2 increases the efficacy of docetaxel in human lung cancer xenografts in mice. Cancer Sci 101:1447–1453

Shi L, Qin N, Hu L et al (2011) Tiliroside-derivatives enhance GLUT4 translocation via AMPK in muscle cells. Diabetes Res Clin Pract 92:e41–e46

Shirwany NA, Zou MH (2014a) AMPK: a cellular metabolic and redox sensor. A minireview. Front Biosci 19:447–474

Shirwany NA, Zou MH (2014b) AMPK: a cellular metabolic and redox sensor. A minireview. Front Biosci (Landmark Ed) 19:447–474

Short JD, Dere R, Houston KD et al (2010) AMPK-mediated phosphorylation of murine p27 at T197 promotes binding of 14-3-3 proteins and increases p27 stability. Mol Carcinog 49:429–439

Snima KS, Pillai P, Cherian AM, Nair SV, Lakshmanan VK (2014) Anti-diabetic drug metformin: challenges and perspectives for cancer therapy. Curr Cancer Drug Targets 14(8):727–736

Sui X, Chen R, Wang Z et al (2013) Autophagy and chemotherapy resistance: a promising therapeutic target for cancer treatment. Cell Death Dis 4, e838

Tao R, Gong J, Luo X, Zang M, Guo W, Wen R, Luo Z (2010) AMPK exerts dual regulatory effects on the PI3K pathway. J Mol Signal 5:1

Tirino V, Desiderio V, Paino F et al (2013) Cancer stem cells in solid tumors: an overview and new approaches for their isolation and characterization. FASEB J 27:13–24

Tripathi DN, Chowdhury R, Trudel LJ et al (2013) Reactive nitrogen species regulate autophagy through ATM-AMPK-TSC2-mediated suppression of mTORC1. Proc Natl Acad Sci U S A 110:E2950–E2957

Venkatesha VA, Parsels LA, Parsels JD et al (2012) Sensitization of pancreatic cancer stem cells to gemcitabine by Chk1 inhibition. Neoplasia 14:519–525

Vucicevic L, Misirkic M, Janjetovic K et al (2009) AMP-activated protein kinase-dependent and -independent mechanisms underlying in vitro antiglioma action of compound C. Biochem Pharmacol 77(11):1684–1693

Vucicevic L, Misirkic M, Janjetovic K et al (2011) Compound C induces protective autophagy in cancer cells through AMPK inhibition-independent blockade of Akt/mTOR pathway. Autophagy 7(1):40–50

Wan Z, Root-McCaig J, Castellani L et al (2014) Evidence for the role of AMPK in regulating PGC-1 alpha expression and mitochondrial proteins in mouse epididymal adipose tissue. Obesity (Silver Spring) 22:730–738

Wang ZY, Loo TY, Shen JG et al (2012) LDH-A silencing suppresses breast cancer tumorigenicity through induction of oxidative stress mediated mitochondrial pathway apoptosis. Breast Cancer Res Treat 131(3):791–800

Wang Z, Wang N, Liu P et al (2014) MicroRNA-25 regulates chemoresistance-associated autophagy in breast cancer cells, a process modulated by the natural autophagy inducer isoliquiritigenin. Oncotarget 5:7013–7026

Wu CA, Chao Y, Shiah SG, Lin WW (2013) Nutrient deprivation induces the Warburg effect through ROS/AMPK-dependent activation of pyruvate dehydrogenase kinase. Biochim Biophys Acta 1833:1147–1156

Xiao B, Sanders MJ, Underwood E et al (2011) Structure of mammalian AMPK and its regulation by ADP. Nature 472:230–233

Xiao X, He Q, Lu C et al (2012) Metformin impairs the growth of liver kinase B1-intact cervical cancer cells. Gynecol Oncol 127:249–255

Xie M, Zhang D, Dyck JR et al (2006) A pivotal role for endogenous TGF-beta-activated kinase-1 in the LKB1/AMP-activated protein kinase energy-sensor pathway. Proc Natl Acad Sci U S A 103:17378–17383

Xin H, Kong Y, Jiang X et al (2013) Multi-drug-resistant cells enriched from chronic myeloid leukemia cells by Doxorubicin possess tumor-initiating-cell properties. J Pharmacol Sci 122:299–304

Xu J, Ji J, Yan XH (2012) Cross-talk between AMPK and mTOR in regulating energy balance. Crit Rev Food Sci Nutr 52:373–381

Yan M, Gingras MC, Dunlop EA et al (2014) The tumor suppressor folliculin regulates AMPK-dependent metabolic transformation. J Clin Invest 124:2640–2650

Yang W, Hong YH, Shen XQ, Frankowski C, Camp HS, Leff T (2001) Regulation of transcription by AMP activated protein kinase: phosphorylation of p300 blocks its interaction with nuclear receptors. J Biol Chem 276(42):38341–38344

Yoon JH, Ahn SG, Lee BH, Jung SH, Oh SH (2012) Role of autophagy in chemoresistance: regulation of the ATM-mediated DNA-damage signaling pathway through activation of DNA-PKcs and PARP-1. Biochem Pharmacol 83:747–757

Yu L, Yang SJ (2010) AMP-activated protein kinase mediates activity-dependent regulation of peroxisome proliferator-activated receptor gamma coactivator-1alpha and nuclear respiratory factor 1 expression in rat visual cortical neurons. Neuroscience 169:23–38

Yu F, Yao H, Zhu P et al (2007) let-7 regulates self renewal and tumorigenicity of breast cancer cells. Cell 131:1109–1123

Zhang L, Wang S, Wangtao et al (2009) Upregulation of GRP78 and GRP94 and its function in chemotherapy resistance to VP-16 in human lung cancer cell line SK-MES-1. Cancer Invest 27 (4):453–458

Zhao M, Klionsky DJ (2011) AMPK-dependent phosphorylation of ULK1 induces autophagy. Cell Metab 13(2):119–120

Zhao Y, Liu H, Liu Z et al (2011) Overcoming trastuzumab resistance in breast cancer by targeting dysregulated glucose metabolism. Cancer Res 71(13):4585–4597

Zhou J, Huang W, Tao R et al (2009) Inactivation of AMPK alters gene expression and promotes growth of prostate cancer cells. Oncogene 28(18):1993–2002

Zhou M, Zhao Y, Ding Y et al (2010) Warburg effect in chemosensitivity: targeting lactate dehydrogenase-A re-sensitizes taxol-resistant cancer cells to taxol. Mol Cancer 9:33

Zou Z, Yuan Z, Zhang Q et al (2012) Aurora kinase A inhibition-induced autophagy triggers drug resistance in breast cancer cells. Autophagy 8(12):1798–1810

Chapter 10
AMPK as a Pro-longevity Target

Kristopher Burkewitz, Heather J.M. Weir, and William B. Mair

Contents

Abstract Chronic, age-associated diseases are already among the leading causes of morbidity and death in the world, a problem exacerbated by the rapidly rising proportion of elderly in the global population. This emergent epidemic represents the next great challenge for biomedical science and public health. Fortunately, decades of studies into the biology of aging have provided a head start by revealing an evolutionarily conserved network of genes that controls the rate and quality of the aging process itself and which can thereby be targeted for protection against age-onset disease. A number of dietary, genetic, and pharmacological interventions, including dietary restriction (DR) and the biguanide metformin, can extend

K. Burkewitz (✉) • H.J.M. Weir • W.B. Mair (✉)
Department of Genetics and Complex Diseases, Harvard T.H. Chan School of Public Health, 665 Huntington Ave, Boston, MA 02115, USA
e-mail: kburkewi@hsph.harvard.edu; wmair@hsph.harvard.edu

M.D. Cordero, B. Viollet (eds.), *AMP-activated Protein Kinase*, Experientia Supplementum 107, DOI 10.1007/978-3-319-43589-3_10

healthy lifespan and reduce the incidence of multiple chronic conditions. Many of these interventions recurrently involve a core network of nutrient sensors: AMP-activated protein kinase (AMPK), mammalian target of rapamycin (mTOR), the insulin/insulin-like growth factor signaling pathway (IIS), and the sirtuins. Here, we will summarize how AMPK acts downstream of these pro-longevity interventions and within this network of nutrient sensors to control the cell and physiological processes important for defining how well we age.

Keywords AMPK • Aging • Longevity • Metabolism • Mitochondria • Sirtuins • TOR • FOXO

10.1 Introduction

Humanity is graying. The elderly population—those 65 and older—is on the verge of outnumbering children for the first time in world history (Population Reference Bureau), while the median age of the world population is expected to rise by 8 % in the next decade, ~22 % by 2050, and >40 % by 2100 (United Nations World Population Prospects 2015). The aging of the population is not restricted to wealthy, developed nations either; in fact, the truth is closer to the converse. While the elderly population is projected to rise by roughly 54 % in high-income nations by 2050, less developed countries are expected to face an increase of over 230 % in that same period. The end result of such dramatic increases in human survival, driven by advances in public health, will be 1.5 billion people over the age of 65 worldwide by 2050 (Fig. 10.1a), more than the current population of China. Historically, demographic shifts of this magnitude have occurred slowly, providing time for societies and economies to adapt, but the present rise in the age of the global population is happening within a single generation (Fig. 10.1a), placing enormous stress on several of the fundamental institutions in our lives, especially retirement and health care.

Reduction of early mortality and increased worldwide survival rates are cause for celebration. However, this shift in population dynamics will trigger an "epidemiological transition" and ensuing public health epidemic for which we are currently unprepared. At least from a biological standpoint, humans are not aging gracefully; our bodies did not evolve to live as long as we now do. As a result, old age carries with it the burden of chronic diseases, including heart disease, cancer, metabolic disease, and neurodegenerative disorders. Unfortunately, this burden only continues to grow as we travel further into old age: two-thirds of the elderly in the United States have multiple chronic diseases (Fig. 10.1b) (CDC 2013). This simple fact means that even a revolutionary cure for one of these chronic diseases would only marginally affect disease-free health span (Goldman et al. 2013). Over the last twenty years, however, a solution has begun to emerge from studies of

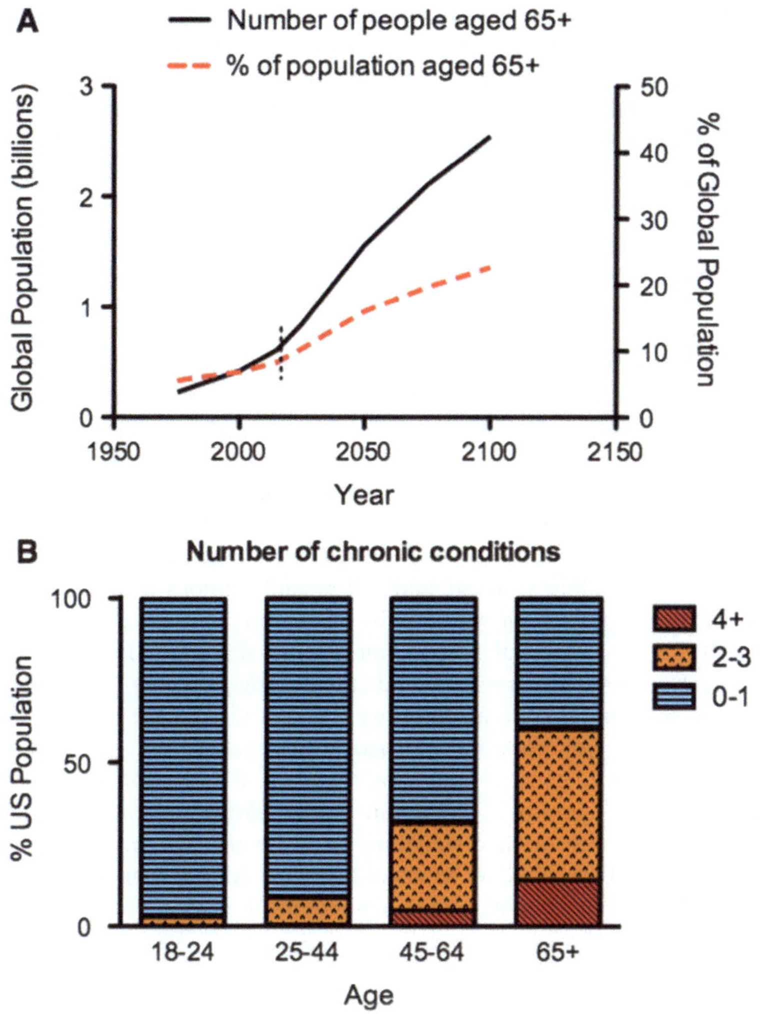

Fig. 10.1 The world population is aging, resulting in an overall increase in the global burden of chronic, age-onset disease. Source data from 2015 Revision of World Population Prospects, United Nations Population Division; esa.un.org/unpd/wpp (**a**); and from Health, United States, 2015; www.cdc.gov/nchs/hus/index.htm (**b**)

model organisms across the evolutionary spectrum. Though it was observed as early as the mid-twentieth century that dietary interventions could extend the lifespan of laboratory animals, the discovery that single gene mutations could dramatically prolong lifespan and maintain animals in a youthful state set off a flurry of research into the genetics of aging (Fontana et al. 2010; Kenyon et al. 1993). The field has since demonstrated unequivocally that the rate and quality of physiological aging are plastic, and thus the molecular mechanisms linking old age to disease risk represent valid therapeutic targets. Many labs have contributed to mapping out entire, evolutionarily conserved pathways with potent effects on animal lifespan, revealing a number of consistent, core longevity factors. This knowledge will ultimately serve in the development of single interventions capable of treating an array of age-onset, chronic pathologies.

A common thread among interventions known to promote healthier aging is an association with energy or nutrient depletion. This theme manifests across virtually all tiers of biological organization, from the molecular level to cells, tissues, and whole-animal physiology. Longevity can result from reducing energy's molecular "currency," adenosine triphosphate (ATP), or from perturbing the function of the cell's energy-producing organelle, the mitochondrion. Reducing systemic, endocrine signals of energy abundance extends lifespan, as does a wide array of dietary interventions all designed to reduce nutrient intake in one form or another. The evolutionarily ancient energy sensor, AMP-activated protein kinase (AMPK), similarly crosses all these organizational boundaries. A heterotrimeric kinase complex composed of a catalytic α subunit and regulatory β and γ subunits, AMPK responds directly to both AMP/ADP:ATP ratios sensed through binding at the γ subunit and to endocrine signals of satiety or hunger. In turn, the kinase can regulate virtually all cell processes—including systemic and cell nonautonomous responses—to maintain energy homeostasis both within cells and throughout an organism. Accordingly, AMPK plays a critical role in transducing signals of low energy into a physiological response that promotes healthier aging, and this is true across the evolutionary spectrum. While neighboring chapters discuss the role of AMPK in the pathogenesis and/or treatment of specific age-related diseases, here we will highlight the dietary, genetic, and pharmacological interventions that require AMPK to modulate aging per se. We will discuss AMPK as a node in a complex network of nutrient sensors and signaling pathways that determines the rate and quality of aging and speculate on which of the many AMPK-regulated cellular and metabolic processes mediate its role as a potent pro-longevity factor.

10.2 Energy Metabolism and Aging Are Linked Through AMPK

Theories of how and why we age have long focused on the idea of limited energy resources forcing trade-offs between self-preservation and growth or reproduction. In harsh, nutrient-scarce environments, organisms enhance cellular defense mechanisms involved in repairing damage and maintaining cell homeostasis, thus slowing or protecting against the onset of pathology that occurs over time. Conversely, in conditions of abundance, organisms seem to generate, store, and spend their energy differently, perhaps more dysfunctionally, to such an extent that one outcome of the current obesity epidemic may be to accelerate the onset of age-related diseases, if not aging itself (Haslam and James 2005; Horvath et al. 2014). Experimentally, Apfeld et al. provided some of the first direct evidence suggesting that energy levels, as reflected by nucleotide charge, could play a direct role in determining lifespan when they revealed that AMP:ATP ratios could be used to predict remaining life expectancy in the nematode *C. elegans* and that hormetic stresses or genetic mutants that extend lifespan also at least temporarily increase AMP:ATP (Apfeld et al. 2004). Consistent with this direct link, interventions that impose conditions of nutrient scarcity or impair energy production represent some of the most evolutionarily robust methods to promote longer life and healthier aging, and AMPK plays an important part in translating energetic deficits into these benefits.

10.2.1 Dietary Restriction

Dietary restriction (DR), defined broadly as a reduction in nutrient intake to an extent that does not cause malnutrition, is one of the first discovered and best understood interventions to extend lifespan (Fontana and Partridge 2015). DR activates a physiological low-energy program which extends the lifespan of virtually all species in which it has been rigorously tested (Mair and Dillin 2008). Beyond longevity, DR regimens protect against diverse and seemingly unrelated age-onset pathologies including neurodegenerative disease, cardiovascular disease, metabolic disease, and cancer (Masoro 2002). DR promises exceptional therapeutic potential, but adherence to such a strict dietary regimen presents a serious challenge. As a result, many investigations have aimed to define the molecular mediators of DR in order to develop pharmacological targets and, ultimately, therapeutics. While the goal of a true DR mimetic is not yet realized, molecular genetic studies in simpler model organisms, e.g., yeast, worm, and fly, have revealed a highly conserved network of nutrient-sensing pathways functioning as the core of this DR low-energy switch (Fig. 10.2).

As a conserved, literal sensor of low-energy stress via AMP/ADP:ATP ratios, AMPK represents perhaps the most appealing mechanistic link between DR, energy status, and lifespan. Given the known upstream activation routes for AMPK, the simplest hypothesis would suggest that DR results in a chronic reduction in cellular energy charge, promoting AMPK activation. Greer et al. performed the first direct test of this idea in *C. elegans*, which feeds on a nonpathogenic *E. coli* lawn. DR can be enforced upon *C. elegans* by dilution of the *E. coli* food source on solid growth media, and employing this protocol, the authors confirmed that DR indeed promotes an increase in the AMP:ATP ratio of worms (Greer et al. 2007a). As the authors hypothesized, lifespan extension on DR in this system requires AAK-2, one of two isoforms of the *C. elegans* AMPK catalytic subunit (Greer et al. 2007a).

Supporting conservation of this hypothesis through evolution, *a Drosophila* study demonstrated that flies harboring mutations in the AMP synthesis pathway exhibit increased AMP- and ADP-to-ATP ratios and are long-lived (Stenesen et al. 2013). Restoring AMP:ATP ratio back to normal levels by supplementing

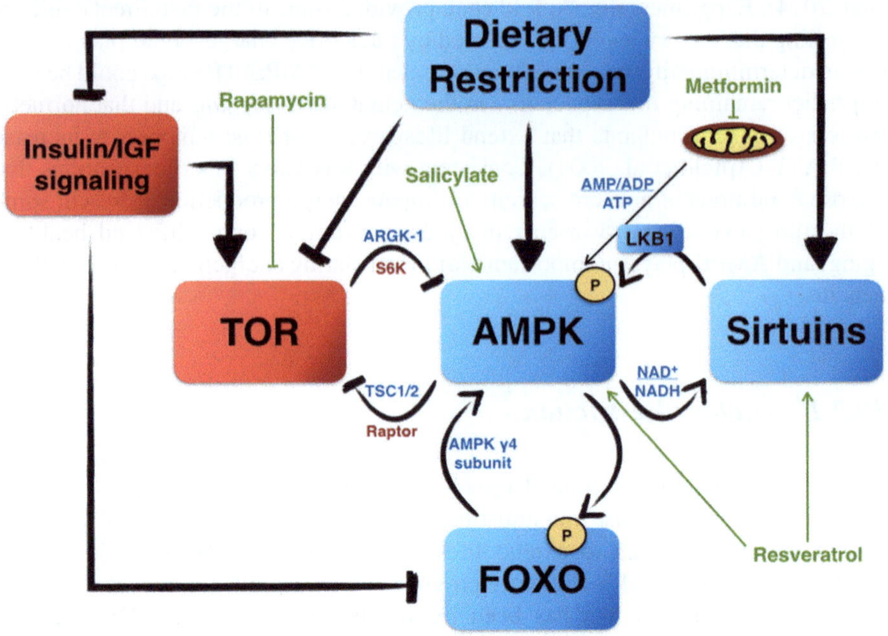

Fig. 10.2 AMPK is a central node in a network of nutrient-sensing pathways regulating longevity. Pro-longevity factors (activity positively correlates with longevity) are colored *blue*, while anti-longevity factors (inhibition correlates with longevity) are *red*

the animals with adenine completely blocks lifespan extension in the AMP synthesis mutants. While this study did not explicitly examine the effects of DR on endogenous AMP or ATP levels in *Drosophila*, adenine supplementation also completely suppressed the lifespan extension in flies undergoing a DR regimen involving reductions in the dietary yeast component from 16 % to 4 %. The lifespan extension induced by AMP synthesis mutants is also suppressed by dominant negative AMPK, suggesting that the kinase mediates the longevity benefits of these high AMP:ATP ratios in fly as well as worm (Stenesen et al. 2013).

Importantly, DR is an umbrella term encompassing a significant variety of nutritional interventions; for example, DR can entail reducing single nutrients in the food source or the total caloric content while keeping relative nutrient proportions equal. In *C. elegans*, where the most direct tests of AMPK's role in DR have been performed, AMPK is not uniformly required for lifespan extension across all these diverse regimens. Contrasting with studies involving bacterial dilution on solid media, *aak-2* is not required for lifespan extension in worms fed diluted bacteria in liquid culture or in *eat-2* mutant worms which possess reduced feeding rates (Curtis et al. 2006; Greer and Brunet 2009; Mair et al. 2009). Futhermore, DR regimens may activate AMPK by mechanisms completely unrelated to cell nucleotide charge, and these likely include altered levels of humoral factors and signaling through alternative nutrient-sensing pathways, some of which are discussed below. In addition, it remains unclear whether dominant negative AMPK blocks lifespan extension by DR in *Drosophila* as it does in AMP synthesis mutants. A great deal more work is required before we can claim to fully understand when and where in the body AMPK becomes activated during DR and the mechanisms involved.

10.2.2 Mitochondrial Inhibition

Mild inhibition of mitochondrial function is another conserved paradigm of longevity. Genetic inhibition of several components of the electron transport chain (ETC) complexes alters developmental timing and extends lifespan (Dell'agnello et al. 2007; Dillin et al. 2002; Houtkooper et al. 2013; Lakowski and Hekimi 1996; Lee et al. 2003; Liu et al. 2005). Generally, declining mitochondrial function and quality is a hallmark of aging (Lopez-Otin et al. 2013), and so the finding that perturbing key functional components of the organelle extends lifespan is somewhat paradoxical. The precise cell and molecular mechanisms by which mitochondrial inhibition promotes longevity remain elusive, but tend to fall in one of two directions: altered signaling and/or cellular damage caused by reactive oxygen species (ROS) from the mitochondria, or enhanced protein homeostasis through the mitochondrial unfolded protein response (UPR^{mt}), which results when the stoichiometry of mitochondrial ETC complexes is disrupted (Haynes et al. 2013). Relatively recent studies have identified noncanonical activation of AMPK by ROS signaling, suggesting that this could be an intriguing connection to pursue in the context of mitochondrial pathways and aging (Hardie et al. 2012). On the other hand, recent work suggests that UPR^{mt} longevity is mediated through epigenetic reprogramming and transcriptional remodeling of the mitochondrial protein homeostasis network (Merkwirth et al. 2016), but whether AMPK plays an important role in these UPR^{mt}-related processes remains mostly unknown. Given the canonical roles of the mitochondrion as the energy-producing organelle and of AMPK as a sensor and transducer of energy stress, it is a simple progression to hypothesize that mitochondrial inhibition inflicts an energy crisis upon the animal to which AMPK responds, but experiments testing this hypothesis in lifespan regulation are relatively scant. Surprisingly, inhibition of ISP-1 (complex III) or CLK-1 (ubiquinone synthesis) still extends lifespan in *C. elegans* lacking AAK-2, though the longevity is partially blunted in this AMPK mutant (Curtis et al. 2006). In this case, involvement of the alternate α subunit, AAK-1, has not been ruled out. Broadly, given the variety of routes by which mitochondrial perturbations could be influencing AMPK function, it naturally remains an attractive candidate to mediate at least some of the associated benefits on aging.

10.3 AMPK Regulates Longevity in Conjunction with Other Energy-Sensing Pathways

Mitochondrial inhibition and especially DR are very broad interventions with far-reaching consequences across cell and animal physiology. Many of the changes they elicit are not coupled to aging, so a great deal of elegant work has sought to identify the downstream molecular mechanisms which are required, and ultimately sufficient, to extend lifespan. These studies found that a number of interventions designed to impair much more specific aspects of energy metabolism also exert potent effects on lifespan. Recurrently, this more targeted work has highlighted multiple energy-sensing genes and pathways—including AMPK—which seem to function as a core, evolutionarily conserved signaling network capable of defining the rate of aging.

10.3.1 Insulin and Growth Factor Signaling

Suppression of the insulin/insulin-like growth factor signaling (IIS) pathway dramatically extends lifespan in multiple systems (Fontana et al. 2010; Kenyon 2011). The IIS pathway responds to nutrient levels in the body to mediate feeding, development, and growth, as well as macromolecular biosynthesis and storage. In a highly conserved pathway, Forkhead box O (FOXO) family transcription factors act downstream of IIS, becoming nuclear and activated only when the IIS pathway is repressed (Matsuzaki et al. 2003). *C. elegans* mutants of the insulin receptor, DAF-2, are completely dependent on the sole FOXO ortholog in worms, DAF-16, for extended lifespan (Kenyon et al. 1993). AMPK/AAK-2 is completely required for lifespan extension in some weaker hypomorphic alleles of *daf-2*, yet only partially suppresses the longer lifespans of stronger reduction-of-function alleles, suggesting the existence of both AMPK-dependent and independent mechanisms of IIS longevity (Apfeld et al. 2004). Though mutants of the IIS pathway extend lifespan in flies, rodents, and possibly humans (Fontana et al. 2010; Pawlikowska et al. 2009; Suh et al. 2008), the requirement of AMPK in these systems requires exploration. Intriguingly, AMPK interacts with these FOXO transcription factors in multiple interesting ways, some of which are evolutionarily conserved. Not only does AMPK phosphorylate and activate FOXO/DAF-16 in *C. elegans* and mammals (Greer et al. 2007a, b), but FOXO/DAF-16 may directly upregulate expression of an atypical γ isoform lacking consensus AMP-binding sites in *C. elegans* (Tullet et al. 2014). This atypical γ is required for maximal lifespan extension of insulin pathway mutants, suggesting that this may represent a natural constitutively active form of the complex mediating the longevity benefits of impaired insulin signaling (Chen et al. 2013; Tullet et al. 2014) This represents just one example out of many where these nutrient-responsive pathways cross-regulate one another.

10.3.2 Sirtuins

Analogous to the increase in AMP/ADP:ATP ratio that promotes AMPK activation, another manifestation of nutrient depletion is a rise in the ratio of NAD^+:NADH, favoring the oxidized form. Sirtuins are a conserved family of protein deacetylaces and deacylases that require NAD^+ for their function, and so similarly to AMPK, their activity also correlates with conditions of nutrient depletion. Early studies in yeast suggested that increased sirtuin function is sufficient to promote healthy aging and longevity as a mediator of DR (Kaeberlein et al. 1999; Lin et al. 2000). These findings were extended to metazoans, and though valid concerns were raised over confounding genetic backgrounds and technical issues, the field has rebounded to demonstrate convincingly that sirtuins can modulate lifespan and regulate age-related pathology in different models, including rodents through overexpression of sirtuin family members SIRT1 and SIRT6 (Burnett et al. 2011; Guarente 2013; Kanfi et al. 2012; Satoh et al. 2013). Like the interaction between FOXO and AMPK, data from multiple studies indicate that sirtuins and AMPK can promote one another's activation. AMPK promotes activation of the mammalian sirtuin homolog SIRT1 indirectly through effects on the NAD^+:NADH ratio in cells (Canto et al. 2010; Fulco et al. 2008). On the other hand, the upstream AMPK kinase LKB1 is a direct target of SIRT1 deacetylation, providing an avenue for SIRT1 to promote AMPK activity (Lan et al. 2008). These two metabolic sensors also require one another to mediate metabolic homeostasis in various contexts, especially as it relates to regulation of mitochondria (see later sections for more detail). Regarding studies directly examining lifespan, *C. elegans* overexpressing the sirtuin family member SIR-2.1 require AMPK/AAK-2 for longevity, but this result should be reconfirmed now that issues with secondary mutations in the background of those strains have been identified (Curtis et al. 2006; Guarente 2013). In rodents, one study suggests that age-dependent declines in NAD^+ levels result in declining SIRT1 activity, ultimately driving aging through a failure to maintain mitochondrial homeostasis (Gomes et al. 2013). In a test of this model, the authors showed that supplementing old mice with NAD^+ reverses age-related decline in skeletal muscle strength, and although they found evidence of altered AMPK activity, it remains unclear whether it plays an important role in these rejuvenating effects (Gomes et al. 2013). Especially in light of these promising links between sirtuins and AMPK, more specific investigation into how AMPK and sirtuins co-regulate lifespan is needed.

10.3.3 TOR

In many respects, TOR and AMPK exist as near-perfect foils, and their mutual antagonism serves to maintain balance in cells. Where AMPK responds to energy depletion to enhance cellular catabolic activities, the TOR kinase complex is activated by nutrient abundance to promote anabolic functions, leading to cell growth and proliferation. Given the close association between nutrient depletion and longevity, it is perhaps unsurprising that components of the TOR signaling pathway are almost universally anti-longevity factors (Johnson et al. 2013). TOR inhibition extends lifespan from yeast to mammals, and AMPK plays an important part in these systems (Fabrizio et al. 2001; Harrison et al. 2009; Kapahi et al. 2004; Vellai et al. 2003). Consistent with their opposing roles, TOR and AMPK share a number of regulatory targets, converging to reciprocally regulate core processes such as lipid metabolism, autophagy, and protein synthesis (detailed further in later sections). TOR and AMPK also directly antagonize one another. AMPK phosphorylates and activates two different inhibitors of the TOR complex, raptor and TSC2 (Gwinn et al. 2008; Inoki et al. 2006), while a canonical TOR effector, S6 kinase (S6K), can directly phosphorylate and inhibit AMPK α (Dagon et al. 2012).

Due to the reciprocal regulation between AMPK and TOR complexes (i.e., activation of one kinase results in inhibition of the other), it is difficult to tease apart which kinase plays a dominant or upstream role in lifespan regulation through genetic epistasis. However, a study by Selman et al. demonstrated that S6K/RSKS-1 mutants are long-lived in both mouse and *C. elegans,* and consistent with the explanation that S6K loss results in AMPK activation, the worm α subunit AAK-2 is required for those longevity benefits (Selman et al. 2009). Carrying this finding further, proteomic approaches aiming to identify mechanisms of TOR pathway longevity in S6K mutant worms revealed that the arginine/creatine kinase, ARGK-1, is 35-fold upregulated in these animals (McQuary et al. 2016). ARGK-1 is both required and sufficient for AMPK α activation in the S6K mutants, and though unconfirmed experimentally, this may occur through the conserved role of this phosphagen kinase family in regulating cellular levels of energy-carrying molecules like creatine (McQuary et al. 2016). Furthermore, ARGK-1 overexpression extends *C. elegans* lifespan and requires AMPK/AAK-2 for this effect. These results more broadly suggest the existence of additional mechanisms by which the TOR pathway may regulate AMPK activity and potentially longevity.

10.3.4 Longevity Signaling Is Highly Interconnected

Though especially apparent from the antagonism between the AMPK and TOR complexes, it is worth noting that many of these longevity pathways are related and/or cross-regulated, resulting in a web-like signaling network that can be difficult to untangle. For example, insulin/IGF1 signaling is reduced in animals undergoing DR (Fontana and Partridge 2015), but while insulin-receptor mutants are completely dependent on FOXO/DAF-16 for longevity, some DR regimens are at least partially independent of this transcription factor (Houthoofd et al. 2003; Panowski et al. 2007). While this suggests that the longevity mechanisms of DR and reduced IIS are separable, an alternative explanation could be that DR benefits can only be replicated by the combined action of multiple nutrient-sensing pathways.

Recent work by Hou et al. elegantly addresses this point (Hou et al. 2016). By analyzing the transcriptomes of nematodes undergoing chronic DR and intermittent fasting and comparing these datasets to established genetic models of longevity, the authors identified that DR-associated gene expression changes seemed to be reprised by the summed effects of 3 distinct genetic pathways: a TOR module, an AMPK/Calcineurin/Xbox-binding protein 1 (XBP1) module, and a FOXO/DAF-16 module. Intriguingly, DR ceased to have an additive effect on lifespan in mutant animals only when all 3 genetic pathways were simultaneously modulated, suggesting that together the combined action of these genetic pathways recapitulates DR. Notably, however, genetically perturbing these pathways can elicit a longevity response much stronger than DR alone. Dual inhibition of the insulin and TOR pathways, via the insulin receptor/DAF-2 and S6K/RSKS-1 in *C. elegans*, promotes an incredible 5-fold extension in lifespan, analogous to humans living for several hundred years (Chen et al. 2013). The delayed aging of these double mutants was completely dependent on FOXO/DAF-16, and the authors exploited this dependency to identify changes in gene expression that mediate this extreme longevity. This analysis ultimately revealed that the AMPK complex was activated both transcriptionally and post-translationally in these animals, and that the AMPK γ subunit, AAKG-4, and α subunit, AAK-2, were required for the synergism between TOR and IIS inhibition (Chen et al. 2013). In the context of this chapter, these studies altogether reveal that AMPK acts as a key node in this longevity network, but also suggest that there are a variety of AMPK-independent routes by which alternative nutrient-sensing pathways can promote healthier aging. While the past two decades of work in the field have focused largely on the concept of single-gene interventions that regulate lifespan, perhaps this recent work indicates that the most potent protection against age-onset disease and morbidity will require combinatorial interventions.

10.4 Pro-longevity Pharmacology and AMPK

The ultimate goal of the aging field is to develop pharmacological interventions that ameliorate the morbidity and disease risk associated with advanced age. In addition to genetic approaches towards the dissection of metabolic pathways involved in longevity, pharmacological studies have also proven fruitful—not only in elucidating longevity pathways through experiments in model organisms but also in providing early evidence that this type of therapeutic approach may be feasible in humans.

Many of these small molecules target one or more of the nutrient-sensing pathways mentioned previously. Because of the interlinked activation status of the nutrient sensors, it can be difficult to determine whether these molecules are "dirty" and bind multiple targets or whether they bind one and initiate a chain reaction of regulation within the network. Resveratrol, for example, is a polyphenol initially thought to extend lifespan in yeast and other invertebrate models through activation of the sirtuin Sir2 and its orthologs, i.e., SIRT1 in mammals (Howitz et al. 2003; Wood et al. 2004). However, resveratrol given to mice on high fat diets dramatically reverses diet-associated pathology and decreased health span by a mechanism associated with decreased insulin/IGF-1 levels and an ability to activate AMPK, even to a similar extent as 5-aminoimidazole-4-carboxamide ribonucleutide (AICAR), an AMP mimetic (Baur et al. 2006). Casting into doubt whether sirtuins are really the primary effectors of resveratrol, a study of mice fed resveratrol revealed that the beneficial effects on most metabolic parameters, including fat mass, glucose homeostasis, and mitochondrial biogenesis, were dependent on AMPK α subunits 1 or 2 and, furthermore, that resveratrol can activate AMPK in cells lacking any SIRT1 (Um et al. 2010). It does appear that resveratrol can directly bind SIRT1, so alternative, dose-dependent mechanisms of resveratrol effects on aging and health span may exist, and future studies will need to address this question (Hubbard et al. 2013; Price et al. 2012).

Notably, resveratrol's ability to influence lifespan and some markers of age-related pathology in flies, rodents, and more recently rhesus monkeys seems achievable primarily in the context of an "unhealthy" diet, which may limit (but not nullify) its therapeutic applications (Baur et al. 2006; Jimenez-Gomez et al. 2013; Mattison et al. 2014; Pearson et al. 2008; Wang et al. 2013). In contrast, the FDA-approved macrolide immunosuppressant rapamycin has repeatedly and robustly extended lifespan in genetically heterogeneous mice fed normal diets, even when initiated late in life (Harrison et al. 2009; Miller et al. 2011, 2014). Rapamycin inhibits TOR complex signaling through binding to its inhibitor FK506-binding protein 12 (FKBP12) (Laplante and Sabatini 2012). Given the reciprocal regulation between AMPK and TOR kinases, this raises the possibility that AMPK plays an important role in rapamycin-induced longevity, but direct tests of this hypothesis are still required. Delineating the downstream mechanisms of rapamycin's effects on aging will be important in order to develop methods to uncouple health span benefits from its established role in immunosuppression.

In addition to targeting the nutrient sensors themselves, some small molecules target key enzymes in energy metabolism, thus indirectly affecting the activity of the core nutrient sensors. One such example arises from chemical inhibitors of glucose metabolism. Supplementing *C. elegans* diet with excess glucose shortens lifespan, while feeding nematodes the glycolytic inhibitor 2-deoxyglucose (2DG) extends lifespan (Schulz et al. 2007). Similarly, feeding worms with D-glucosamine, which—unlike 2DG—is nontoxic in mammalian models, also suppresses glucose metabolism and extends lifespan in both *C. elegans* and mice (Weimer et al. 2014). Intriguingly, the authors in both cases find that the α subunit AAK-2 is required for these lifespan effects in *C. elegans*, and suggest a model whereby reduced glucose availability leads to enhanced, AMPK-dependent mitochondrial biogenesis, respiration, and longevity.

Following a similar line of reasoning, the biguanide metformin has long been prescribed to correct hyperglycemia and related dysregulation of glucose metabolism in diabetic patients. Following the discovery that it exerts protection against cancer (Evans et al. 2005), however, metformin has emerged as one of the most promising compounds to combat age-onset pathology in humans. Due to metformin's status as the most commonly prescribed treatment for type II diabetes worldwide, there is a growing wealth of longitudinal data describing outcomes for patients receiving chronic treatment. Studies following large cohorts of diabetic patients continue to demonstrate that metformin treatment confers protection against cardiovascular disease in addition to multiple forms of cancer and even all-cause mortality, perhaps providing a preliminary proof-of-principle that a single drug is capable of preventing diverse age-dependent pathologies (Milman and Barzilai 2016; Pryor and Cabreiro 2015). Clinical trials are now underway to determine more directly if metformin improves cancer-related outcomes in patients without any confounding effects of diabetes.

Though it was once suspected that metformin could be a direct AMPK agonist, multiple studies ultimately determined that metformin more directly targets mitochondrial function via Complex I (El-Mir et al. 2000; Owen et al. 2000). Regardless, metformin indisputably promotes AMPK activation, likely as a secondary consequence of effects on AMPK's upstream activator LKB1 (Shaw et al. 2005) and/or possibly also through alterations in AMP/ATP levels. Metformin extends lifespan of *C. elegans* (Cabreiro et al. 2013; Onken and Driscoll 2010) and mice (Anisimov et al. 2011; Martin-Montalvo et al. 2013), though notably failed to promote longevity in *Drosophila*, possibly due to toxicity (Slack et al. 2012). Metformin and/or its close relative phenformin potently activates AMPK in all of these systems, and though it was not directly tested in the mouse studies, metformin longevity in *C. elegans* requires intact AMPK (Cabreiro et al. 2013; Onken and Driscoll 2010).

The cellular mechanisms mediating lifespan effects downstream of metformin are not yet well understood, though the working models proposed in these studies converge on enhanced stress defense and detoxification processes. In *C. elegans*, SKN-1/Nrf2, a master regulator of ROS and xenobiotic detoxification, is required in addition to AMPK for metformin to extend lifespan (Onken and Driscoll 2010). Consistent with its effectiveness as a treatment for diabetes, an alternative explanation is that metformin protects metabolic homeostasis, e.g., glucose levels, insulin sensitivity, and blood lipid profiles, later into old age (Martin-Montalvo et al. 2013). Metformin may also represent an interesting tool in dissecting which AMPK-dependent metabolic functions correlate with longevity and which do not, as several of metformin's effects are now known to be AMPK-independent. For example, AMPK is not necessarily required for metformin to inhibit mammalian target of rapamycin complex I (mTORC1) (Kalender et al. 2010) or for effects on glucose metabolism, at least at higher concentrations of the drug (Foretz et al. 2010; Madiraju et al. 2014; Miller et al. 2013).

Efforts to translate some of these experimental successes to humans are already underway; however, a barrier to advancement here is that aging is not recognized as a treatable disorder and, therefore, not amenable to the current design of clinical trials. For this reason, any progress is restricted to molecules, such as rapamycin and metformin, which are already proven safe and effective for specific disease states. In an attempt to transform this paradigm, the Food and Drug Administration recently announced plans to carry forward the Targeting Aging with Metformin (TAME) study. In the TAME study, metformin will be given to participants at risk of cancer, heart disease, or cognitive impairment, and its ability to reduce the risk of death presented by this broad spectrum of age-related diseases will be evaluated. If successful, the results of this study will validate the argument for the ability of a drug to broadly extend health span and inform the design of future clinical trials to identify pharmacological agents capable of doing so.

10.5 AMPK Activation Is Sufficient to Extend Lifespan

Arguably the most important characteristic of a longevity-promoting therapeutic target is that directly modulating its activity is sufficient to extend lifespan. While the previously mentioned drugs and genetic longevity paradigms promote AMPK activation and require its function to extend lifespan, it remains difficult to rule out contributions from off-target (i.e., non-AMPK) effects. However, multiple independent lines of evidence from direct genetic perturbations in both worm and fly highlight the potential of AMPK as just such a target. Methods of constitutive activation of the AMPK complex in vivo have included overexpression of the wild-type α subunit (Apfeld et al. 2004; Stenesen et al. 2013; Ulgherait et al. 2014), expression of mutant forms of the α subunit which enhances its basal activation (Mair et al. 2011), as well as overexpression of mutated γ subunits that mimics AMP-binding (Greer et al. 2007a). In each of these cases, enhanced activation of AMPK is sufficient to extend lifespan even in animals fed ad libitum. While several specific small-molecule activators of AMPK are known (Cool et al. 2006; Li et al. 2013), it remains unclear whether these compounds extend lifespan. Hawley et al. also recently demonstrated that salicylate, the active form of aspirin, directly binds AMPK β and promotes activation of the complex (Hawley et al. 2012). Intriguingly, experiments run as part of the Intervention Testing Program of the National Institute of Health revealed that chronic administration of aspirin significantly extends the lifespan of male mice (Strong et al. 2008), and these effects on aging appear to be evolutionarily conserved, as aspirin also promotes longevity in *C. elegans* (Ayyadevara et al. 2013; Wan et al. 2013). Suggesting that these effects are mediated at least in part through salicylate binding to AMPK, aspirin-treated worms require AMPK/AAK-2 for lifespan extension (Wan et al. 2013). In light of these findings, testing the effects of additional AMPK activators only becomes more compelling.

More recently, studies have begun using AMPK transgenics to address an added layer of complexity that is difficult to tease apart with drug treatments, namely that despite the ubiquity of AMPK expression across metazoan tissues, neither the effects of energy stress nor the roles of AMPK are uniform between cell types.

Furthermore, alternate isoform composition of heterotrimeric AMPK complexes can have differing effects in the same tissue, and specific diseases can be associated with mutations in one variant of a subunit but not another (Ross et al. 2016). This raises the possibility that unique AMPK complex combinations could act specifically in a limited number of tissues to promote lifespan extension. In fact, recent work exploring the outcome of chronic, systemic AMPK activation via mutation of the $\gamma 2$ subunit in mice has revealed a number of adverse consequences, including hyperphagia and obesity (Yavari et al. 2016). Thus, it seems clear that deleterious effects in one tissue may outweigh the benefits of AMPK activation in others, perhaps necessitating a tissue- or subunit-selective approach for therapeutic targeting of the complex.

A collection of studies in both worms and flies have begun to test this possibility—that AMPK action in select tissues can indeed promote longevity—but a clear and consistent model is still emerging. Stenesen et al. overexpressed a wild-type AMPK α subunit in either the fat body or the muscle of adult *Drosophila* and found in both cases that tissue-specific AMPK activation was sufficient to extend lifespan (Stenesen et al. 2013). Similarly, a second study by Ulgherait et al. overexpressed AMPK α in two fly tissues, neurons or intestine, again finding that activation in either tissue was sufficient to extend lifespan (Ulgherait et al. 2014). Two mechanistic explanations for these results are possible: first, in a cell autonomous model, AMPK could be prolonging the health of the neurons or intestine in which it was overexpressed, and prolonging the integrity of individual tissues is sufficient to extend the lifespan of the animal. Alternatively, AMPK could defend against age-related pathology cell nonautonomously. The authors took the next step to address this question, ultimately confirming the latter model by showing that either neuronal or intestinal AMPK overexpression protects against age-related proteotoxicity in distal tissues (e.g., muscle) through a mechanism that involves reduced insulin signaling between tissues and cell-nonautonomous upregulation of autophagy (Ulgherait et al. 2014). In the future, activating AMPK in all individual tissues and comparison to ubiquitous overexpression may help to integrate these observations into a complete model. Taking this systematic approach in *C. elegans* revealed that in contrast to the fly, tissue-specific overexpression of an activated AMPK α, AAK-2, in the worm neurons, muscle, or intestine all failed to extend lifespan, despite robust longevity in animals expressing the same transgene from the native AAK-2 promoter (Burkewitz et al. 2015). Despite the requirement of AMPK activation in multiple tissues, the mechanism of AMPK longevity in the worm still depends on cell nonautonomous signaling, however. Uncoupling AMPK from regulating its conserved downstream substrate, CREB-regulated transcriptional coactivator (CRTC-1), selectively in neurons abolishes lifespan extension. In this model, neuronal AMPK regulates a permissive, central switch, which allows peripheral AMPK action to promote longevity (see later sections for more detail).

10.6 Longevity-Promoting Outputs of AMPK Activation

AMPK modulates a number of cellular processes known to impact aging via posttranslational regulation of signaling, metabolic, and transcriptional effectors. Given that global activation of AMPK is unlikely to be an effective therapeutic strategy, a complete understanding of AMPK's downstream effects may enable a refined approach to harness the beneficial, pro-longevity effects of AMPK (Fig. 10.3).

10.6.1 Autophagy

Macroautophagy—the degradation and recycling of cytoplasmic components via delivery to the lysosome (herein referred to as autophagy)—is primarily thought of as a cell protective process, maintaining cellular quality control via degradation of damaging components, as well as helping to meet bioenergetic demand. A rapidly growing body of literature supports a role for autophagy in aging, which has been reviewed elsewhere (Knuppertz and Osiewacz 2016; Rubinsztein et al. 2011). Given the importance of autophagy in aging and age-related diseases, understanding the precise regulatory mechanisms that govern the process is a rapidly

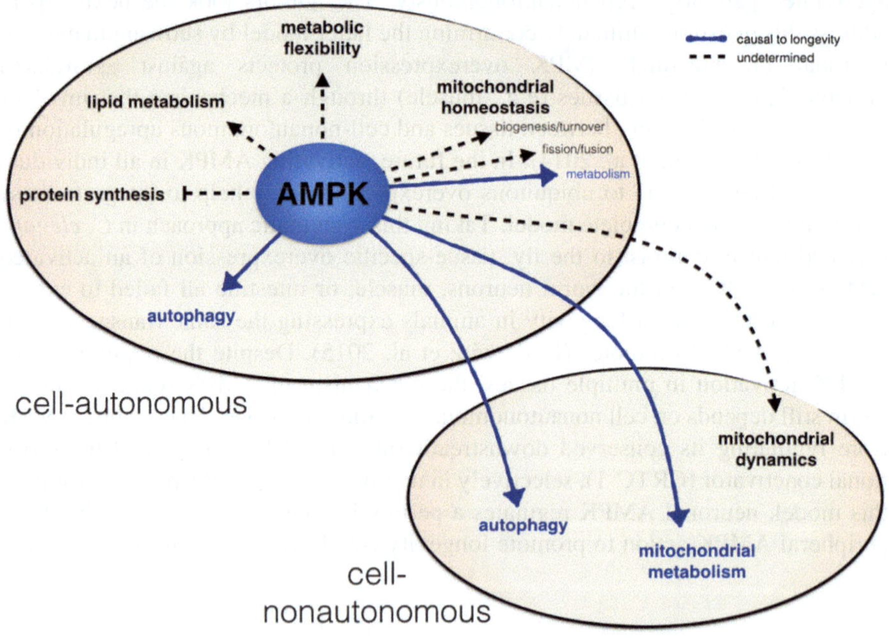

Fig. 10.3 Summary of downstream processes mediating AMPK longevity

expanding area of research. While the molecular mechanisms are undoubtedly complex, there is an emerging role for AMPK, both directly and indirectly, in autophagy induction.

Some of the earliest reports linking autophagy with longevity stemmed from the observation that autophagy is strongly induced by DR in multiple model systems, enabling cells to maintain energy homeostasis during nutrient scarcity via the turnover of cytoplasmic macromolecules (Rubinsztein et al. 2011). Under low nutrient conditions, autophagy induction begins with the formation of autophagosomes, the double membrane vesicles that are delivered to the lysosome. Part of the machinery required for this process is a conserved kinase complex that includes UNC-51-like kinase 1 (ULK-1), a mammalian homologue of yeast Atg1. AMPK, activated by low energy status, directly phosphorylates ULK-1 on at least two residues, resulting in activation of ULK-1 (Egan et al. 2011; Kim et al. 2011). AMPK deficiency in mammalian cells and mouse liver results in autophagy defects, confirming the physiological importance of this AMPK-dependent regulation. The requirement for AMPK to induce autophagy is also conserved across species, as the AMPK ortholog AAK-2 is necessary and sufficient to activate autophagy in *C. elegans* (Egan et al. 2011).

A second step in autophagosome formation was also found to be regulated by AMPK: Vps34, a class III phosphatidylinositol-3 kinase, functions in various cellular processes, including autophagy, by forming multiple different complexes. When complexed with Beclin 1 and other autophagic machinery components, Vps34 controls autophagosome biogenesis and maturation. Kim et al. elucidated an intricate AMPK-dependent mechanism involving differential regulation of Vps34 complexes, showing that under nutrient stress AMPK specifically promotes the pro-autophagy activity of Vps34 by phosphorylating Vps34 and Beclin 1 (Kim and Guan 2013).

Recently, a more specific role for AMPK in regulating the selective clearance of damaged mitochondria by mitophagy was revealed (Toyama et al. 2016). In response to treatment with electron transport chain inhibitors, the mitochondrial network undergoes rapid fragmentation, allowing for degradation of defective mitochondria. AMPK was found to be required for this stress-induced fragmentation, and pharmacological activation of AMPK even in the absence of mitochondrial stress was sufficient to promote acute fragmentation of the network (Toyama et al. 2016). A screen to establish the molecular mechanism responsible revealed a novel role for AMPK in regulating mitochondrial dynamics; AMPK directly phosphorylates mitochondrial fission factor (MFF), an outer mitochondrial membrane protein that recruits the cytoplasmic GTPase dynamin-related protein-1 (DRP1) to sites of fission (Toyama et al. 2016). Interestingly, aged mice lacking skeletal muscle AMPK accumulate p62 and parkin, indicating a defect in mitophagy. This was linked to exacerbation of age-associated myopathy, suggesting that AMPK-mediated induction of mitophagy plays a role in aging (Bujak et al. 2015).

In addition to direct phosphorylation events that lead to the induction of autophagy, AMPK likely plays further roles in regulating the process via crosstalk with other signaling pathways. mTORC1 is a well-established inhibitor of autophagy, and part of AMPK's role in promoting autophagy requires inhibition of mTORC1 under low nutrient conditions to relieve mTORC1-dependent inhibition. mTORC1 disrupts the interaction between AMPK and ULK-1 by phosphorylating ULK-1 at a different site (Kim et al. 2011). When activated by low glucose, AMPK inhibits mTORC1 through phosphorylation of Raptor and TSC, two regulatory components governing mTORC1 activity, which in turn allows for AMPK-dependent phosphorylation of ULK-1 (Gwinn et al. 2008; Inoki et al. 2003; Kim et al. 2011). Inhibition of autophagy by mTORC1 also involves other mechanisms, including regulation of the mammalian transcription factor EB (TFEB). Under nutrient-rich conditions, mTORC1 phosphorylates TFEB, preventing its nuclear localization and thereby inhibiting transcription of its target genes, including multiple autophagy-related and lysosomal genes (Settembre et al. 2012). Interestingly, the *C. elegans* ortholog HLH-30 is required for lifespan extension in multiple longevity pathways, and overexpression extends *C. elegans* lifespan (Lapierre et al. 2013). Since AMPK inhibits mTORC1, it is conceivable that AMPK plays a role in this pathway to permit nuclear translocation of TFEB/HLH-30 when nutrient levels are low; however, this remains to be demonstrated.

It is evident that AMPK participates in the complex, multilevel regulation of autophagy, but until recently it remained unclear whether autophagy played a causal role in AMPK-mediated longevity. Addressing this question in *Drosophila*, Ulgerhait et al. demonstrated that AMPK modulates autophagy in a tissue-specific manner by both cell-autonomous and cell-nonautonomous mechanisms (Ulgherait et al. 2014). Activating AMPK in the nervous system extended lifespan, with a concomitant induction of autophagy in the intestine and autophagy-related gene expression in muscle. Similarly, intestinal activation of AMPK resulted in autophagy induction in the nervous system. Establishing causality, neuronal RNAi knockdown of *Atg1*, a kinase involved in autophagosome formation, suppressed the ability of neuronal AMPK to promote longevity and slow markers of systemic aging. Furthermore, neuronal upregulation of Atg1 was sufficient to extend lifespan and slow aging (Ulgherait et al. 2014). The requirement of autophagy in other longevity pathways closely linked to AMPK further suggest that autophagy has a central role in conferring AMPK-dependent lifespan extension. For example, extension of lifespan by inhibition of calcineurin in *C. elegans* requires autophagy (Dwivedi et al. 2009). AMPK and calcineurin share a direct longevity target, CRTC-1, and have opposing effects on its localization; activation of AMPK or inhibition of calcineurin prevents nuclear localization of CRTC-1, resulting in lifespan extension in *C. elegans*. It will be interesting to determine whether this signaling pathway involving CRTCs converges on the regulation of autophagy.

Further studies are needed to determine precisely how AMPK regulates autophagy in the context of aging. In particular, verifying that the requirement for autophagy in AMPK-mediated longevity is conserved in mammalian systems, and in which tissues the promotion of autophagy is beneficial for slowing aging should be priorities. Another gap in our understanding relates to the identity of specific autophagic cargo that need to be degraded to promote healthy aging. It is plausible that selectivity exists for different cargo according to nutritional and environmental conditions, and so determining the relevant cargo for AMPK-mediated regulation of autophagy will be revealing. For example, since acute activation of AMPK was recently shown to promote mitophagy, it will be interesting to establish if clearance of damaged mitochondria is required for AMPK-mediated longevity.

10.6.2 Mitochondrial Function

Mitochondrial function is central to numerous longevity pathways and has been widely reported to become dysregulated with age. The principal function of AMPK is to maintain energy homeostasis; activated when cellular energy levels are low, AMPK turns on catabolic pathways and inhibits anabolic ATP-consuming pathways to restore ATP levels. Accordingly, AMPK acts as a master regulator of mitochondrial biogenesis and metabolism, and the extent to which it does so in the context of aging is just beginning to be explored.

One way that AMPK increases ATP levels is by inducing mitochondrial biogenesis (Winder et al. 2000; Zong et al. 2002). Peroxisome proliferator-activated receptor γ coactivator-1 α (PGC1α), the master transcriptional regulator of mitochondrial biogenesis, is directly phosphorylated at two residues by AMPK (Jager et al. 2007). Pharmacological activation of AMPK by AICAR induces mitochondrial gene expression but fails to do so in mice lacking PGC1α in skeletal muscle. The mechanism underlying transcriptional induction appears to involve two levels of regulation. First, when phosphorylated by AMPK, PGC1α transcriptionally activates its own promoter (Jager et al. 2007). Secondly, SIRT1 can deacetylate PGC1α to enhance its activity, though this deacetylation first requires priming by AMPK-mediated phosphorylation of PGC1α in order to recruit SIRT1 (Canto et al. 2009). AMPK further enhances mitochondrial biogenesis by increasing cellular NAD+ levels leading to activation of SIRT1 and subsequent activation of PGC1α (Canto et al. 2009). Aging is associated with reductions in mitochondrial function, including reduced mitochondrial biogenesis, which may at least in part be due to a diminished ability to activate AMPK with age. Activation of AMPK, either acutely by AICAR and exercise or chronically by β-guanidinopropionic acid (β-GPA), was blunted in skeletal muscle of aged rats (Reznick et al. 2007). β-GPA treatment also failed to induce mitochondrial biogenesis in old rats, suggesting that age-associated decreases in AMPK activity may be a limiting factor for promoting mitochondrial function in old age (Reznick et al. 2007). The potential for promoting longevity by targeting mitochondrial biogenesis has been demonstrated in flies, where tissue-specific overexpression of PGC1α in intestinal stem cells extends lifespan and delays age-related loss of tissue homeostasis (Rera et al. 2011). It is currently unknown if altered mitochondrial biogenesis is required for AMPK-mediated lifespan extension. A recent study showed that mitochondrial biogenesis and mitophagy are coordinately regulated, and uncoupling of the two processes during aging contributes to dysregulated cellular function in *C. elegans*

(Palikaras et al. 2015). Since AMPK regulates both of these processes, it is tempting to speculate that maintenance of mitochondrial homeostasis and content may be instrumental to AMPK-mediated longevity.

Beyond promoting mitochondrial biogenesis, AMPK specifically regulates systemic mitochondrial function by reprogramming mitochondrial metabolism, which was recently shown to be required for AMPK-mediated lifespan extension in *C. elegans* (Burkewitz et al. 2015). Genetic activation of AAK-2 extends lifespan by inhibiting CRTC-1 via phosphorylation (Mair et al. 2011). Transcriptional and metabolic analyses revealed that suppression of AAK-2 longevity by activating CRTC-1 (rendering it insensitive to AMPK regulation) is associated with broad suppression of mitochondrial metabolic genes and downregulation of TCA cycle intermediates, indicating that AMPK-dependent regulation of mitochondrial function is specifically coupled to its promotion of longevity (Burkewitz et al. 2015). This study further showed that AMPK regulates systemic mitochondrial metabolism via a downstream cell-nonautonomous mechanism: opposing transcriptional effectors CRTC-1/CREB and the PPARα ortholog, NHR-49, act specifically in the nervous system to regulate longevity and systemic metabolism. Notably, activating CRTC-1 in neurons results in fragmentation of mitochondria in muscle cells, demonstrating that cell-nonautonomous remodeling of mitochondrial dynamics is associated with AMPK longevity (Burkewitz et al. 2015). Taken together with the recent finding that AMPK directly phosphorylates MFF to drive fission under mitochondrial stress (Toyama et al. 2016), this highlights an intriguing new role for AMPK in adapting mitochondrial function via remodeling mitochondrial networks.

It is becoming increasingly apparent that AMPK not only regulates mitochondrial function at multiple levels, but that at least some of these are indeed causal to AMPK-mediated longevity. Given the emerging cell-autonomous and nonautonomous mechanisms by which AMPK acts to regulate mitochondrial homeostasis, deciphering the tissue-specific requirements for these effects will be essential in order to selectively harness the pro-longevity effects of AMPK activation.

10.6.3 Lipid Metabolism and Metabolic Flexibility

The capacity to switch between different fuel substrates, defined as metabolic flexibility, enables an organism to adapt to changing nutrient availability. When glucose availability is high, β-oxidation is suppressed in favor of glucose metabolism, whereas in the fasting state a metabolic shift occurs to limit glucose metabolism and instead promote fat metabolism. Impaired metabolic flexibility contributes to dysfunctional energy homeostasis and ultimately metabolic disease and is associated with aging and age-related diseases (Calcada et al. 2014; Gao et al. 2014). Exactly how metabolic function changes with age and the contribution of specific metabolites to the aging process is relatively understudied, but reciprocal changes in phosphoenolpyruvate carboxykinase (PEPCK-C) and glycolytic pyruvate kinase (PK) have been reported in mammalian muscle and liver tissue (Dhahbi et al. 1999). A recent study in *C. elegans* shed more light in this area, employing metabolic flux analyses to show that an age-associated decrease in PEPCK-C and increase in PK is necessary and sufficient to limit lifespan (Yuan et al. 2016). Moreover, PEPCK-C overexpression extends lifespan via promotion of oxidative metabolism and acceleration of the TCA cycle, which requires functional AAK-2 (Yuan et al. 2016). Altered fat metabolism is also associated with aging and linked to several different longevity pathways (Lapierre et al. 2011; Soukas et al. 2009; Wang et al. 2008). Despite clear links between AMPK, metabolic flexibility, and fat metabolism, the contribution of these processes to AMPK-mediated longevity is less well understood.

AMPK is instrumental in the adaptive transcriptional response of cells to altered fuel source availability. Upon sensing energy stress, AMPK acts as a metabolic switch by activating the transcriptional regulators PGC1α and FOXO1 via activation of SIRT1. The net result is an increase in transcription of mitochondrial and fatty acid oxidation genes, shifting fuel source usage from carbohydrate to lipid (Canto et al. 2010). While clearly playing a vital role in cell survival during times of nutrient stress, a causal role in AMPK-mediated longevity remains to be demonstrated. One line of evidence suggests that AMPK controls fuel conservation when nutrient availability is scarce, which could conceivably contribute to prolonged lifespan. *C. elegans* lacking AAK-2 exhibit reduced triglyceride levels basally (Cunningham et al. 2014), but when starved, deplete their lipid stores more rapidly than wild-type worms due to the fact that AMPK limits triglyceride hydrolysis (Narbonne and Roy 2009). When exposed to starvation, *C. elegans* enter an alternative, dauer larval stage through a mechanism that involves AMPK, so confounding effects of larval staging can make interpretation of starvation assays difficult in worms (Cunningham et al. 2014). Nevertheless, in *Drosophila* reduced AMPK function also results in abnormal lipid homeostasis, sensitivity to starvation, and decreased lifespan under starvation (Johnson et al. 2010), and in mice activating AMPK reduces the rate of lipolysis in adipose tissue (Daval et al. 2005). A goal of future studies should be to address whether reduced lipolysis and preservation of lipid stores are causal to AMPK-mediated longevity.

Further enticing links exist between AMPK and the regulation of lipid metabolism that may be suggestive of a role in aging. For example, AMPK phosphorylates and inhibits sterol regulatory element-binding protein (SREBP), the master regulator of lipogenesis (Li et al. 2011). Intriguingly, given the role of CRTC family members as mediators of AMPK longevity (Mair et al. 2011), a recent study demonstrated that CRTC2 in the mammalian liver also regulates SREBP1, but through a mechanism unrelated to its role in transcriptional regulation in the nucleus (Han et al. 2015). Depending on nutrient levels and phosphorylation by mTOR, cytosolic CRTC2 binds COPII components that mediate ER-Golgi trafficking to prevent the maturation of active SREBP1, thus ultimately helping to promote lipid synthesis during feeding and inhibit lipogenesis during fasting (Han et al. 2015). Whether AMPK directly or indirectly plays a role in this mechanism is unknown, but deserves further exploration. More canonically, AMPK promotes β-oxidation via phosphorylation and inactivation of acetyl CoA carboxylase 2 (ACC2). This increases fatty acid uptake into mitochondria, again supporting a role for AMPK in regulating fuel source usage (Hardie et al. 2012). As our understanding of AMPK's role in regulating lipid metabolism moves forward, a priority should be to precisely define which aspects are required for AMPK-dependent promotion of healthy aging.

10.6.4 Protein Synthesis

Pharmacological or genetic inhibition of certain components of the protein synthesis machinery extends lifespan in yeast and *C. elegans* (Chiocchetti et al. 2007; Curran and Ruvkun 2007; Hansen et al. 2007; Henderson et al. 2006; Pan et al. 2007; Steffen et al. 2008; Syntichaki et al. 2007). The most obvious explanation for this phenomenon may be that reduced protein synthesis enhances the ability of cells to maintain protein homeostasis through chaperone action and turnover; but intriguingly, certain mRNAs are also preferentially translated when protein synthesis is inhibited, suggesting that specific changes in gene expression may mediate some of these lifespan phenotypes (Jefferies et al. 1994; Tzamarias et al. 1989). Rogers et al. identified AMPK/AAK-2 as one such preferentially synthesized transcript in *C. elegans* when the translation initiation factor eIF4G/IFG-1 is inhibited (Rogers et al. 2011). Subsequent epistasis analysis revealed that longevity caused by IFG-1 inhibition completely requires intact AAK-2 function, indicating that enhanced AMPK activity may promote at least some of the beneficial effects of ribosome inhibition on aging. AMPK itself is capable of inhibiting protein translation (Hoppe et al. 2009; Horman et al. 2002), suggesting the possibility of a feed-forward interaction or, alternatively, that reduced protein synthesis could also function as a downstream mechanism of AMPK longevity. Additional experiments are required to test these possibilities.

10.7 Concluding Remarks

While a great deal of progress has been made in establishing AMPK as a necessary and important effector of longevity in many paradigms, even more work remains in addressing the mechanisms by which AMPK actually defends against aging and associated pathologies. Does AMPK itself represent a drug target that could protect against an array of age-onset diseases? The effects of manipulating AMPK function in model organisms (e.g., significant growth and fertility defects) suggest that untoward side effects may preclude the benefits of such an approach. Further, the nuanced, differential roles of alternate AMPK heterotrimers in different cell types might limit the therapeutic potential of whole body alterations to global AMPK activity. Identifying the downstream, AMPK-regulated cell processes specific to its role in longevity therefore becomes of utmost importance as the field moves forward, ultimately leading to the design of drugs capable of uncoupling positive and negative AMPK effects. This remains a daunting task given that AMPK regulates virtually all facets of cell behavior. However, recent studies revealing that AMPK action may only be needed in select tissues to extend lifespan suggest that perhaps only a small number of AMPK outputs are required (Stenesen et al. 2013; Ulgherait et al. 2014), and proof-of-principle that longevity can be uncoupled from many unwanted side effects by a single AMPK substrate exists (Burkewitz et al. 2015). Notably, much of this direct evidence for AMPK's roles in aging stems from studies in invertebrate models and remains to be tested in mammals; on the other hand, understanding of AMPK's signaling roles within nutrient-sensing networks are derived almost solely from biochemical studies in mammalian models and extrapolated to invertebrate models through too many unvalidated assumptions. Given the rise of rapid and powerful genome-editing techniques, it will be both fascinating and critically important to integrate genetic models of aging with molecular models of intracellular signaling to precisely define the routes by which AMPK function promotes healthier aging.

References

Anisimov VN, Berstein LM, Popovich IG, Zabezhinski MA, Egormin PA, Piskunova TS, Semenchenko AV, Tyndyk ML, Yurova MN, Kovalenko IG et al (2011) If started early in life, metformin treatment increases life span and postpones tumors in female SHR mice. Aging 3:148–157

Apfeld J, O'Connor G, McDonagh T, DiStefano PS, Curtis R (2004) The AMP-activated protein kinase AAK-2 links energy levels and insulin-like signals to lifespan in C. elegans. Genes Dev 18:3004–3009

Ayyadevara S, Bharill P, Dandapat A, Hu C, Khaidakov M, Mitra S, Shmookler Reis RJ, Mehta JL (2013) Aspirin inhibits oxidant stress, reduces age-associated functional declines, and extends lifespan of Caenorhabditis elegans. Antioxid Redox Signal 18:481–490

Baur JA, Pearson KJ, Price NL, Jamieson HA, Lerin C, Kalra A, Prabhu VV, Allard JS, Lopez-Lluch G, Lewis K et al (2006) Resveratrol improves health and survival of mice on a high-calorie diet. Nature 444:337–342

Bujak AL, Crane JD, Lally JS, Ford RJ, Kang SJ, Rebalka IA, Green AE, Kemp BE, Hawke TJ, Schertzer JD et al (2015) AMPK activation of muscle autophagy prevents fasting-induced hypoglycemia and myopathy during aging. Cell Metab 21:883–890

Burkewitz K, Morantte I, Weir HJ, Yeo R, Zhang Y, Huynh FK, Ilkayeva OR, Hirschey MD, Grant AR, Mair WB (2015) Neuronal CRTC-1 governs systemic mitochondrial metabolism and lifespan via a catecholamine signal. Cell 160:842–855

Burnett C, Valentini S, Cabreiro F, Goss M, Somogyvari M, Piper MD, Hoddinott M, Sutphin GL, Leko V, McElwee JJ et al (2011) Absence of effects of Sir2 overexpression on lifespan in C. elegans and Drosophila. Nature 477:482–485

Cabreiro F, Au C, Leung KY, Vergara-Irigaray N, Cocheme HM, Noori T, Weinkove D, Schuster E, Greene ND, Gems D (2013) Metformin retards aging in C. elegans by altering microbial folate and methionine metabolism. Cell 153:228–239

Calcada D, Vianello D, Giampieri E, Sala C, Castellani G, de Graaf A, Kremer B, van Ommen B, Feskens E, Santoro A et al (2014) The role of low-grade inflammation and metabolic flexibility in aging and nutritional modulation thereof: a systems biology approach. Mech Ageing Dev 136–137:138–147

Canto C, Gerhart-Hines Z, Feige JN, Lagouge M, Noriega L, Milne JC, Elliott PJ, Puigserver P, Auwerx J (2009) AMPK regulates energy expenditure by modulating NAD+ metabolism and SIRT1 activity. Nature 458:1056–1060

Canto C, Jiang LQ, Deshmukh AS, Mataki C, Coste A, Lagouge M, Zierath JR, Auwerx J (2010) Interdependence of AMPK and SIRT1 for metabolic adaptation to fasting and exercise in skeletal muscle. Cell Metab 11:213–219

CDC (2013) The state of aging and health in America 2013. Centers for Disease Control and Prevention, US Dept of Health and Human Services, Atlanta, GA

Chen D, Li PW, Goldstein BA, Cai W, Thomas EL, Chen F, Hubbard AE, Melov S, Kapahi P (2013) Germline signaling mediates the synergistically prolonged longevity produced by double mutations in daf-2 and rsks-1 in C. elegans. Cell Rep 5:1600–1610

Chiocchetti A, Zhou J, Zhu H, Karl T, Haubenreisser O, Rinnerthaler M, Heeren G, Oender K, Bauer J, Hintner H et al (2007) Ribosomal proteins Rpl10 and Rps6 are potent regulators of yeast replicative life span. Exp Gerontol 42:275–286

Cool B, Zinker B, Chiou W, Kifle L, Cao N, Perham M, Dickinson R, Adler A, Gagne G, Iyengar R et al (2006) Identification and characterization of a small molecule AMPK activator that treats key components of type 2 diabetes and the metabolic syndrome. Cell Metab 3:403–416

Cunningham KA, Bouagnon AD, Barros AG, Lin L, Malard L, Romano-Silva MA, Ashrafi K (2014) Loss of a neural AMP-activated kinase mimics the effects of elevated serotonin on fat, movement, and hormonal secretions. PLoS Genet 10, e1004394

Curran SP, Ruvkun G (2007) Lifespan regulation by evolutionarily conserved genes essential for viability. PLoS Genet 3, e56

Curtis R, O'Connor G, DiStefano PS (2006) Aging networks in Caenorhabditis elegans: AMP-activated protein kinase (aak-2) links multiple aging and metabolism pathways. Aging Cell 5:119–126

Dagon Y, Hur E, Zheng B, Wellenstein K, Cantley LC, Kahn BB (2012) p70S6 kinase phosphorylates AMPK on serine 491 to mediate leptin's effect on food intake. Cell Metab 16:104–112

Daval M, Diot-Dupuy F, Bazin R, Hainault I, Viollet B, Vaulont S, Hajduch E, Ferre P, Foufelle F (2005) Anti-lipolytic action of AMP-activated protein kinase in rodent adipocytes. J Biol Chem 280:25250–25257

Dell'agnello C, Leo S, Agostino A, Szabadkai G, Tiveron C, Zulian A, Prelle A, Roubertoux P, Rizzuto R, Zeviani M (2007) Increased longevity and refractoriness to Ca(2+)-dependent neurodegeneration in Surf1 knockout mice. Hum Mol Genet 16:431–444

Dhahbi JM, Mote PL, Wingo J, Tillman JB, Walford RL, Spindler SR (1999) Calories and aging alter gene expression for gluconeogenic, glycolytic, and nitrogen-metabolizing enzymes. Am J Physiol 277:E352–360

Dillin A, Hsu AL, Arantes-Oliveira N, Lehrer-Graiwer J, Hsin H, Fraser AG, Kamath RS, Ahringer J, Kenyon C (2002) Rates of behavior and aging specified by mitochondrial function during development. Science 298:2398–2401

Dwivedi M, Song HO, Ahnn J (2009) Autophagy genes mediate the effect of calcineurin on life span in C. elegans. Autophagy 5:604–607

Egan DF, Shackelford DB, Mihaylova MM, Gelino S, Kohnz RA, Mair W, Vasquez DS, Joshi A, Gwinn DM, Taylor R et al (2011) Phosphorylation of ULK1 (hATG1) by AMP-activated protein kinase connects energy sensing to mitophagy. Science 331:456–461

El-Mir MY, Nogueira V, Fontaine E, Averet N, Rigoulet M, Leverve X (2000) Dimethylbiguanide inhibits cell respiration via an indirect effect targeted on the respiratory chain complex I. J Biol Chem 275:223–228

Evans JM, Donnelly LA, Emslie-Smith AM, Alessi DR, Morris AD (2005) Metformin and reduced risk of cancer in diabetic patients. BMJ 330:1304–1305

Fabrizio P, Pozza F, Pletcher SD, Gendron CM, Longo VD (2001) Regulation of longevity and stress resistance by Sch9 in yeast. Science 292:288–290
Fontana L, Partridge L (2015) Promoting health and longevity through diet: from model organisms to humans. Cell 161:106–118
Fontana L, Partridge L, Longo VD (2010) Extending healthy life span--from yeast to humans. Science 328:321–326
Foretz M, Hebrard S, Leclerc J, Zarrinpashneh E, Soty M, Mithieux G, Sakamoto K, Andreelli F, Viollet B (2010) Metformin inhibits hepatic gluconeogenesis in mice independently of the LKB1/AMPK pathway via a decrease in hepatic energy state. J Clin Invest 120:2355–2369
Fulco M, Cen Y, Zhao P, Hoffman EP, McBurney MW, Sauve AA, Sartorelli V (2008) Glucose restriction inhibits skeletal myoblast differentiation by activating SIRT1 through AMPK-mediated regulation of Nampt. Dev Cell 14:661–673
Gao AW, Canto C, Houtkooper RH (2014) Mitochondrial response to nutrient availability and its role in metabolic disease. EMBO Mol Med 6:580–589
Goldman DP, Cutler D, Rowe JW, Michaud PC, Sullivan J, Peneva D, Olshansky SJ (2013) Substantial health and economic returns from delayed aging may warrant a new focus for medical research. Health Aff 32:1698–1705
Gomes AP, Price NL, Ling AJ, Moslehi JJ, Montgomery MK, Rajman L, White JP, Teodoro JS, Wrann CD, Hubbard BP et al (2013) Declining NAD(+) induces a pseudohypoxic state disrupting nuclear-mitochondrial communication during aging. Cell 155:1624–1638
Greer EL, Brunet A (2009) Different dietary restriction regimens extend lifespan by both independent and overlapping genetic pathways in C. elegans. Aging Cell 8:113–127
Greer EL, Dowlatshahi D, Banko MR, Villen J, Hoang K, Blanchard D, Gygi SP, Brunet A (2007a) An AMPK-FOXO pathway mediates longevity induced by a novel method of dietary restriction in C. elegans. Curr Biol 17:1646–1656
Greer EL, Oskoui PR, Banko MR, Maniar JM, Gygi MP, Gygi SP, Brunet A (2007b) The energy sensor AMP-activated protein kinase directly regulates the mammalian FOXO3 transcription factor. J Biol Chem 282:30107–30119
Guarente L (2013) Calorie restriction and sirtuins revisited. Genes Dev 27:2072–2085
Gwinn DM, Shackelford DB, Egan DF, Mihaylova MM, Mery A, Vasquez DS, Turk BE, Shaw RJ (2008) AMPK phosphorylation of raptor mediates a metabolic checkpoint. Mol Cell 30:214–226
Han J, Li E, Chen L, Zhang Y, Wei F, Liu J, Deng H, Wang Y (2015) The CREB coactivator CRTC2 controls hepatic lipid metabolism by regulating SREBP1. Nature 524:243–246
Hansen M, Taubert S, Crawford D, Libina N, Lee SJ, Kenyon C (2007) Lifespan extension by conditions that inhibit translation in Caenorhabditis elegans. Aging Cell 6:95–110
Hardie DG, Ross FA, Hawley SA (2012) AMPK: a nutrient and energy sensor that maintains energy homeostasis. Nat Rev Mol Cell Biol 13:251–262
Harrison DE, Strong R, Sharp ZD, Nelson JF, Astle CM, Flurkey K, Nadon NL, Wilkinson JE, Frenkel K, Carter CS et al (2009) Rapamycin fed late in life extends lifespan in genetically heterogeneous mice. Nature 460:392–395
Haslam DW, James WP (2005) Obesity. Lancet 366:1197–1209
Hawley SA, Fullerton MD, Ross FA, Schertzer JD, Chevtzoff C, Walker KJ, Peggie MW, Zibrova D, Green KA, Mustard KJ et al (2012) The ancient drug salicylate directly activates AMP-activated protein kinase. Science 336:918–922
Haynes CM, Fiorese CJ, Lin YF (2013) Evaluating and responding to mitochondrial dysfunction: the mitochondrial unfolded-protein response and beyond. Trends Cell Biol 23:311–318
Henderson ST, Bonafe M, Johnson TE (2006) daf-16 protects the nematode Caenorhabditis elegans during food deprivation. J Gerontol A Biol Sci Med Sci 61:444–460
Hoppe S, Bierhoff H, Cado I, Weber A, Tiebe M, Grummt I, Voit R (2009) AMP-activated protein kinase adapts rRNA synthesis to cellular energy supply. Proc Natl Acad Sci U S A 106:17781–17786

Horman S, Browne G, Krause U, Patel J, Vertommen D, Bertrand L, Lavoinne A, Hue L, Proud C, Rider M (2002) Activation of AMP-activated protein kinase leads to the phosphorylation of elongation factor 2 and an inhibition of protein synthesis. Curr Biol 12:1419–1423

Horvath S, Erhart W, Brosch M, Ammerpohl O, von Schonfels W, Ahrens M, Heits N, Bell JT, Tsai PC, Spector TD et al (2014) Obesity accelerates epigenetic aging of human liver. Proc Natl Acad Sci U S A 111:15538–15543

Hou L, Wang D, Chen D, Liu Y, Zhang Y, Cheng H, Xu C, Sun N, McDermott J, Mair WB et al (2016) A Systems Approach to Reverse Engineer Lifespan Extension by Dietary Restriction. Cell Metab 23:529–540

Houthoofd K, Braeckman BP, Johnson TE, Vanfleteren JR (2003) Life extension via dietary restriction is independent of the Ins/IGF-1 signalling pathway in Caenorhabditis elegans. Exp Gerontol 38:947–954

Houtkooper RH, Mouchiroud L, Ryu D, Moullan N, Katsyuba E, Knott G, Williams RW, Auwerx J (2013) Mitonuclear protein imbalance as a conserved longevity mechanism. Nature 497:451–457

Howitz KT, Bitterman KJ, Cohen HY, Lamming DW, Lavu S, Wood JG, Zipkin RE, Chung P, Kisielewski A, Zhang LL et al (2003) Small molecule activators of sirtuins extend Saccharomyces cerevisiae lifespan. Nature 425:191–196

Hubbard BP, Gomes AP, Dai H, Li J, Case AW, Considine T, Riera TV, Lee JE, E SY, Lamming DW et al (2013) Evidence for a common mechanism of SIRT1 regulation by allosteric activators. Science 339:1216–1219

Inoki K, Ouyang H, Zhu T, Lindvall C, Wang Y, Zhang X, Yang Q, Bennett C, Harada Y, Stankunas K et al (2006) TSC2 integrates Wnt and energy signals via a coordinated phosphorylation by AMPK and GSK3 to regulate cell growth. Cell 126:955–968

Inoki K, Zhu T, Guan KL (2003) TSC2 mediates cellular energy response to control cell growth and survival. Cell 115:577–590

Jager S, Handschin C, St-Pierre J, Spiegelman BM (2007) AMP-activated protein kinase (AMPK) action in skeletal muscle via direct phosphorylation of PGC-1alpha. Proc Natl Acad Sci U S A 104:12017–12022

Jefferies HB, Reinhard C, Kozma SC, Thomas G (1994) Rapamycin selectively represses translation of the "polypyrimidine tract" mRNA family. Proc Natl Acad Sci U S A 91:4441–4445

Jimenez-Gomez Y, Mattison JA, Pearson KJ, Martin-Montalvo A, Palacios HH, Sossong AM, Ward TM, Younts CM, Lewis K, Allard JS et al (2013) Resveratrol improves adipose insulin signaling and reduces the inflammatory response in adipose tissue of rhesus monkeys on high-fat, high-sugar diet. Cell Metab 18:533–545

Johnson EC, Kazgan N, Bretz CA, Forsberg LJ, Hector CE, Worthen RJ, Onyenwoke R, Brenman JE (2010) Altered metabolism and persistent starvation behaviors caused by reduced AMPK function in Drosophila. PloS One 5

Johnson SC, Rabinovitch PS, Kaeberlein M (2013) mTOR is a key modulator of ageing and age-related disease. Nature 493:338–345

Kaeberlein M, McVey M, Guarente L (1999) The SIR2/3/4 complex and SIR2 alone promote longevity in Saccharomyces cerevisiae by two different mechanisms. Genes Dev 13:2570–2580

Kalender A, Selvaraj A, Kim SY, Gulati P, Brule S, Viollet B, Kemp BE, Bardeesy N, Dennis P, Schlager JJ et al (2010) Metformin, independent of AMPK, inhibits mTORC1 in a rag GTPase-dependent manner. Cell Metab 11:390–401

Kanfi Y, Naiman S, Amir G, Peshti V, Zinman G, Nahum L, Bar-Joseph Z, Cohen HY (2012) The sirtuin SIRT6 regulates lifespan in male mice. Nature 483:218–221

Kapahi P, Zid BM, Harper T, Koslover D, Sapin V, Benzer S (2004) Regulation of lifespan in Drosophila by modulation of genes in the TOR signaling pathway. Curr Biol 14:885–890

Kenyon C (2011) The first long-lived mutants: discovery of the insulin/IGF-1 pathway for ageing. Philos Trans R Soc Lond B Biol Sci 366:9–16

Kenyon C, Chang J, Gensch E, Rudner A, Tabtiang R (1993) A C. elegans mutant that lives twice as long as wild type. Nature 366:461–464

Kim J, Guan KL (2013) AMPK connects energy stress to PIK3C3/VPS34 regulation. Autophagy 9:1110–1111

Kim J, Kundu M, Viollet B, Guan KL (2011) AMPK and mTOR regulate autophagy through direct phosphorylation of Ulk1. Nat Cell Biol 13:132–141

Knuppertz L, Osiewacz HD (2016) Orchestrating the network of molecular pathways affecting aging: Role of nonselective autophagy and mitophagy. Mech Ageing Dev 153:30–40

Lakowski B, Hekimi S (1996) Determination of life-span in Caenorhabditis elegans by four clock genes. Science 272:1010–1013

Lan F, Cacicedo JM, Ruderman N, Ido Y (2008) SIRT1 modulation of the acetylation status, cytosolic localization, and activity of LKB1. Possible role in AMP-activated protein kinase activation. J Biol Chem 283:27628–27635

Lapierre LR, De Magalhaes Filho CD, McQuary PR, Chu CC, Visvikis O, Chang JT, Gelino S, Ong B, Davis AE, Irazoqui JE et al (2013) The TFEB orthologue HLH-30 regulates autophagy and modulates longevity in Caenorhabditis elegans. Nat Commun 4:2267

Lapierre LR, Gelino S, Melendez A, Hansen M (2011) Autophagy and lipid metabolism coordinately modulate life span in germline-less C. elegans. Curr Biol 21:1507–1514

Laplante M, Sabatini DM (2012) mTOR signaling in growth control and disease. Cell 149:274–293

Lee SS, Lee RY, Fraser AG, Kamath RS, Ahringer J, Ruvkun G (2003) A systematic RNAi screen identifies a critical role for mitochondria in C. elegans longevity. Nat Genet 33:40–48

Li Y, Xu S, Mihaylova MM, Zheng B, Hou X, Jiang B, Park O, Luo Z, Lefai E, Shyy JY et al (2011) AMPK phosphorylates and inhibits SREBP activity to attenuate hepatic steatosis and atherosclerosis in diet-induced insulin-resistant mice. Cell Metab 13:376–388

Li YY, Yu LF, Zhang LN, Qiu BY, Su MB, Wu F, Chen DK, Pang T, Gu M, Zhang W et al (2013) Novel small-molecule AMPK activator orally exerts beneficial effects on diabetic db/db mice. Toxicol Appl Pharmacol 273:325–334

Lin SJ, Defossez PA, Guarente L (2000) Requirement of NAD and SIR2 for life-span extension by calorie restriction in Saccharomyces cerevisiae. Science 289:2126–2128

Liu X, Jiang N, Hughes B, Bigras E, Shoubridge E, Hekimi S (2005) Evolutionary conservation of the clk-1-dependent mechanism of longevity: loss of mclk1 increases cellular fitness and lifespan in mice. Genes Dev 19:2424–2434

Lopez-Otin C, Blasco MA, Partridge L, Serrano M, Kroemer G (2013) The hallmarks of aging. Cell 153:1194–1217

Madiraju AK, Erion DM, Rahimi Y, Zhang XM, Braddock DT, Albright RA, Prigaro BJ, Wood JL, Bhanot S, MacDonald MJ et al (2014) Metformin suppresses gluconeogenesis by inhibiting mitochondrial glycerophosphate dehydrogenase. Nature 510:542–546

Mair W, Dillin A (2008) Aging and survival: the genetics of life span extension by dietary restriction. Annu Rev Biochem 77:727–754

Mair W, Morantte I, Rodrigues AP, Manning G, Montminy M, Shaw RJ, Dillin A (2011) Lifespan extension induced by AMPK and calcineurin is mediated by CRTC-1 and CREB. Nature 470:404–408

Mair W, Panowski SH, Shaw RJ, Dillin A (2009) Optimizing dietary restriction for genetic epistasis analysis and gene discovery in C. elegans. PLoS One 4, e4535

Martin-Montalvo A, Mercken EM, Mitchell SJ, Palacios HH, Mote PL, Scheibye-Knudsen M, Gomes AP, Ward TM, Minor RK, Blouin MJ et al (2013) Metformin improves healthspan and lifespan in mice. Nat Commun 4:2192

Masoro EJ (2002) Caloric restriction: a key to understanding and modulating aging. Elsevier, Amsterdam

Matsuzaki H, Daitoku H, Hatta M, Tanaka K, Fukamizu A (2003) Insulin-induced phosphorylation of FKHR (Foxo1) targets to proteasomal degradation. Proc Natl Acad Sci U S A 100:11285–11290

Mattison JA, Wang M, Bernier M, Zhang J, Park SS, Maudsley S, An SS, Santhanam L, Martin B, Faulkner S et al (2014) Resveratrol prevents high fat/sucrose diet-induced central arterial wall inflammation and stiffening in nonhuman primates. Cell Metab 20:183–190

McQuary PR, Liao CY, Chang JT, Kumsta C, She X, Davis A, Chu CC, Gelino S, Gomez-Amaro RL, Petrascheck M et al (2016) C. elegans S6K Mutants Require a Creatine-Kinase-like Effector for Lifespan Extension. Cell Rep 14:2059–2067

Merkwirth C, Jovaisaite V, Durieux J, Matilainen O, Jordan SD, Quiros PM, Steffen KK, Williams EG, Mouchiroud L, Tronnes SU et al (2016) Two Conserved Histone Demethylases Regulate Mitochondrial Stress-Induced Longevity. Cell 165(5):1209–1223

Miller RA, Chu Q, Xie J, Foretz M, Viollet B, Birnbaum MJ (2013) Biguanides suppress hepatic glucagon signalling by decreasing production of cyclic AMP. Nature 494:256–260

Miller RA, Harrison DE, Astle CM, Baur JA, Boyd AR, de Cabo R, Fernandez E, Flurkey K, Javors MA, Nelson JF et al (2011) Rapamycin, but not resveratrol or simvastatin, extends life span of genetically heterogeneous mice. J Gerontol A Biol Sci Med Sci 66:191–201

Miller RA, Harrison DE, Astle CM, Fernandez E, Flurkey K, Han M, Javors MA, Li X, Nadon NL, Nelson JF et al (2014) Rapamycin-mediated lifespan increase in mice is dose and sex dependent and metabolically distinct from dietary restriction. Aging Cell 13:468–477

Milman S, Barzilai N (2016) Dissecting the Mechanisms Underlying Unusually Successful Human Health Span and Life Span. Cold Spring Harb Perspect Med 6:a025098

Narbonne P, Roy R (2009) Caenorhabditis elegans dauers need LKB1/AMPK to ration lipid reserves and ensure long-term survival. Nature 457:210–214

Onken B, Driscoll M (2010) Metformin induces a dietary restriction-like state and the oxidative stress response to extend C. elegans Healthspan via AMPK, LKB1, and SKN-1. PLoS One 5:8758

Owen MR, Doran E, Halestrap AP (2000) Evidence that metformin exerts its anti-diabetic effects through inhibition of complex 1 of the mitochondrial respiratory chain. Biochem J 348 (Pt 3):607–614

Palikaras K, Lionaki E, Tavernarakis N (2015) Coordination of mitophagy and mitochondrial biogenesis during ageing in C. elegans. Nature 521:525–528

Pan KZ, Palter JE, Rogers AN, Olsen A, Chen D, Lithgow GJ, Kapahi P (2007) Inhibition of mRNA translation extends lifespan in Caenorhabditis elegans. Aging Cell 6:111–119

Panowski SH, Wolff S, Aguilaniu H, Durieux J, Dillin A (2007) PHA-4/Foxa mediates diet-restriction-induced longevity of C. elegans. Nature 447:550–555

Pawlikowska L, Hu D, Huntsman S, Sung A, Chu C, Chen J, Joyner AH, Schork NJ, Hsueh WC, Reiner AP et al (2009) Association of common genetic variation in the insulin/IGF1 signaling pathway with human longevity. Aging Cell 8:460–472

Pearson KJ, Baur JA, Lewis KN, Peshkin L, Price NL, Labinskyy N, Swindell WR, Kamara D, Minor RK, Perez E et al (2008) Resveratrol delays age-related deterioration and mimics transcriptional aspects of dietary restriction without extending life span. Cell Metab 8:157–168

Price NL, Gomes AP, Ling AJ, Duarte FV, Martin-Montalvo A, North BJ, Agarwal B, Ye L, Ramadori G, Teodoro JS et al (2012) SIRT1 is required for AMPK activation and the beneficial effects of resveratrol on mitochondrial function. Cell Metab 15:675–690

Pryor R, Cabreiro F (2015) Repurposing metformin: an old drug with new tricks in its binding pockets. Biochem J 471:307–322

Rera M, Bahadorani S, Cho J, Koehler CL, Ulgherait M, Hur JH, Ansari WS, Lo T Jr, Jones DL, Walker DW (2011) Modulation of longevity and tissue homeostasis by the Drosophila PGC-1 homolog. Cell Metab 14:623–634

Reznick RM, Zong H, Li J, Morino K, Moore IK, Yu HJ, Liu ZX, Dong J, Mustard KJ, Hawley SA et al (2007) Aging-associated reductions in AMP-activated protein kinase activity and mitochondrial biogenesis. Cell Metab 5:151–156

Rogers AN, Chen D, McColl G, Czerwieniec G, Felkey K, Gibson BW, Hubbard A, Melov S, Lithgow GJ, Kapahi P (2011) Life span extension via eIF4G inhibition is mediated by posttranscriptional remodeling of stress response gene expression in C. elegans. Cell Metab 14:55–66

Ross FA, MacKintosh C, Hardie DG (2016) AMP-activated protein kinase: a cellular energy sensor that comes in twelve flavours. FEBS J. doi:10.1111/febs.13698

Rubinsztein DC, Marino G, Kroemer G (2011) Autophagy and aging. Cell 146:682–695

Satoh A, Brace CS, Rensing N, Cliften P, Wozniak DF, Herzog ED, Yamada KA, Imai S (2013) Sirt1 extends life span and delays aging in mice through the regulation of Nk2 homeobox 1 in the DMH and LH. Cell Metab 18:416–430

Schulz TJ, Zarse K, Voigt A, Urban N, Birringer M, Ristow M (2007) Glucose restriction extends Caenorhabditis elegans life span by inducing mitochondrial respiration and increasing oxidative stress. Cell Metab 6:280–293

Selman C, Tullet JM, Wieser D, Irvine E, Lingard SJ, Choudhury AI, Claret M, Al-Qassab H, Carmignac D, Ramadani F et al (2009) Ribosomal protein S6 kinase 1 signaling regulates mammalian life span. Science 326:140–144

Settembre C, Zoncu R, Medina DL, Vetrini F, Erdin S, Erdin S, Huynh T, Ferron M, Karsenty G, Vellard MC et al (2012) A lysosome-to-nucleus signalling mechanism senses and regulates the lysosome via mTOR and TFEB. EMBO J 31:1095–1108

Shaw RJ, Lamia KA, Vasquez D, Koo SH, Bardeesy N, Depinho RA, Montminy M, Cantley LC (2005) The kinase LKB1 mediates glucose homeostasis in liver and therapeutic effects of metformin. Science 310:1642–1646

Slack C, Foley A, Partridge L (2012) Activation of AMPK by the putative dietary restriction mimetic metformin is insufficient to extend lifespan in Drosophila. PLoS One 7, e47699

Soukas AA, Kane EA, Carr CE, Melo JA, Ruvkun G (2009) Rictor/TORC2 regulates fat metabolism, feeding, growth, and life span in Caenorhabditis elegans. Genes Dev 23:496–511

Steffen KK, MacKay VL, Kerr EO, Tsuchiya M, Hu D, Fox LA, Dang N, Johnston ED, Oakes JA, Tchao BN et al (2008) Yeast life span extension by depletion of 60s ribosomal subunits is mediated by Gcn4. Cell 133:292–302

Stenesen D, Suh JM, Seo J, Yu K, Lee KS, Kim JS, Min KJ, Graff JM (2013) Adenosine nucleotide biosynthesis and AMPK regulate adult life span and mediate the longevity benefit of caloric restriction in flies. Cell Metab 17:101–112

Strong R, Miller RA, Astle CM, Floyd RA, Flurkey K, Hensley KL, Javors MA, Leeuwenburgh C, Nelson JF, Ongini E et al (2008) Nordihydroguaiaretic acid and aspirin increase lifespan of genetically heterogeneous male mice. Aging Cell 7:641–650

Suh Y, Atzmon G, Cho MO, Hwang D, Liu B, Leahy DJ, Barzilai N, Cohen P (2008) Functionally significant insulin-like growth factor I receptor mutations in centenarians. Proc Natl Acad Sci U S A 105:3438–3442

Syntichaki P, Troulinaki K, Tavernarakis N (2007) eIF4E function in somatic cells modulates ageing in Caenorhabditis elegans. Nature 445:922–926

Toyama EQ, Herzig S, Courchet J, Lewis TL Jr, Loson OC, Hellberg K, Young NP, Chen H, Polleux F, Chan DC et al (2016) Metabolism. AMP-activated protein kinase mediates mitochondrial fission in response to energy stress. Science 351:275–281

Tullet JM, Araiz C, Sanders MJ, Au C, Benedetto A, Papatheodorou I, Clark E, Schmeisser K, Jones D, Schuster EF et al (2014) DAF-16/FoxO directly regulates an atypical AMP-activated protein kinase gamma isoform to mediate the effects of insulin/IGF-1 signaling on aging in Caenorhabditis elegans. PLoS Genet 10, e1004109

Tzamarias D, Roussou I, Thireos G (1989) Coupling of GCN4 mRNA translational activation with decreased rates of polypeptide chain initiation. Cell 57:947–954

Ulgherait M, Rana A, Rera M, Graniel J, Walker DW (2014) AMPK modulates tissue and organismal aging in a non-cell-autonomous manner. Cell Rep 8:1767–1780

Um JH, Park SJ, Kang H, Yang S, Foretz M, McBurney MW, Kim MK, Viollet B, Chung JH (2010) AMP-activated protein kinase-deficient mice are resistant to the metabolic effects of resveratrol. Diabetes 59:554–563

United Nations, Department of Economic and Social Affairs, Population Division (2015) World population prospects: the 2015 revision, vol I: Comprehensive Tables. ST/ESA/SER.A/379

Vellai T, Takacs-Vellai K, Zhang Y, Kovacs AL, Orosz L, Muller F (2003) Genetics: influence of TOR kinase on lifespan in C. elegans. Nature 426:620

Wan QL, Zheng SQ, Wu GS, Luo HR (2013) Aspirin extends the lifespan of Caenorhabditis elegans via AMPK and DAF-16/FOXO in dietary restriction pathway. Exp Gerontol 48:499–506

Wang C, Wheeler CT, Alberico T, Sun X, Seeberger J, Laslo M, Spangler E, Kern B, de Cabo R, Zou S (2013) The effect of resveratrol on lifespan depends on both gender and dietary nutrient composition in Drosophila melanogaster. Age (Dordr) 35:69–81

Wang MC, O'Rourke EJ, Ruvkun G (2008) Fat metabolism links germline stem cells and longevity in C. elegans. Science 322:957–960

Weimer S, Priebs J, Kuhlow D, Groth M, Priebe S, Mansfeld J, Merry TL, Dubuis S, Laube B, Pfeiffer AF et al (2014) D-Glucosamine supplementation extends life span of nematodes and of ageing mice. Nat Commun 5:3563

Winder WW, Holmes BF, Rubink DS, Jensen EB, Chen M, Holloszy JO (2000). Activation of AMP-activated protein kinase increases mitochondrial enzymes in skeletal muscle. J Appl Physiol 88:2219–2226

Wood JG, Rogina B, Lavu S, Howitz K, Helfand SL, Tatar M, Sinclair D (2004) Sirtuin activators mimic caloric restriction and delay ageing in metazoans. Nature 430:686–689

Yavari A, Stocker CJ, Ghaffari S, Wargent ET, Steeples V, Czibik G, Pinter K, Bellahcene M, Woods A, Martinez de Morentin PB et al (2016) Chronic Activation of gamma2 AMPK induces obesity and reduces beta cell function. Cell Metab 23(5):821–36

Yuan Y, Hakimi P, Kao C, Kao A, Liu R, Janocha A, Boyd-Tressler A, Hang X, Alhoraibi H, Slater E et al (2016) Reciprocal Changes in Phosphoenolpyruvate Carboxykinase and Pyruvate Kinase with Age Are a Determinant of Aging in Caenorhabditis elegans. J Biol Chem 291:1307–1319

Zong H, Ren JM, Young LH, Pypaert M, Mu J, Birnbaum MJ, Shulman GI (2002) AMP kinase is required for mitochondrial biogenesis in skeletal muscle in response to chronic energy deprivation. Proc Natl Acad Sci U S A 99:15983–15987

Chapter 11
Targeting AMPK for the Alleviation of Pathological Pain

Marina N Asiedu, Gregory Dussor, and Theodore J Price

Contents

Abstract Chronic pain is a major clinical problem that is poorly treated with available therapeutics. Adenosine monophosphate-activated protein kinase (AMPK) has recently emerged as a novel target for the treatment of pain with the exciting potential for disease modification. AMPK activators inhibit signaling pathways that are known to promote changes in the function and phenotype of peripheral nociceptive neurons and promote chronic pain. AMPK activators also reduce the excitability of these cells suggesting that AMPK activators may be efficacious for the treatment of chronic pain disorders, like neuropathic pain, where changes in the excitability of nociceptors is thought to be an underlying cause. In agreement with this, AMPK activators have now been shown to alleviate pain in a broad variety of preclinical pain models indicating that this mechanism might be engaged for the treatment of many types of pain in the clinic. A key feature

M.N. Asiedu • G. Dussor • T.J. Price (✉)
University of Texas at Dallas, School of Behavioral and Brain Sciences, JO 4.212, 800 W Campbell Road, Richardson, TX 75080, USA
e-mail: theodore.price@utdallas.edu

M.D. Cordero, B. Viollet (eds.), *AMP-activated Protein Kinase*, Experientia Supplementum 107, DOI 10.1007/978-3-319-43589-3_11

of the effect of AMPK activators in these models is that they can lead to a long-lasting reversal of pain hypersensitivity even long after treatment cessation, indicative of disease modification. Here, we review the evidence supporting AMPK as a novel pain target pointing out opportunities for further discovery that are likely to have an impact on drug discovery efforts centered around potent and specific allosteric activators of AMPK for chronic pain treatment.

Keywords Neuropathic pain • Cancer pain; • AMPK • mTOR • MAPK • Postsurgical pain

11.1 Introduction

Pain is the most common reason that people seek medical attention and is frequently a symptom of underlying injury or disease. Acute pain serves as a protective mechanism that alerts us to damaging or potentially damaging stimuli. Pain can, however, become maladaptive and persist for months or even years after an injury has resolved or emerge in the absence of any apparent injury (Woolf and Salter 2000; Basbaum et al. 2009). This sort of pathological pain is a major clinical and public health problem that adversely affects up to a third of the population of most developed countries and is associated with enormous medical costs and productivity loss in the workplace. (Education 2011). While acute pain can be well controlled with different treatments like nonsteroidal anti-inflammatory drugs or opioids, very few effective therapeutics are currently available for pathological pain and most of them show limited efficacy (Education 2011). A possible reason for this lack of efficacy is that no therapeutics have yet been approved for human use that target underlying mechanisms that cause pain to become pathological. Rather, existing therapeutics (e.g., antiepileptics) only target mechanisms that transiently decrease pathological excitability in the peripheral or central nervous systems [PNS and CNS, respectively, (Baron et al. 2010)]. Hence, disease-modifying therapies to treat pathological pain are greatly needed to tackle the challenges posed by this medical problem.

In order to address how disease-modifying therapies for pathological pain can be developed, we must first elucidate the fundamental molecular events driving pathological pain development and maintenance. Work in this area using preclinical pain models has led to tremendous advancements, yet the molecular underpinnings of pathological pain are still not completely well understood (Mogil 2009). Nonetheless, it is well documented and broadly accepted that plasticity in PNS nociceptors, sensory neurons responsible for the detection of damaging or potentially damaging stimuli, and/or in CNS neurons that receive inputs from these cells promotes pathological pain (Woolf and Walters 1991; Perl 1996; Woolf and Salter 2000; Ji and Woolf 2001; Ji et al. 2003; Basbaum et al. 2009; Latremoliere and Woolf 2009; Price and Geranton 2009; Walters and Moroz 2009; Woolf 2011). This chapter will mainly focus on plasticity in peripheral nociceptors and how this form of plasticity can be targeted with adenosine monophosphate-activated protein kinase (AMPK)-based therapeutics (Price and Dussor 2013). We will showcase the evidence supporting AMPK activators as some of the first disease-modifying agents for pathological pain and highlight the rationale behind this premise as well as the relevant therapeutic strategies that can be utilized to engage AMPK in this therapeutic setting.

11.2 Activity-Dependent Translation as a Key Mechanism Promoting Pathological Pain and Its Negative Regulation by AMPK

Injury to peripheral tissues and/or peripheral nerves changes the sensitivity of nociceptive afferent neurons causing them to become hyperexcitable. The induction of this increase in excitability can occur rapidly after injury and is mediated by a variety of endogenous factors that act on diverse receptors expressed by nociceptors (Woolf and Walters 1991; Perl 1996; Koltzenburg 1999; Basbaum et al. 2009; Price and Geranton 2009; Reichling and Levine 2009; Walters and Moroz 2009). Very rapid changes in excitability are generally attributed to short-term signaling mediated by α subunits of G-protein coupled receptors (GPCRs) or via activation of kinases downstream of tyrosine kinase receptors (Trks) and can largely be attributed to the phosphorylation of ion channels. In some cases, these short-term changes in the sensitivity of nociceptors may resolve once the stimulus causing the signaling events to occur is removed. However, in cases where pain becomes pathological, this may not be the case. One possible explanation for how this occurs is that certain signaling events are capable of causing long-term changes in the function or phenotype of peripheral nociceptors causing potentially permanent alterations in pain sensitivity (Reichling and Levine 2009; Price and Inyang 2015). It is likely that this transition to a pathological pain state is controlled by changes in gene expression, either at the level of transcription (Woolf and Costigan 1999; Ji and Woolf 2001; Obata and Noguchi 2004; Ji et al. 2007, 2009b) or translation changes (Price and Geranton 2009; Obara et al. 2012; Price and Dussor 2013).

Multiple lines of evidence indicate that a critical molecular event for neuronal plasticity is the activity-dependent regulation of translation, often at distinct subcellular locations, for instance at the base of dendritic spines or in axonal terminals (Martin et al. 2000; Giuditta et al. 2002; Klann et al. 2004; Sutton and Schuman 2005; Schuman et al. 2006; Jung et al. 2012). This process of activity-dependent translation has been linked to learning and memory in the central nervous system (CNS) and we will review evidence that it is linked to pathological pain in the peripheral nervous system (PNS). Local translation at distal sites (dendrites or axons) in neurons requires three features: 1) ribosomes must be present at distal sites, 2) mRNAs must be transported from the cell body to these distal sites, and 3) signaling machinery linking extracellular events to intracellular translation must be present and function downstream of cell surface receptors. In the CNS, these criteria are satisfied at the level of dendritic spines where molecular changes involved in learning and memory are thought to occur (Martin et al. 2000; Klann et al. 2004; Sutton and Schuman 2005; Schuman et al. 2006). In the PNS, including in nociceptors, there is now strong evidence that these criteria are also met (Jung et al. 2012), and there is also abundant evidence that local, activity-dependent translation plays a key function in changing the excitability of nociceptors (Price and Geranton 2009; Obara et al. 2012; Price and Dussor 2013).

An early view of translation of new proteins in the PNS was that this solely occurred in the soma of DRG or TG neurons (Koenig and Giuditta 1999). This idea was based on the limited number of ribosomes found in adult DRG axons. It is now well accepted that local translation in DRG neuron axons is an important component of development, especially at growth cones of developing DRG neurons. Following development, there are indications that local translation in axons is decreased; however, these axons still contain ribosomes, mRNAs, mRNA trafficking proteins, and signaling molecules involved in translation control (Koenig and Giuditta 1999; Mohr and Richter 2000; Piper and Holt 2004; Willis and Twiss 2006; Lin and Holt 2008; Jung and Holt 2011; Jung et al. 2012). Work in the area of pain plasticity has focused on signaling molecules with the major focus falling on a kinase called mechanistic target of rapamycin complex 1 (mTORC1). mTORC1 is regulated by upstream activation of Trks and engagement of PI3K/AKT signaling. When mTORC1 is activated, it stimulates phosphorylation of a family of proteins that bind to the 5′m7GTP Cap structure (5′Cap) of mRNAs (Gingras et al. 2004). A key event in this process is the phosphorylation of eukaryotic initiation factor (eIF) binding protein (BP) 4E-BP. When 4E-BP is phosphorylated by mTORC1, 4E-BP dissociates from eIF4E, the 5′ cap binding protein, allowing for more efficient association of eIF4E with eIF4A (a deadbox family helicase) and eIF4G (a scaffolding protein) (Gingras et al. 2004). These three proteins form a functional unit called the eIF4F complex. This eIF4F complex is associated with efficient translation of target mRNAs and is an important regulatory endpoint for mTORC1 signaling (Gingras et al. 1999; Sonenberg 2008).

Augmented mTORC1 activity is unequivocally linked to plasticity in nociceptors. First, rapamycin, a highly selective inhibitor of mTORC1, attenuates plasticity induced by endogenous mediators that act via receptors expressed by nociceptive neurons (Price et al. 2007; Melemedjian et al. 2010). Rapamycin also decreases the sensitivity of a subset of nociceptors thought to play an important role in mechanical hypersensitivity following injury (Jimenez-Diaz et al. 2008; Geranton et al. 2009) and reduces neuropathic mechanical hypersensitivity in rats (Melemedjian et al. 2013c) and mice (Obara et al. 2011). Finally, factors that are well known to sensitize nociceptors, such as nerve growth factor (NGF), and peripheral nerve injury, which causes neuropathic pain, increase mTORC1 activity in DRG neurons and their peripherally projecting axons (Melemedjian et al. 2010, 2011). While much work has focused on mTORC1 in the PNS in pathological pain generation, there is also strong evidence linking increased mTORC1 activity to plasticity in the spinal dorsal horn after peripheral injury (Asante et al. 2009, 2010; Geranton et al. 2009; Norsted Gregory et al. 2010; Xu et al. 2010, 2011; Shih et al. 2012) or even exposure to drugs of abuse (Xu et al. 2014). Here, mTORC1 appears to control changes in synaptic strength between nociceptors and CNS neurons that relay nociceptive information to the brain via changes in postsynaptic expression of plasticity-related genes. However, an alternative interpretation is that presynaptic effects in the dorsal roots, which contain the central projections of DRG neurons, predominate, as there is also evidence that mTORC1 activity in these axons contributes to neuropathic pain (Geranton et al. 2009). While ongoing research is still elucidating the molecular details of how mTORC1 regulates nociceptor and dorsal horn neuron plasticity, it is clear that mTORC1 is a crucial molecular signaling hub underlying pathological pain.

MAPK represents another signaling pathway linking extracellular signals to translation in neurons, including nociceptors. The MAPK isoforms ERK and p38, which play crucial roles in nociceptor excitability and synaptic plasticity in the CNS, phosphorylate eIF4E at Ser 209 via a downstream kinase called MAP kinase interacting kinase [MNK, (Pyronnet et al. 1999)]. Two MNK isoforms are found in mammalian genomes and both are expressed in the nervous system. MNK-mediated phosphorylation of eIF4E is thought to regulate the translation of a subset of mRNAs. These mRNAs have been elucidated using mice where the Ser 209 site of eIF4E has been mutated to Ala rendering the protein unphosphorylatable (Furic et al. 2010). mRNAs whose translation is decreased in the absence of eIF4E phosphorylation encode cytokines, chemokines, and other plasticity-related proteins (Herdy et al. 2012). Among these mRNAs is the matrix metalloprotease isoform 9 (MMP9) (Gkogkas et al. 2014), which is expressed in the DRG, is regulated by injury and/or by chronic opioid exposure, and is involved in pathological pain (Kawasaki et al. 2008; Ji et al. 2009a; Berta et al. 2012; Liu et al. 2012). ERK or p38 signaling to eIF4E through MNK1/2 is also involved in the promotion of pain plasticity by endogenous factors that signal via this pathway through their cognate cell surface receptors. One such example is interleukin 6 (IL6). IL6 exposure to DRG neurons increases ERK and eIF4E phosphorylation and promotes mechanical hypersensitivity in a translation-dependent fashion (Melemedjian et al. 2010, 2013b, 2014; Asiedu et al. 2011).

The data discussed above point to mTORC1 and MAPK, especially ERK, signaling as prime targets for control of pathological pain. While a number of studies have indicated positive effects for the mTORC1 inhibitor rapamycin using an acute drug administration paradigm (Price et al. 2007; Jimenez-Diaz et al. 2008; Asante et al. 2009; Geranton et al. 2009; Melemedjian et al. 2010; Asiedu et al. 2011; Obara et al. 2011; Xu et al. 2014), longer term treatment with rapamycin causes an engagement of feedback signaling through p70 S6 Kinase (p70S6K) in DRG neurons that leads to stimulation of ERK and augmented neuronal excitability (Melemedjian et al. 2013c). Human studies suggest that rapamycin treatment can promote pain hypersensitivity, and rapamycin treatment has been anecdotally linked to devastating human pain disorders such as complex regional pain syndrome (Christopher 2001; Witzig et al. 2005; Molina et al. 2008; Massard et al. 2010; Budde et al. 2011; McCormack et al. 2011). These rodent and human findings likely rule out rapamycin as a pain therapeutic. MAPK inhibitors have been widely investigated as pain therapeutics (Ji et al. 2009b) including some clinical trials for p38 inhibitors that have had mixed success (Anand et al. 2011). While work in that arena will undoubtedly continue, there is a strong scientific rationale for the development of therapeutics for pain that abrogate signaling in both the mTORC1 and MAPK signaling pathways as these are often convergently linked to nociceptor plasticity and pathological pain.

Emerging evidence indicates that AMPK activation negatively regulates activity-dependent protein synthesis and other signaling pathways that drive neuronal plasticity in pathological pain leading to decreased nociceptor excitability (Price and Dussor 2013). AMPK activation also suppresses the feedback signaling mechanism that arises from prolonged blockade of mTORC1 signaling with rapamycin (Melemedjian et al. 2013c). In addition, AMPK activation inhibits the MAPK signaling pathway through phosphorylation of adaptor proteins that regulate tyrosine receptor kinase signaling and through negative regulation of small GTPases that are upstream of MAPK signaling (Melemedjian et al. 2011; Carling et al. 2012; Tillu et al. 2012). Thus, mTORC1 and MAPK signaling pathways are both inhibited by AMPK activation suggesting that pharmacological stimulation of AMPK signaling may interfere with signaling that controls activity-dependent translation in nociceptors and therefore short circuit pain amplification. There is now strong evidence to support this novel role for AMPK as a pain plasticity target, which will be discussed in a variety of experimental contexts below.

11.3 AMPK as a Key Regulator of Ion Channel Activity and Its Role in Pathological Pain

The coupling of ion channel activity to the metabolic state of a cell is logical because ion flow across open cation or anion channels must be balanced by active transporters that consume a great deal of cellular energy. The activation of AMPK under cellular metabolic stress conditions and the well-recognized function of AMPK as a promoter of ATP generating catabolic pathways while turning off ATP-utilizing anabolic pathways (Hardie et al. 2012) posits that AMPK's modulation of ion channel function might result in decreased cellular excitability. Cystic fibrosis transmembrane conductance regulator (CFTR) was discovered by Hallows et al. as one of the earliest channel targets of AMPK. The direct phosphorylation of CFTR by AMPK inhibits the channel's activity by reducing its open probability (Hallows et al. 2000) and also preventing PKA stimulation of the channel (King et al. 2009).

AMPK was later reported to negatively regulate epithelial Na+ channel (ENaC) membrane expression in an indirect manner via phosphorylation of ubiquitin ligase

Nedd4-2 which results in ubiquitination of ENaC and decreased ENaC expression (Carattino et al. 2005; Bhalla et al. 2006). While decreasing Na+ channel expression will result in decreased excitability, increased K+ channel activity will also decrease cellular excitability. In this regard, AMPK has been shown to enhance KATP (Kir6.2) channel activity and membrane expression (Sukhodub et al. 2007; Yoshida et al. 2012). Moreover, in outer hair cells of the ear, AMPK activation increases BK_{Ca} channel expression, which plays a role in protecting these cells from the excitotoxic effect of excessive sound exposure (Foller et al. 2012). On the other hand, AMPK has been shown to contribute to ion channel modulation that consequently increases cellular activity by driving hyperpolarizing shifts in the voltage dependence of Nav1.5 and decreasing current amplitudes of BK_{Ca}, TREK, and TASK channels (Light et al. 2003), as well as reducing the membrane expression of Kv7.1 and Kir2.1 channels (Andersen and Rasmussen 2012). Hence, there is evidence for bidirectional regulation of cellular excitability by AMPK activation. However, as we will argue below, the evidence gleaned from studies of AMPK in neurons overwhelmingly supports a role for AMPK in decreasing excitability in these specialized cells.

Activation of AMPK in rat hippocampal neurons produces a hyperpolarizing shift in the current/voltage relationship of voltage-gated K^+ channels, an effect that was inhibited by introducing antibodies against Kv2.1 into the recording pipette. In addition, direct phosphorylation of Kv2.1 and shifts in channel kinetics similar to those observed in neurons were obtained in cell lines expressing Kv2.1 confirming AMPK's direct effect on regulation of Kv2.1 channels (Ikematsu et al. 2011). This study was among the first to show that AMPK activation results in enhanced K+ channel activity causing a robust decrease in neuronal excitability.

While further studies are needed to elucidate other ion channels that are directly targeted by the kinase, AMPK activation may also modulate channel activity or expression in neurons in an indirect manner similar to what was discussed above for ENaC. Nedd4-2, which is targeted by AMPK increasing its ubiquitin ligase activity, interacts with multiple channel types such as KCNQ channels, CIC-2 chloride channels, and voltage gated Na^+ channels (Bongiorno et al. 2011). Thus, it is formally possible that AMPK-mediated Nedd4-2 activation may be a general channel modulatory mechanism resulting in channel downregulation in neurons, but this idea has not been rigorously assessed. AMPK-mediated interference with other kinase signaling pathways is another potential mechanism through which AMPK can decrease channel activity and therefore excitability in neurons. For instance, ERK has been shown to modulate numerous kinds of channels such as the K^+ channel subtype Kv4.2 that contributes to transient outward or A-type K^+ currents (Adams et al. 2000; Schrader et al. 2006), Cav2.2 channels (Martin et al. 2006), and Nav1.7 channels. The Nav1.7 channel is expressed prominently by nociceptors where it directs the amplification of generator potentials to control action potential firing (Stamboulian et al. 2010). Importantly, Nav1.7 is associated with human pain conditions wherein a loss-of-function produces congenital insensitivity to pain while a gain-in-function results in paroxysmal extreme pain disorder and erythromelalgia (Dib-Hajj et al. 2010). ERK phosphorylates Nav1.7 on specific residues on an intercellular loop of the channel. In sensory neurons stimulated with depolarizing ramp current injection, ERK phosphorylation of Nav1.7 enhances action potential firing and decreases the latency to first action potential (Stamboulian et al. 2010). We have demonstrated that AMPK activators decrease ERK activity in nociceptors and dampen sensory neuron excitability evoked by ramp current injection (Melemedjian et al. 2011). Although a possible direct effect of AMPK on Nav1.7 cannot be completely excluded, the modulation of MAPK activity by AMPK likely orchestrates the observed reduction in neuronal excitability.

The evidence compiled above hints at two primary mechanisms through which AMPK can acutely influence the excitability of nociceptors: via direct phosphorylation events that likely increase the activity of voltage gated K^+ channels and through signaling mechanisms that dampen the activity of other kinases that phosphorylate channels involved in increased excitability. Additional studies are required to more fully grasp how these mechanisms operate in nociceptors. Nevertheless, the existing literature supports a role for AMPK activators for the treatment of pain because this mechanism is likely to both decrease activity-dependent expression of genes that promote excitability and to alter nociceptor excitability through decreased activity in excitatory channels and augmented activity in inhibitory channels. Next, we will discuss the accumulating preclinical evidence that supports AMPK activators as pathological pain therapeutics and their potential for disease modification in specific chronic pain states.

11.4 AMPK as a Target for Disease Modification in Neuropathic Pain

Neuropathic pain is defined as pain arising from injury or disease affecting the peripheral or central nervous system. Neuropathic pain influences a smaller proportion of the population than more common forms of persistent pain; however, it is among the most difficult forms of pathological pain to treat and places a significant burden on patients and healthcare workers. Because diseases that are rising in global prominence (e.g., diabetes) and treatments for other diseases (e.g., cancer chemotherapeutics) can cause peripheral neuropathies that result in unrelenting neuropathic pain, there is increasing emphasis on finding new therapeutic targets for neuropathic pain. Studies in rodents with traumatic injury to peripheral nerves highlight AMPK as a key and unique target for neuropathic pain alleviation (Melemedjian et al. 2011).

As noted above, injury to the PNS due to trauma, metabolic disease, or exposure to drugs, such as chemotherapeutics, is the fundamental etiology of neuropathic pain. The general symptomology of neuropathic pain includes exquisite hypersensitivity to mechanical stimulation, hypersensitivity to cold, and ongoing burning pain (Baron et al. 2010). Peripheral nerve injury (PNI) in rodents stimulates a reorganization of the translation machinery in the PNS characterized by an activation of mTORC1 (Abe et al. 2010; Melemedjian et al. 2011) and MAPK signaling (Obata et al. 2003, 2004a, 2004b; Agthong et al. 2006; Melemedjian et al. 2011) and their downstream signaling counterparts in DRG neurons and their axons. Additionally, there is an enhancement in the expression of RNA binding and transport proteins (Melemedjian et al. 2011) that is presumably involved in the translocation of mRNAs to the axonal compartment following injury (Zheng et al. 2001; Willis et al. 2005, 2011; Thakor et al. 2009; Ruangsri et al. 2011). In effect, these molecular events induce an increase in the rate of translation in peripheral nerves (Melemedjian et al. 2011). Overall, this PNI-induced reorganization process may be linked to regeneration after injury (Willis and Twiss 2006; Jung et al. 2012) as well as changes in excitability observed in injured DRGs (Thakor et al. 2009; Melemedjian et al. 2011).

Systemic administration of AMPK activators (either metformin or A769662) to rodents after peripheral nerve injury decreases mechanical hypersensitivity, an effect that starts 2–3 days after treatment (Melemedjian et al. 2011, 2013c). Strikingly, within 7 days, these compounds produced a total resolution of mechanical hypersensitivity that still persisted following cessation of treatment, supporting the idea that AMPK activators have disease-modifying properties that develop over the course of treatment. Moreover, this disease-modifying effect of treatment with metformin does not seem to depend on when the treatment beings as 7 day systemic treatment reverses PNI-induced mechanical hypersensitivity when the onset of treatment was between 14 and 60 days after injury (Melemedjian et al. 2011, 2013c). In addition, the reversal of mechanical hypersensitivity after metformin treatment is sustained after a prolonged period of treatment cessation (Melemedjian et al. 2013c). Metformin treatment also normalized increased translation rates in injured peripheral nerves (Melemedjian et al. 2011), highlighting that the potential disease-modifying effects of AMPK activators on PNI-induced neuropathic pain are apparent from a behavioral and biochemical perspective. *In vitro* studies using patch-clamp electrophysiology on DRG neurons demonstrate that AMPK activators such as metformin and A769662 also reduce the excitability of these neurons (Melemedjian et al. 2011). Increased excitability of DRG neurons following injury is widely believed to result in ectopic activity causing ongoing burning pain that is a prominent feature of neuropathic pain (Baron et al. 2010). Although it remains to be shown *in vivo*, the profound effect of AMPK activators on DRG excitability suggests that AMPK activators might not only reverse mechanical hypersensitivity but also spontaneous pain associated with persistent neuropathic pain in humans.

An important mechanism of neuropathic pain is activation of glia in the spinal dorsal horn following peripheral nerve injury (Beggs and Salter 2010). This glial activation may lead to the secretion of inflammatory mediators such as cytokines and chemokines that act on neurons in the dorsal horn to enhance their excitability. Another apparent function of this change in glial activity is a decrease in their expression of glutamate transporters that leads to an enhancement of glutamatergic transmission between neurons in the dorsal horn (Nie and Weng 2009, 2010). Using a model of neuropathic pain in rats, Maixner et al. demonstrated that AMPK activation in the spinal cord leads to a reduction in glial secretion of cytokines (e.g., interleukin 1β) and an increase in glutamate transporter expression that restores normal astrocyte-mediated control of synaptic glutamate levels in the dorsal horn after nerve injury (Maixner et al. 2015). A key finding of these studies is that these pharmacological effects are mimicked by genetic manipulation of spinal AMPK α1 subunit expression. In another study, it was shown that blockade of AMPK activity in mice causes an enhancement of NLRP3 inflammasome activity and the generation of neuropathic pain-like behavioral effects (Bullon et al. 2015). This effect was absent in *Nlrp3*$^{-/-}$ mice and could be rescued with subsequent treatment with metformin. Hence, AMPK activator-mediated control of neuropathic pain may have both peripheral and central components that contribute to disease modification. More work is needed to delineate mechanisms underlying these effects.

11.5 A Triple Action for Ampk in Cancer Treatment, Chemotherapy-Induced Neuropathic Pain, and Cancer Pain

A core hallmark of cancer is the dysregulation of cellular energetics resulting in loss of normal regulation of cell growth and proliferation (Martinez-Outschoorn et al. 2010; Pavlides et al. 2010). The mTORC1 and MAPK have been widely implicated in cancer (Gao and Roux 2015; Siddiqui and Sonenberg 2015). A variety of cancer clinical trials have been conducted using mTORC1 or MAPK inhibitors. Clinical trials and preclinical investigations have shown that the use of single kinase inhibitors targeting these mechanisms can lead to the engagement of feedback signaling amplification in these kinase pathways that limits the utility of these therapeutics (Ghosh et al. 2006; Vicier et al. 2013) (Poulikakos and Solit 2011; Chapman 2013; Sale and Cook 2014). Using AMPK activators as a possible alternative approach to circumvent this problem is attractive from a molecular signaling standpoint because AMPK activation negatively influences both the mTORC1 and MAPK signaling pathways and attenuates feedback signaling mechanisms (Jakobsen et al. 2001). There is extensive literature highlighting the tumor suppressor function of AMPK activation. Oral administration of metformin in mice prevents tobacco carcinogen-induced lung tumorigenesis and significantly reduces the development of lung cancers (Memmott et al. 2010). In a model of chemically induced colon cancer, metformin markedly reduces aberrant crypt foci and decreases polyp formation (Hosono et al. 2010). Moreover, in widely cited meta-analyses of numerous diabetic epidemiological studies, there was evidence of a marked reduction in cancer risk and a significant improvement in prognosis in subjects taking metformin compared to those taking other antidiabetic drugs. These risk reductions were particularly associated with cancers of the colon, lung, and liver (Decensi et al. 2010; Hardie 2013). Hence, there is compelling evidence for cancer preventative and cancer therapeutic effects of AMPK activation.

Chemotherapy-induced peripheral neuropathy (CIPN) is a major dose-limiting adverse effect of many chemotherapeutic agents (Balayssac et al. 2011; Ferrier et al. 2013) and can have devastating consequences for patients (Cata et al. 2006). Proposed therapeutics aimed at preventing the neurotoxic effects of chemotherapeutic treatment have had limited success. Additionally, CIPN is resistant to many of the first-line treatments for neuropathic pain (Dworkin et al. 2010; Ferrier et al. 2013). Though the underlying mechanisms of CIPN are not well understood, some of the symptoms and pathophysiology such as epidermal nerve fiber dieback and ectopic activity in DRG neurons parallels that of trauma-induced neuropathic pain (Han and Smith 2013). Simultaneous administration of metformin with chemotherapeutic treatment in mice prevents the loss of tactile sensation and the development mechanical hypersensitivity associated with chemotherapeutics (Mao-Ying et al. 2014). Notably, there is a lack of effect with metformin after the development of CIPN indicating that this therapeutic approach may only be effective when employed during the entire course of chemotherapeutic treatment (Mao-Ying et al. 2014). This is in stark contrast to trauma-induced neuropathic pain where metformin treatment resolved established mechanical hypersensitivity. In spite of this, the active investigation of metformin as an add-on treatment for chemotherapy in cancer trials (Kourelis and Siegel 2012) suggests that this treatment approach may be valuable in the prevention of CIPN during the course of chemotherapy treatment and eliminate the dose-limiting side effect of chemotherapeutic agents.

CIPN is a major cause of pain in cancer patients; if certain cancers progress and metastasize to bone or originate in bone, these can cause severe pain. This cancer-induced pain can be caused by bone destruction induced by the proliferating cancer cells but there is also now strong evidence that the presence of cancer in bone causes alterations in the local microenvironment that alter the function and phenotype of nociceptors that innervate the bone. Many of these factors are cytokines, chemokines, and growth factors, like NGF, that act on sensory neurons to increase mTOR and MAPK signaling. In a rat bone cancer pain model, AICAR and resveratrol, two AMPK activators, cause a significant alleviation of cancer-induced pain that is accompanied by reduced cytokine expression and attenuation of MAPK signaling (Song et al. 2015). These authors also found a reversal of spinal astrocyte and microglial activation with resveratrol treatment suggesting that the CNS immune response induced by persistent nociceptor activation within the cancer-ridden bone is attenuated by AMPK activation (Song et al. 2015). Whether this is mediated by reduced nociceptor excitability in the PNS or a direct effect on CNS cell types remains to be thoroughly investigated.

While more work is clearly needed in this important area, these limited studies point to great opportunities for AMPK activators in the area of CIPN and cancer-induced pain therapeutics. The intense investigation of AMPK activators in cancer prevention and treatment makes it likely that clinical trials in this area will soon lead to opportunities to test these hypotheses in human beings. Investigators designing these trials should be cognizant of the fact that pain is a leading cause of disability and morbidity in cancer patients and survivors and design endpoints for pain measures into studies to examine the potential benefits of these compounds for cancer pain and CIPN in humans.

11.6 AMPK as a Target for Postsurgical Pain Treatment and Prevention

Pain is an unavoidable consequence of surgery. Despite the existence of analgesics aimed at managing acute postsurgical pain, chronic postsurgical pain still develops in up to 50% of surgical patients after certain surgical procedures, and it can be very debilitating in up to 10% of this population (Kehlet et al. 2006). Current therapies for postsurgical pain, which are mainly opioids, have serious side effects, abuse liabilities, and can eventually exacerbate chronic postsurgical pain or increase the chances of chronic postsurgical pain development (Laboureyras et al. 2009; Sehgal et al. 2013). The release of cytokines [e.g., IL6, (Clark et al. 2007; Loram et al. 2007)] and growth factors [e.g., NGF, (Zahn et al. 2004; Banik et al. 2005; Wu et al. 2007)] from the site of incision is one of the many causes of postsurgical pain and is likely linked to the development of chronic postsurgical pain. These endogenous factors, which can act via receptors expressed by nociceptors, perpetuate both phenotypic and functional changes in DRG neurons that result in increased excitability driving pain amplification and chronification (Banik and Brennan 2008; Xu and Brennan 2009).

Pain prior to surgery and/or a previous surgical procedure are the main identified risk factors for chronic postsurgical pain (Kehlet et al. 2006; Pinto et al. 2013). Studies have illustrated that there is an enhanced rate of chronic pain development after a revision to knee replacement surgery compared to an original knee replacement surgery. This points to the idea that the first surgery creates a pathological pain plasticity that alters the response to the second stimulus (the revision surgery) (Kim et al. 2014). The planned nature of surgical interventions presents a unique opportunity for the development of therapeutics targeting the prevention of chronic postsurgical pain development that can be utilized during a surgical procedure. Currently, there are no treatments available targeted at preventing the development of chronic pain after surgery.

The endogenous factors known to promote chronic pain after surgeries predominantly engage the mTORC1 and MAPK signaling pathways [e.g., NGF and IL6, (Zahn et al. 2004; Banik et al. 2005; Wu et al. 2007, 2009)] in nociceptors innervating the incision site to drive nociceptor sensitization pointing to AMPK activators again as a potential therapeutic approach to halt the development of incision-induced plasticity following surgery. Local application of resveratrol, which activates AMPK and decreases ERK and mTORC1 signaling in DRG neurons, at the time of incision in mice, or 24 h following incision, significantly attenuates incision-induced mechanical hypersensitivity (Tillu et al. 2012). Incision to the mouse paw induces a change in the nociceptive system such that a subsequent subthreshold stimulus administered after the incision causes a prolonged and enhanced mechanical hypersensitivity. This subthreshold stimulus shows no effect in naïve mice and is capable of evoking a pain state in "primed" mice even weeks after the incisional injury has healed. Hence, this form of plasticity may model the "second hit" scenario outlined above wherein the incidence of chronic postsurgical pain is very high (Kim et al. 2014). Notably, local resveratrol treatment at the time of the incision completely abolished the development of priming suggesting that AMPK activation may be used as a preventative treatment for chronic postsurgical pain (Tillu et al. 2012). Though this effect on chronic postsurgical pain is yet to be demonstrated with other AMPK activators, recent studies suggest that curcumin inhibits incision-induced pain and prevents incision-induced priming effects (Sahbaie et al. 2014). While the mechanism of action of curcumin was not explored in this study, curcumin activates AMPK in many cell types and likewise attenuates MAPK and mTORC1 signaling (Yu et al. 2008; Kim et al. 2009).

Collectively, these studies indicate that AMPK activation for the treatment of postsurgical pain achieves antinociceptive effects and prevents the development of persistent pain plasticity after incision. The work discussed above warrants further exploration with more specific AMPK activators in preclinical models but suggests that AMPK activators might be employed to reduce postsurgical pain and prevent the development of persistent pain after surgery.

11.7 AMPK Activators for Inflammatory Pain Treatment

The AMPK activators AICAR and metformin were recently evaluated in models of inflammatory nociception in mice (Russe et al. 2013) where both compounds attenuated inflammation-induced pain hypersensitivity and decreased the inflammatory response when compounds were given systemically. AMPK activator treatment also reduced pain-associated activation of MAPKs in the spinal cord and the induction of c-fos expression in dorsal horn neurons. Importantly, in addition to a pharmacological approach, this study also evaluated the contribution of AMPK to inflammatory pain using mouse genetics. Using a global knockout approach for the AMPK α2 subunit, the authors found enhanced inflammatory hyperalgesia and a complete loss of the pain relieving effects of AMPK activators suggesting that the α2 subunit is the key regulatory factor for the antinociceptive effects of AMPK activators (Russe et al. 2013). In addition, these investigators also generated conditional knockout mice lacking the α2 subunit either specifically in sensory neurons or in macrophages and granulocytes. Both of these strains of mice showed enhanced inflammatory nociception suggesting that both sensory afferents and immune cells contribute to AMPK-mediated effects in the context of inflammatory pain.

An interesting new development in the area of therapeutics for inflammatory pain is the discovery that salicylates activate AMPK (Hawley et al. 2012) in addition to their very widely known action as irreversible cyclooxygenase (COX) enzyme inhibitors. The efficacy of salicylates in inflammatory pain has been widely attributed to this irreversible inhibition of COX, but it is formally possible that some of the beneficial actions are due to activation of AMPK with salicylates. Remarkably, other common nonsteroidal anti-inflammatory drugs (NSAIDs) like ibuprofen and diclofenac also activate AMPK suggesting that an acidic structure found in many NSAIDs may also lead to AMPK activation albeit at higher concentrations than are needed to inhibit COX enzymes (King et al. 2015). COX-2-specific inhibitors apparently lack activity at AMPK (King et al. 2015). Given that COX-2-specific inhibitors have been disappointing as next generation inflammatory pain drugs, it is possible that they lose some degree of efficacy due to their inability to activate AMPK.

The discovery that the oldest anti-inflammatory pain medicines known to humans have activity at AMPK (Hawley et al. 2012; King et al. 2015) opens new opportunities for development of AMPK activating compounds for the treatment of inflammatory pain. A large number of COX inhibitors have been described [see, for instance, (Ramalho et al. 2009) for a recent review of compounds described in the patent literature], and screening efforts are underway to understand whether and how these compounds might interact with AMPK (Sisignano et al. 2016). Insofar as these COX inhibitors are clearly efficacious for inflammatory pain but have side effects that preclude their long-term use, these discoveries present great opportunity for rationale drug design for next generation inflammatory pain medicines with dual action at AMPK and COX and possible disease-modifying activities.

11.8 AMPK: Diabetic Neuropathy Prevention and Treatment

For several decades, metformin has been the first-line medication used for type II diabetes. Diabetes can cause irreversible damage to many organ systems when it is left unmanaged. Peripheral neuropathy is a common comorbidity of diabetes. The main symptoms of diabetic neuropathy are tingling and numbness in the distal extremities sometimes accompanied by burning pain (Argoff et al. 2006). Animal and human histochemical studies indicate that diabetes can cause dieback of PNS epidermal endings (Sorensen et al. 2006; Cheng et al. 2013; Cheung et al. 2015; Ebenezer and Polydefkis 2014), an outcome that may not be related to pain (Sorensen et al. 2006). Also, in studies where diabetes is modeled in animals, the development of ectopic activity in DRG neurons (Khan et al. 2002; Serra et al. 2012) has been demonstrated. Clearly, diabetic neuropathy shares a similar pathophysiological profile with trauma-induced neuropathic pain and CIPN though its cause is mediated by metabolic dysfunction.

A current area of investigation is whether pharmacological treatments utilized in the control of glucose levels in type II diabetics are associated with protective effects against diabetic neuropathy. Though no conclusive studies in clinical populations have thus far been conducted, there are some indications that insulin treatment is associated with increased neuropathy incidence while metformin may protect against peripheral neuropathy. In a large study examining coronary artery disease and type II diabetes in a cohort of patients, the rate of neuropathy increased in patients using insulin whereas patients taking metformin had a reduced risk of developing neuropathy. This was independent of the duration of type II diabetes, glycemic index control, and other characteristics of the disease (Pop-Busui et al. 2009). These clinical results were further corroborated by a recent study in rats wherein a high fat diet (HFD) and streptozotocin (STZ) model of diabetes (HFD/STZ) was used to determine metformin's effect on the development of neuropathic pain. These authors found that metformin treatment started 4 days following the induction of diabetes had no effect on blood glucose levels over several weeks of treatment but significantly blocked the development of mechanical hypersensitivity caused by the HFD/STZ treatment (Byrne et al. 2015). This finding has now been corroborated by another study using a similar diabetes model in rats. These authors found that metformin treatment led to a significant reduction in diabetes-induced heat, cold, and mechanical hypersensitivity and reduced levels of glycation end products in blood in these animals without altering blood sugar levels (Ma et al. 2015). Importantly, this study also showed that metformin activates AMPK in sciatic nerves from diabetic rats and that this leads to the induction of AMPK-regulated genes in the PNS. The molecular mechanisms activated by metformin, and possibly other AMPK activators, that protect the peripheral nervous system during metabolic disease should be further explored. One possible mechanism for the neuroprotective effect of AMPK activation is via the induction of apolipoprotein E expression, a well-known neuroprotective factor, that is induced in the PNS by metformin treatment and is further enhanced by metformin treatment after nerve injury (Melemedjian et al. 2013a).

Additionally, AMPK may be directly and intimately linked to the development and maintenance of diabetic neuropathy. Hyperglycemia is a proposed contributing factor to nerve dysfunction in diabetes. It has been shown that hyperglycemia impairs bioenergetics in both DRG neurons and their mitochondria resulting in the downregulation of AMPK expression, phosphorylation, and activity (Roy Chowdhury et al. 2012). Elevating endogenous AMPK expression levels with adenoviral vectors or with augmenting AMPK activity using the AMPK positive modulator resveratrol restored the impaired biogenetic profile of cultured DRG neurons from streptozotocin-diabetic rats. In addition, resveratrol significantly reversed established deficits in thermal sensitivity and attenuated epidermal innervation density deficiency in STZ-treated rodents (Roy Chowdhury et al. 2012). These results suggest that deficits in AMPK may be a major driver of diabetic neuropathy development. Taken together, the findings discussed here create a strong rationale for targeting AMPK as a promising therapeutic for diabetic neuropathy treatment.

11.9 AMPK and Morphine/Opioid Tolerance

Opioids such as morphine are the gold standard treatment for severe pain management (Manchikanti et al. 2011; Sehgal et al. 2013). Despite their efficacy in providing acute pain relief, persistent and long-term use of opioids produces a reduction in their analgesic efficacy, also known as tolerance, necessitating dose escalation to manage pain. This dose escalation heightens the risk for adverse side effects such as sedation, respiratory depression, and constipation associated with opioid use and drastically reduces the quality of life of patients (Xu et al. 2014). The mechanisms underlying opioid tolerance are complex and not well understood. Recent studies implicate opioid-induced microglia activation and the resulting pro-inflammatory cytokine production as key pathogenic players in opioid tolerance. Chronic morphine exposure activates spinal and cortical glial cells (Cui et al. 2008; Mika 2008) to induce microglia and astrocyte changes (Horvath et al. 2010) resulting in the enhanced secretion of many substances such as pro-inflammatory neuromodulators including interleukin-1β (IL-1β), IL-6, and tumor necrosis factor-α (TNF-α), CCL2, CXCL1, ATP, nitric oxide (NO) and excitatory amino acids which counteract the analgesic effects of morphine and accelerate the development of tolerance (Hameed et al. 2010; Zhao et al. 2012; Sun et al. 2014). Pharmacological inhibition of microglia reduces morphine-induced cytokine release and attenuates morphine tolerance (Mika 2008). Administration of microglial inhibitors, which suppress microglia activation and cytokine release, significantly attenuates the development of tolerance after chronic intrathecal or systemic opioid administration (Cui et al. 2008; Sun et al. 2012; Fukagawa et al. 2013).

The MAPK signaling pathway, including extracellular signal-regulated protein kinase (ERK), p38 MAPK, c-Jun N-terminal kinase, and extracellular signal-regulated protein kinase 5 (ERK5), has been implicated in morphine-induced neuroinflammation and tolerance (Chen and Sommer 2009). There is also indication that opioid-induced tolerance and neuropathic pain share some similar mechanism profiles (Watkins et al. 2005). Based on the evidence that AMPK activation inhibits MAPK signaling and that AMPK activators alleviate acute and neuropathic pain (Melemedjian et al. 2011; Tillu et al. 2012), Han et al. investigated the effect of AMPK activation on morphine-induced microglial activation and tolerance. They demonstrated that resveratrol, an AMPK activator, suppresses morphine-induced p38/NF-kB signaling in microglia, and this effect was dependent on AMPK activation. They also showed that resveratrol blocks acute and chronic morphine tolerance (Han et al. 2014). This suggests that the engagement of AMPK signaling using AMPK activators represent a novel pharmacological method to attenuate morphine tolerance via suppression of morphine-induced microglial activation. Since opioids remain the most important drug class for the treatment of severe pain, this work creates a rationale for further exploration of AMPK activators as adjuvants to opioid therapy to reduce development of side effects that limit drug efficacy.

11.10 What Do We Know (and, More Importantly, Not Know) About AMPK Activators for Pain?

While there are a growing number of investigators actively working on AMPK as a novel target for pain therapeutics, we still have an enormous amount to learn about the target. In our view, the opportunity for disease modification via AMPK targeting presents the greatest opportunity for clinical benefit in patients suffering from chronic neuropathic pain. Neuropathic pain is perhaps the area of greatest unmet medical need in the clinical pain arena, and it is also heavily reliant on drugs that achieve little benefit in most patients (Baron et al. 2010). Having said this, it is still not known precisely how AMPK activators achieve this effect in rodent models of trauma-induced neuropathic pain, making it very difficult to make predictions of whether similar effects could be expected in human patients or even if trials should be designed with these endpoints in mind. Future work in this area should define the effects of AMPK activators on injured sensory afferents that drive persistent input to the CNS after nerve injury and on changes in microglia and other cell types in the CNS after injury (summarized in Fig. 11.1). While there is now substantial evidence that AMPK activation can influence both of these endpoints (Melemedjian et al. 2011, 2013c; Maixner et al. 2015), it is not clear which is responsible for the disease-modifying properties of AMPK activation in neuropathic pain. It is also, of course, possible that it is a different mechanism altogether (either AMPK-dependent or AMPK-independent) but investigations honing in on these two well prominent neuropathic pain mechanisms should yield preliminary insight into the problem.

Another major issue in the field is the reliance on imprecise pharmacological tools. This fact obscures whether AMPK is responsible for many of the behavioral observations that have been reported above. Before getting into how to combat this problem in future studies, two important points should be highlighted. First, metformin has been used widely in many of the preclinical studies described above. This is well justified because metformin is safe, widely available, and has the potential for immediate clinical translation. In fact, a retrospective study of metformin in low back pain patients suggests that the drug may be effective for sciatic nerve pain (Taylor et al. 2013), and a small prospective trial showed positive effects of metformin in fibromyalgia (Bullon et al. 2015). However, another retrospective study failed to find a positive effect for metformin on acute pain (Smith and Ang 2015). Metformin is widely thought to act via AMPK although this is still an area of great controversy and other mechanisms have been proposed (Zhou et al. 2001; Ouyang et al. 2011; Forslund et al. 2015). In agreement with this, a recent study investigating the effects of methylglyoxal, an important end glycation product produced in diabetes patients (Bierhaus et al. 2012), on diabetic neuropathic pain found that metformin blocked the effects of methylglyoxal via a direct reactive oxygen species sequestration mechanism that is independent of any intracellular signaling pathways (Huang et al. 2016). Second, while genetic tools have been employed in this area, their use is limited by the important function of

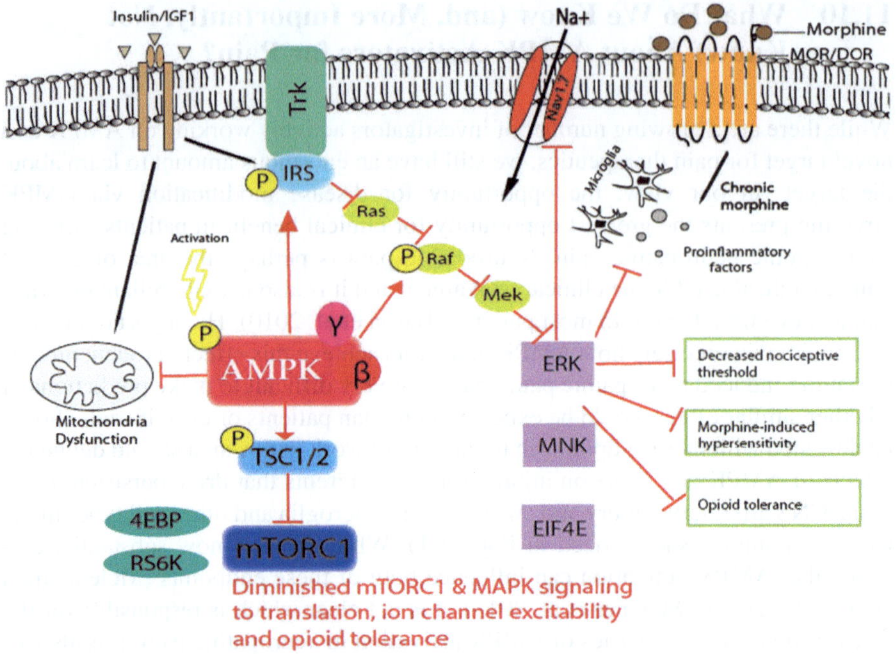

Fig. 11.1 Impact of AMPK activation on pathological pain-related signaling cascades. Factors associated with pathological pain employ signaling mechanisms that converge onto the mTORC1 and MAPK pathways to induce and maintain pain plasticity in nociceptors. Activation of AMPK abrogates these effects by negatively regulating these signaling events via phosphorylation of key effector molecules. AMPK activation phosphorylates IRS, an adaptor protein of the Trk receptor, at Ser789 to inhibit signaling via Trk and Insulin/IGF1 receptors. AMPK activation dampens mTORC1 signaling by phosphorylating TSC1/2 at Ser 1227 and 1345. The phosphorylation of Braf (Raf), a small GTPase linking receptors to MAPK signaling, by AMPK attenuates downstream ERK signaling. These signaling events produce several pathological pain attenuation outcomes such as: (1) a decrease in the phosphorylation of voltage-gated sodium channels such as $Na_v1.7$ leading to blunted neuronal excitability, (2) a reduction in microglial activation and microglia-related release of pro-inflammatory factors potentially resulting in reduced neuropathic pain, opioid-induced hypersensitivity, and diminished opioid tolerance, and (3) AMPK activation also normalizes dysregulation in mitochondrial and neuronal bioenergetics that is likely associated with diabetic neuropathic pain

AMPK in the development of the nervous system (Spasic et al. 2009). As noted above, genetic ablation of AMPK α2 subunits leads to an exacerbation of inflammatory pain and nullifies the antihyperalgesic effect of metformin (Russe et al. 2013). This finding strongly supports a mechanistic role for AMPK in modulation of inflammatory pain but similar experimental strategies have not yet been undertaken in the setting of neuropathic or cancer pain models. Coming back to the issue of using imprecise pharmacological tools, two things can immediately be done in this area. First, investigators should look carefully at readily available data on the specificity of the AMPK blocker compound C and more carefully consider this in their interpretations. While compound C does inhibit AMPK kinase activity, it is one of the least specific kinase inhibitors available and should only be used under very specific experimental conditions where the use of the tool can lead to valid interpretations (Arrowsmith et al. 2015). Second, the past two or three years have seen a major proliferation in the availability of positive allosteric modulators of AMPK as well as an improvement in their potency and specificity (Giordanetto and Karis 2012; Tripodi et al. 2012; Xiao et al. 2013; Zhang et al. 2013; Liu et al. 2014; Dokla et al. 2015). While it is always possible that off target effects will be found for these compounds in the future, many of them possess different allosteric binding sites on AMPK and have distinct chemical structures. Carefully choosing pharmacological tools for activation of AMPK in experimental pain studies can lead to significant mechanistic insight and may open the door to new pharmacological strategies for AMPK activation in a broad number of chronic pain conditions. For instance, compounds that allosterically enhance AMPK activity via binding to the β1 subunit [e.g., A760662, (Cool et al. 2006)] vs. those that protect against dephosphorylation via binding to the α subunit of the kinase [e.g., PT1, (Pang et al. 2008)] may have differential effects under different cellular circumstances but so far no work has been done on this area in the PNS or CNS. Findings along these lines would greatly enhance drug discovery efforts around AMPK as a pain target.

Work on AMPK as a pain target began based on mechanisms related to AMPK's role in regulating activity-dependent translation (Melemedjian et al. 2011; Price and Dussor 2013; Price et al. 2016) but it is still far from clear that this is the end of the story for AMPK-mediated control of nociceptor excitability, much less for other roles of AMPK activation in pain relief (e.g., modulation of microglial function). While it is clear that AMPK activation results in mTOR and MAPK inhibition in DRG neurons and a reduction in activity-dependent translation (Melemedjian et al. 2011; Price and Dussor 2013; Price et al. 2016), AMPK also regulates other cellular functions that may ultimately also be linked to inhibition of nociceptor excitability. For instance, AMPK is a core regulator of phospholipid biosynthesis and metabolism, but this aspect of AMPK function has been almost entirely ignored in the context of neuronal excitability. This area of investigation offers great opportunity for new discovery because it is well known that lipids are key regulators of channel, G-protein coupled receptor, and kinase function in nociceptors (Piomelli et al. 2014; Loo et al. 2015) and other neurons. Here, we propose that basic cell biology studies on the role of AMPK in neuronal excitability can greatly enhance our understanding of the target and how it can be manipulated to achieve effects that are relevant for pain therapeutics.

11.11 Closing Remarks

In this chapter, we have laid out the scientific rationale for targeting AMPK for the alleviation of pathological pain states and presented evidence in support of this therapeutic approach. While the existing evidence is compelling, we have tried to make the case that we are only now scratching the surface of the potential for this novel target, and continued discovery in this arena of investigation is likely to lead to a clearer picture of how AMPK can best be targeted for the alleviation of different types of pain. What is already clear is that AMPK activators offer an opportunity for disease modification for some of the most challenging forms of chronic pain and that this target has the potential to lead to treatments that will greatly benefit patients who are terribly underserved by the available pharmacological armamentarium.

Acknowledgements This work was supported by NIH grants R01NS065926 (TJP), R01GM102575 (TJP and GD), and The University of Texas STARS program (TJP and GD).

The authors declare no conflicts of interest.

References

Abe N, Borson SH, Gambello MJ, Wang F, Cavalli V (2010) Mammalian target of rapamycin (mTOR) activation increases axonal growth capacity of injured peripheral nerves. J Biol Chem 285:28034–28043

Adams JP, Anderson AE, Varga AW, Dineley KT, Cook RG, Pfaffinger PJ, Sweatt JD (2000) The A-type potassium channel Kv4.2 is a substrate for the mitogen-activated protein kinase ERK. J Neurochem 75:2277–2287

Agthong S, Kaewsema A, Tanomsridejchai N, Chentanez V (2006) Activation of MAPK ERK in peripheral nerve after injury. BMC Neurosci 7:45

Anand P, Shenoy R, Palmer JE, Baines AJ, Lai RY, Robertson J, Bird N, Ostenfeld T, Chizh BA (2011) Clinical trial of the p38 MAP kinase inhibitor dilmapimod in neuropathic pain following nerve injury. Eur J Pain 15:1040–1048

Andersen MN, Rasmussen HB (2012) AMPK: a regulator of ion channels. Commun Integr Biol 5:480–484

Argoff CE, Backonja MM, Belgrade MJ, Bennett GJ, Clark MR, Cole BE, Fishbain DA, Irving GA, McCarberg BH, McLean MJ (2006) Consensus guidelines: treatment planning and options. Diabetic peripheral neuropathic pain. Mayo Clinic Proc 81:S12–S25

Arrowsmith CH et al (2015) The promise and peril of chemical probes. Nat Chem Biol 11:536–541

Asante CO, Wallace VC, Dickenson AH (2009) Formalin-induced behavioural hypersensitivity and neuronal hyperexcitability are mediated by rapid protein synthesis at the spinal level. Mol Pain 5:27

Asante CO, Wallace VC, Dickenson AH (2010) Mammalian target of rapamycin signaling in the spinal cord is required for neuronal plasticity and behavioral hypersensitivity associated with neuropathy in the rat. J Pain 11(12):1356–1367

Asiedu MN, Tillu DV, Melemedjian OK, Shy A, Sanoja R, Bodell B, Ghosh S, Porreca F, Price TJ (2011) Spinal protein kinase M zeta underlies the maintenance mechanism of persistent nociceptive sensitization. J Neurosci 31:6646–6653

Balayssac D, Ferrier J, Descoeur J, Ling B, Pezet D, Eschalier A, Authier N (2011) Chemotherapy-induced peripheral neuropathies: from clinical relevance to preclinical evidence. Expert Opin Drug Saf 10:407–417

Banik RK, Brennan TJ (2008) Sensitization of primary afferents to mechanical and heat stimuli after incision in a novel in vitro mouse glabrous skin-nerve preparation. Pain 138:380–391

Banik RK, Subieta AR, Wu C, Brennan TJ (2005) Increased nerve growth factor after rat plantar incision contributes to guarding behavior and heat hyperalgesia. Pain 117:68–76

Baron R, Binder A, Wasner G (2010) Neuropathic pain: diagnosis, pathophysiological mechanisms, and treatment. Lancet Neurol 9:807–819

Basbaum AI, Bautista DM, Scherrer G, Julius D (2009) Cellular and molecular mechanisms of pain. Cell 139:267–284

Beggs S, Salter MW (2010) Microglia-neuronal signalling in neuropathic pain hypersensitivity 2.0. Curr Opin Neurobiol 20:474–480

Berta T, Liu T, Liu YC, Xu ZZ, Ji RR (2012) Acute morphine activates satellite glial cells and up-regulates IL-1beta in dorsal root ganglia in mice via matrix metalloprotease-9. Mol Pain 8:18

Bhalla V, Oyster NM, Fitch AC, Wijngaarden MA, Neumann D, Schlattner U, Pearce D, Hallows KR (2006) AMP-activated kinase inhibits the epithelial Na+ channel through functional regulation of the ubiquitin ligase Nedd4-2. J Biol Chem 281:26159–26169

Bierhaus A et al (2012) Methylglyoxal modification of Nav1.8 facilitates nociceptive neuron firing and causes hyperalgesia in diabetic neuropathy. Nat Med 18:926–933

Bongiorno D, Schuetz F, Poronnik P, Adams DJ (2011) Regulation of voltage-gated ion channels in excitable cells by the ubiquitin ligases Nedd4 and Nedd4-2. Channels (Austin) 5:79–88

Budde K, Becker T, Arns W, Sommerer C, Reinke P, Eisenberger U, Kramer S, Fischer W, Gschaidmeier H, Pietruck F (2011) Everolimus-based, calcineurin-inhibitor-free regimen in recipients of de-novo kidney transplants: an open-label, randomised, controlled trial. Lancet 377:837–847

Bullon P, Alcocer-Gomez E, Carrion AM, Garrido-Maraver J, Marin-Aguilar F, Roman-Malo L, Ruiz-Cabello J, Culic O, Ryffel B, Apetoh L, Ghiringhelli F, Battino M, Sanchez-Alcazar JA, Cordero MD (2015) AMPK phosphorylation modulates pain by activation of NLRP3-inflammasome. Antioxidants & redox signaling. Antioxid Redox Signal 24(3):157–170

Byrne F, Cheetham SC, Vickers SP, Chapman VR (2015) Characterisation of pain responses in the high fat diet/streptozotocin model of diabetes, and the analgesic effects of antidiabetic treatments. J Diabetes Res 2015:752481

Carattino MD, Edinger RS, Grieser HJ, Wise R, Neumann D, Schlattner U, Johnson JP, Kleyman TR, Hallows KR (2005) Epithelial sodium channel inhibition by AMP-activated protein kinase in oocytes and polarized renal epithelial cells. J Biol Chem 280:17608–17616

Carling D, Thornton C, Woods A, Sanders MJ (2012) AMP-activated protein kinase: new regulation, new roles? Biochem J 445:11–27

Cata JP, Weng HR, Lee BN, Reuben JM, Dougherty PM (2006) Clinical and experimental findings in humans and animals with chemotherapy-induced peripheral neuropathy. Minerva Anestesiol 72:151–169

Chapman PB (2013) Mechanisms of resistance to RAF inhibition in melanomas harboring a BRAF mutation. Am Soc Clin Oncol Educ Book. http://meetinglibrary.asco.org/content/158-132

Chen Y, Sommer C (2009) The role of mitogen-activated protein kinase (MAPK) in morphine tolerance and dependence. Mol Neurobiol 40:101–107

Cheng HT, Dauch JR, Porzio MT, Yanik BM, Hsieh W, Smith AG, Singleton JR, Feldman EL (2013) Increased axonal regeneration and swellings in intraepidermal nerve fibers characterize painful phenotypes of diabetic neuropathy. J Pain 14:941–947

Cheung A, Podgorny P, Martinez JA, Chan C, Toth C (2015) Epidermal axonal swellings in painful and painless diabetic peripheral neuropathy. Muscle Nerve 51(4):505–513

Christopher JEW (2001) Sirolimus (rapamycin) in clinical transplantation. Transplant Rev 15:165–177

Clark JD, Shi X, Li X, Qiao Y, Liang D, Angst MS, Yeomans DC (2007) Morphine reduces local cytokine expression and neutrophil infiltration after incision. Mol Pain 3:28

Cool B, Zinker B, Chiou W, Kifle L, Cao N, Perham M, Dickinson R, Adler A, Gagne G, Iyengar R, Zhao G, Marsh K, Kym P, Jung P, Camp HS, Frevert E (2006) Identification and characterization of a small molecule AMPK activator that treats key components of type 2 diabetes and the metabolic syndrome. Cell Metab 3:403–416

Cui Y, Liao XX, Liu W, Guo RX, Wu ZZ, Zhao CM, Chen PX, Feng JQ (2008) A novel role of minocycline: attenuating morphine antinociceptive tolerance by inhibition of p38 MAPK in the activated spinal microglia. Brain Behav Immun 22:114–123

Decensi A, Puntoni M, Goodwin P, Cazzaniga M, Gennari A, Bonanni B, Gandini S (2010) Metformin and cancer risk in diabetic patients: a systematic review and meta-analysis. Cancer Prev Res (Phila) 3:1451–1461

Dib-Hajj SD, Cummins TR, Black JA, Waxman SG (2010) Sodium channels in normal and pathological pain. Annu Rev Neurosci 33:325–347

Dokla EM, Fang CS, Lai PT, Kulp SK, Serya RA, Ismail NS, Abouzid KA, Chen CS (2015) Development of potent adenosine monophosphate activated protein kinase (AMPK) activators. ChemMedChem 10:1915–1923

Dworkin RH et al (2010) Recommendations for the pharmacological management of neuropathic pain: an overview and literature update. Mayo Clin Proc 85:S3–S14

Ebenezer G, Polydefkis M (2014) Epidermal innervation in diabetes. Handb Clin Neurol 126:261–274

Education IoMCoAPRCa (2011) Relieving pain in America: a blueprint for transforming prevention, care, education, and research. Washington, DC

Ferrier J, Pereira V, Busserolles J, Authier N, Balayssac D (2013) Emerging trends in understanding chemotherapy-induced peripheral neuropathy. Curr Pain Headache Rep 17:364

Foller M, Jaumann M, Dettling J, Saxena A, Pakladok T, Munoz C, Ruth P, Sopjani M, Seebohm G, Ruttiger L, Knipper M, Lang F (2012) AMP-activated protein kinase in BK-channel regulation and protection against hearing loss following acoustic overstimulation. FASEB J 26:4243–4253

Forslund K et al (2015) Disentangling type 2 diabetes and metformin treatment signatures in the human gut microbiota. Nature 528:262–266

Fukagawa H, Koyama T, Kakuyama M, Fukuda K (2013) Microglial activation involved in morphine tolerance is not mediated by toll-like receptor 4. J Anesth 27:93–97

Furic L, Rong L, Larsson O, Koumakpayi IH, Yoshida K, Brueschke A, Petroulakis E, Robichaud N, Pollak M, Gaboury LA, Pandolfi PP, Saad F, Sonenberg N (2010) eIF4E phosphorylation promotes tumorigenesis and is associated with prostate cancer progression. Proc Natl Acad Sci U S A 107:14134–14139

Gao B, Roux PP (2015) Translational control by oncogenic signaling pathways. Biochim Biophys Acta 1849:753–765

Geranton SM, Jimenez-Diaz L, Torsney C, Tochiki KK, Stuart SA, Leith JL, Lumb BM, Hunt SP (2009) A rapamycin-sensitive signaling pathway is essential for the full expression of persistent pain states. J Neurosci 29:15017–15027

Ghosh S, Tergaonkar V, Rothlin CV, Correa RG, Bottero V, Bist P, Verma IM, Hunter T (2006) Essential role of tuberous sclerosis genes TSC1 and TSC2 in NF-kappaB activation and cell survival. Cancer Cell 10:215–226

Gingras AC, Raught B, Sonenberg N (1999) eIF4 initiation factors: effectors of mRNA recruitment to ribosomes and regulators of translation. Annu Rev Biochem 68:913–963

Gingras AC, Raught B, Sonenberg N (2004) mTOR signaling to translation. Curr Top Microbiol Immunol 279:169–197

Giordanetto F, Karis D (2012) Direct AMP-activated protein kinase activators: a review of evidence from the patent literature. Expert Opin Ther Pat 22:1467–1477

Giuditta A, Kaplan BB, van Minnen J, Alvarez J, Koenig E (2002) Axonal and presynaptic protein synthesis: new insights into the biology of the neuron. Trends Neurosci 25:400–404

Gkogkas CG, Khoutorsky A, Cao R, Jafarnejad SM, Prager-Khoutorsky M, Giannakas N, Kaminari A, Fragkouli A, Nader K, Price TJ, Konicek BW, Graff JR, Tzinia AK, Lacaille JC, Sonenberg N (2014) Pharmacogenetic inhibition of eIF4E-dependent Mmp9 mRNA translation reverses fragile X syndrome-like phenotypes. Cell Rep 9:1742–1755

Hallows KR, Raghuram V, Kemp BE, Witters LA, Foskett JK (2000) Inhibition of cystic fibrosis transmembrane conductance regulator by novel interaction with the metabolic sensor AMP-activated protein kinase. J Clin Invest 105:1711–1721

Hameed H, Hameed M, Christo PJ (2010) The effect of morphine on glial cells as a potential therapeutic target for pharmacological development of analgesic drugs. Curr Pain Headache Rep 14:96–104

Han Y, Smith MT (2013) Pathobiology of cancer chemotherapy-induced peripheral neuropathy (CIPN). Front Pharmacol 4:156

Han Y, Jiang C, Tang J, Wang C, Wu P, Zhang G, Liu W, Jamangulova N, Wu X, Song X (2014) Resveratrol reduces morphine tolerance by inhibiting microglial activation via AMPK signalling. Eur J Pain 18:1458–1470

Hardie DG (2013) AMPK: a target for drugs and natural products with effects on both diabetes and cancer. Diabetes 62:2164–2172

Hardie DG, Ross FA, Hawley SA (2012) AMPK: a nutrient and energy sensor that maintains energy homeostasis. Nat Rev Mol Cell Biol 13:251–262

Hawley SA, Fullerton MD, Ross FA, Schertzer JD, Chevtzoff C, Walker KJ, Peggie MW, Zibrova D, Green KA, Mustard KJ, Kemp BE, Sakamoto K, Steinberg GR, Hardie DG (2012) The ancient drug salicylate directly activates AMP-activated protein kinase. Science 336:918–922

Herdy B et al (2012) Translational control of the activation of transcription factor NF-kappaB and production of type I interferon by phosphorylation of the translation factor eIF4E. Nat Immunol 13:543–550

Horvath RJ, Romero-Sandoval EA, De Leo JA (2010) Inhibition of microglial P2X4 receptors attenuates morphine tolerance, Iba1, GFAP and mu opioid receptor protein expression while enhancing perivascular microglial ED2. Pain 150:401–413

Hosono K, Endo H, Takahashi H, Sugiyama M, Uchiyama T, Suzuki K, Nozaki Y, Yoneda K, Fujita K, Yoneda M, Inamori M, Tomatsu A, Chihara T, Shimpo K, Nakagama H, Nakajima A (2010) Metformin suppresses azoxymethane-induced colorectal aberrant crypt foci by activating AMP-activated protein kinase. Mol Carcinog 49:662–671

Huang Q, Chen Y, Gong N, Wang Y (2016) Methylglyoxal mediates streptozotocin-induced diabetic neuropathic pain via activation of the peripheral TRPA1 and Nav1.8 channels. Metabolism 65(4):463–474

Ikematsu N, Dallas ML, Ross FA, Lewis RW, Rafferty JN, David JA, Suman R, Peers C, Hardie DG, Evans AM (2011) Phosphorylation of the voltage-gated potassium channel Kv2.1 by AMP-activated protein kinase regulates membrane excitability. Proc Natl Acad Sci U S A 108:18132–18137

Jakobsen SN, Hardie DG, Morrice N, Tornqvist HE (2001) 5′-AMP-activated protein kinase phosphorylates IRS-1 on Ser-789 in mouse C2C12 myotubes in response to 5-aminoimidazole-4-carboxamide riboside. J Biol Chem 276:46912–46916

Ji RR, Woolf CJ (2001) Neuronal plasticity and signal transduction in nociceptive neurons: implications for the initiation and maintenance of pathological pain. Neurobiol Dis 8:1–10

Ji RR, Kohno T, Moore KA, Woolf CJ (2003) Central sensitization and LTP: do pain and memory share similar mechanisms? Trends Neurosci 26:696–705

Ji RR, Kawasaki Y, Zhuang ZY, Wen YR, Zhang YQ (2007) Protein kinases as potential targets for the treatment of pathological pain. Handb Exp Pharmacol:359–389

Ji RR, Xu ZZ, Wang X, Lo EH (2009a) Matrix metalloprotease regulation of neuropathic pain. Trends Pharmacol Sci 30:336–340

Ji RR, Gereau RW, Malcangio M, Strichartz GR (2009b) MAP kinase and pain. Brain Res Rev 60:135–148

Jimenez-Diaz L, Geranton SM, Passmore GM, Leith JL, Fisher AS, Berliocchi L, Sivasubramaniam AK, Sheasby A, Lumb BM, Hunt SP (2008) Local translation in primary afferent fibers regulates nociception. PLoS One 3, e1961

Jung H, Holt CE (2011) Local translation of mRNAs in neural development. Wiley Interdiscip Rev RNA 2:153–165
Jung H, Yoon BC, Holt CE (2012) Axonal mRNA localization and local protein synthesis in nervous system assembly, maintenance and repair. Nat Rev Neurosci 13:308–324
Kawasaki Y, Xu ZZ, Wang X, Park JY, Zhuang ZY, Tan PH, Gao YJ, Roy K, Corfas G, Lo EH, Ji RR (2008) Distinct roles of matrix metalloproteases in the early- and late-phase development of neuropathic pain. Nat Med 14:331–336
Kehlet H, Jensen TS, Woolf CJ (2006) Persistent postsurgical pain: risk factors and prevention. Lancet 367:1618–1625
Khan GM, Chen SR, Pan HL (2002) Role of primary afferent nerves in allodynia caused by diabetic neuropathy in rats. Neuroscience 114:291–299
Kim T, Davis J, Zhang AJ, He X, Mathews ST (2009) Curcumin activates AMPK and suppresses gluconeogenic gene expression in hepatoma cells. Biochem Biophys Res Commun 388:377–382
Kim MH, Nahm FS, Kim TK, Chang MJ, Do SH (2014) Comparison of postoperative pain in the first and second knee in staged bilateral total knee arthroplasty: clinical evidence of enhanced pain sensitivity after surgical injury. Pain 155:22–27
King JD Jr, Fitch AC, Lee JK, McCane JE, Mak DO, Foskett JK, Hallows KR (2009) AMP-activated protein kinase phosphorylation of the R domain inhibits PKA stimulation of CFTR. Am J Physiol Cell Physiol 297:C94–C101
King TS, Russe OQ, Moser CV, Ferreiros N, Kynast KL, Knothe C, Olbrich K, Geisslinger G, Niederberger E (2015) AMP-activated protein kinase is activated by non-steroidal anti-inflammatory drugs. Eur J Pharmacol 762:299–305
Klann E, Antion MD, Banko JL, Hou L (2004) Synaptic plasticity and translation initiation. Learn Mem 11:365–372
Koenig E, Giuditta A (1999) Protein-synthesizing machinery in the axon compartment. Neuroscience 89:5–15
Koltzenburg M (1999) The changing sensitivity in the life of the nociceptor. Pain Suppl 6:S93–S102
Kourelis TV, Siegel RD (2012) Metformin and cancer: new applications for an old drug. Med Oncol 29:1314–1327
Laboureyras E, Chateauraynaud J, Richebe P, Simonnet G (2009) Long-term pain vulnerability after surgery in rats: prevention by nefopam, an analgesic with antihyperalgesic properties. Anesth Analg 109:623–631
Latremoliere A, Woolf CJ (2009) Central sensitization: a generator of pain hypersensitivity by central neural plasticity. J Pain 10:895–926
Light PE, Wallace CH, Dyck JR (2003) Constitutively active adenosine monophosphate-activated protein kinase regulates voltage-gated sodium channels in ventricular myocytes. Circulation 107:1962–1965
Lin AC, Holt CE (2008) Function and regulation of local axonal translation. Curr Opin Neurobiol 18:60–68
Liu YC, Berta T, Liu T, Tan PH, Ji RR (2012) Acute morphine induces matrix metalloproteinase-9 up-regulation in primary sensory neurons to mask opioid-induced analgesia in mice. Mol Pain 8:19
Liu J, Chen D, Liu P, He M, Li J, Li J, Hu L (2014) Discovery, synthesis, and structure-activity relationships of 20(S)-protopanaxadiol (PPD) derivatives as a novel class of AMPKalpha2beta1gamma1 activators. Eur J Med Chem 79:340–349
Loo L, Wright BD, Zylka MJ (2015) Lipid kinases as therapeutic targets for chronic pain. Pain 156 (Suppl 1):S2–S10
Loram LC, Themistocleous AC, Fick LG, Kamerman PR (2007) The time course of inflammatory cytokine secretion in a rat model of postoperative pain does not coincide with the onset of mechanical hyperalgesia. Can J Physiol Pharmacol 85:613–620

Ma J, Yu H, Liu J, Chen Y, Wang Q, Xiang L (2015) Metformin attenuates hyperalgesia and allodynia in rats with painful diabetic neuropathy induced by Streptozotocin. Eur J Pharmacol 764:599–606

Maixner DW, Yan X, Gao M, Yadav R, Weng HR (2015) Adenosine monophosphate-activated protein kinase regulates interleukin-1 beta expression and glial glutamate transporter function in rodents with neuropathic pain. Anesthesiology 122:1401–1413

Manchikanti L, Vallejo R, Manchikanti KN, Benyamin RM, Datta S, Christo PJ (2011) Effectiveness of long-term opioid therapy for chronic non-cancer pain. Pain Physician 14:E133–E156

Mao-Ying QL, Kavelaars A, Krukowski K, Huo XJ, Zhou W, Price TJ, Cleeland C, Heijnen CJ (2014) The anti-diabetic drug metformin protects against chemotherapy-induced peripheral neuropathy in a mouse model. PLoS One 9, e100701

Martin KC, Barad M, Kandel ER (2000) Local protein synthesis and its role in synapse-specific plasticity. Curr Opin Neurobiol 10:587–592

Martin SW, Butcher AJ, Berrow NS, Richards MW, Paddon RE, Turner DJ, Dolphin AC, Sihra TS, Fitzgerald EM (2006) Phosphorylation sites on calcium channel alpha1 and beta subunits regulate ERK-dependent modulation of neuronal N-type calcium channels. Cell Calcium 39:275–292

Martinez-Outschoorn UE, Trimmer C, Lin Z, Whitaker-Menezes D, Chiavarina B, Zhou J, Wang C, Pavlides S, Martinez-Cantarin MP, Capozza F, Witkiewicz AK, Flomenberg N, Howell A, Pestell RG, Caro J, Lisanti MP, Sotgia F (2010) Autophagy in cancer associated fibroblasts promotes tumor cell survival: Role of hypoxia, HIF1 induction and NFkappaB activation in the tumor stromal microenvironment. Cell Cycle 9:3515–3533

Massard C, Fizazi K, Gross-Goupil M, Escudier B (2010) Reflex sympathetic dystrophy in patients with metastatic renal cell carcinoma treated with Everolimus. Investig New Drugs 28:879–881

McCormack FX et al (2011) Efficacy and safety of Sirolimus in Lymphangioleiomyomatosis. N Engl J Med 364(17):1595–1606

Melemedjian OK, Asiedu MN, Tillu DV, Peebles KA, Yan J, Ertz N, Dussor GO, Price TJ (2010) IL-6- and NGF-induced rapid control of protein synthesis and nociceptive plasticity via convergent signaling to the eIF4F complex. J Neurosci 30:15113–15123

Melemedjian OK, Asiedu MN, Tillu DV, Sanoja R, Yan J, Lark A, Khoutorsky A, Johnson J, Peebles KA, Lepow T, Sonenberg N, Dussor G, Price TJ (2011) Targeting adenosine monophosphate-activated protein kinase (AMPK) in preclinical models reveals a potential mechanism for the treatment of neuropathic pain. Mol Pain 7:70

Melemedjian OK, Khoutorsky A, Sorge RE, Yan J, Asiedu MN, Valdez A, Ghosh S, Dussor G, Mogil JS, Sonenberg N, Price TJ (2013a) mTORC1 inhibition induces pain via IRS-1-dependent feedback activation of ERK. Pain 154(7):1080–1091

Melemedjian OK, Yassine HN, Shy A, Price TJ (2013b) Proteomic and functional annotation analysis of injured peripheral nerves reveals ApoE as a protein upregulated by injury that is modulated by metformin treatment. Mol Pain 9:14

Melemedjian OK, Tillu DV, Asiedu MN, Mandell EK, Moy JK, Blute VM, Taylor CJ, Ghosh S, Price TJ (2013c) BDNF regulates atypical PKC at spinal synapses to initiate and maintain a centralized chronic pain state. Mol Pain 9:12

Melemedjian OK, Tillu DV, Moy JK, Asiedu MN, Mandell EK, Ghosh S, Dussor G, Price TJ (2014) Local translation and retrograde axonal transport of CREB regulates IL-6-induced nociceptive plasticity. Mol Pain 10:45

Memmott RM, Mercado JR, Maier CR, Kawabata S, Fox SD, Dennis PA (2010) Metformin prevents tobacco carcinogen--induced lung tumorigenesis. Cancer Prev Res (Phila) 3:1066–1076

Mika J (2008) Modulation of microglia can attenuate neuropathic pain symptoms and enhance morphine effectiveness. Pharmacol Rep 60:297–307

Mogil JS (2009) Animal models of pain: progress and challenges. Nat Rev Neurosci 10:283–294

Mohr E, Richter D (2000) Axonal mRNAs: functional significance in vertebrates and invertebrates. J Neurocytol 29:783–791
Molina MG, Diekmann F, Burgos D, Cabello M, Lopez V, Oppenheimer F, Navarro A, Campistol J (2008) Sympathetic dystrophy associated with sirolimus therapy. Transplantation 85:290–292. doi:10.1097/TP.1090b1013e3181601230
Nie H, Weng HR (2009) Glutamate transporters prevent excessive activation of NMDA receptors and extrasynaptic glutamate spillover in the spinal dorsal horn. J Neurophysiol 101:2041–2051
Nie H, Weng HR (2010) Impaired glial glutamate uptake induces extrasynaptic glutamate spillover in the spinal sensory synapses of neuropathic rats. J Neurophysiol 103:2570–2580
Norsted Gregory E, Codeluppi S, Gregory JA, Steinauer J, Svensson CI (2010) Mammalian target of rapamycin in spinal cord neurons mediates hypersensitivity induced by peripheral inflammation. Neuroscience 169:1392–1402
Obara I, Tochiki KK, Geranton SM, Carr FB, Lumb BM, Liu Q, Hunt SP (2011) Systemic inhibition of the mammalian target of rapamycin (mTOR) pathway reduces neuropathic pain in mice. Pain 152:2582–2595
Obara I, Geranton SM, Hunt SP (2012) Axonal protein synthesis: a potential target for pain relief? Curr Opin Pharmacol 12:42–48
Obata K, Noguchi K (2004) MAPK activation in nociceptive neurons and pain hypersensitivity. Life Sci 74:2643–2653
Obata K, Yamanaka H, Dai Y, Tachibana T, Fukuoka T, Tokunaga A, Yoshikawa H, Noguchi K (2003) Differential activation of extracellular signal-regulated protein kinase in primary afferent neurons regulates brain-derived neurotrophic factor expression after peripheral inflammation and nerve injury. J Neurosci 23:4117–4126
Obata K, Yamanaka H, Dai Y, Mizushima T, Fukuoka T, Tokunaga A, Noguchi K (2004a) Differential activation of MAPK in injured and uninjured DRG neurons following chronic constriction injury of the sciatic nerve in rats. Eur J Neurosci 20:2881–2895
Obata K, Yamanaka H, Kobayashi K, Dai Y, Mizushima T, Katsura H, Fukuoka T, Tokunaga A, Noguchi K (2004b) Role of mitogen-activated protein kinase activation in injured and intact primary afferent neurons for mechanical and heat hypersensitivity after spinal nerve ligation. J Neurosci 24:10211–10222
Ouyang J, Parakhia RA, Ochs RS (2011) Metformin activates AMP kinase through inhibition of AMP deaminase. J Biol Chem 286:1–11
Pang T, Zhang ZS, Gu M, Qiu BY, Yu LF, Cao PR, Shao W, Su MB, Li JY, Nan FJ, Li J (2008) Small molecule antagonizes autoinhibition and activates AMP-activated protein kinase in cells. J Biol Chem 283:16051–16060
Pavlides S, Tsirigos A, Migneco G, Whitaker-Menezes D, Chiavarina B, Flomenberg N, Frank PG, Casimiro MC, Wang C, Pestell RG, Martinez-Outschoorn UE, Howell A, Sotgia F, Lisanti MP (2010) The autophagic tumor stroma model of cancer: Role of oxidative stress and ketone production in fueling tumor cell metabolism. Cell Cycle 9:3485–3505
Perl ER (1996) Cutaneous polymodal receptors: characteristics and plasticity. Prog Brain Res 113:21–37
Pinto PR, McIntyre T, Ferrero R, Almeida A, Araujo-Soares V (2013) Risk factors for moderate and severe persistent pain in patients undergoing total knee and hip arthroplasty: a prospective predictive study. PLoS One 8, e73917
Piomelli D, Hohmann AG, Seybold V, Hammock BD (2014) A lipid gate for the peripheral control of pain. J Neurosci 34:15184–15191
Piper M, Holt C (2004) RNA translation in axons. Annu Rev Cell Dev Biol 20:505–523
Pop-Busui RLJ, Lopes N, Jones TL, BARI 2D Investigators (2009) Prevalence of diabetic peripheral neuropathy and relation to glycemic control therapies at baseline in the BARI 2D cohort. J Peripher Nerv Syst 14:1–13
Poulikakos PI, Solit DB (2011) Resistance to MEK inhibitors: should we co-target upstream? Sci Signal 4:pe16

Price TJ, Dussor G (2013) AMPK: An emerging target for modification of injury-induced pain plasticity. Neurosci Lett 557 Pt A:9–18
Price TJ, Geranton SM (2009) Translating nociceptor sensitivity: the role of axonal protein synthesis in nociceptor physiology. Eur J Neurosci 29:2253–2263
Price TJ, Inyang KE (2015) Commonalities between pain and memory mechanisms and their meaning for understanding chronic pain. In: Price TJ, Dussor G (eds) Molecular biology of pain, progress in molecular and translational science. Elsevier: Amsterdam. Prog Mol Biol Transl Sci.131:409–34
Price TJ, Rashid MH, Millecamps M, Sanoja R, Entrena JM, Cervero F (2007) Decreased nociceptive sensitization in mice lacking the fragile X mental retardation protein: role of mGluR1/5 and mTOR. J Neurosci 27:13958–13967
Price TJ, Das V, Dussor G (2016) Adenosine monophosphate-activated protein kinase (AMPK) activators for the prevention, treatment and potential reversal of pathological pain. Curr Drug Targets 17(8):908–920
Pyronnet S, Imataka H, Gingras AC, Fukunaga R, Hunter T, Sonenberg N (1999) Human eukaryotic translation initiation factor 4G (eIF4G) recruits mnk1 to phosphorylate eIF4E. EMBO J 18:270–279
Ramalho TC, Rocha M, da Cunha EF, Freitas MP (2009) The search for new COX-2 inhibitors: a review of 2002–2008 patents. Expert Opin Ther Pat 19:1193–1228
Reichling DB, Levine JD (2009) Critical role of nociceptor plasticity in chronic pain. Trends Neurosci 32:611–618
Roy Chowdhury SK, Smith DR, Saleh A, Schapansky J, Marquez A, Gomes S, Akude E, Morrow D, Calcutt NA, Fernyhough P (2012) Impaired adenosine monophosphate-activated protein kinase signalling in dorsal root ganglia neurons is linked to mitochondrial dysfunction and peripheral neuropathy in diabetes. Brain 135:1751–1766
Ruangsri S, Lin A, Mulpuri Y, Lee K, Spigelman I, Nishimura I (2011) Relationship of axonal voltage-gated sodium channel 1.8 (NaV1.8) mRNA accumulation to sciatic nerve injury-induced painful neuropathy in rats. J Biol Chem 286:39836–39847
Russe OQ, Moser CV, Kynast KL, King TS, Stephan H, Geisslinger G, Niederberger E (2013) Activation of the AMP-activated protein kinase reduces inflammatory nociception. J Pain 14:1330–1340
Sahbaie P, Sun Y, Liang DY, Shi XY, Clark JD (2014) Curcumin treatment attenuates pain and enhances functional recovery after incision. Anesth Analg 118:1336–1344
Sale MJ, Cook SJ (2014) Intrinsic and acquired resistance to MEK1/2 inhibitors in cancer. Biochem Soc Trans 42:776–783
Schrader LA, Birnbaum SG, Nadin BM, Ren Y, Bui D, Anderson AE, Sweatt JD (2006) ERK/MAPK regulates the Kv4.2 potassium channel by direct phosphorylation of the pore-forming subunit. Am J Physiol Cell Physiol 290:C852–C861
Schuman EM, Dynes JL, Steward O (2006) Synaptic regulation of translation of dendritic mRNAs. J Neurosci 26:7143–7146
Sehgal N, Colson J, Smith HS (2013) Chronic pain treatment with opioid analgesics: benefits versus harms of long-term therapy. Expert Rev Neurother 13:1201–1220
Serra J, Bostock H, Sola R, Aleu J, Garcia E, Cokic B, Navarro X, Quiles C (2012) Microneurographic identification of spontaneous activity in C-nociceptors in neuropathic pain states in humans and rats. Pain 153:42–55
Shih MH, Kao SC, Wang W, Yaster M, Tao YX (2012) Spinal cord NMDA receptor-mediated activation of mammalian target of rapamycin is required for the development and maintenance of bone cancer-induced pain hypersensitivities in rats. J Pain 13:338–349
Siddiqui N, Sonenberg N (2015) Signalling to eIF4E in cancer. Biochem Soc Trans 43:763–772
Sisignano M, Parnham MJ, Geisslinger G (2016) Drug repurposing for the development of novel analgesics. Trends Pharmacol Sci 37(3):172–183
Smith B, Ang D (2015) Metformin: potential analgesic? Pain Med 16(12):2256–2260
Sonenberg N (2008) eIF4E, the mRNA cap-binding protein: from basic discovery to translational research. Biochem Cell Biol 86:178–183

Song H, Han Y, Pan C, Deng X, Dai W, Hu L, Jiang C, Yang Y, Cheng Z, Li F, Zhang G, Wu X, Liu W (2015) Activation of adenosine monophosphate-activated protein kinase suppresses neuroinflammation and ameliorates bone cancer pain: involvement of inhibition on mitogen-activated protein kinase. Anesthesiology 123:1170–1185

Sorensen L, Molyneaux L, Yue DK (2006) The relationship among pain, sensory loss, and small nerve fibers in diabetes. Diabetes Care 29:883–887

Spasic MR, Callaerts P, Norga KK (2009) AMP-activated protein kinase (AMPK) molecular crossroad for metabolic control and survival of neurons. Neuroscientist 15:309–316

Stamboulian S, Choi JS, Ahn HS, Chang YW, Tyrrell L, Black JA, Waxman SG, Dib-Hajj SD (2010) ERK1/2 mitogen-activated protein kinase phosphorylates sodium channel Na(v)1.7 and alters its gating properties. J Neurosci 30:1637–1647

Sukhodub A, Jovanovic S, Du Q, Budas G, Clelland AK, Shen M, Sakamoto K, Tian R, Jovanovic A (2007) AMP-activated protein kinase mediates preconditioning in cardiomyocytes by regulating activity and trafficking of sarcolemmal ATP-sensitive K(+) channels. J Cell Physiol 210:224–236

Sun J, Liu S, Mata M, Fink DJ, Hao S (2012) Transgene-mediated expression of tumor necrosis factor soluble receptor attenuates morphine tolerance in rats. Gene Ther 19:101–108

Sun Y, Sahbaie P, Liang D, Li W, Clark JD (2014) Opioids enhance CXCL1 expression and function after incision in mice. J Pain 15:856–866

Sutton MA, Schuman EM (2005) Local translational control in dendrites and its role in long-term synaptic plasticity. J Neurobiol 64:116–131

Taylor A, Westveld AH, Szkudlinska M, Guruguri P, Annabi E, Patwardhan A, Price TJ, Yassine HN (2013) The use of metformin is associated with decreased lumbar radiculopathy pain. J Pain Res 6:755–763

Thakor DK, Lin A, Matsuka Y, Meyer EM, Ruangsri S, Nishimura I, Spigelman I (2009) Increased peripheral nerve excitability and local NaV1.8 mRNA up-regulation in painful neuropathy. Mol Pain 5:14

Tillu DV, Melemedjian OK, Asiedu MN, Qu N, De Felice M, Dussor G, Price TJ (2012) Resveratrol engages AMPK to attenuate ERK and mTOR signaling in sensory neurons and inhibits incision-induced acute and chronic pain. Mol Pain 8:5

Tripodi F, Pagliarin R, Fumagalli G, Bigi A, Fusi P, Orsini F, Frattini M, Coccetti P (2012) Synthesis and biological evaluation of 1,4-diaryl-2-azetidinones as specific anticancer agents: activation of adenosine monophosphate activated protein kinase and induction of apoptosis. J Med Chem 55:2112–2124

Vicier C, Dieci MV, Andre F (2013) New strategies to overcome resistance to mammalian target of rapamycin inhibitors in breast cancer. Curr Opin Oncol 25:587–593

Walters ET, Moroz LL (2009) Molluscan memory of injury: evolutionary insights into chronic pain and neurological disorders. Brain Behav Evol 74:206–218

Watkins LR, Hutchinson MR, Johnston IN, Maier SF (2005) Glia: novel counter-regulators of opioid analgesia. Trends Neurosci 28:661–669

Willis DE, Twiss JL (2006) The evolving roles of axonally synthesized proteins in regeneration. Curr Opin Neurobiol 16:111–118

Willis D, Li KW, Zheng JQ, Chang JH, Smit AB, Kelly T, Merianda TT, Sylvester J, van Minnen J, Twiss JL (2005) Differential transport and local translation of cytoskeletal, injury-response, and neurodegeneration protein mRNAs in axons. J Neurosci 25:778–791

Willis DE, Xu M, Donnelly CJ, Tep C, Kendall M, Erenstheyn M, English AW, Schanen NC, Kirn-Safran CB, Yoon SO, Bassell GJ, Twiss JL (2011) Axonal localization of transgene mRNA in mature PNS and CNS neurons. J Neurosci 31:14481–14487

Witzig TE, Geyer SM, Ghobrial I, Inwards DJ, Fonseca R, Kurtin P, Ansell SM, Luyun R, Flynn PJ, Morton RF, Dakhil SR, Gross H, Kaufmann SH (2005) Phase II trial of single-agent temsirolimus (CCI-779) for relapsed mantle cell lymphoma. J Clin Oncol 23:5347–5356

Woolf CJ (2011) Central sensitization: implications for the diagnosis and treatment of pain. Pain 152:S2–S15

Woolf CJ, Costigan M (1999) Transcriptional and posttranslational plasticity and the generation of inflammatory pain. Proc Natl Acad Sci U S A 96:7723–7730
Woolf CJ, Salter MW (2000) Neuronal plasticity: increasing the gain in pain. Science 288:1765–1769
Woolf CJ, Walters ET (1991) Common patterns of plasticity contributing to nociceptive sensitization in mammals and Aplysia. Trends Neurosci 14:74–78
Wu C, Boustany L, Liang H, Brennan TJ (2007) Nerve growth factor expression after plantar incision in the rat. Anesthesiology 107:128–135
Wu C, Erickson MA, Xu J, Wild KD, Brennan TJ (2009) Expression profile of nerve growth factor after muscle incision in the rat. Anesthesiology 110:140–149
Xiao B, Sanders MJ, Carmena D, Bright NJ, Haire LF, Underwood E, Patel BR, Heath RB, Walker PA, Hallen S, Giordanetto F, Martin SR, Carling D, Gamblin SJ (2013) Structural basis of AMPK regulation by small molecule activators. Nat Commun 4:3017
Xu J, Brennan TJ (2009) Comparison of skin incision vs. skin plus deep tissue incision on ongoing pain and spontaneous activity in dorsal horn neurons. Pain 144:329–339
Xu JT, Zhao X, Yaster M, Tao YX (2010) Expression and distribution of mTOR, p70S6K, 4E-BP1, and their phosphorylated counterparts in rat dorsal root ganglion and spinal cord dorsal horn. Brain Res 1336:46–57
Xu Q, Fitzsimmons B, Steinauer J, Neill AO, Newton AC, Hua XY, Yaksh TL (2011) Spinal phosphinositide 3-kinase-akt-Mammalian target of rapamycin signaling cascades in inflammation-induced hyperalgesia. J Neurosci 31:2113–2124
Xu JT, Zhao JY, Zhao X, Ligons D, Tiwari V, Atianjoh FE, Lee CY, Liang L, Zang W, Njoku D, Raja SN, Yaster M, Tao YX (2014) Opioid receptor-triggered spinal mTORC1 activation contributes to morphine tolerance and hyperalgesia. J Clin Invest 124:592–603
Yoshida H, Bao L, Kefaloyianni E, Taskin E, Okorie U, Hong M, Dhar-Chowdhury P, Kaneko M, Coetzee WA (2012) AMP-activated protein kinase connects cellular energy metabolism to KATP channel function. J Mol Cell Cardiol 52:410–418
Yu S, Shen G, Khor TO, Kim JH, Kong AN (2008) Curcumin inhibits Akt/mammalian target of rapamycin signaling through protein phosphatase-dependent mechanism. Mol Cancer Ther 7:2609–2620
Zahn PK, Subieta A, Park SS, Brennan TJ (2004) Effect of blockade of nerve growth factor and tumor necrosis factor on pain behaviors after plantar incision. J Pain 5:157–163
Zhang LN, Xu L, Zhou HY, Wu LY, Li YY, Pang T, Xia CM, Qiu BY, Gu M, Dong TC, Li JY, Shen JK, Li J (2013) Novel small-molecule AMP-activated protein kinase allosteric activator with beneficial effects in db/db mice. PLoS One 8, e72092
Zhao CM, Guo RX, Hu F, Meng JL, Mo LQ, Chen PX, Liao XX, Cui Y, Feng JQ (2012) Spinal MCP-1 contributes to the development of morphine antinociceptive tolerance in rats. Am J Med Sci 344:473–479
Zheng JQ, Kelly TK, Chang B, Ryazantsev S, Rajasekaran AK, Martin KC, Twiss JL (2001) A functional role for intra-axonal protein synthesis during axonal regeneration from adult sensory neurons. J Neurosci 21:9291–9303
Zhou G, Myers R, Li Y, Chen Y, Shen X, Fenyk-Melody J, Wu M, Ventre J, Doebber T, Fujii N, Musi N, Hirshman MF, Goodyear LJ, Moller DE (2001) Role of AMP-activated protein kinase in mechanism of metformin action. J Clin Invest 108:1167–1174

Chapter 12
AMPK in Pathogens

Inês Mesquita, Diana Moreira, Belém Sampaio-Marques, Mireille Laforge, Anabela Cordeiro-da-Silva, Paula Ludovico, Jérôme Estaquier, and Ricardo Silvestre

Contents

Inês Mesquita and **Diana Moreira** are contributed equally with all other contributors.

I. Mesquita • B. Sampaio-Marques • P. Ludovico • R. Silvestre (✉)
Life and Health Sciences Research Institute (ICVS), School of Health Sciences, University of Minho, Braga, Portugal

ICVS/3Bs-PT Government Associate Laboratory, Guimarães, Braga, Portugal
e-mail: ricardosilvestre@ecsaude.uminho.pt

D. Moreira • A. Cordeiro-da-Silva
i3S-Instituto de Investigação e Inovação em Saúde, Universidade do Porto, Porto, Portugal

IBMC-Instituto de Biologia Molecular e Celular, Universidade do Porto, Porto, Portugal

Departamento de Ciências Biológicas, Faculdade de Farmácia, Universidade do Porto, Porto, Portugal

M. Laforge
CNRS FR 3636, Université Paris Descartes, Paris, France

J. Estaquier
CNRS FR 3636, Université Paris Descartes, Paris, France

Centre de Recherche du CHU de Québec, Université Laval, Québec, QC, Canada

M.D. Cordero, B. Viollet (eds.), *AMP-activated Protein Kinase*, Experientia Supplementum 107, DOI 10.1007/978-3-319-43589-3_12

Abstract During host–pathogen interactions, a complex web of events is crucial for the outcome of infection. Pathogen recognition triggers powerful cellular signaling events that is translated into the induction and maintenance of innate and adaptive host immunity against infection. In opposition, pathogens employ active mechanisms to manipulate host cell regulatory pathways toward their proliferation and survival. Among these, subversion of host cell energy metabolism by pathogens is currently recognized to play an important role in microbial growth and persistence. Extensive studies have documented the role of AMP-activated protein kinase (AMPK) signaling, a central cellular hub involved in the regulation of energy homeostasis, in host–pathogen interactions. Here, we highlight the most recent advances detailing how pathogens hijack cellular metabolism by suppressing or increasing the activity of the host energy sensor AMPK. We also address the role of lower eukaryote AMPK orthologues in the adaptive process to the host microenvironment and their contribution for pathogen survival, differentiation, and growth. Finally, we review the effects of pharmacological or genetic AMPK modulation on pathogen growth and persistence.

Keywords SNF1/AMPK • Host-pathogen interactions • Infection • Metabolism • Microbial auxotrophy • Bioenergetics

12.1 Introduction

Adenosine 5′ monophosphate-activated protein kinase (AMPK) is a heterotrimeric serine/threonine kinase consisting of a catalytic subunit (α) and two regulatory subunits (β and γ) (Xiao et al. 2011). AMPK is considered a pivotal regulator of cellular metabolism known as a metabolic "master regulator," switching on catabolic pathways that generate ATP, while switching off anabolic pathways that consume ATP. Therefore, upon activation, AMPK downregulates several anabolic enzymes through phosphorylation leading to the inhibition of both translation initiation by restraining mammalian target of rapamycin complex 1 (mTORC1) activity, as well as translation elongation through the inactivation of Eukaryotic Elongation Factor 2 (eEF2), which will ultimately decrease cellular ATP consumption. In opposition, AMPK activates a catabolic state by inducing oxidative pathways generating energy through the activation of glucose uptake (via activation of both glucose transporter 1 (GLUT1) and GLUT4) (Barnes et al. 2002; Holmes et al. 1999), glycolysis (via phosphorylation and activation of two of four isoforms of 6-phosphofructo-2-kinase) (Marsin et al. 2002), fatty acid uptake (via translocation of the fatty acid transporter FAT/CD36) (Bonen et al. 2007), and fatty acid oxidation (via phosphorylation of the isoform 2 of acetyl-CoA carboxylase (ACC2)) (Winder and Hardie 1996). Classically, the canonical upstream activator that phosphorylates AMPK on Thr 172 (Stein et al. 2000) is the constitutively active tumor suppressor liver kinase B1 (LKB1) accompanied by two accessory subunits, sterile 20 protein-related adaptor (STRAD) and mouse protein 25 (MO25) (Hawley et al. 2003; Woods et al. 2003), but additional activators such as the Ca^{2+}-calmodulin-dependent kinase kinase (CaMKK) has been also identified (Woods et al. 2005; Hawley et al. 2005).

AMPK is capable of sensing changes in the energy status of the cell, and therefore it is not surprising that it plays a pivotal role during infections (Moreira et al. 2015b). Nutritional immunity has been described as a defence mechanism employed by host cells to prevent the acquisition of essential nutrients by intracellular pathogens, thus preventing pathogen scavenging and further utilization for replication and survival (Hood and Skaar 2012). Furthermore, recent advances have underlined the role of AMPK in macrophage, T cell, and DC functions in distinct settings, providing a molecular link between bioenergetics homeostasis, viability, and effector functions of adaptive and innate immune cells (Blagih et al. 2015; Kelly and O'Neill 2015; Yang and Chi 2015). These concepts are the core of host–pathogen dynamics. Herein, we review the contribution of AMPK sensor in regulating viral, bacterial, parasitic, and fungal infections. Furthermore, given the high degree of conservation among eukaryotes, we intend to discuss the crucial role of AMPK in parasites and fungi. The most recent findings concerning the AMPK role in the parasite and fungal biology during host–pathogen interaction will be also addressed.

12.2 AMPK: A Regulator in Viral Infection

12.2.1 Modulation of AMPK Activity by Viruses

Simian virus 40 (SV40) is a polyoma virus that encodes for small and large T antigens implicated in tumor formation. SV40 small T antigen has been shown to protect human cells in glucose deprivation. This survival advantage for cancer cells was demonstrated to be correlated with an increased phosphorylation of AMPK and consequent downregulation of mTORC1, which ultimately triggered autophagy (Kumar and Rangarajan 2009). In vitro infection with avian reovirus (ARV) resulted in an increase in AMPK phosphorylation on Thr172 (Ji et al. 2009) and a concomitant increase in mitogen-activated protein kinase (MAPK) p38 phosphorylation. AMPK phosphorylation has also been described to be increased early after infection with Rift Valley fever virus (RVFV), an important reemerging arthropod-borne human pathogen (Moser et al. 2012).

On the contrary, infections with Hepatitis C virus (HCV), Human cytomegalovirus (HCMV), Herpes simplex virus (HSV), Epstein–Barr virus (EBV), and human immunodeficiency virus-1 (HIV-1) have been reported to inhibit AMPK activity. In this context, cells infected with HCV or harboring an HCV subgenomic replicon inhibited the phosphorylation of AMPK at Thr172. This was consistent with the concomitant reduction on AMPK activity and increased hepatic lipid accumulation required for virus replication (Mankouri et al. 2010). Inhibition of Thr172 phosphorylation was associated with an increased phosphorylation of AMPK at an alternative site, Ser485, which is phosphorylated by protein kinase B (PKB/Akt) (Horman et al. 2006). Interestingly, both HCV NS4B and NS5A proteins have been shown to activate the protein kinase AKT (Park et al. 2009; Street et al. 2004). In addition to HCV, HCMV, a beta herpes virus that establishes chronic infections, activates mTOR signaling by inhibiting AMPK Thr172 phosphorylation at early time points postinfection (Kudchodkar et al. 2007). Interestingly, at later time points of infection, a kinome RNAi screen performed in HCMV-infected MRC5 fibroblasts identified 106 cellular kinases that influenced the growth of the virus, including multiple elements of the AMPK pathway (Terry et al. 2012). The effector proteins responsible for the AMPK inhibition still remain elusive, although the inhibition of CaMKK was reported to block HCMV-mediated AMPK activation (McArdle et al. 2011, 2012). Furthermore, it has been shown that HCMV blocked the function of the mitochondrial trifunctional protein (TFP), a key enzyme during fatty acid-β-oxidation (Seo and Cresswell 2013), through the redistribution of viperin to the mitochondria. The resulting decrease in cellular ATP levels and consequent increase in AMP-activated AMPK. The latent membrane protein 1 (LMP1) encoded by EBV, which is associated with the development of nasopharyngeal carcinoma, inhibited the phosphorylation of LKB1 at serine 428 leading to the inhibition of AMPK phosphorylation (Plummer et al. 2013).

Sirtuin 1 (SIRT1), a nicotinamide adenine dinucleotide-dependent class III protein deacetylase, and AMPK are activated during the cellular response to nutrient deprivation and exhibit a reciprocal regulation (Fulco et al. 2008). In the context of HCV core protein expression in HepG2 cells, it has been shown that the activity of SIRT1 along with AMPK is decreased (Yu et al. 2013). In the context of HIV infection, it has been proposed that the transactivator Tat protein mediates its effect on AMPK via the inhibition of SIRT1 (Zhang and Wu 2009). However, a certain complexity is associated with the fact that SIRT1 deacetylates Tat (Pagans et al. 2005) and Tat inhibits SIRT1 (Kwon et al. 2008). More recently, it has been proposed that curcumin reversed Tat-mediated reduction in AMPK activation and downstream ACC activation (Zhang et al. 2011). Therefore, the relationship between HIV infection and AMPK modulation merits to be further explored in the context of viral infection.

12.2.2 Role of AMPK in Virus Entry and Replication

Since some viruses manipulate host AMPK toward immune evasion, a valuable therapeutic antiviral strategy may pass through the pharmacological manipulation of AMPK activity. In this context, Vaccinia virus infects a wide range of host cells, and all three subunits of AMPK facilitated virus entry (Moser et al. 2010). AMPK is involved in poxvirus entry in a manner that is independent of its role as a metabolic regulator. The authors point to a novel role of AMPK in promoting macropinocytosis and cellular motility by regulating actin dynamics, independently of LKB1 or CaMKK (Moser et al. 2010). AMPK deficiency attenuated vaccinia infection by interfering with viral entry. This supports the development of selective AMPK inhibitors or other inhibitors of macropinocytosis against poxviruses, as well as for other viruses that hijack this endocytic route for their entry mechanism. In vitro treatment of hepatoma cells with the AMPK activator 5-aminoimidazole-4-carboxamide 1-β-D-ribofuranoside (AICAR) was shown to suppress HCV replication (Nakashima et al. 2011; Mankouri et al. 2010). The importance of AMPK in HCMV replication is supported by the observation that the inhibition of AMPK kinase with compound C attenuates early and late HCMV replication (Hutterer et al. 2013; McArdle et al. 2012; Terry et al. 2012). Moreover, the addition of 2-octynoic acid (2-OA) to HCV-infected cells inhibited viral replication in a process involving the activation of AMPK and the inhibition of ACC, the first rate-limiting enzyme in fatty acid synthesis (Yang et al. 2013). Importantly, it has been shown that AMPK and its upstream activator LKB1 are essential for the control of RVFV infection, which is associated with the phosphorylation and inhibition of ACC. Treatment with the fatty acid palmitate bypasses this restriction, demonstrating that AMPK restricts RVFV infection through its inhibition of fatty acid biosynthesis (Moser et al. 2012). Furthermore, the authors showed that AMPK restricts the growth of multiple arboviruses from disparate families, including the Flavivirus Kunjin virus (KUNV), the Togavirus Sindbis virus (SINV), and the Rhabdovirus Vesicular stomatitis virus (VSV) (Moser et al. 2012). Therefore, this process represents a novel mechanism of interest regarding viruses' control mechanisms. Because many viruses require complex and unique interactions with cellular lipid metabolism through both synthesis and degradation pathways, it cannot be excluded that viruses inhibit AMPK activation in order to stimulate lipid synthesis within the infected cells. Thus, AMPK activation is broadly antiviral and may provide a novel antiviral therapeutic target.

Finally, it has been proposed that AMPK is involved in the reactivation of latent viruses, such as HIV-1, through bryostatin-induced activation, a protein kinase C activator (Mehla et al. 2010). This was further confirmed by the use of compound C that partially reduced HIV-1 reactivation. In the same study, the authors showed that metformin, an oral antidiabetic drug that activates AMPK indirectly through mitochondrial complex I inhibition (Zhou et al. 2001), inhibits HIV-1 replication. These results underline differential effects either on the reactivation of HIV-1 or the productive viral replication (Mehla et al. 2010). Furthermore, metformin also inhibits in vitro HCV replication (Huang et al. 2013). The addition of metformin to current HCV treatment regimens had promising, albeit modest, effects on reducing patient viral loads (Romero-Gomez et al. 2009). Moreover, several AMPK activating drugs have been shown to reduce morbidity and mortality during lethal influenza infection in mice (Moseley et al. 2010).

12.2.3 AMPK and Autophagy During Viral Infections

Autophagy becomes particularly important during cellular starvation as a means to recycle amino acids and cellular components for use as catabolic fuels. mTOR forms two functionally distinct complexes in mammals, mTORC1 and mTORC2. In particular, mTORC1 senses nutrient deprival. This is an essential information to take into account when addressing host–pathogen interactions, given that AMPK is an established negative regulator of the mTOR signaling cascade (Inoki et al. 2003; Shaw et al. 2004). AMPK can directly phosphorylate Raptor, a positive regulatory subunit of the mTORC1 complex, and, in addition, it has been shown that AMPK phosphorylates the protein kinase UNC-51-like kinase 1 (ULK1), which is a key signaling complex required for autophagosome formation (Egan et al. 2011; Kim et al. 2011). During the last two decades, autophagy has been analyzed more in detail and described to regulate host–pathogen interactions. Growing evidences indicate that autophagy acts both in the antiviral and proviral pathway (Rey-Jurado et al. 2015; Santarelli et al. 2015) Viral infections may result in the inhibition of host cell metabolism, thus leading to a concomitant reduction in energy demands. This could happen either by blocking host cell protein synthesis, while in the opposite way, by placing increased energetic demands upon infection, which might result in an activation of AMPK. Further studies are necessary to assess the contributing role of the AMPK pathway in regulating autophagy in the context of viral infections. Therefore, it cannot be excluded that AMPK is a critical regulator of cell autophagy in the context of host–pathogen interactions that need to be further addressed.

12.2.4 AMPK and Immune Effector Functions Cells During Viral Infection

Immune response against pathogens and in particular viruses leads to a rapid modification of cellular metabolism. It has been initially identified that signal transduction pathways, which modulate T-cell metabolism, involve LKB1 and AMPK as key regulators (Blagih et al. 2015; Blagih et al. 2012). Thus, LKB1-deficient T cells exhibited defects in cell proliferation and viability, accompanied by altered glycolytic and lipid metabolism. Interestingly, loss of LKB1 promoted an increase in T cell activation and inflammatory cytokine production by both $CD4^+$ and $CD8^+$ T cells (MacIver et al. 2011). However, although AMPKα1 activity is dispensable for proliferation and differentiation of cytotoxic T lymphocytes (CTLs), AMPK knockout (KO) mice show a striking defect in their ability to generate memory $CD8^+$ T-cell responses in vivo, along with a lower survival of CTLs following withdrawal of immune stimulation (Rolf et al. 2013). The role of metformin has also been addressed in arthritis. It was recently shown that the attenuation of the disease in AMPK KO mice could result from decreased levels of pro-inflammatory cytokines. Interestingly, the impact of AMPK on the differentiation of T cells into T helper 17 (Th17) cells upon stimulation was also demonstrated (Kang et al. 2013). Recently, it has been demonstrated that AMPKα1 is essential for Th1 and Th17 cell development and primary T cell responses to viral and bacterial infections. A reduction of both Th1 and Th17 responses (Estaquier et al. 1996; Estaquier et al. 1995; Raffatellu et al. 2008; Brenchley et al. 2008; Cecchinato et al. 2008; Campillo-Gimenez et al. 2010) were associated with a higher propensity of the cells to die by apoptosis observed during the course of HIV-1 infection (Estaquier et al. 1994; Monceaux et al. 2003; Hurtrel et al. 2005; Cumont et al. 2007). This originates the question as whether or not metformin could restore functional immune effector T cells through AMPK activation and if this approach could have potential interest in the development of therapeutic strategy against acquired immunodeficiency syndrome (AIDS). Thus, given the critical role played by $CD4^+$ and $CD8^+$ T cells in controlling viruses, regulation of metabolic homeostasis represents a quite remarkable field to explore in the context T cell immunity.

In addition to adaptive immunity, a role for AMPK in regulating innate immunity has been proposed. AMPK activation promotes macrophage polarization toward an anti-inflammatory M2 phenotype (Sag et al. 2008) and modulates inflammatory gene expression through activation of SIRT1 (Yang et al. 2010b). However, one important effect of AMPK is the suppression of signal transducers and activators of transcription 1 (STAT1) signaling induced by interferon-gamma (IFN-γ), which inhibits inflammation and chemokines in primary astrocytes and microglia (Meares et al. 2013). This could be particularly important in the context of neuronal infections because inflammation in the central nervous system contributes to neurologic disorders. Furthermore, type I IFN-derived immune responses, which are essential for controlling virus spread, decreases the phosphorylation of AMPK. Consistently, metformin enhances the antiviral effect of IFN-β in Coxsackievirus B3 virus infection (Burke et al. 2014). It has been also proposed that 2-OA antiviral activity against HCV is associated with AMPK activation, through induction of interferon stimulated genes (ISGs) and the inhibition of miR-122 expression (Yang et al. 2013). In the context of RVFV, however, the authors have excluded a role of type I IFN in regulating virus infection downstream AMPK (Moser et al. 2012). Therefore, AMPK appears to be a central player in regulating innate and adaptive immune responses, which are essential in controlling viral infections. Figure 12.1 illustrates the differential regulation of AMPK activity by viruses.

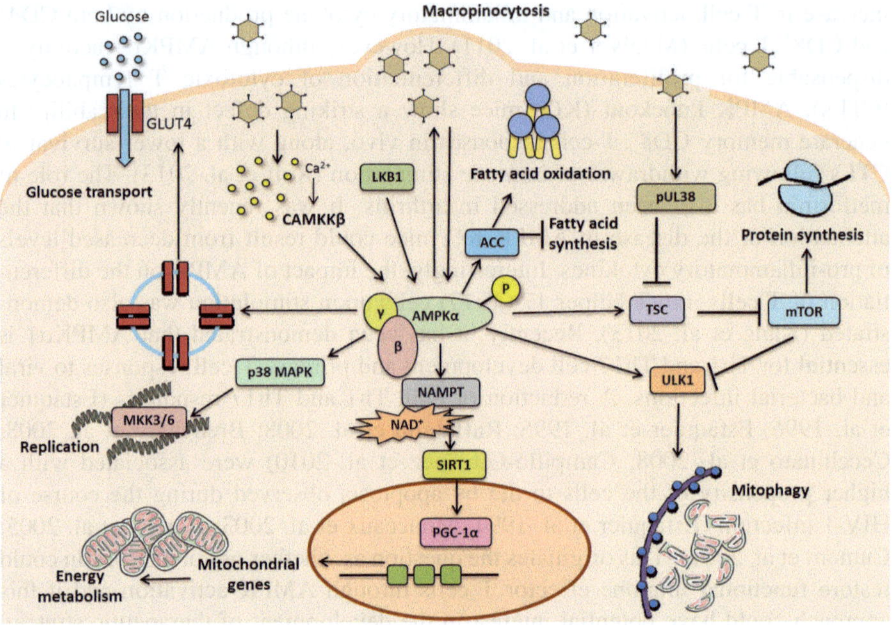

Fig. 12.1 Viral infection induces a differential regulation of AMPK activity. Host AMPK may be modulated in order to promote and ease virus entry (e.g., by macropinocytosis) and immune evasion. AMPK modulation results in the activation of several downstream targets and consequent host metabolism reprograming. Catabolic processes, as fatty acid oxidation and glucose metabolism, are activated through phosphorylation of ACC and increased expression of GLUT4, respectively. On the other side, anabolic processes, as protein synthesis and fatty acid synthesis, are inhibited. NAMPT-induced activation of SIRT1 originates increased PGC1α-induced transcription of mitochondrial genes. Mitophagy may also be induced by AMPK activation, although its role in viral infections has yet to be addressed. *ACC* acetyl-coA carboxylase, *AMPK* AMP-activated protein kinase, *CaMKKβ* calcium/calmodulin-dependent protein kinase kinase β, *GLUT4* glucose transporter 4, *NAD* nicotinamide adenine dinucleotide, *Nampt* nicotinamide phosphoribosyltransferase, *MKK3/6* MAPK kinase 3/6, *mTOR* mechanistic target of rapamycin, *LKB1* liver kinase B1, *p38 MAPK* p38 mitogen-activated protein kinase, *PGC1α* peroxisome proliferator-activated receptor γ co-activator-1α, *SIRT1* Sirtuin 1, *TSC* tuberous sclerosis, *ULK1* Unc-51-like kinase 1

12.3 AMPK in the Control of Bacterial Infection

Pathogenic intracellular bacteria infect their hosts by exploiting its cytoplasmic milieu toward survival, growth, and dissemination. During infection, intracellular bacteria must compete with the host in order to obtain the necessary energy and carbon sources. The specific scavenge of nutritional sources by the bacteria may increase their survival odds and consequently can be detrimental to the survival of the permissive host. In parallel, the host cell is forced to respond vigorously to the internal threat by using specific microbicidal mechanisms at the same time that need to maintain bioenergetic homeostatic levels to prevent failure and death. Yet, AMPK beneficial or prejudicial role during bacterial infections appears to be pathogen-specific (Fig. 12.2).

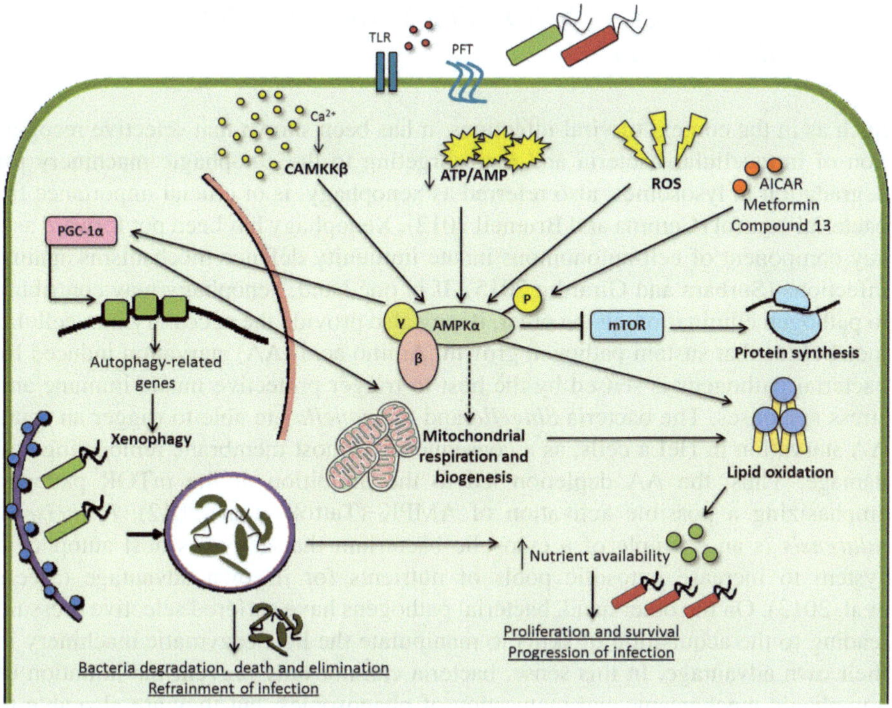

Fig. 12.2 AMPK signaling pathways during host response to bacterial infection. The release of bacterial products such as TLR agonists and PFT or the infectious process itself may lead to AMPK activation. This occurs through alteration of the intracellular calcium concentration and consequent CAMKKβ activation, decrease of ATP/AMP ratios, or ROS production. Additionally, AMPK may also be activated by chemical compounds such as AICAR, metformin, and compound. The phosphorylation of AMPK in the threonine residue 172 (Thr172) during bacterial infections has multiple effects in energy metabolism and autophagy that ultimately contribute to the resolution or progression of the infection. AMPK is a negative regulator of mTOR, so, upon Thr172 phosphorylation, mTOR is inhibited, as well as protein synthesis. Furthermore, mitochondrial metabolism is impacted during infection: mitochondrial respiration and biogenesis may be indirectly increased during bacterial-induced AMPK activation. Furthermore, the transcriptional co-activator PGC-1α is directly modulated by AMPK and originates the increase in transcription of autophagy-related genes, thus leading to increased xenophagy. However, the role of xenophagy in the disclosure of a bacterial infection is still controversial. On one hand, this antibacterial process may cause bacterial death and elimination, through acidification of the intracellular niches and consequent degradation of pathogen. On the other hand, it may increase nutrient availability, thus contributing to pathogen survival. Similarly, the increase in mitochondrial mass may cause an increase in lipid oxidation, which contributes to increased intracellular nutrient availability and, consequently, to proliferation and survival of the pathogen. *AICAR* 5-aminoimidazole-4-carboxamide-1-β-D-ribofuranoside, *AMP* adenosine monophosphate, *AMPK* AMP-activated protein kinase, *ATP* adenosine triphosphate, *CaMKKβ* calcium/calmodulin-dependent protein kinase kinase β, *mTOR* mechanistic target of rapamycin, *PFT* pore-forming toxins, *PGC1α* peroxisome proliferator-activated receptor γ co-activator-1α, *ROS* reactive species of oxygen, *TLR* Toll-like receptor

12.3.1 Autophagy: Is Self-Eating Saving or Killing the Bacteria?

Such as in the context of viral infections, it has been shown that selective recognition of intracellular bacteria and their targeting to the autophagic machinery for degradation in lysosomes, also referred as xenophagy, is of crucial importance for bacterial control (Cemma and Brumell 2012). Xenophagy has been put forward as a key component of cell-autonomous innate immunity defence mechanisms against infections (Sorbara and Girardin 2015). If in one hand, xenophagy may contribute to pathogen elimination, in the other, it may also provide the necessary intracellular metabolites that sustain pathogen growth. Amino acid (AA) starvation induced by bacterial pathogens is sensed by the host to trigger protective innate immune and stress responses. The bacteria *Shigella* and *Salmonella* are able to trigger an acute AA starvation in HeLa cells, as a consequence of host membrane remodeling and damage. Thus, the AA depletion led to the inhibition of the mTOR pathway emphasizing a possible activation of AMPK (Tattoli et al. 2012). *Francisella tularensis* is an example of a cytosolic bacterium that activates host autophagic system to increase cytosolic pools of nutrients for its own advantage (Steele et al. 2013). On the other hand, bacterial pathogens have suffered selective pressure leading to the acquisition of skills to manipulate the host enzymatic machinery in their own advantage. In this sense, bacteria can not only prevent the initiation of autophagic mechanisms and maturation of phagosomes, but they are also able to use some components of xenophagy toward its replication and survival (Levine et al. 2011; Deretic et al. 2013). *Mycobacterium tuberculosis* (*Mtb*) and *Yersinia pseudotuberculosis* represent well this last group. *Y. pseudotuberculosis* can replicate inside autophagosomes, which are blocked in their maturation and thus unable to fuse with lysosomes (Moreau et al. 2010). *M. tuberculosis* is capable of evading autophagy in macrophages through the inhibition of xenophagic pathways (Kumar et al. 2010). In this case, treatment with Isoniazid, a first-line drug against tuberculosis, during the first 24 h of infection is capable to restore autophagy in *M. tuberculosis*-infected bone marrow-derived macrophages (BMDMs) in an AMPK-dependent manner leading to pathogen elimination (Kim et al. 2012).

Nonetheless, the replenishment of essential substrates for the pathogen may be assured through modulation of AMPK activity (Brunton et al. 2013). *Mycobacterium tuberculosis* is capable of utilizing fatty acids as the major energy source (Rhee et al. 2011). While fatty acids and cholesterol are viewed as the predominant carbon source throughout infection (Lee et al. 2013; McKinney et al. 2000; Munoz-Elias and McKinney 2005; Pandey and Sassetti 2008), triglycerides could represent an energetic reservoir during bacteria dormancy (Daniel et al. 2004). *M. tuberculosis* is capable of inhibiting AMPK and activating mTOR by inducing lipid synthesis. It was recently shown that *Mtb* accumulated intracellular triacylglycerides (TAG) whose composition is nearly identical to host TAG. Accordingly, it was demonstrated that *Mtb* TAG were synthesized using free fatty acids released from host macrophages. This lipid scavenging created an energetic reservoir that ultimately allows the bacteria to enter in a latent stage (Daniel et al. 2011). Recently, these kinases arose as a potential trigger of autophagy during infection. *Escherichia coli* (ETEC) exploits host's autophagic machinery through the inhibition of mTOR pathway and ERK1/2 and AMPK phosphorylation (Tang et al. 2014). Similarly, the peroxisome proliferator-activated receptor-gamma coactivator 1α (PPARGC1A)-dependent activation of AMPK in *M. tuberculosis* infection results on increased transcription of autophagy-related genes via CCAAT/enhancer binding protein β (CEBPB). This activation was also associated with increased mitochondrial respiration and biogenesis (Yang et al. 2014). TLR 1/2 has been associated with the initiation of autophagic mechanisms during mycobacterial infections (Yuk et al. 2009). This was shown to be mediated by the mycobacterial lipoprotein LqpH, which activates TLR1/2 inducing a rapid and transient increase in intracellular calcium concentration leading to AMPK activation, possibly through upstream activation of CaMKK. Furthermore, AMPK activation was shown to be essential for autophagy, resulting in the phosphorylation of p38-MAPK (Shin et al. 2010). Similarly to LqpH, pore-forming toxins (PFT) are bacterial virulence factors that have been implicated in the prompting of autophagy, although the underlining mechanisms remain elusive (Gonzalez et al. 2008). Kloft and colleagues observed that several PFT (streptolysin O, *Vibrio cholera* cytolysin, *Staphylococcus aureus* α-toxin, and *Escherichia coli* haemolysin A) cause a drop in intracellular ATP levels leading to AMPK activation (Kloft et al. 2010). It has been hypothesized that the formation of membrane pores leads to the activation of downstream targets, as the AMPK signalling pathway, which may contribute for the maintenance of cellular homeostasis.

Further studies suggest that AMPK signaling may regulate and facilitate the intracellular replication of this pathogen. Although xenophagy may be presented as an antibacterial mechanism, it is important to reckon that other evasion mechanisms driven by a chronic activation of AMPK in the infected cells can be used by invading bacteria, such as *Salmonella*, to subvert the immune response (Tattoli et al. 2012). Likewise, *Legionella pneumophila*, the causative agent of Legionnaires' disease, is capable of inhibiting the phagosome-lysosome fusion, in order to subvert host cell immune response (Horwitz 1983). Macrophage mitochondria and endoplasmic reticulum-derived vesicles are recruited to the vicinity of the phagosome within the first minutes of *L. pneumophila* infection (Tilney et al. 2001). The consequent encapsulation of *Legionella*-containing phagosomes increases the survival odds of this bacteria, since it prevents its fusion with lysosomes or acidifying vesicles. This subservient mechanism clearly suggests the utilization of mitochondria and other organelles for support in early stages of infection. Francione et al. (2009) using a *Dictyostelium discoideum* model for mitochondrial disease observed a faster growth and higher bacterial burden in mitochondrial disease strains. AMPK activation may also be beneficial for *Helicobacter pylori* and *Neisseria meningitidis* infections. These gram-negative bacteria colonize the gastric mucosa causing several local distresses as gastritis, gastric ulcers, or gastric cancer. AMPK activation during gastric epithelial cells

(GEC) cell infection is induced by transforming growth factor-β (TGF-β)-activated kinase 1 (TAK1), resulting in decreased apoptosis of GEC cells and consequent bacterial survival (Lv et al. 2014). The decreased level of apoptosis originates persistence of infection and GEC proliferation, contributing to the development of carcinogenesis. Accordingly with these results, administration of compound 13, a novel α1-selective activator of AMPK, was shown to decrease *H. pylori*-induced GEC apoptosis through reactive oxygen species (ROS) scavenging and activation of the AMPK-heme oxygenase (HO-1) pathway (Zhao et al. 2015). Similarly, activation and overexpression of AMPK improve human brain microvascular endothelial cells (HBMECs) permeability after a challenge with lipopolysaccharide (LPS) through suppression of the induction of NAD(P)H oxidase-derived ROS (Zhao et al. 2014). AMPK was activated at 24 h postinfection, while PKB-Akt protein analysis showed dephosphorylation and thus inactivation (Schubert-Unkmeir et al. 2007). In spite of being demonstrated that *N. meningitidis* regulates gene transcription, translation, and cell metabolism, it is necessary to understand the virulence factors involved.

12.3.2 Bacterial Host Immune Response: A Double-Edge Sword Controlled by AMPK

Nutrient availability is an important factor when evaluating effector T cell function, since their dysfunction can impair the appropriate responses that culminate with pathogen elimination. Mice with T cell-specific deletion of AMPKα1 infected with an overexpressing OVA *Listeria monocytogenes* showed a decreased number of total $CD4^+$ and $CD8^+$ T cells at 7 days postinfection. Not only their numbers were reduced as their effector function, since *ex vivo* stimulation of infected splenocytes in the conditional KO mice displayed lower levels of $IFN\gamma^+$ (Blagih et al. 2015). This work highlights the importance of AMPK in allowing the proper function of T cells in response to pathogens, which is ultimately connected to the metabolic pathways established in these cells. ROS production is one of the hallmarks of innate host response against intracellular pathogens. Concomitant administration of the AMPK inhibitor compound C with LPS originated an impaired immune response to endotoxemia via the decrease of nuclear-factor kB (NF-kB) activation pathway. This was correlated with a decreased chemotaxis of macrophages and neutrophils to the liver, with decreased ROS production and TNFα levels in the serum (Guo et al. 2014). Furthermore, upon AMPK activation, neutrophils showed an enhancement in *E. coli* and *S. aureus* uptake and consequent killing. These observations were supported in vivo using murine models for peritonitis-induced sepsis where metformin-treated mice were found to have fewer viable bacteria (Park et al. 2013). In opposition, in tuberculosis (TB) patients, AMPK-expressing neutrophils were shown to be able to secrete higher quantities of metalloproteinase-8 (MMP-8), in a NF-kB-dependent fashion, leading to matrix destruction and collagen. This study demonstrated that AMPK activation has a role in the secretion of MMP-8 contributing to lung immunopathology during *M. tuberculosis* infection (Ong et al. 2015). In a very elegant work, Chakrabarti et al. showed that the disruption of *Drosophila* flies gut homeostasis upon *Pseudomonas entomophila* infection was caused by the impairment of barrier repair pathways. The mechanism of subversion of gut immune response was mediated by an activation of AMPK-Tuberous sclerosis complex (TSC) stress pathways, which are responsible for TOR inhibition and consequent repression of host translation mechanisms. Ultimately, this contributes to the entrance of infected cells in a quiescent state, more prone for tissue repair but that also allows *P. entomophila* survival (Chakrabarti et al. 2012).

12.3.3 Modulation of Host AMPK as a Pharmacological Approach to Target Bacterial Infection

Metformin treatment restricts the intracellular growth of multidrug-resistant *Mycobacterium tuberculosis* strains in THP-1 and human monocyte-derived macrophages through mitochondrial ROS induction. Moreover, metformin administration ameliorates lung pathology and reduces chronic inflammation while enhancing the efficacy of conventional anti-tuberculosis drugs (isoniazid and ethionamide) in mice models of acute and chronic tuberculosis. In this retrospective study, metformin treatment was associated with improved control of infection and decreased disease severity (Singhal et al. 2014). Moreover, administration of metformin suppresses the inflammatory effects of LPS via induction of activating transcription factor-3 (ATF-3) and AMPK activation. Thus, mice with LPS-induced endotoxemia treated with metformin displayed lower levels of IL-6 and TNF-α, thus decreasing inflammation in vivo (Kim et al. 2014). Furthermore, metformin-induced activation of AMPK inhibited the release of high mobility group box 1 (HMGB1), a protein involved in severe sepsis. This inhibition originated the observed anti-inflammatory effect in LPS-treated RAW264.7 macrophages and in endotoxemic mice, characterized by decreased levels of IL-6, TNF-α, IL-1β, prostaglandin E_2 (PGE_2), and nitric oxide (NO) (Tsoyi et al. 2011). The activation of AMPK can also be obtained by treatment with AICAR. Similarly, in vivo treatment with this drug resulted in decreased lipoteichoic acid (LTA)-driven lung inflammation (Hoogendijk et al. 2013). Using cecal ligation and puncture (CPL) as a model of sepsis, it was shown that AICAR prevented CLP-induced liver and kidney damage. Additionally, AMPK inhibition in hepatic and renal cells with compound c exacerbated cytokine production and prevented autophagy, thus culminating in increased tissue injury (Escobar et al. 2015). The role of a new bioactive lignan, sauchinone, in AMPK modulation has been studied. The activation of LKB1-AMPK pathway by this anti-inflammatory molecule has been shown to decrease liver toxicity through prevention of iron accumulation (Kim et al. 2009). Recently, Jeong and colleagues explored the role of sauchinone-induced AMPK activation in macrophage phagocytosis (Jeong et al. 2014). The authors demonstrated that sauchinone increased phosphorylation of AMPK and p38 MAPK, which correlated with increased *E. coli* phagocytic uptake by macrophages in mice lungs.

12.4 Interaction Between AMPK and Parasites

12.4.1 Host–Parasite Nutrient Sensing

All the organisms have the ability to sense their surroundings in search for nutrients through a variety of strategies that have evolved to suit the particular needs of each living form. In particular, protozoan parasites need to adapt to more than one host to survive and multiply. Therefore, a dynamic interplay between the host and the parasite is necessary to efficiently coordinate the various parasite developmental stages. The best example would be the maintenance of limited parasite density inside the host or in a distinct host niche that will be responsible for initiating a specific response by ensuring space and nutrients to the entire parasite community (Mony and Matthews 2015; van Zandbergen et al. 2010). This is quite remarkable in African trypanosomes, *Trypanosoma brucei* spp., where a morphological and molecular alteration into a "slender" form without proliferative capacity is critical to restrain parasitemia and also to have an efficient transmission to the insect host (Gjini et al. 2010; Vassella et al. 1997; Vickerman 1985). Moreover, the confirmed presence of programmed cell death (PCD) in parasites such as *Tetrahymena thermophila, Leishmania spp., Trypanosoma cruzi, Plasmodium spp., Trypanosome brucei, Giardia lamblia, Dictyostelium discoideum, Trichomonas vaginalis, Peridinium gatunense, and Blastocystis hominis* (Al-Olayan et al. 2002; Ameisen et al. 1995; Pollitt et al. 2011; Rousset and Roze 2007; Zangger et al. 2002) is another mechanism to control parasite density. As a result, the premature death of host or vectors is prevented and the right number of parasites are then able to dampen the host immune system and avoid the limitation of nutrients and resources (Ameisen et al. 1995; Reece et al. 2011). Thus, nutrient availability is a key factor for parasite fitness during infection.

These mechanisms will ultimately ensure a physical space and a nutrient-rich niche that support parasite survival and high growth rate, while maintaining the viability of its host reservoir. During the evolutionary process, unicellular organisms developed strategies to sense extracellular nutrients and intracellular metabolite concentrations since they are often exposed to relevant variations. Some nutrient sensing pathways, as chemoreceptors and PII proteins in bacteria, MEP2 sensor, SPS pathway, and Snf3/Rgt2 sensors in fungi and PII proteins in plants, are unique for these organisms. Such pathways respond to nitrogen, amino acids, ribose, galactose, glucose, and ammonium fluctuations (Chantranupong et al. 2015). Three important pathways, highly conserved from yeast to man, are responsible to sense different nutrient pools. These are general amino acid control non-derepressible 2 (GCN2), TOR kinase and AMPK. Amino-acid levels are sensed by GCN2 and mTOR, while glucose is regulated by mTOR and AMPK and energetic fluctuations are uniquely controlled by AMPK (Chantranupong et al. 2015). Thus, while host and parasites can actually share some nutrient sensors, this section will focus only in AMPK.

12.4.2 AMPK in the Context of Parasite Infection: Host–Parasite Nutrient Dynamics

The exquisite host–parasite interaction is the result of a coevolution network that has been established continuously since the very first ancient association. The parasite and host coevolution has been developed in parallel throughout time although parasites possess an advantage due to their shorter generation time and to the higher growth rate. To preserve its higher proliferative capacity, parasites have to acquire nutrients from the host, its nutritional requirement being one of the key features in the host–parasite interaction. Within host parasite this dynamic, a nutrient competition is established between the manipulative parasite trying to obtain usable energy and metabolites and the host attempting to sequester the same precursors from the pathogen. The heavy metabolic pressure established dictates the parasite auxotrophy to several nutrients, the host being the only viable source for acquisition. Interestingly, parasites are auxotrophic for several amino acids that are essential for the host. The nutritional pressure on extracellular and intracellular parasites is differently exerted. Extracellular parasites can salvage the nutrients directly from the extracellular fluids, while intracellular parasites need to import nutrients across two or three host membrane systems. Indeed, intracellular parasites need to gain access to nutrients present in the cytosol or inside the intravacuolar structures where the threshold levels of amino acids do not match with the high demands of these organisms for carbon, nitrogen, and energy to support its high proliferative rate (Abu Kwaik and Bumann 2013). However, despite their localization, parasites developed several strategies to acquire sufficient nutrients in response to their biological needs. *Plasmodium* spp. (intracellular), *T. brucei* (extracellular), and *Leishmania* spp. (intracellular) acquire nutrients from their hosts by employing different transport proteins, named permeases, located inside their plasma membrane. The nutrients imported from the host include hexoses, purines, iron, polyamines, carboxylates, and amino acids. The different niches occupied by these parasites in their hosts dictate ultimately the mode of nutrient acquisition (Landfear 2011). As an example, during *L. major* amastigotes persistence in the macrophage phagolysosomes, the parasite exploits the macrophage function on extracellular matrix turnover and remodeling in terms of internalization and degradation of glycosaminoglycans (hyaluronan). This mechanism constitutes another process to exploit additional nutrients such as amino sugars as a carbon source. The dependency of *Leishmania* on a higher nutrient variety suggests a nutritional diversity inside phagolysosomes that could ultimately represent an advantage to colonize such harsh environment (Naderer et al. 2015). The response of the host against the diversion of nutrients by the parasite is through the restriction of parasite access to nutrient sources. Thus, the process of nutritional immunity limits the availability of iron, zinc, and manganese from the invading pathogens (Crawford and Wilson 2015). An iron-depleted niche is established during *L. major* infection through the transporter NRAMP1 located at the phagolysosomal compartment. Indeed, NRAMP1 mutation renders the host macrophages more susceptible to *Leishmania* infections (Appelberg 2006; Forbes and Gros 2001). The strategies developed by the host and parasite in terms of nutrients dynamic will impact the infection outcome.

12.4.3 AMPK on the Core of Host–Parasite Metabolic Coupling

The nutritional flow established within host–parasite communities will impact the host metabolic background. Several reports have become to identify the pathways involved in the metabolic manipulation of host metabolism required for intracellular pathogen growth. Recent studies with *Toxoplasma gondii* (MacRae et al. 2012), *Trypanosoma brucei* (Wang et al. 2008), *Leishmania* spp. (Moreira et al. 2015a; Rabhi et al. 2012), *Schistosoma mansoni* (Wang et al. 2008), and *Plasmodium berghei* (Li et al. 2008) have paved the way to the understanding on the molecular mechanisms used by parasites to take advantage of the nutritive host resources. At the core of these metabolic dynamic interactions, AMPK was identified as having a crucial role to balance the energetic status of the host with a positive or negative impact on parasite survival. Nevertheless, the role of AMPK during the metabolic host–parasite crosstalk still remains largely unexplored.

A microarray analysis performed on mice livers of experimental *P. berghei* infection revealed significant variations within genes related to carbohydrate and energetic metabolisms. This approach demonstrated an upregulation of gluconeogenesis pathways while glycolysis appears downregulated on liver extracts from mice infected with *P. berghei*. Interestingly, these modifications were accompanied by an increase in the energetic sensors, *Prkaa2 (AMPKα)* and *Prkag2 (AMPKγ)* transcripts (Sales-Dias 2011). Analyzing AMPK translational levels will be important to clarify the actual role during *P. berghei* liver stage infection. A potential role of AMPK during an experimental model of cerebral malaria (ECM) was recently suggested (Gordon et al. 2015). The mTOR inhibitor rapamycin protects against ECM when administered within the first 4 days of infection. Rapamycin increased survival, blocked the breakdown of the blood–brain barrier and brain hemorrhaging, and decreased the influx of both $CD4^+$ and $CD8^+$ T cells into the brain and the accumulation of parasitized red blood cells in the brain. The impact of mTOR-controlled metabolic pathways and the recent knowledge regarding the activation of metabolic pathways in T cells upon antigen recognition led the authors to suggest a possible impact of some metabolic pathways, such as AMPK, in the control of this disease.

An intense inflammatory reaction accompanies infection with *Trypanosoma cruzi*, the etiologic agent of Chagas disease. *T. cruzi* targets the adipose tissue leading to a release of inflammatory cytokines as well as a decrease in adiponectin and PPAR-γ contributing to the establishment of an inflammatory niche. Interestingly, given that adiponectin activates AMPK, it was suggested that a downregulation of the former could be a evasion mechanism to keep AMPK in check and favor parasite growth (Nagajyothi et al. 2008). A recent work confirmed a protective role for AMPK during *T. cruzi* infection (Caradonna et al. 2013). A genome-wide RNA interference screen identified that a sustained AKT-mTORC1 pathway regulate intracellular *T. cruzi* growth. The maintenance of cellular ATP/ADP ratios at higher levels provide a distinct advantage for the parasite-limiting AMPK activity. Further, the acute silencing of AMPK catalytic (*Prkaa1*) or the regulatory subunit (*Prkab1*) in vitro provides a more favorable growth environment for intracellular *T. cruzi*. Although it has not been confirmed in vivo, AMPK inhibition is suggested to contribute for *T. cruzi* survival.

Recently, we demonstrated that AMPK is, in contrast, crucial for the establishment of a microenvironment more prone for *L. infantum* survival in macrophages (Moreira et al. 2015a). Previous analysis on the transcriptomic signature of *L. major*-infected macrophages revealed that carbohydrate and lipid metabolism were among the most altered pathways during infection. Increased mRNA levels of glucose transporters as well as key glycolytic enzymes encoding genes, such as hexokinases (*Hk*), pyruvate kinase M2 (*Pkm2*), and lactate dehydrogenase a (*Ldha*), were induced in the presence of live but not heat-killed *L. major* promastigotes. *L. major* also induced a downregulation of a number of genes implicated in the tricarboxylic acid (TCA) cycle and oxidative phosphorylation suggesting that infected macrophages mainly rely on increased glycolytic flow for energy production. On the other hand, *L. major* led to cholesterol and triglycerides accumulation on infected macrophages by enhancing the expression of scavenger receptors involved in the uptake of low-density lipoprotein (LDL), inhibiting cholesterol efflux and increasing the synthesis of triacylglycerides (Rabhi et al. 2012). The accumulation of lipid droplets in close proximity to parasitophorous vacuoles advocates the former structure as a potential high-energy substrate source for the intracellular parasite. We further observed that following *L. infantum* infection, macrophages switch from an early glycolytic to an oxidative metabolism, in a process requiring SIRT1 and LKB1/AMPK. In the absence of SIRT1 or LKB1, infected macrophages are not able to induce AMPK activation leading to an impairment of the metabolic switch. In that sense, AICAR-induced AMPK activation contributes to parasite survival while inhibition of AMPK using compound C resulted in lower parasite numbers in vitro. This phenotype was further corroborated in vivo as shown by a significantly reduced parasite burden in mice with a myeloid-specific AMPK deficiency (Moreira et al. 2015a).

Gastrointestinal (GI) nematodes also affect profoundly host metabolism. The infection of CD11c-specific AMPK$\alpha^{-/-}$ (DC-AMPK$^{-/-}$) mice with the gastrointestinal roundworm *Nippostrongylus brasiliensis* led to a dysregulation of Th2 immune response concomitantly with a failure to regenerate tissue damage mediated by the pathogen. Imbalanced responses generated in DC AMPK$^{-/-}$ mice were associated with increased Type-1 responses, greater numbers of Th17 cells, and defects in the generation of alternatively activated macrophages. Therefore, AMPK activity in myeloid cells was shown to regulate host protection against this GI parasite (Nieves et al. 2014). The manipulation of AMPK activities by distinct parasites are summarized in Fig. 12.3. Although still insufficient, these examples highlight the strategies used by the parasites to explore host resources shedding some light on cellular metabolic subversion mechanisms induced by microbe infections within the host. In this context, the modulation of AMPK activity has been put forward as a possible therapeutic target against parasitic diseases.

12.5 Conservation of the AMPK Machinery in Eukaryote Pathogens

12.5.1 SNF1/AMPK Pathways in Parasites

The AMPK family of protein kinases is highly conserved among eukaryotes. This protein has been extensively studied in mammals and yeast, where it was demonstrated to have a huge similarity from a structural and functional point of view. The conserved kinase family present in yeast and plants is denominated as sucrose non-fermenting 1 (SNF1) and SNF1-related protein kinase 1 (SnRK1), respectively (Hardie 2007; Polge and Thomas 2007). Genes encoding orthologues of the three domains (α, β, and γ subunits) are described in all eukaryotic parasite species and even in the primitive *Giardia lamblia*, which lacks mitochondria (Adam 2000; Hardie et al. 2003). The exception is the obligate intracellular parasite *Encephalitozoon cuniculi* that does not possess an identifiable AMPK orthologue (Miranda-Saavedra et al. 2007).

SNF1 and SnRK1 homologs develop similar functions in what concerns the surveillance of the metabolic status in response to nutrient and environmental stress through the induction of catabolic processes and a general repression of anabolism. Similarly to these organisms, some reports have defined the presence of AMPK/SNF1 protein kinases in eukaryote parasites. The first evidence of the presence of an SNF1 homologue in parasites was described in apicomplexa phylum for *Plasmodium falciparum* (Bracchi et al. 1996). In this work, an *SNF1* homologous gene defined as *PfKIN* was found to be increased, at transcriptional level, in the gametocyte stage which is involved in the transmission and adaptation of the malaria

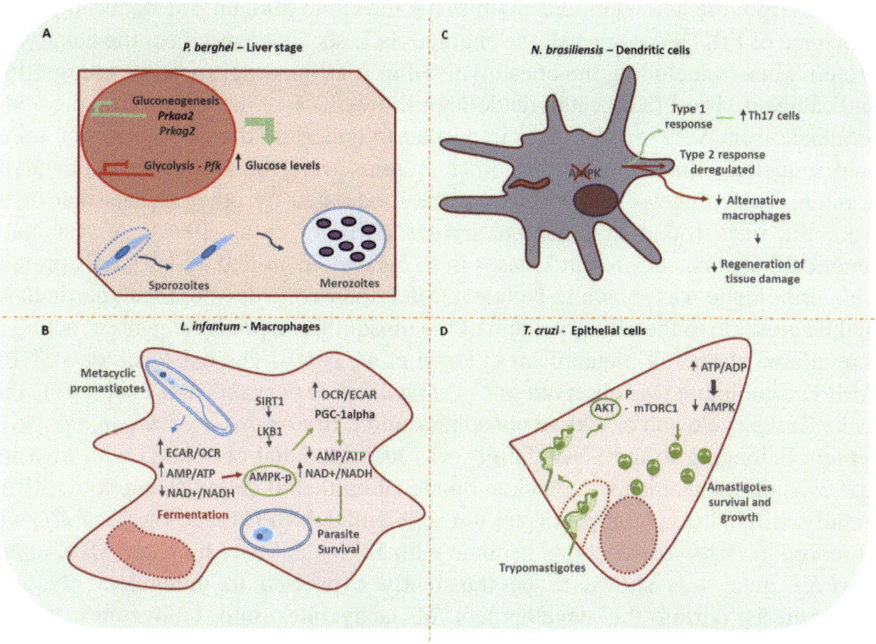

Fig. 12.3 Host AMPK dictates parasite infection outcome. (**a**) In a context of malaria liver stage infection, *Plasmodium berghei* modulates the transcriptional program of the host hepatocytes. Within the host cells, *P. berghei* increases gluconeogenesis enzymes and *Prkaa2* (AMPKα), *Prkag2* (AMPKγ) transcripts, decreasing in parallel the transcription of *Pfk* glycolytic enzyme. At this stage, an increase of glucose levels is established generating a permissive niche for sporozoites survival, which are highly dependent on glucose. (**b**) During *Leishmania infantum* infection, the promastigote form induces a metabolic alteration toward a glycolytic environment with a decrease of energetic (ATP/AMP) and redox status of the host. The increased AMP levels triggers the activation of AMPK that will be phosphorylated by SIRT1-LKB1 pathway which, in this context, is considered an AMPK upstream activator. Afterwards, a metabolic rearrangement occurs through the increase of respiration and PGC-1α activation. At this point, the energetic and redox pools are recovered leading ultimately to the parasite survival. (**c**) A detrimental effect of AMPK was defined for *Nippostrongylus brasiliensis* persistence within the host. The specific ablation of AMPK in dendritic cells deregulated the type 2 response establishing a type 1 through the augmentation of Th17 cells. Simultaneously, the ability to alternatively activate macrophages is dampened, concomitantly with the decrease capacity to regenerate tissue damage induced by the pathogen. (**d**) AMPK in a context of *T. cruzi* infection has also a deleterious effect for the pathogen survival and growth. Inside the host, trypomastigotes supports AKT-mTORC1 pathway activation maintaining in parallel, a high energetic level that consequently ablates AMPK activity. Ultimately, the established niche becomes permissive for the amastigotes survival and growth inside the host. *Pfk* phosphofructokinase 1, *ECAR* extracellular acidification rate, *OCR* oxygen consumption, *SIRT1* silent mating type information regulation 2 homolog 1, *LKB1* liver kinase B1, *AMPK-p* AMP-activated protein kinase α-phosphorylated, *PGC-1alpha* peroxisome proliferator-activated receptor gamma coactivator 1-alpha and mTORC1, mammalian target of rapamycin complex 1

parasite from the human bloodstream to the mosquito midgut. The deduced protein sequence of PfKIN contains all the characteristic sequence motifs of the eukaryotic protein kinases including the ones involved in ATP-binding, substrate recognition, and catalysis. The PfKIN catalytic domain shows 40 % of homology with the SNF1 sequence from *S. cerevisiae*. It is important to underline that no significant amino acid sequence similarities to any other proteins were found outside the catalytic domain. This suggests that PfKIN can be modulated by other signals that differ from the ones found in other eukaryotes (Bracchi et al. 1996). A systematic functional analysis of protein kinases in *P. berghei* identified a SNF1/KIN protein. This orthologue was shown to be relevant for sporozoite development, particularly in the egression to the salivary gland of the mosquito *Anopheles stephensi*, acting as a regulator of energy metabolism (Tewari et al. 2010). The existence of a SNF1/AMPK orthologue was observed in *Cryptosporidium parvum* (Artz et al. 2011) that lacks Krebs cycle and oxidative phosphorylation. In *Eimeria acervulina*, a putative protein belonging to the SNF1 family was identified and correlated with invasion and evasion mechanisms in chicken duodenal epithelial cells (Zhang et al. 2015). Finally, still in the apicomplexa phylum, a potential AMPK homolog (*ToxPK1*) was observed in *Toxoplasma gondii* genome with 58 % identity to human AMPK alpha. *ToxPK1* gene was shown to be transiently expressed to upregulate glycogen biosynthesis during the development of tachyzoites into bradyzoites (Ghosh et al. 2012; Ng et al. 1995, 1997).

A comparative analysis of the kinomes of representative members of pathogenic trypanosomatids, namely *Trypanosoma brucei*, *Trypanosoma cruzi*, and *Leishmania major*, have highlighted that AMPK homologues are relatively poorly represented within trypanosomatid genomes as compared to humans. Yet, these are predicted to be active (Parsons et al. 2005). In *T. brucei*, the fine-tuned control of transient alteration from the "slender" (proliferative) to "stumpy" (quiescent) forms is performed by intracellular signaling pathways triggered by the "stumpy induction factor" (SIF). The slender state is retained until SIF accumulation reaches threshold levels triggering the nutritional stress-like response, which leads to concomitant repression of the "slender retainers" and activation of "stumpy inducers." This dynamic transition leads to cell cycle arrest and prepares the cell for its next life cycle stage, in the insect, by reactivating mitochondrial functions required for oxidative phosphorylation (Mony and Matthews 2015). A potential role for AMPK in this developmental transition can be speculated. The signaling components that drives stumpy formation evaluated in a genome-wide RNA interference library screen identified a AMPK/SNF1/KIN11 homologue. Interestingly, the authors suggest that AMPK homologue, which acts as a SIF, could be a potential inhibitor of trypanosomes TORC4 (Mony et al. 2014). TORC4 activity is proposed to prevent stumpy formation, since the knockdown of TOR4 drove the cells to develop features characteristic of stumpy forms (Barquilla et al. 2012). TOR4 is one of the *T. brucei* TOR paralogues that retain the classical structure of mTOR kinase domains, yet displaying new features not described in other TORs (Barquilla et al. 2008). *T. brucei* procyclic forms are able to monitor glucose levels, through the surface molecule expression of procyclins, which are crucial for the parasite survive in the tsetse fly vector (Clemmens et al. 2009). Procyclins are glycoproteins that cover the surface of procyclic forms that are suspected to have a protective role against the action of insect proteases (Acosta-Serrano et al. 2001). Two major procyclin proteins EP and GPEET were found to be regulated by the beta and gamma subunits (*TbAMPK beta* and *TbAMPK gamma*, respectively), since the silencing of these genes leads to their upregulation. Interestingly, the latter procyclin is highly expressed in glucose-deficient environment within tsetse fly. Moreover, the localization of the scaffold beta subunit (glycosomes and flagellum) suggests a central positioning between surface molecule expression and glycolysis providing thus a molecular connection between these two mechanisms (Clemmens et al. 2009).

It is important to stress out that the role of AMPK/SNF1 protein kinase family are quite similar for all the organisms (extracellular or intracellular), playing a central positioning in its most relevant pathways. The described AMPK/SNF1 homologues have been implicated in the parasite developmental stages that occur within the same host or in different hosts in order to recover from a nutrition/energetic depletion status. Parasites have to deal with different microenvironments during proliferation and differentiation recurring to sensing mechanisms as the AMPK/SNF1 protein kinases that will have definitely a huge impact on parasite fitness.

12.5.2 SNF1/AMPK Pathways in Fungi

Over the past years, the incidence of fungal infections has increased in several countries. Compared to bacteria, infections caused by fungi are less frequent; nevertheless, their treatment is still complicated for systemic infections particularly due to the close functional similarity between fungal cells and the host mammalian cells (Navarro-Garcia et al. 2001). Human pathogenic fungi encounter a broad range of stress conditions in their natural environment as well as in the host during infection that challenge their ability to grow. The fungi ability to adapt to a variety of conditions has contributed to the ubiquitous nature in the environment and to the success of them as pathogens. In fact, adaptation to nutrient fluctuations is crucial for survival. Human pathogenic fungi must adapt to oxidant, pH, or nutritional stress, otherwise they are eliminated by the host defense system. Fungi survival, both in the environment or within the human host, requires the activation of signal transduction pathways that sense environmental or host stress cues (Hohmann 2002). During host nutritional deprivation, fungi pathogens are able to use alternative carbon sources, such mechanism being considered a virulence trait crucial to adapt to stress conditions (Lorenz and Fink 2001). Therefore, it is not surprising that genes that control several metabolic pathways may have a central role in fungal virulence (Navarro-Garcia et al. 2001).

12.5.3 Regulation and Function of Fungi SNF1 Protein Kinase

Snf1 protein kinase was initially identified by a screening performed with the budding yeast *Saccharomyces cerevisiae* (Celenza and Carlson 1986). Snf1 was identified as the yeast homolog of the mammalian AMPK acting as an energy sensor. This kinase, in fungi, is also able to reprogram the cellular metabolism through the energetic balance which is essential to sustain cell metabolism and to support the development of a stress response (Sampaio-Marques et al. 2014). *S. cerevisiae* Snf1 is a serine/threonine protein kinase (α subunit) complexed with other proteins, namely γ activating subunit Snf4 and β subunits Sip1, Sip2, or Gal83 depending on Snf1 localization (Celenza and Carlson 1989; Celenza et al. 1989; Estruch et al. 1992; Vincent and Carlson 1999). AMPK/Snf1 activation in fungi could be dependent or not on AMP/ATP cellular fluctuations. In the former, the allosteric regulation by AMP is not crucial for Snf1 activation while in the latter Snf1 function is regulated by cAMP-dependent protein kinase (PKA) (Ferretti et al. 2012).

A crucial activity on glucose balance has been also attributed to Snf1 kinase. This is of particular importance since glucose is one of the major carbon and energy source for the eukaryotic cells. Under carbon stress conditions, fungal cells required the activity of the Snf1 kinase in order to adapt to this harsh environment. Furthermore, Snf1 activity is requested for transcription of glucose repressed genes and increases with aging, even when glucose is abundant (Ashrafi et al. 2000; Hedbacker and Carlson 2008). Taken together, Snf1 allows cells to adapt to limitations on glucose availability using instead alternative carbon sources, such as sucrose, galactose and ethanol (Carlson 1999; Gancedo 1998). Additionally, Snf1 is highly important in many other cellular stress conditions helping to establish a response against sodium ion stress, heat shock, alkaline pH, and oxidative and genotoxic stress (Hedbacker and Carlson 2008). Other crucial biological impact of Snf1 has been described for glycogen, sterol, fatty acid biosynthesis, fatty acid-β oxidation, peroxisome biogenesis, and ultimately in the fungi sporulation developmental stage (Sanz 2003). The main functions of Snf1 kinase in pathogenic fungi are resumed in Fig. 12.4.

12.5.4 SNF1 Protein Kinase in Pathogenic Fungi

In *Candida* spp., namely *Candida glabrata*, the importance of Snf1 kinase for the regulation of nutrient sources availability was already demonstrated. Snf1 ablation resulted in the loss of its capacity to utilize trehalose, which together with glucose are the carbon sources preferentially used by this organism (Petter and Kwon-Chung 1996). In the yeast *Cryptococcus neoformans*, a pathogenic fungus that infects via respiratory tract, particularly immunocompromised individual

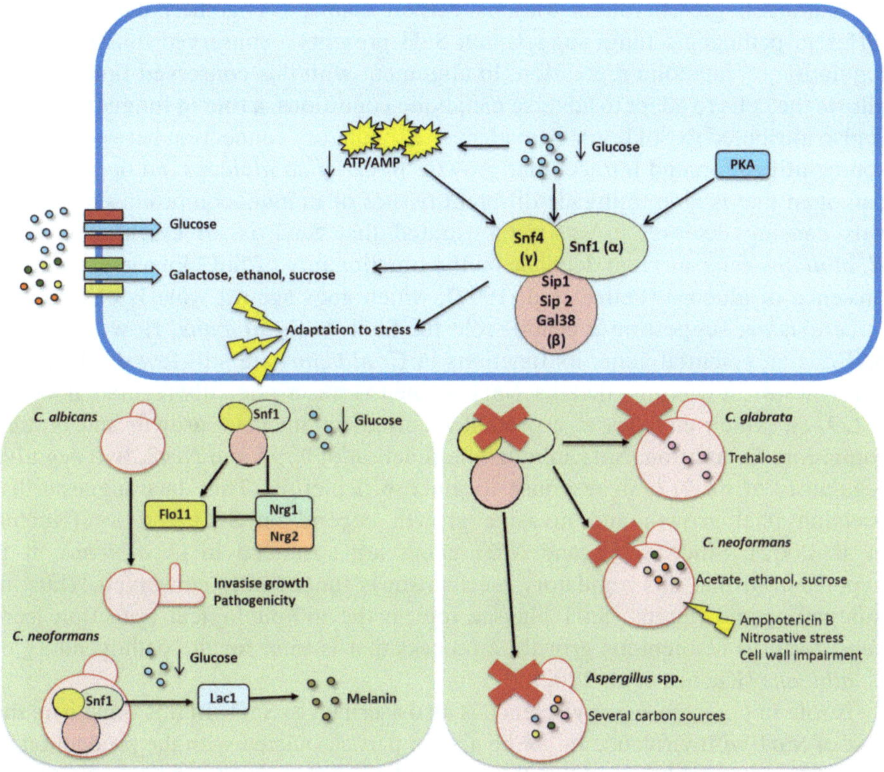

Fig. 12.4 SNF1 protein kinase in pathogenic fungi. Snf1, AMPK homolog in fungi, functions as an energy sensor and allows for a proper response and adaptation to environmental stress. This kinase is activated during low ATP/AMP ratio and also by PKA, in an independent fashion. Snf1 activation allows the adaption to limitations on glucose availability and prompts the use of other alternative carbon sources, like sucrose, galactose, and ethanol. Snf1 ablation results in metabolic impairment and the loss of inability to use different carbon sources, with possible consequences in fungi pathogenicity, namely, regarding susceptibility to antifungal therapy and cell wall constitution. Furthermore, in glucose deprivation conditions, Snf1 promotes transcription of *FLO11* gene, which encodes to a protein required for invasive growth and biofilms, thus impacting *C. albicans* invasion and consequent pathogenicity. Snf1 is also responsible for the regulation of melanin production by *C. neoformans*, which is an important cell wall-associated virulence factor. *AMP* adenosine monophosphate, *AMPK* AMP-activated protein kinase, *ATP* adenosine triphosphate, *PKA* cAMP-dependent protein kinase

(Buchanan and Murphy 1998; Hull and Heitman 2002; Mitchell and Perfect 1995), the *SNF1* abrogation leads to the incapacity to use alternative carbon sources such as acetate, ethanol, and sucrose. Furthermore, an increase on sensitivity to antifungal drugs such as amphotericin B and to nitrosative stress was also detected, manifested particularly at the host temperature of 37 °C (Hu et al. 2008; Yang et al. 2010a). In other fungi species such as *Aspergillus* spp., the absence of SnfA1 (Snf1 homologue) causes a defect on glucose-depression mechanisms and impairs the pathogen growth under various carbon sources. Together, data from the different pathogenic fungi suggest that Snf1 presents a conserved function in the regulation of catabolic repression. In alignment with this conserved function that allows the cells to adapt to adverse metabolic conditions, a role in fungus virulence is also attributed to Snf1, which evidences a regulatory connection between carbon source utilization and intracellular growth. In *Candida albicans*, an opportunistic pathogen that is able to invade different tissues of immunocompromised individuals causing disease, it was demonstrated that Snf1 is an essential protein. *C. albicans* cells survival depends on the functionality of Snf1 kinase even in the presence of glucose (Petter et al. 1997), which goes against what is observed in *S. cerevisiae*, suggesting a central role for Snf1 in *C. albicans*. However, since *SNF1* is an essential gene, its functions in *C. albicans* are still few disclosed. In *S. cerevisiae*, it was unveiled that Snf1 is able to transcriptionally regulate flocullin (*FLO11*), which encodes to a protein required for invasive growth and biofilm formation. In addition, Snf1 inhibits the function of *Nrg1* and *Nrg2*, two negative regulators of *FLO11*, in response to glucose depletion. This data suggests that pseudohyphal growth and invasive growth depend on Snf1 activity (Kuchin et al. 2002). Since *SNF1* and *NRG* genes are sustained in *C. albicans*, it is hypothesized that this regulatory mechanism is functionally conserved. Thus, in this pathogenic fungus, Snf1 plays a role in the morphological transition from yeast form to filamentous growth, a process that is vital for the pathogenicity of *C. albicans* (Kuchin et al. 2002).

A role in *C. neoformans* virulence is also ascribed to Snf1. In this organism, the link of Snf1 with virulence might be also in part associated with the production of the pigment melanin, a cell wall-associated virulence factor, since *snf1* mutant strains presented reduced melanin production at 37 °C (Hu et al. 2008). Melanin production is regulated by the activity of the copper-containing polyphenolic oxidase laccase, majority promoted by the *LAC1* gene. *LAC1* transcription is enhanced by glucose starvation, yet by undisclosed mechanisms; nevertheless, it is suggested that Snf1 participates in the *LAC1* derepression by glucose depletion (Yang et al. 2010a).

Adhesion capacity is another important feature for fungus pathogenicity because it is able to mediate the interaction between colonies as well as host–pathogen interactions (Ramsook et al. 2010). Concerning this pathogenicity feature, it was demonstrated, in vitro, that *SNF1* abrogation in *C. neoformans* impairs the cells ability to adhere to agar in comparison with control strains (Yang et al. 2011). Furthermore, the same study also demonstrated that *snf1* mutant cells presented a phonotype compared with cell wall defects, since they are sensitive to SDS and Congo red (Yang et al. 2011). Thus, Snf1 apparently has also a contribution concerning the modulation of the cell wall integrity that functions as essential barrier to protect cells from harmful effects caused by the extreme conditions frequently face by fungal cells (Bermejo et al. 2008; Levin 2005; Waterman et al. 2007). Together, the data suggests that the observed cell wall impairment observed in the absence of Snf1 might cause changes of the cellular surface components such as glycoproteins which could lead to the loss of adhesion capacity; nevertheless, the adhesion regulation pathways in *C. neoformans* are yet undisclosed.

The kinase Snf1 functions are well characterized in the yeast *S. cerevisiae*, a model organism in which it was demonstrated that this kinase has a central metabolic function. At the same regulation level, several studies in important human pathogenic fungi showed that Snf1 homologs presented a conserved function, evidencing that this protein can be a potential antifungal target. More interestingly, several observations point out Snf1 as a regulator of virulence factors as well as a virulence factor. Concerning this aspect, Snf1 is able to modulate crucial fungal pathogenic mechanisms, as above discussed; nevertheless and due to its crucial role in the cells viability, new approaches must be developed in an attempt to disclose the precise pathogenic mechanism regulated by Snf1. Together, all the pieces of information highlighted that Snf1 might represent a promising alternative target of interest to drug development.

12.6 Final Remarks

Host–pathogen interactions are highly dynamic processes where both compete to acquire a myriad of essential nutrients, crucial to survive inside an infection niche. Although it seems well established that pathogens manipulate host metabolism in its own advantage, the preferred nutrient sources and the triggered metabolic pathways are still to be completely understood. In this context, the manipulation of host AMPK activity seems to be a valuable approach for controlling pathogen burden and infection progression.

The positive or negative impact on host AMPK protein during an infection could reflect the existence of distinct biological and nutritional needs for the distinct pathogens. Overall, two main factors contribute for the different responses observed concerning host AMPK modulation during infection. First, in vivo, microbial pathogens can find distinct niche/cells with different nutritional/immunological status. Secondly, an heterogeneity in host–pathogen interactions (various pathogens subsets) could occur simultaneously within the same host tissue reflecting disparate outcomes (Bumann 2015). We can then speculate that pathogen survival will depend on the ability to sense and find the most permissive cells without discarding the overall contribution of the different pathogen subsets within the same tissue. This vast heterogeneity during in vivo infections set the nutrient sensors, as AMPK, in a more privileged core to impact on the disease outcome. Moreover, an additional diversity in terms of AMPK/SNF1 orthologues function was described within parasites and fungal organisms. The particularities of each parasite or fungus life cycle and the pressure exerted by the different hosts could guide the evolutionary process of these orthologues in distinct branches of the phylogenetic tree that could account ultimately for their variability inside these organisms. In that sense, the parasite and fungal AMPK/SNF1 members could participate in the intracellular sensing of the nutritional status of the host. Thus, a rearrangement of their metabolic pathways may occur in order to acquire the sufficient amount of nutrients and energy from the host to survive. Nevertheless, the role for AMPK/SNF1 orthologues during the infectious process still remains uncomprehend.

The quite unexplored field of nutrient sensors open a new avenue to exploit AMPK as a possible drug target against microbial pathogens. The design of new drugs to activate or inhibit this kinase in the host during infection could be an interesting domain to explore. In fact, there are drugs currently used to treat metabolic diseases that could be considered an alternative approach to treat infectious diseases. A classic example is metformin, which is able to activate AMPK through mitochondria complex I inhibition that is currently the most common treatment for diabetes type 2. Although due to the large variation described in terms of niche/cells and the presence of various pathogen subsets during in vivo infections, several concerns are being raised. The described disparities may account for the unsuccessful treatment or resistance development found in conventional therapeutics. The possible addition/synergy of AMPK activators or inhibitors to the conventional treatment could represent a viable and important therapeutic alternative to overcome this infection issue. Despite the highly conservation among eukaryotes, AMPK orthologues described in parasites and fungi could be also explored as potential drug targets. To develop selective activators or inhibitors to these orthologues, the homology to mammals should be low, although relevant divergences in the catalytic domain or differences in the protein weight/conformation should be considered. As an example, we can speculate that PfKIN from *P. falciparum* could be considered a promising drug target due to the fact that its catalytic domain presents 40 % of homology with the SNF1 sequence of *S. cerevisiae*, and no significant amino acid sequence similarities to any other proteins were found outside the catalytic domain. AMPK orthologues have been more studied in *Plasmodium*; however, further work will be needed to characterize and define this protein as a possible drug target as well as in other pathogens.

Overall, the nutritional status of the host and the infection outcome can influence each other occupying the microbial pathogens a central position in this dynamic balance. The hallmark of the host metabolic status and infection outcome is the gut microbial communities, defined as microbiota, which is currently considered a hot research topic. In a healthy state, they contribute nutrients and energy to the host, and a balance is maintained with the host's metabolism and immune system. On the other side, intestinal dysbiosis can act as a source of inflammation and in-fection and possible contributions to diabetes mellitus and obesity (Flint et al. 2012).This striking regulation could be essential to improve or trace new antimicrobial strategies through the modulation of host metabolism.

Funding Statement This work was supported by grants to JE from the Agence Nationale de Recherches sur le Sida et les Hépatites Virales (ANRS) and from The Canadian HIV Cure Enterprise Team Grant HIG-13305 from the Canadian Institutes of Health Research (CIHR) in partnership with CANFAR and IAS. JE acknowledges the support of the Canada Research Chair program. It was also supported by FCT—Fundação para a Ciência e a Tecnologia/MEC—Ministério da Educação e Ciência através de fundos nacionais e quando aplicável cofinanciado pelo FEDER, no âmbito do Acordo de Parceria PT2020 referente à unidade de investigação n° 4293. DM is supported by SFRH/BD/91543/2012. RS is supported by the Fundação para a Ciência e Tecnologia (FCT) (IF/00021/2014).

References

Abu Kwaik Y, Bumann D (2013) Microbial quest for food in vivo: 'nutritional virulence' as an emerging paradigm. Cell Microbiol 15(6):882–890. doi:10.1111/cmi.12138

Acosta-Serrano A, Vassella E, Liniger M, Kunz Renggli C, Brun R, Roditi I, Englund PT (2001) The surface coat of procyclic Trypanosoma brucei: programmed expression and proteolytic cleavage of procyclin in the tsetse fly. Proc Natl Acad Sci U S A 98(4):1513–1518. doi:10.1073/pnas.041611698

Adam RD (2000) The Giardia lamblia genome. Int J Parasitol 30(4):475–484

Al-Olayan EM, Williams GT, Hurd H (2002) Apoptosis in the malaria protozoan, Plasmodium berghei: a possible mechanism for limiting intensity of infection in the mosquito. Int J Parasitol 32(9):1133–1143

Ameisen JC, Idziorek T, Billaut-Mulot O, Loyens M, Tissier JP, Potentier A, Ouaissi A (1995) Apoptosis in a unicellular eukaryote (Trypanosoma cruzi): implications for the evolutionary origin and role of programmed cell death in the control of cell proliferation, differentiation and survival. Cell Death Differ 2(4):285–300

Appelberg R (2006) Macrophage nutriprive antimicrobial mechanisms. J Leukoc Biol 79 (6):1117–1128. doi:10.1189/jlb.0206079

Artz JD, Wernimont AK, Allali-Hassani A, Zhao Y, Amani M, Lin YH, Senisterra G, Wasney GA, Fedorov O, King O, Roos A, Lunin VV, Qiu W, Finerty P Jr, Hutchinson A, Chau I, von Delft F, MacKenzie F, Lew J, Kozieradzki I, Vedadi M, Schapira M, Zhang C, Shokat K, Heightman T, Hui R (2011) The Cryptosporidium parvum kinome. BMC Genomics 12:478. doi:10.1186/1471-2164-12-478

Ashrafi K, Lin SS, Manchester JK, Gordon JI (2000) Sip2p and its partner snf1p kinase affect aging in S. cerevisiae. Genes Dev 14(15):1872–1885

Barnes K, Ingram JC, Porras OH, Barros LF, Hudson ER, Fryer LG, Foufelle F, Carling D, Hardie DG, Baldwin SA (2002) Activation of GLUT1 by metabolic and osmotic stress: potential involvement of AMP-activated protein kinase (AMPK). J Cell Sci 115(Pt 11):2433–2442

Barquilla A, Crespo JL, Navarro M (2008) Rapamycin inhibits trypanosome cell growth by preventing TOR complex 2 formation. Proc Natl Acad Sci U S A 105(38):14579–14584. doi:10.1073/pnas.0802668105

Barquilla A, Saldivia M, Diaz R, Bart JM, Vidal I, Calvo E, Hall MN, Navarro M (2012) Third target of rapamycin complex negatively regulates development of quiescence in Trypanosoma brucei. Proc Natl Acad Sci U S A 109(36):14399–14404. doi:10.1073/pnas.1210465109

Bermejo C, Rodriguez E, Garcia R, Rodriguez-Pena JM, Rodriguez de la Concepcion ML, Rivas C, Arias P, Nombela C, Posas F, Arroyo J (2008) The sequential activation of the yeast HOG and SLT2 pathways is required for cell survival to cell wall stress. Mol Biol Cell 19 (3):1113–1124. doi:10.1091/mbc.E07-08-0742

Blagih J, Krawczyk CM, Jones RG (2012) LKB1 and AMPK: central regulators of lymphocyte metabolism and function. Immunol Rev 249(1):59–71. doi:10.1111/j.1600-065X.2012.01157.x

Blagih J, Coulombe F, Vincent EE, Dupuy F, Galicia-Vazquez G, Yurchenko E, Raissi TC, van der Windt GJ, Viollet B, Pearce EL, Pelletier J, Piccirillo CA, Krawczyk CM, Divangahi M, Jones RG (2015) The energy sensor AMPK regulates T cell metabolic adaptation and effector responses in vivo. Immunity 42(1):41–54. doi:10.1016/j.immuni.2014.12.030

Bonen A, Han XX, Habets DD, Febbraio M, Glatz JF, Luiken JJ (2007) A null mutation in skeletal muscle FAT/CD36 reveals its essential role in insulin- and AICAR-stimulated fatty acid metabolism. Am J Physiol Endocrinol Metab 292(6):E1740–E1749. doi:10.1152/ajpendo.00579.2006

Bracchi V, Langsley G, Thelu J, Eling W, Ambroise-Thomas P (1996) PfKIN, an SNF1 type protein kinase of Plasmodium falciparum predominantly expressed in gametocytes. Mol Biochem Parasitol 76(1–2):299–303

Brenchley JM, Paiardini M, Knox KS, Asher AI, Cervasi B, Asher TE, Scheinberg P, Price DA, Hage CA, Kholi LM, Khoruts A, Frank I, Else J, Schacker T, Silvestri G, Douek DC (2008) Differential Th17 CD4 T-cell depletion in pathogenic and nonpathogenic lentiviral infections. Blood 112(7):2826–2835. doi:10.1182/blood-2008-05-159301

Brunton J, Steele S, Ziehr B, Moorman N, Kawula T (2013) Feeding uninvited guests: mTOR and AMPK set the table for intracellular pathogens. PLoS Pathog 9(10):e1003552. doi:10.1371/journal.ppat.1003552

Buchanan KL, Murphy JW (1998) What makes Cryptococcus neoformans a pathogen? Emerg Infect Dis 4(1):71–83. doi:10.3201/eid0401.980109

Bumann D (2015) Heterogeneous host-pathogen encounters: act locally, think globally. Cell Host Microbe 17(1):13–19. doi:10.1016/j.chom.2014.12.006

Burke JD, Platanias LC, Fish EN (2014) Beta interferon regulation of glucose metabolism is PI3K/Akt dependent and important for antiviral activity against coxsackievirus B3. J Virol 88 (6):3485–3495. doi:10.1128/JVI.02649-13

Campillo-Gimenez L, Cumont MC, Fay M, Kared H, Monceaux V, Diop O, Muller-Trutwin M, Hurtrel B, Levy Y, Zaunders J, Dy M, Leite-de-Moraes MC, Elbim C, Estaquier J (2010) AIDS progression is associated with the emergence of IL-17-producing cells early after simian immunodeficiency virus infection. J Immunol 184(2):984–992. doi:10.4049/jimmunol.0902316

Caradonna KL, Engel JC, Jacobi D, Lee CH, Burleigh BA (2013) Host metabolism regulates intracellular growth of Trypanosoma cruzi. Cell Host Microbe 13(1):108–117. doi:10.1016/j.chom.2012.11.011

Carlson M (1999) Glucose repression in yeast. Curr Opin Microbiol 2(2):202–207. doi:10.1016/S1369-5274(99)80035-6

Cecchinato V, Trindade CJ, Laurence A, Heraud JM, Brenchley JM, Ferrari MG, Zaffiri L, Tryniszewska E, Tsai WP, Vaccari M, Parks RW, Venzon D, Douek DC, O'Shea JJ, Franchini G (2008) Altered balance between Th17 and Th1 cells at mucosal sites predicts AIDS progression in simian immunodeficiency virus-infected macaques. Mucosal Immunol 1 (4):279–288. doi:10.1038/mi.2008.14

Celenza JL, Carlson M (1986) A yeast gene that is essential for release from glucose repression encodes a protein kinase. Science 233(4769):1175–1180

Celenza JL, Carlson M (1989) Mutational analysis of the Saccharomyces cerevisiae SNF1 protein kinase and evidence for functional interaction with the SNF4 protein. Mol Cell Biol 9 (11):5034–5044

Celenza JL, Eng FJ, Carlson M (1989) Molecular analysis of the SNF4 gene of Saccharomyces cerevisiae: evidence for physical association of the SNF4 protein with the SNF1 protein kinase. Mol Cell Biol 9(11):5045–5054

Cemma M, Brumell JH (2012) Interactions of pathogenic bacteria with autophagy systems. Curr Biol 22(13):R540–R545. doi:10.1016/j.cub.2012.06.001

Chakrabarti S, Liehl P, Buchon N, Lemaitre B (2012) Infection-induced host translational blockage inhibits immune responses and epithelial renewal in the Drosophila gut. Cell Host Microbe 12(1):60–70. doi:10.1016/j.chom.2012.06.001

Chantranupong L, Wolfson RL, Sabatini DM (2015) Nutrient-sensing mechanisms across evolution. Cell 161(1):67–83. doi:10.1016/j.cell.2015.02.041

Clemmens CS, Morris MT, Lyda TA, Acosta-Serrano A, Morris JC (2009) Trypanosoma brucei AMP-activated kinase subunit homologs influence surface molecule expression. Exp Parasitol 123(3):250–257. doi:10.1016/j.exppara.2009.07.010

Crawford A, Wilson D (2015) Essential metals at the host-pathogen interface: nutritional immunity and micronutrient assimilation by human fungal pathogens. FEMS Yeast Res 15(7):pii: fov071. doi:10.1093/femsyr/fov071

Cumont MC, Monceaux V, Viollet L, Lay S, Parker R, Hurtrel B, Estaquier J (2007) TGF-beta in intestinal lymphoid organs contributes to the death of armed effector CD8 T cells and is associated with the absence of virus containment in rhesus macaques infected with the simian immunodeficiency virus. Cell Death Differ 14(10):1747–1758. doi:10.1038/sj.cdd.4402192

Daniel J, Deb C, Dubey VS, Sirakova TD, Abomoelak B, Morbidoni HR, Kolattukudy PE (2004) Induction of a novel class of diacylglycerol acyltransferases and triacylglycerol accumulation in Mycobacterium tuberculosis as it goes into a dormancy-like state in culture. J Bacteriol 186 (15):5017–5030. doi:10.1128/JB.186.15.5017-5030.2004

Daniel J, Maamar H, Deb C, Sirakova TD, Kolattukudy PE (2011) Mycobacterium tuberculosis uses host triacylglycerol to accumulate lipid droplets and acquires a dormancy-like phenotype in lipid-loaded macrophages. PLoS Pathog 7(6):e1002093. doi:10.1371/journal.ppat.1002093

Deretic V, Saitoh T, Akira S (2013) Autophagy in infection, inflammation and immunity. Nat Rev Immunol 13(10):722–737. doi:10.1038/nri3532

Egan DF, Shackelford DB, Mihaylova MM, Gelino S, Kohnz RA, Mair W, Vasquez DS, Joshi A, Gwinn DM, Taylor R, Asara JM, Fitzpatrick J, Dillin A, Viollet B, Kundu M, Hansen M, Shaw RJ (2011) Phosphorylation of ULK1 (hATG1) by AMP-activated protein kinase connects energy sensing to mitophagy. Science 331(6016):456–461. doi:10.1126/science.1196371

Escobar DA, Botero-Quintero AM, Kautza BC, Luciano J, Loughran P, Darwiche S, Rosengart MR, Zuckerbraun BS, Gomez H (2015) Adenosine monophosphate-activated protein kinase activation protects against sepsis-induced organ injury and inflammation. J Surg Res 194 (1):262–272. doi:10.1016/j.jss.2014.10.009

Estaquier J, Idziorek T, de Bels F, Barre-Sinoussi F, Hurtrel B, Aubertin AM, Venet A, Mehtali M, Muchmore E, Michel P, Mouton Y, Girard M, Ameisen JC (1994) Programmed cell death and AIDS: significance of T-cell apoptosis in pathogenic and nonpathogenic primate lentiviral infections. Proc Natl Acad Sci U S A 91(20):9431–9435

Estaquier J, Idziorek T, Zou W, Emilie D, Farber CM, Bourez JM, Ameisen JC (1995) T helper type 1/T helper type 2 cytokines and T cell death: preventive effect of interleukin 12 on activation-induced and CD95 (FAS/APO-1)-mediated apoptosis of CD4+ T cells from human immunodeficiency virus-infected persons. J Exp Med 182(6):1759–1767

Estaquier J, Tanaka M, Suda T, Nagata S, Golstein P, Ameisen JC (1996) Fas-mediated apoptosis of CD4+ and CD8+ T cells from human immunodeficiency virus-infected persons: differential in vitro preventive effect of cytokines and protease antagonists. Blood 87(12):4959–4966

Estruch F, Treitel MA, Yang X, Carlson M (1992) N-terminal mutations modulate yeast SNF1 protein kinase function. Genetics 132(3):639–650

Ferretti AC, Larocca MC, Favre C (2012) Nutritional stress in eukaryotic cells: oxidative species and regulation of survival in time of scarceness. Mol Genet Metab 105(2):186–192. doi:10.1016/j.ymgme.2011.11.007

Flint HJ, Scott KP, Louis P, Duncan SH (2012) The role of the gut microbiota in nutrition and health. Nat Rev Gastroenterol Hepatol 9(10):577–589. doi:10.1038/nrgastro.2012.156

Forbes JR, Gros P (2001) Divalent-metal transport by NRAMP proteins at the interface of host-pathogen interactions. Trends Microbiol 9(8):397–403

Francione L, Smith PK, Accari SL, Taylor PE, Bokko PB, Bozzaro S, Beech PL, Fisher PR (2009) Legionella pneumophila multiplication is enhanced by chronic AMPK signalling in mitochondrially diseased Dictyostelium cells. Dis Model Mech 2(9–10):479–489. doi:10.1242/dmm.003319

Fulco M, Cen Y, Zhao P, Hoffman EP, McBurney MW, Sauve AA, Sartorelli V (2008) Glucose restriction inhibits skeletal myoblast differentiation by activating SIRT1 through AMPK-mediated regulation of Nampt. Dev Cell 14(5):661–673. doi:10.1016/j.devcel.2008.02.004

Gancedo JM (1998) Yeast carbon catabolite repression. Microbiol Mol Biol Rev 62(2):334–361

Ghosh D, Walton JL, Roepe PD, Sinai AP (2012) Autophagy is a cell death mechanism in Toxoplasma gondii. Cell Microbiol 14(4):589–607. doi:10.1111/j.1462-5822.2011.01745.x

Gjini E, Haydon DT, Barry JD, Cobbold CA (2010) Critical interplay between parasite differentiation, host immunity, and antigenic variation in trypanosome infections. Am Nat 176 (4):424–439. doi:10.1086/656276

Gonzalez MR, Bischofberger M, Pernot L, van der Goot FG, Freche B (2008) Bacterial pore-forming toxins: the (w)hole story? Cell Mol Life Sci 65(3):493–507. doi:10.1007/s00018-007-7434-y

Gordon EB, Hart GT, Tran TM, Waisberg M, Akkaya M, Skinner J, Zinocker S, Pena M, Yazew T, Qi CF, Miller LH, Pierce SK (2015) Inhibiting the Mammalian target of rapamycin blocks the development of experimental cerebral malaria. MBio 6(3):e00725. doi:10.1128/mBio.00725-15

Guo Y, Zhang Y, Hong K, Luo F, Gu Q, Lu N, Bai A (2014) AMPK inhibition blocks ROS-NFkappaB signaling and attenuates endotoxemia-induced liver injury. PLoS One 9(1): e86881. doi:10.1371/journal.pone.0086881

Hardie DG (2007) AMP-activated/SNF1 protein kinases: conserved guardians of cellular energy. Nat Rev Mol Cell Biol 8(10):774–785. doi:10.1038/nrm2249

Hardie DG, Scott JW, Pan DA, Hudson ER (2003) Management of cellular energy by the AMP-activated protein kinase system. FEBS Lett 546(1):113–120

Hawley SA, Boudeau J, Reid JL, Mustard KJ, Udd L, Makela TP, Alessi DR, Hardie DG (2003) Complexes between the LKB1 tumor suppressor, STRAD alpha/beta and MO25 alpha/beta are upstream kinases in the AMP-activated protein kinase cascade. J Biol 2(4):28. doi:10.1186/1475-4924-2-28

Hawley SA, Pan DA, Mustard KJ, Ross L, Bain J, Edelman AM, Frenguelli BG, Hardie DG (2005) Calmodulin-dependent protein kinase kinase-beta is an alternative upstream kinase for AMP-activated protein kinase. Cell Metab 2(1):9–19. doi:10.1016/j.cmet.2005.05.009

Hedbacker K, Carlson M (2008) SNF1/AMPK pathways in yeast. Front Biosci 13:2408–2420

Hohmann S (2002) Osmotic stress signaling and osmoadaptation in yeasts. Microbiol Mol Biol Rev 66(2):300–372

Holmes BF, Kurth-Kraczek EJ, Winder WW (1999) Chronic activation of 5′-AMP-activated protein kinase increases GLUT-4, hexokinase, and glycogen in muscle. J Appl Physiol 87 (5):1990–1995

Hood MI, Skaar EP (2012) Nutritional immunity: transition metals at the pathogen-host interface. Nat Rev Microbiol 10(8):525–537. doi:10.1038/nrmicro2836

Hoogendijk AJ, Pinhancos SS, van der Poll T, Wieland CW (2013) AMP-activated protein kinase activation by 5-aminoimidazole-4-carbox-amide-1-beta-D-ribofuranoside (AICAR) reduces lipoteichoic acid-induced lung inflammation. J Biol Chem 288(10):7047–7052. doi:10.1074/jbc.M112.413138

Horman S, Vertommen D, Heath R, Neumann D, Mouton V, Woods A, Schlattner U, Wallimann T, Carling D, Hue L, Rider MH (2006) Insulin antagonizes ischemia-induced Thr172 phosphorylation of AMP-activated protein kinase alpha-subunits in heart via hierarchical phosphorylation of Ser485/491. J Biol Chem 281(9):5335–5340. doi:10.1074/jbc.M506850200

Horwitz MA (1983) The Legionnaires' disease bacterium (Legionella pneumophila) inhibits phagosome-lysosome fusion in human monocytes. J Exp Med 158(6):2108–2126

Hu G, Cheng PY, Sham A, Perfect JR, Kronstad JW (2008) Metabolic adaptation in Cryptococcus neoformans during early murine pulmonary infection. Mol Microbiol 69(6):1456–1475. doi:10.1111/j.1365-2958.2008.06374.x

Huang H, Kang R, Wang J, Luo G, Yang W, Zhao Z (2013) Hepatitis C virus inhibits AKT-tuberous sclerosis complex (TSC), the mechanistic target of rapamycin (MTOR) pathway, through endoplasmic reticulum stress to induce autophagy. Autophagy 9(2):175–195. doi:10.4161/auto.22791

Hull CM, Heitman J (2002) Genetics of Cryptococcus neoformans. Annu Rev Genet 36:557–615. doi:10.1146/annurev.genet.36.052402.152652

Hurtrel B, Petit F, Arnoult D, Muller-Trutwin M, Silvestri G, Estaquier J (2005) Apoptosis in SIV infection. Cell Death Differ 12(Suppl 1):979–990. doi:10.1038/sj.cdd.4401600

Hutterer C, Wandinger SK, Wagner S, Muller R, Stamminger T, Zeittrager I, Godl K, Baumgartner R, Strobl S, Marschall M (2013) Profiling of the kinome of cytomegalovirus-infected cells reveals the functional importance of host kinases Aurora A, ABL and AMPK. Antiviral Res 99(2):139–148. doi:10.1016/j.antiviral.2013.04.017

Inoki K, Zhu T, Guan KL (2003) TSC2 mediates cellular energy response to control cell growth and survival. Cell 115(5):577–590

Jeong KM, Choi JI, Lee SH, Lee HJ, Son JK, Seo CS, Song SW, Kwak SH, Bae HB (2014) Effect of sauchinone, a lignan from Saururus chinensis, on bacterial phagocytosis by macrophages. Eur J Pharmacol 728:176–182. doi:10.1016/j.ejphar.2014.01.039

Ji WT, Lee LH, Lin FL, Wang L, Liu HJ (2009) AMP-activated protein kinase facilitates avian reovirus to induce mitogen-activated protein kinase (MAPK) p38 and MAPK kinase 3/6 signalling that is beneficial for virus replication. J Gen Virol 90(Pt 12):3002–3009. doi:10.1099/vir.0.013953-0

Kang KY, Kim YK, Yi H, Kim J, Jung HR, Kim IJ, Cho JH, Park SH, Kim HY, Ju JH (2013) Metformin downregulates Th17 cells differentiation and attenuates murine autoimmune arthritis. Int Immunopharmacol 16(1):85–92. doi:10.1016/j.intimp.2013.03.020

Kelly B, O'Neill LA (2015) Metabolic reprogramming in macrophages and dendritic cells in innate immunity. Cell Res 25(7):771–784. doi:10.1038/cr.2015.68

Kim YW, Lee SM, Shin SM, Hwang SJ, Brooks JS, Kang HE, Lee MG, Kim SC, Kim SG (2009) Efficacy of sauchinone as a novel AMPK-activating lignan for preventing iron-induced oxidative stress and liver injury. Free Radic Biol Med 47(7):1082–1092. doi:10.1016/j.freeradbiomed.2009.07.018

Kim J, Kundu M, Viollet B, Guan KL (2011) AMPK and mTOR regulate autophagy through direct phosphorylation of Ulk1. Nat Cell Biol 13(2):132–141. doi:10.1038/ncb2152

Kim JJ, Lee HM, Shin DM, Kim W, Yuk JM, Jin HS, Lee SH, Cha GH, Kim JM, Lee ZW, Shin SJ, Yoo H, Park YK, Park JB, Chung J, Yoshimori T, Jo EK (2012) Host cell autophagy activated by antibiotics is required for their effective antimycobacterial drug action. Cell Host Microbe 11(5):457–468. doi:10.1016/j.chom.2012.03.008

Kim J, Kwak HJ, Cha JY, Jeong YS, Rhee SD, Kim KR, Cheon HG (2014) Metformin suppresses lipopolysaccharide (LPS)-induced inflammatory response in murine macrophages via activating transcription factor-3 (ATF-3) induction. J Biol Chem 289(33):23246–23255. doi:10.1074/jbc.M114.577908

Kloft N, Neukirch C, Bobkiewicz W, Veerachato G, Busch T, von Hoven G, Boller K, Husmann M (2010) Pro-autophagic signal induction by bacterial pore-forming toxins. Med Microbiol Immunol 199(4):299–309. doi:10.1007/s00430-010-0163-0

Kuchin S, Vyas VK, Carlson M (2002) Snf1 protein kinase and the repressors Nrg1 and Nrg2 regulate FLO11, haploid invasive growth, and diploid pseudohyphal differentiation. Mol Cell Biol 22(12):3994–4000

Kudchodkar SB, Del Prete GQ, Maguire TG, Alwine JC (2007) AMPK-mediated inhibition of mTOR kinase is circumvented during immediate-early times of human cytomegalovirus infection. J Virol 81(7):3649–3651. doi:10.1128/JVI.02079-06

Kumar SH, Rangarajan A (2009) Simian virus 40 small T antigen activates AMPK and triggers autophagy to protect cancer cells from nutrient deprivation. J Virol 83(17):8565–8574. doi:10.1128/JVI.00603-09

Kumar D, Nath L, Kamal MA, Varshney A, Jain A, Singh S, Rao KV (2010) Genome-wide analysis of the host intracellular network that regulates survival of Mycobacterium tuberculosis. Cell 140(5):731–743. doi:10.1016/j.cell.2010.02.012

Kwon HS, Brent MM, Getachew R, Jayakumar P, Chen LF, Schnolzer M, McBurney MW, Marmorstein R, Greene WC, Ott M (2008) Human immunodeficiency virus type 1 Tat protein inhibits the SIRT1 deacetylase and induces T cell hyperactivation. Cell Host Microbe 3 (3):158–167. doi:10.1016/j.chom.2008.02.002

Landfear SM (2011) Nutrient transport and pathogenesis in selected parasitic protozoa. Eukaryot Cell 10(4):483–493. doi:10.1128/EC.00287-10

Lee W, VanderVen BC, Fahey RJ, Russell DG (2013) Intracellular Mycobacterium tuberculosis exploits host-derived fatty acids to limit metabolic stress. J Biol Chem 288(10):6788–6800. doi:10.1074/jbc.M112.445056

Levin DE (2005) Cell wall integrity signaling in Saccharomyces cerevisiae. Microbiol Mol Biol Rev 69(2):262–291. doi:10.1128/MMBR.69.2.262-291.2005

Levine B, Mizushima N, Virgin HW (2011) Autophagy in immunity and inflammation. Nature 469 (7330):323–335. doi:10.1038/nature09782

Li JV, Wang Y, Saric J, Nicholson JK, Dirnhofer S, Singer BH, Tanner M, Wittlin S, Holmes E, Utzinger J (2008) Global metabolic responses of NMRI mice to an experimental Plasmodium berghei infection. J Proteome Res 7(9):3948–3956. doi:10.1021/pr800209d

Lorenz MC, Fink GR (2001) The glyoxylate cycle is required for fungal virulence. Nature 412 (6842):83–86. doi:10.1038/35083594

Lv G, Zhu H, Zhou F, Lin Z, Lin G, Li C (2014) AMP-activated protein kinase activation protects gastric epithelial cells from Helicobacter pylori-induced apoptosis. Biochem Biophys Res Commun 453(1):13–18. doi:10.1016/j.bbrc.2014.09.028

MacIver NJ, Blagih J, Saucillo DC, Tonelli L, Griss T, Rathmell JC, Jones RG (2011) The liver kinase B1 is a central regulator of T cell development, activation, and metabolism. J Immunol 187(8):4187–4198. doi:10.4049/jimmunol.1100367

MacRae JI, Sheiner L, Nahid A, Tonkin C, Striepen B, McConville MJ (2012) Mitochondrial metabolism of glucose and glutamine is required for intracellular growth of Toxoplasma gondii. Cell Host Microbe 12(5):682–692. doi:10.1016/j.chom.2012.09.013

Mankouri J, Tedbury PR, Gretton S, Hughes ME, Griffin SD, Dallas ML, Green KA, Hardie DG, Peers C, Harris M (2010) Enhanced hepatitis C virus genome replication and lipid accumulation mediated by inhibition of AMP-activated protein kinase. Proc Natl Acad Sci U S A 107 (25):11549–11554. doi:10.1073/pnas.0912426107

Marsin AS, Bouzin C, Bertrand L, Hue L (2002) The stimulation of glycolysis by hypoxia in activated monocytes is mediated by AMP-activated protein kinase and inducible 6-phosphofructo-2-kinase. J Biol Chem 277(34):30778–30783. doi:10.1074/jbc.M205213200

McArdle J, Schafer XL, Munger J (2011) Inhibition of calmodulin-dependent kinase kinase blocks human cytomegalovirus-induced glycolytic activation and severely attenuates production of viral progeny. J Virol 85(2):705–714. doi:10.1128/JVI.01557-10

McArdle J, Moorman NJ, Munger J (2012) HCMV targets the metabolic stress response through activation of AMPK whose activity is important for viral replication. PLoS Pathog 8(1): e1002502. doi:10.1371/journal.ppat.1002502

McKinney JD, Honer zu Bentrup K, Munoz-Elias EJ, Miczak A, Chen B, Chan WT, Swenson D, Sacchettini JC, Jacobs WR Jr, Russell DG (2000) Persistence of Mycobacterium tuberculosis in macrophages and mice requires the glyoxylate shunt enzyme isocitrate lyase. Nature 406 (6797):735–738. doi:10.1038/35021074

Meares GP, Qin H, Liu Y, Holdbrooks AT, Benveniste EN (2013) AMP-activated protein kinase restricts IFN-gamma signaling. J Immunol 190(1):372–380. doi:10.4049/jimmunol.1202390

Mehla R, Bivalkar-Mehla S, Zhang R, Handy I, Albrecht H, Giri S, Nagarkatti P, Nagarkatti M, Chauhan A (2010) Bryostatin modulates latent HIV-1 infection via PKC and AMPK signaling but inhibits acute infection in a receptor independent manner. PLoS One 5(6):e11160. doi:10.1371/journal.pone.0011160

Miranda-Saavedra D, Stark MJ, Packer JC, Vivares CP, Doerig C, Barton GJ (2007) The complement of protein kinases of the microsporidium Encephalitozoon cuniculi in relation to those of Saccharomyces cerevisiae and Schizosaccharomyces pombe. BMC Genomics 8:309. doi:10.1186/1471-2164-8-309

Mitchell TG, Perfect JR (1995) Cryptococcosis in the era of AIDS—100 years after the discovery of Cryptococcus neoformans. Clin Microbiol Rev 8(4):515–548

Monceaux V, Estaquier J, Fevrier M, Cumont MC, Riviere Y, Aubertin AM, Ameisen JC, Hurtrel B (2003) Extensive apoptosis in lymphoid organs during primary SIV infection predicts rapid progression towards AIDS. AIDS 17(11):1585–1596. doi:10.1097/01.aids.0000076286.54156.d6

Mony BM, Matthews KR (2015) Assembling the components of the quorum sensing pathway in African trypanosomes. Mol Microbiol 96(2):220–232. doi:10.1111/mmi.12949

Mony BM, MacGregor P, Ivens A, Rojas F, Cowton A, Young J, Horn D, Matthews K (2014) Genome-wide dissection of the quorum sensing signalling pathway in Trypanosoma brucei. Nature 505(7485):681–685. doi:10.1038/nature12864

Moreau K, Lacas-Gervais S, Fujita N, Sebbane F, Yoshimori T, Simonet M, Lafont F (2010) Autophagosomes can support Yersinia pseudotuberculosis replication in macrophages. Cell Microbiol 12(8):1108–1123. doi:10.1111/j.1462-5822.2010.01456.x

Moreira D, Rodrigues V, Abengozar M, Rivas L, Rial E, Laforge M, Li X, Foretz M, Viollet B, Estaquier J, Cordeiro da Silva A, Silvestre R (2015a) Leishmania infantum modulates host macrophage mitochondrial metabolism by hijacking the SIRT1-AMPK axis. PLoS Pathog 11 (3):e1004684. doi:10.1371/journal.ppat.1004684

Moreira D, Silvestre R, Cordeiro-da-Silva A, Estaquier J, Foretz M, Viollet B (2015b) AMP-activated protein kinase as a target for pathogens: friends or foes? Curr Drug Targets. doi:10.1007/978-3-319-43589-3 [pii]

Moseley CE, Webster RG, Aldridge JR (2010) Peroxisome proliferator-activated receptor and AMP-activated protein kinase agonists protect against lethal influenza virus challenge in mice. Influenza Other Respi Viruses 4(5):307–311. doi:10.1111/j.1750-2659.2010.00155.x

Moser TS, Jones RG, Thompson CB, Coyne CB, Cherry S (2010) A kinome RNAi screen identified AMPK as promoting poxvirus entry through the control of actin dynamics. PLoS Pathog 6(6):e1000954. doi:10.1371/journal.ppat.1000954

Moser TS, Schieffer D, Cherry S (2012) AMP-activated kinase restricts Rift Valley fever virus infection by inhibiting fatty acid synthesis. PLoS Pathog 8(4):e1002661. doi:10.1371/journal.ppat.1002661

Munoz-Elias EJ, McKinney JD (2005) Mycobacterium tuberculosis isocitrate lyases 1 and 2 are jointly required for in vivo growth and virulence. Nat Med 11(6):638–644. doi:10.1038/nm1252 [pii]

Naderer T, Heng J, Saunders EC, Kloehn J, Rupasinghe TW, Brown TJ, McConville MJ (2015) Intracellular survival of leishmania major depends on uptake and degradation of extracellular matrix glycosaminoglycans by macrophages. PLoS Pathog 11(9):e1005136. doi:10.1371/journal.ppat.1005136

Nagajyothi F, Desruisseaux MS, Thiruvur N, Weiss LM, Braunstein VL, Albanese C, Teixeira MM, de Almeida CJ, Lisanti MP, Scherer PE, Tanowitz HB (2008) Trypanosoma cruzi infection of cultured adipocytes results in an inflammatory phenotype. Obesity (Silver Spring) 16(9):1992–1997

Nakashima K, Takeuchi K, Chihara K, Hotta H, Sada K (2011) Inhibition of hepatitis C virus replication through adenosine monophosphate-activated protein kinase-dependent and -independent pathways. Microbiol Immunol 55(11):774–782. doi:10.1111/j.1348-0421.2011.00382.x

Navarro-Garcia F, Sanchez M, Nombela C, Pla J (2001) Virulence genes in the pathogenic yeast Candida albicans. FEMS Microbiol Rev 25(2):245–268

Ng HC, Singh M, Jeyaseelan K (1995) Identification of two protein serine/threonine kinase genes and molecular cloning of a SNF1 type protein kinase gene from Toxoplasma gondii. Biochem Mol Biol Int 35(1):155–165

Ng HC, Singh M, Jeyaseelan K (1997) Nucleotide sequence of ToxPK1 gene from Toxoplasma gondii. DNA Seq 7(3-4):179–191

Nieves W, Oniskey T, Hung L, Herbert D (2014) Metabolic regulation of type 2 immunity controls tissue repair. In: 63rd annual ASTMH meeting, New Orleans, USA, 2–6 November 2014

Ong CW, Elkington PT, Brilha S, Ugarte-Gil C, Tome-Esteban MT, Tezera LB, Pabisiak PJ, Moores RC, Sathyamoorthy T, Patel V, Gilman RH, Porter JC, Friedland JS (2015) Neutrophil-derived MMP-8 drives AMPK-dependent matrix destruction in human pulmonary tuberculosis. PLoS Pathog 11(5):e1004917. doi:10.1371/journal.ppat.1004917

Pagans S, Pedal A, North BJ, Kaehlcke K, Marshall BL, Dorr A, Hetzer-Egger C, Henklein P, Frye R, McBurney MW, Hruby H, Jung M, Verdin E, Ott M (2005) SIRT1 regulates HIV transcription via Tat deacetylation. PLoS Biol 3(2):e41. doi:10.1371/journal.pbio.0030041

Pandey AK, Sassetti CM (2008) Mycobacterial persistence requires the utilization of host cholesterol. Proc Natl Acad Sci U S A 105(11):4376–4380. doi:10.1073/pnas.0711159105

Park CY, Jun HJ, Wakita T, Cheong JH, Hwang SB (2009) Hepatitis C virus nonstructural 4B protein modulates sterol regulatory element-binding protein signaling via the AKT pathway. J Biol Chem 284(14):9237–9246. doi:10.1074/jbc.M808773200

Park DW, Jiang S, Tadie JM, Stigler WS, Gao Y, Deshane J, Abraham E, Zmijewski JW (2013) Activation of AMPK enhances neutrophil chemotaxis and bacterial killing. Mol Med 19:387–398. doi:10.2119/molmed.2013.00065

Parsons M, Worthey EA, Ward PN, Mottram JC (2005) Comparative analysis of the kinomes of three pathogenic trypanosomatids: Leishmania major, Trypanosoma brucei and Trypanosoma cruzi. BMC Genomics 6:127. doi:10.1186/1471-2164-6-127

Petter R, Kwon-Chung KJ (1996) Disruption of the SNF1 gene abolishes trehalose utilization in the pathogenic yeast Candida glabrata. Infect Immun 64(12):5269–5273

Petter R, Chang YC, Kwon-Chung KJ (1997) A gene homologous to Saccharomyces cerevisiae SNF1 appears to be essential for the viability of Candida albicans. Infect Immun 65 (12):4909–4917

Plummer R, Lorigan P, Steven N, Scott L, Middleton MR, Wilson RH, Mulligan E, Curtin N, Wang D, Dewji R, Abbattista A, Gallo J, Calvert H (2013) A phase II study of the potent PARP inhibitor, Rucaparib (PF-01367338, AG014699), with temozolomide in patients with metastatic melanoma demonstrating evidence of chemopotentiation. Cancer Chemother Pharmacol 71(5):1191–1199. doi:10.1007/s00280-013-2113-1

Polge C, Thomas M (2007) SNF1/AMPK/SnRK1 kinases, global regulators at the heart of energy control? Trends Plant Sci 12(1):20–28. doi:10.1016/j.tplants.2006.11.005

Pollitt LC, Mideo N, Drew DR, Schneider P, Colegrave N, Reece SE (2011) Competition and the evolution of reproductive restraint in malaria parasites. Am Nat 177(3):358–367. doi:10.1086/658175

Rabhi I, Rabhi S, Ben-Othman R, Rasche A, Daskalaki A, Trentin B, Piquemal D, Regnault B, Descoteaux A, Guizani-Tabbane L, Sysco C (2012) Transcriptomic signature of Leishmania infected mice macrophages: a metabolic point of view. PLoS Negl Trop Dis 6(8):e1763. doi:10.1371/journal.pntd.0001763

Raffatellu M, Santos RL, Verhoeven DE, George MD, Wilson RP, Winter SE, Godinez I, Sankaran S, Paixao TA, Gordon MA, Kolls JK, Dandekar S, Baumler AJ (2008) Simian immunodeficiency virus-induced mucosal interleukin-17 deficiency promotes Salmonella dissemination from the gut. Nat Med 14(4):421–428. doi:10.1038/nm1743

Ramsook CB, Tan C, Garcia MC, Fung R, Soybelman G, Henry R, Litewka A, O'Meally S, Otoo HN, Khalaf RA, Dranginis AM, Gaur NK, Klotz SA, Rauceo JM, Jue CK, Lipke PN (2010)

Yeast cell adhesion molecules have functional amyloid-forming sequences. Eukaryot Cell 9 (3):393–404. doi:10.1128/EC.00068-09
Reece SE, Pollitt LC, Colegrave N, Gardner A (2011) The meaning of death: evolution and ecology of apoptosis in protozoan parasites. PLoS Pathog 7(12):e1002320. doi:10.1371/journal.ppat.1002320
Rey-Jurado E, Riedel CA, Gonzalez PA, Bueno SM, Kalergis AM (2015) Contribution of autophagy to antiviral immunity. FEBS Lett 589(22):3461–3470. doi:10.1016/j.febslet.2015.07.047
Rhee KY, de Carvalho LP, Bryk R, Ehrt S, Marrero J, Park SW, Schnappinger D, Venugopal A, Nathan C (2011) Central carbon metabolism in Mycobacterium tuberculosis: an unexpected frontier. Trends Microbiol 19(7):307–314. doi:10.1016/j.tim.2011.03.008
Rolf J, Zarrouk M, Finlay DK, Foretz M, Viollet B, Cantrell DA (2013) AMPKalpha1: a glucose sensor that controls CD8 T-cell memory. Eur J Immunol 43(4):889–896. doi:10.1002/eji.201243008
Romero-Gomez M, Diago M, Andrade RJ, Calleja JL, Salmeron J, Fernandez-Rodriguez CM, Sola R, Garcia-Samaniego J, Herrerias JM, De la Mata M, Moreno-Otero R, Nunez O, Olveira A, Duran S, Planas R, Spanish Treatment of Resistance to Insulin in Hepatitis C Genotype 1 Group (2009) Treatment of insulin resistance with metformin in naive genotype 1 chronic hepatitis C patients receiving peginterferon alfa-2a plus ribavirin. Hepatology 50 (6):1702–1708. doi:10.1002/hep.23206
Rousset F, Roze D (2007) Constraints on the origin and maintenance of genetic kin recognition. Evolution 61(10):2320–2330. doi:10.1111/j.1558-5646.2007.00191.x
Sag D, Carling D, Stout RD, Suttles J (2008) Adenosine 5′-monophosphate-activated protein kinase promotes macrophage polarization to an anti-inflammatory functional phenotype. J Immunol 181(12):8633–8641. doi:10.4049/jimmunol.181.12.8633 [pii]
Sales-Dias J (2011) The role of glucose in Plasmodium liver infection. Biochemistry, University of Lisbon, Lisbon
Sampaio-Marques B, Burhans WC, Ludovico P (2014) Longevity pathways and maintenance of the proteome: the role of autophagy and mitophagy during yeast ageing. Microb Cell 1 (4):118–127
Santarelli R, Granato M, Faggioni A, Cirone M (2015) Interference with the autophagic process as a viral strategy to escape from the immune control: lesson from gamma herpesviruses. J Immunol Res 2015:546063. doi:10.1155/2015/546063
Sanz P (2003) Snf1 protein kinase: a key player in the response to cellular stress in yeast. Biochem Soc Trans 31(Pt 1):178–181. doi:10.1042/bst0310178
Schubert-Unkmeir A, Sokolova O, Panzner U, Eigenthaler M, Frosch M (2007) Gene expression pattern in human brain endothelial cells in response to Neisseria meningitidis. Infect Immun 75 (2):899–914. doi:10.1128/IAI.01508-06
Seo JY, Cresswell P (2013) Viperin regulates cellular lipid metabolism during human cytomegalovirus infection. PLoS Pathog 9(8):e1003497. doi:10.1371/journal.ppat.1003497
Shaw RJ, Bardeesy N, Manning BD, Lopez L, Kosmatka M, DePinho RA, Cantley LC (2004) The LKB1 tumor suppressor negatively regulates mTOR signaling. Cancer Cell 6(1):91–99. doi:10.1016/j.ccr.2004.06.007
Shin DM, Yuk JM, Lee HM, Lee SH, Son JW, Harding CV, Kim JM, Modlin RL, Jo EK (2010) Mycobacterial lipoprotein activates autophagy via TLR2/1/CD14 and a functional vitamin D receptor signalling. Cell Microbiol 12(11):1648–1665. doi:10.1111/j.1462-5822.2010.01497.x
Singhal A, Jie L, Kumar P, Hong GS, Leow MK, Paleja B, Tsenova L, Kurepina N, Chen J, Zolezzi F, Kreiswirth B, Poidinger M, Chee C, Kaplan G, Wang YT, De Libero G (2014) Metformin as adjunct antituberculosis therapy. Sci Transl Med 6(263):263ra159. doi:10.1126/scitranslmed.3009885
Sorbara MT, Girardin SE (2015) Emerging themes in bacterial autophagy. Curr Opin Microbiol 23:163–170. doi:10.1016/j.mib.2014.11.020

Steele S, Brunton J, Ziehr B, Taft-Benz S, Moorman N, Kawula T (2013) Francisella tularensis harvests nutrients derived via ATG5-independent autophagy to support intracellular growth. PLoS Pathog 9(8):e1003562. doi:10.1371/journal.ppat.1003562

Stein SC, Woods A, Jones NA, Davison MD, Carling D (2000) The regulation of AMP-activated protein kinase by phosphorylation. Biochem J 345(Pt 3):437–443

Street A, Macdonald A, Crowder K, Harris M (2004) The Hepatitis C virus NS5A protein activates a phosphoinositide 3-kinase-dependent survival signaling cascade. J Biol Chem 279 (13):12232–12241. doi:10.1074/jbc.M312245200

Tang Y, Li F, Tan B, Liu G, Kong X, Hardwidge PR, Yin Y (2014) Enterotoxigenic Escherichia coli infection induces intestinal epithelial cell autophagy. Vet Microbiol 171(1-2):160–164. doi:10.1016/j.vetmic.2014.03.025

Tattoli I, Sorbara MT, Vuckovic D, Ling A, Soares F, Carneiro LA, Yang C, Emili A, Philpott DJ, Girardin SE (2012) Amino acid starvation induced by invasive bacterial pathogens triggers an innate host defense program. Cell Host Microbe 11(6):563–575. doi:10.1016/j.chom.2012.04.012

Terry LJ, Vastag L, Rabinowitz JD, Shenk T (2012) Human kinome profiling identifies a requirement for AMP-activated protein kinase during human cytomegalovirus infection. Proc Natl Acad Sci U S A 109(8):3071–3076. doi:10.1073/pnas.1200494109

Tewari R, Straschil U, Bateman A, Bohme U, Cherevach I, Gong P, Pain A, Billker O (2010) The systematic functional analysis of Plasmodium protein kinases identifies essential regulators of mosquito transmission. Cell Host Microbe 8(4):377–387. doi:10.1016/j.chom.2010.09.006

Tilney LG, Harb OS, Connelly PS, Robinson CG, Roy CR (2001) How the parasitic bacterium Legionella pneumophila modifies its phagosome and transforms it into rough ER: implications for conversion of plasma membrane to the ER membrane. J Cell Sci 114(Pt 24):4637–4650

Tsoyi K, Jang HJ, Nizamutdinova IT, Kim YM, Lee YS, Kim HJ, Seo HG, Lee JH, Chang KC (2011) Metformin inhibits HMGB1 release in LPS-treated RAW 264.7 cells and increases survival rate of endotoxaemic mice. Br J Pharmacol 162(7):1498–1508. doi:10.1111/j.1476-5381.2010.01126.x

van Zandbergen G, Luder CG, Heussler V, Duszenko M (2010) Programmed cell death in unicellular parasites: a prerequisite for sustained infection? Trends Parasitol 26(10):477–483. doi:10.1016/j.pt.2010.06.008

Vassella E, Reuner B, Yutzy B, Boshart M (1997) Differentiation of African trypanosomes is controlled by a density sensing mechanism which signals cell cycle arrest via the cAMP pathway. J Cell Sci 110(Pt 21):2661–2671

Vickerman K (1985) Developmental cycles and biology of pathogenic trypanosomes. Br Med Bull 41(2):105–114

Vincent O, Carlson M (1999) Gal83 mediates the interaction of the Snf1 kinase complex with the transcription activator Sip4. EMBO J 18(23):6672–6681. doi:10.1093/emboj/18.23.6672

Wang Y, Utzinger J, Saric J, Li JV, Burckhardt J, Dirnhofer S, Nicholson JK, Singer BH, Brun R, Holmes E (2008) Global metabolic responses of mice to Trypanosoma brucei brucei infection. Proc Natl Acad Sci U S A 105(16):6127–6132. doi:10.1073/pnas.0801777105

Waterman SR, Hacham M, Panepinto J, Hu G, Shin S, Williamson PR (2007) Cell wall targeting of laccase of Cryptococcus neoformans during infection of mice. Infect Immun 75(2):714–722. doi:10.1128/IAI.01351-06

Winder WW, Hardie DG (1996) Inactivation of acetyl-CoA carboxylase and activation of AMP-activated protein kinase in muscle during exercise. Am J Physiol 270(2 Pt 1):E299–E304

Woods A, Johnstone SR, Dickerson K, Leiper FC, Fryer LG, Neumann D, Schlattner U, Wallimann T, Carlson M, Carling D (2003) LKB1 is the upstream kinase in the AMP-activated protein kinase cascade. Curr Biol 13(22):2004–2008

Woods A, Dickerson K, Heath R, Hong SP, Momcilovic M, Johnstone SR, Carlson M, Carling D (2005) Ca2+/calmodulin-dependent protein kinase kinase-beta acts upstream of AMP-activated protein kinase in mammalian cells. Cell Metab 2(1):21–33. doi:10.1016/j.cmet.2005.06.005

Xiao B, Sanders MJ, Underwood E, Heath R, Mayer FV, Carmena D, Jing C, Walker PA, Eccleston JF, Haire LF, Saiu P, Howell SA, Aasland R, Martin SR, Carling D, Gamblin SJ (2011) Structure of mammalian AMPK and its regulation by ADP. Nature 472(7342):230–233. doi:10.1038/nature09932

Yang K, Chi H (2015) AMPK helps T cells survive nutrient starvation. Immunity 42(1):4–6. doi:10.1016/j.immuni.2014.12.029

Yang J, Li D, Liu X, Pan J, Yan B, Zhu X (2010a) Regulation of virulence factors, carbon utilization and virulence by SNF1 in Cryptococcus neoformans JEC21 and divergent actions of SNF1 between cryptococcal strains. Fungal Genet Biol 47(12):994–1000. doi:10.1016/j.fgb.2010.08.002

Yang Z, Kahn BB, Shi H, Xue BZ (2010b) Macrophage alpha1 AMP-activated protein kinase (alpha1AMPK) antagonizes fatty acid-induced inflammation through SIRT1. J Biol Chem 285 (25):19051–19059. doi:10.1074/jbc.M110.123620

Yang J, Li D, Pan J, Zhu X (2011) Snf1/AMPK protein kinase modulates cell wall integrity in the human pathogenic yeast Cryptococcus neoformans. Wei Sheng Wu Xue Bao 51(6):740–746

Yang D, Xue B, Wang X, Yu X, Liu N, Gao Y, Liu C, Zhu H (2013) 2-octynoic acid inhibits hepatitis C virus infection through activation of AMP-activated protein kinase. PLoS One 8(5): e64932. doi:10.1371/journal.pone.0064932

Yang CS, Kim JJ, Lee HM, Jin HS, Lee SH, Park JH, Kim SJ, Kim JM, Han YM, Lee MS, Kweon GR, Shong M, Jo EK (2014) The AMPK-PPARGC1A pathway is required for antimicrobial host defense through activation of autophagy. Autophagy 10(5):785–802. doi:10.4161/auto.28072

Yu JW, Sun LJ, Liu W, Zhao YH, Kang P, Yan BZ (2013) Hepatitis C virus core protein induces hepatic metabolism disorders through down-regulation of the SIRT1-AMPK signaling pathway. Int J Infect Dis 17(7):e539–e545. doi:10.1016/j.ijid.2013.01.027

Yuk JM, Shin DM, Lee HM, Yang CS, Jin HS, Kim KK, Lee ZW, Lee SH, Kim JM, Jo EK (2009) Vitamin D3 induces autophagy in human monocytes/macrophages via cathelicidin. Cell Host Microbe 6(3):231–243. doi:10.1016/j.chom.2009.08.004

Zangger H, Mottram JC, Fasel N (2002) Cell death in Leishmania induced by stress and differentiation: programmed cell death or necrosis? Cell Death Differ 9(10):1126–1139. doi:10.1038/sj.cdd.4401071

Zhang HS, Wu MR (2009) SIRT1 regulates Tat-induced HIV-1 transactivation through activating AMP-activated protein kinase. Virus Res 146(1–2):51–57. doi:10.1016/j.virusres.2009.08.005

Zhang HS, Ruan Z, Sang WW (2011) HDAC1/NFkappaB pathway is involved in curcumin inhibiting of Tat-mediated long terminal repeat transactivation. J Cell Physiol 226 (12):3385–3391. doi:10.1002/jcp.22691

Zhang Z, Wang S, Huang J, Liu L, Lu M, Li M, Sui Y, Xu L, Yan R, Song X, Li X (2015) Proteomic analysis of Eimeria acervulina sporozoite proteins interaction with duodenal epithelial cells by shotgun LC-MS/MS. Mol Biochem Parasitol 202(2):29–33. doi:10.1016/j.molbiopara.2015.09.006

Zhao Z, Hu J, Gao X, Liang H, Liu Z (2014) Activation of AMPK attenuates lipopolysaccharide-impaired integrity and function of blood-brain barrier in human brain microvascular endothelial cells. Exp Mol Pathol 97(3):386–392. doi:10.1016/j.yexmp.2014.09.006

Zhao H, Zhu H, Lin Z, Lin G, Lv G (2015) Compound 13, an alpha1-selective small molecule activator of AMPK, inhibits Helicobacter pylori-induced oxidative stresses and gastric epithelial cell apoptosis. Biochem Biophys Res Commun 463(4):510–517. doi:10.1016/j.bbrc.2015.05.059

Zhou G, Myers R, Li Y, Chen Y, Shen X, Fenyk-Melody J, Wu M, Ventre J, Doebber T, Fujii N, Musi N, Hirshman MF, Goodyear LJ, Moller DE (2001) Role of AMP-activated protein kinase in mechanism of metformin action. J Clin Invest 108(8):1167–1174. doi:10.1172/JCI13505

Part III
Pharmacology of AMPK

Chapter 13
Targeting AMPK: From Ancient Drugs to New Small-Molecule Activators

Bruno Guigas and Benoit Viollet

Contents

Abstract The AMP-activated protein kinase (AMPK) is an evolutionary conserved and ubiquitously expressed serine/threonine kinase mainly acting as a key regulator of cellular energy homeostasis. AMPK is a heterotrimeric protein

B. Guigas (✉)
Department of Molecular Cell Biology, Leiden University Medical Center, Leiden, The Netherlands

Department of Parasitology, Leiden University Medical Center, P.O. Box 9600, Postzone L40-Q, 2300 RC Leiden, The Netherlands
e-mail: b.g.a.guigas@lumc.nl

B. Viollet
INSERM, U1016, Institut Cochin, Paris, France

CNRS, UMR8104, Paris, France

Université Paris Descartes, Sorbonne Paris Cité, France

M.D. Cordero, B. Viollet (eds.), *AMP-activated Protein Kinase*, Experientia Supplementum 107, DOI 10.1007/978-3-319-43589-3_13

complex, consisting of a catalytic α subunit and two regulatory β and γ subunits, whose activity is tightly regulated by changes in adenine nucleotides and several posttranslational modifications. Once activated in response to energy deficit, AMPK concomitantly inhibits ATP-consuming anabolic processes and promotes ATP-generating catabolic pathways via direct phosphorylation of multiple downstream effectors, leading to restoration of cellular energy balance. A growing number of energy/nutrient-independent functions of AMPK are also regularly reported, progressively expanding its role to regulation of non-metabolic cellular processes. Historically, AMPK as a therapeutic target has attracted much of interest due to its potential impact on metabolic disorders, such as obesity and type 2 diabetes, but has also recently received considerable renewed attention in the framework of cancer studies, highlighting the persistent need for selective, reversible, potent, and tissue-specific activators. In this chapter, we review the most recent advances in the understanding of the mechanism(s) of action of the current portfolio of AMPK activators, including plant-derived natural compounds and newly discovered small-molecule agonists directly targeting various AMPK subunits.

Keywords AMPK • Metformin • Salicylate • A-769662 • Compound-13 • 991

13.1 AMPK: A Cellular Energy Sensor with Pleiotropic Functions

The AMP-activated protein kinase (AMPK) is a serine/threonine protein kinase expressed in all eukaryotic cells that mainly acts as a sensor of cellular energy status, although its role is currently being expanded and not anymore strictly restricted to the regulation of energy homeostasis (Hardie et al. 2016). AMPK is an heterotrimeric complex consisting of a catalytic α subunit and two regulatory β and γ subunits that contain a glycogen-binding domain (carbohydrate-binding module, CBM) and four adenylate nucleotide-binding sites (cystathionine-β-synthase, CBS), respectively (Carling et al. 2012; Hardie et al. 2016; Oakhill et al. 2012). Each of these subunits has several isoforms encoded by different genes (α1, α2, β1, β2, γ1, γ2, γ3), leading to multiple tissue- and species-specific heterotrimeric combinations with eventually different subcellular localizations, regulations, and functions (See Chap. 1).

AMPK functions as an adenylate charge-regulated kinase which constantly senses the cellular energy status by monitoring intracellular AMP, ADP, and ATP levels (Oakhill et al. 2012). A better understanding of the mechanism(s) by which adenine nucleotides regulate AMPK activity came notably from the step-by-step improvement in the elucidation of the crystal structure of mammalian AMPK complex (Calabrese et al. 2014; Chen et al. 2013; Li et al. 2015; Xiao et al. 2007, 2011, 2013). In response to energy deficit characterized by rapid depletion of cellular ATP, the concomitant rise in AMP owing to adenylate kinase leads to a direct but modest allosteric activation of AMPK. However, the binding of AMP to the CBS domains of the AMPK γ subunit also promotes the phosphorylation on the Thr172 residue within the activation loop of the α-catalytic subunit by the upstream AMPK kinases (AMPKK) liver kinase B (LKB1) and/or Ca^{2+}/calmodulin-dependent protein kinase kinase beta (CaMKKβ), considerably amplifying AMPK activity. Although not allosterically activating the kinase, ADP has also been shown to bind instead of ATP to some of the CBS domains of the γ subunit and to enhance AMPKK-mediated Thr172 phosphorylation (Oakhill et al. 2011). AMP and ADP also sustain AMPK activation through inhibition of Thr172 dephosphorylation mediated by yet unidentified AMPK-specific protein phosphatase(s) (Gowans et al. 2013; Sanders et al. 2007b; Xiao et al. 2011). Of note, AMPK complexes containing different γ-subunit isoforms respond differently to changes in adenine nucleotides, suggesting that these variations in affinity might result in heterotrimer-specific functional consequences (Ross et al. 2016). In addition to adenine nucleotides, glycogen has also been shown to regulate AMPK by a mechanism dependent on its binding to the carbohydrate-binding loop of the CBM (Bieri et al. 2012; Koay et al. 2010; Oligschlaeger et al. 2015). Finally, although the translational and posttranscriptional mechanisms involved in the regulation of AMPK subunits expression remain mostly unknown to date, a variety of posttranslational modifications other than the abovementioned Thr172 phosphorylation have been reported, such as multisite phosphorylation [α: Ser485/491, Thr479; β: Ser108, Ser182, Thr148; for review Carling et al. (2012)], myristoylation (Liang et al. 2015; Oakhill et al. 2010; Warden et al. 2001), acetylation (Lin et al. 2012), sumoylation (Rubio et al. 2013; Yan et al. 2015), and O-GlcNAcylation (Bullen et al. 2014), adding another layer of complexity to the already intricate regulation of AMPK activity.

Once activated in the context of energy crisis, AMPK inhibits ATP-consuming anabolic processes and promotes the rates of ATP-generating catabolic pathways, leading to restoration of cellular energy balance (Carling et al. 2011; Hardie 2014b; Oakhill et al. 2012). This implicates direct phosphorylation by AMPK of a broad range of downstream effectors [for reviews, see Hardie (2014a) and Steinberg and Kemp (2009)] that are involved in the regulation of multiple cellular and whole-body metabolic processes (Andris and Leo 2015; Bijland et al. 2013; Daskalopoulos et al. 2016; Hardie 2014a; Marcinko and Steinberg 2014; Mounier et al. 2015; O'Neill and Hardie 2013; Qi and Young 2015; Steinberg and Kemp 2009). Furthermore, a growing number of energy/nutrient-independent functions of AMPK have been recently reported, progressively expanding its roles to regulation of non-metabolic cellular processes [for recent review, see Hardie et al. (2016)].

The first report of a pharmacological AMPK activator came from an international patent filled in 1994 where the cell-permeable 5-Aminoimidazole-4-carboxamide-1-β-D-ribofuranoside (AICA riboside) was described as a ZMP-generating agent mimicking the allosteric effects of AMP on the AMPK system owing to its structural analogy with the adenine nucleotide. AICA riboside was later confirmed to cause intracellular accumulation of ZMP and AMPK activation in metabolic cells (Corton et al. 1995; Henin et al. 1995; Sullivan et al. 1994) and next extensively used during more than a decade for investigating AMPK functions both in vitro and in vivo [for review, see Guigas et al. (2009)]. In the wake of AICA riboside, a large variety of compounds found to activate AMPK have

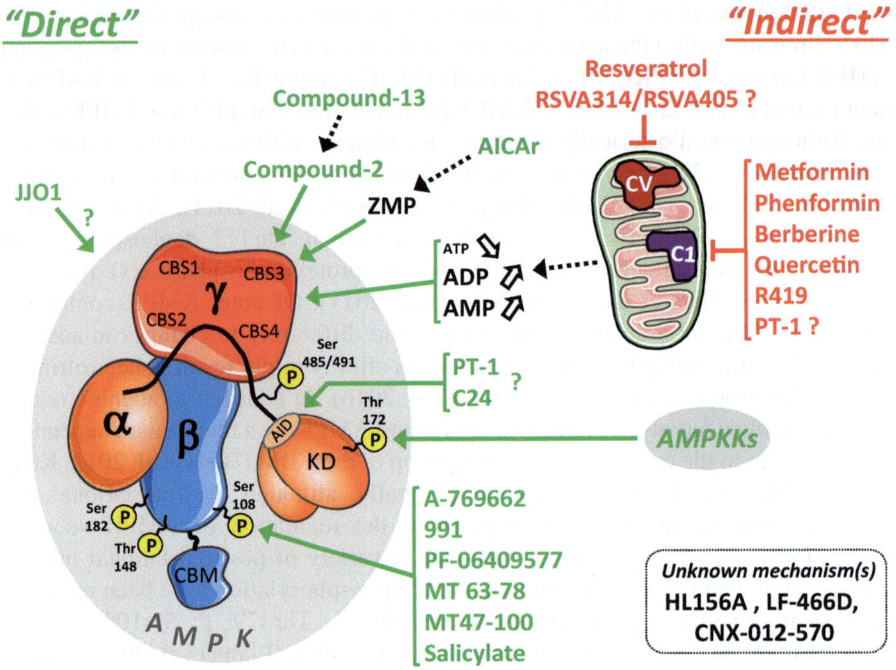

Fig. 13.1 Mechanisms of action of the current portfolio of AMPK activators. *CBS* cystathionine-β-synthase, *CBM* carbohydrate-binding module, *KD* kinase domain, *AID* auto-inhibitory domain

been reported to alter cellular energy status via inhibition of mitochondrial function, leading to decreased ATP synthesis and consequent rise in cellular AMP levels. All these compounds cause activation of AMPK by increasing cellular AMP:ATP and ADP:ATP ratios and are termed indirect AMPK activators. Their mechanism of action can be clearly demonstrated by using cells expressing AMPK complexes containing AMP-insensitive AMPK-γ mutants (Hawley et al. 2010; Jensen et al. 2015). While off-target effects for these compounds have been clearly demonstrated (Cameron et al. 2016; Foretz et al. 2010; Guigas et al. 2006, 2007; Lim et al. 2016; Liu et al. 2014; Vincent et al. 2015; Rao et al. 2016), development of potent and specific AMPK activators was a long-awaited milestone for the AMPK community. In the last decade, many high-throughput screens have been conducted to identify novel AMPK activators (Kim et al. 2015; Sinnett and Brenman 2014). So far, more than 10 different classes of direct AMPK activators have been published in patent and research articles (Giordanetto and Karis 2012; Yun and Ha 2011; Rana et al. 2015) and have been instrumental for further elucidation of physiological responses caused by increased AMPK activity. In this chapter, we provide a brief synopsis of key indirect and direct AMPK activators (Fig. 13.1), notably focusing on recent molecular breakthroughs in the understanding of their mechanisms of action.

13.2 AMPK Activation by Plant-Derived Natural Compounds

13.2.1 Metformin

Metformin (Dimethybiguanide) is a synthetic derivative of a natural plant-derived product extracted from the so-called French lilac *Galega officinalis* (Witters 2001). Metformin is the most prescribed antidiabetic agents and is currently used as a first-line drug for treatment of type 2 diabetes for its potent anti-hyperglycemic action presumably mediated by AMPK-independent inhibition of hepatic gluconeogenesis (Foretz et al. 2010; Miller et al. 2013). This 60-year-old last therapeutic survivor of the biguanide family (phenformin and buformin were withdrawn due to frequent lactic acidosis) has also recently attracted renewed attention as potential antineoplastic agent for the treatment of various cancers (Morales and Morris 2015). An exhaustive overview of the pleiotropic effects and up-to-date underlying molecular mechanisms of metformin is currently out of the scope of this chapter and has been recently done elsewhere (Foretz et al. 2014). However, the first demonstration in 2001 that metformin can activate AMPK in primary mouse hepatocytes was an important step that has shed a complete new light on the drug and also later boosted the effort to elucidate its exact mechanism of action (Zhou et al. 2001). After a decade of controversy, the mechanism by which metformin activates AMPK starts to reach a consensus. It is indeed now well accepted that activation of AMPK is not mediated by direct interaction of the drug with the kinase but results from a specific and mild inhibition of the mitochondrial respiratory-chain complex 1 by metformin (El-Mir et al. 2000; Owen et al. 2000), leading to an increase in ADP:ATP and AMP:ATP ratios (Foretz et al. 2014). The key role played by mitochondria in AMPK activation by metformin was first demonstrated in bovine aortic endothelial cells depleted of mitochondria where the effect of the drug on kinase activation was abolished (Zou et al. 2004). Furthermore, it has been shown that bypassing the inhibition of mitochondrial respiratory-chain complex 1 by using methyl succinate, a substrate of the respiratory-chain complex 2, allowed to counteract both the metformin-induced alteration of cellular energy status and AMPK activation in pancreatic MIN6 β-cells, supporting a causal relationship between the specific mitochondrial action of the drug and AMPK activation (Hinke et al. 2007). Although the exact mechanism(s) by which metformin inhibits complex 1 still remains to be elucidated, this indirect and adenine nucleotide-dependent mechanism involving mitochondria as primary cellular target (Leverve et al. 2003) was further supported by the demonstration that the drug failed to activate AMPK in cells expressing AMP-insensitive AMPKγ2 mutant (Hawley et al. 2010). Not surprisingly, R419, a pharmacological inhibitor of the mitochondrial respiratory-chain complex 1, was also recently shown to promote indirect activation of AMPK both in vitro and in vivo (Marcinko et al. 2015; Jenkins et al. 2013).

13.2.2 Resveratrol

Resveratrol is a plant-derived natural polyphenol famous for being present in significant amount in the skin of grapes and, therefore, in red wine. In recent years, although still controversial, resveratrol has emerged as a compound that might protect against metabolic, cardiovascular, and other age-related complications, including neurodegeneration and cancer (Kulkarni and Canto 2015). A landmark study published in 2006 have reported that resveratrol improved whole-body insulin sensitivity and promoted healthy aging in high-fat diet (HFD)-fed mice, an effect associated with increased AMPK activity in various metabolic tissues (Baur et al. 2006). Remarkably, these beneficial effects are lost in AMPKα1- or -α2-deficient mice where treatment with resveratrol failed to reduce fat mass and improve whole-body insulin sensitivity and glucose tolerance, confirming that AMPK is a central target for the metabolic effects of the plant-derived polyphenol (Um et al. 2010). A randomized double-blind crossover study in humans also reported that treatment with resveratrol for 30 days improved metabolic homeostasis in obese men, an effect associated with enhanced AMPK Thr172 phosphorylation in skeletal muscle (Timmers et al. 2011). Activation of AMPK by resveratrol was confirmed in various ex vivo/in vitro cellular models, including primary neurons (Dasgupta and Milbrandt 2007; Vingtdeux et al. 2010), C2C12 and L6 myotubes (Park et al. 2007; Breen et al. 2008), rat cardiomyocytes (Chan et al. 2008), human umbilical vein endothelial cells (Xu et al. 2009), primary stromal vascular cells (Wang et al. 2015), or human colorectal tissues (Cai et al. 2015), although the underlying mechanism(s) involved remained unclear. Finally, an important breakthrough came from the use of a cell line stably expressing AMPK complexes containing an AMP-insensitive γ2 variant (Hawley et al. 2010) for clarifying the adenine nucleotide-dependent or -independent mechanism(s) by which known natural compounds and small-molecule agonists activate AMPK. This elegant study undoubtedly demonstrated that AMPK activation by resveratrol was secondary to increased ADP:ATP ratio and presumably due to inhibition of mitochondrial oxidative phosphorylation by the drug at the level of ATP synthase, as previously suggested (Zheng and Ramirez 2000). Similarly, other polyphenols used in Chinese ancestral pharmacopeia, such as berberine or quercetin, were also reported to activate AMPK by an indirect adenine nucleotide-dependent mechanism involving inhibition of the mitochondrial machinery and subsequent alteration of cellular energetics (Turner et al. 2008; Hawley et al. 2010).

Two synthetic small molecules selected for their structural similarity with resveratrol, RSVA314 and RSVA405, were later found to activate AMPK with a potency nearly 40 times higher than resveratrol (Vingtdeux et al. 2011). Not surprisingly, these compounds did not activate recombinant AMPK in vitro but rather led to a dose-dependent decrease in cellular ATP levels (Vingtdeux et al. 2011), suggesting a similar mechanism of action than resveratrol, i.e. an adenine nucleotide-mediated activation of AMPK secondary to inhibition of mitochondrial oxidative phosphorylation.

13.2.3 Salicylate

Salicylate is a plant-derived product mostly extracted from willow bark which belongs to the ancestral human pharmacopeia, some Egyptian pharaonic papyri tracking back its first mention as medicinal drug for reducing fever to the second millennium BC and medieval herbalists relied on salicylate-containing extracts for their palliative properties (Hedner and Everts 1998). Synthetic derivatives, such as aspirin (acetylsalicylate) and salsalate, are now widely used for their anti-coagulation and anti-inflammatory properties, and both of them are rapidly metabolized to salicylate by various esterases (Higgs et al. 1987). Strikingly, concentrations of salicylate in the upper therapeutic range for humans were recently reported to activate AMPK by a mechanism mostly independent of changes in cellular AMP or ADP levels but involving direct interaction of the drug with the AMPK-β1 subunit (Hawley et al. 2012). Indeed, salicylate increased AMPK activity by promoting both direct allosteric activation of the kinase and inhibition of Thr172 dephosphorylation by protein phosphatase, an effect that is abolished by S108A substitution in the β1 subunit of the α1β1γ1 complex and not observed in β2-containing complex (Hawley et al. 2012). Furthermore, in vivo administration of salicylate increased AMPK activity in liver, soleus muscle, and adipose tissue, reduced circulating non-esterified fatty acid levels, and promoted fatty acid oxidation in wild-type mice but not in whole-body AMPKβ1$^{-/-}$ mice, confirming the involvement of β1-containing AMPK complexes in the mechanism of action of the drug (Hawley et al. 2012). Of note, salicylate also improved metabolic homeostasis in high-fat diet-fed insulin-resistant mice (Hawley et al. 2012) by a yet unclear AMPK-independent mechanism(s) that might involve activation of brown adipose tissue by the drug (van Dam et al. 2015).

More recently, a series of studies from the same group also reported that salicylate can activate AMPK by directly interacting with the β1-subunit drug-binding site in bone marrow-derived mouse macrophages (BMDMs) (Fullerton et al. 2015), mouse livers and primary human hepatocytes (Ford et al. 2015), and ex vivo prostate and lung cancer cells (O'Brien et al. 2015). Indeed, direct activation of AMPK by salicylate in BMDMs reduced foam cell formation by promoting cholesterol efflux to HDL, suggesting that pharmacological activation of β1-containing AMPK complex in macrophages might be beneficial during the early stages of atherosclerosis (Fullerton et al. 2015). In addition, the AMPK β1-mediated inhibition of lipogenesis induced by therapeutic concentrations of salicylate in mouse liver, human hepatocytes, and cancer cells has been shown to promote insulin sensitivity and cell death, respectively, an effect that synergizes with the one of metformin (Ford et al. 2015; O'Brien et al. 2015). Altogether, these data indicate that a combination of salicylate and metformin may have a better therapeutic efficacy than each of these mechanistically divergent AMPK activators used separately (see also Sect. 13.4.1). Of note, a recent study has reported that some of the anti-neoplastic properties of various AMPK activators in tumor cells, including those of salicylate, are actually AMPK independent (Vincent et al. 2015), underlining the need for systematic validation of any effects observed with these agonists using AMPK-deficient cellular models.

13.3 AMPK Activation by Small Drug Molecules

13.3.1 A-769662 (Thienopyridone)

The development of direct AMPK activators has been pioneered by Abbott Laboratories. By using a microarrayed compound screening (μARCS) technology, the screen of a chemical library of over 700,000 compounds was performed (Anderson et al. 2004; Cool et al. 2006). In μARCS AMPK assays, activity of partially purified AMPKαβγ complex from rat liver was monitored by phosphorylation of SAMS peptide substrate (HMRSAMSGLHLVKRR) after incubation with the chemical library arrayed on polystyrene sheets. A non-nucleoside thienopyridone compound A-592107 was initially identified and served as a structural template for the development of the more potent A-769662 AMPK activator (Fig. 13.2), a thienopyridone scaffold with a phenylphenol substituent (Cool et al. 2006). The specificity of A-769662 (EC50: 0.8 μM) was tested in a cell-free assay against a

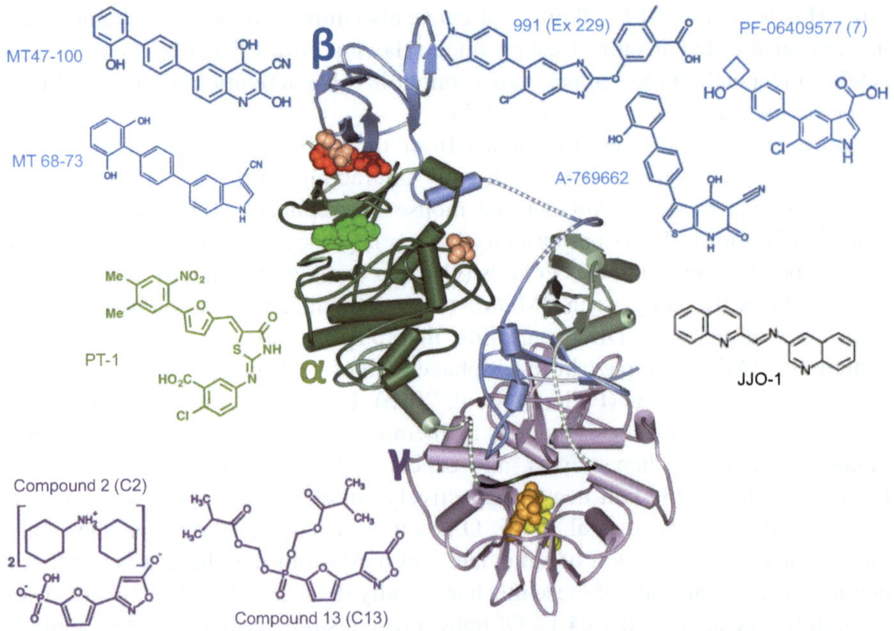

Fig. 13.2 Chemical structures of the main small-molecule AMPK activators

panel of 76 protein kinases, including members from the AMPK-related kinase family, and it was found that the majority of them were not significantly affected (Goransson et al. 2007).

Early studies indicated that A-769662-induced AMPK activation occurs by reversible binding (Cool et al. 2006) and revealed that, similar to AMP, A-769662 activates AMPK both allosterically and by inhibiting AMPK-α-subunit Thr172 dephosphorylation by inactivating protein phosphatase (Goransson et al. 2007; Sanders et al. 2007a). Kinetic studies have also suggested that A-769662 exerts its effects by lowering the K_m for the SAMS peptide substrate (Calabrese et al. 2014). Importantly, A-769662 did not induce significant inhibition of mitochondrial oxygen consumption rate, nor increase cellular AMP:ATP and ADP:ATP ratios in intact cells (Foretz et al. 2010; Guigas et al. 2009; Cool et al. 2006), and validation studies further confirmed that A-769662 does not act as an AMP mimetic. Indeed, high concentrations of A-769662 had no effect on the activity of AMP-sensitive enzymes, such as fructose 1,6-biphosphatase and glycogen phosphorylase (Cool et al. 2006). In addition, A-769662 activates AMPK complexes containing AMP-insensitive AMPK-γ1R298G or AMPK-γ2R531Q mutants in vitro and in vivo, respectively (Hawley et al. 2010; Sanders et al. 2007a), indicating a mechanism of action that differs from AMP. In line with these observations, combination studies showed that A-769662 increased AMPK activity in the presence of saturating concentration of AMP (Cool et al. 2006) and failed to displace labeled AMP bound to a GST fusion of the four CBS motifs from human γ2 subunit (Goransson et al. 2007). These data support the notion that A769662 binds to an alternate allosteric site. Several studies provided compelling evidence that the CBM domain of the AMPK-β subunit, and in particular phosphorylation of AMPK-β Ser108 (an autophosphorylation site within the CBM), is crucial for the allosteric effect of A-769662 (Scott et al. 2008, 2014; Sanders et al. 2007a; Xiao et al. 2013; Calabrese et al. 2014; Hawley et al. 2012). However, it is unlikely that A-769662 binds to the glycogen-binding site or alternate site located within the CBM as demonstrated by NMR and the absence of an obvious binding pocket on the structure of isolated CBS motif (Polekhina et al. 2005; Calabrese et al. 2014; Sanders et al. 2007a). Furthermore, A-769662 does act by binding to the AMPK-α subunit or by relieving inhibition of the kinase domain by the auto-inhibitory domain (AID), suggesting that it must utilize a novel allosteric binding site (Goransson et al. 2007). A clue to the A769662 binding site came from hydrogen-deuterium exchange mass spectrometry studies showing that A-769662 causes conformational changes around the N-lobe of the kinase domain (KD) on the AMPK-α subunit (Landgraf et al. 2013). A-769662 is likely to bind at the interface of the AMPK-β CBM and the AMPK-α KD for functional activation and protects AMPK-α Thr172 from phosphatases through conformational changes in the activation loop of the AMPK-α KD. The presence of this novel allosteric binding pocket has been recently confirmed by the crystal structures of AMPK heterotrimeric α2β1γ1 and α2β1γ1 complexes bound to A-769662 (Calabrese et al. 2014; Xiao et al. 2013) and is made by a three-helical bundle formed between an α-helix at the C-terminal portion of AMPK-β CBM and C-α-helices from the

AMPK-α KD. Of note, A769662 preferentially activates AMPK complexes containing the AMPK-β1 over the AMPK-β2-subunit isoform (Hawley et al. 2012; Rajamohan et al. 2016; Scott et al. 2008). The estimated K_D for A-769662 to the α1β1γ1 heterotrimer is around 30-48 nM, whereas the binding to α1β2γ1 is negligible (Calabrese et al. 2014; Rajamohan et al. 2016). Furthermore, the potency of allosteric modulation and protection of AMPK-α Thr172 from phosphatase of AMPK-β2 containing heterotrimeric complexes is evident only at relatively high concentrations in comparison to AMPK-β1 containing heterotrimeric complexes (Rajamohan et al. 2016). These data indicate that the allosteric binding interface between AMPK-α KD and AMPK-β2 CBM is substantially different from the AMPK-α KD and AMPK-β1 CBM interface, resulting in a reduced binding affinity for A-769662. Swapping of non-conserved amino acids in the CBM from AMPK-β1 and AMPK-β2 was sufficient to change the potency of A-769662 and alter ligand specificity toward the AMPK-β isoforms, showing their importance in the topology of the CBM-KD interface (Calabrese et al. 2014). These studies highlight an unexpected opportunity for the development of isoform-specific small-molecule activators that can target cell- and tissue-specific AMPK complexes. However, analysis of tissue distribution of the various AMPK-β isoforms in different species has revealed that the predominant AMPK-β isoform in human skeletal muscle and liver is AMPK-β2 (Wu et al. 2013; Stephenne et al. 2011; Birk and Wojtaszewski 2006). Thus, these observations raise concerns about the clinical utility of small-molecule AMPK activators with restricted effect to AMPK-β1 containing complexes, such as A-769662.

In vitro incubation of primary rodent hepatocytes and mouse embryonic fibroblasts resulted in a dose-dependent increase in AMPK activity and phosphorylation of the well-established downstream AMPK target acetyl-CoA carboxylase (ACC), effects that are completely abolished in AMPK-α$1^{-/-}$α$2^{-/-}$ and AMPK-β$1^{-/-}$ cells (Hawley et al. 2012; Scott et al. 2008; Foretz et al. 2010; Cool et al. 2006; Guigas et al. 2009; Goransson et al. 2007). Intriguingly, the effects of A-769662 on AMPK phosphorylation were quite small in comparison to the effects on ACC phosphorylation, raising some doubt about the requirement of AMPK-α activation loop phosphorylation for A-769662 action (Foretz et al. 2010; Goransson et al. 2007). Consistent with this observation, it was recently demonstrated that A-769662 mediates allosteric activation of AMPK independently of AMPK-α Thr172 phosphorylation, provided AMPK-β Ser108 is phosphorylated (Scott et al. 2014; Viollet et al. 2014). In rodent primary hepatocytes, A-769662-induced AMPK activation was associated with a decrease in fatty acid synthesis (Cool et al. 2006) and enhanced fatty acid oxidation (Hawley et al. 2012). In contrast, no significant effects were reported on basal and cAMP-stimulated gluconeogenic gene expression and glucose production rates (Foretz et al. 2010), while another study indicated that high concentrations of A-769662 lowered basal hepatic glucose production by an AMPK-independent mechanism (Scott et al. 2008). When administrated acutely in Sprague Dawley rats (Cool et al. 2006) or wild type but not in AMPK-β$1^{-/-}$ mice (Hawley et al. 2012) at a dose of 30 mg/kg (0.905 nmol/g), A-769662 induced a rapid shift from carbohydrate to fat oxidation, as measured by indirect calorimetry, consistent with the inactivation of ACC and reduction of malonyl CoA levels observed in the liver. Furthermore, administration of A-769662 (3, 10, and 30 mg/kg) for 5 days in diabetic and obese *ob*/*ob* mice showed a significant and dose-dependent lowering of plasma triglyceride and hepatic triglyceride levels, concomitant with a reduction in ACC activity in the liver (Cool et al. 2006). This was associated with a modest reduction in body weight gain and asignificant decrease infed plasma glucose only in *ob*/*ob* mice treated with high-dose A-769662 (30 mg/kg). Based on the observations that A-769662 has a poor oral bioavailability and reached the highest concentration in liver (with much lower levels in skeletal muscle), it is likely that the beneficial effects on lipid metabolism are primarily due to stimulation of AMPK, leading to increased oxidation and decreased synthesis of fatty acids in liver (Cool et al. 2006). Therefore, by decreasing hepatic lipid accumulation, A-769662 will subsequently improve insulin sensitivity and overall glucose homeostasis.

Although the contribution of AMPK activation in skeletal muscle cannot be excluded, it has been reported that A-769662 is unable to stimulate glucose uptake in mouse skeletal muscle (Scott et al. 2008). In addition to the beneficial effects on lipid metabolism, treatment with A-769662 has recently emerged as an effective strategy to protect the heart against ischemia–reperfusion injury (Kim et al. 2011). Activation of AMPK pathway prior to ischemia by A-769662 pretreatment replicated the potent protection seen with ischemic preconditioning. A-769662-induced AMPK activation during ischemia preserved myocardial energy charge by improving the balance between energy generation and utilization pathways (Kim et al. 2011). AMPK activation limits energy expenditure by activating eEF2 kinase, which phosphorylates and inactivates eEF2, an important regulator of protein translation. A-769662-induced AMPK activation also promoted cardioprotection through the phosphorylation and activation of endothelial nitric oxide synthase (eNOS), which plays several protective effects following ischemia (Kim et al. 2011). The cardioprotective effect of A-769662 is abrogated in transgenic mice that express a kinase dead AMPK-α2 isoform (K45R mutation) in the heart, indicating that A-769662 preconditioning action is AMPK dependent (Kim et al. 2011). This is an important finding as it is now beginning to emerge that A769662 can interfere with various biological pathways unrelated to AMPK through multiple off-target effects (Benziane et al. 2009; Moreno et al. 2008; Treebak et al. 2009).

13.3.2 991 (ex229)

A series of patents published by Merck Sharp & Dohme Corporation and Metabasis Therapeutics have claimed novel cyclic benzimidazole derivatives as relevant AMPK activators (Giordanetto and Karis 2012). Among these, ex229 (also later referred to as compound 991, Fig. 13.2) was found to have half-maximal effective concentrations (EC50 relative to maximal activation by AMP) of less than 5 μM and activation effect relative to maximal activation by AMP greater than 200 % using recombinant human α1β1γ1 complex. Compound 991 was five- to tenfold more potent than A-769662 in activating AMPK, as assessed by allosteric activation and protection against Thr172 dephosphorylation (Xiao et al. 2013). Similarly to A-769662, 991 does not activate AMPK complexes lacking the AMPK-β CBM and preferentially activates those containing the AMPK-β1 isoform (Xiao et al. 2013). Circular dichroism data revealed that there is a single tight site for 991 binding. Interestingly, A-769662 competed for 991 binding to AMPK, suggesting that the two compounds bind to a common site. Indeed, the recent crystal structure of full-length human AMPK-α2β1γ1 complex bound to 991 showed that the compound binds at the interface between the N-terminal lobe of AMPK-α KD and the AMPK-β CBM, the same binding site as A-769662 (Xiao et al. 2013). Compound 991 efficiently activated AMPK in isolated rodent skeletal muscle without altering cellular AMP:ATP and ADP:ATP ratios and stimulated both glucose uptake and fatty acid oxidation by a mechanism that was demonstrated to be AMPK dependent by using AMPK-$\alpha1^{-/-}\alpha2^{-/-}$ myotubes (Lai et al. 2014).

13.3.3 PF-06409577 (Compound 7)

Recently, another small-molecule screening conducted by Pfizer using a novel time-resolved fluorescence resonance energy transfer activation/protection assay identified PF-06409577 (also called Compound 7, Fig. 13.2) as a specific and indole acid-based direct AMPK activator (Cameron et al. 2016). PF-06409577 was shown to be a potent allosteric activator of both human and rat recombinant AMPK isoforms that contain the β 1 subunit (α1β 1γ1), and to also protect against Thr172 dephosphorylation by protein phosphatase PP2A. In term of specificity, a broad screening against a panel of receptors, channels, PDEs and kinases showed that PF-06409577 exhibited minimal off-target effects. At the mechanistic level, a crystallographic approach using recombinant α1β 1γ1 AMPK coupled to computational modelling indicated that PF-06409577 binds at the interface between the α- and β-subunits. Interestingly, despite possessing a different bicyclic core ring, this indazole acid compound therefore shared a common binding site with A-769662 and 991 (Cameron et al. 2016). Remarkably, treatment with PF-06409577 promoted a dosedependent activation of AMPK in whole kidney tissue and improved renal functions in ZSF-1 obese rat, a model of diabetic nephropathy (Cameron et al. 2016).

13.3.4 Compound-2/Compound-13

Metabasis Therapeutics screened a focused library of AMP mimetics against both human and rat AMPK by monitoring the phosphorylation of the SAMS peptide and identified 5-(5-hydroxy-isoxazol-3-yl)-furan-2-phosphonic acid (Compound-2, C2) that showed high potency for AMPK activation (Fig. 13.2) (Gomez-Galeno et al. 2010). C2 bears little structural resemblance to AMP and does not affect the activity of several AMP-regulated enzymes (Gomez-Galeno et al. 2010; Hunter et al. 2014). In addition, C2 had no effect on a panel of 138 protein kinases, including members of the AMPK-related family and known upstream kinases of AMPK (Hunter et al. 2014). This compound shows an improved specificity toward AMPK and is a potent allosteric activator of the kinase (EC_{50} of 10–30 nM). It is structurally distinct from the prototypical non-nucleotide AMPK activator, A-769662, and does not require the presence of the AMPK-β CBM for activation. It is likely that C2 exerts its allosteric effects by exploiting the same binding site on the AMPK-γ subunits as AMP. It has been reported that C2 failed to stimulate AMPK complexes containing AMPK-γ2R531G mutant, which renders AMPK complexes insensitive to AMP. Moreover, C2 and AMP displaced a GST-AMPK-γ2 subunit fusion from ATP-γ-Sepharose to the same extent, indicating that both ligands compete for the same site. However, the recent crystal structure of full-length α2β1γ1 isoform co-crystallized with C2 and AMP revealed two C2-binding sites in the AMPK-γ subunit distinct from nucleotide-binding sites (Langendorf et al. 2016). The two C2 molecules bind the AMPK-γ at the interface between the CBS-binding sites 1, 3, and 4, with the phosphate groups of both C2 molecules overlapping the phosphate binding sites of AMP in sites 1 and 4.

Interestingly, C2 shows a preference for AMPK-α1 containing complexes and, like AMP, protects against AMPK-α1 Thr172 dephosphorylation. In contrast, C2 is a partial agonist of AMPK-α2 containing complexes and fails to protect against dephosphorylation (Hunter et al. 2014). Using chimeric AMPK-α2, in which the α-regulatory subunit-interacting motif-2 (α-RIM2) from AMPK-α1 was used to replace the equivalent region, full allosteric activation and protection against Thr172 dephosphorylation by C2 could be fully restored (Hunter et al. 2014). These data indicate the importance of the different sequences of AMPK-α1 and AMPK-α2 in the α-RIM2 region for the selectivity of C2 toward AMPK-α1 versus AMPK-α2 containing complexes. This was confirmed by structural comparison of the interaction between AMPK-γ subunit and AMPK-α1 and α2 isoform α-RIM2, showing that α1-RIM2 interacts more strongly than α2-RIM2 upon C2 binding (Langendorf et al. 2016).

To overcome poor cellular permeability, an esterase-sensitive phosphonate prodrug termed Compound-13 (C13) was designed to evaluate the metabolic outcome in vitro and in vivo (Fig. 13.2) (Gomez-Galeno et al. 2010). In primary hepatocytes C13 dose dependently increased ACC phosphorylation without any significant change in adenine nucleotide levels at concentrations up to 100 μM (Hunter et al. 2014). Consistent with the phosphorylation of ACC, C13 potently inhibited hepatic lipogenesis in primary mouse hepatocytes (Hunter et al. 2014) and whole-body lipogenesis in mice (Gomez-Galeno et al. 2010). Regulation of lipid synthesis by C13 was blunted in primary hepatocytes from liver-specific AMPK-α1$^{-/-}$AMPK-α2$^{-/-}$ mice but not from AMPK-β1$^{-/-}$ or AMPK-β2$^{-/-}$ mice, showing that C13 action is independent of a specific AMPK-β-subunit isoform (Hunter et al. 2014).

13.3.5 PT-1

Next to the kinase domain of the AMPK-α catalytic subunit, the auto-inhibitory domain (AID) maintains the kinase in an inactive form through binding to the N- and C-lobes of AMPK-α KD. In contrast, upon activation, conformational changes in AMPK-α induce dissociation of the AMPK-α AID from the AMPK-α KD, relieving the auto-inhibitory conformation. A library of 3600 organic compounds was screened to identify small molecules that affect conformational change in AMPK-α and antagonize the interaction between AMPK-α AID and AMPK-α KD. The small-molecule PT-1 was identified through screening with a truncated form of AMPK-α ($\alpha1_{1\text{-}394}$) containing only the AMPK-α AID and AMPK-α KD, which is repressed under basal condition (Fig. 13.2) (Pang et al. 2008). PT-1 activates the inactive AMPK-$\alpha1_{1\text{-}394}$ construct as well as the AMPK-α1β1γ1 heterotrimeric complex in cell-free essays and in intact cells. It was proposed that PT-1 antagonizes the auto-inhibitory conformation of AMPK-α by binding in the cleft between AMPK-α AID and AMPK-α KD (Pang et al. 2008). C24, another small-molecule activator issued from optimization of PT1 by the same group, was also reported to allosterically stimulate inactive AMPK-α-subunit truncations and activate AMPK heterotrimers by antagonizing autoinhibition (Li et al. 2013). However, a recent study has revisited the mechanism of action of PT-1 and showed that instead of binding directly to the AMPK-α1 subunit, as previously suggested, it indirectly activates AMPK by inhibiting the respiratory chain and thus increasing cellular AMP levels (Jensen et al. 2015). Consistently, PT-1 failed to activate AMPK complexes containing AMP-insensitive AMPK-γ1R299G mutant, suggesting it functions as an indirect activator (Jensen et al. 2015).

13.3.6 MT 63-78 (Debio 0930)

A small-molecule screening using human recombinant AMPK-α1β1γ1 performed in the framework of a partnership between Mercury Therapeutics and CreaGen Biosciences identified MT 68-73 (also called Debio 0930, Fig. 13.2) as a specific and potent direct AMPK activator (Zadra et al. 2014). It was shown that MT68-78 allosterically activates AMPK and protects from AMPK-α Thr172 dephosphorylation by protein phosphatase PP2C. Specificity of MT 63-78 was tested against a panel of 93 protein kinases. At 25 μM, two kinases were slightly activated and six kinases were marginally inhibited, whereas at 5 μM the majority of the kinases were not affected, including human members of the AMPK-related kinase family (Zadra et al. 2014). Functional analyses revealed that MT 63-78 selectively activates complexes containing the AMPK-β1 isoform in cell-free essays. At high doses, MT 68-73 is able to activate AMPK-β2 subunits, especially when complexed with AMPK-α2. This was confirmed in a cell-based context where AMPK-β1 subunit but not AMPK-β2 subunit silencing abolished MT 63-78-mediated activation of

AMPK and phosphorylation of ACC (Zadra et al. 2014). These results indicate that MT 68-73 is likely specific to the class of small-molecule activators whose binding involves the AMPK-β subunit. Interestingly, MT 63-78 showed antitumor effects in a number of different cancer cell types and inhibition of cell growth was strictly dependent on AMPK activity (Zadra et al. 2014).

13.3.7 MT47-100

MT47-100 identified by Mercury Therapeutics is structurally similar to A-769662 but possesses a dihydroxyquinoline core instead of the thienopyridone core of A-769662 (Fig. 13.2). This compound has a unique feature, being simultaneously a direct activator and inhibitor of AMPK complexes containing AMPK-β1 or AMPK-β2 subunit, respectively (Scott et al. 2015). Like A-769662, activation of AMPK-β1 containing complexes by MT47-100 is dependent on the presence of AMPK-β CBM and AMPK-β Ser108 (Scott et al. 2015). In addition, synergistic activation in combination with AMP has been demonstrated. The inhibitory effect of MT47-100 on AMPK-β2 containing complexes is also dependent on the presence of AMPK-β CBM but not AMPK-β Ser108 phosphorylation (Scott et al. 2015). The residue determining the agonistic/antagonistic properties of MT47-100 is non-conserved residue located within the AMPK-β CBM at position Phe82, Tyr92, and Leu93 in AMPK-β1 and the corresponding residues Ile81, Phe91, and Ile92 in AMPK-β2. A more complete understanding of the mechanism by which these three specific AMPK-β2 residues inactivate AMPK may facilitate the design of additional isoform-specific AMPK allosteric inhibitors.

13.3.8 JJO1

By screening the Diverse Compound Library Set 1 from the National Cancer Institute using recombinant AMPK-α1β1γ1, Scott et al. identified a bi-quinoline compound JJO-1 (Fig. 13.2) as a direct AMPK activator but only at low ATP concentrations (Scott et al. 2014). This compound activates all possible AMPK-αβγ combinations except those containing the AMPK-γ3 isoform. Unlike A-769662, JJO-1 has no effect on the protection against AMPK-α Thr172 dephosphorylation by protein phosphatases and does not synergize with AMP on the allosteric activation of AMPK-α1β1γ1. Further functional analyses revealed that the mechanism of action of JJO-1 is distinct from the various mechanisms demonstrated to date for the abovementioned small-molecule activators. Intriguingly, JJO-1 allosterically activates AMPK independently of all identified effector domains, AMPK-αAID, AMPK-β CBM, and AMPK-γ nucleotide-binding sites (Scott et al. 2015). Further studies are required to gain insight into the mechanisms and the alternate allosteric site by which this small molecule activates AMPK.

13.4 Future Directions

13.4.1 Combinations of Direct and Indirect AMPK Activators

As described above, A-769662 and AMP can activate AMPK complexes by distinct molecular mechanisms through binding at distinct allosteric binding sites on the AMPK heterotrimeric complex (Calabrese et al. 2014). Interestingly, when A-769662 is combined with AMP, a synergistic allosteric activation of "naïve" AMPK complexes that are not phosphorylated is observed (Scott et al. 2014; Viollet et al. 2014). These data reveal that AMPK can be activated via a purely allosteric mechanism, bypassing the requirement of AMPK-α Thr172 and AMPK-β Ser108 phosphorylation, considered as major steps for AMPK activity and A-769662-mediated AMPK activation, respectively. These findings are highly relevant from a pharmaceutical point of view for the development of combined treatment of AMPK agonists in pathophysiological circumstances where AMPK-α Thr172 (and/or AMPK-β Ser108) phosphorylation is altered (Scott et al. 2014; Viollet et al. 2014). The demonstration of synergism between activating binding sites in intact cells has been first provided on enhanced AMPK activation and phosphorylation of downstream targets by the combination of A-769662 with AMP-dependent indirect AMPK activators such as AICAR, phenformin, or oligomycin (Ducommun et al. 2014; Foretz et al. 2010; Timmermans et al. 2014). Consistent with enhanced AMPK activation by co-treatment, inhibition of hepatic lipogenesis (Ducommun et al. 2014) and activation of cardiac glucose transport (Timmermans et al. 2014) were greatly improved compared with A-769662 alone. Similarly, salicylate [which binds to the A-769662 site (Hawley et al. 2012)] and metformin [which increases cellular AMP (Foretz et al. 2010)] act synergistically to activate AMPK and inhibit lipogenesis in primary mouse and human hepatocytes as well as in prostate and lung cancer cells (O'Brien et al. 2015; Ford et al. 2015). This mechanism of synergy may be also exploited for compounds that bind at the A-769662 site and activate AMPK-β2 complexes. Thus, these results reinforce the view that combinatorial treatments would be of value to enhance AMPK activation in patients and could help to reduce the amount of drugs administrated for a better tolerability and efficacy.

13.4.2 Identification of New Small-Molecule AMPK Activators

AMPK still holds promise as pharmacological target for various metabolic and non-metabolic diseases (Dasgupta and Chhipa 2016; Fullerton et al. 2013; Ruderman et al. 2013; Salminen et al. 2011; Wang et al. 2012; Viollet et al. 2006), emphasizing the continuous need for identification and development of novel direct and heterotrimer/cell/tissue-specific AMPK modulators. Recently, new molecules, such as CNX-012-570 (Anil et al. 2014), HL156A (Ju et al. 2016), and YLF-466D (Liu et al. 2015), were added to the currently available portfolio of AMPK activators, but the mechanism(s) by which they modulate the kinase activity still remain elusive and would require further studies. The elucidation of the molecular mechanisms of action of these molecules and of the newly discovered ones might allow structural refining and, as a virtuous cycle, ultimately leads to even more potent and specific AMPK activators (or inhibitors). Among the pioneering tools that can contribute to fasten the identification of new leads, computer-based drug design seems to offer a number of opportunities by permitting the virtual screening of theoretically infinite numbers of hypothetical molecules for isoform-specific small-molecule AMPK activators and next by prioritizing the selection of the compounds that are the most suitable for further in vitro and in vivo testing (Miglianico et al. 2016). Importantly, chronic and sustained tissue nonselective AMPK activation by means of constitutive activating mutation in the $\gamma 2$ AMPK subunit has been recently reported to impair whole-body metabolic homeostasis in mice (Yavari et al. 2016), emphasizing the critical requirement for both tissue-specific and reversible pharmacological AMPK activation by small-molecule agonists in the framework of strategies seeking to target the kinase, notably for the treatment of metabolic disorders.

Duality of Interest The authors declare that they have no conflict of interest.

References

Anderson SN, Cool BL, Kifle L, Chiou W, Egan DA, Barrett LW, Richardson PL, Frevert EU, Warrior U, Kofron JL, Burns DJ (2004) Microarrayed compound screening (microARCS) to identify activators and inhibitors of AMP-activated protein kinase. J Biomol Screen 9:112–121

Andris F, Leo O (2015) AMPK in lymphocyte metabolism and function. Int Rev Immunol 34:67–81

Anil TM, Harish C, Lakshmi MN, Harsha K, Onkaramurthy M, Sathish Kumar V, Shree N, Geetha V, Balamurali GV, Gopala AS, Madhusudhan Reddy B, Govind MK, Anup MO, Moolemath Y, Venkataranganna MV, Jagannath MR, Somesh BP (2014) CNX-012-570, a direct AMPK activator provides strong glycemic and lipid control along with significant reduction in body weight; studies from both diet-induced obese mice and db/db mice models. Cardiovasc Diabetol 13:27

Baur JA, Pearson KJ, Price NL, Jamieson HA, Lerin C, Kalra A, Prabhu VV, Allard JS, Lopez-Lluch G, Lewis K, Pistell PJ, Poosala S, Becker KG, Boss O, Gwinn D, Wang M, Ramaswamy S, Fishbein KW, Spencer RG, Lakatta EG, Le Couteur D, Shaw RJ, Navas P, Puigserver P, Ingram DK, de Cabo R, Sinclair DA (2006) Resveratrol improves health and survival of mice on a high-calorie diet. Nature 444:337–342

Benziane B, Bjornholm M, Lantier L, Viollet B, Zierath JR, Chibalin AV (2009) AMP-activated protein kinase activator A-769662 is an inhibitor of the Na(+)-K(+)-ATPase. Am J Physiol Cell Physiol 297:C1554–C1566

Bieri M, Mobbs JI, Koay A, Louey G, Mok YF, Hatters DM, Park JT, Park KH, Neumann D, Stapleton D, Gooley PR (2012) AMP-activated protein kinase beta-subunit requires internal motion for optimal carbohydrate binding. Biophys J 102:305–314

Bijland S, Mancini SJ, Salt IP (2013) Role of AMP-activated protein kinase in adipose tissue metabolism and inflammation. Clin Sci (Lond) 124:491–507

Birk JB, Wojtaszewski JF (2006) Predominant alpha2/beta2/gamma3 AMPK activation during exercise in human skeletal muscle. J Physiol 577:1021–1032

Breen DM, Sanli T, Giacca A, Tsiani E (2008) Stimulation of muscle cell glucose uptake by resveratrol through sirtuins and AMPK. Biochem Biophys Res Commun 374:117–122

Bullen JW, Balsbaugh JL, Chanda D, Shabanowitz J, Hunt DF, Neumann D, Hart GW (2014) Cross-talk between two essential nutrient-sensitive enzymes: O-GlcNAc transferase (OGT) and AMP-activated protein kinase (AMPK). J Biol Chem 289:10592–10606

Cai H, Scott E, Kholghi A, Andreadi C, Rufini A, Karmokar A, Britton RG, Horner-Glister E, Greaves P, Jawad D, James M, Howells L, Ognibene T, Malfatti M, Goldring C, Kitteringham N, Walsh J, Viskaduraki M, West K, Miller A, Hemingway D, Steward WP, Gescher AJ, Brown K (2015) Cancer chemoprevention: Evidence of a nonlinear dose response for the protective effects of resveratrol in humans and mice. Sci Transl Med 7:298ra117

Calabrese MF, Rajamohan F, Harris MS, Caspers NL, Magyar R, Withka JM, Wang H, Borzilleri KA, Sahasrabudhe PV, Hoth LR, Geoghegan KF, Han S, Brown J, Subashi TA, Reyes AR, Frisbie RK, Ward J, Miller RA, Landro JA, Londregan AT, Carpino PA, Cabral S, Smith AC, Conn EL, Cameron KO, Qiu X, Kurumbail RG (2014) Structural basis for AMPK activation: natural and synthetic ligands regulate kinase activity from opposite poles by different molecular mechanisms. Structure 22:1161–1172

Cameron AR, Logie L, Patel K, Bacon S, Forteath C, Harthill J, Roberts A, Sutherland C, Stewart D, Viollet B, Sakamoto K, McDougall G, Foretz M, Rena G (2016) Investigation of salicylate hepatic responses in comparison with chemical analogues of the drug. Acta Biochim Biophys 1862(8):1412–1422

Cameron KO, Kung DW, Kalgutkar AS, Kurumbail RG, Miller R, Salatto CT, Ward J, Withka JM, Bhattacharya SK, Boehm M, Borzilleri KA, Brown JA, Calabrese M, Caspers NL, Cokorinos E, Conn EL, Dowling MS, Edmonds DJ, Eng H, Fernando DP, Frisbie R, Hepworth D, Landro J, Mao Y, Rajamohan F, Reyes AR, Rose CR, Ryder T, Shavnya A, Smith AC, Tu M, Wolford AC, Xiao J (2016) Discovery and preclinical characterization of 6-Chloro-5-[4-(1-hydroxycyclobutyl)phenyl]-1H-indole-3-carboxylic Acid (PF-06409577), a direct activator of adenosine monophosphate-activated protein kinase (AMPK), for the potential treatment of diabetic nephropathy. J Med Chem (in press). doi:10.1021/acs.jmedchem.6b00866

Carling D, Mayer FV, Sanders MJ, Gamblin SJ (2011) AMP-activated protein kinase: nature's energy sensor. Nat Chem Biol 7:512–518

Carling D, Thornton C, Woods A, Sanders MJ (2012) AMP-activated protein kinase: new regulation, new roles? Biochem J 445:11–27

Chan AY, Dolinsky VW, Soltys CL, Viollet B, Baksh S, Light PE, Dyck JR (2008) Resveratrol inhibits cardiac hypertrophy via AMP-activated protein kinase and Akt. J Biol Chem 283:24194–24201

Chen L, Xin FJ, Wang J, Hu J, Zhang YY, Wan S, Cao LS, Lu C, Li P, Yan SF, Neumann D, Schlattner U, Xia B, Wang ZX, Wu JW (2013) Conserved regulatory elements in AMPK. Nature 498:E8–E10

Cool B, Zinker B, Chiou W, Kifle L, Cao N, Perham M, Dickinson R, Adler A, Gagne G, Iyengar R, Zhao G, Marsh K, Kym P, Jung P, Camp HS, Frevert E (2006) Identification and characterization of a small molecule AMPK activator that treats key components of type 2 diabetes and the metabolic syndrome. Cell Metab 3:403–416

Corton JM, Gillespie JG, Hawley SA, Hardie DG (1995) 5-aminoimidazole-4-carboxamide ribonucleoside. A specific method for activating AMP-activated protein kinase in intact cells? Eur J Biochem 229:558–565

Dasgupta B, Chhipa RR (2016) Evolving lessons on the complex role of AMPK in normal physiology and cancer. Trends Pharmacol Sci 37:192–206

Dasgupta B, Milbrandt J (2007) Resveratrol stimulates AMP kinase activity in neurons. Proc Natl Acad Sci USA 104:7217–7222

Daskalopoulos EP, Dufeys C, Bertrand L, Beauloye C, Horman S (2016) AMPK in cardiac fibrosis and repair: actions beyond metabolic regulation. J Mol Cell Cardiol 91:188–200

Ducommun S, Ford RJ, Bultot L, Deak M, Bertrand L, Kemp BE, Steinberg GR, Sakamoto K (2014) Enhanced activation of cellular AMPK by dual-small molecule treatment: AICAR and A769662. Am J Physiol Endocrinol Metab 306:E688–E696

El-Mir MY, Nogueira V, Fontaine E, Averet N, Rigoulet M, Leverve X (2000) Dimethylbiguanide inhibits cell respiration via an indirect effect targeted on the respiratory chain complex I. J Biol Chem 275:223–228

Ford RJ, Fullerton MD, Pinkosky SL, Day EA, Scott JW, Oakhill JS, Bujak AL, Smith BK, Crane JD, Blumer RM, Marcinko K, Kemp BE, Gerstein HC, Steinberg GR (2015) Metformin and salicylate synergistically activate liver AMPK, inhibit lipogenesis and improve insulin sensitivity. Biochem J 468:125–132

Foretz M, Hebrard S, Leclerc J, Zarrinpashneh E, Soty M, Mithieux G, Sakamoto K, Andreelli F, Viollet B (2010) Metformin inhibits hepatic gluconeogenesis in mice independently of the LKB1/AMPK pathway via a decrease in hepatic energy state. J Clin Invest 120:2355–2369

Foretz M, Guigas B, Bertrand L, Pollak M, Viollet B (2014) Metformin: from mechanisms of action to therapies. Cell Metab 20:953–966

Fullerton MD, Steinberg GR, Schertzer JD (2013) Immunometabolism of AMPK in insulin resistance and atherosclerosis. Mol Cell Endocrinol 366:224–234

Fullerton MD, Ford RJ, McGregor CP, LeBlond ND, Snider SA, Stypa SA, Day EA, Lhotak S, Schertzer JD, Austin RC, Kemp BE, Steinberg GR (2015) Salicylate improves macrophage cholesterol homeostasis via activation of Ampk. J Lipid Res 56:1025–1033

Giordanetto F, Karis D (2012) Direct AMP-activated protein kinase activators: a review of evidence from the patent literature. Expert Opin Ther Pat 22:1467–1477

Gomez-Galeno JE, Dang Q, Nguyen TH, Boyer SH, Grote MP, Sun Z, Chen M, Craigo WA, van Poelje PD, MacKenna DA, Cable EE, Rolzin PA, Finn PD, Chi B, Linemeyer DL, Hecker SJ, Erion MD (2010) A potent and selective AMPK activator that inhibits de novo lipogenesis. ACS Med Chem Lett 1:478–482

Goransson O, McBride A, Hawley SA, Ross FA, Shpiro N, Foretz M, Viollet B, Hardie DG, Sakamoto K (2007) Mechanism of action of A-769662, a valuable tool for activation of AMP-activated protein kinase. J Biol Chem 282:32549–32560

Gowans GJ, Hawley SA, Ross FA, Hardie DG (2013) AMP is a true physiological regulator of AMP-activated protein kinase by both allosteric activation and enhancing net phosphorylation. Cell Metab 18:556–566

Guigas B, Bertrand L, Taleux N, Foretz M, Wiernsperger N, Vertommen D, Andreelli F, Viollet B, Hue L (2006) 5-Aminoimidazole-4-carboxamide-1-beta-D-ribofuranoside and metformin inhibit hepatic glucose phosphorylation by an AMP-activated protein kinase-independent effect on glucokinase translocation. Diabetes 55:865–874

Guigas B, Taleux N, Foretz M, Detaille D, Andreelli F, Viollet B, Hue L (2007) AMP-activated protein kinase-independent inhibition of hepatic mitochondrial oxidative phosphorylation by AICA riboside. Biochem J 404:499–507

Guigas B, Sakamoto K, Taleux N, Reyna SM, Musi N, Viollet B, Hue L (2009) Beyond AICA riboside: in search of new specific AMP-activated protein kinase activators. IUBMB Life 61:18–26

Hardie DG (2014a) AMP-activated protein kinase: maintaining energy homeostasis at the cellular and whole-body levels. Annu Rev Nutr 34:31–55

Hardie DG (2014b) AMPK—sensing energy while talking to other signaling pathways. Cell Metab 20:939–952

Hardie DG, Schaffer BE, Brunet A (2016) AMPK: an energy-sensing pathway with multiple inputs and outputs. Trends Cell Biol 26:190–201

Hawley SA, Ross FA, Chevtzoff C, Green KA, Evans A, Fogarty S, Towler MC, Brown LJ, Ogunbayo OA, Evans AM, Hardie DG (2010) Use of cells expressing gamma subunit variants to identify diverse mechanisms of AMPK activation. Cell Metab 11:554–565

Hawley SA, Fullerton MD, Ross FA, Schertzer JD, Chevtzoff C, Walker KJ, Peggie MW, Zibrova D, Green KA, Mustard KJ, Kemp BE, Sakamoto K, Steinberg GR, Hardie DG

(2012) The ancient drug salicylate directly activates AMP-activated protein kinase. Science 336:918–922

Hedner T, Everts B (1998) The early clinical history of salicylates in rheumatology and pain. Clin Rheumatol 17:17–25

Henin N, Vincent MF, Gruber HE, Van den Berghe G (1995) Inhibition of fatty acid and cholesterol synthesis by stimulation of AMP-activated protein kinase. FASEB J 9:541–546

Higgs GA, Salmon JA, Henderson B, Vane JR (1987) Pharmacokinetics of aspirin and salicylate in relation to inhibition of arachidonate cyclooxygenase and antiinflammatory activity. Proc Natl Acad Sci USA 84:1417–1420

Hinke SA, Martens GA, Cai Y, Finsi J, Heimberg H, Pipeleers D, Van de Casteele M (2007) Methyl succinate antagonises biguanide-induced AMPK-activation and death of pancreatic beta-cells through restoration of mitochondrial electron transfer. Br J Pharmacol 150:1031–1043

Hunter RW, Foretz M, Bultot L, Fullerton MD, Deak M, Ross FA, Hawley SA, Shpiro N, Viollet B, Barron D, Kemp BE, Steinberg GR, Hardie DG, Sakamoto K (2014) Mechanism of action of compound-13: an alpha1-selective small molecule activator of AMPK. Chem Biol 21:866–879

Jenkins Y, Sun TQ, Markovtsov V, Foretz M, Li W, Nguyen H, Li Y, Pan A, Uy G, Gross L, Baltgalvis K, Yung SL, Gururaja T, Kinoshita T, Owyang A, Smith IJ, McCaughey K, White K, Godinez G, Alcantara R, Choy C, Ren H, Basile R, Sweeny DJ, Xu X, Issakani SD, Carroll DC, Goff DA, Shaw SJ, Singh R, Boros LG, Laplante MA, Marcotte B, Kohen R, Viollet B, Marette A, Payan DG, Kinsella TM, Hitoshi Y (2013) AMPK activation through mitochondrial regulation results in increased substrate oxidation and improved metabolic parameters in models of diabetes. PLoS One 8:e81870

Jensen TE, Ross FA, Kleinert M, Sylow L, Knudsen JR, Gowans GJ, Hardie DG, Richter EA (2015) PT-1 selectively activates AMPK-gamma1 complexes in mouse skeletal muscle, but activates all three gamma subunit complexes in cultured human cells by inhibiting the respiratory chain. Biochem J 467:461–472

Ju KD, Kim HJ, Tsogbadrakh B, Lee J, Ryu H, Cho EJ, Hwang YH, Kim K, Yang J, Ahn C, Oh KH (2016) HL156A, a novel AMP-activated protein kinase activator, is protective against peritoneal fibrosis in an in vivo and in vitro model of peritoneal fibrosis. Am J Physiol Renal Physiol 310:F342–F350

Kim AS, Miller EJ, Wright TM, Li J, Qi D, Atsina K, Zaha V, Sakamoto K, Young LH (2011) A small molecule AMPK activator protects the heart against ischemia-reperfusion injury. J Mol Cell Cardiol 51:24–32

Kim J, Shin J, Ha J (2015) Screening methods for AMP-activated protein kinase modulators: a patent review. Expert Opin Ther Pat 25:261–277

Koay A, Woodcroft B, Petrie EJ, Yue H, Emanuelle S, Bieri M, Bailey MF, Hargreaves M, Park JT, Park KH, Ralph S, Neumann D, Stapleton D, Gooley PR (2010) AMPK beta subunits display isoform specific affinities for carbohydrates. FEBS Lett 584:3499–3503

Kulkarni SS, Canto C (2015) The molecular targets of resveratrol. Biochim Biophys Acta 1852:1114–1123

Lai YC, Kviklyte S, Vertommen D, Lantier L, Foretz M, Viollet B, Hallen S, Rider MH (2014) A small-molecule benzimidazole derivative that potently activates AMPK to increase glucose transport in skeletal muscle: comparison with effects of contraction and other AMPK activators. Biochem J 460:363–375

Landgraf RR, Goswami D, Rajamohan F, Harris MS, Calabrese MF, Hoth LR, Magyar R, Pascal BD, Chalmers MJ, Busby SA, Kurumbail RG, Griffin PR (2013) Activation of AMP-activated protein kinase revealed by hydrogen/deuterium exchange mass spectrometry. Structure 21:1942–1953

Langendorf CG, Ngoei KR, Scott JW, Ling NX, Issa SM, Gorman MA, Parker MW, Sakamoto K, Oakhill JS, Kemp BE (2016) Structural basis of allosteric and synergistic activation of AMPK by furan-2-phosphonic derivative C2 binding. Nat Commun 7:10912

Leverve XM, Guigas B, Detaille D, Batandier C, Koceir EA, Chauvin C, Fontaine E, Wiernsperger NF (2003) Mitochondrial metabolism and type-2 diabetes: a specific target of metformin. Diabetes Metab 29:6S88–6S94

Li YY, Yu LF, Zhang LN, Qiu BY, Su MB, Wu F, Chen DK, Pang T, Gu M, Zhang W, Ma WP, Jiang HW, Li JY, Nan FJ, Li J (2013) Novel small-molecule AMPK activator orally exerts beneficial effects on diabetic db/db mice. Toxicol Appl Pharmacol 273:325–334

Li X, Wang L, Zhou XE, Ke J, de Waal PW, Gu X, Tan MH, Wang D, Wu D, Xu HE, Melcher K (2015) Structural basis of AMPK regulation by adenine nucleotides and glycogen. Cell Res 25:50–66

Liang J, Xu ZX, Ding Z, Lu Y, Yu Q, Werle KD, Zhou G, Park YY, Peng G, Gambello MJ, Mills GB (2015) Myristoylation confers noncanonical AMPK functions in autophagy selectivity and mitochondrial surveillance. Nat Commun 6:7926

Lim EG, Kim GT, Lee SH, Kim SY, Kim YM (2016) Apoptotic effects of extract from Cnidium monnieri (L.) Cusson by adenosine monosphosphate-activated protein kinase-independent pathway in HCT116 colon cancer cells. Mol Med Rep 13:4681–4688

Lin YY, Kiihl S, Suhail Y, Liu SY, Chou YH, Kuang Z, Lu JY, Khor CN, Lin CL, Bader JS, Irizarry R, Boeke JD (2012) Functional dissection of lysine deacetylases reveals that HDAC1 and p300 regulate AMPK. Nature 482:251–255

Liu X, Chhipa RR, Pooya S, Wortman M, Yachyshin S, Chow LM, Kumar A, Zhou X, Sun Y, Quinn B, McPherson C, Warnick RE, Kendler A, Giri S, Poels J, Norga K, Viollet B, Grabowski GA, Dasgupta B (2014) Discrete mechanisms of mTOR and cell cycle regulation by AMPK agonists independent of AMPK. Proc Natl Acad Sci USA 111:E435–E444

Liu Y, Park JM, Oh SJ, Chang KH, Lee MY (2015) Antiplatelet effect of a newly developed AMP-activated protein kinase activator YLF-466D. Eur J Pharmacol 760:81–87

Marcinko K, Steinberg GR (2014) The role of AMPK in controlling metabolism and mitochondrial biogenesis during exercise. Exp Physiol 99:1581–1585

Marcinko K, Bujak AL, Lally JS, Ford RJ, Wong TH, Smith BK, Kemp BE, Jenkins Y, Li W, Kinsella TM, Hitoshi Y, Steinberg GR (2015) The AMPK activator R419 improves exercise capacity and skeletal muscle insulin sensitivity in obese mice. Mol Metab 4:643–651

Miglianico M, Nicolaes GA, Neumann D (2016) Pharmacological targeting of AMP-activated protein kinase and opportunities for computer-aided drug design. J Med Chem 59:2879–2893

Miller RA, Chu Q, Xie J, Foretz M, Viollet B, Birnbaum MJ (2013) Biguanides suppress hepatic glucagon signalling by decreasing production of cyclic AMP. Nature 494:256–260

Morales DR, Morris AD (2015) Metformin in cancer treatment and prevention. Annu Rev Med 66:17–29

Moreno D, Knecht E, Viollet B, Sanz P (2008) A769662, a novel activator of AMP-activated protein kinase, inhibits non-proteolytic components of the 26S proteasome by an AMPK-independent mechanism. FEBS Lett 582:2650–2654

Mounier R, Theret M, Lantier L, Foretz M, Viollet B (2015) Expanding roles for AMPK in skeletal muscle plasticity. Trends Endocrinol Metab 26:275–286

O'Brien AJ, Villani LA, Broadfield LA, Houde VP, Galic S, Blandino G, Kemp BE, Tsakiridis T, Muti P, Steinberg GR (2015) Salicylate activates AMPK and synergizes with metformin to reduce the survival of prostate and lung cancer cells ex vivo through inhibition of de novo lipogenesis. Biochem J 469:177–187

O'Neill LA, Hardie DG (2013) Metabolism of inflammation limited by AMPK and pseudo-starvation. Nature 493:346–355

Oakhill JS, Chen ZP, Scott JW, Steel R, Castelli LA, Ling N, Macaulay SL, Kemp BE (2010) Beta-Subunit myristoylation is the gatekeeper for initiating metabolic stress sensing by AMP-activated protein kinase (AMPK). Proc Natl Acad Sci USA 107:19237–19241

Oakhill JS, Steel R, Chen ZP, Scott JW, Ling N, Tam S, Kemp BE (2011) AMPK is a direct adenylate charge-regulated protein kinase. Science 332:1433–1435

Oakhill JS, Scott JW, Kemp BE (2012) AMPK functions as an adenylate charge-regulated protein kinase. Trends Endocrinol Metab 23:125–132

Oligschlaeger Y, Miglianico M, Chanda D, Scholz R, Thali RF, Tuerk R, Stapleton DI, Gooley PR, Neumann D (2015) The recruitment of AMP-activated protein kinase to glycogen is regulated by autophosphorylation. J Biol Chem 290:11715–11728

Owen MR, Doran E, Halestrap AP (2000) Evidence that metformin exerts its anti-diabetic effects through inhibition of complex 1 of the mitochondrial respiratory chain. Biochem J 348 (Pt 3):607–614

Pang T, Zhang ZS, Gu M, Qiu BY, Yu LF, Cao PR, Shao W, Su MB, Li JY, Nan FJ, Li J (2008) Small molecule antagonizes autoinhibition and activates AMP-activated protein kinase in cells. J Biol Chem 283:16051–16060

Park CE, Kim MJ, Lee JH, Min BI, Bae H, Choe W, Kim SS, Ha J (2007) Resveratrol stimulates glucose transport in C2C12 myotubes by activating AMP-activated protein kinase. Exp Mol Med 39:222–229

Polekhina G, Gupta A, van Denderen BJ, Feil SC, Kemp BE, Stapleton D, Parker MW (2005) Structural basis for glycogen recognition by AMP-activated protein kinase. Structure 13:1453–1462

Qi D, Young LH (2015) AMPK: energy sensor and survival mechanism in the ischemic heart. Trends Endocrinol Metab 26:422–429

Rajamohan F, Reyes AR, Frisbie RK, Hoth LR, Sahasrabudhe P, Magyar R, Landro JA, Withka JM, Caspers NL, Calabrese MF, Ward J, Kurumbail RG (2016) Probing the enzyme kinetics, allosteric modulation and activation of alpha1- and alpha2-subunit-containing AMP-activated protein kinase (AMPK) heterotrimeric complexes by pharmacological and physiological activators. Biochem J 473:581–592

Rana S, Blowers EC, Natarajan A (2015) Small molecule adenosine 5′-monophosphate activated protein kinase (AMPK) modulators and human diseases. J Med Chem 58:2–29

Rao E, Zhang Y, Li Q, Hao J, Egilmez NK, Suttles J, Li B (2016) AMPK-dependent and independent effects of AICAR and compound C on T-cell responses. Oncotarget, (in press). doi:10.18632/oncotarget.9277

Ross FA, Jensen TE, Hardie DG (2016) Differential regulation by AMP and ADP of AMPK complexes containing different gamma subunit isoforms. Biochem J 473:189–199

Rubio T, Vernia S, Sanz P (2013) Sumoylation of AMPKbeta2 subunit enhances AMP-activated protein kinase activity. Mol Biol Cell 24(1801–11):S1–S4

Ruderman NB, Carling D, Prentki M, Cacicedo JM (2013) AMPK, insulin resistance, and the metabolic syndrome. J Clin Invest 123:2764–2772

Salminen A, Kaarniranta K, Haapasalo A, Soininen H, Hiltunen M (2011) AMP-activated protein kinase: a potential player in Alzheimer's disease. J Neurochem 118:460–474

Sanders MJ, Ali ZS, Hegarty BD, Heath R, Snowden MA, Carling D (2007a) Defining the mechanism of activation of AMP-activated protein kinase by the small molecule A-769662, a member of the thienopyridone family. J Biol Chem 282:32539–32548

Sanders MJ, Grondin PO, Hegarty BD, Snowden MA, Carling D (2007b) Investigating the mechanism for AMP activation of the AMP-activated protein kinase cascade. Biochem J 403:139–148

Scott JW, van Denderen BJ, Jorgensen SB, Honeyman JE, Steinberg GR, Oakhill JS, Iseli TJ, Koay A, Gooley PR, Stapleton D, Kemp BE (2008) Thienopyridone drugs are selective activators of AMP-activated protein kinase beta1-containing complexes. Chem Biol 15:1220–1230

Scott JW, Ling N, Issa SM, Dite TA, O'Brien MT, Chen ZP, Galic S, Langendorf CG, Steinberg GR, Kemp BE, Oakhill JS (2014) Small molecule drug A-769662 and AMP synergistically activate naive AMPK independent of upstream kinase signaling. Chem Biol 21:619–627

Scott JW, Galic S, Graham KL, Foitzik R, Ling NX, Dite TA, Issa SM, Langendorf CG, Weng QP, Thomas HE, Kay TW, Birnberg NC, Steinberg GR, Kemp BE, Oakhill JS (2015) Inhibition of AMP-activated protein kinase at the allosteric drug-binding site promotes islet insulin release. Chem Biol 22:705–711

Sinnett SE, Brenman JE (2014) Past strategies and future directions for identifying AMP-activated protein kinase (AMPK) modulators. Pharmacol Ther 143:111–118
Steinberg GR, Kemp BE (2009) AMPK in health and disease. Physiol Rev 89:1025–1078
Stephenne X, Foretz M, Taleux N, van der Zon GC, Sokal E, Hue L, Viollet B, Guigas B (2011) Metformin activates AMP-activated protein kinase in primary human hepatocytes by decreasing cellular energy status. Diabetologia 54:3101–3110
Sullivan JE, Carey F, Carling D, Beri RK (1994) Characterisation of 5′-AMP-activated protein kinase in human liver using specific peptide substrates and the effects of 5′-AMP analogues on enzyme activity. Biochem Biophys Res Commun 200:1551–1556
Timmermans AD, Balteau M, Gelinas R, Renguet E, Ginion A, de Meester C, Sakamoto K, Balligand JL, Bontemps F, Vanoverschelde JL, Horman S, Beauloye C, Bertrand L (2014) A-769662 potentiates the effect of other AMP-activated protein kinase activators on cardiac glucose uptake. Am J Physiol Heart Circ Physiol 306:H1619–H1630
Timmers S, Konings E, Bilet L, Houtkooper RH, van de Weijer T, Goossens GH, Hoeks J, van der Krieken S, Ryu D, Kersten S, Moonen-Kornips E, Hesselink MK, Kunz I, Schrauwen-Hinderling VB, Blaak EE, Auwerx J, Schrauwen P (2011) Calorie restriction-like effects of 30 days of resveratrol supplementation on energy metabolism and metabolic profile in obese humans. Cell Metab 14:612–622
Treebak JT, Birk JB, Hansen BF, Olsen GS, Wojtaszewski JF (2009) A-769662 activates AMPK beta1-containing complexes but induces glucose uptake through a PI3-kinase-dependent pathway in mouse skeletal muscle. Am J Physiol Cell Physiol 297:C1041–C1052
Turner N, Li JY, Gosby A, To SW, Cheng Z, Miyoshi H, Taketo MM, Cooney GJ, Kraegen EW, James DE, Hu LH, Li J, Ye JM (2008) Berberine and its more biologically available derivative, dihydroberberine, inhibit mitochondrial respiratory complex I: a mechanism for the action of berberine to activate AMP-activated protein kinase and improve insulin action. Diabetes 57:1414–1418
Um JH, Park SJ, Kang H, Yang S, Foretz M, McBurney MW, Kim MK, Viollet B, Chung JH (2010) AMP-activated protein kinase-deficient mice are resistant to the metabolic effects of resveratrol. Diabetes 59:554–563
van Dam AD, Nahon KJ, Kooijman S, van den Berg SM, Kanhai AA, Kikuchi T, Heemskerk MM, van Harmelen V, Lombes M, van den Hoek AM, de Winther MP, Lutgens E, Guigas B, Rensen PC, Boon MR (2015) Salsalate activates brown adipose tissue in mice. Diabetes 64:1544–1554
Vincent EE, Coelho PP, Blagih J, Griss T, Viollet B, Jones RG (2015) Differential effects of AMPK agonists on cell growth and metabolism. Oncogene 34:3627–3639
Vingtdeux V, Giliberto L, Zhao H, Chandakkar P, Wu Q, Simon JE, Janle EM, Lobo J, Ferruzzi MG, Davies P, Marambaud P (2010) AMP-activated protein kinase signaling activation by resveratrol modulates amyloid-beta peptide metabolism. J Biol Chem 285:9100–9113
Vingtdeux V, Chandakkar P, Zhao H, d'Abramo C, Davies P, Marambaud P (2011) Novel synthetic small-molecule activators of AMPK as enhancers of autophagy and amyloid-beta peptide degradation. FASEB J 25:219–231
Viollet B, Foretz M, Guigas B, Horman S, Dentin R, Bertrand L, Hue L, Andreelli F (2006) Activation of AMP-activated protein kinase in the liver: a new strategy for the management of metabolic hepatic disorders. J Physiol 574:41–53
Viollet B, Foretz M, Schlattner U (2014) Bypassing AMPK phosphorylation. Chem Biol 21:567–569
Wang S, Song P, Zou MH (2012) AMP-activated protein kinase, stress responses and cardiovascular diseases. Clin Sci (Lond) 122:555–573
Wang S, Liang X, Yang Q, Fu X, Rogers CJ, Zhu M, Rodgers BD, Jiang Q, Dodson MV, Du M (2015) Resveratrol induces brown-like adipocyte formation in white fat through activation of AMP-activated protein kinase (AMPK) alpha1. Int J Obes (Lond) 39:967–976
Warden SM, Richardson C, O'Donnell J Jr, Stapleton D, Kemp BE, Witters LA (2001) Post-translational modifications of the beta-1 subunit of AMP-activated protein kinase affect enzyme activity and cellular localization. Biochem J 354:275–283

Witters LA (2001) The blooming of the French lilac. J Clin Invest 108:1105–1107

Wu J, Puppala D, Feng X, Monetti M, Lapworth AL, Geoghegan KF (2013) Chemoproteomic analysis of intertissue and interspecies isoform diversity of AMP-activated protein kinase (AMPK). J Biol Chem 288:35904–35912

Xiao B, Heath R, Saiu P, Leiper FC, Leone P, Jing C, Walker PA, Haire L, Eccleston JF, Davis CT, Martin SR, Carling D, Gamblin SJ (2007) Structural basis for AMP binding to mammalian AMP-activated protein kinase. Nature 449:496–500

Xiao B, Sanders MJ, Underwood E, Heath R, Mayer FV, Carmena D, Jing C, Walker PA, Eccleston JF, Haire LF, Saiu P, Howell SA, Aasland R, Martin SR, Carling D, Gamblin SJ (2011) Structure of mammalian AMPK and its regulation by ADP. Nature 472:230–233

Xiao B, Sanders MJ, Carmena D, Bright NJ, Haire LF, Underwood E, Patel BR, Heath RB, Walker PA, Hallen S, Giordanetto F, Martin SR, Carling D, Gamblin SJ (2013) Structural basis of AMPK regulation by small molecule activators. Nat Commun 4:3017

Xu Q, Hao X, Yang Q, Si L (2009) Resveratrol prevents hyperglycemia-induced endothelial dysfunction via activation of adenosine monophosphate-activated protein kinase. Biochem Biophys Res Commun 388:389–394

Yan Y, Ollila S, Wong IP, Vallenius T, Palvimo JJ, Vaahtomeri K, Makela TP (2015) SUMOylation of AMPKalpha1 by PIAS4 specifically regulates mTORC1 signalling. Nat Commun 6:8979

Yavari A, Stocker CJ, Ghaffari S, Wargent ET, Steeples V, Czibik G, Pinter K, Bellahcene M, Woods A, Martinez de Morentin PB, Cansell C, Lam BY, Chuster A, Petkevicius K, Nguyen-Tu MS, Martinez-Sanchez A, Pullen TJ, Oliver PL, Stockenhuber A, Nguyen C, Lazdam M, O'Dowd JF, Harikumar P, Toth M, Beall C, Kyriakou T, Parnis J, Sarma D, Katritsis G, Wortmann DD, Harper AR, Brown LA, Willows R, Gandra S, Poncio V, de Oliveira Figueiredo MJ, Qi NR, Peirson SN, McCrimmon RJ, Gereben B, Tretter L, Fekete C, Redwood C, Yeo GS, Heisler LK, Rutter GA, Smith MA, Withers DJ, Carling D, Sternick EB, Arch JR, Cawthorne MA, Watkins H, Ashrafian H (2016) Chronic activation of gamma2 AMPK induces obesity and reduces beta cell function. Cell Metab 23:821–836

Yun H, Ha J (2011) AMP-activated protein kinase modulators: a patent review (2006–2010). Expert Opin Ther Pat 21:983–1005

Zadra G, Photopoulos C, Tyekucheva S, Heidari P, Weng QP, Fedele G, Liu H, Scaglia N, Priolo C, Sicinska E, Mahmood U, Signoretti S, Birnberg N, Loda M (2014) A novel direct activator of AMPK inhibits prostate cancer growth by blocking lipogenesis. EMBO Mol Med 6:519–538

Zheng J, Ramirez VD (2000) Inhibition of mitochondrial proton F0F1-ATPase/ATP synthase by polyphenolic phytochemicals. Br J Pharmacol 130:1115–1123

Zhou G, Myers R, Li Y, Chen Y, Shen X, Fenyk-Melody J, Wu M, Ventre J, Doebber T, Fujii N, Musi N, Hirshman MF, Goodyear LJ, Moller DE (2001) Role of AMP-activated protein kinase in mechanism of metformin action. J Clin Invest 108:1167–1174

Zou MH, Kirkpatrick SS, Davis BJ, Nelson JS, Wiles WG 4th, Schlattner U, Neumann D, Brownlee M, Freeman MB, Goldman MH (2004) Activation of the AMP-activated protein kinase by the anti-diabetic drug metformin in vivo. Role of mitochondrial reactive nitrogen species. J Biol Chem 279: 43940–43951

Part IV
AMPK Non-mammalian Systems

Chapter 14
AMPK in Yeast: The SNF1 (Sucrose Non-fermenting 1) Protein Kinase Complex

Pascual Sanz, Rosa Viana, and Maria Adelaida Garcia-Gimeno

Contents

Abstract In yeast, SNF1 protein kinase is the orthologue of mammalian AMPK complex. It is a trimeric complex composed of Snf1 protein kinase (orthologue of AMPKα catalytic subunit), Snf4 (orthologue of AMPKγ regulatory subunit),

P. Sanz (✉) • R. Viana
Instituto de Biomedicina de Valencia, CSIC and Centro de Investigación Biomédica en Red de Enfermedades Raras (CIBERER-ISCiii), Jaime Roig 11, 46010 Valencia, Spain
e-mail: sanz@ibv.csic.es

M.A. Garcia-Gimeno
Department of Biotecnología, Escuela Técnica Superior de Ingeniería Agronómica y del Medio Natural (ETSIAMN), Universitat Politécnica de Valencia, Valencia, Spain

M.D. Cordero, B. Viollet (eds.), *AMP-activated Protein Kinase*, Experientia Supplementum 107, DOI 10.1007/978-3-319-43589-3_14

and a member of the Gal83/Sip1/Sip2 family of proteins (orthologues of AMPKβ subunit) that act as scaffolds and also regulate the subcellular localization of the complex. In this chapter, we review the recent literature on the characteristics of SNF1 complex subunits, the structure and regulation of the activity of the SNF1 complex, its role at the level of transcriptional regulation of relevant target genes and also at the level of posttranslational modification of targeted substrates. We also review the crosstalk of SNF1 complex activity with other key protein kinase pathways such as cAMP–PKA, TORC1, and PAS kinase.

Keywords Snf1 • Snf4 • Gal83 • Sip1 • Sip2 • Glucose repression • Transcriptional regulation • Protein structure analysis • Response to cellular stress • Posttranslational regulation • Energy metabolism • Protein kinase

List of Abbreviations

AID	Autoinhibitory domain
ASC	Association with Snf1 complex
AMPK	AMP-activated protein kinase
CBS	Cystathionine-β-synthase domain
CSRE	Carbon source-responsive element
eIF2α	Eukaryotic initiation factor 2α
GBD	Glycogen-binding domain
KD	Catalytic domain
RD	Regulatory domain
Snf1	sucrose non-fermenting 1
TORC1	Target of rapamycin complex 1
PAS kinase	Protein kinase with a PAS domain
PKA	Protein kinase A

14.1 Introduction

The yeast *Saccharomyces cerevisiae* adapts its metabolism to the quality of the nutrients in the surrounding media. In the presence of preferred carbon sources (glucose), this yeast represses the transcription of a large number of genes, including those involved in the utilization of alternative carbon sources, gluconeogenesis, and respiration [see Santangelo (2006), Conrad et al. (2014), and Kayikci and Nielsen (2015) for review]. Biochemical and genetic studies have identified the Snf1 (sucrose non-fermenting 1) protein kinase as a master player in the glucose repression pathway (Celenza and Carlson 1984). Snf1 is a serine/threonine protein kinase that, in the absence of glucose, promotes respiratory metabolism, glycogen accumulation, and gluconeogenesis, by inhibiting transcriptional repressors (i.e., Mig1) or by stimulating transcriptional activators (i.e., Cat8 and Sip4). Snf1 is the yeast orthologue of the catalytic α-subunit of the mammalian AMP-activated protein kinase (AMPK) (Hardie et al. 1998) and works as a part of a heterotrimeric complex containing the regulatory subunit Snf4 (orthologue of mammalian AMPKγ subunit) and one member of the Sip1/Sip2/Gal83 family of proteins (orthologues of mammalian AMPKβ subunit) that act as scaffolds and also regulate the subcellular localization of the complex (Jiang and Carlson 1997; Vincent et al. 2001; Hedbacker and Carlson 2006). In this chapter, we name the heterotrimeric yeast complex as SNF1.

In addition to its role in the de-repression of genes controlled by glucose, SNF1 complex participates in a wide variety of physiological processes that range from meiosis and spore formation (Honigberg and Lee 1998), aging (Ashrafi et al. 2000; Yao et al. 2015), autophagy (Wang et al. 2001), filamentous and invasive growth (Cullen and Sprague 2000), and biofilm formation (Kuchin et al. 2002). It also participates in the response to environmental stresses, such as heat shock (Hahn and Thiele 2004), endoplasmic reticulum stress (Ferrer-Dalmau et al. 2015), alkaline pH (Casamayor et al. 2012), and salt stress (Hong and Carlson 2007). Since the function of yeast SNF1 and mammalian AMPK complexes is evolutionally conserved (Hardie et al. 1998), unraveling the function and regulation of yeast SNF1 complex has been very valuable to understand the same principles in its mammalian orthologue AMPK. In this chapter, we review the recent literature on the regulation of the activity of the SNF1 complex and its involvement in energy metabolism and environmental stresses.

14.2 The SNF1 Complex

14.2.1 Snf1 Catalytic Subunit

Snf1 protein kinase is the catalytic subunit of the SNF1 complex. It is a protein of 633 aa with a molecular weight of 72 kDa. It contains two different characteristic domains: an N-terminal catalytic kinase domain (KD), which spans from residue 55–306, and a C-terminal regulatory domain (RD), which spans from residue 392–633 (Fig. 14.1). In this Snf1-RD domain is where the rest of SNF1 subunits (Snf4 and Sip1/Sip2/Gal83) bind, the interaction of Sip1/Sip2/Gal83 being constitutive, but the binding of Snf4 is negatively regulated by glucose (see below).

The Snf1-KD is conserved throughout evolution, sharing a 47 % identity (61 % homology) with mammalian AMPKα subunit (Hardie et al. 1998). Its structure shares strong similarities to other protein kinases, with a small N-terminal lobe (with five β-sheets and one α-helix) and a large C-terminal lobe mostly helical (6 α-helices), the regulatory Thr210 residue (see below) being presumably located on the top of the cleft between the N-terminal and C-terminal lobes (Rudolph et al. 2005; Nayak et al. 2006).

The Snf1-RD is divergent (only 32 % homology with mammalian AMPKα subunit) (Hardie et al. 1998). This domain contains a region called AutoInhibitory Domain (AID; residues 392–488). Under high glucose conditions, the Snf1-KD interacts with AID, resulting in an autoinhibited inactive SNF1 complex. Under low

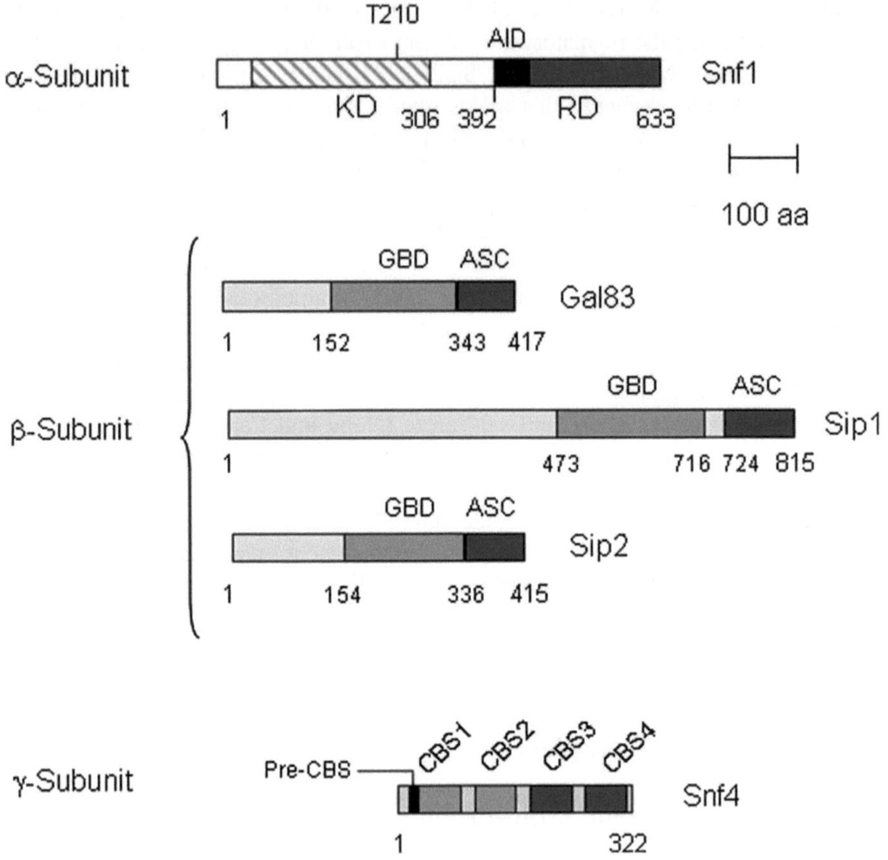

Fig. 14.1 Diagram of yeast SNF1 subunits. *KD* kinase domain, *RD* regulatory domain, *AID* autoinhibitory domain, *GBD* glycogen-binding domain, *ASC* association with Snf1 complex domain, *CBS* cystathionine-β-synthase domain. Numbers refer to the position of the corresponding amino acid residue

glucose conditions, Snf4 binds to the Snf1-RD domain forcing the release of the binding of the Snf1-KD domain from the AID, adopting Snf1 in this way an active open conformation (Jiang and Carlson 1996) (Fig. 14.2). Removal of the autoinhibitory domain in Snf1 renders a protein that does not require Snf4 for activation under low glucose conditions; thus, it is constitutively active. In addition, it has been described that mutations clustered near the catalytic site of Snf1 (L183I, K192R, Y167H, and I241N) reduce the interaction of Snf1-KD with the autoinhibitory domain (AID) and makes SNF1 complex independent of Snf4 for activation (Leech et al. 2003). All these results confirmed the autoinhibited conformation of Snf1 protein kinase in the presence of glucose.

Activation of SNF1 complex under low glucose conditions involves two steps, one that requires phosphorylation of the Snf1 catalytic subunit and another that is

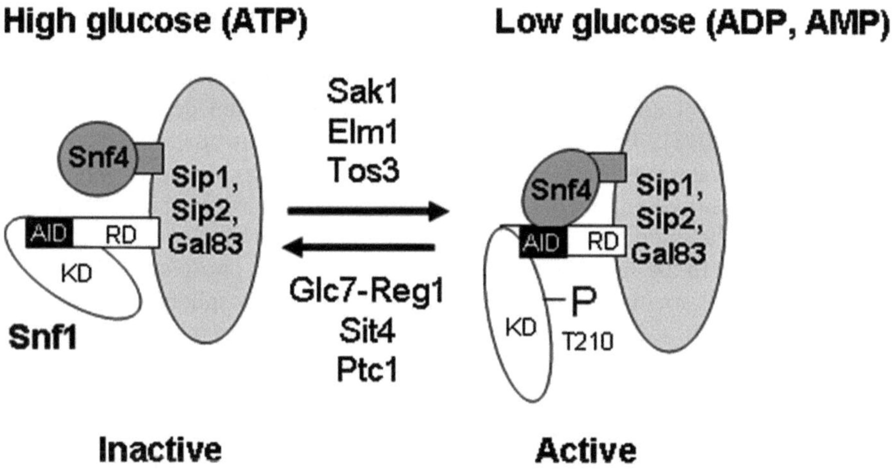

Fig. 14.2 Glucose regulates the activity of the SNF1 complex by a two-step process: phosphorylation by upstream kinases (Sak1, Elm1, and Tos3) and dephosphorylation by phosphatases (Gcl7-Reg1, Sit4, and Ptc1) and Snf4-dependent rearrangement within the SNF1 complex. *KD* kinase domain, *RD* regulatory domain, *AID* autoinhibitory domain in Snf1. See text for details

mediated by the interaction between Snf4 and Snf1 (Jiang and Carlson 1996). Phosphorylation occurs at residue Thr210 in Snf1 located in the activation loop of the KD domain, and although it can occur independently of the Snf4 step, the complex has to undergo a Snf4-dependent rearrangement to achieve full activity (McCartney and Schmidt 2001). Phosphorylation of Thr210 is performed by three partially redundant upstream kinases: Sak1, Elm1, and Tos3 (Fig. 14.2). Among them, Sak1 has been shown to be the most important upstream kinase in the glucose repression pathway (Hedbacker et al. 2004a; McCartney et al. 2005). Sak1, Elm1, and Tos3 are members of the ELM family of protein kinases (Hunter and Plowman 1997). While none of them had a clear mammalian orthologue, their kinase domains were most closely related to those of the protein kinase LKB1 and the calmodulin-dependent protein kinase kinase CaMKKβ. This information on the initial characterization of yeast Snf1 upstream kinases triggered the rapid identification of mammalian AMPK upstream kinases (Hawley et al. 2003; Woods et al. 2003, 2005).

Since in yeast the activity of the three yeast upstream kinases is constitutive, the regulation of the phosphorylation status of Snf1 must depend on the phosphatase involved in the dephosphorylation of residue Thr210. In this sense, some reports indicate that yeast protein phosphatase type 1 (Glc7) and its regulatory subunit Reg1 play a key role in the dephosphorylation of Thr210 in response to glucose addition (Ludin et al. 1998; Sanz et al. 2000) (Fig. 14.2). In the absence of Reg1, Glc7 cannot perform its function, so Snf1 is constitutively phosphorylated and active, even in the presence of glucose (Ludin et al. 1998; Sanz et al. 2000). Since the activity of the Glc7-Reg1 complex does not change depending on the presence of glucose (Rubenstein et al. 2008), it has been hypothesized that the accessibility of Snf1 for dephosphorylation is the key factor in the dephosphorylation event (Rubenstein et al. 2008). In this regard, it has been recently described that a conformational change in Snf4 due to the presence of ADP bound to the corresponding CBS domains (see below) protects Snf1 from dephosphorylation (Mayer et al. 2011). Therefore, a plausible hypothesis is that the Glc7-Reg1 phosphatase holoenzyme is always bound to SNF1 complex, but it is only accessible to dephosphorylate Thr210 when a conformational change occurs at Snf4 due presumably to the interchange of ADP by ATP under high glucose conditions.

Recently, the PP2A-like phosphatase Sit4 and the PP2C phosphatase Ptc1 have been defined as alternative phosphatases that regulate the phosphorylation status of Snf1 (Ruiz et al. 2011, 2013) (Fig. 14.2). As phosphorylation of mammalian AMPK complex can also be regulated by protein phosphatase 1 [and its regulatory subunit R6 (PPP1R3D)] and members of the PP2A and PP2C protein phosphatase families (Garcia-Haro et al. 2010; Joseph et al. 2015; Chida et al. 2013), these results indicate a conserved mechanism of regulation of the phosphorylation status of SNF1/AMPK complexes throughout evolution.

14.2.2 Snf4 Regulatory Subunit

Snf4 is the regulatory subunit of the SNF1 complex. It is a protein of 322 aa with a molecular weight of 36 kDa. Snf4 contains an N-terminal domain (pre-CBS) that allows its interaction with Snf1 and Gal83 (Viana et al. 2007) (Fig. 14.1). In addition, it contains four tandem repeats of a structural module called CBS, described initially in cystathionine-β-synthase, which is involved in binding to AMP, ADP, and ATP (Fig. 14.1). However, and contrary to what is observed in its mammalian orthologue (AMPKγ subunit), although AMP binds to the Snf4-CBS domains, it does not act as an allosteric activator of the SNF1 complex (Wilson et al. 1996). In yeast, it seems that binding of ADP to Snf4-CBS domains ensures a direct link between energy metabolism and the activity of the SNF1 complex (Mayer et al. 2011). As indicated above, binding of ADP to Snf4 promotes a conformational change that protects SNF1 complex from dephosphorylation by exogenous phosphatases (Mayer et al. 2011).

As described above, Snf4 is required for the full activation of SNF1 complex under low glucose conditions because it interacts with Snf1-RD and releases Snf1-KD from its binding to the AID domain (Fig. 14.2). In addition, it has been recently described that interaction of Snf4 with Gal83 (a member of the yeast AMPKβ subunit orthologues) plays a negative role in the regulation of the activity of the SNF1 complex. Mutations in residues of Snf4-CBS2 forming part of a hydrophobic patch or close to it (i.e., C136Y, G145E, R169A, and N177Y) abolished the interaction with the glycogen-binding domain (GBD) of Gal83 and strongly relieved glucose inhibition of SNF1 complex (Momcilovic et al. 2008). This suggests a negative role of the interaction of Snf4 with the GBD of Gal83/Sip1/

Sip2, which adds an extra layer of complexity in the regulation of the activity of SNF1 complex (see below) (Fig. 14.3).

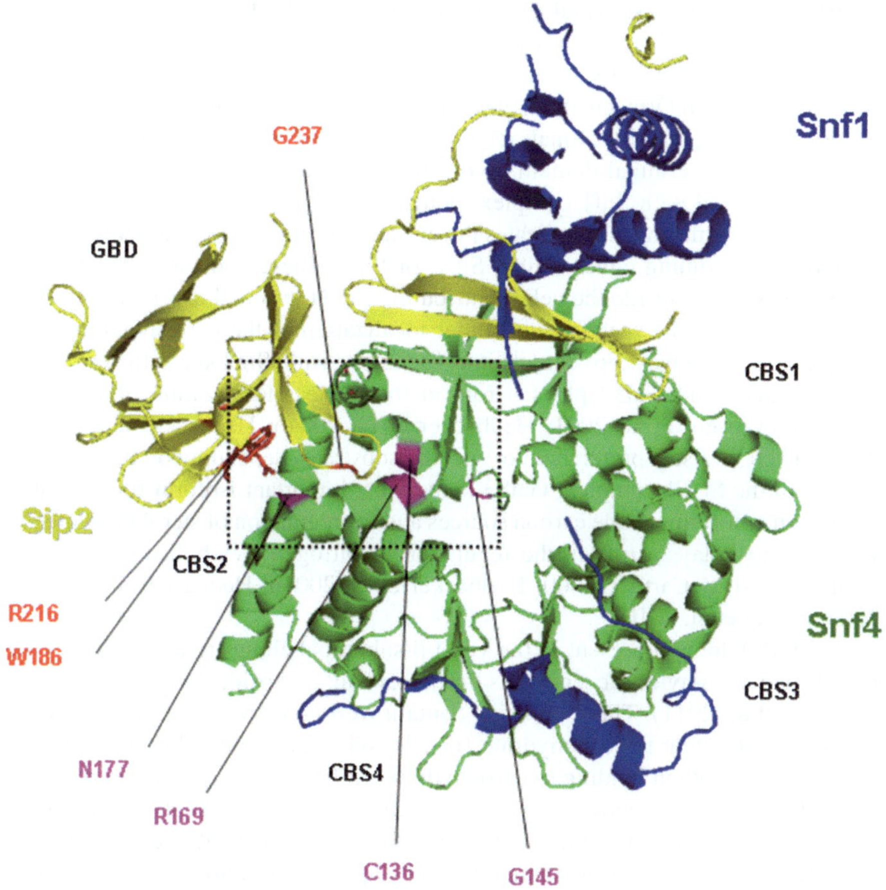

Fig. 14.3 Structural model of the heterotrimeric SNF1 complex. Crystal structure of the heterotrimeric core of the *Saccharomyces cerevisiae* SNF1 complex (PDB, 2QLV) (Amodeo et al. 2007) was modeled using the PyMol program (DeLano Scientific LLC, San Carlos, CA). The C-terminal part of Snf1 (residues 398–633) is colored in blue, the C-terminal part of Sip2 (residues 154–415) is colored in *yellow*, and the full-length Snf4 is colored in *green*. Critical residues forming part of the hydrophobic patch present in the CBS2 of Snf4 are indicated in *magenta* (C136, G145, R169, and N177), and those critical residues of the glycogen-binding domain of Sip2 [W186, R216, and G237 (G235 in Gal83)] are indicated in *red*. The *dotted line* indicates the area of interaction between the GBD of Sip2 and the CBS2 of Snf4

14.2.3 Gal83, Sip1, and Sip2 Scaffolding Subunits

These subunits (Gal83, 417 aa, 46.6 kDa; Sip1, 815 aa, 91 kDa; Sip2, 415 aa, 46.4 kDa) are the yeast orthologues of mammalian AMPKβ subunits (Fig. 14.1). They act as scaffolds to allow the interaction between Snf1 and Snf4. They also regulate the subcellular localization of the SNF1 complex and contribute to substrate specification (Vincent et al. 2001; Hedbacker et al. 2004b). These β-subunits contain a divergent N-terminal region, a conserved glycogen-binding domain (GBD), and a C-terminal domain involved in the interaction with Snf1 and Snf4 (ASC, associated with Snf1 complex, domain) (Fig. 14.1). The presence of three different β-subunits in yeast indicates the existence of three different pools of SNF1 complexes containing either Gal83, Sip1, or Sip2. These pools are distributed in different locations inside the cell, contributing the N-terminal divergent region of each subunit into the specific subcellular localization of the corresponding SNF1 complex. In this sense, Sip1 directs it to vacuoles, while Sip2 keeps the enzyme in the cytoplasm and Gal83 plays a role in the nuclear localization of the SNF1 complex (Vincent et al. 2001; Hedbacker et al. 2004b; Hedbacker and Carlson 2006). There is also a preference for specific substrates depending on the β-subunit present in the SNF1 complex. Gal83 is the most important isoform for regulating growth on non-fermentable carbon sources and for regulation of sterol biosynthesis, whereas Sip1 plays a role in the regulation of nitrogen metabolism and meiosis (Schmidt and McCartney 2000; Hedbacker et al. 2004b; Hedbacker and Carlson 2006; Zhang et al. 2010).

The GBD domain present in the yeast β-subunits belongs to the CBM48 type, found also in amylolytic enzymes from several glycoside hydrolase families (Janecek et al. 2011). These domains contain well-conserved consensus residues involved in binding to carbohydrate (i.e., W184, R214, and G235 in Gal83). In addition to its role in binding to carbohydrates, it has been recently demonstrated that GBDs play a negative role in the regulation of the activity of the SNF1 complex. It has been shown that deletion of the GBD from Gal83 relieves glucose inhibition of SNF1: Snf1 maintains its Thr210 phosphorylation status even in the presence of glucose (Momcilovic et al. 2008). Since alterations in the levels of glycogen do not affect the regulation of SNF1 activity by glucose, these results indicate that it is not the binding to glycogen which inhibits SNF1 but rather the GBD itself which contributes to glucose inhibition of SNF1 complex through its interaction with the rest of the SNF1 heterotrimer (Wiatrowski et al. 2004; Momcilovic et al. 2008). This hypothesis was confirmed when it was demonstrated that substitutions in residues of the GBD that abolished the interaction of Gal83 with Snf4 also relieved glucose inhibition of SNF1 complex: Gal83-W184A, -R214Q, and -G235R mutants showed an increase in the activity of the SNF1 complex in cells growing in the presence of glucose (Fig. 14.3, with the numbers W186, R216, and G237 corresponding to the same conserved residues in Sip2). These results were complementary to the ones presented above on mutations in residues of the Snf4 CBS2 domain that prevented the binding to Gal83 and increased the activity of the SNF1 complex (Momcilovic et al. 2008) and suggest a negative role of Gal83-GBD in regulating SNF1 complex activity. Surprisingly, when the structure of a form of the heterotrimeric SNF1 complex containing the GBD of Sip2 and the entire Snf4 became available, it was noticed that the patches of critical residues present in the GBD of Sip2 and those present in the CBS2 domain of Snf4 were spatially close, suggesting a direct interaction between these residues (Amodeo et al. 2007). If the interaction between these two modules is perturbed, then the SNF1 complex would become constitutively activated.

It has also been recently described that the GBD domain of the yeast β-subunits participates in the binding of Reg1 (Mangat et al. 2010), a Glc7 protein phosphatase regulatory subunit involved in the regulation of the phosphorylation status of SNF1 complex (see above). This opens the possibility that in the absence of a functional GBD, the Glc7-Reg1 phosphatase holoenzyme would not have access to the SNF1 complex, so it would be constitutively phosphorylated even in the presence of glucose. In other words, the GBD domain would be essential for recruiting the Glc7-Reg1 phosphatase holoenzyme to SNF1 complex for regulation.

14.3 Structure of the SNF1 Complex

The structural characterization of yeast SNF1 complex was one of the first achievements to understand the complexity of the interaction among the three SNF1 subunits. The first released heterotrimeric structure corresponding to *Saccharomyces cerevisiae* SNF1 complex was composed by a truncated form of Snf1 (residues 398–633), a truncated form of Sip2 (residues 154–415, but containing the GBD), and the entire Snf4 (residues 1–322) (PDB, 2QLV) (Amodeo et al. 2007) (Fig. 14.3). This structure showed a triangular form with a base formed by Snf4 which associates with the C-terminal domains of Snf1 and Sip2. Since this structure contained larger fragments from Snf1 and Sip2 in comparison with alternative structures either from *Schizosaccharomyces pombe* (PDB, 2OOX) or from *Rattus novergicus* (2V8Q), it allowed for the first time the definition of a close interaction between the GBD of Sip2 (residues 154–247) and an α-helix of CBS2 of Snf4, explaining why alterations in some residues of yeast GBD or in residues from Snf4-CBS2 can lead to an activation of the SNF1 complex (see above) (Fig. 14.3, region with a dotted line). The structure also indicated that residues 460–495 of Snf1, corresponding to the Snf1-RD, had well-defined interactions with Snf4 (residues 270–275). This demonstrates a direct interaction between Snf1 and Snf4 within SNF1 complex, which is fully supported by earlier biochemical and genetic data (Amodeo et al. 2007). Again, these initial studies in yeast promoted the work in mammalian AMPK complex that ended with a thorough description of the structure of this complex in mammalian cells. These studies in mammalian AMPK showed extraordinary similarity with the previously described structure of the SNF1 complex (Xiao et al. 2011, 2013; Calabrese et al. 2014; Li et al. 2015).

14.4 Regulation of the SNF1 Complex

As indicated above, in yeast, regulation of the activity of the SNF1 complex involves two steps, one that requires the phosphorylation of the Snf1 catalytic subunit (at residue Thr210 located within its activation loop) by an upstream kinase (see above) and another one that is mediated by the interaction between Snf4 and Snf1. After these two steps, phosphorylated Snf1 kinase opens its active site, making it accessible to substrates.

The activity of the SNF1 complex is also regulated by posttranslational modification of any of its subunits. In this sense, it has been reported that acetylation of Sip2 stabilizes its interaction with the Snf1 catalytic subunit, thereby inhibiting the activity of the SNF1 complex (Lin et al. 2009). Sip2 is also myristoylated, allowing its localization at the plasma membrane, sequestering in this way Snf4 and decreasing SNF1 activity (Lin et al. 2003). Sip1 is also myristoylated, triggering its relocalization from the cytosol to the vacuolar membrane in response to different stress conditions (Hedbacker et al. 2004b).

On the other hand, in the presence of glucose, Snf1 is the target of the E3-SUMO ligase Mms21 which catalyzes the attachment of a SUMO1 moiety to residue K549 in the C-terminal regulatory domain of Snf1. Sumoylation inhibits SNF1 activity and targets Snf1 to ubiquitination via the SUMO-targeted E3 ubiquitin ligase Slx5–Slx8, leading to the degradation of Snf1. As this modification only takes place in the presence of glucose, it can be envisioned as a response to glucose sensing (Simpson-Lavy and Johnston 2013). The same authors indicated that this sumoylation could be reverted by the expression of the SUMO-protease Ulp1 (Simpson-Lavy and Johnston 2013).

Ubiquitination also modulates negatively Snf1 stability and catalytic activity. Although the identity of the E3-ubiquitin ligases involved in this reaction is still unknown, it has been described that deubiquitination of Snf1 is carried out at least by Ubp8 deubiquitinase, improving the stability and phosphorylation status of Snf1 and activation of the SNF1 complex (Wilson et al. 2011). On the other hand, deletion of the *UBP8* gene causes a decrease in SNF1 activity, probably due to hyperubiquitination (Wilson et al. 2011). Recently, it has been described that Ubp10 deubiquitinase also plays a role in deubiquitinating SNF1 complex. In fact in a double mutant lacking *UBP8* and *UBP10*, the levels of Snf1 protein were dramatically decreased due to hyperubiquitination and degradation. However, the Snf1 species still present in this double mutant were hyperactivated upon stress, which may compensate for the loss of the Snf1 protein and protect cells in this way against stress (Hsu et al. 2015).

14.5 Transcriptional Effects of SNF1 on Energy Metabolism and Cellular Stress

As indicated above, SNF1 complex activates transcription by inhibiting transcriptional repressors (i.e., Mig1) or by stimulating transcriptional activators (i.e., Cat8 and Sip4). In this section, we will focus on the action of the SNF1 complex on the transcription of relevant genes related to energy metabolism and cellular stress:

14.5.1 Glucose-Repressed Genes

In the presence of glucose, the expression of a set of genes [such as *SUC2, MAL, GAL*, etc., mainly related to the utilization of alternative carbon sources (sucrose, maltose, galactose)] is repressed by the action of the Mig1 transcriptional repressor. This repressor is a major target of the action of SNF1 complex. Under low glucose conditions, SNF1 phosphorylates Mig1 allowing its dissociation from the co-repressors Ssn6 and Tup1 and its release from DNA promoters (Treitel and Carlson 1995). Phosphorylated Mig1 is then exported from the nucleus to the cytoplasm, thereby permitting gene expression of glucose-repressed genes (*SUC2, MAL, GAL,* etc.) [reviewed in Kayikci and Nielsen (2015) and Shashkova et al. (2015)]. In addition to Ssn6 and Tup1 co-repressors, Mig1 interacts inside the nucleus with Hxk2 (hexokinase PII), a "moonlighting enzyme" that in addition to its role in glycolysis (it phosphorylates glucose to render glucose-6P) has a nuclear role in the regulation of the transcription of glucose-repressed genes (Ahuatzi et al. 2007). Binding of Hxk2 to Mig1 prevents the phosphorylation of the repressor by SNF1 complex and hence removal of Mig1 from the nucleus. Under low glucose conditions, SNF1 complex phosphorylates Hxk2 and prevents its nuclear localization; thus, Mig1 becomes available for phosphorylation by nuclear SNF1 complex, thus triggering its exit from the nucleus (Fernandez-Garcia et al. 2012). Phosphorylation of Hxk2 and the induced translocation of the phosphoenzyme to the cytosol can occur alternatively by the action of Ymr291w/Tda1 protein kinase, which adds an extra contribution of a protein kinase to glucose signaling (Kaps et al. 2015).

14.5.2 Gluconeogenic Genes

Glucose also represses the expression of gluconeogenic genes such as *PCK1* (phosphoenolpyruvate carboxykinase 1), *FBP1* (fructose 1,6 bisphosphatase), *MLS1* (malate synthase 1), *ACS1* (acetylCoA synthetase 1), *MDH2* (malate dehydrogenase 2), *SFC1* (succinate/fumarate transporter), *CAT2* (carnitine acetyltransferase), *IDP2* (NADP-dependent isocitrate dehydrogenase), and *JEN1* (carboxylic acid transporter protein homologue) (Turcotte et al. 2010). All these genes contain in their promoters a carbon source-responsive element (CSRE) with a consensus sequence YCCRTTNRNCGG that responds to zinc cluster transcriptional activators such as Cat8, Sip4, and Rds2 (Roth et al. 2004). When glucose is absent, SNF1 complex phosphorylates and activates Cat8, Sip4, and Rds2 transcription factors allowing the expression of gluconeogenic genes (Vincent and Carlson 1999; Roth et al. 2004). Phosphorylation of a single serine residue in Cat8 by SNF1 complex is responsible for the activation of this factor (Charbon et al. 2004). Similarly, phosphorylation of Sip4 by SNF1 correlates with an increase in its transcriptional activity (Lesage et al. 1996) and hyperphosphorylation of Rds2 in ethanol is Snf1-dependent (Soontorngun et al. 2007).

Another target of SNF1 action is Adr1 transcriptional activator, which mediates the expression of genes related to the utilization of ethanol and fatty acid metabolism. Adr1 coordinates metabolic activities important for the generation of acetylCoA and NADH from alternative carbon sources such as lipids (Young et al. 2003). Adr1 also binds to CSRE-containing promoters as in the case of Cat8. In fact, they both target common CSRE genes for derepression (i.e., *ADH2*, alcohol dehydrogenase; *ACS1*, acetylCoA synthetase 1) (Tachibana et al. 2005). However, the action of SNF1 complex is indirect in this case as it mediates the dephosphorylation of Adr1, which allows the binding of Adr1 to specific promoters (Turcotte et al. 2010).

14.5.3 Chromatin Remodeling

In addition to the role of SNF1 complex on regulating the activity of transcriptional activators (i.e., Cat8, Sip4, Rds2, Adr1) and repressors (i.e., Mig1), SNF1 has a role in chromatin remodeling for proper expression of glucose-repressed genes. SNF1 complex mediates chromatin remodeling by directly interacting and phosphorylating histone acetyltransferase Gcn5, which triggers its histone-modifying activity (Lo et al. 2001; Abate et al. 2012). Another general effect of SNF1 on transcription is by a direct interaction with RNAII-polymerase holoenzyme. In response to glucose limitation, SNF1 complex phosphorylates this holoenzyme which enhances the transcription process (Kuchin et al. 2000).

14.5.4 Salt Stress

One of the main systems that allow yeast to grow in the presence of high NaCl concentrations is the extrusion of cytosolic Na^+ through the action of P-type ATPases (i.e., Ena1 and Pmr2). The expression of *ENA1* shows a dual mode of regulation. It is induced by salt stress by a mechanism involving Hog1 and calcineurin (PP2B phosphatase) pathways, and it is repressed by glucose. The effect of glucose is independent of the Hog1 or calcineurin (PP2B phosphatase) pathways. Glucose repression involves the action of Mig1 and Ssn6 repressors, whose activity is regulated by SNF1 complex. Regulation of *ENA1* expression by Snf1 kinase could be part of a general mechanism through which yeast cells respond to carbon source starvation by activating protective systems against different types of stress (Alepuz et al. 1997; Ye et al. 2008). In the absence of a functional SNF1 complex, the expression of *ENA1* is compromised and this results in a higher sensitivity to salt stress. In fact, the growth of a *snf1* mutant is severely inhibited in the presence of 1.2 M NaCl and 0.3 M LiCl and *snf4* mutants also showed the same salt sensitivity as *snf1* cells. This effect is not a consequence of the high osmolarity of the medium since these mutants grow as wild type in 1.2 M KCl, 200 mM $CaCl_2$ or 1.8 M sorbitol (Alepuz et al. 1997). Therefore, low SNF1 activity leads to salt sensitivity and, on the contrary, high SNF1 activity, as in the case of *reg1* mutants, leads to salt resistance even in glucose growing cells.

14.5.5 Heat Shock

When yeast cells are exposed to heat shock conditions, they enhance the expression of a specific set of genes (those expressing heat shock proteins, among others) that allow cells to cope with the heat stress. The expression of this set of genes is governed by the heat shock factor 1 (Hsf1) transcriptional activator. It has been observed that under low glucose conditions the SNF1 complex phosphorylates Hsf1 causing its trimerization and activation (Hahn and Thiele 2004). The presence of glucose inactivates Hsf1 via inactivation of SNF1 complex and results in a higher sensitivity to heat shock. This explains why the growth of a *snf1* mutant is more sensitive to heat stress (Hahn and Thiele 2004).

14.6 Posttranslational Effects of SNF1 on Energy Metabolism

SNF1 complex is a sensor of energy charge. When activated it slows down anabolic processes that consume ATP whereas it stimulates catabolic processes that produce ATP in order to restore energy homeostasis. In this way, SNF1 complex adjusts energy metabolism to energetic demand. In this section, we will review the posttranslational role of SNF1 complex in inactivating energetically expensive cellular processes such as protein synthesis and lipid metabolism.

14.6.1 Protein Synthesis

In order to coordinate carbon and nitrogen metabolism, under amino acid limitation but in the presence of glucose, Snf1 interacts with and activates Gcn2 protein kinase, promoting the formation of phosphorylated eIF2α, which slows down translation but stimulates the synthesis of Gcn4, a transcription factor that enhances the expression of genes involved in amino acid biosynthesis and nitrogen utilization [reviewed in Cherkasova et al. (2010) and Conrad et al. (2014)]. However, under conditions of glucose deprivation, SNF1 complex downregulates both transcription and translation of *GCN4* (Shirra et al. 2008). In this way, SNF1 complex balances amino acid synthesis to carbon metabolism and cellular energy balance (Kayikci and Nielsen 2015).

14.6.2 Lipid Metabolism

SNF1 complex also affects lipid metabolism. It phosphorylates and inactivates acetyl CoA carboxylase (*ACC1*) under low energy conditions (Woods et al. 1994). The inactivation of Acc1 results in a decrease in the levels of malonyl-CoA, which is the first intermediate of fatty acid synthesis. So, activation of SNF1 complex promotes inhibition of fatty acid synthesis by decreasing the levels of the first building block, malonyl CoA. At the same time, low levels of malonyl-CoA stimulate the entry of fatty acids into the mitochondria, enhancing β-oxidation. In this way, cell obtains ATP to balance energy demand (Woods et al. 1994).

Under conditions of elevated Acc1 activity (i.e., high glucose), the activity of inositol 1-phosphate synthase (*INO1*; that mediates the conversion of glucose-6P to inositol-1P) is reduced, affecting in this way inositol-containing phospholipid biosynthesis. For this reason, *snf1* mutants, which cannot inhibit Acc1, do not produce enough inositol-1P and require exogenous inositol for growth (Shirra et al. 2001).

14.7 Crosstalk Between SNF1 Complex and Other Protein Kinase Pathways

14.7.1 cAMP–PKA

Protein kinase A is a master regulator in yeast cell physiology since it affects a large number of different cellular processes, including stress tolerance, fermentation, and cell proliferation [see Santangelo (2006), Smets et al. (2010), and Conrad et al. (2014) for review]. Early genetic studies in yeast defined that SNF1 and

PKA were members of antagonistic and parallel pathways: *snf1* mutant phenotypes resembled those associated with an overactivation of the cAMP–PKA pathway and mutations in components of the PKA pathway that decrease the activity of PKA moderated the *snf1* phenotype (Thompson-Jaeger et al. 1991). In this sense, a double *snf1 bcy1* mutant (in which SNF1 is inactive and PKA is overactive) shows extreme phenotypes, simply because of the additive effects of the absence of Snf1 and overactivation of PKA (Thompson-Jaeger et al. 1991). Later studies have confirmed the inversed relationship between the SNF1 and the PKA pathways: on one hand, Sak1 (a member of the Snf1 upstream kinases) contains putative PKA phosphorylation sites that would lead to its inactivation, and on the other hand, phosphorylation of Bcy1 (a regulatory subunit of PKA) by SNF1 complex enhances its binding to PKA, decreasing in this way its activity [reviewed in Conrad et al. (2014) and Shashkova et al. (2015)]. This negative relationship between SNF1 and PKA pathways is also present in mammalian cells where it has been described that PKA phosphorylates AMPKα subunit at different sites and this triggers the inactivation of the AMPK complex (Djouder et al. 2010).

14.7.2 TORC1 Kinase Complex

TOR (target of rapamacin) kinase complex 1 (TORC1), a complex formed by TOR kinase and a specific set of regulatory subunits, regulates cell growth and proliferation in response to a variety of signals, including nitrogen, amino acids, and stress factors [see Porta et al. (2014) and Shimobayashi and Hall (2014) for reviews]. In yeast, TORC1 senses the amount of available nitrogen sources and regulates the expression of nitrogen catabolite repressible (NCR) genes by regulating the subcellular localization of the transcription factors Gln3 and Gat1. In the presence of available nitrogen sources, TORC1 is active and phosphorylates Gln3 and Gat1, maintaining these transcription factors anchored to the scaffold protein Ure2 in the cytosol. Upon inactivation of TORC1 (either by nutrient depletion or by rapamycin treatment), Gln3 is released from Ure2 and imported to the nucleus, where it activates the expression of NCR genes. Glucose also regulates Gln3 phosphorylation and subcellular localization by a process which is mediated by SNF1 complex. Snf1 interacts and phosphorylates Gln3 and promotes its nuclear internalization (in this case, SNF1 phosphorylation of Gln3 causes an opposite effect to the one made by TORC1). Snf1 only participates in the induction of Gln3 activation upon carbon starvation, having no effect on nitrogen starvation or rapamycin induction of Gln3 activation. Therefore, nitrogen and glucose signaling pathways converge onto Gln3, which may be critical for both nutrient-sensing and starvation responses (Bertram et al. 2002). Gat1, another transcriptional activator of NCR genes, is also phosphorylated by Snf1 under low glucose conditions (Kulkarni et al. 2006), adding another piece of regulation of SNF1 complex on TORC1-related transcriptional activators.

An additional proof of the relationship between SNF1 and TORC1 kinase pathways is the fact that *snf1* and *snf4* mutants are resistant to rapamycin, whereas *reg1* mutant (with an hyperactive Snf1) displays hypersensitivity to the drug (Bertram et al. 2002). These results suggest that SNF1 complex negatively participates in a cellular process that also involves TORC1 (Bertram et al. 2002). Recently, it has been demonstrated that SNF1 complex affects directly the activity of the TORC1 complex. SNF1 complex promotes the phosphorylation of Kog1 and Tco89 subunits of the TORC1 complex, influencing negatively the activity of this complex [reviewed in Shashkova et al. (2015)]. This negative relationship is also present in mammalian cells where it has been demonstrated that AMPK regulates negatively mTOR function (Hardie and Ashford 2014).

14.7.3 PAS Kinases

PAS (Per-Arnt-Sim domain containing) kinases are a family of protein kinases that play a role in sugar metabolism and storage as well as enhancement of protein synthesis, maintaining in this way cell wall integrity [see Wilson et al. (2005) and Hao et al. (2007) for review]. In yeast, there are two PAS kinase homologues, Psk1 and Psk2, which play a key role in the regulation of respiration and glycogen storage (Grose et al. 2007, 2009; DeMille et al. 2014). Recently, it has been demonstrated that Psk1 interacts with and becomes phosphorylated by SNF1 complex containing Gal83 as β-subunit (DeMille et al. 2015). This modification leads to the activation of Psk1, which phosphorylates and activates poly(A)-binding protein (Pab) binding protein 1 (Pbp1), which then inhibits TOR kinase complex 1 (TORC1) through its sequestration at stress granules. These data demonstrate an alternative layer of negative regulation of TORC1 by SNF1, since Snf1-deficient cells display increased TORC1 activity, whereas cells containing hyperactive Snf1 display a PAS kinase-dependent decrease in TORC1 activity. These results suggest a close relationship between nutrient-sensing kinases such as SNF1, Psk1, and TORC1 that allow the reprogramming of metabolic pathways in order to maintain energy homeostasis in response to nutrient status (DeMille et al. 2015).

14.8 Perspectives

The yeast Snf1 protein kinase was first described back in 1984 (Celenza and Carlson 1984). Since then, a lot of information on the function and regulation of the activity of Snf1 protein kinase has been gained by biochemical and genetic methods. This initial work carried out in yeast paved the way to the understanding of the function of mammalian AMPK. In fact, most of the breakthroughs on the mammalian AMPK field came from previous studies in the yeast system (i.e., regulation of the activity by glucose, identification of upstream kinases, crystallographic structural studies of the heterotrimeric complex, etc.). However, there are still areas that need to be addressed, such as the role of different phosphatases in regulating the phosphorylation status of SNF1 complex in response to different growth conditions, the determinants that target the different SNF1 complexes to different subcellular locations, the role of SNF1 complex in different metabolic pathways, etc. As the function of SNF1/AMPK complexes is conserved throughout evolution, in our opinion the knowledge that is gained day by day on both mammalian AMPK and yeast SNF1 systems benefits each other and will end with a better understanding of the function of these metabolic sensors.

Acknowledgements This study was supported by grants from the Spanish Ministry of Education and Science SAF2014-54604-C3-1-R and Generalitat Valenciana (PrometeoII/2014/029) to P.S.

References

Abate G, Bastonini E, Braun KA, Verdone L, Young ET, Caserta M (2012) Snf1/AMPK regulates Gcn5 occupancy, H3 acetylation and chromatin remodelling at S. cerevisiae ADY2 promoter. Biochim Biophys Acta 1819(5):419–427

Ahuatzi D, Riera A, Pelaez R, Herrero P, Moreno F (2007) Hxk2 regulates the phosphorylation state of Mig1 and therefore its nucleocytoplasmic distribution. J Biol Chem 282(7):4485–4493

Alepuz PM, Cunningham KW, Estruch F (1997) Glucose repression affects ion homeostasis in yeast through the regulation of the stress-activated ENA1 gene. Mol Microbiol 26(1):91–98

Amodeo GA, Rudolph MJ, Tong L (2007) Crystal structure of the heterotrimer core of Saccharomyces cerevisiae AMPK homologue SNF1. Nature 449(7161):492–495

Ashrafi K, Lin SS, Manchester JK, Gordon JI (2000) Sip2p and its partner snf1p kinase affect aging in S. cerevisiae. Genes Dev 14(15):1872–1885

Bertram PG, Choi JH, Carvalho J, Chan TF, Ai W, Zheng XF (2002) Convergence of TOR-nitrogen and Snf1-glucose signaling pathways onto Gln3. Mol Cell Biol 22 (4):1246–1252

Calabrese MF, Rajamohan F, Harris MS, Caspers NL, Magyar R, Withka JM, Wang H, Borzilleri KA, Sahasrabudhe PV, Hoth LR, Geoghegan KF, Han S, Brown J, Subashi TA, Reyes AR, Frisbie RK, Ward J, Miller RA, Landro JA, Londregan AT, Carpino PA, Cabral S, Smith AC, Conn EL, Cameron KO, Qiu X, Kurumbail RG (2014) Structural basis for AMPK activation: natural and synthetic ligands regulate kinase activity from opposite poles by different molecular mechanisms. Structure 22(8):1161–1172

Casamayor A, Serrano R, Platara M, Casado C, Ruiz A, Arino J (2012) The role of the Snf1 kinase in the adaptive response of Saccharomyces cerevisiae to alkaline pH stress. Biochem J 444 (1):39–49

Celenza JL, Carlson M (1984) Cloning and genetic mapping of SNF1, a gene required for expression of glucose-repressible genes in Saccharomyces cerevisiae. Mol Cell Biol 4 (1):49–53

Conrad M, Schothorst J, Kankipati HN, Van Zeebroeck G, Rubio-Texeira M, Thevelein JM (2014) Nutrient sensing and signaling in the yeast Saccharomyces cerevisiae. FEMS Microbiol Rev 38 (2):254–299

Cullen PJ, Sprague GF Jr (2000) Glucose depletion causes haploid invasive growth in yeast. Proc Natl Acad Sci USA 97(25):13619–13624

Charbon G, Breunig KD, Wattiez R, Vandenhaute J, Noel-Georis I (2004) Key role of Ser562/661 in Snf1-dependent regulation of Cat8p in Saccharomyces cerevisiae and Kluyveromyces lactis. Mol Cell Biol 24(10):4083–4091

Cherkasova V, Qiu H, Hinnebusch AG (2010) Snf1 promotes phosphorylation of the alpha subunit of eukaryotic translation initiation factor 2 by activating Gcn2 and inhibiting phosphatases Glc7 and Sit4. Mol Cell Biol 30(12):2862–2873

Chida T, Ando M, Matsuki T, Masu Y, Nagaura Y, Takano-Yamamoto T, Tamura S, Kobayashi T (2013) N-Myristoylation is essential for protein phosphatases PPM1A and PPM1B to dephosphorylate their physiological substrates in cells. Biochem J 449(3):741–749

DeMille D, Badal BD, Evans JB, Mathis AD, Anderson JF, Grose JH (2015) PAS kinase is activated by direct SNF1-dependent phosphorylation and mediates inhibition of TORC1 through the phosphorylation and activation of Pbp1. Mol Biol Cell 26(3):569–582

DeMille D, Bikman BT, Mathis AD, Prince JT, Mackay JT, Sowa SW, Hall TD, Grose JH (2014) A comprehensive protein-protein interactome for yeast PAS kinase 1 reveals direct inhibition of respiration through the phosphorylation of Cbf1. Mol Biol Cell 25(14):2199–2215

Djouder N, Tuerk RD, Suter M, Salvioni P, Thali RF, Scholz R, Vaahtomeri K, Auchli Y, Rechsteiner H, Brunisholz RA, Viollet B, Makela TP, Wallimann T, Neumann D, Krek W (2010) PKA phosphorylates and inactivates AMPKalpha to promote efficient lipolysis. EMBO J 29(2):469–481

Fernandez-Garcia P, Pelaez R, Herrero P, Moreno F (2012) Phosphorylation of yeast hexokinase 2 regulates its nucleocytoplasmic shuttling. J Biol Chem 287(50):42151–42164

Ferrer-Dalmau J, Randez-Gil F, Marquina M, Prieto JA, Casamayor A (2015) Protein kinase Snf1 is involved in the proper regulation of the unfolded protein response in Saccharomyces cerevisiae. Biochem J 468(1):33–47

Garcia-Haro L, Garcia-Gimeno MA, Neumann D, Beullens M, Bollen M, Sanz P (2010) The PP1-R6 protein phosphatase holoenzyme is involved in the glucose-induced dephosphorylation and inactivation of AMP-activated protein kinase, a key regulator of insulin secretion, in MIN6 beta cells. FASEB J 24(12):5080–5091

Grose JH, Smith TL, Sabic H, Rutter J (2007) Yeast PAS kinase coordinates glucose partitioning in response to metabolic and cell integrity signaling. EMBO J 26(23):4824–4830

Grose JH, Sundwall E, Rutter J (2009) Regulation and function of yeast PAS kinase: a role in the maintenance of cellular integrity. Cell Cycle 8(12):1824–1832

Hahn JS, Thiele DJ (2004) Activation of the Saccharomyces cerevisiae heat shock transcription factor under glucose starvation conditions by Snf1 protein kinase. J Biol Chem 279 (7):5169–5176

Hao HX, Cardon CM, Swiatek W, Cooksey RC, Smith TL, Wilde J, Boudina S, Abel ED, McClain DA, Rutter J (2007) PAS kinase is required for normal cellular energy balance. Proc Natl Acad Sci USA 104(39):15466–15471

Hardie DG, Ashford ML (2014) AMPK: regulating energy balance at the cellular and whole body levels. Physiology (Bethesda) 29(2):99–107

Hardie DG, Carling D, Carlson M (1998) The AMP-activated/SNF1 protein kinase subfamily: metabolic sensors of the eukaryotic cell? Annu Rev Biochem 67:821–855

Hawley SA, Boudeau J, Reid JL, Mustard KJ, Udd L, Makela TP, Alessi DR, Hardie DG (2003) Complexes between the LKB1 tumor suppressor, STRADalpha/beta and MO25alpha/beta are upstream kinases in the AMP-activated protein kinase cascade. J Biol 2(4):28

Hedbacker K, Carlson M (2006) Regulation of the nucleocytoplasmic distribution of Snf1-Gal83 protein kinase. Eukaryot Cell 5(12):1950–1956

Hedbacker K, Hong SP, Carlson M (2004a) Pak1 protein kinase regulates activation and nuclear localization of Snf1-Gal83 protein kinase. Mol Cell Biol 24(18):8255–8263

Hedbacker K, Townley R, Carlson M (2004b) Cyclic AMP-dependent protein kinase regulates the subcellular localization of Snf1-Sip1 protein kinase. Mol Cell Biol 24(5):1836–1843

Hong SP, Carlson M (2007) Regulation of snf1 protein kinase in response to environmental stress. J Biol Chem 282(23):16838–16845

Honigberg SM, Lee RH (1998) Snf1 kinase connects nutritional pathways controlling meiosis in Saccharomyces cerevisiae. Mol Cell Biol 18(8):4548–4555

Hsu HE, Liu TN, Yeh CS, Chang TH, Lo YC, Kao CF (2015) Feedback control of Snf1 protein and its phosphorylation is necessary for adaptation to environmental stress. J Biol Chem 290 (27):16786–16796

Hunter T, Plowman GD (1997) The protein kinases of budding yeast: six score and more. Trends Biochem Sci 22(1):18–22

Janecek S, Svensson B, Macgregor EA (2011) Structural and evolutionary aspects of two families of non-catalytic domains present in starch and glycogen binding proteins from microbes, plants and animals. Enzym Microb Technol 49(5):429–440

Jiang R, Carlson M (1996) Glucose regulates protein interactions within the yeast SNF1 protein kinase complex. Genes Dev 10:3105–3115

Jiang R, Carlson M (1997) The Snf1 protein kinase and its activating subunit, Snf4, interact with distinct domains of the Sip1/Sip2/Gal83 component in the kinase complex. Mol Cell Biol 17:2099–2106

Joseph BK, Liu HY, Francisco J, Pandya D, Donigan M, Gallo-Ebert C, Giordano C, Bata A, Nickels JT Jr (2015) Inhibition of AMP kinase by the protein phosphatase 2A heterotrimer, PP2APpp2r2d. J Biol Chem 290(17):10588–10598

Kaps S, Kettner K, Migotti R, Kanashova T, Krause U, Rodel G, Dittmar G, Kriegel TM (2015) Protein kinase Ymr291w/Tda1 is essential for glucose signaling in saccharomyces cerevisiae on the level of hexokinase isoenzyme ScHxk2 phosphorylation*. J Biol Chem 290 (10):6243–6255

Kayikci O, Nielsen J (2015) Glucose repression in Saccharomyces cerevisiae. FEMS Yeast Res 15 (6):fov068. doi:10.1093/femsyr/fov068

Kuchin S, Treich I, Carlson M (2000) A regulatory shortcut between the Snf1 protein kinase and RNA polymerase II holoenzyme. Proc Natl Acad Sci USA 97(14):7916–7920

Kuchin S, Vyas VK, Carlson M (2002) Snf1 protein kinase and the repressors Nrg1 and Nrg2 regulate FLO11, haploid invasive growth, and diploid pseudohyphal differentiation. Mol Cell Biol 22(12):3994–4000

Kulkarni A, Buford TD, Rai R, Cooper TG (2006) Differing responses of Gat1 and Gln3 phosphorylation and localization to rapamycin and methionine sulfoximine treatment in Saccharomyces cerevisiae. FEMS Yeast Res 6(2):218–229

Leech A, Nath N, McCartney RR, Schmidt MC (2003) Isolation of mutations in the catalytic domain of the snf1 kinase that render its activity independent of the snf4 subunit. Eukaryot Cell 2(2):265–273

Lesage P, Yang X, Carlson M (1996) Yeast SNF1 protein kinase interacts with SIP4, a C6 zinc cluster transcriptional activator: a new role for SNF1 in the glucose response. Mol Cell Biol 16 (5):1921–1928

Li X, Wang L, Zhou XE, Ke J, de Waal PW, Gu X, Tan MH, Wang D, Wu D, Xu HE, Melcher K (2015) Structural basis of AMPK regulation by adenine nucleotides and glycogen. Cell Res 25 (1):50–66

Lin SS, Manchester JK, Gordon JI (2003) Sip2, an N-myristoylated beta subunit of Snf1 kinase, regulates aging in Saccharomyces cerevisiae by affecting cellular histone kinase activity, recombination at rDNA loci, and silencing. J Biol Chem 278(15):13390–13397

Lin YY, Lu JY, Zhang J, Walter W, Dang W, Wan J, Tao SC, Qian J, Zhao Y, Boeke JD, Berger SL, Zhu H (2009) Protein acetylation microarray reveals that NuA4 controls key metabolic target regulating gluconeogenesis. Cell 136(6):1073–1084

Lo WS, Duggan L, Emre NC, Belotserkovskya R, Lane WS, Shiekhattar R, Berger SL (2001) Snf1—a histone kinase that works in concert with the histone acetyltransferase Gcn5 to regulate transcription. Science 293(5532):1142–1146

Ludin K, Jiang R, Carlson M (1998) Glucose-regulated interaction of a regulatory subunit of protein phosphatase 1 with the Snf1 protein kinase in Saccharomyces cerevisiae. Proc Natl Acad Sci USA 95(11):6245–6250

Mangat S, Chandrashekarappa D, McCartney RR, Elbing K, Schmidt MC (2010) Differential roles of the glycogen-binding domains of beta subunits in regulation of the Snf1 kinase complex. Eukaryot Cell 9(1):173–183

Mayer FV, Heath R, Underwood E, Sanders MJ, Carmena D, McCartney RR, Leiper FC, Xiao B, Jing C, Walker PA, Haire LF, Ogrodowicz R, Martin SR, Schmidt MC, Gamblin SJ, Carling D (2011) ADP regulates SNF1, the Saccharomyces cerevisiae homolog of AMP-activated protein kinase. Cell Metab 14(5):707–714

McCartney RR, Rubenstein EM, Schmidt MC (2005) Snf1 kinase complexes with different beta subunits display stress-dependent preferences for the three Snf1-activating kinases. Curr Genet 47(6):335–344

McCartney RR, Schmidt MC (2001) Regulation of Snf1 kinase. Activation requires phosphorylation of threonine 210 by an upstream kinase as well as a distinct step mediated by the Snf4 subunit. J Biol Chem 276(39):36460–36466

Momcilovic M, Iram SH, Liu Y, Carlson M (2008) Roles of the glycogen-binding domain and Snf4 in glucose inhibition of SNF1 protein kinase. J Biol Chem 283(28):19521–19529

Nayak V, Zhao K, Wyce A, Schwartz MF, Lo WS, Berger SL, Marmorstein R (2006) Structure and dimerization of the kinase domain from yeast Snf1, a member of the Snf1/AMPK protein family. Structure 14(3):477–485

Porta C, Paglino C, Mosca A (2014) Targeting PI3K/Akt/mTOR signaling in cancer. Front Oncol 4:64

Roth S, Kumme J, Schuller HJ (2004) Transcriptional activators Cat8 and Sip4 discriminate between sequence variants of the carbon source-responsive promoter element in the yeast Saccharomyces cerevisiae. Curr Genet 45(3):121–128

Rubenstein EM, McCartney RR, Zhang C, Shokat KM, Shirra MK, Arndt KM, Schmidt MC (2008) Access denied: Snf1 activation loop phosphorylation is controlled by availability of the phosphorylated threonine 210 to the PP1 phosphatase. J Biol Chem 283(1):222–230

Rudolph MJ, Amodeo GA, Bai Y, Tong L (2005) Crystal structure of the protein kinase domain of yeast AMP-activated protein kinase Snf1. Biochem Biophys Res Commun 337(4):1224–1228

Ruiz A, Xu X, Carlson M (2011) Roles of two protein phosphatases, Reg1-Glc7 and Sit4, and glycogen synthesis in regulation of SNF1 protein kinase. Proc Natl Acad Sci USA 108 (16):6349–6354

Ruiz A, Xu X, Carlson M (2013) Ptc1 protein phosphatase 2C contributes to glucose regulation of SNF1/AMP-activated protein kinase (AMPK) in Saccharomyces cerevisiae. J Biol Chem 288 (43):31052–31058

Santangelo GM (2006) Glucose signaling in Saccharomyces cerevisiae. Microbiol Mol Biol Rev 70(1):253–282

Sanz P, Alms GR, Haystead TA, Carlson M (2000) Regulatory interactions between the Reg1-Glc7 protein phosphatase and the Snf1 protein kinase. Mol Cell Biol 20(4):1321–1328

Schmidt MC, McCartney RR (2000) beta-subunits of Snf1 kinase are required for kinase function and substrate definition. EMBO J 19(18):4936–4943

Shashkova S, Welkenhuysen N, Hohmann S (2015) Molecular communication: crosstalk between the Snf1 and other signaling pathways. FEMS Yeast Res 15(4):fov026

Shimobayashi M, Hall MN (2014) Making new contacts: the mTOR network in metabolism and signalling crosstalk. Nat Rev Mol Cell Biol 15(3):155–162

Shirra MK, McCartney RR, Zhang C, Shokat KM, Schmidt MC, Arndt KM (2008) A chemical genomics study identifies Snf1 as a repressor of GCN4 translation. J Biol Chem 283 (51):35889–35898

Shirra MK, Patton-Vogt J, Ulrich A, Liuta-Tehlivets O, Kohlwein SD, Henry SA, Arndt KM (2001) Inhibition of acetyl coenzyme A carboxylase activity restores expression of the INO1 gene in a snf1 mutant strain of Saccharomyces cerevisiae. Mol Cell Biol 21(17):5710–5722

Simpson-Lavy KJ, Johnston M (2013) SUMOylation regulates the SNF1 protein kinase. Proc Natl Acad Sci USA 110(43):17432–17437

Smets B, Ghillebert R, De Snijder P, Binda M, Swinnen E, De Virgilio C, Winderickx J (2010) Life in the midst of scarcity: adaptations to nutrient availability in Saccharomyces cerevisiae. Curr Genet 56(1):1–32

Soontorngun N, Larochelle M, Drouin S, Robert F, Turcotte B (2007) Regulation of gluconeogenesis in Saccharomyces cerevisiae is mediated by activator and repressor functions of Rds2. Mol Cell Biol 27(22):7895–7905

Tachibana C, Yoo JY, Tagne JB, Kacherovsky N, Lee TI, Young ET (2005) Combined global localization analysis and transcriptome data identify genes that are directly coregulated by Adr1 and Cat8. Mol Cell Biol 25(6):2138–2146

Thompson-Jaeger S, Francois J, Gaughran JP, Tatchell K (1991) Deletion of SNF1 affects the nutrient response of yeast and resembles mutations which activate the adenylate cyclase pathway. Genetics 129(3):697–706

Treitel MA, Carlson M (1995) Repression by SSN6-TUP1 is directed by MIG1, a repressor/activator protein. Proc Natl Acad Sci USA 92(8):3132–3136

Turcotte B, Liang XB, Robert F, Soontorngun N (2010) Transcriptional regulation of nonfermentable carbon utilization in budding yeast. FEMS Yeast Res 10(1):2–13

Viana R, Towler MC, Pan DA, Carling D, Viollet B, Hardie DG, Sanz P (2007) A conserved sequence immediately N-terminal to the Bateman domains in AMP-activated protein kinase gamma subunits is required for the interaction with the beta subunits. J Biol Chem 282 (22):16117–16125

Vincent O, Carlson M (1999) Gal83 mediates the interaction of the Snf1 kinase complex with the transcription activator Sip4. EMBO J 18(23):6672–6681

Vincent O, Townley R, Kuchin S, Carlson M (2001) Subcellular localization of the Snf1 kinase is regulated by specific beta subunits and a novel glucose signaling mechanism. Genes Dev 15 (9):1104–1114

Wang Z, Wilson WA, Fujino MA, Roach PJ (2001) Antagonistic controls of autophagy and glycogen accumulation by Snf1p, the yeast homolog of AMP-activated protein kinase, and the cyclin- dependent kinase Pho85p. Mol Cell Biol 21(17):5742–5752

Wiatrowski HA, Van Denderen BJ, Berkey CD, Kemp BE, Stapleton D, Carlson M (2004) Mutations in the gal83 glycogen-binding domain activate the snf1/gal83 kinase pathway by a glycogen-independent mechanism. Mol Cell Biol 24(1):352–361

Wilson MA, Koutelou E, Hirsch C, Akdemir K, Schibler A, Barton MC, Dent SY (2011) Ubp8 and SAGA regulate Snf1 AMP kinase activity. Mol Cell Biol 31(15):3126–3135

Wilson WA, Hawley SA, Hardie DG (1996) Glucose repression/derepression in budding yeast: SNF1 protein kinase is activated by phosphorylation under derepressing conditions, and this correlates with a high AMP:ATP ratio. Curr Biol 6(11):1426–1434

Wilson WA, Skurat AV, Probst B, de Paoli-Roach A, Roach PJ, Rutter J (2005) Control of mammalian glycogen synthase by PAS kinase. Proc Natl Acad Sci USA 102(46):16596–16601

Woods A, Dickerson K, Heath R, Hong SP, Momcilovic M, Johnstone SR, Carlson M, Carling D (2005) C(Ca2+)/calmodulin-dependent protein kinase kinase-beta acts upstream of AMP-activated protein kinase in mammalian cells. Cell Metab 2(1):21–33

Woods A, Johnstone SR, Dickerson K, Leiper FC, Fryer LG, Neumann D, Schlattner U, Wallimann T, Carlson M, Carling D (2003) LKB1 is the upstream kinase in the AMP-activated protein kinase cascade. Curr Biol 13(22):2004–2008

Woods A, Munday MR, Scott J, Yang X, Carlson M, Carling D (1994) Yeast SNF1 is functionally related to mammalian AMP-activated protein kinase and regulates acetyl-CoA carboxylase in vivo. J Biol Chem 269(30):19509–19515

Xiao B, Sanders MJ, Carmena D, Bright NJ, Haire LF, Underwood E, Patel BR, Heath RB, Walker PA, Hallen S, Giordanetto F, Martin SR, Carling D, Gamblin SJ (2013) Structural basis of AMPK regulation by small molecule activators. Nat Commun 4:3017

Xiao B, Sanders MJ, Underwood E, Heath R, Mayer FV, Carmena D, Jing C, Walker PA, Eccleston JF, Haire LF, Saiu P, Howell SA, Aasland R, Martin SR, Carling D, Gamblin SJ (2011) Structure of mammalian AMPK and its regulation by ADP. Nature 472(7342):230–233

Yao Y, Tsuchiyama S, Yang C, Bulteau AL, He C, Robison B, Tsuchiya M, Miller D, Briones V, Tar K, Potrero A, Friguet B, Kennedy BK, Schmidt M (2015) Proteasomes, Sir2, and Hxk2 form an interconnected aging network that impinges on the AMPK/Snf1-regulated transcriptional repressor Mig1. PLoS Genet 11(1), e1004968

Ye T, Elbing K, Hohmann S (2008) The pathway by which the yeast protein kinase Snf1p controls acquisition of sodium tolerance is different from that mediating glucose regulation. Microbiology 154(Pt 9):2814–2826

Young ET, Dombek KM, Tachibana C, Ideker T (2003) Multiple pathways are co-regulated by the protein kinase Snf1 and the transcription factors Adr1 and Cat8. J Biol Chem 278 (28):26146–26158

Zhang J, Olsson L, Nielsen J (2010) The beta-subunits of the Snf1 kinase in Saccharomyces cerevisiae, Gal83 and Sip2, but not Sip1, are redundant in glucose derepression and regulation of sterol biosynthesis. Mol Microbiol 77(2):371–383

Chapter 15
5′-AMP-Activated Protein Kinase Signaling in *Caenorhabditis elegans*

Moloud Ahmadi and Richard Roy

Contents

Abstract AMP-activated protein kinase (AMPK) is one of the central regulators of cellular and organismal metabolism in eukaryotes. Once activated by decreased energy levels, it induces ATP production by promoting catabolic pathways while conserving ATP by inhibiting anabolic pathways. AMPK plays a crucial role in various aspects of cellular function such as regulating growth, reprogramming metabolism, autophagy, and cell polarity. In this chapter, we focus on how recent breakthroughs made using the model organism *Caenorhabditis elegans* have contributed to our understanding of AMPK function and how it can be utilized in the future to elucidate hitherto unknown aspects of AMPK signaling.

Keywords AMPK • AAK-2 • *Caenorhabditis elegans* • Diapause • Dauer • Germline stem cell quiescence • Lifespan • Longevity

M. Ahmadi • R. Roy (✉)
Department of Biology, McGill University, 1205 Avenue Docteur Penfield, Montreal, QC, Canada H3A 1B1
e-mail: richard.roy@mcgill.ca

M.D. Cordero, B. Viollet (eds.), *AMP-activated Protein Kinase*, Experientia Supplementum 107, DOI 10.1007/978-3-319-43589-3_15

15.1 Introduction

5′-AMP-activated protein kinase (AMPK) is a metabolic sensor that plays a central role in regulating fuel homeostasis. Like other organisms, in *Caenorhabditis elegans* AMPK targets a wide spectrum of substrates involved in diverse cellular processes that all contribute to appropriate adaptation during conditions of stress, particularly energy stress. The *C. elegans* genome encodes two catalytic α subunits, *aak-1* and *aak-2*; two β regulatory subunits, *aakb-1* and *aakb-2*; and five γ regulatory subunits called *aakg-1 aakg-2*, *aakg-3*, *aakg-4*, *and aakg-5* (Apfeld et al. 2004; Beale 2008). As is the case in most metazoa, AMPK activation is mainly achieved through phosphorylation of Thr^{243} by PAR-4, the LKB-1 orthologue in *C. elegans* (Narbonne and Roy 2006). AMPK has been shown to enhance lifespan and autophagy and to modify the feeding and foraging behavior of starved animals (Mair et al. 2011; Egan et al. 2011; Mihaylova and Shaw 2011; Lee et al. 2008; Lemieux et al. 2015; Cunningham et al. 2012).

The α subunits of *C. elegans* AMPK are striking in that 30% of the residues are identical and 78% are similar to the human AMPK catalytic subunits. Notably, the β subunit alignment also shows 30% identity and 84% similarity to human AMPK β subunit. Interestingly, the *C. elegans* β subunits are most similar to human γ 1 subunits with the maximum homology in the AMP/ATP binding Bateman domains. Because of this high degree of homology shared between the AMPK subunits of *C. elegans* and humans, and that *C. elegans* mutants that lack all AMPK signaling are viable, many laboratories have turned to *C. elegans* as a model to identify and characterize the molecular targets and the associated cellular roles of AMPK in animal development and physiology (Beale 2008).

A functional AMPK/GFP fusion protein is expressed in several tissues during hermaphrodite development including the pharynx, the head and ventral cord neurons, hermaphrodite-specific neuron (HSN), body wall muscles, the vulva, the excretory canal, distal tip cells, spermatheca, sheath cells, and intestine (Lee et al. 2008). This broad expression pattern suggests that AMPK may have unforeseen roles in the correct execution of several important signaling pathways that range from organogenesis and tissue maintenance to stem cell integrity, metabolism, and behavior.

In this chapter, we highlight the value of *C. elegans* as a genetically tractable metazoan to elaborate on the known and previously uncharacterized AMPK signaling pathways, their targets in specific tissues, and how modification of these processes may be useful for discovery of new therapeutic targets. By using an unbiased genetic approach, novel unpredicted targets can be identified that could help advance our treatment of many diseases that impinge upon the broad spectrum of AMPK-dependent processes, while also providing a means of developing novel diagnostic tools.

15.2 An Overview on *C. elegans* as a Model Organism

One of the most striking features of *C. elegans* is the reproductive plasticity the animal possesses that is intrinsic to its life cycle. This facilitates dispersal and delays reproduction during environmental duress, enhancing their ability to occupy niches of varying extreme. Prior to a commitment to reproductive development, *C. elegans* can undergo a series of reversible developmental arrests or "diapauses" at several points throughout their developmental progression. For example, if upon hatching the environment is resource depleted, *C. elegans* hatchlings will arrest in the first larval stage (L1) and execute a diapause. This period of quiescence is reversible so when appropriate food resources become available this period of quiescence has little to no consequence on later fitness. In addition, if the animals sense suboptimal growth conditions such as continued reduced resources, high growth temperature or population density later after they have initiated postembryonic development, the animals can opt for an alternative developmental trajectory referred to as "dauer" that allows them to survive under very harsh conditions where other organisms could never thrive (Wood et al. 1980). Population density increases the levels of a constitutively synthesized pheromone signal that induces dauer arrest while also blocking dauer recovery. Likewise, guanylyl cyclase, TGFB-like, insulin-like, and steroid/nuclear receptor signaling pathways together constitute an evolutionarily conserved signal transduction module that regulates dauer formation in both an independent and collective manner (Hu 2005).

The onset of dauer development is associated with numerous morphological and physiological changes that can protect the free-living and the parasitic larva from its environment. Both in *C. elegans* or in parasitic nematodes, these dauer-associated adaptive responses greatly enhance fitness in environments that are inherently variable. But these advantages come at a price, namely, long-term survival is counterbalanced by a suspension of reproduction until conditions improve (Hubbard et al. 2013; Fielenbach and Antebi 2008; Bargmann 2006).

Despite the apparent quiescence of dauer larvae, several thousand genes continue to be expressed and are critical for their survival. If environmental conditions improve, dauer larvae will recover and continue their life cycle and grow to reproductive adulthood. Similar changes are triggered in some mammals faced with environmental and nutrient stress which eventually leads to quiescence (Finch 1994).

15.3 Gonadal Function of AMPK

C. elegans germline is supported by two gonadal arms in which germline stem cells (GSCs) proliferate at the distal ends and differentiate more proximally into mature gametes. The entire reproductive apparatus arises from a gonadal primordium that consists of two somatic gonad precursors and two primordial germ cells (PGCs) present at the first larval (L1) stage (Hirsh et al. 1976; Kimble and Hirsh 1979). The PGCs are born during embryogenesis at the 100 cell stage and remain quiescent thereafter until the animal emerges from the egg.

Upon exposure to food, these four cells that have remained quiescent for much of embyrogenesis resume mitotic cell divisions and initiate development (Hirsh et al. 1976; Kimble and Hirsh 1979). Upon initiation of postembryonic development, cell divisions will begin but will arrest again in the L1 stage and then resume proliferation and differentiation later in the L3 stage (Killian and Hubbard 2004).

Upon the initiation of differentiation, the proliferative zone of the germ line is restricted to the area distal to the mitosis/meiosis border. All the germ cells proximal to this border are differentiated (in prophase of meiosis I) (Hubbard et al. 2013). Of note, many factors have been revealed to influence the expansion of the germline proliferative zone, making these physiologically sensitive progenitor cells an interesting model for understanding developmental and physiological control of cell proliferation (Killian and Hubbard 2004, 2005).

The germ line is subjected to environmentally regulated diapause points that allow *C. elegans* to survive the environmental conditions ranging from deserts to the Antarctic. AMPK signaling has been revealed as one of the conserved mechanisms that mediates the sensitivity of germline development in response to such challenges.

Dauer formation results in the establishment of a G2 phase cell cycle arrest and ultimately the inhibition of germline proliferation (Narbonne and Roy 2006). In a genetic screen conducted to identify factors required for germline stem cell quiescence, AMPK signaling was identified as one of the genetic mechanisms that limits the capacity of germline proliferation. Notably, AMPK function was found to be crucial for the establishment of GSC quiescence as the removal of any one of the AMPK subunit orthologues, namely, *aak-1*, *aak-2*, or *aakb*-1 leads to supernumerary germ cell divisions during the dauer stage. AMPK therefore ensures that GSC proliferation is restricted during periods when inadequate resources are available to fuel critical cellular processes (Narbonne and Roy 2006; Narbonne et al. 2010).

Like most organisms, the optimal activation of AMPK in *C. elegans* requires PAR-4(LKB1)/STRD-1-dependent phosphorylation at Thr^{243}. In the dauer germ line, AMPK acts downstream of PAR-4 to coordinate proliferation with somatic development in response to all three dauer signaling pathways, although compound mutant analysis suggests that PAR-4/LKB1 may have additional AMPK-independent effects on germline quiescence (Narbonne and Roy 2006). Studies performed in mammalian systems suggest that the TSC1/2/mTOR (target of rapamycin) pathway may contribute to quiescence in various tissues (Avruch et al. 2006). However, based on bioinformatic analyses, there is no clear TSC1/2 orthologue annotated in the *C. elegans* genome, suggesting that alternative mechanisms must exist that control quiescence downstream of AMPK in the GSCs during the dauer stage in *C. elegans* that are independent of TSC1/2 (Gwinn et al. 2008).

Like the GSCs that escape quiescence in the dauer stage, compromise of either of the two catalytic subunits results in an abnormal PGC proliferation defect and reduced viability during L1 diapause. AMPK signaling is therefore also required to maintain quiescence and presumably to reallocate energy resources in response to food deprivation upon hatching (Fukuyama et al. 2012). Similarly, many of these phenotypes are mirrored by the *C. elegans* PTEN orthologue *daf-18* (Baugh and Sternberg 2006; Sigmond et al. 2008). Although intestinal and pan-neuronal promoters restored the survival of AMPK mutant animals, neither of these tissue-specific transgenes could suppress the abnormal proliferation of PGCs during the L1 diapause consistent with AMPK acting in distinct tissues for survival and to establish and maintain PGC quiescence during the L1 diapause (Fukuyama et al. 2012).

The mechanisms by which AMPK regulates germline quiescence remain unknown (Fukuyama et al. 2012). Previous findings from mammalian studies suggest that AMPK may function to promote the activity of p53 homologue *cep-1* or the cyclin-dependent kinase inhibitor orthologues *cki-1* and *cki-2* to ultimately modulate cell cycle arrest (Tobin and Saito 2012). However, since both *cep-1*/p53/ and *cki-1*/*-2* modulate their targets predominantly at the G1-S transition, it is difficult to account for their role in maintaining the typical G2 arrest that is observed in the PGCs and the GSCs during periods of quiescence based on their canonical mode of action. To elucidate the mechanism(s) by which AMPK regulates cell cycle quiescence, the identification of additional AMPK substrates that may ultimately link energy stress and cell cycle arrest will be essential.

15.4 AMPK Regulates Fat Rationing During the Dauer Stage

In situations of cyclical climate change, many organisms will accumulate fuel resources during the periods of abundance that precede a long cold or dry season. By altering its metabolism accordingly prior to long periods of resource scarcity, the *C. elegans* dauer larva can augment its triglyceride stockpile to act as an energy source during periods when the animal cannot feed. The utilization of this resource must therefore be tightly managed as premature exhaustion of the fuel stockpile could jeopardize survival.

Unlike most organisms, animals that lack all AMPK signaling are viable in *C. elegans*. However, AMPK must affect critical pathways to ensure L1 survival through phosphorylation of targets that have yet to be determined. On the other hand, AMPK mutant dauer larvae die prematurely despite the transient accumulation of fat upon dauer entry. AMPK mutant dauers hydrolyze and deplete their accumulated fat reservoirs in less than 48 h after dauer formation (Narbonne and Roy 2009). LKB1/AMPK signaling mediates the survival of dauer animals by functioning within both the skin-like hypodermis, which is a major site of *C. elegans* fat storage, and in the kidney-like excretory cell. At the time of dauer entry, *aak-2* mutants display abundant fat accumulation and normal osmoregulation. However, animals quickly deplete their fat reservoirs due to the increased activity of the lipase, ATGL-1, the orthologue of mammalian adipose triglyceride lipase. Moreover, they also become osmosensitive resulting in a rapid decline in viability. Taken together, AMPK ensures long-term dauer survival by limiting fat utilization by downregulation of ATGL-1 activity all the while ensuring appropriate osmoregulation (Narbonne and Roy 2009).

AMPK directly phosphorylates serine 303 on adipose triglyceride lipase (ATGL-1), to release a single free fatty acid (Narbonne and Roy 2009; Cunningham and Ashrafi 2009). It therefore regulates the rate-limiting step in triglyceride hydrolysis and hence energy production from this source. This ATGL-1 inactivation following AMPK phosphorylation is triggered by the generation of 14-3-3 binding sites on ATGL-1, which are recognized by the *C. elegans* 14-3-3 protein orthologue PAR-5, resulting in sequestration of ATGL-1 away from the lipid droplets and subsequent proteasome-mediated degradation (Xie and Roy 2015).

This AMPK-dependent block in lipid hydrolysis is somewhat counterintuitive considering its role in mammalian cells that have been subjected to energy stress. Activation of AMPK is reflected by a direct phospho-inhibition of ACC1/2 which affects both lipid biosynthesis indirectly by relieving the malonyl CoA-mediated inhibition of CPT. The result of this is an increase in fatty acid transport to the mitochondria, subsequent β-oxidation, and increased energy production from lipid (Mihaylova and Shaw 2011). Although the differences are striking, they may be justified based on the timescales associated with each of the two physiological responses: one being activated on a very short timescale (minutes/h) following starvation and the other dominating during the comparatively longer timescales typical of dauer survival (several months amounting to 10 times the normal lifespan of the animal).

The survival of AMPK mutant dauers is highly dependent on generating novel energy sources. One such mechanism includes regulation of the abundance and the nature of the fatty acid content by increasing the HIF-1-dependent expression of the enzymes that are critical for fatty acid biosynthesis. This phenomenon can occur downstream of reduced catalase activity which allows cellular levels of hydrogen peroxide to progressively increase during dauer. This metabolic by-product is often vilified because of its ROS-generating capacity although at low levels it ensures readjustment of lipid biosynthetic capacity downstream of HIF-1 activation to compensate for cellular energy deficiencies in the AMPK mutants (Xie and Roy 2012). Similar metabolic readjustments can also occur in tumor cells and are often associated with the Warburg effect, where energy-consuming pathways are modified toward a growth condition of these rogue cells, while consequently compromising their energetic efficiency (Deberardinis et al. 2008).

If tumor cells can generate their own energy through activation of HIF-1, they can become less dependent on nutrient delivery. This property could potentially enhance their capacity for unscheduled growth. As AMPK has been shown to play an important role in blocking cell growth in response to nutrient/energy stress or during oxidative stress and in certain cell types loss of AMPK does lead to increased HIF levels (Shackelford et al. 2009), it is paramount to investigate whether activation of HIF-1 is associated with the loss of AMPK or its upstream kinase LKB1 in various tumors, or in patients with tumor predisposing diseases such as Peutz–Jeghers Syndrome.

15.5 AMPK: A Modulator of Antiaging Signaling Network

The lifespan of most animals is organism specific and it appears to be selected for some optimal time duration that does not necessarily correspond to the maximal capacity of the organism in question. Through studies in model organisms, it has since become clear that our lifespan is by no means finite in length, and it can be extended under specific genetic and metabolic conditions. This can be achieved by impinging on a number of genetic pathways. First, by limiting caloric intake or by altering the activity of single genes within the defined pathways that modulate metabolism, mutant organisms can live significantly longer than wild-type animals (Schulz et al. 2007; Hansen et al. 2007; Mair et al. 2011). Various regimens of calorie restriction in which macronutrient sources of protein, carbohydrate, and lipid are scarce can consequently extend lifespan through mechanisms that are as of yet unclear. As a consequence of calorie restriction, physical aging is slowed and there is an increased resistance to disease (Longo and Finch 2003; Bordone and Guarente 2005). Alternatively, modification of the insulin signaling pathway also results in significant lengthening of lifespan, while mitochondrial homeostasis also influences lifespan, although whether this arises through increased or decreased generation of ROS is still somewhat controversial (Yee et al. 2014).

C. elegans has proven to be an excellent model to test many of these concepts, while its genetic tractability has provided particularly valuable information concerning the molecular genetic mechanisms that underlie our genetically encoded lifespan. Given the importance of energy homeostasis in the genetic regulation of aging, it is not surprising that AMPK has surfaced as a key regulator of lifespan determination by converging on an integrated signaling network that intersects with multiple genetic pathways including insulin-like signaling, enhanced SIR-2.1 deacetylase activity, and/or mitochondrial homeostasis.

15.5.1 Insulin/IGF-1/FOXO Signaling

Mutations in the *C. elegans* dauer formation abnormal gene *daf-2*/InR provided the first genetically characterized long-lived mutants where loss-of-function mutations conferred a greater than twofold extension to the wild-type lifespan (Kenyon et al. 1993). Molecular analyses revealed that many of the genes that extended lifespan were indeed implicated in insulin signaling: DAF-2 was identified as the *C. elegans* orthologue of the mammalian insulin/IGF-1 receptor (IGFR), AGE-1 *C. elegans* orthologue of PI 3-kinase (PI3K), and DAF-16/*C. elegans* orthologue of forkhead box protein O (FOXO) and together they form the components of a conserved insulin-like signaling pathway (Morris et al. 1996; Kimura et al. 1997; Lin et al. 1997; Ogg et al. 1997; Paradis and Ruvkun 1998).

Insulin/IGF-1 signaling is one of the key regulators of metabolism and energy homeostasis stimulating glucose uptake and growth signaling pathways, thereby coordinating food intake and cellular energy homeostasis by promoting anabolic processes. The DAF-16/FOXO transcription factor modulates growth, metabolism, and aging in *C. elegans*. The serine/threonine protein kinases AKT-1/2 and SGK-1 respond to activation of the insulin receptor DAF-2 and phosphorylate and inactivate the DAF-16/FOXO by preventing its translocation to the nucleus and thereby blocking transcription of its target genes. As expected, abolished insulin/IGF-1 signaling (e.g., using mutations in *daf-2*/InR) relieves this block resulting in DAF-16/FOXO translocation into the nucleus where it induces the expression of genes that are involved in resistance to various stresses that ultimately affect lifespan determination (Paradis and Ruvkun 1998; Hertweck et al. 2004).

Exposure of *daf-2*/InR mutants and wild type animals to starvation, heat stress or mitochondrial poisoning increases the intracellular AMP to ATP ratios, resembling a condition that activates AMPK (Apfeld et al. 2004). Consistent with this, *aak-2* functions both in a linear pathway and in parallel with *daf-16*/FOXO to extend lifespan of *daf-2*/InR mutants, at least partially through inhibition of the CREB-regulated transcriptional coactivator CRTC-1 (Apfeld et al. 2004; Mair et al. 2011). Interestingly, it was revealed that *crtc-1* specifically mediates the longevity output of AMPK through the direct phosphorylation of CRTC-1 in the nervous system resulting in 14-3-3 binding, cytosolic sequestration, and subsequent inactivation, leading to longevity and systemic changes in metabolic transcription (Burkewitz et al. 2015).

15.5.2 TOR Signaling

The TOR pathway plays a key role in the regulation of growth and reproduction in response to nutrient and energy availability. As such, blocking its activity has been shown to extend lifespan in a variety of species (Kapahi et al. 2010). Inhibition of TOR signaling and reduced insulin/IGF-1 signaling both converge on common transcriptional targets and non-transcriptional effectors and these likely function together since blocking TOR signaling does not further extend the increased lifespan of *daf-2*/InR mutants (Vellai et al. 2003; Hansen et al. 2007).

In mammals, TOR exists in the form of TORC1 and TORC2 complexes, which contain the DAF-15/Raptor and RICT-1/Rictor coactivators, respectively. These two complexes act in independent genetic pathways to affect distinct cellular processes; TORC1 controls cell growth and proliferation in response to nutrient signals largely by regulating protein synthesis, whereas TORC2 modulates cell morphology. In *C. elegans*, TORC1 modulates longevity at least partially via the GTPases RAGA-1/RAGC-1 (Robida-Stubbs et al. 2012; Schreiber et al. 2010), RHEB-1/Rheb (Honjoh et al. 2009), and DAF-15/Raptor (Jia et al. 2004). In addition, TOR may also exert its role in the modulation of longevity by non-transcriptional mechanisms such as TORC1-mediated phosphorylation of S6 kinase (S6K), which is one of the principal growth-promoting targets of TOR through its role in promoting translation. Interestingly, the combination of mutations in the insulin signaling *daf-2*/InR and the *rsks-1* (S6K) pathways in *C. elegans* produces a synergistic lifespan extension that requires the positive feedback regulation of DAF-16/FOXO which is AMPK dependent (Greer et al. 2007; Chen et al. 2013). This positive feedback is most likely achieved by the direct phosphorylation of DAF-16/FOXO via AMPK (Greer et al. 2007), which eventually leads to regulation of *aakg-4* (a gene encoding an atypical γ subunit of AMPK) by DAF-16/FOXO promoting AMPK activation (Tullet et al. 2014).

15.5.3 Autophagy

Autophagy has recently been identified as a significant player in the genetic determination of lifespan and it is also affected by activated AMPK in *C. elegans*. Autophagy is a self-degradative process that has emerged as a modulator of longevity in both the TOR and the insulin-IGF-1 signaling pathways as well as in the dietary-restriction paradigm (Lapierre et al. 2011). *daf-2*/InR mutants display enhanced autophagy that is AMPK dependent and which contributes significantly to its effects on lifespan extension (Egan et al. 2011). Although somewhat speculative, autophagy may extend lifespan through the inhibition of protein aggregation, thereby delaying the collapse of proteostasis observed during aging (Florez-McClure et al. 2007; Ben-Zvi et al. 2009). In mammals, AMPK directly phosphorylates the autophagy initiating kinase, UNC-51 (ATG-1/ULK-1), which consequently induces autophagy. Consistent with this, both AMPK and ULK1 are necessary for inducing autophagy upon reduced insulin signaling in *C. elegans* (Egan et al. 2011).

15.5.4 Dietary Restriction

Dietary restriction (DR) is a conserved lifespan-extending paradigm which increases longevity in a manner that is independent of insulin signaling (Lapierre et al. 2011). Intriguingly, AMPK controls longevity in the diverse food deprivation regimens via distinct signaling pathways. It is demonstrated that the lifespan extension can also be acquired by direct interaction of AMPK and DAF-16/FOXO signaling independent of *daf-2*/InR signaling upon feeding restrictive amounts of bacteria which resemble DR resulting in increased AMP/ATP ratios (Greer et al. 2007). By contrast, AMPK modulates longevity independent of *daf-2*/*daf-16* signaling when treated with Metformin which induces DR-like state. This effect is achieved through the AMPK-dependent induction of nuclear localization of the oxidative stress-responsive transcription factor SKN-1/Nrf2 which eventually confers Metformin-dependent lifespan extension (Onken and Driscoll 2010). Taken together, these findings suggest an important energy-sensing AMPK–FOXO pathway that mediates lifespan extension. However, AMPK also functions independently of DAF-16/FOXO to elicit lifespan extension by resveratrol, a natural polyphenol that mimics some aspects of DR, suggesting that activation of AMPK is not always necessarily coupled to DAF-16/FOXO in order for it to extend lifespan (Greer and Brunet 2009; Parker et al. 2005).

15.5.5 Mitochondrial Control of Lifespan

Metabolic stress and ROS signaling have been proposed as causative factors in determining longevity through involvement of mitochondria (Hekimi and Guarente 2003; Schriner et al. 2005). AMPK signaling plays a key role in the mitochondrial regulation of lifespan. Accordingly, AMPK activation under reduced insulin signaling has been shown to stimulate L-proline catabolism that eventually generates a transient ROS signal in the mitochondria that results in an adaptive induction of endogenous stress defenses to extend lifespan (Zarse et al. 2012). Similarly, impairment of glycolysis using inhibitors increases lifespan in an *aak-2*-dependent manner due to enhanced mitochondrial respiration that causes ROS-dependent induction of stress resistance (Haesa et al. 2014). By contrast, compound-based inhibition of mitochondrial complex I using rotenone induces mitochondrial ROS signal leading to lifespan extension that is independent of AMPK/AAK-2 signaling, suggesting that AMPK is dispensable for sensing ATP depletion that occurs upon inhibition of complex I in *C. elegans* (Schmeisser et al. 2013).

The effect of AMPK signaling on aging has received much attention in recent years. Given the conserved AMPK signaling and targets in *C. elegans*, *Drosophila*, and humans, testing the role of AMPK and its targets in longevity will yield valuable insight into the biology of aging.

15.6 Concluding Remarks

AMPK modulates cellular energy metabolism by its direct and indirect regulation of gene transcription and key metabolic enzymes. AMPK-mediated regulation of lipid metabolism occurs at multiple points in the complex reactions that control lipid homeostasis, while the same is true in its regulation of enzyme function in glucose metabolism. But even though AMPK has been well characterized as a key regulator of metabolism, there is now accumulating evidence that suggests the involvement of AMPK in almost all aspects of cellular function, including cell polarity, cell growth and proliferation, autophagy, maintenance of mitochondrial homeostasis, and gene expression. It has been shown that both tumor cells and viruses develop mechanisms that lead to downregulation of AMPK which allow them to escape from its restraining influences on growth and biosynthesis (Hardie 2011). At the whole-body level, AMPK function is required for modulation of many physiological processes such as appetite, energy expenditure, as well as substrate utilization in response to exercise, cytokines, and nutrients (Hardie 2011).

Various diseases have been linked to altered AMPK signaling. However, despite the obvious requirement of AMPK signaling under a number of different conditions and treatments, the quantitative role of AMPK in the modulation of physiological responses has been less extensive, potentially because only a minority of the critical AMPK substrates have been identified and characterized biochemically. By using in vivo genetic models of AMPK loss of function, greater understanding could be achieved regarding the significance of AMPK signaling to the etiology of type 2 diabetes as well as its implication in obesity treatment, cancer therapy, or potentially even in developing improvements in the quality of life in the elderly.

Given the accessibility and depth of knowledge accrued by the community of *C. elegans* researchers, this extraordinary model organism will continue to provide an exceptional wealth of knowledge for future studies to help clarify the role of AMPK in development and in the modulation of metabolism and physiology in response to stress.

References

Apfeld J, O'Connor G, McDonagh T, DiStefano PS, Curtis R (2004) The AMP-activated protein kinase AAK-2 links energy levels and insulin-like signals to lifespan in C. elegans. Genes Dev 15:3004–9

Avruch J, Hara K, Lin Y, Liu M, Long X, Ortiz-Vega S, Yonezawa K (2006) Insulin and amino-acid regulation of mTOR signalling and kinase activity through the Rheb GTPase. Oncogene 16:6361–6372

Bargmann CI (2006) Chemosensation in C. elegans. Worm Book

Baugh LR, Sternberg PW (2006) DAF-16/FOXO regulates transcription of cki-1/Cip/Kip and repression of lin-4 during C. elegans L1 arrest. Curr Biol 16:780–785

Beale EG (2008) 5′-AMP-activated protein kinase signalling in Caenorhabditis elegans. Exp Biol Med (Maywood) 233:12–20

Ben-Zvi A, Miller EA, Morimoto RI (2009) Collapse of proteostasis represents an early molecular event in Caenorhabditis elegans aging. Proc Natl Acad Sci 106:14914–14919

Bordone L, Guarente L (2005) Calorie restriction, SIRT1 and metabolism: understanding longevity. Nat Rev Mol Cell Biol 6:298–305

Burkewitz K, Morantte I, Weir HJM, Yeo R, Zhang Y, Huynh FK, Ilkayeva OR, Hirschey MD, Grant AR, Mair WB (2015) Neuronal CRTC-1 governs systemic mitochondrial metabolism and lifespan via a catecholamine signal. Cell 160:842–855

Chen D, Li PW, Goldstein BA, Cai W, Thomas EL, Chen F, Hubbard AE, Melov S, Kapahi P (2013) Germline signalling mediates the synergistically prolonged longevity produced by double mutations in daf-2 and rsks-1 in C. elegans. Cell Rep 5:1600–1610

Cunningham KA, Ashrafi K (2009) Fat rationing in dauer times. Cell Metab 9:113–114

Cunningham KA, Hua Z, Srinivasan S, Liu J, Lee BH, Edwards RH, Ashrafi K (2012) AMP-activated kinase links serotonergic signaling to glutamate release for regulation of feeding behavior in C. elegans. Cell Metab 16:113–121

DeBerardinis RJ, Lum JJ, Hatzivassiliou G, Thompson CB (2008) The biology of cancer: metabolic reprogramming fuels cell growth and proliferation. Cell Metab 7:11–20
Egan DF, Shackelford DB, Mihaylova MM, Gelino S, Kohnz RA, Mair W, Vasquez DS, Joshi A, Gwinn DM, Taylor R, Asara JM, Fitzpatrick J, Dillin A, Viollet B, Kundu M, Hansen M, Shaw RJ (2011) Phosphorylation of ULK1 (hATG1) by AMP activated protein kinase connects energy sensing to mitophagy. Science 331:456–461
Fielenbach N, Antebi AC (2008) C. elegans dauer formation and the molecular basis of plasticity. Genes Dev 22:2149–2165
Finch CE (1994) Longevity, senescence and the genome. University of Chicago Press, Chicago
Florez-McClure ML, Hohsfield LA, Fonte G, Bealor MT, Link CD (2007) Decreased insulin-receptor signalling promotes the autophagic degradation of beta-amyloid peptide in C. elegans. Autophagy 3:569–580
Fukuyama M, Sakuma K, Park R, Kasuga H, Nagaya R, Atsumi Y, Shimomura Y, Takahashi S, Kajiho H, Rougvie A, Kontani K, Katada T (2012) C. elegans AMPKs promote survival and arrest germline development during nutrient stress. Biol Open 1:929–936
Greer EL, Brunet A (2009) Different dietary restriction regimens extend lifespan by both independent and overlapping genetic pathways in C. elegans. Aging Cell 8:113–127
Greer EL, Dowlatshahi D, Banko MR, Villen J, Hoang K, Blanchard D, Gygi SP, Brunet A (2007) An AMPK-FOXO pathway mediates longevity induced by a novel method of dietary restriction in C. elegans. Curr Biol 17:1646–1656
Gwinn DM, Shackelford DB, Egan DF, Mihaylova MM, Mery A, Vasquez DS, Turk BE, Shaw RJ (2008) AMPK phosphorylation of raptor mediates a metabolic checkpoint. Mol Cell 30:214–226
Haesa WD, Frooninckxa L, Asschea RV, Smoldersb A, Depuydta G, Billenc J, Braeckmanb BP, Schoofsa L, Temmermana L (2014) Metformin promotes lifespan through mitohormesis via the peroxiredoxin PRDX-2. Proc Natl Acad Sci 111:2501–2509
Hansen M, Taubert S, Crawford D, Libina N, Lee SJ, Kenyon C (2007) Lifespan extension by conditions that inhibit translation in Caenorhabditis elegans. Aging Cell 6:95–110
Hardie DG (2011) AMP-activated protein kinase—an energy sensor that regulates all aspects of cell function. Genes Dev 25:1895–1908
Hekimi S, Guarente L (2003) Genetics and the specificity of the aging process. Science 299:1351–1354
Hertweck M, Göbel C, Baumeister RC (2004) C. elegans SGK-1 is the critical component in the Akt/PKB kinase complex to control stress response and life span. Dev Cell 6:577–88
Hirsh D, Oppenheim D, Klass M (1976) Development of the reproductive system of Caenorhabditis elegans. Dev Biol 49:200–219
Honjoh S, Yamamoto T, Uno M, Nishida E (2009) Signalling through RHEB-1 mediates intermittent fasting-induced longevity in C. elegans. Nature 457:726–730
Hu PJ (2005) The online review of C. elegans biology. Worm Book
Hubbard EJ, Korta DZ, Dalfó D (2013) Physiological control of germline development. Adv Exp Med Biol 757:101–131
Jia K, Chen D, Riddle DL (2004) The TOR pathway interacts with the insulin signalling pathway to regulate C. elegans larval development, metabolism and life span. Development 131:3897–3906
Kapahi P, Chen D, Rogers AN, Katewa SD, Li PW, Thomas EL, Kockel L (2010) With TOR, less is more: a key role for the conserved nutrient-sensing TOR pathway in aging. Cell Metab 11:453–465
Kenyon C, Chang J, Gensch E, Rudner A, Tabtiang R (1993) A C. elegans mutant that lives twice as long as wild type. Nature 366:461–464
Killian DJ, Hubbard EJA (2004) C. elegans pro-1 activity is required for soma/germline interactions that influence proliferation and differentiation in the germ line. Development 131:1267–1278

Killian DJ, Hubbard EJA (2005) Caenorhabditis elegans germline patterning requires coordinated development of the somatic gonadal sheath and the germ line. Dev Biol 279:322–335

Kimble J, Hirsh D (1979) The postembryonic cell lineages of the hermaphrodite and male gonads in Caenorhabditis elegans. Dev Biol 70:396–417

Kimura KD, Tissenbaum HA, Liu Y, Ruvkun G (1997) daf-2, an insulin receptor-like gene that regulates longevity and diapause in Caenorhabditis elegans. Science 277:942–946

Lapierre LR, Gelino S, Meléndez A, Hansen M (2011) Autophagy and lipid metabolism coordinately modulate life span in germline-less C. elegans. Curr Biol 21:1507–1514

Lee H, Cho JS, Lambacher N, Lee J, Lee S, Lee TH, Gartner A, Koo H (2008) The Caenorhabditis elegans AMP-activated protein kinase AAK-2 is phosphorylated by LKB1 and is required for resistance to oxidative stress and for normal motility and foraging behavior. J Biol Chem 283:14988–14993

Lemieux GA, Cunningham KA, Lin L, Mayer F, Werb Z, Ashrafi K (2015) Kynurenic acid is a nutritional cue that enables behavioral plasticity. Cell 160:119–131

Lin K, Dorman JB, Rodan A, Kenyon C (1997) daf-16: An HNF-3/forkhead family member that can function to double the life-span of Caenorhabditis elegans. Science 278:1319–22

Longo VD, Finch CE (2003) Evolutionary medicine: from dwarf model systems to healthy centenarians. Science 299:1342–1346

Mair W, Morantte I, Rodrigues AP, Manning G, Montminy M, Shaw RJ, Dillin A (2011) A Lifespan extension induced by AMPK and calcineurin is mediated by CRTC-1 and CREB. Nature 470:404–408

Mihaylova MM, Shaw RJ (2011) The AMP-activated protein kinase (AMPK) signalling pathway coordinates cell growth, autophagy, & metabolism. Nat Cell Biol 13:1016–1023

Morris JZ, Tissenbaum HA, Ruvkun G (1996) A phosphatidylinositol-3-OH kinase family member regulating longevity and diapause in Caenorhabditis elegans. Nature 382:536–9

Narbonne P, Roy R (2006) Inhibition of germline proliferation during C. elegans dauer development requires PTEN, LKB1 and AMPK signalling. Development 133:61–619

Narbonne P, Roy R (2009) Caenorhabditis elegans dauers need LKB1/AMPK to ration lipid reserves and ensure long-term survival. Nature 457:210–214

Narbonne P, Hyenne V, Li S, Labbé J, Roy R (2010) Differential requirements for STRAD in LKB1-dependent functions in C. elegans. Development 137:661–670

Ogg S, Paradis S, Gottlieb S, Patterson GI, Lee L, Tissenbaum HA, Ruvkun G (1997) The Fork head transcription factor DAF-16 transduces insulin-like metabolic and longevity signals in C. elegans. Nature 389:994–999

Onken B, Driscoll M (2010) Metformin induces a dietary restriction-like state and the oxidative stress response to extend C. elegans healthspan via AMPK, LKB1, and SKN-1. PLoS One 5, e8758

Paradis S, Ruvkun G (1998) Caenorhabditis elegans Akt/PKB transduces insulin receptor-like signals from AGE-1 PI3 kinase to the DAF-16 transcription factor. Genes Dev 12:2488–2498

Parker JA, Arango M, Abderrahmane S, Lambert E, Tourette C, Catoire H, Néri C (2005) Resveratrol rescues mutant polyglutamine cytotoxicity in nematode and mammalian neurons. Nat Genet 37:349–50

Robida-Stubbs S, Glover-Cutter K, Lamming DW, Mizunuma M, Narasimhan SD, Neumann-Haefelin E, Sabatini DM, Blackwell TK (2012) TOR signalling and rapamycin influence longevity by regulating SKN-1/Nrf and DAF-16/FoxO. Cell Metab 15:713–724

Schmeisser S, Priebe S, Groth M, Monajembashi S, Hemmerich P, Guthke R, Platzer M, Ristow M (2013) Neuronal ROS signalling rather than AMPK/sirtuin-mediated energy sensing links dietary restriction to lifespan extension. Mol Metab 2:92–102

Schreiber MA, Pierce-Shimomura JT, Chan S, Parry D, McIntire SL (2010) Manipulation of behavioral decline in Caenorhabditis elegans with the Rag GTPase raga-1. PLoS Genet 6, e1000972

Schriner SE, Linford NJ, Martin GM, Treuting P, Ogburn CE, Emond M, Coskun PE, Ladiges W, Wolf N, Van Remmen H, Wallace DC, Rabinovitch PS (2005) Extension of murine life span by overexpression of catalase targeted to mitochondria. Science 308:1909–11

Schulz TJ, Zarse K, Voigt A, Urban N, Birringer M, Ristow M (2007) Glucose restriction extends Caenorhabditis elegans life span by inducing mitochondrial respiration and increasing oxidative stress. Cell Metab 6:280–293

Shackelford DB, Vasquez DS, Corbeil J, Wu S, Leblanc M, Wu CL, Vera DR, Shaw RJ (2009) mTOR and HIF-1alpha-mediated tumor metabolism in an LKB1 mouse model of Peutz-Jeghers syndrome. Proc Natl Acad Sci 106:11137–11142

Sigmond T, Barna J, Tóth ML, Takács-Vellai K, Pásti G, Kovács AL, Vellai T (2008) Autophagy in *Caenorhabditis elegans*. Methods Enzymol 451:521–540

Tobin DV, Saito RM (2012) Developmental decisions: Balancing genetics and the environment by C. elegans. Cell Cycle 11:1666–1671

Tullet JM, Araiz C, Sanders MJ, Au C, Benedetto A, Papatheodorou I, Clark E, Schmeisser K, Jones D, Schuster EF, Thornton JM, Gems D (2014) DAF-16/FoxO directly regulates an atypical AMP-activated protein kinase gamma isoform to mediate the effects of Insulin/IGF-1 signalling on aging in Caenorhabditis elegans. PLoS Genet 10, e1004109

Vellai T, Takacs-Vellai K, Zhang Y, Kovacs AL, Orosz L, Müller F (2003) Genetics: influence of TOR kinase on lifespan in C. elegans. Nature 426:620

Wood WB, Hecht R, Carr S, Vanderslice R, Wolf N, Hirsh D (1980) Parental effects and phenotypic characterization of mutations that affect early development in Caenorhabditis elegans. Dev Biol 74:446–469

Xie M, Roy R (2012) Increased levels of hydrogen peroxide induce a HIF-1-dependent modification of lipid metabolism in AMPK compromised C. elegans dauer larvae. Cell Metab 16:322–35

Xie M, Roy R (2015) AMP-Activated kinase regulates lipid droplet localization and stability of adipose triglyceride lipase in C. elegans dauer larvae. Plos One 10:e0130480

Yee C, Yang W, Hekimi S (2014) The pro-longevity response to mitochondrial ROS in C. elegans is mediated by the intrinsic apoptosis pathway. Cell 157:897–909

Zarse K, Schmeisser S, Groth M, Priebe S, Beuster G, Kuhlow D, Guthke R, Platzer M, Kahn CR, Ristow M (2012) Impaired insulin/IGF1 signalling extends life span by promoting mitochondrial L-proline catabolism to induce a transient ROS signal. Cell Metab 15:451–65

Chapter 16
The Role of AMPK in *Drosophila melanogaster*

Sarah E. Sinnett and Jay E. Brenman

Contents

Abstract In the fruit fly, *Drosophila melanogaster*, mono-allelic expression of AMPK-α, -β, and -γ yields a single heterotrimeric energy sensor that regulates cellular and whole-body energetic homeostasis. The genetic simplicity of *Drosophila*, with only a single gene for each subunit, makes the fruit fly an appealing organism for elucidating the effects of AMPK mutations on signaling pathways and phenotypes. In addition, *Drosophila* presents researchers with an opportunity to use straightforward genetic approaches to elucidate metabolic signaling pathways that contain a level of complexity similar to that observed in mammalian pathways. Just as in mammals, however, the regulatory realm of AMPK function extends beyond metabolic rates and lipid metabolism. Indeed, experiments using *Drosophila* have shown that AMPK may exert protective effects with regard to life span and neurodegeneration. This chapter addresses a few of the research areas in which *Drosophila* has been used to elucidate the physiological functions of AMPK. In doing so, this chapter provides a primer for basic *Drosophila* nomenclature, thereby

S.E. Sinnett (✉)
Gene Therapy Center, University of North Carolina (UNC) at Chapel Hill, Chapel Hill, NC, USA
e-mail: ssinnett@email.unc.edu

J.E. Brenman
Department of Cell Biology and Physiology, Neuroscience Center, UNC Chapel Hill School of Medicine, Chapel Hill, NC, USA

M.D. Cordero, B. Viollet (eds.), *AMP-activated Protein Kinase*, Experientia Supplementum 107, DOI 10.1007/978-3-319-43589-3_16

eliminating a communication barrier that persists for AMPK researchers trained in mammalian genetics.

Keywords Drosophila melanogaster • AMPK • LKB1 • Gal4 • neurodegeneration

16.1 Introduction

16.1.1 Overview of AMPK Signaling in Drosophila

In the fruit fly, mono-allelic expression of AMPK-α, -β, and -γ yields a single heterotrimeric energy sensor that regulates cellular and whole-body energetic homeostasis (Johnson et al. 2010; Pan and Hardie 2002). In contrast to the streamlined fly genome, the more complex mammalian genome encodes multiple AMPK subunit isoforms ($\alpha_{1\text{-}2}$, $\beta_{1\text{-}2}$, and $\gamma_{1\text{-}3}$) that can theoretically form 12 unique heterotrimers (Iseli et al. 2005; Pan and Hardie 2002). Thus, the genetic simplicity of *Drosophila*, with only a single gene for each subunit, makes the fruit fly an appealing organism for elucidating the effects of AMPK mutations on signaling pathways and phenotypes.

Understandably, the phylogenetic separation between *D. melanogaster* and *H. sapiens* may make some AMPK researchers reluctant to explore the elegant genetic tools (e.g., temporally and/or spatially driven gene expression systems) that are easily available to fly geneticists. Two lines of evidence, however, point to the utility of *Drosophila* as a useful model organism for studying AMPK. First, functionally critical amino acids in AMPK subunits are either conserved or replaced by biochemically similar residues across species (Figs. 16.1 and 16.2) (Chen et al. 2012; Kazgan et al. 2010; Pan and Hardie 2002; Spasic et al. 2008; Xiao et al. 2007). These conserved amino acid residues include those that regulate AMPK localization, subunit interactions, and activity. Several examples are highlighted here: (1) Cytoplasmic localization of AMPK-α_2 is controlled by a C-terminal export signal that is conserved in mammals and flies (Kazgan et al. 2010). Deletion of this export signal has been shown to restrict localization of *Drosophila* AMPK (dAMPK)-α to nuclei (Kazgan et al. 2010). (2) The scaffolding β_1 subunit relies on its C-terminal residues (amino acids 186–270 in rat) to bind AMPK-α_1 and -γ_1 in vitro (Iseli et al. 2005). Although N-terminal AMPK-β residues have diverged among species, the C-terminal residues in dAMPK-β are almost perfectly identical to those of human and rat AMPK-β_1 (Fig. 16.1). (3) Finally, amino acid residues that mediate binding of AMP at two regulatory sites in AMPK-γ_1 (i.e., the allosteric activation site and dephosphorylation inhibition site) are conserved in *Drosophila* (Fig. 16.2) (Chen et al. 2012; Xiao et al. 2007). In addition, key AMPK-γ amino acid residues that have been shown

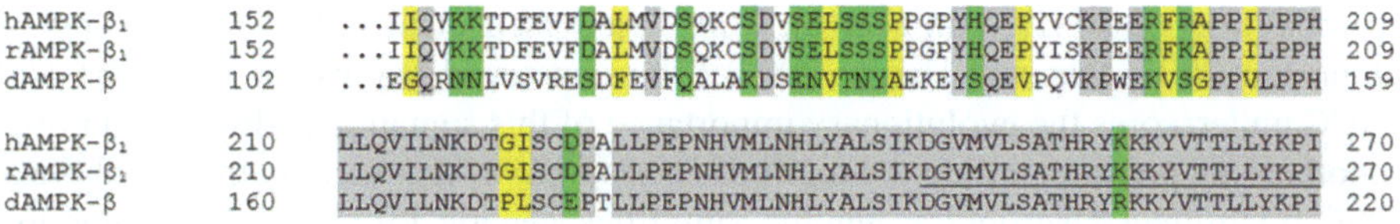

Fig. 16.1 *Conservation of carboxy-terminus residues in AMPK-β*. Most amino acids in the C-terminus of human AMPK-β_1 (NM_006253.4), rat AMPK-β_1 (GenBank: X95577.1), and *Drosophila* AMPK-β (NP_995783.1) are either invariant (*gray*) or have residues with conserved hydrophilicity (*green*) or hydrophobicity (*yellow*). Amino acids 246–270 (*underlined*) in rAMPK-β_1 are required for binding to AMPK-γ; amino acids 186–270 in rAMPK-β_1 are required for heterotrimer formation (Iseli et al. 2005)

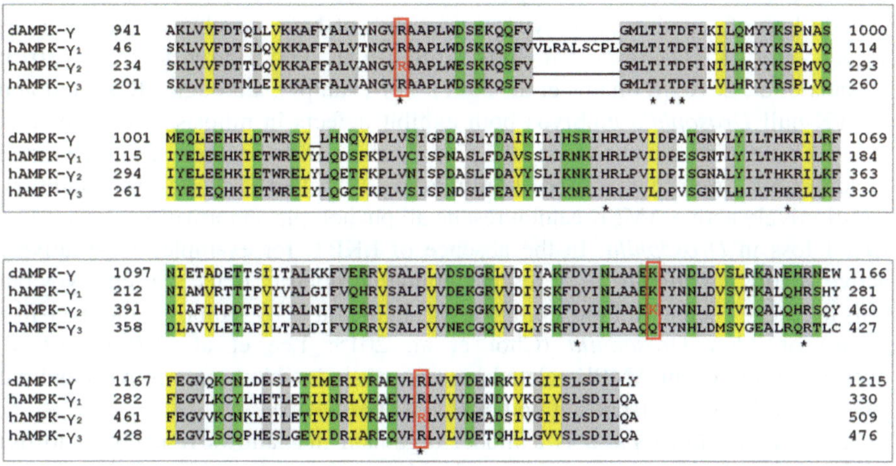

Fig. 16.2 *Alignment of human and Drosophila AMPK-γ*. CBS domains of AMPK are highly conserved between *H. sapiens* and *D. melanogaster*. Most amino acids in the CBS domains of *Drosophila* AMPK-γ (NP_732598.1) and human AMPK-$\gamma_{1\text{-}3}$ (γ_1, NP_001193638.1; γ_2, NP_001035723.1; γ_3, NP_059127.2) are either invariant (*gray*) or have residues with conserved hydrophilicity (*green*) or hydrophobicity (*yellow*). In SNF4 (dAMPK-γ), the first pair of CBS domains spans from amino acids 941–1069, while the second pair spans from amino acids 1097–1215 (Marchler-Bauer et al. 2015). Human AMPK-γ_2 amino acids highlighted in red represent conserved residues that, when mutated, can lead to cardiomyopathy in humans (Burwinkel et al. 2005; Liu et al. 2013; Moffat and Harper 2010). *Stars* indicate some of the conserved amino acid residues that have been shown to mediate nucleotide binding at regulatory sites in crystal structures containing a specific transcript variant of rat AMPK-γ_1 (Chen et al. 2012; Xiao et al. 2007). *Note*: Amino acid numbering will vary among transcript variants; the numbering listed in the figure corresponds to the provided accession numbers

to alter cardiac AMPK activity and cause human cardiomyopathies (when mutated) are conserved in both human and fly genomes (Fig. 16.2) (Burwinkel et al. 2005; Liu et al. 2013; Moffat and Harper 2010).

A second line of evidence pointing to the utility of *Drosophila* is the conserved expression of the AMPK kinase LKB1 and the AMPK substrate acetyl CoA carboxylase (ACC-1) (Amin et al. 2009; Andersen et al. 2012; Martin and St Johnston 2003; Pan and Hardie 2002). The conserved expression of LKB1 and ACC underscores the evolutionary importance of this signaling pathway and raises another key point regarding the merit of *Drosophila* as a model organism: whereas the streamlined *Drosophila* genome simplifies genetic loss- or gain-of-function studies in vivo, the complexity of LKB1 signaling—with both AMPK-*dependent* and AMPK-*independent* effects—is also conserved in both mammals and flies (Amin et al. 2009; Andersen et al. 2012; Choi et al. 2015). This conserved signaling complexity means that researchers can use *Drosophila* to help identify (or eliminate) the role of AMPK with respect to various LKB1-dependent developmental and physiological processes. In mice, flies, and even nematodes, LKB1 genetic loss-of-function phenotypes can be more severe than those of AMPK genetic loss-of-function phenotypes (yeast do not have an obvious homologous sequence for LKB1) (Amin et al. 2009; Apfeld et al. 2004; Hardie and Alessi 2013; Nakada et al. 2010; Williams et al. 2011). For example, although LKB1-null and AMPK-null *Drosophila* embryos both exhibit defects in mitosis, loss of AMPK (in eye tissue) does not phenocopy the pitted surface or rhabdomere fusion observed in adult LKB1-null fly eye tissue (Amin et al. 2009; Lee et al. 2007). In addition, constitutively active AMPK cannot rescue all phenotypic abnormalities induced by LKB1 loss in *Drosophila*. In the absence of LKB1, for example, constitutively active AMPK (AMPK-α^{T184D}) rescues epithelial polarity defects in embryos but fails to rescue decreased lipid storage levels in the fat body (adipose- and liver-like tissue) of adult *Drosophila* (Choi et al. 2015; Lee et al. 2007). Instead, overexpression of an AMPK-related kinase, salt-inducible kinase 3, rescues lipid storage levels in LKB1-null fat bodies (Choi et al. 2015).

In summary, *Drosophila* is a model organism that allows researchers to use straightforward genetic approaches to elucidate conserved signaling pathways that can be characterized by a level of complexity similar to that of mammalian pathways. For an overview of *Drosophila* AMPK's (dAMPK's) role as a metabolic regulator—particularly in the context of dietary restriction—we invite readers to review the biochemical and behavioral experiments published by Johnson et al. (2010). AMPK-dependent feeding behaviors published by Johnson and colleagues include hyperphagia triggered by genetically reduced dAMPK function as well as a normally pro-survival "foraging" behavior induced in starved flies with genetically reduced AMPK function (Johnson et al. 2010). Just as in mammals, however, the regulatory realm of AMPK extends beyond metabolic rates and lipid metabolism. Indeed, as Sect. 2 will show, experiments using *Drosophila* have shown that AMPK may exert protective effects with regard to life span and neurodegeneration (Ng et al. 2012; Pimenta de Castro et al. 2012; Ulgherait et al. 2014). This chapter addresses a few of the research areas—many of which translate to higher organisms—in which *Drosophila* has been used to elucidate the physiological functions of AMPK. In doing so, this chapter will also function as a primer for basic *Drosophila* nomenclature, thereby eliminating a communication barrier that persists between AMPK researchers trained in either mammalian genetics or *Drosophila* genetics.

16.1.2 Overview of Tools Available to Drosophila Geneticists

Many readers of this chapter may have limited or no experience working with *Drosophila* and the genetic tools that are uniquely available to fly geneticists. These tools include P elements, the Gal4-upstream activating sequence (UAS) gene expression system, and genome-wide availability of *Drosophila* transgenic RNAi approaches. Understandably, researchers with expertise in mammalian AMPK may have difficulty mining *Drosophila* literature for insights into cellular signaling pathways. Thus, we have illustrated the technical approaches to help readers understand the ways in which *Drosophila* has contributed to AMPK research (Fig. 16.3). The Gal4-UAS expression system is an especially noteworthy and commonly used tool, as it allows geneticists to manipulate the AMPK signaling pathway spatially and/or temporally (Brand and Perrimon 1993; McGuire et al. 2004). Essentially, this approach relies on a yeast transcription activator (Gal4) (that has no known function in *Drosophila*) to precisely control the expression of any gene placed after the yeast UAS promoter within transgenic *Drosophila*. A detailed description of this system has been published by McGuire et al. (2004). In short, Gal4 binds to UAS targets upstream of a transgene that has been inserted into the fly genome. To achieve either ubiquitous or spatially restricted transgene expression, researchers can select from a number of freely available fly lines (from stock centers) that express Gal4 drivers under various well-characterized enhancers (Table 16.1). To temporally regulate transgene expression, researchers can use two derivatives of the classical Gal4-UAS system: a pharmacologically inducible Gene-Switch system and a temperature-controllable TARGET system (McGuire et al. 2004; Osterwalder et al. 2001). Mechanistic explanations of these systems are beyond the scope of this paper but are provided alongside excellent illustrations in the above-mentioned review (McGuire et al. 2004). For this chapter, it is sufficient for readers to simply know that fly geneticists have established tools for silencing (via Gal4-driven RNAi expression) or overexpressing AMPK subunits both spatially and temporally.

16.2 Selected Disease Models Studied in *Drosophila*

16.2.1 Overview of Disease Models

Before beginning a discussion of translational research in *Drosophila*, we would like to first acknowledge an excellent review that explains how *Drosophila* has helped to expand our understanding of human neurodegenerative diseases (Lessing and Bonini 2009). "Maintaining the Brain: Insight into Human Neurodegeneration From *Drosophila* Mutants," by Lessing and Bonini, touches upon research areas that are also described herein (e.g., Parkinson's disease (PD) and Alzheimer's Disease (AD)). We recommend this published review to readers and will expand

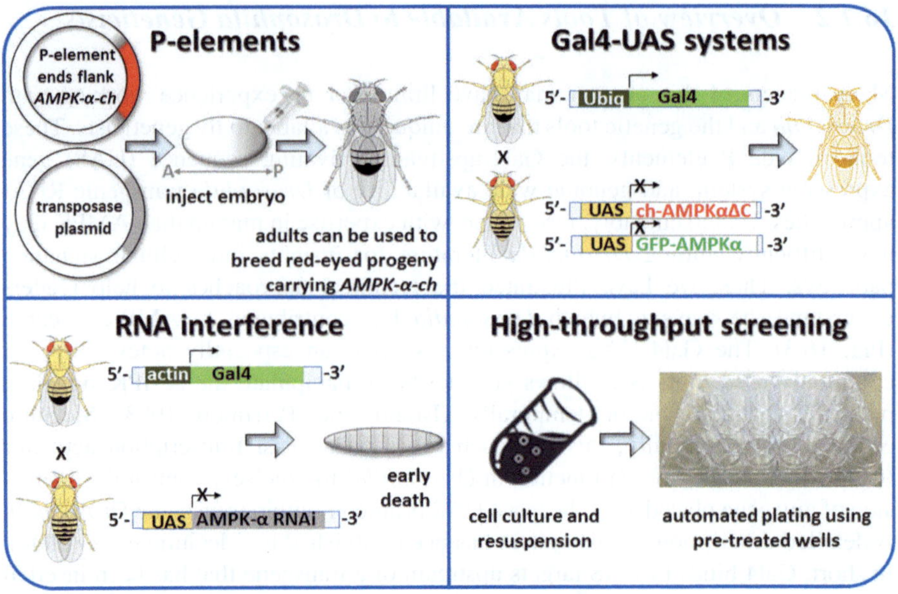

Fig. 16.3 *Tools used in Drosophila research*. Techniques used by fly geneticists are illustrated. (Clockwise from *top left*) *P-elements*: A transgene plasmid (e.g., encoding cherry-tagged AMPK-α) and helper plasmid (encoding transposase) can be mixed and injected into *Drosophila* embryos (Bachmann and Knust 2008; Beall and Rio 1997). The inverted terminal repeats flanking the transgene are recognized by transposase, which randomly inserts the transgene into genomic DNA (Bachmann and Knust 2008; Beall and Rio 1997). Embryos will then develop into adults in which some germ cells carry a transgene insertion. Adult males can be crossed to females to generate offspring that retain the insertion in every cell (Bachmann and Knust 2008; Majumdar and Rio 2015). *Drosophila* offspring carrying the P-element insertion can be identified by a marker (typically a gene for red eyes) that is also encoded separately in the transgene plasmid (Bachmann and Knust 2008; Majumdar and Rio 2015). *Gal4-UAS gene expression systems*: Gal4 and UAS-transgene parental lines can be crossed to generate offspring in which transgene expression is driven by Gal4 (Brand and Perrimon 1993). Kazgan et al., for example, used Ubiq-Gal4 to drive ubiquitous (Ubiq) expression of both full-length GFP-tagged dAMPK-α and mCherry (ch)-tagged truncated dAMPK-α in offspring, thereby allowing them to identify a nuclear export signal in dAMPK-α (Kazgan et al. 2010). The figure depicting Kazgan et al.' s technical approach is adapted from the earliest publication describing the Gal4-UAS system (Brand and Perrimon 1993). *RNA interference*: Gal4 can also be used to drive the expression of RNA (Perkins et al. 2015). In the example shown, knockdown of AMPK-α can be used to help identify changes in AMPK-dependent signaling pathways in larvae (Onyenwoke et al. 2012). *High-throughput screening*: *Drosophila* cell lines allow researchers to take advantage of *Drosophila*'s streamlined genome without having to breed stable fly lines. In the example shown, cells can be rapidly plated onto wells that have been pretreated with an RNAi library. Moser and colleagues have used this approach to screen for host factors (such as AMPK) that regulate viral infection (Moser et al. 2010a, b)

upon Lessing's and Bonini's work by discussing dAMPK-related research published during or after 2010. As stated earlier, we will focus on recent studies that demonstrate the protective role of dAMPK with regard to life span and neurodegeneration (Lin et al. 2014; Ng et al. 2012; Pimenta de Castro et al. 2012; Ulgherait et al. 2014).

Table 16.1 Use of Gal4-UAS system in AMPK research

Expression	Gal4 driven	Transgene	Study
Ubiquitous	Tubulin[1]	AMPK-α-DN AMPK-RNAi (α, γ) AMPK-α	Effect of AMPK on lethality
	Actin[1, 6]	AMPK-α-DN AMPK-RNAi (α, γ) AMPK-α	Effect of AMPK on lethality
		AMPK-α-RNAi	Effect of AMPK on NDPK activity
	Ubiquitin [1,5]	AMPK-α-DN AMPK-γ-RNAi	Effect of AMPK on life span and feeding behavior
		AMPK-α (± mCh) truncated AMPK-α (± mCh)	Identification of nuclear export signal in AMPK-α
	Hsp70[1,a]	AMPK-α-DN AMPK-RNAi (α, γ)	Effect of AMPK on life span during starvation
Pan-neuronal	ELAV[4,b]	AMPK-γ truncated AMPK-γ	Effect of WT or mutated AMPK-γ on vacuolar pathology in the brain
	ELAV GS[3,c]	AMPK-α (± mCh)	Effect of neuronal AMPK overexpression on life span (± starvation), intestinal aging, and the expression of autophagy related genes (ATG)
Intestinal	TIGS-2 GS[3, c]	mCh-AMPK-α	Effect of intestinal AMPK overexpression on intestinal aging, life span
Fat body, digestive system	S106 GS[2,7, c]	AMPK-α-RNAi	Role of AMPK in mediating life span extension induced by β-sitosterol
Sensory neurons	109(2)80[1,5]	Actin-GFP, AMPK-α-DN AMPK-RNAi (α, γ) truncated AMPK-α	Effect of AMPK on neuronal morphology

[a]Heat shock protein (Hsp)70-Gal4 is temperature controllable (McGuire et al. 2004)
[b]ELAV, embryonic lethal abnormal vision promoter (McGuire et al. 2004) is pan-neuronal
[c]*GS* Gene-Switch Gal4s that are expressed after feeding flies with RU486 (Osterwalder et al. 2001). *mCh* mCherry fluorescent tag. (1) Johnson et al. (2010), (2) Lin et al. (2014), (3) Ulgherait et al. (2014), (4) Tschape et al. (2002), (5) Kazgan et al. (2010), (6) Onyenwoke et al. (2012), (7) Poirier et al. (2008)

16.2.2 Role of AMPK in Longevity

With its short life span, *Drosophila melanogaster* is a convenient model organism for longevity studies in the lab (Linford et al. 2013). Multiple publications have explored the role of AMPK in regulating life span. To begin, the Gal4-UAS system has been used to show that—depending on the strength of the Gal4 driver—a decrease in dAMPK activity may be either lethal during the larval stage or may shorten the life span of adult transgenic animals (Gal4 was used to drive the near ubiquitous overexpression of either AMPK RNAi or a kinase-dead AMPK mutant) (Johnson et al. 2010). In agreement with this observation, a second lab has shown that *increased* pan-neuronal dAMPK activity *extends* adult female life span (Ulgherait et al. 2014). Furthermore, genetic knockdown of dAMPK-α (in the fat body) has been shown to prevent the increased life span induced by β-sitosterol, a natural product that promotes AMPK phosphorylation in *Drosophila* S2 cell culture (Lin et al. 2014). In light of these 3 lines of evidence, it may be tempting to simply conclude that life span positively correlates with dAMPK activity in flies. During starvation, however, this correlation may not always be observed. For example, overexpression of AMPK in neurons has been shown to *shorten* the life span of starved transgenics, while overexpression of dominant-negative AMPK in neuroendocrine cells has been shown to *extend* the life span of starved transgenic animals (Braco et al. 2012; Ulgherait et al. 2014). Interestingly, decreased triglyceride levels—an indicator of altered metabolism—have been documented for both fed mutants with reduced (ubiquitous) AMPK function and starved mutants with increased neuronal AMPK function (Johnson et al. 2010; Ulgherait et al. 2014). Together, these findings highlight the complex relationships among AMPK activity, diet, and life span. In addition, these findings indicate that ubiquitous and tissue-specific modulation of AMPK activity may have different consequences on life span (Braco et al. 2012; Johnson et al. 2010).

Finally, before continuing to the next section, we would be remiss in omitting a revelatory finding described in Ulgherait et al.'*s* lifespan studies. Just as increased pan-neuronal dAMPK activity extends adult life span, the reciprocal experiment (increased intestinal dAMPK activity) also extends *Drosophila* life span *while upregulating autophagy in the brain* (Ulgherait et al. 2014). Importantly, Poirier and colleagues have confirmed that the intestinal Gal4 driver TIGS-2 permits UAS-transgene expression solely within the digestive system of male and female *Drosophila*—without detectable levels of UAS-transgene expression in the brain (Poirier et al. 2008). The non-cell autonomous effects demonstrated by Ulgherait and colleagues are particularly timely in light of an increasingly widespread interest regarding the interdependence of brain and intestinal health (Duca et al. 2015; Reardon 2014; Ulgherait et al. 2014).

16.2.3 Role of AMPK in Neurodegenerative Disorders

Many experts in mammalian AMPK signaling may be familiar with the regulation or alteration of autophagy by AMPK, as well as the physiological relevance of autophagy in proteinopathic diseases (Cai et al. 2012; Vingtdeux et al. 2011). In cell cultures, for example, activation of AMPK by resveratrol has been shown to promote autophagy and—as a result—decrease levels of amyloid-β (Aβ), a secreted peptide that contributes to the development of Alzheimer's disease (AD) (Vingtdeux et al. 2010). In agreement with in vitro findings, dietary resveratrol has been shown to significantly increase AMPK phosphorylation and decrease Aβ deposition in the brains of mice modeling AD (Vingtdeux et al. 2010).

More recently, *Drosophila* researchers have used genetic tools to address a different and perhaps lesser-known facet of proteinopathic diseases: *mitochondrial* pathology. Researchers have generated genetically distinct *Drosophila* models of PD-related mitochondrial pathology. This chapter addresses two *Drosophila* models: one in which transgenic animals express misfolded ornithine transcarbamylase (dOTC) and another in which transgenic *Drosophila* express mutated leucine-rich repeat kinase 2 (LRRK2) (Ng et al. 2012; Pimenta de Castro et al. 2012). The former model was motivated by the observation that mutated PTEN-induced putative kinase 1 (PINK1) is accompanied by increased levels of misfolded mitochondrial protein (in both PD patients and fly mutants) (Pimenta de Castro et al. 2012). By mutating the mitochondrial matrix protein OTC instead of PINK1, researchers were able to examine the effects of *organelle-restricted* protein misfolding (Pimenta de Castro et al. 2012). Misfolded mitochondrial protein was accompanied by an accumulation of light chain 3-II (LC3-II; an autophagy marker) in flies expressing dOTC but not in flies expressing dOTC with Gal4-driven AMPK-α-RNAi (Pimenta de Castro et al. 2012). Pimenta de Castro and colleagues have speculated that AMPK-dependent autophagy may exert a protective role in this fly model of protein misfolding (2012). Notably, shortly after the *Drosophila* model of dOTC was published, researchers published a mouse model of PD where dOTC expression is driven by the tyrosine hydroxylase promoter (Moisoi et al. 2014). This progression from early experiments in flies to ensuing studies in mice underscores the translational merit of *Drosophila* research.

As alluded to earlier, a second *Drosophila* model of PD-related mitochondrial pathology was achieved by using the Gal4-UAS system to express mutated LRRK2 (a disease gene mutated in PD) in muscle tissue (Ng et al. 2012). In addition to exhibiting mitochondrial pathology, these flies exhibited locomotor dysfunction which was rescued by ectopic expression of parkin (mutations in parkin can cause PD) (Ng et al. 2012). Researchers demonstrated that ectopic expression of constitutively active AMPK rescued both mitochondrial and locomotor abnormalities in LRRK2 G2019S mutants as well as in *parkin* mutants (Ng et al. 2012).

Finally, no discussion of the role of dAMPK in the brain would be complete without mentioning *loechrig* (*loe*) (also reviewed by Lessing and Bonini 2009). *Loe* mutants have reduced ("hypomorphic") dAMPK-γ expression in the nervous system (Tschape et al. 2002). This reduced AMPK function allows adult flies to develop past the earlier lethality found in AMPK-null mutations, but results in a severe vacuolar pathology and widespread neuronal death in the adult fly brain (Kazgan et al. 2010; Tschape et al. 2002). More recently, Kretzschmar and colleagues have demonstrated that *loe* mutants have increased prenylation of Rho1 (Rho1 is orthologous to mammalian RhoA) (Cook et al. 2012). Although *loechrig* does not model the etiology of AD, Cook et al. stress the translational value of their research by discussing the link between activated RhoA/Rho-associated protein kinase (ROCK) signaling and AD (Cook et al. 2012). Pharmacological inhibition of ROCK, for example, has been shown to decrease Aβ42 levels in vivo (Cole and Vassar 2006; Zhou et al. 2003).

16.3 Summary

Drosophila is a model organism that allows researchers to use straightforward genetic approaches to elucidate conserved, AMPK-dependent signaling pathways. Fly geneticists have used traditional loss-of-function or hypomorphic AMPK subunit alleles and the versatile Gal4-UAS gene-expression system to elucidate the physiological functions of AMPK in starvation and longevity studies as well as in in vivo models of neurodegenerative disorders (Cook et al. 2012; Johnson et al. 2010; Ng et al. 2012; Pimenta de Castro et al. 2012; Tschape et al. 2002; Ulgherait et al. 2014). Importantly, translational research in *Drosophila* may help provide new insights into the etiology of mammalian disease models, as demonstrated by recently published models of PD in transgenic flies and mice (Moisoi et al. 2014; Pimenta de Castro et al. 2012).

Acknowledgments We would like to thank Dr. Rob Onyenwoke and Dr. Nevzat Kazgan for critiquing this manuscript prior to submission. This work was funded by NS080108 to J.B.

References

Amin N, Khan A, St Johnston D, Tomlinson I, Martin S, Brenman J, McNeill H (2009) LKB1 regulates polarity remodeling and adherens junction formation in the Drosophila eye. Proc Natl Acad Sci US A 106:8941–8946. doi:10.1073/pnas.0812469106

Andersen RO, Turnbull DW, Johnson EA, Doe CQ (2012) Sgt1 acts via an LKB1/AMPK pathway to establish cortical polarity in larval neuroblasts. Dev Biol 363:258–265. doi:10.1016/j.ydbio.2011.12.047

Apfeld J, O'Connor G, McDonagh T, DiStefano PS, Curtis R (2004) The AMP-activated protein kinase AAK-2 links energy levels and insulin-like signals to lifespan in C. elegans. Genes Dev 18:3004–3009. doi:10.1101/gad.1255404

Bachmann A, Knust E (2008) The use of P-element transposons to generate transgenic flies. Methods Mol Biol 420:61–77. doi:10.1007/978-1-59745-583-1_4

Beall EL, Rio DC (1997) Drosophila P-element transposase is a novel site-specific endonuclease. Genes Dev 11:2137–2151

Braco JT, Gillespie EL, Alberto GE, Brenman JE, Johnson EC (2012) Energy-dependent modulation of glucagon-like signaling in Drosophila via the AMP-activated protein kinase. Genetics 192:457–466. doi:10.1534/genetics.112.143610

Brand AH, Perrimon N (1993) Targeted gene expression as a means of altering cell fates and generating dominant phenotypes. Development 118:401–415

Burwinkel B et al (2005) Fatal congenital heart glycogenosis caused by a recurrent activating R531Q mutation in the gamma 2-subunit of AMP-activated protein kinase (PRKAG2), not by phosphorylase kinase deficiency. Am J Hum Genet 76:1034–1049. doi:10.1086/430840

Cai Z, Yan LJ, Li K, Quazi SH, Zhao B (2012) Roles of AMP-activated protein kinase in Alzheimer's disease. Neuromol Med 14:1–14. doi:10.1007/s12017-012-8173-2

Chen L et al (2012) AMP-activated protein kinase undergoes nucleotide-dependent conformational changes. Nat Struct Mol Biol 19:716–718. doi:10.1038/nsmb.2319

Choi S, Lim DS, Chung J (2015) Feeding and fasting signals converge on the LKB1-SIK3 pathway to regulate lipid metabolism in Drosophila. PLoS Genet 11, e1005263. doi:10.1371/journal.pgen.1005263

Cole SL, Vassar R (2006) Isoprenoids and Alzheimer's disease: a complex relationship. Neurobiol Dis 22:209–222. doi:10.1016/j.nbd.2005.11.007

Cook M, Mani P, Wentzell JS, Kretzschmar D (2012) Increased RhoA prenylation in the loechrig (loe) mutant leads to progressive neurodegeneration. PLoS One 7, e44440. doi:10.1371/journal.pone.0044440

Duca FA, Cote CD, Rasmussen BA, Zadeh-Tahmasebi M, Rutter GA, Filippi BM, Lam TK (2015) Metformin activates a duodenal Ampk-dependent pathway to lower hepatic glucose production in rats. Nat Med 21:506–511. doi:10.1038/nm.3787

Hardie DG, Alessi DR (2013) LKB1 and AMPK and the cancer-metabolism link—ten years after. BMC Biol 11:36. doi:10.1186/1741-7007-11-36

Iseli TJ et al (2005) AMP-activated protein kinase beta subunit tethers alpha and gamma subunits via its C-terminal sequence (186-270). J Biol Chem 280:13395–13400. doi:10.1074/jbc.M412993200

Johnson EC et al (2010) Altered metabolism and persistent starvation behaviors caused by reduced AMPK function in Drosophila. PLoS One 5. doi:10.1371/journal.pone.0012799

Kazgan N, Williams T, Forsberg LJ, Brenman JE (2010) Identification of a nuclear export signal in the catalytic subunit of AMP-activated protein kinase. Mol Biol Cell 21:3433–3442. doi:10.1091/mbc.E10-04-0347

Lee JH et al (2007) Energy-dependent regulation of cell structure by AMP-activated protein kinase. Nature 447:1017–1020. doi:10.1038/nature05828

Lessing D, Bonini NM (2009) Maintaining the brain: insight into human neurodegeneration from Drosophila melanogaster mutants. Nat Rev Genet 10:359–370. doi:10.1038/nrg2563

Lin WS et al (2014) The anti-aging effects of Ludwigia octovalvis on Drosophila melanogaster and SAMP8 mice. Age (Dordr) 36:689–703. doi:10.1007/s11357-013-9606-z

Linford NJ, Bilgir C, Ro J, Pletcher SD (2013) Measurement of lifespan in Drosophila melanogaster. J Vis Exp 71. doi:10.3791/50068

Liu Y et al (2013) Identification of a novel de novo mutation associated with PRKAG2 cardiac syndrome and early onset of heart failure. PLoS One 8, e64603. doi:10.1371/journal.pone.0064603

Majumdar S, Rio DC (2015) P transposable elements in Drosophila and other eukaryotic organisms. Microbiol Spectr 3:MDNA3-0004-2014. doi:10.1128/microbiolspec.MDNA3-0004-2014

Marchler-Bauer A et al (2015) CDD: NCBI's conserved domain database. Nucleic Acids Res 43: D222–D226. doi:10.1093/nar/gku1221
Martin SG, St Johnston D (2003) A role for Drosophila LKB1 in anterior-posterior axis formation and epithelial polarity. Nature 421:379–384. doi:10.1038/nature01296
McGuire SE, Roman G, Davis RL (2004) Gene expression systems in Drosophila: a synthesis of time and space. Trends Genet 20:384–391. doi:10.1016/j.tig.2004.06.012
Moffat C, Harper ME (2010) Metabolic functions of AMPK: aspects of structure and of natural mutations in the regulatory gamma subunits. IUBMB Life 62:739–745. doi:10.1002/iub.387
Moisoi N, Fedele V, Edwards J, Martins LM (2014) Loss of PINK1 enhances neurodegeneration in a mouse model of Parkinson's disease triggered by mitochondrial stress. Neuropharmacology 77:350–357. doi:10.1016/j.neuropharm.2013.10.009
Moser TS, Jones RG, Thompson CB, Coyne CB, Cherry S (2010a) A kinome RNAi screen identified AMPK as promoting poxvirus entry through the control of actin dynamics. PLoS Pathog 6, e1000954. doi:10.1371/journal.ppat.1000954
Moser TS, Sabin LR, Cherry S (2010b) RNAi screening for host factors involved in Vaccinia virus infection using Drosophila cells. J Vis Exp. doi:10.3791/2137
Nakada D, Saunders TL, Morrison SJ (2010) Lkb1 regulates cell cycle and energy metabolism in haematopoietic stem cells. Nature 468:653–658. doi:10.1038/nature09571
Ng CH et al (2012) AMP kinase activation mitigates dopaminergic dysfunction and mitochondrial abnormalities in Drosophila models of Parkinson's disease. J Neurosci 32:14311–14317. doi:10.1523/JNEUROSCI.0499-12.2012
Onyenwoke RU, Forsberg LJ, Liu L, Williams T, Alzate O, Brenman JE (2012) AMPK directly inhibits NDPK through a phosphoserine switch to maintain cellular homeostasis. Mol Biol Cell 23:381–389. doi:10.1091/mbc.E11-08-0699
Osterwalder T, Yoon KS, White BH, Keshishian H (2001) A conditional tissue-specific transgene expression system using inducible GAL4. Proc Natl Acad Sci USA 98:12596–12601. doi:10.1073/pnas.221303298
Pan DA, Hardie DG (2002) A homologue of AMP-activated protein kinase in Drosophila melanogaster is sensitive to AMP and is activated by ATP depletion. Biochem J 367:179–186. doi:10.1042/BJ20020703
Perkins LA et al. (2015) The transgenic RNAi project at Harvard Medical School: Resources and Validation Genetics. doi:10.1534/genetics.115.180208
Pimenta de Castro I et al (2012) Genetic analysis of mitochondrial protein misfolding in Drosophila melanogaster. Cell Death Differ 19:1308–1316. doi:10.1038/cdd.2012.5
Poirier L, Shane A, Zheng J, Seroude L (2008) Characterization of the Drosophila gene-switch system in aging studies: a cautionary tale. Aging Cell 7:758–770. doi:10.1111/j.1474-9726.2008.00421.x
Reardon S (2014) Gut-brain link grabs neuroscientists. Nature 515:175–177. doi:10.1038/515175a
Spasic MR, Callaerts P, Norga KK (2008) Drosophila alicorn is a neuronal maintenance factor protecting against activity-induced retinal degeneration. J Neurosci 28:6419–6429. doi:10.1523/JNEUROSCI.1646-08.2008
Tschape JA, Hammerschmied C, Muhlig-Versen M, Athenstaedt K, Daum G, Kretzschmar D (2002) The neurodegeneration mutant lochrig interferes with cholesterol homeostasis and Appl processing. EMBO J 21:6367–6376
Ulgherait M, Rana A, Rera M, Graniel J, Walker DW (2014) AMPK modulates tissue and organismal aging in a non-cell-autonomous manner. Cell Rep 8:1767–1780. doi:10.1016/j.celrep.2014.08.006
Vingtdeux V et al (2010) AMP-activated protein kinase signaling activation by resveratrol modulates amyloid-beta peptide metabolism. J Biol Chem 285:9100–9113. doi:10.1074/jbc.M109.060061

Vingtdeux V, Chandakkar P, Zhao H, d'Abramo C, Davies P, Marambaud P (2011) Novel synthetic small-molecule activators of AMPK as enhancers of autophagy and amyloid-beta peptide degradation. FASEB J 25:219–231. doi:10.1096/fj.10-167361

Williams T, Courchet J, Viollet B, Brenman JE, Polleux F (2011) AMP-activated protein kinase (AMPK) activity is not required for neuronal development but regulates axogenesis during metabolic stress. Proc Natl Acad Sci U S A 108:5849–5854. doi:10.1073/pnas.1013660108

Xiao B et al (2007) Structural basis for AMP binding to mammalian AMP-activated protein kinase. Nature 449:496–500. doi:10.1038/nature06161

Zhou Y et al (2003) Nonsteroidal anti-inflammatory drugs can lower amyloidogenic Abeta42 by inhibiting Rho. Science 302:1215–1217. doi:10.1126/science.1090154

Chapter 17
Plant SnRK1 Kinases: Structure, Regulation, and Function

Leonor Margalha, Concetta Valerio, and Elena Baena-González

Contents

Abstract SnRK1 is an evolutionarily conserved protein kinase complex that regulates energy homeostasis in plants. In doing so, it promotes tolerance to adverse environmental conditions and influences a large array of growth and developmental processes. SnRK1 shares clear structural and functional similarities with its orthologs, yeast SNF1 and mammalian AMPK, but has evolved unique features that presumably provide a better adaptation to an autotrophic lifestyle. In this

L. Margalha • C. Valerio • E. Baena-González (✉)
Instituto Gulbenkian de Ciência, Rua da Quinta Grande 6, 2780-156 Oeiras, Portugal
e-mail: ebaena@igc.gulbenkian.pt

M.D. Cordero, B. Viollet (eds.), *AMP-activated Protein Kinase*, Experientia Supplementum 107, DOI 10.1007/978-3-319-43589-3_17

chapter, we review current knowledge on SnRK1, an atypical member of the SNF1/AMPK family, providing insight into its structure, regulation, and functions.

Keywords SNF1-related protein kinase (SnRK1) • Energy signaling • Homeostasis • Stress • Carbon metabolism • Carbon allocation • Plant

17.1 Introduction

During photosynthesis, plants convert light into chemical energy, i.e., sugars, in order to sustain their growth and development. In addition to being metabolic substrates, sugars are also potent signaling molecules, and therefore, sugar production in photosynthetic source organs and their distribution and storage in sink organs are tightly coordinated to ensure proper growth and development (Lastdrager et al. 2014). This regulation of sugar distribution becomes particularly relevant in situations of stress, when ATP production through photosynthesis and/or respiration may be compromised, severely limiting plant growth and crop productivity (De Block and Van Lijsebettens 2011). The adequate allocation of nutrients during normal development and upon environmental challenges is poorly understood, but an increasing body of evidence implicates the Snf1-related protein kinase 1 (SnRK1) in the regulation of these processes. SnRK1 is the plant ortholog of yeast Sucrose non-fermenting1 (SNF1) and mammalian AMP-activated protein kinase (AMPK). Similarly to SNF1 and AMPK, SnRK1 undergoes activation when energy levels decline, triggering vast metabolic and transcriptional changes that promote stress tolerance and survival (Baena-González et al. 2007; Baena-González and Sheen 2008). Besides orchestrating an efficient stress response, the SnRK1 energy sensor influences overall plant growth and nutrient remobilization processes such as seed filling and senescence (Lastdrager et al. 2014; Tsai and Gazzarrini 2014).

Despite its centrality, little is known about how SnRK1 functions and how its activity is regulated. In this chapter, we will review our current understanding of SnRK1 complex composition and its regulation and function. Sharing clear structural and functional similarities with SNF1 and AMPK, the plant kinase has also evolved unique features that presumably provide a better adaptation to an autotrophic lifestyle. The atypical nature of the plant kinase and its relevance for virtually all aspects of plant life highlight the need for understanding the SnRK1 system through dedicated research.

17.2 First Findings About SnRK1

SnRK1 research was initiated with the identification of a Snf1-related protein kinase in the endosperm of rye (RKIN1) (Alderson et al. 1991). This kinase showed remarkable sequence similarity with yeast Snf1 and mammalian AMPK, suggesting the existence of an evolutionarily conserved family of protein kinases (Carling et al. 1994). This discovery was followed by the isolation of related cDNAs from multiple plant sources such as Arabidopsis, barley, and tobacco, among others. Rye *RKIN1* (Alderson et al. 1991), as well as tobacco *NPK5* (Muranaka et al. 1994), complemented the yeast *snf1* mutant, suggesting a conserved function for the plant kinases in the control of carbon metabolism. The first *in planta* evidence for such a role came from the antisense-mediated silencing of *SnRK1α* in potato (Halford and Hardie 1998; Purcell et al. 1998) and wheat embryos (Laurie et al. 2003), which resulted in decreased expression of the sucrose synthase (*SUS4*, sucrose-inducible) and α-amylase (α-*AMY2*, glucose-repressible) genes, respectively.

Several biochemical studies showed the phosphorylation and inactivation of key metabolic enzymes of isoprenoid biosynthesis (3-hydroxy-3-methylglutaryl CoA reductase, HMGR), sucrose synthesis (sucrose phosphate synthase, SPS), and nitrogen assimilation (nitrate reductase, NR) in cauliflower (Mackintosh et al. 1992; Ball et al. 1994; Dale et al. 1995a), spinach (McMichael et al. 1995; Douglas et al. 1997), and Arabidopsis (Dale et al. 1995a; Douglas et al. 1997). Later on, it was realized that the kinases responsible for these different phosphorylation events belong to the SnRK1 family (Sugden et al. 1999b). SnRK1 shares similar substrate recognition motifs with SNF1 and AMPK (Dale et al. 1995b) and has been shown to phosphorylate and inactivate mammalian HMGR and acetyl-CoA carboxylase (ACC, involved in lipid biosynthesis) in the same sites as AMPK (Mackintosh et al. 1992). However, the plant ACC is not phosphorylated by SnRK1, where possibly different regulatory mechanisms have evolved (Mackintosh et al. 1992).

17.3 SnRK1 Complex Composition

The SNF1/AMPK/SnRK1 kinases show remarkable sequence conservation and structural similarities (Polge and Thomas 2007; Hedbacker and Carlson 2008; Carling et al. 2012). As first demonstrated by the purification of mammalian AMPK (Davies et al. 1994; Mitchelhill et al. 1994), all these kinases function as heterotrimeric complexes, composed by a catalytic α subunit and two regulatory subunits, β and γ (Fig. 17.1).

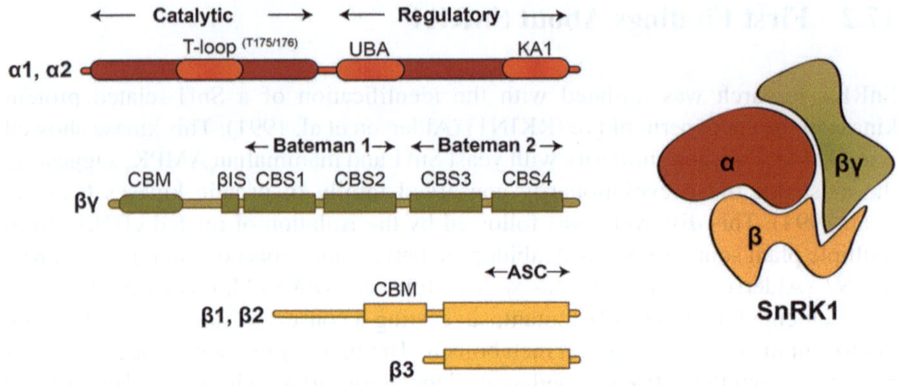

Fig. 17.1 SnRK1 heterotrimeric complex composition. The α subunit (in *red*) is composed of a catalytic domain, harboring the conserved T-loop, and a regulatory domain, which encompasses an ubiquitin-associated (UBA) domain, and a kinase-associated 1 (KA1) domain for binding the β and γ subunits. The plant-specific SnRK1βγ (in *green*) appears to be the only canonical γ-type subunit in plants. In addition to two "Bateman" domains, each containing two cystathionine-β-synthase (CBS) motifs, and a β-interacting sequence (βIS), SnRK1βγ unconventionally harbors a carbohydrate-binding module (CBM). The β subunit (in *yellow*) comprises an N-terminal variable extension, a CBM, and an association with the SNF1 complex (ASC) domain, containing sites of interaction with α and γ. The plant-specific SnRK1β3 is an atypical β subunit, as it does not possess the N-terminal variable extension or the CBM

17.3.1 α-Catalytic Subunit

In plants, the genes encoding the α-catalytic subunits are part of a large SnRK superfamily that comprises the SnRK1, SnRK2, and SnRK3 groups (Hrabak et al. 2003). Whereas the SnRK1 group (with 3 members in Arabidopsis, SnRK1α1/KIN10, SnRK1α2/KIN11, and SnRK1α3/KIN12) is closely related to the yeast Snf1 and mammalian AMPKα, SnRK2 (with 10 members) and SnRK3 (with 25 members) belong to more divergent plant-specific groups of kinases that cannot functionally replace SnRK1 (Hrabak et al. 2003) and have been implicated in abscisic acid (ABA) signaling as well as osmotic and salt stress responses (Coello et al. 2011). In cereals, the SnRK1 group is further divided into two subgroups, SnRK1a and SnRK1b, with SnRK1a being more closely related to dicot SnRK1 and SnRK1b being cereal specific and mostly expressed in the seed (Halford and Hardie 1998).

The SnRK1α-catalytic subunit comprises an N-terminal kinase domain and a C-terminal regulatory domain. It has the highest degree of cross-species conservation among the three subunits, especially in the kinase domain of the protein (Carling et al. 1994; Halford et al. 2003). The kinase domain displays a canonical fold (Hanks and Hunter 1995) and contains the activation loop (T-loop), in which the phosphorylation of the conserved threonine (T; SnRK1α1^{T175}/SnRK1α2^{T176}) by upstream kinases is essential for activity (Stein et al. 2000; Baena-González et al. 2007; Shen et al. 2009; Crozet et al. 2010, 2014). The C-terminal regulatory region harbors an ubiquitin-associated (UBA) domain that may mediate noncovalent interactions with ubiquitylated proteins (Hofmann and Bucher 1996; Farrás et al. 2001) and a kinase-associated 1 (KA1) domain, responsible for the interaction with regulatory subunits and upstream phosphatases, as well as with other interacting proteins (Bhalerao et al. 1999; Kleinow et al. 2000; Farrás et al. 2001; Chiang et al. 2013; Rodrigues et al. 2013).

The α-catalytic subunits can form dimers in yeast and mammals, as observed in crystals in vitro and in co-immunoprecipitation assays in vivo (Rudolph et al. 2005; Nayak et al. 2006; Scholz et al. 2009), and the same might occur with the plant α subunits (Nunes et al. 2013b; Coello and Martínez-Barajas 2014b). However, the physiological relevance of such a level of organization is still unclear.

17.3.2 β- and γ-Regulatory Subunits

Sequence homology searches also allowed the identification of conserved β- and γ-regulatory subunits and prompted studies demonstrating physical interactions with the yeast and among the plant subunits (Bouly et al. 1999; Kleinow et al. 2000; Gissot et al. 2004). Atypical plant-specific β- and γ-regulatory subunits were also identified (Lumbreras et al. 2001; Gissot et al. 2004; Hardie 2007; Ghillebert et al. 2011).

Arabidopsis harbors two highly conserved SnRK1β1/AKINβ1 and SnRK1β2/AKINβ2 subunits, with a high degree of identity (49 %) between them (Bouly et al. 1999). The β1 and β2 subunits contain a carbohydrate-binding motif (CBM) in the middle of the protein as well as a C-terminal association with SNF1 complex (ASC) domain. The CBM domain is lacking in the SnRK1β3/AKINβ3 protein, which nevertheless is, likewise the β1 and β2 subunits, able to interact with the α and γ subunits and complement a yeast mutant devoid of all three β subunits (Gissot et al. 2004; Polge et al. 2008). The primordial function of the regulatory β subunits is therefore to act as a scaffold between the α and γ subunits. The CBM domain binds glycogen in yeast and mammals (Polekhina et al. 2003; Wiatrowski et al. 2004; Polekhina et al. 2005; McBride et al. 2009). In plants, recent evidence suggests that the CBM of the β or the atypical βγ subunits does not mediate SnRK1 binding to carbohydrates, including starch, the plant carbon storage analogue to glycogen (Emanuelle et al. 2015), although an opposite conclusion was reached in a previous study (Avila-Castañeda et al. 2014). The CBM domain largely overlaps with the previously assigned kinase interaction sequence (KIS) domain, required for the interaction with the α-catalytic subunit of yeast SNF1 and plant SnRK1 (Jiang and Carlson 1997; Bouly et al. 1999). Besides the conserved domains, β subunits have an N-terminal variable extension (Warden et al. 2001; Hedbacker et al. 2004; Hedbacker and Carlson 2006, 2008; Polge et al. 2008; Oakhill et al. 2010, 2011) that in the Arabidopsis SnRK1β1 subunit is important for the interaction with NR (Polge et al. 2008).

The SNF1/AMPK/SnRK1 γ subunit harbors two Bateman domains (Bateman 1997), comprising four cystathionine-beta-synthase (CBS) repeats that in mammals bind adenylates, providing a mechanism for the allosteric regulation of AMPK (Carling et al. 1987; Scott et al. 2004). The CBS motifs are preceded by a pre-CBS domain in the N-terminus that is highly divergent (Viana et al. 2007). In plants, the γ subunit is highly atypical, as it harbors a CBM sequence present in the β subunits fused to the N-terminus of a γ subunit (Lumbreras et al. 2001; Gissot et al. 2006). This plant-specific SnRK1βγ is able to complement the yeast *snf4* mutant and to interact with α and β subunits, forming SnRK1 complexes exclusive to plants (Kleinow et al. 2000; Gissot et al. 2006; López-Paz et al. 2009; Bitrián et al. 2011; Ramon et al. 2013; Emanuelle et al. 2015). Similarly to AMPKγ (Viana et al. 2007), the pre-CBS sequence seems to be essential for SnRK1βγ function, as deletion of this region results in loss of binding to the AKINβ2 subunit and the inability to complement the yeast *snf4* mutant (Ramon et al. 2013). On the contrary, the SnRK1γ subunit is unable to rescue *snf4* and does not interact with α and β subunits in vivo (Ramon et al. 2013; Emanuelle et al. 2015). In addition to SnRK1γ, an extended family of γ-subunit-like proteins is present in plants (Gissot et al. 2006; Robaglia et al. 2012). However, phylogenetic analyses place these genes outside of the monophyletic clade to which SnRK1βγ and yeast and animal SNF4/AMPKγ-like genes belong (Ramon et al. 2013). Furthermore, genes representative of these more distant clades are unable to complement the yeast *snf4* mutant, altogether indicating that in higher plants only the βγ-type proteins have retained the canonical γ-subunit function (Ramon et al. 2013). Interestingly, SnRK1βγ is able to homodimerize through the CBM and the CBS domains (López-Paz et al. 2009). Given that dimeric associations through a KIS/CBM-related domain of glucoamylase are critical for binding and hydrolyzing starch in *Aspergillus niger* (Jørgensen et al. 2008), SnRK1βγ homodimers could potentially interact with carbohydrates to modulate the activity of the SnRK1 complex (López-Paz et al. 2009).

The number of SnRK1 complexes that can theoretically be formed varies significantly between plants. In Arabidopsis, six different SnRK1 αβγ heterotrimers assemble through various combinations of α1/α2 [*SnRK1α3* is poorly expressed in most plant tissues; (Baena-González et al. 2007)], β1/β2/β3, and βγ (Bouly et al. 1999; Gissot et al. 2004, 2006; Ramon et al. 2013; Emanuelle et al. 2015). However, the number of possible heterotrimers increases if alternative transcription initiation or splicing events are considered (Gissot et al. 2006; Steinberg and Kemp 2009; Ghillebert et al. 2011). In addition, hormones and metabolic and environmental cues differentially regulate the expression of specific SnRK1 subunits (Bradford et al. 2003; Buitink et al. 2003; Schwachtje et al. 2006; Cernadas et al. 2008; Polge et al. 2008; Li et al. 2009; Radchuk et al. 2010). SnRK1 complex composition may also be unique to particular subcellular compartments, tissues, or developmental stages. For instance, Arabidopsis *SnRK1β1* has specific patterns of expression throughout plant development and the expression of *SnRK1β1* is induced by ammonium and dark and repressed by sugar (Polge et al. 2008; Li et al. 2009). SnRK1β1-containing complexes interact specifically with NR, promoting its phosphorylation-dependent inhibition with a consequent downregulation of nitrogen assimilation (Polge et al. 2008). In a study using bean seeds, where SnRK1 complexes were separated by size-exclusion chromatography (Coello and Martínez-Barajas 2014a), the differential distribution of SnRK1 activity per size/fraction throughout seed development supported the idea that multiple types of SnRK1 complexes may exist. Different SnRK1 complexes may possess distinct substrate specificities and differ in their susceptibility to inhibitory metabolites (Nunes et al. 2013b; Coello and Martínez-Barajas 2014a).

It was also suggested that SnRK1 could form heterodimers composed of α and βγ subunits and devoid of a β-regulatory subunit (Lumbreras et al. 2001; Bitrián et al. 2011), but recent studies do not support this hypothesis (Ramon et al. 2013; Emanuelle et al. 2015).

17.4 Regulation of the SnRK1 Complex

In addition to the subunit composition of the complex, SNF1/AMPK/SnRK1 kinases are controlled by numerous other mechanisms that have an impact on kinase activity, stability, and/or subcellular localization. However, compared to its yeast and mammalian orthologs, several studies have highlighted differences in how SnRK1 is activated in response to energy stress.

17.4.1 Post-translational Modifications

Post-translational modifications allow rapid and dynamic changes on the proteome that do not rely on translation, which is one of the most energy-demanding processes in the cell (Lane and Martin 2010; Chantranupong et al. 2015). Besides the essential role of T-loop phosphorylation on the catalytic subunit of SNF1/AMPK/SnRK1 complexes for activity, several other post-translational modifications have been characterized and implicated in the regulation of this conserved family of protein kinases (Crozet et al. 2014) (Table 17.1).

17.4.2 Phosphorylation

The phosphorylation of a conserved threonine (T) residue (Snf1^{T210}; AMPKα1^{T183}/AMPKα2^{T172}; SnRK1α1^{T175}/SnRK1α2^{T176}) in the catalytic subunit is essential for SNF1/AMPK/SnRK1 activity (Stein et al. 2000; McCartney and Schmidt 2001; Baena-González et al. 2007; Shen et al. 2009; Crozet et al. 2010). In yeast and mammalian systems, a clear correlation is observed between kinase activation under low glucose/energy and T-loop phosphorylation (Hedbacker and Carlson

Table 17.1 List of post-translational modifications reported for SnRK1 subunits and factors (potentially) influencing these modifications

Event	Effect	System	References
T-Loop Phosphorylation			
SnRK1α1[T175]/ SnRK1α2[T176]		In vitro	(1)
Regulators			
SnAK1/2 (GRIK2/1)	SnRK1 phosphorylation/activation	In vitro, Y2H, yeast	(2–5)
ABI1/PP2CA phosphatases	SnRK1 dephosphorylation/ repression	In vitro, In vivo	(6)
Mammalian PP2C and PP2A	SnRK1 dephosphorylation/ inactivation	In vitro on SnRK1 partly purified from plants	(1)
Mammalian PP2C, PP1. λPP	No effect	In vitro on recombinant trimer	(7)
AMP	Indirect protection from dephosphorylation	In vitro on SnRK1 partly purified from plants	(1)
AMP/ADP	No effect	In vitro on recombinant heterotrimers	(7)
Other Phosphorylation			
α1, α2, β1, β2, βγ	Unknown	In vitro	(8–14)
Ubiquitylation			
α1, α2, βγ	Unknown	In vivo	(10–17)
Regulators			
5PTase13	Interaction/SnRK1 stability/ activity	In vitro, Y2H, In vivo	(18)
PRL1	Interaction/SnRK1 stability/ activity	In vitro, Y2H, In vivo	(16, 19, 20)
ASK1/SKP1	Interaction	In vitro, Y2H, In vivo	(20)
Myristoylation			
β1, β2	SnRK1 localization to PM/inhibition	In vivo	(21)
Sumoylation			
α1, β1, β2	Attenuation of SnRK1 signaling	In vitro, In vivo	(22)
Regulators			
SCE1	Interaction	Y2H	(23)
ESD4	Interaction	Y2H	(23)
SIZ1	Hyperactivation of SnRK1 signaling in *siz1-2*	In vitro, In vivo	(22)

(1) Sugden, C. et al., Plant J 19, 433–439 (1999); (2) Crozet, P. et al., J Biol Chem 285, 12071–12077 (2010); (3) Hey, S. et al., J Biol Chem 282, 10472–10479 (2007); (4) Shen, W., Hanley-Bowdoin, L., Plant Physiol 142, 1642–1655 (2006); (5) Shen, W. et al., Plant Physiol 150, 996–1005 (2009); (6) Rodrigues, A. et al., The Plant Cell 25, 3871–3884 (2013); (7) Emanuelle, S. et al., Plant J 82, 183–192 (2015); (8) Choudhary, M. K. et al., Mol Cell Proteomics 14, 2243–2260 (2015); (9) Engelsberger, W. R., Schulze, W. X., Plant J 69, 978–995 (2012); (10) Heazlewood, J. L. et al., Nucleic Acids Res 36, D1015–1021 (2008); (11) Nakagami, H. et al., Plant Physiol 153, 1161–1174 (2010); (12) Umezawa, T. et al., Sci Signal 6, rs8 (2013); (13) Wang, P. et al., Proc Natl Acad Sci USA 110, 11205–11210 (2013); (14) Yao, Q. et al., Front. Plant Sci. 3, 206 (2012); (15) Kim, D.-Y. et al., Plant Cell 25, 1523–1540 (2013); (16) Lee, J.-H. et al., Plant Cell 20, 152–167 (2008); (17) Maor, R. et al., Mol Cell Proteomics 6, 601–610 (2007); (18) Ananieva, E. A. et al., Plant Physiol 148, 1868–1882 (2008); (19) Bhalerao, R. P. et al., Proc Natl Acad Sci USA 96, 5322–5327 (1999); (20) Farrás, R. et al., EMBO J 20, 2742–2756 (2001); (21) Pierre, M. et al., Plant Cell 19, 2804–2821 (2007); (22) Crozet, P. et al., Plant J 85, 120–133 (2016); (23) Elrouby, N., Coupland, G., Proc Natl Acad Sci USA 107, 17415–17420 (2010)

2008; Steinberg and Kemp 2009). Unlike the yeast and mammalian counterparts, equal phosphorylation levels of SnRK1α1^{T175}/SnRK1α2^{T176} are detected in control and low-energy (SnRK1 activating) conditions as well as in *pp2c* mutant plants that display deficient repression of the SnRK1 pathway (Baena-González et al. 2007; Fragoso et al. 2009; Coello et al. 2012; Rodrigues et al. 2013). It is possible that a finer level of regulation, with tissue or subcellular specificity, may remain undetected with analyses of total plant cell extracts. Indeed, size fractionation analyses from Arabidopsis seedlings revealed stronger SnRK1α1^{T175}/SnRK1α2^{T176} phosphorylation in fractions with maximal SnRK1 activity and of approximately 100 kDa in size, suggesting that phosphorylation may be higher when the catalytic subunit is incorporated into a heterotrimeric complex (Nunes et al. 2013b). It is also possible that SnRK1 is regulated by other phosphorylation events (Heazlewood et al. 2008; Nakagami et al. 2010; Engelsberger and Schulze 2012; Yao et al. 2012; Umezawa et al. 2013; Wang et al. 2013) or mechanisms (Table 17.1).

Several kinases are known to activate Snf1 (Sak1, Elm1, and Tos3) and AMPK (LKB1, CaMKKβ, TAK1) by phosphorylation of the T-loop (Hedbacker and Carlson 2008; Steinberg and Kemp 2009). Based on sequence similarity, yeast mutant complementation, and phosphorylation assays, SnRK1 activating kinases 1 and 2 (SnAK1/2; also called Geminivirus Rep-interacting kinase 2/1, GRIK2/1) were identified as upstream kinases of SnRK1 (Shen and Hanley-Bowdoin 2006; Hey et al. 2007; Shen et al. 2009; Crozet et al. 2010). SnAKs appear to be constitutively active and are insensitive to Ca^{2+}, to AMP, and to the CaMKK-specific inhibitor STO-609, when assayed in vitro (Shen et al. 2009). SnAKs autophosphorylate on SnAK1^{T153}/SnAK2^{T154}, and this phosphorylation is required for their activity (Crozet et al. 2010). SnRK1 cross-phosphorylates and inactivates SnAK1^{S260}/SnAK2^{S261} in vitro, probably as part of a negative feedback loop to tightly control SnRK1 activation (Crozet et al. 2010). However, along with the lack of *in planta* evidence for SnAKs' functional relevance on SnRK1 signaling, SnAKs/GRIKs are only detected in geminivirus-infected mature leaves and in actively proliferating tissues that do not cover the range of detection of SnRK1 phosphorylation (Shen et al. 2009). Proteasomal degradation of SnAKs may justify the undetectably low levels of these proteins in several tissues, but it is also possible that other upstream kinases exist (Shen and Hanley-Bowdoin 2006; Shen et al. 2009). Mammalian CaMKK can phosphorylate SnRK1 purified from spinach in vitro (Sugden et al. 1999a), and the CIPK15-SnRK1A-MYBS1 circuitry, required for rice seedlings to tolerate hypoxic conditions during flooding (Lee et al. 2009), suggests that endogenous Ca^{2+}-dependent kinases like CIPKs or others may also serve as SnRK1 upstream kinases (Crozet et al. 2014).

Conversely, several phosphatases act on Snf1 and AMPK to counteract phosphorylation when glucose/energy levels are restored (Hedbacker and Carlson 2008; Steinberg and Kemp 2009). Mammalian PP2A and PP2C phosphatases were reported to dephosphorylate and inactivate plant SnRK1 in vitro (Sugden et al. 1999a). Recently, clade A PP2Cs, ABI1 and PP2CA, were reported to interact and dephosphorylate Arabidopsis SnRK1α1 (Rodrigues et al. 2013). PP2Cs inactivate SnRK1 in response to sugars, and conversely, repression of PP2Cs by the ABA receptors allows the activation of SnRK1 by this phytohormone (Rodrigues et al. 2013). Other phosphatases were reported to interact with SnRK1α in vitro and in yeast two-hybrid assays, namely PP2C74 (Tsugama et al. 2012) and the dual-specificity protein tyrosine phosphatase PTPKIS1 (Fordham-Skelton et al. 2002), although the functional relevance of these interactions is still unknown. PTPKIS1 binds starch through a CBM domain (Fordham-Skelton et al. 2002; Kerk et al. 2006) and was identified as the component responsible for starch overaccumulation in the *sex4* mutant (Niittylä et al. 2006), suggesting that it may regulate the initial steps of starch degradation at the granule surface (Niittylä et al. 2006; Kötting et al. 2009).

Searches in public phosphoproteomics databases retrieve several phosphoresidues in the SnRK1α catalytic subunits in addition to the conserved T-loop threonine, as well as several phosphoresidues in the regulatory subunits (Heazlewood et al. 2008; Nakagami et al. 2010; Engelsberger and Schulze 2012; Yao et al. 2012; Umezawa et al. 2013; Wang et al. 2013; Choudhary et al. 2015), but no particular function has yet been ascribed for these phosphorylation events. Interestingly, the pathogen-induced cell death suppressor Adi3 interacts with SnRK1α1 in tomato, phosphorylating the Gal83 β subunit and thereby inhibiting SnRK1 activity (Avila et al. 2012). Whether this or similar mechanisms connect cell death to SnRK1 control and whether this plays a role in other types of environmental, metabolic, and hormonal cues awaits further investigation.

17.4.3 Ubiquitylation

The activity of protein kinases commonly initiates their downregulation through the ubiquitin-proteasome system (Lu and Hunter 2009). In agreement with this view, the inactive kinases SnRK1α1^{K48M} and SnRK1α1^{T175A} (impaired in phosphotransfer activity and in T-loop phosphorylation, respectively) accumulate to much higher levels than the active SnRK1α1 protein in Arabidopsis cells (Baena-González et al. 2007), suggesting a connection between SnRK1 activity and stability. Large proteomics studies to identify ubiquitylated proteins *in planta* detected ubiquitylation of SnRK1α1, SnRK1α2, and SnRK1βγ (Maor et al. 2007; Kim et al. 2013). The extent of ubiquitylation increased substantially upon MG132 treatment, in line with other studies reporting the proteasomal degradation of SnRK1 subunits (Bhalerao et al. 1999; Ananieva et al. 2008; Lee et al. 2008; Kim et al. 2013; Crozet et al. 2016).

Remarkably, SnRK1α1 immunoprecipitates contain vast amounts of ubiquitin conjugates, and the levels of these are greatly reduced in the *siz1-2* E3 SUMO ligase mutant, suggesting a cooperative interplay between sumoylation and ubiquitylation in SnRK1 degradation (Crozet et al. 2016). Such cross talk has also been identified in yeast, where sumoylated Snf1 is ubiquitylated via the SUMO-targeted ubiquitin ligase Slx5–Slx8 and subsequently degraded (Simpson-Lavy and Johnston 2013).

In addition to SIZ1-mediated sumoylation, other factors have been reported to influence SnRK1α1 stability. SnRK1α1 is targeted for proteasomal degradation in a 5PTase13-dependent manner under low nutrient conditions (Ananieva et al. 2008). SnRK1α1 degradation may also be mediated by the DDB1-CUL4-ROC1-PRL1 E3 ubiquitin ligase, in which PRL1 is the putative substrate receptor of the complex (Lee et al. 2008). SnRK1α1 interacts with PRL1 (Bhalerao et al. 1999; Farrás et al. 2001), and its proteasomal degradation is slower in *prl1* and *cul4cs* extracts than in the wild type, accumulating to a higher extent in these mutants (Lee et al. 2008). On the other hand, PRL1 was reported to compete with SKP1/ASK1 for binding SnRK1α1 and SnRK1α2 (Farrás et al. 2001). SKP1/ASK1 is a component of SCF E3 ubiquitin ligases, and although SnRK1 participates in the assembly of a proteasomal complex with this E3 ligase, there is also the possibility that SnRK1 degradation is mediated by either the SCF complex or the CUL4-DDB1 machinery (Farrás et al. 2001; Lee et al. 2008).

17.4.4 Sumoylation

Arabidopsis SnRK1α1 was shown to interact with the E2 SUMO conjugating enzyme SCE1 and with the SUMO protease ESD4 in a yeast two-hybrid screen (Elrouby and Coupland 2010). Furthermore, in the same study, SnRK1α1 was found to be sumoylated with both SUMO1 and SUMO3 isoforms in a high-throughput heterologous assay in *E. coli*. Recently, SnRK1 sumoylation was shown to occur *in planta*, seemingly as a coordinated event at the level of the whole complex, and SIZ1 was identified as the E3 SUMO ligase responsible for this modification (Crozet et al. 2016). Sumoylation represses SnRK1 signaling in vivo, promoting degradation of the kinase through the ubiquitin-proteasome system. Interestingly, SIZ1-dependent degradation of SnRK1 is strictly dependent on the activity of the kinase, suggesting this as a negative feedback mechanism to attenuate SnRK1 signaling and prevent detrimental overactivation of the pathway (Castro et al. 2015; Crozet et al. 2016).

Sumoylation has recently emerged as an important mechanism also in the regulation of Snf1 (Simpson-Lavy and Johnston 2013; Simpson-Lavy et al. 2015) and AMPK (Rubio et al. 2013; Hendriks et al. 2014). Although differences with the regulation of the plant kinase have been reported in both cases, the fact that AMPKα sumoylation increases in response to heat shock or proteasome inhibition (Hendriks et al. 2014) and the fact that sumoylated Snf1 is ubiquitylated and targeted for degradation (Simpson-Lavy and Johnston 2013) also suggest some common principles.

17.4.5 Myristoylation

Myristoylation of AMPK, SNF1, and SnRK1 modulates kinase activity and subcellular localization (Crozet et al. 2014). Myristoylation of SnRK1 was found to be critical for differentiation of the shoot apical meristem and overall plant development. In the myristoylation-deficient *nmt1-1* mutant, SnRK1 kinase activity and *SnRK1β1* transcript levels were fivefold and twofold higher, respectively, than in wild-type plants (Pierre et al. 2007). In the same study, GFP fusions of either SnRK1β1 or SnRK1β2 localized to the plasma membrane, whereas a G2A substitution preventing myristoylation of the β subunits re-localized SnRK1β1 to the nucleus and SnRK1β2 to the cytoplasm, mimicking what happens in the *nmt1-1* line. It was hence suggested that myristoylation of β1 or β2 subunits sequesters the complex to the plasma membrane, acting as a negative regulator of the SnRK1 pathway, in accordance with the SnRK1 overactivation and glucose hypersensitivity of the *nmt1-1* mutant (Pierre et al. 2007; Traverso et al. 2008).

17.4.6 Adenylates

The high AMP/ATP and ADP/ATP ratios present in metabolically stressed cells activate Snf1 and AMPK, promoting phosphorylation by upstream kinases and conferring protection from dephosphorylation by upstream phosphatases (Hardie 2014). Furthermore, in mammals, AMP binding to the CBS motifs in the γ-regulatory subunit of AMPK triggers its allosteric activation (Oakhill et al. 2010, 2011; Xiao et al. 2011; Carling et al. 2012; Oakhill et al. 2012; Chandrashekarappa et al. 2013; Gowans et al. 2013). In contrast, SnRK1 regulation by AMP was only associated with protection from dephosphorylation in an assay with recombinant mammalian PP2C and purified SnRK1 complex from spinach leaves (Sugden et al. 1999a). A more recent work has confirmed the unique nature of the plant kinase and the lack of direct effects by AMP or ADP on kinase activity using a recombinant trimer (Emanuelle et al. 2015). Interestingly, modeling of the SnRK1βγ CBS motifs revealed that all of the three sites equivalent to the AMPKγ nucleotide-binding sites lack residues that are important for accommodating adenine nucleotides, providing an explanation for the lack of allosteric regulation of SnRK1 by adenylates (Sugden et al. 1999a; Emanuelle et al. 2015).

17.4.7 Sugars

Several lines of evidence implicated SnRK1 in sugar sensing and signaling. On one hand, the activation of SnRK1 by several stresses that converge as a common energy-deficit signal can be blocked by glucose or sucrose supply (Baena-González et al. 2007). Consistent with this, the transcriptional profile induced by SnRK1 pathway activation largely overlaps with profiles obtained under a variety of starvation conditions (Contento et al. 2004; Lin and Wu 2004; Thimm et al. 2004; Buchanan-Wollaston et al. 2005; Baena-González et al. 2007) and is opposite to profiles obtained under sucrose or glucose feeding (Palenchar et al. 2004; Price et al. 2004). On the other hand, sugars modulate SnRK1 signaling at several levels. Transcriptionally, sucrose deprivation stimulated *Vicia faba SnRK1* promoter activity in Arabidopsis protoplasts (Radchuk et al. 2010), and glucose depletion induced *SnRK1β1* expression in germinating *Medicago truncatula* seeds (Buitink et al. 2003). Sugars can also modify the transcript levels of C/S1-class bZIP transcription factors, downstream effectors of SnRK1 signaling (Rook et al. 1998; Wiese et al. 2005; Weltmeier et al. 2009). *bZIP11* expression is induced by sugars, but its translation is repressed by sucrose (Ma et al. 2011). Other studies reported the downregulation of *bZIP63* by mRNA decay in response to sugar and ABA signals (Matiolli et al. 2011) and, conversely, a change in the dimerization properties of bZIP63 during starvation, induced by its SnRK1-dependent phosphorylation (Mair et al. 2015). Glucose deprivation activated SnRK1A post-transcriptionally in rice embryos and suspension cells, and this activation was dependent on the calcineurin B-like-interacting protein kinase 15 (CIPK15) kinase (Lee et al. 2009). SnRK1A protein accumulation was higher after 24 h of glucose deprivation, and in the absence of *SnRK1A* transcript changes, a twofold increase in kinase activity was achieved in comparison to glucose-fed conditions (Lu et al. 2007). Despite all these observations, it must be noted that the ultimate stress-derived signal(s) activating SnRK1, as well as the sugar feeding-derived signal(s) inhibiting SnRK1, remains unidentified.

A more direct effect of specific sugars has been demonstrated with partially purified SnRK1 complexes on which the addition of trehalose-6-phosphate (T6P), glucose-6-phosphate (G6P), or glucose-1-phosphate (G1P) has an inhibitory effect (Toroser et al. 2000; Zhang et al. 2009; Nunes et al. 2013b). T6P is a precursor of trehalose and plays an important signaling role in plant carbohydrate metabolism, growth, and development (Tsai and Gazzarrini 2014). Increased T6P accumulation generally occurs when sucrose levels are high and is thus considered an indicator of the sugar status of the plant (Lunn et al. 2014). T6P represses SnRK1 activity in physiological (μM) amounts through an unknown intermediary factor only present in actively growing young tissues (Schluepmann et al. 2003, 2004; Gomez et al. 2006; Paul 2007; Paul et al. 2008; Schluepmann and Paul 2009; Zhang et al. 2009; Debast et al. 2011; Martínez-Barajas et al. 2011). Other sugars such as G6P and G1P also inhibit SnRK1 in young tissues *via* an intermediary protein factor that is separable from SnRK1 (Nunes et al. 2013b). Since maximal inhibition by G6P and G1P was found on SnRK1 complexes of the same size as for T6P (~174 kDa) (Nunes et al. 2013b), it is likely that the SnRK1 complexes inhibited in the three cases are similar, possibly requiring the same intermediary factor.

17.4.8 Hormones

The SnRK1 pathway is intimately associated with plant hormone signaling, and several hormones, in particular abscisic acid (ABA), have been shown to control SnRK1 at different levels. In tomato, expression of the SnRK1γ-regulatory subunit, *LeSNF4,* was induced in response to ABA or dehydration and was repressed by gibberellin (GA) (Bradford et al. 2003). Consistent with this, an inhibitor of ABA synthesis repressed *SNF4b* gene expression in *Medicago truncatula* (Buitink et al. 2003). In another study, auxin and ABA stimulated *Vicia faba SnRK1* promoter activity in Arabidopsis and tobacco protoplasts (Radchuk et al. 2010).

Genetic screens to dissect sugar signaling have often retrieved components of ABA metabolism or signaling, revealing an intimate interplay between sugar and ABA (Rolland et al. 2006). ABA controls SnRK1 activity posttranscriptionally in a 2C-type protein phosphatase (PP2C)-dependent manner. Two known repressors of the ABA pathway, the PP2Cs ABI1 and PP2CA, interact with and dephosphorylate SnRK1α1 and are important for SnRK1 repression in response to sugars (Rodrigues et al. 2013). ABI1 and PP2CA interact with SnRK2 kinases and are blocked by the ABA receptors upon ABA binding (Cutler et al. 2010). Thereby, ABA inhibits PP2C action, resulting in the activation not only of ABA signaling but also of the SnRK1 pathway. Consequently, SnRK1 and ABA induce largely overlapping transcriptional responses (Rodrigues et al. 2013). On the other hand, SnRK1 gain- and loss-of-function approaches result in ABA-related phenotypes that may suggest additional nodes of cross talk between the two pathways downstream of PP2Cs. For example, localized *SnRK1* silencing in pea seeds via antisense expression causes defects in seed storage, maturation, and dormancy together with transcriptional changes that are reminiscent of ABA insensitivity (Radchuk et al. 2006; 2010). Conversely, overexpression of SnRK1α1 in Arabidopsis results in delayed germination and in ABA hypersensitivity (Jossier et al. 2009; Tsai and Gazzarrini 2012).

Some studies report a negative regulation of ABA on SnRK1 signaling, possibly due to the use of different (autotrophic vs. heterotrophic) plant material or conditions that may determine different SnRK1 complex composition (Coello et al. 2012; Lin et al. 2014).

17.4.9 Scaffold Proteins

The binding of adaptor proteins can further increase the versatility of SNF1/AMPK/SnRK1 complexes providing specificity toward distinct regulators or substrates under certain conditions and also by turning the complex susceptible to different metabolites. This appears to be the case for the SnRK1 inhibition by T6P, G1P, and G6P through a yet unknown intermediary factor (Zhang et al. 2009; Nunes et al. 2013b). A more recent study reported the inhibition of SnRK1 by a large protein factor (>30 kDa), although in this case the inhibition was not shown to depend on specific sugars (Emanuelle et al. 2015). On the other hand, proteins containing the domain of unknown function 581 (DUF581) were proposed to function as auxiliary adaptor modules of SnRK1 (Nietzsche et al. 2014). DUF581 proteins interact with Arabidopsis SnRK1α1 and α2 catalytic subunits and their co-expression re-locates SnRK1 to different subnuclear loci. For example, SnRK1 catalytic subunits and DUF581-18 interact with the transcriptional regulator storekeeper related 1 (STKR1) (Nietzsche et al. 2014), homologous to a transcription factor implicated in sucrose-regulated gene expression in potato (Zourelidou et al. 2002).

17.4.10 Oligomerization

The formation of higher order oligomeric complexes can provide an additional layer of kinase regulation (Pellicena and Kuriyan 2006). The expected molecular weights of SnRK1 heterotrimers were largely exceeded in size-exclusion chromatography analyses (Nunes et al. 2013b), suggesting the existence of higher order SnRK1 oligomers and/or complex assembly with other proteins. On the other hand, dimerization was observed with maize SnRK1βγ subunits (López-Paz et al. 2009), but whether this property is related to the potential oligomerization of the heterotrimeric complex is not known. It was proposed that oligomerization may be particularly important in specific subcellular loci where the kinases are highly concentrated, since dimerization appears to be a reversible and concentration-dependent process (Riek et al. 2008; Scholz et al. 2009). Higher order oligomers were associated with an inactive state of the AMPK complex (Scholz et al. 2009), and it was also speculated that upon dimerization Snf1 becomes inaccessible for T-loop phosphorylation, rendering the oligomer inactive (Nayak et al. 2006). A similar outcome could result from SnRK1 oligomerization, since size fractionation studies revealed a weaker phospho-signal associated with higher order complexes and a stronger phospho-signal in fractions with the average heterotrimer molecular weight (Nunes et al. 2013b).

17.4.11 Subcellular Localization

Using a BAC-recombineering strategy to tag endogenous SnRK1 subunits in Arabidopsis, SnRK1βγ and SnRK1α1 were shown to be most abundant in young organs, such as leaf primordia, and in the meristematic zones of roots and shoots (Bitrián et al. 2011). SnRK1βγ and SnRK1α1 co-localized in these tissues to the plasma membrane, although SnRK1βγ was preferentially localized to the nucleus and SnRK1α1 to the cytoplasm and ring-shaped structures around the nuclei. In other tissues, co-localization of SnRK1βγ and SnRK1α1 in the nucleus depended on the cell type and/or developmental stage and was observed, e.g., in stomata cells and in ovules (Bitrián et al. 2011). The nuclear localization of SnRK1 subunits

(Gissot et al. 2006; Pierre et al. 2007; Bitrián et al. 2011; Tsai and Gazzarrini 2012; Williams et al. 2014; O'Brien et al. 2015) is in accordance with the central role of SnRK1 in transcriptional reprogramming (Baena-González et al. 2007). Both nuclear localization (NLS) and nuclear export signal (NES) motifs were characterized in AMPKα (Suzuki et al. 2007; Kazgan et al. 2010) and seem to be conserved in the SnRK1α1 and SnRK1α2 catalytic subunits. It was hypothesized that the transcription factors FUS3 and PTL may retain SnRK1α1 in the nucleus by binding to its C-terminal part and masking the conserved NES in this region (O'Brien et al. 2015). SnRK1β1 and β2 localize to the plasma membrane as a consequence of their myristoylation, supporting the localization of SnRK1 heterotrimeric complexes also at this subcellular site (Pierre et al. 2007; Bitrián et al. 2011). However, it cannot be excluded the presence of SnRK1β2 and β1 in nuclear SnRK1 complexes, since SnRK1β2 can co-localize with βγ in the nucleus (Gissot et al. 2006) and deletion of the N-terminal myristoylation signal results in nuclear accumulation of SnRK1β1 complexes (Ferrando et al. 2001; Gissot et al. 2006). The atypical SnRK1β3 subunit lacks the conserved N-terminal and CBM domains and does not present a glycine in position 2, precluding its myristoylation and membrane association (Gissot et al. 2004, 2006; Pierre et al. 2007). Transient co-expression of tagged maize proteins revealed the co-localization of SnRK1βγ, SnRK1α, and SnRK1β also in cytoplasmic speckles, in agreement with a role in the regulation of key metabolic cytoplasmic enzymes (López-Paz et al. 2009). SnRK1 was further observed in the Golgi (O'Brien et al. 2015) and in chloroplasts (Fragoso et al. 2009), although it is not clear why such localization was not detected in other studies.

17.5 SnRK1 Function

SnRK1 is at the interface of several stress, hormone, and metabolic signaling pathways (Baena-González et al. 2007; Baena-González and Sheen 2008; Halford and Hey 2009). In this section, we will provide an overview of the processes in which SnRK1 has been implicated, ranging from carbon metabolism and allocation to various growth and developmental aspects and stress responses (Fig. 17.2).

17.5.1 Energy and Carbon Metabolism

Early work on SnRK1 suggested a conserved and ancestral role in metabolic control through enzyme phosphorylation and transcriptional regulation (Halford and Hardie 1998; Purcell et al. 1998; Sugden et al. 1999b). Although the phosphorylation of HMGR, NR, and SPS by SnRK1 has been demonstrated only in vitro (Sugden et al. 1999b), several observations support that this is the case also in vivo. SPS and NR are rapidly inactivated upon transition from light to dark, coinciding with their strong phosphorylation (McMichael et al. 1995). On the other

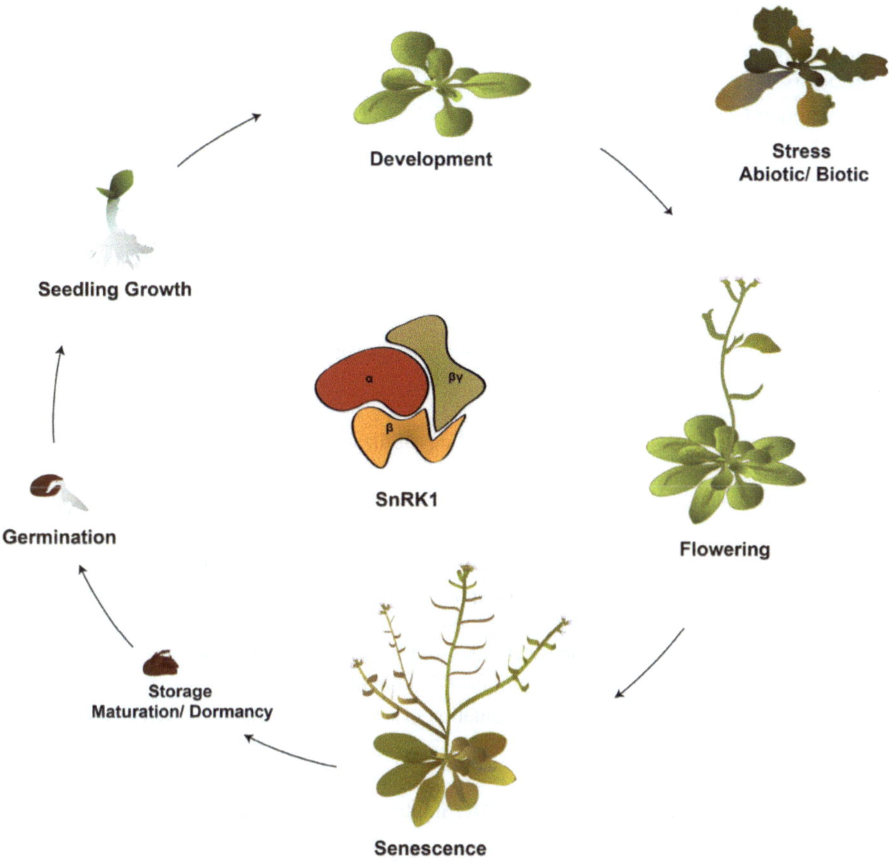

Fig. 17.2 SnRK1 influences the stress response and a wide range of growth and developmental processes. SnRK1 is important for balancing energy use and storage and has been implicated in fundamental growth and developmental processes in plants, including cell cycle regulation, seed development (storage, maturation, dormancy), germination, seedling growth, vegetative development, flowering, and senescence. In most cases, the underlying molecular details are unknown, but increasing mechanistic insight is being provided by the identification of targets and other interacting proteins (Table 17.2)

hand, their phosphorylation is prevented by G6P, which in turn is known to repress SnRK1 activity (Toroser et al. 2000; Zhang et al. 2009; Nunes et al. 2013b). Members of the class II T6P synthase (TPS) proteins are also phosphorylated by SnRK1 (Glinski and Weckwerth 2005; Harthill et al. 2006). Class II TPSs are catalytically inactive and are thought to play a regulatory role in carbon metabolism (Ramon et al. 2009).

Other additional putative targets of phosphorylation are (i) the cytosolic pyruvate kinase (PKc), key enzyme of glycolysis (Beczner et al. 2010); (ii) the nonphosphorylating glyceraldehyde-3-phosphate dehydrogenase (np-Ga3PDHase), a cytosolic unconventional glycolytic enzyme specifically phosphorylated in

Table 17.2 List of SnRK1 targets and other identified SnRK1 interactors

Targets/Other interactors	Effect	System	References
Transcription factors			
bZIP63	Interaction/phosphorylation/altered bZIP dimerization	Y2H, In vivo	(1)
FUS3	Interaction/phosphorylation/FUS3 stabilization	In vitro, Y2H, In vivo	(2)
PTL	Interaction	In vitro, Y2H, In vivo	(3)
ATAF1	Interaction	In vitro, Y2H	(4)
IDD8	Interaction/phosphorylation/IDD8 inactivation	In vitro, In vivo	(5)
STKR1	Interaction	Y2H	(6)
MYC2	Interaction/MYC2 destabilization	In vitro; Y2H; In vivo	(7)
Metabolic enzymes			
HMGR	Phosphorylation/inactivation	In vitro	(8–11)
SPS	Phosphorylation/inactivation	In vitro	(11, 12)
NR	Interaction/phosphorylation/14-3-3 binding/inactivation	In vitro; In vivo	(10–15)
F2KP	Phosphorylation/14-3-3 binding	In vitro	(16)
TPS5	Phosphorylation/14-3-3 binding	In vitro	(17)
PKc	Interaction/regulation of activity	In vitro; Y2H	(18)
np-Ga3PDHase	Phosphorylation/14-3-3 binding	In vitro	(19, 20)
Biotic stress interactors			
Geminivirus AL2 and L2 proteins	Interaction/SnRK1 inhibition	In vitro, Y2H, In vivo	(21)
Geminivirus AL2/C2 proteins	Phosphorylation/delay of viral infection	In vitro, In vivo	(22)
Geminivirus βC1 protein	Interaction/phosphorylation/delay of viral infection	In vitro, In vivo	(23)
Bacterial AvrBsT protein	Interaction	Y2H	(24)
Plant Adi3 cell death-supressor	Interaction/phosphorylation of SnRK1β/SnRK1 inhibition	In vitro; Y2H	(25)
Plant HSPRO1/2 proteins	Interaction	Y2H	(26)
Abiotic stress interactors			
BHSP17 (class I heat shock protein)	Phosphorylation	In vitro	(27)
SKIN1 and 2 kinases	Interaction/inhibition of SnRK1A signaling	Y2H	(28)
CPK15 kinase	Interaction	In vivo	(29)

(continued)

Table 17.2 (continued)

Targets/Other interactors	Effect	System	References
Others			
SnAK1/2 (GRIK2/1) kinases	Phosphorylation/SnAK inactivation	In vitro	(30)
PP2C74, phosphatase	Interaction	In vitro; Y2H	(31)
PTPKIS1, phosphatase	Interaction	In vitro; Y2H	(32)
KRP6 and KRP7, cell cycle regulators	Phosphorylation/KRP6 inhibition	In vitro; In vivo	(33)
CDKE1, cyclin-dependent kinase	Interaction	In vivo	(34)
SKD1, AAA-type ATPase	Interaction/phosphorylation	In vitro, Y2H, In vivo	(35)
DUF581 proteins	Interaction	Y2H, In vivo	(6)

(1) Mair, A. et al., eLife 4 (2015); (2) Tsai, A. Y.-L., Gazzarrini, S., Plant J 69, 809–821 (2012); (3) O'Brien, M. et al., J Exp Bot 66, 2475–2485 (2015); (4) Kleinow, T. et al., Plant Sci 177, 360–370 (2009); (5) Jeong, E.-Y. et al., BMC Plant Biol 15, 110 (2015); (6) Nietzsche, M. et al., Front Plant Sci 5, 54 (2014); (7) Im, J. H. et al., Plant Cell Environ 37, 2303–2312 (2014); (8) Ball, K. L. et al., Eur J Biochem 219, 743–750 (1994); (9) Dale, S. et al., Eur J Biochem 233, 506–513 (1995); (10) Douglas, P. et al., Biochem J 325 (Pt 1), 101–109 (1997); (11) Sugden, C. et al., Plant Physiol 120, 257–274 (1999); (12) McMichael, R. W. et al., Plant Physiol 108, 1077–1082 (1995); (13) Jossier, M. et al., Plant J 59, 316–328 (2009); (14) Li, X.-F. et al., J Integr Plant Biol 51, 513–520 (2009); (15) Polge, C. et al., Plant Physiol 148, 1570–1582 (2008); (16) Kulma, A. et al., Plant J 37, 654–667 (2004); (17) Harthill, J. E. et al., Plant J 47, 211–223 (2006); (18) Beczner, F. et al., J Plant Physiol 167, 1046–1051 (2010); (19) Bustos, D. M., Iglesias, A. A., Plant Physiol 133, 2081–2088 (2003); (20) Piattoni, C. V. et al., Plant Physiol 156, 1337–1350 (2011); (21) Hao, L. et al., Plant Cell 15, 1034–1048 (2003); (22) Shen, W. et al., J Virol 88, 10598–10612 (2014); (23) Shen, Q. et al., Plant Physiol 157, 1394–1406 (2011); (24) Szczesny, R. et al., New Phytol 187, 1058–1074 (2010); (25) Avila, J. et al., Plant Physiol 159, 1277–1290 (2012); (26) Gissot, L. et al., Plant Physiol 142, 931–944 (2006); (27) Slocombe, S. P. et al., Plant Physiol Biochem 42, 111–116 (2004); (28) Lin, C.-R. et al., Plant Cell 26, 808–827 (2014); (29) Lee, K.-W. et al., Sci Signal 2, ra61 (2009); (30) Crozet, P. et al., J Biol Chem 285, 12071–12077 (2010); (31) Tsugama, D. et al., FEBS Lett 586, 693–698 (2012); (32) Fordham-Skelton, A. P. et al., Plant J 29, 705–715 (2002); (33) Guérinier, T. et al., Plant J 75, 515–525 (2013); (34) Ng, S. et al., J Biol Chem 288, 3449–3459 (2013); (35) Chiang, C.-P. et al., J Exp Bot 64, 2385–2400 (2013)

heterotrophic tissues (Piattoni et al. 2011); and (iii) the bifunctional enzyme 6-phosphofructo-2-kinase/fructose-2-6-bisphosphatase (F2KP) (Kulma et al. 2004), which produces and degrades fructose-2-6-bisphosphate (Fru-2,6bP). Fru-2,6bP inhibits cytosolic fructose-1,6-bisphosphatase (FBPase), which catalyzes the first committed step in the pathway of sucrose production. In this way, Fru-2,6bP coordinates the rates of sucrose production and carbon assimilation and controls partitioning of photoassimilate between sucrose and starch (Stitt 1987). This capacity to modulate photoassimilate partitioning appears to be an important determinant of growth and fitness in highly fluctuating natural environments (Mccormick and Kruger 2015).

In the case of NR, F2KP, np-Ga3PDHase, and TPS5, phosphorylation by SnRK1 results in the recruitment of 14-3-3 proteins, and, at least for NR and np-Ga3PDHase, binding of 14-3-3 s contributes to enzyme inactivation (Moorhead et al. 1996; Ikeda et al. 2000; Bustos and Iglesias 2003; Kulma et al. 2004; Harthill et al. 2006; Piattoni et al. 2011).

In addition to direct enzyme regulation, SnRK1 controls energy homeostasis also by triggering a vast transcriptional reprogramming (Baena-González et al. 2007; Baena-González and Sheen 2008). Part of these transcriptional changes occur through S1- and C-class bZIP transcription factors, which can form homo- and heterodimers (Lastdrager et al. 2014). As recently shown for the C-class bZIP63, phosphorylation by SnRK1 changes the dimerization properties of the transcription factor (Mair et al. 2015), therefore generating a large and complex network of bZIP regulators. In the last years, miRNAs have also been implicated in the regulation of SnRK1 targets (Confraria et al. 2013).

SnRK1 induces more than 300 genes involved in various remobilization processes, including amino acid catabolism; cell wall, sucrose, starch, and polysaccharide hydrolysis; lipid mobilization; and beta-oxidation (Baena-González et al. 2007; Baena-González and Sheen 2008). One major class of genes induced by the SnRK1-bZIP axis relates to proline and branched-chain amino acids (Mair et al. 2015) which may serve as alternative energy sources during stress (Szal and Podgorska 2012). Interestingly, several of the class II TPS genes are differentially regulated by SnRK1 and in response to carbon supply. Expression of *TPS5* is induced by sugars and repressed by SnRK1 activation, whereas *TPS8-9-10* are repressed by sugars and induced by SnRK1 and starvation (Price et al. 2004; Blasing et al. 2005; Baena-González et al. 2007; Osuna et al. 2007). Concomitantly, SnRK1 represses more than 300 genes involved in biosynthetic processes like amino acid, cell wall, nucleotide, protein, and carbohydrate synthesis. Transcriptional regulators, chromatin remodeling factors, and signal transduction components are also transcriptionally regulated by SnRK1 (Baena-González and Sheen 2008).

The connection to metabolism is further supported by the phenotypes of plants with an altered SnRK1 pathway. *SnRK1* silencing causes male sterility in barley (Zhang et al. 2001) and defective seed storage, maturation, and dormancy in pea (Radchuk et al. 2006). Presumably, pollen grains and seeds fail to enter the maturation stage because of their inability to convert sucrose to starch and storage products. In the moss *Physcomitrella patens,* depletion of SnRK1 in a *snf1a snf1b* null mutant impairs starch accumulation and growth in a normal day/night cycle. The mutant is only viable if grown under constant illumination or with exogenous sugar supply (Thelander et al. 2004). In contrast, in Arabidopsis, decreased starch accumulation was reported for plants overexpressing SnRK1α1 (Jossier et al. 2009). Full SnRK1 depletion seems to be lethal in higher plants (Baena-González et al. 2007), but when *SnRK1α* is transiently and systemically silenced, starch degradation is impaired leading to starch accumulation in the end of the night (Baena-González et al. 2007). This is accompanied by growth arrest and premature senescence, which cannot be reverted by constant illumination or sugar provision.

Conversely, SnRK1α1 overexpression in Arabidopsis allows plants to better tolerate conditions of limited photosynthesis and carbohydrate supply (Baena-González et al. 2007).

In cereals, SnRK1 induces α-amylase genes in seeds and consequently promotes energy mobilization during seed germination and seedling growth (Laurie et al. 2003; Lu et al. 2007). In potato tubers, SnRK1 regulates Sucrose-Synthase (SuSy) and ADP-glucose pyrophosphorylase (AGPase), the two major enzymes determining sink strength and partitioning of sucrose into starch. SnRK1 mediates the sucrose induction of *SuSy* gene expression (Purcell et al. 1998) and the sucrose/T6P-induced redox activation of the AGPase activity (Tiessen et al. 2003; Kolbe et al. 2005).

Altogether, SnRK1 appears to promote on one hand starch degradation and consequently carbon availability in source tissues (Arabidopsis leaves and endosperm of germinating cereal seeds) and, on the other, starch synthesis and partitioning of nutrient to long-term storages in sink tissues (barley pollen, potato tubers, and pea seeds). The molecular basis of such a differential outcome of SnRK1 activity in different organs and cell types has not been clarified yet, but it is likely to involve differences in the composition of the SnRK1 complex and, in consequence, differences in the way it is regulated and it controls downstream processes.

17.5.2 Growth and Development

SnRK1 conveys the metabolic status of the plant toward important developmental decisions, and accordingly, manipulation of the SnRK1 pathway has an impact on plant growth and architecture. In potato, antisense transcript repression of the α-subunit *PKIN1* (Halford et al. 2003) or of the β-subunit *StubGAL83* (Lovas et al. 2003) affects tuber and root development. Conversely, engineered reduction of T6P levels in potato tubers activates SnRK1 and induces downregulation of genes associated to cell proliferation and growth, and upregulation of an inhibitor of cell cycle progression (Debast et al. 2011). T6P- and SnRK1-mediated signaling is also involved in the repression of new fruit growth by preexisting fruits in cucumber plants (Zhang et al. 2015), whereas fruits of transgenic tomato plants overexpressing SnRK1 ripened earlier than the wild type (Wang et al. 2012). In *Medicago truncatula* seeds, repression of *SNF4b* resulted in reduced seed longevity (Rosnoblet et al. 2007).

SnRK1A was shown to promote seed germination and seedling growth and to modulate development in rice (Lu et al. 2007; Cho et al. 2012). SnRK1A-Interacting Negative Regulators (SKINs) counteracted SnRK1A function, inhibiting starch and nutrient mobilization from the endosperm and in consequence seed germination and growth (Lin et al. 2014). Antisense repression of *SnRK1* in pea seeds caused defects in seed storage and maturation reminiscent of ABA insensitivity. This included downregulation of genes related to cell proliferation, meristem maintenance and differentiation, leaf formation, and polarity, through interaction with both ABA-dependent and independent pathways (Radchuk et al. 2006, 2010). Additionally, *SnRK1* was identified as a very early marker gene of developing leaf primordia in tomato (Pien et al. 2001).

In Arabidopsis, transient *SnRK1α* silencing caused dramatic growth defects, aberrant leaf shape, strong accumulation of anthocyanins, and early senescence (Baena-González et al. 2007). Less severe silencing allowed the formation of inflorescences, although they were rudimentary and unviable. The growth defects of the Arabidopsis SnRK1 loss-of-function plants could not be rescued by constant illumination or exogenous supply of sucrose, suggesting fundamental roles in normal vegetative and reproductive growth beyond a purely metabolic regulation. Conversely, SnRK1α1 overexpression in Arabidopsis caused alterations in inflorescence architecture, as well as delayed flowering and senescence under long-day conditions (Baena-González et al. 2007; Williams et al. 2014; Jeong et al. 2015).

AMPK was shown to regulate fundamental cellular functions, such as cell polarity and cell division (Williams and Brenman 2008). Phosphorylation of the cyclin-dependent kinase inhibitor $p27_{KIP1}$ by AMPK causes its stabilization and thereby cell cycle arrest, autophagy, or apoptosis (Liang et al. 2007; Short et al. 2008; Short et al. 2010). Similarly, SnRK1 phosphorylates the Arabidopsis $p27_{KIP1}$ homologs, KRP6 and KRP7, but this phosphorylation seems to promote cell division (Guérinier et al. 2013). This seemingly counterintuitive function of SnRK1 in promoting cell cycle progression could be unrelated to its energy-sensing function, ensuring proper cell division in the absence of stress. It may also serve a coordinating function, allowing an orderly cell cycle arrest in the ensuing G1 phase.

In a study in Arabidopsis, SnRK1α1 was shown to interact, phosphorylate, and stabilize the FUSCA3 (FUS3) transcription factor. SnRK1α1 and FUS3 interact genetically and functionally as positive regulators of ABA signaling, antagonize embryonic-to-vegetative and vegetative-to-reproductive phase transitions, and regulate lateral organ development in Arabidopsis (Tsai and Gazzarrini 2012). SnRK1 interacts also with the PETAL LOSS (PTL) transcription factor that represses growth in inter-sepal zones of Arabidopsis (O'Brien et al. 2015) and with the plant-specific NAC TF family member ATAF1. ATAF subfamily members are positive regulators of plant development, and, accordingly, *ATAF* downregulation leads to dwarfism, sterility, and reduced or absent flower initiation (Kleinow et al. 2009). Recently, SnRK1α1 was shown to phosphorylate and repress the indeterminate domain 8 (IDD8) TF, with a consequent delay in flowering (Jeong et al. 2015).

17.5.3 Stress

Adverse environmental conditions, such as drought, extreme temperatures, extended darkness, pollutants, and flooding, result in declining energy levels that trigger SnRK1 signaling (Baena-González and Sheen 2008). One of the first observations connecting environmental stress and SnRK1 was the increased salt sensitivity of potato plants silenced for the *SnRK1β* (*StubGal83*) subunit (Lovas et al. 2003). Concomitantly, it was also discovered in the leguminous plant *Medicago truncatula* that the expression of regulatory β (*MtAKINb2*) and γ subunits (*MtSNF4b*) of the SnRK1 complex is induced by desiccation and osmotic treatment in germinating seeds and developing seedlings (Buitink et al. 2003).

SnRK1 promotes tolerance to submergence in rice and Arabidopsis by inducing hypoxia-responsive genes (Baena-González et al. 2007; Lee et al. 2009; Cho et al. 2012; Im et al. 2014). In rice, the O_2 deficiency signal is transmitted by CIPK15 to the SnRK1A/MYBS1 axis to stimulate carbohydrate catabolism and promote growth under floodwater (Lee et al. 2009). Under other abiotic stress conditions, such as salt or cold stress, ABA counteracts the growth-promoting function of SnRK1A (Lu et al. 2007; Lin et al. 2014). The ABA-induced SKIN repressors confine both SnRK1A and MYBS1 in the cytoplasm, preventing SnRK1A/MYBS1 nuclear localization and thus the consequent mobilization of nutrient reserves from source tissues. Interestingly, in the case of coupled salt and hypoxia stress caused by submerging plants in saltwater, SnRK1 antagonizes ABA action. Stimulated by hypoxia, SnRK1 phosphorylates the ABA-inducible transcription factor AtMYC2 and enhances its proteasomal degradation. This results in the disruption of the tolerance to salt induced by ABA (Im et al. 2014).

On the contrary, in some cases SnRK1 and ABA cooperate to induce stress tolerance. In salt-stressed Arabidopsis roots, ABA induces general defense-related functions and SnRK1 completes the stress response supporting a substantial metabolic reprogramming through the coordinated activation of bZIP transcription factors (Hartmann et al. 2015).

As observed in normal growth conditions, SnRK1 displays a differential mode of regulation in different organs also under stress. In maize, under salt stress, SnRK1 activity is increased and sensitive to T6P in the kernels after pollination, whereas in leaves it appears to be uncoupled and independent from T6P. For example, class II TPS genes are induced in salt-stressed maize leaves, but probably not by SnRK1 activity, which is very low and unaffected by T6P in this case (Henry et al. 2015).

SnRK1 has also been implicated in response to phosphate starvation (Fragoso et al. 2009). Under these limited conditions, SnRK1α2/AKIN11 is degraded, while SnRK1α1/AKIN10 activity is increased, potentially playing a role in response to phosphate deficiency. SnRK1 has also been associated with responses to temperature stress. Its ability to interact with and phosphorylate the barley heat shock protein 17 (HSP17) suggests that it may be important for the heat stress response (Slocombe et al. 2004). On the other hand, under sink limitation conditions, such as low temperature, the inhibitory effect of T6P on SnRK1 activation is necessary to enable plant recovery after the relief of stress conditions (Nunes et al. 2013a).

An increasing number of studies implicate SnRK1 also in the plant response to biotic stress. SnRK1 can be a component of the plant defense system, potentially activated by metabolic and energy imbalances derived from pathogen attack, and can therefore become a target upon infection. In tobacco, SnRK1 increases resistance against tomato golden mosaic virus and beet curly top virus. However, the viral AL2 and L2 proteins interact and inactivate SnRK1, thereby counteracting the defense response (Hao et al. 2003). AL2 viral proteins can also repress adenosine kinase (ADK) with a consequent decrease in AMP levels. This may increase the susceptibility of SnRK1 to dephosphorylation and inactivation and may be an indirect mechanism for SnRK1 downregulation (Hao et al. 2003; Wang et al. 2003). AC2/C2 geminivirus proteins inhibit Arabidopsis SnRK1α2. The infection results in broad effects, namely on RNA silencing and autophagy, likely components of antiviral immunity (Liu et al. 2014). Begomovirus and curtovirus AL1 and AL2/C2 proteins interact with the host geminivirus Rep-interacting kinases (GRIKs), which are upstream activating kinases of SnRK1 (Shen and Hanley-Bowdoin 2006; Shen et al. 2009, 2014). Moreover, SnRK1 phosphorylates the AL2/C2 proteins in vitro. It is suggested that through this mechanism SnRK1 delays viral DNA accumulation and symptom development during Arabidopsis infection (Shen et al. 2014). SnRK1 also interacts and phosphorylates the βC1 protein encoded by a Geminivirus β-Satellite, as part of the host defense in tomato plants (Shen et al. 2011, 2012).

Regarding bacterial pathogens, infection of sweet orange plants by *Xanthomonas axonopodis pv. aurantifolii* increased the expression of trehalose biosynthesis enzymes and SnRK1 kinases, likely associated with a defense response (Cernadas et al. 2008). The AvrBsT protein of *Xanthomonas campestris pv. vesicatoria* targets SnRK1 and suppresses the hypersensitive response elicited by the effector protein AvrBs1 in pepper. Intriguingly, SnRK1 is required for AvrBs1-induced hypersensitive response (Szczesny et al. 2010). In another case, the plant AvrPto-dependent Pto-interacting protein 3 (Adi3) interacts and phosphorylates the SnRK1β subunit Gal83, with a consequent suppression of SnRK1 activity. Adi3 is a cell death suppressor with an important role in the resistance of tomato plants to *Pseudomonas syringae pv tomato* (Avila et al. 2012).

Furthermore, the plant-specific SnRK1βγ subunit interacts with HSPRO1/2 proteins involved in plant defense and possibly in the leaf senescence program (Gissot et al. 2006). HSPRO was shown to participate in a whole-plant change in growth physiology associated with SnRK1 signaling, when *Nicotiana attenuata* seedlings interact with the fungus *Piriformospora indica* (Schuck et al. 2012, 2013). Interestingly, herbivore attack results in downregulation of *SnRK1β* in *Nicotiana attenuata* source leaves and thereby in increased assimilate transport to the roots. Plants with enhanced root reserves prolong reproduction and tolerate better herbivory by allocating reserves to a less vulnerable location within the plant (Schwachtje et al. 2006).

17.6 Concluding Remarks

Eukaryotic SNF1/AMPK/SnRK1 kinases are part of a sophisticated signaling network that allows organisms to adapt their physiology and behavior to the availability of nutrients. Accordingly, one major role of these kinases is the regulation of energy and carbon metabolism, a function conserved also in plants. SnRK1 kinases have been implicated in numerous nutrient management processes both at the cellular and whole-plant levels, ranging from sucrose, starch, and amino acid metabolism to senescence, seed filling, and tuber growth.

Research on SnRK1 has been largely driven by the underlying assumption that the mechanisms regulating the plant kinase are also conserved. However, the increasing realization that the structure and mode of regulation of SnRK1 are unique should encourage the use of less biased approaches in the future. The presence of atypical subunits may explain the evolution of adenylate and T-loop-independent regulatory mechanisms that render SnRK1 activity susceptible to plant-specific signals, such as sugar phosphates, unique protein regulators, and phytohormones.

Understanding SnRK1 heterogeneity represents a major challenge, especially considering the dual nature of plants as autotrophic and heterotrophic organisms. Reconstitution of the various SnRK1 complexes in vitro and in vivo has recently been achieved providing a powerful tool in this direction. Further elucidation of SnRK1 complex composition in different conditions, tissues, and developmental stages will be key to understanding how SnRK1 is regulated and how different downstream targets are controlled.

Many of the processes governed by the SnRK1 pathway are highly relevant for crop performance under an increasingly variable climate as well as for various other aspects of agricultural production. A detailed comprehension of how SnRK1 operates will therefore be essential for the targeted manipulation of these processes and ultimately for understanding the diversity of roles that sugar signals play in plant growth and development.

Acknowledgments We thank Pierre Crozet and Ana Confraria for criticism and suggestions on the manuscript and Nuno Margalha for valuable help with the figures. This work was supported by Fundação para a Ciência e a Tecnologia projects PTDC/BIA-PLA/3937/2012 and UID/Multi/04551/2013.

References

Alderson A, Sabelli PA, Dickinson JR, Cole D, Richardson M, Kreis M, Shewry PR, Halford NG (1991) Complementation of snf1, a mutation affecting global regulation of carbon metabolism in yeast, by a plant protein kinase cDNA. Proc Natl Acad Sci USA 88:8602–8605

Ananieva EA, Gillaspy GE, Ely A, Burnette RN, Erickson FL (2008) Interaction of the WD40 domain of a myoinositol polyphosphate 5-phosphatase with SnRK1 links inositol, sugar, and stress signaling. Plant Physiol 148:1868–1882

Avila J, Gregory OG, Su D, Deeter TA, Chen S, Silva-Sanchez C, Xu S, Martin GB, Devarenne TP (2012) The β-subunit of the SnRK1 complex is phosphorylated by the plant cell death suppressor Adi3. Plant Physiol 159:1277–1290

Avila-Castañeda A, Gutiérrez-Granados N, Ruiz-Gayosso A, Sosa-Peinado A, Martínez-Barajas E, Coello P (2014) Structural and functional basis for starch binding in the SnRK1 subunits AKINβ2 and AKINβγ. Front Plant Sci 5:199

Baena-González E, Sheen J (2008) Convergent energy and stress signaling. Trends Plant Sci 13:474–482

Baena-González E, Rolland F, Thevelein JM, Sheen J (2007) A central integrator of transcription networks in plant stress and energy signalling. Nature 448:938–942

Ball KL, Dale S, Weekes J, Hardie DG (1994) Biochemical characterization of two forms of 3-hydroxy-3-methylglutaryl-CoA reductase kinase from cauliflower (Brassica oleracia). Eur J Biochem 219:743–750

Bateman A (1997) The structure of a domain common to archaebacteria and the homocystinuria disease protein. Trends Biochem Sci 22:12–13

Beczner F, Dancs G, Sos-Hegedus A, Antal F, Banfalvi Z (2010) Interaction between SNF1-related kinases and a cytosolic pyruvate kinase of potato. J Plant Physiol 167:1046–1051

Bhalerao RP, Salchert K, Bakó L, Okrész L, Szabados L, Muranaka T, Machida Y, Schell J, Koncz C (1999) Regulatory interaction of PRL1 WD protein with Arabidopsis SNF1-like protein kinases. Proc Natl Acad Sci U S A 96:5322–5327

Bitrián M, Roodbarkelari F, Horváth M, Koncz C (2011) BAC-recombineering for studying plant gene regulation: developmental control and cellular localization of SnRK1 kinase subunits. Plant J 65:829–842

Blasing OE, Gibon Y, Gunther M, Hohne M, Morcuende R, Osuna D, Thimm O, Usadel B, Scheible WR, Stitt M (2005) Sugars and circadian regulation make major contributions to the global regulation of diurnal gene expression in Arabidopsis. Plant Cell 17:3257–3281

Bouly JP, Gissot L, Lessard P, Kreis M, Thomas M (1999) Arabidopsis thaliana proteins related to the yeast SIP and SNF4 interact with AKINalpha1, an SNF1-like protein kinase. Plant J 18:541–550

Bradford KJ, Downie AB, Gee OH, Alvarado V, Yang H, Dahal P (2003) Abscisic acid and gibberellin differentially regulate expression of genes of the SNF1-related kinase complex in tomato seeds. Plant Physiol 132:1560–1576

Buchanan-Wollaston V, Page T, Harrison E, Breeze E, Lim PO, Nam HG, Lin J-F, Wu S-H, Swidzinski J, Ishizaki K et al (2005) Comparative transcriptome analysis reveals significant differences in gene expression and signalling pathways between developmental and dark/starvation-induced senescence in Arabidopsis. Plant J 42:567–585

Buitink J, Thomas M, Gissot L, Leprince O (2003) Starvation, osmotic stress and desiccation tolerance lead to expression of different genes of the regulatory b and g subunits of the SnRK1 complex in germinating seeds of Medicago truncatula. Plant Cell Environ 27:55–67

Bustos DM, Iglesias AA (2003) Phosphorylated non-phosphorylating glyceraldehyde-3-phosphate dehydrogenase from heterotrophic cells of wheat interacts with 14-3-3 proteins. Plant Physiol 133:2081–2088

Carling D, Zammit VA, Hardie DG (1987) A common bicyclic protein kinase cascade inactivates the regulatory enzymes of fatty acid and cholesterol biosynthesis. FEBS Lett 223:217–222

Carling D, Aguan K, Woods A, Verhoeven AJ, Beri RK, Brennan CH, Sidebottom C, Davison MD, Scott J (1994) Mammalian AMP-activated protein kinase is homologous to yeast and plant protein kinases involved in the regulation of carbon metabolism. J Biol Chem 269:11442–11448

Carling D, Thornton C, Woods A, Sanders MJ (2012) AMP-activated protein kinase: new regulation, new roles? Biochem J 445:11–27

Castro PH, Verde N, Lourenco T, Magalhaes AP, Tavares RM, Bejarano ER, Azevedo H (2015) SIZ1-dependent post-translational modification by SUMO modulates sugar signaling and metabolism in Arabidopsis thaliana. Plant Cell Physiol 56(12):2297–2311

Cernadas RA, Camillo LR, Benedetti CE (2008) Transcriptional analysis of the sweet orange interaction with the citrus canker pathogens Xanthomonas axonopodis pv. citri and Xanthomonas axonopodis pv. aurantifolii. Mol Plant Pathol 9:609–631

Chandrashekarappa DG, McCartney RR, Schmidt MC (2013) Ligand binding to the AMP-activated protein kinase active site mediates protection of the activation loop from dephosphorylation. J Biol Chem 288:89–98

Chantranupong L, Wolfson Rachel L, Sabatini David M (2015) Nutrient-sensing mechanisms across evolution. Cell 161:67–83

Chiang C-P, Li C-H, Jou Y, Chen Y-C, Lin Y-C, Yang F-Y, Huang N-C, Yen HE (2013) Suppressor of K+ transport growth defect 1 (SKD1) interacts with RING-type ubiquitin ligase and sucrose non-fermenting 1-related protein kinase (SnRK1) in the halophyte ice plant. J Exp Bot 64:2385–2400

Cho Y-H, Hong J-W, Kim E-C, Yoo S-D (2012) Regulatory functions of SnRK1 in stress-responsive gene expression and in plant growth and development. Plant Physiol 158:1955–1964

Choudhary MK, Nomura Y, Wang L, Nakagami H, Somers DE (2015) Quantitative circadian phosphoproteomic analysis of Arabidopsis reveals extensive clock control of key components in physiological, metabolic and signaling pathways. Mol Cell Proteomics 14:2243–2260

Coello P, Martínez-Barajas E (2014a) The activity of SnRK1 is increased in Phaseolus vulgaris seeds in response to a reduced nutrient supply. Front Plant Sci 5:196

Coello P, Martínez-Barajas E (2014b) SnRK1 is differentially regulated in the cotyledon and embryo axe of bean (Phaseolus vulgaris L) seeds. Plant Physiol Biochem 80:153–159

Coello P, Hey SJ, Halford NG (2011) The sucrose non-fermenting-1-related (SnRK) family of protein kinases: potential for manipulation to improve stress tolerance and increase yield. J Exp Bot 62:883–893

Coello P, Hirano E, Hey SJ, Muttucumaru N, Martinez-Barajas E, Parry MAJ, Halford NG (2012) Evidence that abscisic acid promotes degradation of SNF1-related protein kinase (SnRK) 1 in wheat and activation of a putative calcium-dependent SnRK2. J Exp Bot 63:913–924

Confraria A, Martinho C, Elias A, Rubio-Somoza I, Baena-Gonzalez E (2013) miRNAs mediate SnRK1-dependent energy signaling in Arabidopsis. Front Plant Sci 4:197

Contento AL, Kim S-J, Bassham DC (2004) Transcriptome profiling of the response of Arabidopsis suspension culture cells to Suc starvation. Plant Physiol 135:2330–2347

Crozet P, Jammes F, Valot B, Ambard-Bretteville F, Nessler S, Hodges M, Vidal J, Thomas M (2010) Cross-phosphorylation between Arabidopsis thaliana sucrose nonfermenting 1-related protein kinase 1 (AtSnRK1) and its activating kinase (AtSnAK) determines their catalytic activities. J Biol Chem 285:12071–12077

Crozet P, Margalha L, Confraria A, Rodrigues A, Martinho C, Adamo M, Elias CA, Baena-González E (2014) Mechanisms of regulation of SNF1/AMPK/SnRK1 protein kinases. Front Plant Sci 5:190

Crozet P, Margalha L, Butowt R, Fernandes N, Elias A, Orosa B, Tomanov K, Teige M, Bachmair A, Sadanandom A et al (2016) SUMOylation represses SnRK1 signaling in Arabidopsis. Plant J 85:120–133

Cutler SR, Rodriguez PL, Finkelstein RR, Abrams SR (2010) Abscisic acid: emergence of a core signaling network. Annu Rev Plant Biol 61:651–679

Dale S, Arro M, Becerra B, Morrice NG, Boronat A, Hardie DG, Ferrer A (1995a) Bacterial expression of the catalytic domain of 3-hydroxy-3-methylglutaryl-CoA reductase (isoform HMGR1) from Arabidopsis thaliana, and its inactivation by phosphorylation at Ser577 by Brassica oleracea 3-hydroxy-3-methylglutaryl-CoA reductase kinase. Eur J Biochem 233:506–513

Dale S, Wilson WA, Edelman AM, Hardie DG (1995b) Similar substrate recognition motifs for mammalian AMP-activated protein kinase, higher plant HMG-CoA reductase kinase-A, yeast SNF1, and mammalian calmodulin-dependent protein kinase I. FEBS Lett 361:191–195

Davies SP, Hawley SA, Woods A, Carling D, Haystead TA, Hardie DG (1994) Purification of the AMP-activated protein kinase on ATP-gamma-sepharose and analysis of its subunit structure. Eur J Biochem 223:351–357

De Block M, Van Lijsebettens M (2011) Energy efficiency and energy homeostasis as genetic and epigenetic components of plant performance and crop productivity. Curr Opin Plant Biol 14:275–282

Debast S, Nunes-Nesi A, Hajirezaei MR, Hofmann J, Sonnewald U, Fernie AR, Börnke F (2011) Altering trehalose-6-phosphate content in transgenic potato tubers affects tuber growth and alters responsiveness to hormones during sprouting. Plant Physiol 156:1754–1771

Douglas P, Pigaglio E, Ferrer A, Halford NG, MacKintosh C (1997) Three spinach leaf nitrate reductase-3-hydroxy-3-methylglutaryl-CoA reductase kinases that are regulated by reversible phosphorylation and/or Ca2+ ions. Biochem J 325(Pt 1):101–109

Elrouby N, Coupland G (2010) Proteome-wide screens for small ubiquitin-like modifier (SUMO) substrates identify Arabidopsis proteins implicated in diverse biological processes. Proc Natl Acad Sci USA 107:17415–17420

Emanuelle S, Hossain MI, Moller IE, Pedersen HL, van de Meene AML, Doblin MS, Koay A, Oakhill JS, Scott JW, Willats WGT et al (2015) SnRK1 from Arabidopsis thaliana is an atypical AMPK. Plant J 82:183–192

Engelsberger WR, Schulze WX (2012) Nitrate and ammonium lead to distinct global dynamic phosphorylation patterns when resupplied to nitrogen-starved Arabidopsis seedlings. Plant J 69:978–995

Farrás R, Ferrando A, Jásik J, Kleinow T, Okrész L, Tiburcio A, Salchert K, del Pozo C, Schell J, Koncz C (2001) SKP1-SnRK protein kinase interactions mediate proteasomal binding of a plant SCF ubiquitin ligase. EMBO J 20:2742–2756

Ferrando A, Koncz-Kalman Z, Farras R, Tiburcio A, Schell J, Koncz C (2001) Detection of in vivo protein interactions between Snf1-related kinase subunits with intron-tagged epitope-labelling in plants cells. Nucleic Acids Res 29:3685–3693

Fordham-Skelton AP, Chilley P, Lumbreras V, Reignoux S, Fenton TR, Dahm CC, Pages M, Gatehouse JA (2002) A novel higher plant protein tyrosine phosphatase interacts with SNF1-related protein kinases via a KIS (kinase interaction sequence) domain. Plant J 29:705–715

Fragoso S, Espíndola L, Páez-Valencia J, Gamboa A, Camacho Y, Martínez-Barajas E, Coello P (2009) SnRK1 isoforms AKIN10 and AKIN11 are differentially regulated in Arabidopsis plants under phosphate starvation. Plant Physiol 149:1906–1916

Ghillebert R, Swinnen E, Wen J, Vandesteene L, Ramon M, Norga K, Rolland F, Winderickx J (2011) The AMPK/SNF1/SnRK1 fuel gauge and energy regulator: structure, function and regulation. FEBS J 278:3978–3990

Gissot L, Polge C, Bouly J-P, Lemaitre T, Kreis M, Thomas M (2004) AKINbeta3, a plant specific SnRK1 protein, is lacking domains present in yeast and mammals non-catalytic beta-subunits. Plant Mol Biol 56:747–759

Gissot L, Polge C, Jossier M, Girin T, Bouly J-P, Kreis M, Thomas M (2006) AKINbetagamma contributes to SnRK1 heterotrimeric complexes and interacts with two proteins implicated in plant pathogen resistance through its KIS/GBD sequence. Plant Physiol 142:931–944

Glinski M, Weckwerth W (2005) Differential multisite phosphorylation of the trehalose-6-phosphate synthase gene family in Arabidopsis thaliana: a mass spectrometry-based process for multiparallel peptide library phosphorylation analysis. Mol Cell Proteomics 4:1614–1625

Gomez LD, Baud S, Gilday A, Li Y, Graham IA (2006) Delayed embryo development in the ARABIDOPSIS TREHALOSE-6-PHOSPHATE SYNTHASE 1 mutant is associated with altered cell wall structure, decreased cell division and starch accumulation. Plant J 46:69–84

Gowans GJ, Hawley SA, Ross FA, Hardie DG (2013) AMP is a true physiological regulator of AMP-activated protein kinase by both allosteric activation and enhancing net phosphorylation. Cell Metab 18:556–566

Guérinier T, Millan L, Crozet P, Oury C, Rey F, Valot B, Mathieu C, Vidal J, Hodges M, Thomas M et al (2013) Phosphorylation of p27(KIP) (1) homologs KRP6 and 7 by SNF1-related protein kinase-1 links plant energy homeostasis and cell proliferation. Plant J 75:515–525

Halford NG, Hardie DG (1998) SNF1-related protein kinases: global regulators of carbon metabolism in plants? Plant Mol Biol 37:735–748

Halford NG, Hey SJ (2009) Snf1-related protein kinases (SnRKs) act within an intricate network that links metabolic and stress signalling in plants. Biochem J 419:247–259

Halford NG, Hey S, Jhurreea D, Laurie S, McKibbin RS, Paul M, Zhang Y (2003) Metabolic signalling and carbon partitioning: role of Snf1-related (SnRK1) protein kinase. J Exp Bot 54:467–475

Hanks SK, Hunter T (1995) Protein kinases 6. The eukaryotic protein kinase superfamily: kinase (catalytic) domain structure and classification. FASEB J 9:576–596

Hao L, Wang H, Sunter G, Bisaro DM (2003) Geminivirus AL2 and L2 proteins interact with and inactivate SNF1 kinase. Plant Cell 15:1034–1048

Hardie DG (2007) AMP-activated/SNF1 protein kinases: conserved guardians of cellular energy. Nat Rev Mol Cell Biol 8:774–785

Hardie DG (2014) AMP-activated protein kinase: maintaining energy homeostasis at the cellular and whole-body levels. Annu Rev Nutr 34:31–55

Harthill JE, Meek SEM, Morrice N, Peggie MW, Borch J, Wong BHC, Mackintosh C (2006) Phosphorylation and 14-3-3 binding of Arabidopsis trehalose-phosphate synthase 5 in response to 2-deoxyglucose. Plant J 47:211–223

Hartmann L, Pedrotti L, Weiste C, Fekete A, Schierstaedt J, Göttler J, Kempa S, Krischke M, Dietrich K, Mueller MJ et al (2015) Crosstalk between two bZIP signaling pathways orchestrates salt-induced metabolic reprogramming in Arabidopsis roots. Plant Cell 27:2244–2260

Heazlewood JL, Durek P, Hummel J, Selbig J, Weckwerth W, Walther D, Schulze WX (2008) PhosPhAt: a database of phosphorylation sites in Arabidopsis thaliana and a plant-specific phosphorylation site predictor. Nucleic Acids Res 36:D1015–1021

Hedbacker K, Carlson M (2006) Regulation of the nucleocytoplasmic distribution of Snf1-Gal83 protein kinase. Eukaryotic Cell 5:1950–1956

Hedbacker K, Carlson M (2008) SNF1/AMPK pathways in yeast. Front Biosci 13:2408–2420

Hedbacker K, Townley R, Carlson M (2004) Cyclic AMP-dependent protein kinase regulates the subcellular localization of Snf1-Sip1 protein kinase. Mol Cell Biol 24:1836–1843

Hendriks IA, D'souza RCJ, Yang B, Verlaan-de Vries M, Mann M, Vertegaal ACO (2014) Uncovering global SUMOylation signaling networks in a site-specific manner. Nat Struct Mol Biol 53:1053–1066

Henry C, Bledsoe SW, Griffiths CA, Kollman A, Paul MJ, Sakr S, Lagrimini LM (2015) Differential role for trehalose metabolism in salt-stressed maize. Plant Physiol 169:1072–1089

Hey S, Mayerhofer H, Halford NG, Dickinson JR (2007) DNA sequences from Arabidopsis, which encode protein kinases and function as upstream regulators of Snf1 in yeast. J Biol Chem 282:10472–10479

Hofmann K, Bucher P (1996) The UBA domain: a sequence motif present in multiple enzyme classes of the ubiquitination pathway. Trends Biochem Sci 21:172–173

Hrabak EM, Chan CWM, Gribskov M, Harper JF, Choi JH, Halford N, Kudla J, Luan S, Nimmo HG, Sussman MR et al (2003) The Arabidopsis CDPK-SnRK superfamily of protein kinases. Plant Physiol 132:666–680

Ikeda Y, Koizumi N, Kusano T, Sano H (2000) Specific binding of a 14-3-3 protein to autophosphorylated WPK4, an SNF1-related wheat protein kinase, and to WPK4-phosphorylated nitrate reductase. J Biol Chem 275, 41528

Im JH, Cho Y-H, Kim G-D, Kang G-H, Hong J-W, Yoo S-D (2014) Inverse modulation of the energy sensor Snf1-related protein kinase 1 on hypoxia adaptation and salt stress tolerance in Arabidopsis thaliana. Plant Cell Environ 37:2303–2312

Jeong E-Y, Seo PJ, Woo JC, Park C-M (2015) AKIN10 delays flowering by inactivating IDD8 transcription factor through protein phosphorylation in Arabidopsis. BMC Plant Biol 15:110

Jiang R, Carlson M (1997) The Snf1 protein kinase and its activating subunit, Snf4, interact with distinct domains of the Sip1/Sip2/Gal83 component in the kinase complex. Mol Cell Biol 17:2099–2106

Jørgensen AD, Nøhr J, Kastrup JS, Gajhede M, Sigurskjold BW, Sauer J, Svergun DI, Svensson B, Vestergaard B (2008) Small angle X-ray studies reveal that Aspergillus niger glucoamylase has a defined extended conformation and can form dimers in solution. J Biol Chem 283:14772–14780

Jossier M, Bouly J-P, Meimoun P, Arjmand A, Lessard P, Hawley S, Grahame Hardie D, Thomas M (2009) SnRK1 (SNF1-related kinase 1) has a central role in sugar and ABA signalling in Arabidopsis thaliana. Plant J 59:316–328

Kazgan N, Williams T, Forsberg LJ, Brenman JE (2010) Identification of a nuclear export signal in the catalytic subunit of AMP-activated protein kinase. Mol Biol Cell 21:3433–3442

Kerk D, Conley TR, Rodriguez FA, Tran HT, Nimick M, Muench DG, Moorhead GBG (2006) A chloroplast-localized dual-specificity protein phosphatase in Arabidopsis contains a phylogenetically dispersed and ancient carbohydrate-binding domain, which binds the polysaccharide starch. Plant J 46:400–413

Kim D-Y, Scalf M, Smith LM, Vierstra RD (2013) Advanced proteomic analyses yield a deep catalog of ubiquitylation targets in Arabidopsis. Plant Cell 25:1523–1540

Kleinow T, Bhalerao R, Breuer F, Umeda M, Salchert K, Koncz C (2000) Functional identification of an Arabidopsis snf4 ortholog by screening for heterologous multicopy suppressors of snf4 deficiency in yeast. Plant J 23:115–122

Kleinow T, Himberta S, Krenza B, Jeskea H, Koncz C (2009) NAC domain transcription factor ATAF1 interacts with SNF1-related kinases and silencing of its subfamily causes severe developmental defects in Arabidopsis. Plant Sci 177:360–370

Kolbe A, Tiessen A, Schluepmann H, Paul M, Ulrich S, Geigenberger P (2005) Trehalose 6-phosphate regulates starch synthesis via posttranslational redox activation of ADP-glucose pyrophosphorylase. Proc Natl Acad Sci U S A 102:11118–11123

Kötting O, Santelia D, Edner C, Eicke S, Marthaler T, Gentry MS, Comparot-Moss S, Chen J, Smith AM, Steup M et al (2009) STARCH-EXCESS4 is a laforin-like Phosphoglucan phosphatase required for starch degradation in Arabidopsis thaliana. Plant Cell 21:334–346

Kulma A, Villadsen D, Campbell DG, Meek SEM, Harthill JE, Nielsen TH, Mackintosh C (2004) Phosphorylation and 14-3-3 binding of Arabidopsis 6-phosphofructo-2-kinase/fructose-2,6-bisphosphatase. Plant J 37:654–667

Lane N, Martin W (2010) The energetics of genome complexity. Nature 467:929–934

Lastdrager J, Hanson J, Smeekens S (2014) Sugar signals and the control of plant growth and development. J Exp Bot 65:799–807

Laurie S, McKibbin RS, Halford NG (2003) Antisense SNF1-related (SnRK1) protein kinase gene represses transient activity of an alpha-amylase (alpha-Amy2) gene promoter in cultured wheat embryos. J Exp Bot 54:739–747

Lee JH, Terzaghi W, Gusmaroli G, Charron JB, Yoon HJ, Chen H, He YJ, Xiong Y, Deng XW (2008) Characterization of Arabidopsis and rice DWD proteins and their roles as substrate receptors for CUL4-RING E3 ubiquitin ligases. Plant Cell 20:152–167

Lee K-W, Chen P-W, Lu C-A, Chen S, Yu S-M, Ho T-HD (2009) Coordinated responses to oxygen and sugar deficiency allow rice seedlings to tolerate flooding. Sci Signal 2:ra61

Li X-F, Li Y-J, An Y-H, Xiong L-J, Shao X-H, Wang Y, Sun Y (2009) AKINbeta1 is involved in the regulation of nitrogen metabolism and sugar signaling in Arabidopsis. J Integr Plant Biol 51:513–520

Liang J, Shao SH, Xu Z-X, Hennessy B, Ding Z, Larrea M, Kondo S, Dumont DJ, Gutterman JU, Walker CL et al (2007) The energy sensing LKB1-AMPK pathway regulates p27(kip1) phosphorylation mediating the decision to enter autophagy or apoptosis. Nat Cell Biol 9:218–224

Lin J-F, Wu S-H (2004) Molecular events in senescing Arabidopsis leaves. Plant J 39:612–628

Lin C-R, Lee K-W, Chen C-Y, Hong Y-F, Chen J-L, Lu C-A, Chen K-T, Ho T-HD YS-M (2014) SnRK1A-interacting negative regulators modulate the nutrient starvation signaling sensor SnRK1 in source-sink communication in cereal seedlings under abiotic stress. Plant Cell 26:808–827

Liu L, Chung HY, Lacatus G, Baliji S, Ruan J, Sunter G (2014) Altered expression of Arabidopsis genes in response to a multifunctional geminivirus pathogenicity protein. BMC Plant Biol 14:302

López-Paz C, Vilela B, Riera M, Pagès M, Lumbreras V (2009) Maize AKINbetagamma dimerizes through the KIS/CBM domain and assembles into SnRK1 complexes. FEBS Lett 583:1887–1894

Lovas A, Bimbó A, Szabó L, Bánfalvi Z (2003) Antisense repression of StubGAL83 affects root and tuber development in potato. Plant J 33:139–147
Lu Z, Hunter T (2009) Degradation of activated protein kinases by ubiquitination. Annu Rev Biochem 78:435–475
Lu C-A, Lin C-C, Lee K-W, Chen J-L, Huang L-F, Ho S-L, Liu H-J, Hsing Y-I, Yu S-M (2007) The SnRK1A protein kinase plays a key role in sugar signaling during germination and seedling growth of rice. Plant Cell 19:2484–2499
Lumbreras V, Alba MM, Kleinow T, Koncz C, Pagès M (2001) Domain fusion between SNF1-related kinase subunits during plant evolution. EMBO Rep 2:55–60
Lunn JE, Delorge I, Figueroa CM, Van Dijck P, Stitt M (2014) Trehalose metabolism in plants. Plant J 79:544–567
Ma J, Hanssen M, Lundgren K, Hernández L, Delatte T, Ehlert A, Liu C-M, Schluepmann H, Dröge-Laser W, Moritz T et al (2011) The sucrose-regulated Arabidopsis transcription factor bZIP11 reprograms metabolism and regulates trehalose metabolism. New Phytol 191:733–745
Mackintosh RW, Davies SP, Clarke PR, Weekes J, Gillespie JG, Gibb BJ, Hardie DG (1992) Evidence for a protein kinase cascade in higher plants. 3-Hydroxy-3-methylglutaryl-CoA reductase kinase. Eur J Biochem 209:923–931
Mair A, Pedrotti L, Wurzinger B, Anrather D, Simeunovic A, Weiste C, Valerio C, Dietrich K, Kirchler T, Nägele T et al (2015) SnRK1-triggered switch of bZIP63 dimerization mediates the low-energy response in plants. eLife 4
Maor R, Jones A, Nühse TS, Studholme DJ, Peck SC, Shirasu K (2007) Multidimensional protein identification technology (MudPIT) analysis of ubiquitinated proteins in plants. Mol Cell Proteomics 6:601–610
Martínez-Barajas E, Delatte T, Schluepmann H, de Jong GJ, Somsen GW, Nunes C, Primavesi LF, Coello P, Mitchell RAC, Paul MJ (2011) Wheat grain development is characterized by remarkable trehalose 6-phosphate accumulation pregrain filling: tissue distribution and relationship to SNF1-related protein kinase1 activity. Plant Physiol 156:373–381
Matiolli CC, Tomaz JP, Duarte GT, Prado FM, Del Bem LEV, Silveira AB, Gauer L, Corrêa LGG, Drumond RD, Viana AJC et al (2011) The Arabidopsis bZIP gene AtbZIP63 is a sensitive integrator of transient ABA and glucose signals. Plant Physiol 157:692–705
McBride A, Ghilagaber S, Nikolaev A, Hardie DG (2009) The glycogen-binding domain on the AMPK beta subunit allows the kinase to act as a glycogen sensor. Cell Metab 9:23–34
McCartney RR, Schmidt MC (2001) Regulation of Snf1 kinase. Activation requires phosphorylation of threonine 210 by an upstream kinase as well as a distinct step mediated by the Snf4 subunit. J Biol Chem 276:36460–36466
Mccormick AJ, Kruger NJ (2015) Lack of fructose 2,6-bisphosphate compromises photosynthesis and growth in Arabidopsis in fluctuating environments. Plant J 81:670–683
McMichael RW, Bachmann M, Huber SC (1995) Spinach leaf sucrose-phosphate synthase and nitrate reductase are phosphorylated/inactivated by multiple protein kinases in vitro. Plant Physiol 108:1077–1082
Mitchelhill KI, Stapleton D, Gao G, House C, Michell B, Katsis F, Witters LA, Kemp BE (1994) Mammalian AMP-activated protein kinase shares structural and functional homology with the catalytic domain of yeast Snf1 protein kinase. J Biol Chem 269:2361–2364
Moorhead G, Douglas P, Morrice N, Scarabel M, Aitken A, MacKintosh C (1996) Phosphorylated nitrate reductase from spinach leaves is inhibited by 14-3-3 proteins and activated by fusicoccin. Curr Biol 6:1104–1113
Muranaka T, Banno H, Machida Y (1994) Characterization of tobacco protein kinase NPK5, a homolog of Saccharomyces cerevisiae SNF1 that constitutively activates expression of the glucose-repressible SUC2 gene for a secreted invertase of S. cerevisiae. Mol Cell Biol 14:2958–2965
Nakagami H, Sugiyama N, Mochida K, Daudi A, Yoshida Y, Toyoda T, Tomita M, Ishihama Y, Shirasu K (2010) Large-scale comparative phosphoproteomics identifies conserved phosphorylation sites in plants. Plant Physiol 153:1161–1174

Nayak V, Zhao K, Wyce A, Schwartz MF, Lo W-S, Berger SL, Marmorstein R (2006) Structure and dimerization of the kinase domain from yeast Snf1, a member of the Snf1/AMPK protein family. Structure 14:477–485

Nietzsche M, Schießl I, Börnke F (2014) The complex becomes more complex: protein-protein interactions of SnRK1 with DUF581 family proteins provide a framework for cell- and stimulus type-specific SnRK1 signaling in plants. Front Plant Sci 5:54

Niittylä T, Comparot-Moss S, Lue W-L, Messerli G, Trevisan M, Seymour MDJ, Gatehouse JA, Villadsen D, Smith SM, Chen J et al (2006) Similar protein phosphatases control starch metabolism in plants and glycogen metabolism in mammals. J Biol Chem 281:11815–11818

Nunes C, O'Hara LE, Primavesi LF, Delatte TL, Schluepmann H, Somsen GW, Silva AB, Fevereiro PS, Wingler A, Paul MJ (2013a) The trehalose 6-phosphate/SnRK1 signaling pathway primes growth recovery following relief of sink limitation. Plant Physiol 162:1720–1732

Nunes C, Primavesi LF, Patel MK, Martinez-Barajas E, Powers SJ, Sagar R, Fevereiro PS, Davis BG, Paul MJ (2013b) Inhibition of SnRK1 by metabolites: tissue-dependent effects and cooperative inhibition by glucose 1-phosphate in combination with trehalose 6-phosphate. Plant Physiol Biochem 63:89–98

O'Brien M, Kaplan-Levy RN, Quon T, Sappl PG, Smyth DR (2015) PETAL LOSS, a trihelix transcription factor that represses growth in Arabidopsis thaliana, binds the energy-sensing SnRK1 kinase AKIN10. J Exp Bot 66:2475–2485

Oakhill JS, Chen Z-P, Scott JW, Steel R, Castelli LA, Ling N, Macaulay SL, Kemp BE (2010) β-Subunit myristoylation is the gatekeeper for initiating metabolic stress sensing by AMP-activated protein kinase (AMPK). Proc Natl Acad Sci USA 107:19237–19241

Oakhill JS, Steel R, Chen Z-P, Scott JW, Ling N, Tam S, Kemp BE (2011) AMPK is a direct adenylate charge-regulated protein kinase. Science 332:1433–1435

Oakhill JS, Scott JW, Kemp BE (2012) AMPK functions as an adenylate charge-regulated protein kinase. Trends Endocrinol Metab 23:125–132

Osuna D, Usadel B, Morcuende R, Gibon Y, Blasing OE, Hohne M, Gunter M, Kamlage B, Trethewey R, Scheible WR et al (2007) Temporal responses of transcripts, enzyme activities and metabolites after adding sucrose to carbon-deprived Arabidopsis seedlings. Plant J 49:463–491

Palenchar PM, Kouranov A, Lejay LV, Coruzzi GM (2004) Genome-wide patterns of carbon and nitrogen regulation of gene expression validate the combined carbon and nitrogen (CN)-signaling hypothesis in plants. Genome Biol 5:R91

Paul M (2007) Trehalose 6-phosphate. Curr Opin Plant Biol 10:303–309

Paul MJ, Primavesi LF, Jhurreea D, Zhang Y (2008) Trehalose metabolism and signaling. Annu Rev Plant Biol 59:417–441

Pellicena P, Kuriyan J (2006) Protein-protein interactions in the allosteric regulation of protein kinases. Curr Opin Struct Biol 16:702–709

Piattoni CV, Bustos DM, Guerrero SA, Iglesias AA (2011) Nonphosphorylating glyceraldehyde-3-phosphate dehydrogenase is phosphorylated in wheat endosperm at serine-404 by an SNF1-related protein kinase allosterically inhibited by ribose-5-phosphate. Plant Physiol 156:1337–1350

Pien S, Wyrzykowska J, Fleming AJ (2001) Novel marker genes for early leaf development indicate spatial regulation of carbohydrate metabolism within the apical meristem. Plant J 25:663–674

Pierre M, Traverso JA, Boisson B, Domenichini S, Bouchez D, Giglione C, Meinnel T (2007) N-Myristoylation regulates the SnRK1 pathway in Arabidopsis. Plant Cell 19:2804–2821

Polekhina G, Gupta A, Michell BJ, Van Denderen B, Murthy S, Feil SC, Jennings IG, Campbell DJ, Witters LA, Parker MW et al (2003) AMPK beta subunit targets metabolic stress sensing to glycogen. Curr Biol 13:867–871

Polekhina G, Gupta A, Van Denderen BJW, Feil SC, Kemp BE, Stapleton D, Parker MW (2005) Structural basis for glycogen recognition by AMP-activated protein kinase. Structure 13:1453–1462

Polge C, Thomas M (2007) SNF1/AMPK/SnRK1 kinases, global regulators at the heart of energy control? Trends Plant Sci 12:20–28

Polge C, Jossier M, Crozet P, Gissot L, Thomas M (2008) Beta-subunits of the SnRK1 complexes share a common ancestral function together with expression and function specificities; physical interaction with nitrate reductase specifically occurs via AKINbeta1-subunit. Plant Physiol 148:1570–1582

Price J, Laxmi A, St Martin SK, Jang J-C (2004) Global transcription profiling reveals multiple sugar signal transduction mechanisms in Arabidopsis. Plant Cell 16:2128–2150

Purcell P, Smith AM, Halford NG (1998) Antisense expression of sucrose non-fermenting-1-related protein kinase sequence in potato results in decreased expression of sucrose synthase in tubers and loss of sucrose-inducibility of sucrose synthase transcripts in leaves. Plant J 14:195–203

Radchuk R, Radchuk V, Weschke W, Borisjuk L, Weber H (2006) Repressing the expression of the SUCROSE NONFERMENTING-1-RELATED PROTEIN KINASE gene in pea embryo causes pleiotropic defects of maturation similar to an abscisic acid-insensitive phenotype. Plant Physiol 140:263–278

Radchuk R, Emery RJN, Weier D, Vigeolas H, Geigenberger P, Lunn JE, Feil R, Weschke W, Weber H (2010) Sucrose non-fermenting kinase 1 (SnRK1) coordinates metabolic and hormonal signals during pea cotyledon growth and differentiation. Plant J 61:324–338

Ramon M, De Smet I, Vandesteene L, Naudts M, Leyman B, Van Dijck P, Rolland F, Beeckman T, Thevelein JM (2009) Extensive expression regulation and lack of heterologous enzymatic activity of the class II trehalose metabolism proteins from Arabidopsis thaliana. Plant Cell Environ 32:1015–1032

Ramon M, Ruelens P, Li Y, Sheen J, Geuten K, Rolland F (2013) The hybrid four-CBS-domain KINβγ subunit functions as the canonical γ subunit of the plant energy sensor SnRK1. Plant J 75:11–25

Riek U, Scholz R, Konarev P, Rufer A, Suter M, Nazabal A, Ringler P, Chami M, Müller SA, Neumann D et al (2008) Structural properties of AMP-activated protein kinase: dimerization, molecular shape, and changes upon ligand binding. J Biol Chem 283:18331–18343

Robaglia C, Thomas M, Meyer C (2012) Sensing nutrient and energy status by SnRK1 and TOR kinases. Curr Opin Plant Biol 15:301–307

Rodrigues A, Adamo M, Crozet P, Margalha L, Confraria A, Martinho C, Elias A, Rabissi A, Lumbreras V, González-Guzmán M et al (2013) ABI1 and PP2CA phosphatases are negative regulators of Snf1-related protein kinase1 signaling in Arabidopsis. Plant Cell 25:3871–3884

Rolland F, Baena-Gonzalez E, Sheen J (2006) Sugar sensing and signaling in plants: conserved and novel mechanisms. Annu Rev Plant Biol 57:675–709

Rook F, Gerrits N, Kortstee A, van Kampen M, Borrias M, Weisbeek P, Smeekens S (1998) Sucrose-specific signalling represses translation of the Arabidopsis ATB2 bZIP transcription factor gene. Plant J 15:253–263

Rosnoblet C, Aubry C, Leprince O, Vu BL, Rogniaux H, Buitink J (2007) The regulatory gamma subunit SNF4b of the sucrose non-fermenting-related kinase complex is involved in longevity and stachyose accumulation during maturation of Medicago truncatula seeds. Plant J 51:47–59

Rubio T, Vernia S, Sanz P (2013) Sumoylation of AMPKβ2 subunit enhances AMP-activated protein kinase activity. Mol Biol Cell 24:1801–1811

Rudolph MJ, Amodeo GA, Bai Y, Tong L (2005) Crystal structure of the protein kinase domain of yeast AMP-activated protein kinase Snf1. Biochem Biophys Res Commun 337:1224–1228

Schluepmann H, Paul M (2009) Trehalose metabolites in Arabidopsis-elusive, active and central. Arabidopsis Book 7, e0122

Schluepmann H, Pellny T, van Dijken A, Smeekens S, Paul M (2003) Trehalose 6-phosphate is indispensable for carbohydrate utilization and growth in Arabidopsis thaliana. Proc Natl Acad Sci U S A 100:6849–6854

Schluepmann H, van Dijken A, Aghdasi M, Wobbes B, Paul M, Smeekens S (2004) Trehalose mediated growth inhibition of Arabidopsis seedlings is due to trehalose-6-phosphate accumulation. Plant Physiol 135:879–890

Scholz R, Suter M, Weimann T, Polge C, Konarev PV, Thali RF, Tuerk RD, Viollet B, Wallimann T, Schlattner U et al (2009) Homo-oligomerization and activation of AMP-activated protein kinase are mediated by the kinase domain alphaG-helix. J Biol Chem 284:27425–27437

Schuck S, Camehl I, Gilardoni PA, Oelmueller R, Baldwin IT, Bonaventure G (2012) HSPRO controls early Nicotiana attenuata seedling growth during interaction with the fungus Piriformospora indica. Plant Physiol 160:929–943

Schuck S, Baldwin IT, Bonaventure G (2013) HSPRO acts via SnRK1-mediated signaling in the regulation of Nicotiana attenuata seedling growth promoted by Piriformospora indica. Plant Signal Behav 8, e23537

Schwachtje J, Minchin PEH, Jahnke S, van Dongen JT, Schittko U, Baldwin IT (2006) SNF1-related kinases allow plants to tolerate herbivory by allocating carbon to roots. Proc Natl Acad Sci U S A 103:12935–12940

Scott JW, Hawley SA, Green KA, Anis M, Stewart G, Scullion GA, Norman DG, Hardie DG (2004) CBS domains form energy-sensing modules whose binding of adenosine ligands is disrupted by disease mutations. J Clin Invest 113:274–284

Shen W, Hanley-Bowdoin L (2006) Geminivirus infection up-regulates the expression of two Arabidopsis protein kinases related to yeast SNF1- and mammalian AMPK-activating kinases. Plant Physiol 142:1642–1655

Shen W, Reyes MI, Hanley-Bowdoin L (2009) Arabidopsis protein kinases GRIK1 and GRIK2 specifically activate SnRK1 by phosphorylating its activation loop. Plant Physiol 150:996–1005

Shen Q, Liu Z, Song F, Xie Q, Hanley-Bowdoin L, Zhou X (2011) Tomato SlSnRK1 protein interacts with and phosphorylates betaC1, a pathogenesis protein encoded by a geminivirus beta-satellite. Plant Physiol 157:1394–1406

Shen Q, Bao M, Zhou X (2012) A plant kinase plays roles in defense response against geminivirus by phosphorylation of a viral pathogenesis protein. Plant Signal Behav 7:888–892

Shen W, Dallas MB, Goshe MB, Hanley-Bowdoin L (2014) SnRK1 phosphorylation of AL2 delays cabbage leaf curl virus infection in Arabidopsis. J Virol 88:10598–10612

Short JD, Houston KD, Dere R, Cai S-L, Kim J, Johnson CL, Broaddus RR, Shen J, Miyamoto S, Tamanoi F et al (2008) AMP-activated protein kinase signaling results in cytoplasmic sequestration of p27. Cancer Res 68:6496–6506

Short JD, Dere R, Houston KD, Cai S-L, Kim J, Bergeron JM, Shen J, Liang J, Bedford MT, Mills GB et al (2010) AMPK-mediated phosphorylation of murine p27 at T197 promotes binding of 14-3-3 proteins and increases p27 stability. Mol Carcinog 49:429–439

Simpson-Lavy KJ, Johnston M (2013) SUMOylation regulates the SNF1 protein kinase. Proc Natl Acad Sci U S A 110:17432–17437

Simpson-Lavy KJ, Bronstein A, Kupiec M, Johnston M (2015) Cross-talk between carbon metabolism and the DNA damage response in S. cerevisiae. Cell Rep 12:1865–1875

Slocombe SP, Beaudoin F, Donaghy PG, Hardie DG, Dickinson JR, Halford NG (2004) SNF1-related protein kinase (snRK1) phosphorylates class I heat shock protein. Plant Physiol Biochem 42:111–116

Stein SC, Woods A, Jones NA, Davison MD, Carling D (2000) The regulation of AMP-activated protein kinase by phosphorylation. Biochem J 345:437–443

Steinberg GR, Kemp BE (2009) AMPK in health and disease. Physiol Rev 89:1025–1078

Stitt M (1987) Fructose 2,6-bisphosphate and plant carbohydrate metabolism. Plant Physiol 84:201–204

Sugden C, Crawford RM, Halford NG, Hardie DG (1999a) Regulation of spinach SNF1-related (SnRK1) kinases by protein kinases and phosphatases is associated with phosphorylation of the T loop and is regulated by 5′-AMP. Plant J 19:433–439

Sugden C, Donaghy PG, Halford NG, Hardie DG (1999b) Two SNF1-related protein kinases from spinach leaf phosphorylate and inactivate 3-hydroxy-3-methylglutaryl-coenzyme A reductase, nitrate reductase, and sucrose phosphate synthase in vitro. Plant Physiol 120:257–274

Suzuki A, Okamoto S, Lee S, Saito K, Shiuchi T, Minokoshi Y (2007) Leptin stimulates fatty acid oxidation and peroxisome proliferator-activated receptor alpha gene expression in mouse C2C12 myoblasts by changing the subcellular localization of the alpha2 form of AMP-activated protein kinase. Mol Cell Biol 27:4317–4327

Szal B, Podgorska A (2012) The role of mitochondria in leaf nitrogen metabolism. Plant Cell Environ 35:1756–1768

Szczesny R, Büttner D, Escolar L, Schulze S, Seiferth A, Bonas U (2010) Suppression of the AvrBs1-specific hypersensitive response by the YopJ effector homolog AvrBsT from Xanthomonas depends on a SNF1-related kinase. New Phytol 187:1058–1074

Thelander M, Olsson T, Ronne H (2004) Snf1-related protein kinase 1 is needed for growth in a normal day–night light cycle. EMBO J 23:1900–1910

Thimm O, Bläsing O, Gibon Y, Nagel A, Meyer S, Krüger P, Selbig J, Müller LA, Rhee SY, Stitt M (2004) Mapman: a user-driven tool to display genomics data sets onto diagrams of metabolic pathways and other biological processes. Plant J 37:914–939

Tiessen A, Prescha K, Branscheid A, Palacios N, Mckibbin R, Halford NG, Geigenberger P (2003) Evidence that SNF1-related kinase and hexokinase are involved in separate sugar-signalling pathways modulating post-translational redox activation of ADP-glucose pyrophosphorylase in potato tubers. Plant J 35:490–500

Toroser D, Plaut Z, Huber SC (2000) Regulation of a plant SNF1-related protein kinase by glucose-6-phosphate. Plant Physiol 123:403–412

Traverso JA, Meinnel T, Giglione C (2008) Expanded impact of protein N-myristoylation in plants. Plant Signal Behav 3:501–502

Tsai AY-L, Gazzarrini S (2012) AKIN10 and FUSCA3 interact to control lateral organ development and phase transitions in Arabidopsis. Plant J 69:809–821

Tsai AY, Gazzarrini S (2014) Trehalose-6-phosphate and SnRK1 kinases in plant development and signaling: the emerging picture. Front Plant Sci 5:119

Tsugama D, Liu S, Takano T (2012) A putative myristoylated 2C-type protein phosphatase, PP2C74, interacts with SnRK1 in Arabidopsis. FEBS Lett 586:693–698

Umezawa T, Sugiyama N, Takahashi F, Anderson JC, Ishihama Y, Peck SC, Shinozaki K (2013) Genetics and phosphoproteomics reveal a protein phosphorylation network in the abscisic acid signaling pathway in Arabidopsis thaliana. Sci Signal 6:rs8

Viana R, Towler MC, Pan DA, Carling D, Viollet B, Hardie DG, Sanz P (2007) A conserved sequence immediately N-terminal to the Bateman domains in AMP-activated protein kinase gamma subunits is required for the interaction with the beta subunits. J Biol Chem 282:16117–16125

Wang H, Hao L, Shung CY, Sunter G, Bisaro DM (2003) Adenosine kinase is inactivated by geminivirus AL2 and L2 proteins. Plant Cell 15:3020–3032

Wang X, Peng F, Li M, Yang L, Li G (2012) Expression of a heterologous SnRK1 in tomato increases carbon assimilation, nitrogen uptake and modifies fruit development. J Plant Physiol 169:1173–1182

Wang P, Xue L, Batelli G, Lee S, Hou Y-J, Van Oosten MJ, Zhang H, Tao WA, Zhu J-K (2013) Quantitative phosphoproteomics identifies SnRK2 protein kinase substrates and reveals the effectors of abscisic acid action. Proc Natl Acad Sci U S A 110:11205–11210

Warden SM, Richardson C, O'Donnell J, Stapleton D, Kemp BE, Witters LA (2001) Post-translational modifications of the beta-1 subunit of AMP-activated protein kinase affect enzyme activity and cellular localization. Biochem J 354:275–283

Weltmeier F, Rahmani F, Ehlert A, Dietrich K, Schutze K, Wang X, Chaban C, Hanson J, Teige M, Harter K et al (2009) Expression patterns within the Arabidopsis C/S1 bZIP transcription factor network: availability of heterodimerization partners controls gene expression during stress response and development. Plant Mol Biol 69:107–119

Wiatrowski HA, Van Denderen BJW, Berkey CD, Kemp BE, Stapleton D, Carlson M (2004) Mutations in the gal83 glycogen-binding domain activate the snf1/gal83 kinase pathway by a glycogen-independent mechanism. Mol Cell Biol 24:352–361

Wiese A, Elzinga N, Wobbes B, Smeekens S (2005) Sucrose-induced translational repression of plant bZIP-type transcription factors. Biochem Soc Trans 33:272–275

Williams T, Brenman JE (2008) LKB1 and AMPK in cell polarity and division. Trends Cell Biol 18:193–198

Williams SP, Rangarajan P, Donahue JL, Hess JE, Gillaspy GE (2014) Regulation of sucrose non-fermenting related kinase 1 genes in Arabidopsis thaliana. Front Plant Sci 5:324

Xiao B, Sanders MJ, Underwood E, Heath R, Mayer FV, Carmena D, Jing C, Walker PA, Eccleston JF, Haire LF et al (2011) Structure of mammalian AMPK and its regulation by ADP. Nature 472:230–233

Yao Q, Bollinger C, Gao J, Xu D, Thelen JJ (2012) P(3)DB: an integrated database for plant protein phosphorylation. Front Plant Sci 3:206

Zhang Y, Shewry PR, Jones H, Barcelo P, Lazzeri PA, Halford NG (2001) Expression of antisense SnRK1 protein kinase sequence causes abnormal pollen development and male sterility in transgenic barley. Plant J 28:431–441

Zhang Y, Primavesi LF, Jhurreea D, Andralojc PJ, Mitchell RAC, Powers SJ, Schluepmann H, Delatte T, Wingler A, Paul MJ (2009) Inhibition of SNF1-related protein kinase1 activity and regulation of metabolic pathways by trehalose-6-phosphate. Plant Physiol 149:1860–1871

Zhang Z, Deng Y, Song X, Miao M (2015) Trehalose-6-phosphate and SNF1-related protein kinase 1 are involved in the first-fruit inhibition of cucumber. J Plant Physiol 177:110–120

Zourelidou M, de Torres-Zabala M, Smith C, Bevan MW (2002) Storekeeper defines a new class of plant-specific DNA-binding proteins and is a putative regulator of patatin expression. Plant J 30:489–497

Part V
Methods of Study in AMPK

Chapter 18
Animal Models to Study AMPK

Benoit Viollet and Marc Foretz

Contents

Abstract AMPK is an evolutionary conserved energy sensor involved in the regulation of energy metabolism. Based on biochemical studies, AMPK has brought much of interest in recent years due to its potential impact on metabolic disorders. Suitable animal models are therefore essential to promote our understanding of the molecular and functional roles of AMPK but also to bring novel information for the development of novel therapeutic strategies. The organism systems include pig (*Sus scrofa*), mouse (*Mus musculus*), fly (*Drosophila melanogaster*), worm (*Caenorhabditis elegans*), and fish (*Danio rerio*) models. These animal models have provided reliable experimental evidence demonstrating the crucial role of AMPK in the regulation of metabolism but also of cell polarity, autophagy, and oxidative stress. In this chapter, we update the new development in the generation and application of animal models for the study of AMPK biology.

B. Viollet (✉) • M. Foretz
INSERM U1016, Institut Cochin, Paris, France

CNRS UMR 8104, Paris, France

Université Paris Descartes, Sorbonne Paris Cité, Paris, France
e-mail: benoit.viollet@inserm.fr

M.D. Cordero, B. Viollet (eds.), *AMP-activated Protein Kinase*, Experientia Supplementum 107, DOI 10.1007/978-3-319-43589-3_18

We also discuss recent breakthroughs from studies in mice, flies, and worms showing how AMPK has a primary role in initiating or promoting pathological or beneficial impact on health.

Keywords AMPK-activated protein kinase • Animal models • Transgenic animals • Energy metabolism • Pharmacological drugs • Therapeutics

18.1 Introduction

AMP-activated protein kinase (AMPK) is widely accepted as a sensor of cellular energy balance (Hardie 2014). At the cellular level, AMPK promotes ATP-producing catabolic pathways, while simultaneously inhibiting ATP-consuming anabolic pathways. At the organismal level, AMPK integrates stress responses such as exercise as well as nutrient and hormonal signals to control whole-body energy expenditure and substrate utilization. As such a potent regulator of cellular and whole-body metabolism, AMPK has become the focus of a great deal of attention and appeared as an obvious target for the treatment of metabolic disorders such as obesity and type 2 diabetes (Winder and Hardie 1999). In several rodents models of diabetes and obesity, pharmacological activation of AMPK results in the remodeling of a wide range of metabolic pathways and has led to substantial improvement of disease outcome (Buhl et al. 2002; Halseth et al. 2002; Song et al. 2002; Cool et al. 2006; Fullerton et al. 2013). In parallel with rapid scientific discovery in cell-free and cellular systems, the development of animal models has been instrumental to the expanding field of AMPK (Viollet et al. 2009). Over the last decade, knockout (KO) mouse models for the different AMPK subunit isoforms as well as transgenic mouse models overexpressing loss-of-function or gain-of-function AMPK mutants have been developed to better understand the impact of AMPK on metabolic health and disease and have largely contributed to expand our knowledge on the relevance of AMPK in human disease (Fig. 18.1). Conditional targeting approaches are now at the forefront of mechanistic studies to investigate the relevance and specificity of AMPK in the homeostasis of multiple organs. These mouse models are also critically important for preclinical translational studies and early-stage clinical investigation to progress. In addition, the use of model organisms such as *Saccharomyces cerevisiae* (yeast) [see chapter AMPK in yeast: the SNF1 (sucrose non-fermenting 1) protein kinase complex], *Drosophila melanogaster* (fly) [see chapter The role of AMPK in *Drosophila melanogaster*], and *Caenorhabditis elegans* (roundworm) [see Chap. 5′-AMP-Activated Protein Kinase Signaling in *Caenorhabditis elegans*] has played a key role in delineating the physiological role of AMPK pathway and has been instrumental in providing novel information on the biology of AMPK system (Fig. 18.1). These various models helped to expand the paradigm of AMPK as a metabolic sensor and showed that AMPK has broad effects on cellular function, regulating cell growth, autophagy, oxidative stress, and cell polarity.

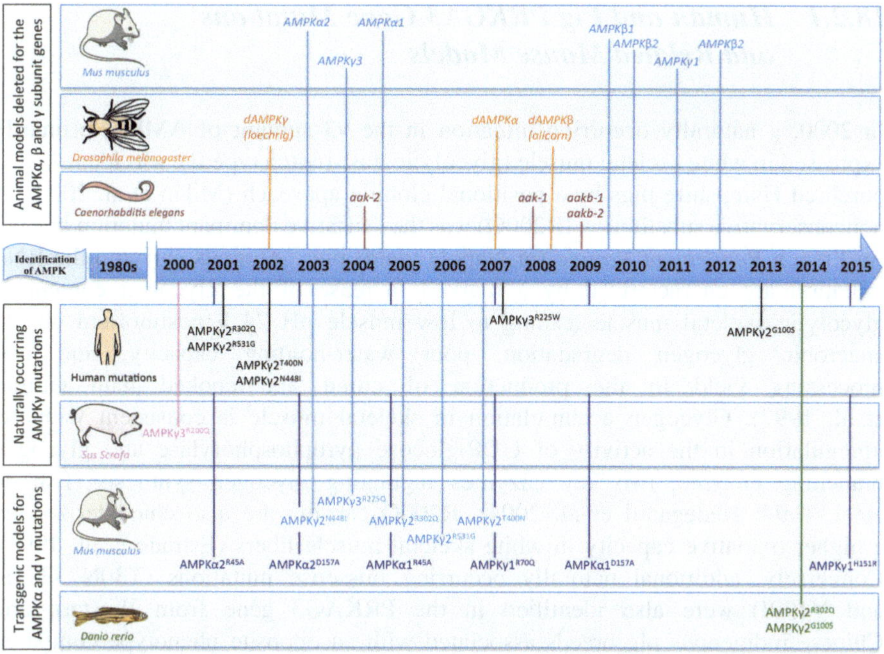

Fig. 18.1 Timeline of animal models generated for the study of AMPK biology. Deletion of the AMPKα, β, and γ genes in *Caenorhabditis elegans* (*aak-1*, *aak-2*, *aakb-1,* and *aakb-2*), *Drosophila melanogaster* (*dAMPKα, dAMPKβ/Loechrig,* and *dAMPKγ/Alicorn*), and *mus musculus* (*AMPKαl*, *AMPKα*, *AMPKβ1*, *AMPKβ2*, *AMPKγ1*, and *AMPKγ3*) is shown in the upper panel. Naturally occurring mutations in the human and porcine AMPKγ gene are shown in the middle panel. Transgenic mouse and fish models expressing AMPKα and γ mutations are shown in the lower panel. Refer to Tables 18.1, 18.3, and 18.5 for references

18.2 Naturally Occurring Mutations

Many genetically engineered animal models have been generated to investigate the physiopathological role of AMPK, but understanding of AMPK function has been also greatly advanced by studies of naturally occurring mutations in AMPK genes (Fig. 18.1). These mutations are apparently fairly rare but can result in pronounced pathological changes (Milan et al. 2000; Blair et al. 2001; Gollob et al. 2001a; Arad et al. 2002). Naturally occurring mutations have been initially characterized in the pig *PRKAG3* gene and the human *PRKAG2* gene (encoding the γ3 and γ2 subunit of AMPK, respectively) and are associated with abnormally high glycogen accumulation that results in pathological changes to skeletal and cardiac muscle (Milan et al. 2000; Blair et al. 2001; Gollob et al. 2001a; Arad et al. 2002). Later, a mutation in human PRKAG3 gene has been identified in association with an increase in skeletal muscle glycogen content and a decrease in intramuscular triglyceride (Costford et al. 2007).

18.2.1 Human and Pig PRKGA3 Gene Mutations and Related Mouse Models

In 2000, a naturally occurring mutation in the γ3 subunit of AMPK, primarily expressed in white skeletal muscle (glycolytic, fast-twitch type II), was identified in purebred Hampshire pigs by a positional cloning approach (Milan et al. 2000). A nonconservative substitution (R200Q) was the causative dominant mutation in RN^{-} (in French *Rendement Napole* for Napole yield) pigs. Animals carrying the RN^{-} mutation are characterized by a marked increase of the glycogen content in glycolytic skeletal muscle leading to low muscle pH 24 h postmortem due to anaerobic glycogen degradation, poor water-holding capacity, and low processing yield in the production of cured and cooked ham (Enfalt et al. 1997). Glycogen accumulation in skeletal muscle is consistent with the upregulation in the activity of UDP-glucose pyrophosphorylase and glycogen branching enzyme, two key enzymes regulating glycogen synthesis (Estrade et al. 1994; Hedegaard et al. 2004). R200Q carriers are also characterized by a higher oxidative capacity in white skeletal muscle fibers (Estrade et al. 1994). Conversely, additional naturally occurring missense mutations (T30N, G52S, and V199I) were also identified in the PRKAG3 gene from Western and Chinese indigenous pig breeds associated with an opposite phenotype compared to the RN^{-} pigs, resulting in reduced skeletal muscle glycogen content and additive effect on meat quality traits (Ciobanu et al. 2001; Huang et al. 2004).

Strong support for the causative nature of the R200Q mutation in the elevated skeletal muscle glycogen content has been provided by transgenic mouse models overexpressing primarily in white skeletal muscle the mouse γ3 R200Q mutant form (in these studies, the R200Q missense mutation is designated R225Q as the start methionine codon proposed by Cheung et al. (2000) is used) (Table 18.1). Muscle-specific transgenic mouse lines overexpressing the AMPKγ3 R225Q mutant under the control of either the myosin light chain or the muscle creatine kinase promoter/enhancer were generated and resulted in significant increases in skeletal muscle glycogen, replicating the pig RN^{-} phenotype (Barnes et al. 2004; Yu et al. 2006) (Table 18.2).

The R200Q missense mutation occurs within the cystathionine *β*-synthase domain 1 (CBS1) of the regulatory AMPKγ3 subunit, which is involved in the binding of the allosteric activator of the kinase, AMP. Originally, there was great controversy over the impact of the R200Q mutation on AMPK activity. It was initially reported that the activity of AMPK was reduced in muscle extracts from RN^{-} pigs both in the presence and absence of AMP (Milan et al. 2000). In resting muscle from AMPKγ3 R225Q mutant mice, AMPK activity was found to be decreased (Yu et al. 2006) or unaltered (Barnes et al. 2004). However, reduced AMPK activity could reflect the potential feedback inhibition of glycogen overload on AMPK activation (Milan et al. 2000; Jorgensen et al. 2004a). Thus, to evaluate the impact of CBS domain mutation, activity of $\alpha2\beta2\gamma3^{R225Q}$ trimer was measured from heterotrimeric complexes purified from transiently transfected COS cells and

Table 18.1 Genetically modified mouse models for the study of AMPK signaling pathway

Targeted gene	Genetic modification	Viability	References
Knockout and floxed mice			
AMPKα1	$AMPK\alpha 1^{-/-}$	Viable	Jorgensen et al. (2004b)
AMPKα2	$AMPK\alpha 2^{-/-}$	Viable	Viollet et al. (2003)
AMPKα1/α2	$AMPK\alpha 1^{-/-}\alpha 2^{-/-}$	Embryonic lethal	unpublished data
AMPKα1	$AMPK\alpha 1^{fl/fl}$	Viable	Nakada et al. (2010), Lantier et al. (2014)
AMPKα2	$AMPK\alpha 2^{fl/fl}$	Viable	Viollet et al. (2003), Nakada et al. (2010)
AMPKβ1	$AMPK\beta 1^{-/-}$	Viable/postnatal death	Dasgupta and Milbrandt (2009), Dzamko et al. (2010)
AMPKβ2	$AMPK\beta 2^{-/-}$	Viable	Steinberg et al. (2010), Dasgupta et al. (2012)
AMPKβ1/β2	$AMPK\beta 1^{-/-}\beta 2^{-/-}$	Embryonic lethal	O'Neill et al. (2011)
AMPKβ1	$AMPK\beta 1^{fl/fl}$	Viable	O'Neill et al. (2011)
AMPKβ2	$AMPK\beta 2^{fl/fl}$	Viable	O'Neill et al. (2011)
AMPKγ1	$AMPK\gamma 1^{-/-}$	Viable	Foretz et al. (2011)
AMPKγ1	$AMPK\gamma 1^{fl/fl}$	Viable	Foretz et al. (2011)
AMPKγ3	$AMPK\gamma 3^{-/-}$	Viable	Barnes et al. (2004)

Targeted gene	Genetic modification	Phenotype	References
Knock-in mice			
AMPKγ2	$AMPK\gamma 2^{R299Q}$	Obesity, hyperphagia, and impaired insulin secretion	Yavari et al. (2016)

Transgene	Targeted tissue	Promoter	References
Transgenic mice			
$AMPK\alpha 1^{K45R}$	Skeletal and cardiac muscle cells	MCK promoter/enhancer	Fujii et al. (2005)
$AMPK\alpha 1^{D157A}$	Skeletal muscle	HSA promoter	Miura et al. (2009)
$AMPK\alpha 1^{1\text{-}312}$	Vascular endothelium	VE–cadherin promoter	Li et al. (2012)
$AMPK\alpha 1_{1\text{-}312/T172D}$	Hepatocytes	Human ApoE promoter	Yang et al. (2008)
$AMPK\alpha 1_{1\text{-}312/T172D}$	Pancreatic β cells	RIP2 promoter	Sun et al. (2010)
AMPKα2	Ubiquitous	CMV promoter	Gong et al. (2011)
$AMPK\alpha 2^{K45R}$	Skeletal and cardiac muscle cells	MCK promoter/enhancer	Mu et al. (2001)
$AMPK\alpha 2^{D157A}$	Skeletal and cardiac muscle cells	MCK promoter/enhancer	Fujii et al. (2005)
$AMPK\alpha 2^{D157A}$	Pancreatic β cells	RIP2 promoter	Sun et al. (2010)

(continued)

Table 18.1 (continued)

Transgene	Targeted tissue	Promoter	References
$AMPK\alpha2^{D157A}$	Cardiomyocytes	α-MHC promoter	Xing et al. (2003)
$AMPK\gamma1^{R70Q}$	Skeletal muscle cells	α-Actin promoter	Barre et al. (2007)
$AMPK\gamma1^{H151R}$	Skeletal muscle cells	MLC1 promoter/ enhancer	Schonke et al. (2015)
$AMPK\gamma2$	Cardiomyocytes	α-MHC promoter	Arad et al. (2003), Sidhu et al. (2005)
$AMPK\gamma2^{R302Q}$	Cardiomyocytes	α-MHC promoter	Sidhu et al. (2005), Folmes et al. (2009)
$AMPK\gamma2^{N488I}$	Cardiomyocytes	α-MHC promoter α-MHC promoter/ tTA (Tet-off)	Arad et al. (2003), Luptak et al. (2007) Wolf et al. (2008)
$AMPK\gamma2^{T400N}$	Cardiomyocytes	α-MHC promoter	Banerjee et al. (2007)
$AMPK\gamma2^{R531G}$	Cardiomyocytes	α-MHC promoter	Davies et al. (2006)
$AMPK\gamma3$	Skeletal and cardiac muscle cells	MCK promoter/ enhancer MLC1 promoter/ enhancer	Yu et al. (2006) Barnes et al. (2004)
$AMPK\gamma3^{R225Q}$	Skeletal and cardiac muscle cells	MCK promoter/ enhancer MLC1 promoter/ enhancer	Yu et al. (2006) Barnes et al. (2004)
Targeted gene	**Targeted tissue**	**Cre-driven recombination**	**References**
Tissue-specific deletion of AMPK subunits			
$AMPK\alpha1$	T cells	CD4-Cre	Rolf et al. (2013)
	Myeloid cells	LysM-Cre	Mounier et al. (2013)
	Chondrocytes	Col2α1-Cre	Yang et al. (2016)
	White adipose tissue stromal vascular cells (in vitro deletion)	Gt(ROSA)26Sor-Cre tamoxifen inducible	Wang et al. (2015)
	Central nervous system	hGFAP-Cre	Maixner et al. (2015)
	Muscle satellite cells	Pax7-Cre tamoxifen inducible	Fu et al. (2015, 2016)
	Adipocytes	Adipoq-Cre	Wu et al. (2015)
$AMPK\alpha2$	Hepatocytes	Alfp-Cre	Andreelli et al. (2006)
	Hypothalamus AgRP neurons	AgRP-Cre	Claret et al. (2007)
	Hypothalamus POMC neurons	POMC-Cre	Claret et al. (2007)
	Endothelial cells	VE-cadherin-Cre	Kohlstedt et al. (2013)
	Adipocytes	Adipoq-Cre	Wu et al. (2015)

(continued)

Table 18.1 (continued)

Targeted gene	Targeted tissue	Cre-driven recombination	References
AMPKα1/ α2	Hypothalamus POMC neurons	POMC-Cre	Claret et al. (2007)
	Skeletal muscle cells	HSA-Cre	Lantier et al. (2014)
	Hepatocytes	Alfp-Cre	Guigas et al. (2006), Foretz et al. (2010)
	Pancreatic β cells	RIP2-Cre	Sun et al. (2010)
	Pancreatic β cells	mIns1-Cre	Kone et al. (2014)
	Pancreatic α cells	PPG-Cre	Sun et al. (2015)
	Catecholaminergic cells	TH-Cre	Mahmoud et al. (2015)
	Muscle satellite cells (in vitro deletion)	Gt(ROSA)26Sor-Cre (tamoxifen inducible)	Fu et al. (2013)
	Hematopoietic stem cells	Mx1-Cre (pIpC inducible)	Nakada et al. (2010)
	Proglucagon-expressing cells	Proglucagon-Cre	Sayers et al. (2016)
AMPKβ1	Skeletal and heart muscle cells	MCK-Cre	O'Neill et al. (2011), Sung et al. (2015)
AMPKβ2	Skeletal and heart muscle cells	MCK-Cre	O'Neill et al. (2011), Sung et al. (2015)
AMPKβ1/ β2	Skeletal and heart muscle cells	MCK-Cre	O'Neill et al. (2011), Sung et al. (2015)

Adipoq adiponectin, *AgRP* agouti-related protein, *Alfp* albumin/α-fetoprotein, *ApoE* apolipoprotein E, *CMV* cytomegalovirus, *Col2α1*, collagen type 2 α1, *hGFAP* human glial fibrillary acidic protein, *HSA* human α-skeletal actin, *Ins1* mouse insulin 1, *LysM* lysozyme M, *MCK* muscle creatine kinase, *MLC1* myosin light chain 1, *α-MHC* α-myosin heavy chain, *Mx1* myxovirus resistance protein 1, *Pax7* paired box-containing 7, *pIpC* polyinosine-polycytidine, *POMC* pro-opiomelanocortin, *PPG* preproglucagon, *RIP2* rat insulin promoter 2, *TH* tyrosine hydroxylase, *tTA* tetracycline-controlled transcriptional activator system, *VE* vascular endothelial

resulted in a higher basal AMPK activity and loss of AMP dependence (Barnes et al. 2004). Nevertheless, question remains on the mechanisms regulating AMP-independent activity of the AMPKγ3 R225Q mutant (Lindgren et al. 2007), indicating that more detailed functional studies are needed to precisely define the nature of this mutation in the CBS domain of AMPKγ3. Interestingly, a homologous mutation to the RN$^-$ mutation found in pigs has been identified in the human PRKAG3 gene from genetic studies of lean and obese human populations (Costford et al. 2007). Subjects bearing the AMPKγ3 R225W mutation exhibit increased skeletal muscle glycogen content, indicating conserved AMPKγ3 function across mammalian species (Costford et al. 2007). The AMPK γ3 R225W mutation is associated with increased activity of AMPK in differentiated muscle cells derived from *vastus lateralis* biopsies (Costford et al. 2007). Furthermore, it should be noted that introduction of the equivalent mutation R70Q in AMPKγ1 and R302Q in AMPKγ2 also caused a marked increase in AMPK activity (Gollob et al. 2001a; Hamilton et al. 2001; Yavari et al. 2016). More recently, studies with exercise-

Table 18.2 Metabolic pathways dysregulated in transgenic mice overexpressing AMPKα and γ mutations in skeletal and cardiac muscles

Organ	Pathway/function	Transgenic mice	References
Heart	Glycogen storage cardiomyopathy	AMPKγ2^{R302Q} AMPKγ2^{N488I} AMPKγ2^{R531G} AMPKγ2^{T400N}	Sidhu et al. (2005) Arad et al. (2003), Luptak et al. (2007) Davies et al. (2006) Banerjee et al. (2007)
	Ischemic/postischemic tolerance	AMPKα2^{R45A} AMPKα2^{D157A}	Russell et al. (2004) Xing et al. (2003)
	Glucose transport, glycogen metabolism, glycolysis	AMPKα2^{R45A} AMPKα2^{D157A}	Xing et al. (2003), Russell et al. (2004)
	Mitochondrial biogenesis/function	AMPKα2^{D157A}	Gundewar et al. (2009)
Skeletal muscle	Glucose transport	AMPKα^{2R45A} AMPKα2^{D157A}	Mu et al. (2001) Xing et al. (2003)
	Contraction/exercise tolerance	AMPKα2^{R45A} AMPKα2^{D157A} AMPKγ1^{R70Q} AMPKγ3^{R225Q}	Mu et al. (2001) Fujii et al. (2007), Rockl et al. (2007) Barre et al. (2007), Rockl et al. (2007) Barnes et al. (2004, 2005)
	Glycogen metabolism	AMPKγ3^{R225Q} AMPKγ1^{R70Q}	Barnes et al. (2004), Yu et al. (2006) Barre et al. (2007)
	Mitochondrial biogenesis/ function	AMPKα2^{R45A}	Zong et al. (2002)

trained RN$^-$ pigs have shifted toward the recognition that γ3 R200Q mutation is a gain-of-function mutation, which results in hyperaccumulation of glycogen due to increased influx of glucose and enhanced endurance exercise capacity (Granlund et al. 2010, 2011). Increased glycogen level is not due to impaired glycogen utilization as glycogen breakdown is similar after exercise in carriers and non-carriers of the RN$^-$ mutation (Granlund et al. 2010). These results are consistent with the finding that pigs carrying the R200Q mutation show increased signaling response of Akt and phosphorylation of its substrate, AS160, a higher capacity for phosphorylation of glucose, and faster muscle glycogen resynthesis after exercise (Granlund et al. 2010; Essen-Gustavsson et al. 2011; Granlund et al. 2011). However, slightly reduced basal and contraction-stimulated rates of glucose uptake were evidenced in transgenic AMPKγ3 R225Q mutant mice (Barnes et al. 2004; Yu et al. 2006), suggesting that enhanced glycogen accumulation is not due to a single mechanism. The possibility that increases in total glycogen synthase activity contribute to elevated glycogen concentrations cannot be excluded (Yu et al. 2006). Other factors that could influence glycogen synthesis are the difference in the gene regulatory responses that facilitate metabolic adaptations. It has been shown that overexpression of AMPKγ3 R225Q coordinates the transcription of genes important for glycolytic and oxidative metabolism (Nilsson et al. 2006) and leads to change in the adaptive metabolic response of glycolytic skeletal muscle by inducing mitochondrial biogenesis and function (Garcia-Roves et al. 2008). Accordingly, transgenic AMPKγ3 R225Q mutant mice display a greater reliance on lipid oxidation and are protected from high-fat diet-induced insulin resistance in skeletal muscle through increased fat oxidation (Barnes et al. 2004, 2005). One consequence of increased fatty acid oxidation in AMPKγ3 R225Q skeletal muscle could be the reduction of the demand for glucose oxidation via the Randle effect (Hue and Taegtmeyer 2009), inducing a glucose-sparing effect that drives glucose toward glycogen synthesis.

18.2.2 Human PRKGA2 Gene Mutations and Related Animal Models

Point mutations in the AMPKγ2 subunit, encoded by the *Prkag2* gene, have been associated with a rare autosomal-dominant genetic disease of the heart in humans (Blair et al. 2001; Gollob et al. 2001a, b). Genetic defects in the *Prkag2* gene are characterized by a cardiac glycogen overload, ventricular preexcitation (Wolff–Parkinson–White syndrome), and cardiac hypertrophy. One of the first human mutation identified occurred at residue 302, resulting in change of arginine to glutamine (R302Q), which is homologous to the R200Q *Prkag3* gene mutation in pigs (Milan et al. 2000; Gollob et al. 2001a). Subsequently, identification of additional dominant mutations of PRKAG2 gene has been reported in families coupled with congenital hypertrophic cardiomyopathy and familial preexcitation syndrome. However, distinct clinical onset and variability in clinical manifestations exist between the various PRKAG2 mutations (Porto et al. 2016). This phenotypic disparity may be a result of the specific effects of individual mutations on AMPKγ2 activity and function. To date, missense mutations include G100S (Zhang et al. 2013), R302Q (Gollob et al. 2001b; Arad et al. 2002), H383R (Blair et al. 2001), R384T (Akman et al. 2007), T400N (Arad et al. 2002), K485E (Liu et al. 2013), Y487H (Arad et al. 2005), N488I (Arad et al. 2002), E506K (Bayrak et al. 2006), E506Q (Kelly et al. 2009), H530R (Morita et al. 2008), R531G (Gollob et al. 2001b), R531Q (Burwinkel et al. 2005), S548P (Laforet et al. 2006), and insertion of an additional leucine residue L351Ins (Blair et al. 2001). These mutations occur very selectively in strategic positions within the CBS motifs or in linker sequences between these motifs but not in other parts of the AMPKγ2 subunit. One interesting exception is the G100S mutation mapped into a non-CBS domain. However, it remains unclear if a mutation outside the CBS domains can indirectly change the binding ability of CBS domains to AMP and ATP. Collectively, these data strongly suggest a specific connection between AMP and ATP binding and the different PRKAG2 mutations, supporting the notion that dysregulation of AMPK activity (gain or loss of function) contributes to the development of the cardiomyopathy.

Consistent with a causative role of PRKAG2 in Wolff–Parkinson–White syndrome, heterozygous mice overexpressing mutant PRKAG2 alleles (R302Q, T400N, N488I, R531G) under the cardiac-specific α-myosin heavy chain (α-MHC) promoter develop pathological cardiac changes resembling the human PRKAG2 cardiomyopathy with significant glycogen accumulation, ventricular preexcitation, and cardiac hypertrophy (Arad et al. 2002, 2003; Sidhu et al. 2005; Davies et al. 2006; Banerjee et al. 2007) (Table 18.2). In addition, transgenic zebrafish with cardiac-specific expression of the G100S and R302Q mutations in PRKAG2 exhibit typical features of the human disease with cardiac hypertrophy and increased glycogen storage in the heart (Zhang et al. 2014). The PRKAG2 cardiac phenotype has been attributed to alterations in AMPK activity resulting from the AMPKγ2 mutations. However, the action of the different PRKAG2 mutations on AMPK activity and sensitivity to AMP appears to be partially different if not opposite (Daniel and Carling 2002; Scott et al. 2004). Transgenic mouse model with cardiac-specific overexpression of the AMPK$\gamma 2^{N448I}$ and AMPK$\gamma 2^{T400N}$ mutations has been reported to increase AMPK basal activity (Arad et al. 2003; Banerjee et al. 2007), whereas the AMPK$\gamma 2^{R302Q}$ and AMPKγ2 ^{R531G} mutations result in the inhibition of AMPK (Sidhu et al. 2005; Davies et al. 2006). These discrepant findings could be related to a biphasic response of AMPK activity in response to the overexpression of PRKAG2 mutations (Banerjee et al. 2007; Folmes et al. 2009). Alteration in cardiac AMPK activity is possibly due to a feedback inhibition of the kinase activity by glycogen accumulation (Davies et al. 2006; Folmes et al. 2009), masking the consequences of human PRKAG2 mutations on AMPK activity. To confirm that increased AMPK activity is responsible for PRKAG2 cardiomyopathy, transgenic mice overexpressing AMPK $\gamma 2^{N448I}$ and $\gamma 2^{T400N}$ were crossbred with transgenic mice expressing a dominant-negative form of the AMPKα2 subunit, AMPK$\alpha 2^{D157A}$ (Table 18.1), in the heart to alter AMPK activity (Ahmad et al. 2005; Banerjee et al. 2007). Compound heterozygous mice have reduced cardiac AMPK activity and minimal cardiac dysfunction, demonstrating that activation of α2-containing AMPK heterotrimeric complexes rather than α1-containing complexes is responsible for the cardiac phenotype (Ahmad et al. 2005; Banerjee et al. 2007). Furthermore, studies using a transgenic mouse model expressing a human AMPK$\gamma 2^{N488I}$ mutant protein under transcriptional control of a tetracycline-repressible α-myosin heavy chain promoter (Table 18.1) provide evidence that clinical manifestations of PRKAG2 cardiomyopathy are significantly reversed by the suppression of mutant AMPK activity after the onset of the disease (Wolf et al. 2008). This finding suggests that the appropriate pharmacological targeting of AMPK or its downstream effectors may help to improve the phenotypic expression of the disease. Compelling evidence now exists that aberrant increase of basal AMPK activity by PRKAG2 mutations results in global remodeling of the metabolic network in favor of glycogen storage (Zou et al. 2005; Luptak et al. 2007) (Table 18.2). Thus, the metabolic consequences of chronic activation of AMPK in the absence of energy deficiency are distinct from those previously reported during stress conditions. In heart carrying a mutant PRKAG2 allele, inappropriate activation of AMPK triggers an increase in both glucose uptake and fatty acid oxidation, inducing, via the Randle effect, an inhibition of glucose oxidation leading to the storage of the exceeding glucose into glycogen. The mechanism of increased glucose uptake has been attributed to the upregulation of the sodium-dependent glucose co-transporter (SGLT) isoform SGLT1 but not facilitated-diffusion glucose transporter 1 (GLUT1) or GLUT4 in AMPK $\gamma 2^{T400N}$ transgenic mice (Banerjee et al. 2010b). Confirmation of the role of SGLT1 in the phenotypic features of the PRKAG2 cardiomyopathy has been demonstrated by the knockdown of cardiac SGLT1 in transgenic mice overexpressing the *PRKAG2* T400N mutation, which attenuates the cardiomyopathy phenotype (Ramratnam et al. 2014). Due to increased glucose uptake, the transgenic mice overexpressing the AMPK $\gamma 2^{N448I}$ mutation manifest high intracellular glucose-6-phosphate (G6P) levels, which contribute to the allosteric activation of glycogen synthase (GS) and enhanced glycogen synthesis (Luptak et al. 2007). By genetic inhibition of G6P-stimulated glycogen synthase activity, the pathological glycogen storage phenotype was rescued, providing definitive evidence for extensive remodeling of substrate metabolism and the causative role for high intracellular G6P in *Prkag2* cardiomyopathy (Kim et al. 2014). Surprisingly, elimination of excessive glycogen accumulation eliminated the ventricular preexcitation but not the cardiac hypertrophy phenotype, indicating that the abnormal cardiac growth is regulated by separate mechanisms. Recent studies indicate that AMPK$\gamma 2^{N448I}$ and AMPK $\gamma 2^{T400N}$ mutations stimulate hypertrophic signaling with activation of the transcription factors nuclear factor kB (NF-kB) and forkhead box O transcription factor (FoxO) and the mammalian target of rapamycin (mTOR) signaling pathway (Banerjee et al. 2010a; Kim et al. 2014).

18.3 Genetically Modified Mouse Models

Murine models have been widely used in biomedical research. Extensive similarities in anatomy, physiology, and genetics have allowed numerous inferences about human biology to be drawn from mouse models. The recent development of conditional targeting approaches and the availability of numerous genetically modified mouse models greatly facilitate functional studies. Important progress has been made during the last decade in the understanding of the pathophysiological function of AMPK, partly due to the generation of whole-body and conditional KO mouse models as well as tissue-specific transgenic mice (Table 18.1). These mouse models have made it possible to decipher the distinct physiopathological functions of the multiple AMPK isoforms and AMPK heterotrimer combinations.

18.3.1 AMPK Knockout Mouse Models

Generation of whole-body deletion of each catalytic or regulatory AMPK subunits results in viable mice (Table 18.1). Of note, a mouse model of AMPKβ1 deletion using a gene trap approach resulted in severe brain developmental defect leading to postnatal death (Dasgupta and Milbrandt 2009). These mice express a fusion protein containing a AMPKβ1 N-terminal fragment (2–224) fused to β-galactosidase and may explain the abnormal brain development and not the loss of the AMPKβ1 subunit (Dzamko et al. 2010). Interestingly, targeted disruption of AMPK subunits is associated with distinct phenotypic abnormalities (Viollet et al. 2003; Barnes et al. 2004; Jorgensen et al. 2004a; Dzamko et al. 2010; Steinberg et al. 2010; Foretz et al. 2011; Dasgupta et al. 2012). These diverse phenotypes could simply reflect isoform-preferred substrate phosphorylation between the combination of AMPKα1- and AMPKα2-containing heterotrimeric complexes, but a recent phosphoproteomic approach in cancer cells fails to identify specific targets for AMPKα1 or AMPKα2 (Schaffer et al. 2015). These data are rather suggestive of distinct tissue-specific contribution of the different isoforms in the control of AMPK activity and function. This is in line with the description of specific pattern of expression for AMPKα, AMPKβ, and AMPKγ isoforms between tissues (Stapleton et al. 1996; Thornton et al. 1998; Cheung et al. 2000). Indeed, the expression of AMPKβ1 and AMPKβ2 isoforms differs in a number of tissues, with AMPKβ1 highly expressed in liver and weakly expressed in skeletal muscle, whereas the opposite pattern is observed for AMPKβ2 (Thornton et al. 1998). Studies from AMPKβ1$^{-/-}$ and AMPKβ2$^{-/-}$ mice have highlighted the relative importance of these two isoforms in the liver and skeletal muscle, respectively (Dzamko et al. 2010; Steinberg et al. 2010). Another striking example of restricted expression of AMPK isoforms comes from reports showing exclusive expression of AMPKα1, but not AMPKα2, in erythrocytes, macrophages, and T cells (Tamas et al. 2006; Sag et al. 2008; Foretz et al. 2010). Mice lacking AMPKα1 are anemic and had markedly enlarged spleens (Foretz et al. 2010). Similar phenotypes are observed in mice lacking AMPKγ1, the only AMPKγ isoform expressed in murine erythrocytes (Foretz et al. 2011). Studies of AMPKα1$^{-/-}$ and AMPKγ1$^{-/-}$ mice revealed that AMPK is required for erythrocyte homeostasis by regulating erythrocyte membrane elasticity (Foretz et al. 2010, 2011) and autophagy-dependent mitochondrial clearance during erythrocyte differentiation (Wang et al. 2010). In addition, with the use of bone marrow-derived macrophages from AMPKα1$^{-/-}$ and AMPKγ1$^{-/-}$ mice, the regulatory role of AMPK in the macrophagic differentiation of monocytes (Obba et al. 2015) and the polarization of macrophages to an anti-inflammatory phenotype (Zhu et al. 2015) has been demonstrated. Support for a critical role of AMPK in the regulation of the pro-inflammatory/anti-inflammatory balance in macrophage has been provided in previous studies using adoptive transfer of AMPKβ1$^{-/-}$ bone marrow in control recipient mice (Galic et al. 2011). It has been also established that AMPK is a key determinant of T cell effector responses by controlling T cell metabolism. AMPKα1-deficient T cells display reduced metabolic plasticity in vitro as well as in vivo during viral and bacterial infections (Blagih et al. 2015). However, with the exception of these studies on erythrocytes, macrophages, and T cells, analysis of the consequence of genetic deletion of one catalytic isoform should be done with caution, since compensatory increase in expression and activity of the remaining isoform can mask a particular phenotype. This has been clearly evidenced in primary culture of mouse proximal tubules from AMPKα1$^{-/-}$ and AMPKα2$^{-/-}$ mice, where adaptive upregulation of the other AMPKα isoform fully compensates and can substitute for the other for ameliorating the response to metabolic stress (Lieberthal et al. 2013). Similarly, it has been reported that AMPKα1 activity is higher in AMPKα2$^{-/-}$ muscle as compared to control muscles, suggesting a compensatory effect (Jorgensen et al. 2004b) (see below). In addition, overexpression of AMPKα2 compensated for the loss of AMPKα1 and may explain the lack of phenotype in chondrocyte-specific ablation of AMPKα1 (Yang et al. 2016). In an attempt to generate full KO of AMPK, AMPKα1$^{-/-}$ and AMPKα2$^{-/-}$ or AMPKβ1$^{-/-}$ and AMPKβ2$^{-/-}$ mice have been crossed, but combined disruption of AMPK isoforms is incompatible with life, indicating that AMPK is required during embryogenesis (Viollet et al. 2009; O'Neill et al. 2011). AMPK α1$^{-/-}$α2$^{-/-}$ double knockout embryos die at ~10.5 days post-conception (B. Viollet, unpublished data). The recent development of tissue-specific and conditional knockout technology has now allowed the generation of animal models completely lacking AMPK activity in a specific tissue (Table 18.1). Among the first tissues to be targeted, skeletal muscle-specific deletion of both AMPKα1/AMPKα2 and AMPKβ1/AMPKβ2 has been obtained by crossing AMPKα1$^{fl/fl}$α2$^{fl/fl}$ and AMPKβ$^{fl/fl}$β2$^{fl/fl}$ mice with transgenic mice expressing the Cre recombinase under the control of the human skeletal actin and muscle creatine kinase promoter (O'Neill et al. 2011; Lantier et al. 2014).

18.3.2 Insight from Mouse Models into AMPK-Dependent Stimulation of Glucose Uptake in Skeletal Muscle

Skeletal muscle contraction is associated with a dramatic increase in energy turnover rates that represents a major metabolic challenge. Since the first report on acute skeletal muscle AMPK activation in response to physical exercise in rodents (Winder and Hardie 1996) and later in humans (Chen et al. 2000; Fujii et al. 2000; Wojtaszewski et al. 2000), the role of AMPK in the adaptive changes in skeletal muscle has been a subject of intense research. During exercise, contracting skeletal muscle rapidly increases glucose uptake in an intensity-dependent manner to sustain the energy demand caused by increased ATP turnover. The idea that AMPK was involved in regulating glucose transport in skeletal muscle was supported by the observations that stimulation of glucose transport was achieved upon AMPK activation by AICAR (Merrill et al. 1997) and later by other pharmacological AMPK agonists such as Ex229/991 (Lai et al. 2014). The importance of AMPK in skeletal muscle glucose uptake has been investigated in genetically modified mouse models, including transgenic mice expressing naturally occurring mutation in the AMPKγ3 isoform (Table 18.1). Muscles from knockout mice (AMPKα2$^{-/-}$, AMPKβ2$^{-/-}$, and AMPKγ3$^{-/-}$) or transgenic mice expressing a dominant-negative form of AMPKα2 in skeletal muscle (AMPKα2-KD and AMPKα2i) completely abolished ex vivo AICAR- and EX229/991-stimulated glucose uptake (Mu et al. 2001; Barnes et al. 2004; Jorgensen et al. 2004a; Fujii et al. 2005; Steinberg et al. 2010; Lai et al. 2014). From these data, it has been suggested that contraction-stimulated glucose uptake is dependent on AMPKα2β2γ3, the major heterotrimer activated by exercise in human muscle. However, the role of AMPK in contraction-stimulated glucose uptake remains controversial. In AMPKα$2^{-/-}$, AMPKβ$2^{-/-}$, AMPKγ$3^{-/-}$, AMPKα2-KD, and AMPKα2i mice ex vivo contraction-stimulated glucose uptake is normal or only moderately reduced (Mu et al. 2001; Barnes et al. 2004; Jorgensen et al. 2004a; Fujii et al. 2005; Steinberg et al. 2010). Similarly, the role of AMPK in exercise-induced skeletal muscle glucose transport in vivo is not clear. While one study reported no alterations in exercise-induced glucose transport (Maarbjerg et al. 2009), a clear reduction was shown in another study (Lee-Young et al. 2009). These discrepancies are probably due to redundancy of signaling coming from residual AMPK activity in these mouse models where only a single AMPK isoform is genetically altered/deleted and is probably sufficient to increase glucose uptake during muscle contractions. AMPKα1 subunit is still present in AMPKα$2^{-/-}$, and AMPKα1 activity can be detected in AMPKα2-KD mice and may sustain alone the coordination of muscle metabolism and adaptation to exercise. To circumvent this problem, conditional muscle-specific knockout of both AMPKα and AMPKβ subunits has been generated. Deletion of AMPKβ1 and AMPKβ2 isoforms inhibited both contraction- (ex vivo) and exercise- (in vivo) induced glucose transport in skeletal muscle (O'Neill et al. 2011), while the deletion of both AMPKα1 and α2 showed reduced contraction-stimulated glucose uptake in soleus but not EDL muscle (Lantier et al. 2014). Consistent with the idea of functional redundancy between isoforms, it should be noted that mice lacking the AMPKβ2 subunit specifically in skeletal muscle had normal glucose uptake despite reductions in AMPK activity of more than 90 % (O'Neill et al. 2011). Altogether, these findings support the notion that AMPK is necessary to achieve the effects of exercise on muscle glucose transport. The mechanism underlying AMPK-dependent contraction-induced stimulation of glucose uptake has been described to act through the phosphorylation of the Rab GTPase-activating protein Tre-2/ BUB2/ cdc 1 domain (TBC1D) 1. TBC1D1 has several common contraction and AICAR responsive phosphorylation sites that are blunted in AMPKα$2^{-/-}$, AMPKγ$3^{-/-}$, AMPKα2-KD, and AMPKα2i mice (Treebak et al. 2006; Pehmoller et al. 2009; Vichaiwong et al. 2010). This has been further supported by recent findings showing a reduction in TBC1D1 phosphorylation in contracted muscle from skeletal muscle-specific AMPKβ1β2 KO mice concomitantly with decreased glucose transport (O'Neill et al. 2011).

18.4 Genetically Modified Fly Models

AMPK system is highly conserved between insects and mammals. *Drosophila* AMPK contains three protein subunits, α, β, and γ, which are encoded, in contrast to mammalian AMPK, by single genes named dAMPKα, alicorn, and loechrig (also known as SNF4Aγ), respectively (Pan and Hardie 2002) (Fig. 18.2), making Drosophila an attractive animal model to study AMPK functions in vivo. However, Drosophila expresses several isoforms for each subunit generated from alternative splicing and differential transcription initiation (Tschape et al. 2002) and can presumably form a high number of different heterotrimeric complexes. Like mammalian AMPK, drosophila AMPK is allosterically activated by AMP and by treatments that depleted cellular ATP. This is associated with phosphorylation of Thr-184 within the activation loop of dAMPKα and of acetyl-CoA carboxylase at a homologous site also conserved between mammals and insects (Pan and Hardie 2002). Although mice have been the most commonly used animal model to decipher the biology of AMPK, genetically engineered fruit fly *Drosophila melanogaster* (Tables 18.3 and 18.4) has been particularly important in demonstrating the role of AMPK in the regulation of cell polarity during energy stress and in neuronal survival (Tschape et al. 2002; Lee et al. 2007; Mirouse et al. 2007; Spasic et al. 2008). Identification and generation of mutant AMPK drosophila has been realized by forward genetic screens using ethylmethanesulfonate (EMS) mutagenesis (Medina et al. 2006) or by the P-element transposon technology (Rubin and Spradling 1982). Germline transformation of drosophila can be realized

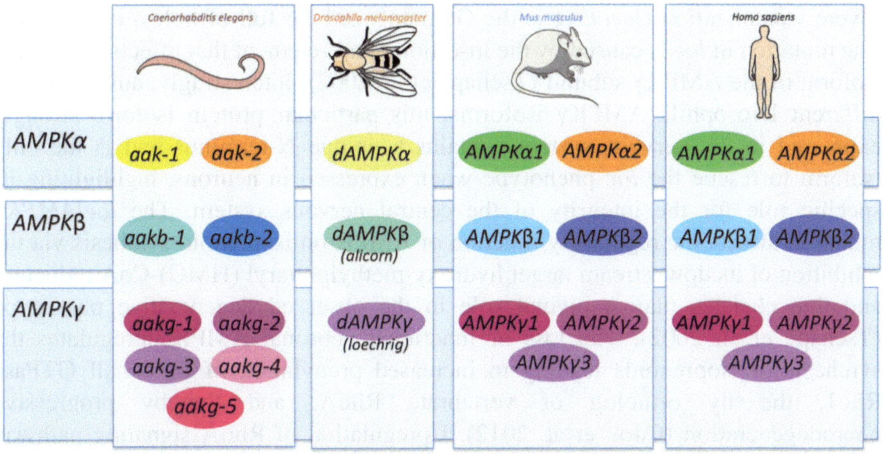

Fig. 18.2 Diversity in the genes encoding the AMPK subunits between animal species. Schematic representation of the different genes encoding the AMPKα, AMPKβ, and AMPKγ subunits in *Caenorhabditis elegans* (aak-1/2, aakb-1/2, and aakg-1/2/3/4/5), *Drosophila melanogaster* (dAMPKα, dAMPKβ/Loechrig and dAMPKγ/Alicorn), *mus musculus* (AMPKα1/2, AMPKβ1/2, and AMPKγ1/2/3), and *Homo sapiens* (AMPKα1/2, AMPKβ1/2, and AMPKγ1/2/3)

Table 18.3 AMPK mutations in *Drosophila*

Gene	Allele	Description	Mutagenesis	References
dAMPKα	AMPKα^{D1} AMPKα^{D2} AMPKα^{1} AMPKα^{2}	Deletion 1,268,785–1,270,743 bp Deletion 1,269,080–1,270,246 bp Amino acid replacement (Y141) Amino acid replacement (S211L)	Imprecise excision P-element Imprecise excision P-element EMS mutagenesis EMS mutagenesis	Lee et al. (2007), Schertel et al. (2013) Mirouse et al. (2007) www.flybase.org www.flybase.org
dAMPKβ alicorn (alc)	*Δ12.125* *l(2)45Ad^{2}*	Deletion of 1718 bp (nonsense point mutation replacing codon 233/112 in *alc* transcripts RA/RB)	Imprecise excision P-element EMS mutagenesis	Spasic et al. (2008) Dockendorff et al. (2000)
dAMPKγ loechrig (loe)	*loechrig*	P-element insertion	P-element insertion	Deak et al. (1997), Tschape et al. (2002)

via insertion of a single P transposable element or by imprecise mobilization of the P-elements (excision) to generate null mutants. In addition, creation of transgenic flies using different approaches, including the UAS/Gal4 and "tet-on" systems (Tower 2000), has been instrumental in complementing more traditional genetics.

The Drosophila *loechrig* (*loe*) mutant is characterized by degeneration and severe vacuolization (*loechrig* is the German word for full of holes) in the brain. The mutation in *loe* is caused by the insertion of a P-element that affects a neuronal isoform of the AMPKγ subunit (Tschape et al. 2002). Interestingly, among the six different Drosophila AMPKγ isoforms, this particular protein isoform strongly expressed in the nervous system contains a unique N-terminus and is the only isoform to rescue the *loe* phenotype when expressed in neurons, highlighting its specific role for the integrity of the central nervous system. The *loe/AMPKγ* mutation affects the regulatory function of AMPK on isoprenoid synthesis via the inhibition of its downstream target hydroxy-methylglutaryl (HMG)-CoA reductase and that changes play a pivotal role in the observed degenerative phenotype (Tschape et al. 2002). The loss of functional neuronal AMPK upregulates the synthesis of isoprenoids leading to increased prenylation of the small GTPase Rho1, the fly ortholog of vertebrate RhoA, and thereby progressive neurodegeneration (Cook et al. 2012). Upregulation of RhoA signaling pathway induces changes in the actin cytoskeleton through an increase in phosphorylated cofilin and accumulation of F-actin, resulting in deleterious consequences on neuronal growth and impaired axonal integrity (Cook et al. 2014). Additional evidence for a role of AMPK in neuroprotective processes comes from the study of a P-element insertion line found to target *alicorn* (*alc*), encoding the Drosophila

Table 18.4 Transgenic *Drosophila* for the study of AMPK signaling pathway

Gene	Allele	Description	References
dAMPKα	AMPKα$^{WT.Scer\backslash UAS}$	Wild-type AMPKα	Lee et al. (2007), Schertel et al. (2013)
	AMPKα$^{KR.Scer\backslash UAS}$	Kinase dead form of AMPKα	Lee et al. (2007)
	AMPKα$^{K56R.Scer\backslash UAS}$	Kinase dead version of AMPKα with a K56R amino acid substitution	Bland et al. (2010)
	AMPKα$^{K57A.Scer\backslash UAS}$	Kinase dead version of AMPKα with a K57A amino acid substitution	Johnson et al. (2010), Stenesen et al. (2013)
	AMPKα$^{TD.Scer\backslash UAS}$	Constitutively active form of AMPKα	Lee et al. (2007)
	AMPKα$^{T184D.Scer\backslash UAS}$	Constitutively active form of AMPKα	Mirouse et al. (2007)
	Scer\UAS- AMPKα RNAi	AMPKα RNAi	Dietzl et al. (2007)
	AMPKα Scer\UAS.T:Avic\GFP	AMPKα GFP tagged at the N-terminal end	Kazgan et al. (2010)
	AMPKα Scer\UAS.ORF.T:Ivir\HA1	Full-length AMPKα tagged at the C-terminal end with three copies of HA	Schertel et al. (2013)
	AMPKα ΔC.Scer\UAS.T:Disc\RFP-mCherry	Truncated form of AMPKα lacking the C-terminal 22 amino acid residues (a stop codon has been introduced after Pro561) tagged at the N-terminal end with mCherry	Mirouse et al. (2007)
	AMPKα Scer\UAS.T:Disc\RFP-mCherry	AMPKα tagged at the N-terminal end with mCherry	Kazgan et al. (2010)
AMPKγ loechrig (loe)	Scer\UAS- AMPKγ^{RNAi}	Scer\UAS- AMPKγ^{RNAi}	Johnson et al. (2010)

homolog of the regulatory AMPKβ subunit. Disruption of the *alc*/AMPKβ gene causes early-onset progressive retinal degeneration, characterized by extensive vacuolization and general structural disorganization (Spasic et al. 2008). The mechanism of progressive neuronal death observed as a consequence of loss of AMPK does not involve apoptosis (Tschape et al. 2002; Poels et al. 2012). It has been suggested that AMPK could contribute to the protection of neurons from increased metabolic activity by the regulation of the autophagic process. Accordingly, the *loechrig*/ *SNF4Aγ* gene has been identified in a genetic screen to select P-element insertion that affects autophagy in the larval fat body (Lippai et al. 2008). Surprisingly, in *alc* mutants, a severe induction of autophagy was reported (Poels et al. 2012). This is most likely a consequence of the absence of the negative regulatory feedback loop mediated by AMPK under condition of excessive autophagic induction (Loffler et al. 2011).

Parkinson's disease is one of the most common neurodegenerative diseases in the aging population. Recently, a *Drosophila* Parkinson's disease model was used as an initial system to evaluate the therapeutic potential of a number of candidate compounds (Ng et al. 2012). Epigallocatechin gallate (EGCG), a green tea-derived catechin, was found to provide the best protection against the loss of dopaminergic neurons and mitochondrial dysfunction, the essential pathological phenotypes of human Parkinson's patients. Importantly, the protective effects of EGCG are abolished when AMPK is knocked down, and loss of AMPK activity exacerbates neuronal loss and associated phenotypes (Ng et al. 2012), suggestive of a pathological role of AMPK in neurodegenerative diseases. In addition, it was demonstrated that genetic activation of AMPK also protects against neuronal loss and reproduces EGCG's protective effects, indicating that targeting AMPK will be useful therapeutically in the treatment of neurodegenerative disorders. Indeed, as mitochondrial dysfunction is currently widely accepted to be a key driver of neurodegeneration, activation of AMPK could preserve neuronal function by preserving the energy balance and by restoring the clearance of damaged mitochondria via induction of mitophagy through the phosphorylation of the autophagy initiator autophagy-related gene 1 (Atg1)/ Unc-51-like kinase 1 (ULK1).

In agreement, with a role of AMPK in neural maintenance, lethal mutations in AMPKα, identified by forward genetic screen with EMS mutagenesis or generated by imprecise excision of P-elements, confirmed the importance for the kinase in the maintenance of cell integrity (Lee et al. 2007; Swick et al. 2013). AMPKα-null fly embryos showed severe abnormalities in cell polarity and disorganization of epithelial structures and lead to embryonic lethality (Lee et al. 2007). Moreover, AMPK-null embryos contained defective mitotic divisions with lagging or polyploid chromosomes. It was established that AMPK functions in mitosis and epithelial polarity by targeting myosin II regulatory light chain (MRLC), since phosphomimetic mutant of MRLC rescued the AMPK-null defects in cell polarity and mitosis (Lee et al. 2007). These findings uncovered a link between energy status and cell structures, revealing new pathophysiological functions for AMPK signaling pathway in the regulation of cellular structures.

Drosophila has been used as a model system to dissect possible roles of AMPK in aging and lifespan determination. Dietary restriction, a reduction in total food intake, has been shown to increase lifespan by modulating nutrient-sensing pathways in flies as well as in a wide range of organisms including nematodes and mammals (Fontana et al. 2010). However, question remains regarding the importance of cellular energy homeostasis and AMPK signaling as an evolutionary conserved determinant of lifespan control. For this purpose, to circumvent the lethality caused by AMPKα deletion in late larval stages (Lee et al. 2007), the consequences of tissue-specific knockdown of AMPKα and transgenic expression of AMPKα using the inducible GeneSwitch system (Poirier et al. 2008), which provided both temporal and spatial control, have been examined to influence drosophila lifespan. Different approaches were used including RNAi-mediated knockdown of AMPKα (Tohyama and Yamaguchi 2010) or AMPKγ (Johnson, Kazgan et al. 2010) and expression of a dominant-negative construct (AMPKα^{K57A}), in which the catalytic domain of the AMPKα subunit is inactive (Johnson et al. 2010) (Table 18.4). Consistent with a causal role for AMPK in regulating energy homeostasis, reduced AMPK signaling leads to hypersensitivity to starvation conditions (Bland et al. 2010; Johnson et al. 2010; Tohyama and Yamaguchi 2010). Interestingly, a similar sensitivity to starvation conditions was observed with a selective loss of AMPK signaling in muscle (Tohyama and Yamaguchi 2010) and was associated with a reduction in lifespan (Bland et al. 2010; Stenesen et al. 2013), suggesting potential tissue-specific requirements for AMPK-mediated lifespan extension. To study the impact of tissue-restricted upregulation of AMPK, pro-longevity effects were investigated in transgenic flies overexpressing AMPK in a subset of metabolic tissues. It was found that localized AMPK activation (e.g., by overexpressing wild-type AMPK in muscle, fat body, brain, or intestinal epithelium) extends lifespan in a non-cell-autonomous manner (Bland et al. 2010; Stenesen et al. 2013; Ulgherait et al. 2014). Activation of AMPK produces a non-cell autonomous induction of autophagy due to upregulation of Atg1/ULK1 signaling, which appears to be necessary and sufficient to slow systemic aging (Ulgherait et al. 2014). A recent study demonstrated that dietary application of β-guanidinopropionic acid (β-GPA), a creatine analog, can extend lifespan through AMPK-dependent induction of autophagy (Yang et al. 2015). Such findings may provide significant insight for pharmaceutical strategies to manipulate AMPK function in single tissue to prolong lifespan in mammals.

18.5 Genetically Modified Worm Models

The nematode *Caenorhabditis elegans* has many excellent advantages as an in vivo model for detailed molecular analyses and functional genomics. It is a small nematode with a life cycle of 3.5 days and a lifespan of about 2–3 weeks. In addition, *C. elegans* displays conserved developmental programs, genetic tractability, and a fully sequenced genome, making an ideal model system to understand biological processes. To identify the biological function of specific genes in *C. elegans*, different methods have been developed using reverse and forward genetics or transgenesis techniques (Baylis and Vazquez-Manrique 2011; Boulin and Hobert 2012). *C. elegans* has many genes which are similar to those of other higher eukaryotes, suggesting many similar functions in cellular and molecular mechanisms. The *C. elegans* genome encodes the *aak-1* and *aak-2* genes, which are homologs of the catalytic α subunits of mammalian AMPK, the *aakb-1* and *aakb-2* genes, which are homologs of the regulatory β subunits of mammalian AMPK, and the aakg-1, aakg-2, aakg-3, aakg-4, and aakg-5 genes, which are homologs of the regulatory γ subunits of mammalian AMPK (Apfeld et al. 2004) (Fig. 18.2). Hence, a potential of 20 alternative heterotrimeric complexes can be formed in *C. elegans*.

Table 18.5 AMPK mutant strains for the study of AMPK signaling pathway in *C. elegans*

Gene	Allele	description	References
aak	*aak-1 (tm1944)*	Deletion of 618 bp	Lee et al. (2008)
	aak-2 (ok524)	Deletion of 408 bp (protein truncated after amino acid 164 lacking complete kinase domain as well as the inhibitory and AMPKβγ-binding domains)	Apfeld et al. (2004)
	aak-2 (rr48)	Point mutation H208Y predicted to disrupt the catalytic activity of the α subunit	Narbonne and Roy (2006)
	aak-2 (gt33)	Deletion of 606 bp generated by random UV-activated trimethyl-psoralen mutagenesis	Jansen et al. (1997)
	Paak-2::gfp	Promoter GFP fusion	Tullet et al. (2014)
aakb	aakb-1 *(tm2658)*	Deletion of 146 bp (part of the glycogen-binding domain is missing)	Narbonne and Roy (2009)
	Paakb-1::gfp	Promoter GFP fusion	Tullet et al. (2014)
	aakb-2 (rr88)	C262T transition introducing a premature stop codon	Narbonne and Roy (2009)
aakg	aakg-4 (tm5097)	Deletion of 231 bp + insertion of 24 bp	www.wormbase.org
	aakg-4 (tm5269)	Deletion of 272 bp	www.wormbase.org
	aakg-4 (tm5539)	Deletion of 405 bp (region encoding portions of both CBS3 and CBS4 domains)	Tullet et al. (2014)
	Paakg-4::gfp	Promoter GFP fusion	Tullet et al. (2014)
	Paakg-5::gfp	Promoter GFP fusion	Tullet et al. (2014)

During the last decade, *C. elegans* has been at the forefront as a model system to understand biological processes linking energetics to longevity. Limiting energy availability through dietary restriction confers lifespan extension in organisms as diverse as yeasts, nematodes, and rodents (Fontana et al. 2010). It has been found that increases in cellular AMP/ATP ratio are associated with age in *C. elegans*, suggesting a link between genes known to affect lifespan and the control of energy metabolism (Apfeld et al. 2004). Thus, it is not surprising that AMPK has emerged as a key regulator of lifespan determination (Burkewitz et al. 2014). The use of *aak-2* mutants (Table 18.5) has highlighted the role of AMPK in the regulation of lifespan in response to environmental stress and insulin-like signaling (Apfeld et al. 2004; Narbonne and Roy 2006). *C. elegans* lacking *aak-2* failed to extend lifespan in response to dietary restriction and low-energy conditions (Greer et al. 2007; Schulz et al. 2007; Lee et al. 2008; Narbonne and Roy 2009; Fukuyama et al. 2012). Conversely, overexpression of aak-2 significantly promotes lifespan extension and mimics dietary restriction in well-fed wild-type animals (Apfeld et al. 2004). Similar increase in lifespan is observed in transgenic worms expressing an active form of the AMPKα (Mair et al. 2011) or the AMPKγ (Greer et al. 2007) subunits. *C. elegans* fed with the AMPK agonist metformin also display an increase in lifespan that is dependent on *aak-2* (Onken and Driscoll 2010). The effect of metformin on extension of *C. elegans* lifespan can be direct by promoting resistance to biguanide toxicity through AMPK activation and indirect by impairing folate metabolism of *E. coli*, its trophic microbial partner (Cabreiro et al. 2013).

Under condition of energetic stress, the aak-2 subunit becomes phosphorylated at threonine 243 (Thr243), equivalent to Thr172 in the mammalian ortholog, and primarily takes part in its activity to phosphorylate and inhibits the CREB-regulated transcriptional coactivator CRTC-1, a cofactor involved in diverse physiological processes including energy homeostasis (Mair et al. 2011). Modulation of CRTC-1 phosphorylation status by AMPK activation in the central nervous system is required to locally and cell autonomously promote remodeling of mitochondrial metabolic networks and increase longevity (Burkewitz et al. 2015). Upon reduced AMPK activity, CRTC-1 modulates AMPK-mediated longevity cell non-autonomously via regulation of the neurotransmitter/hormone octopamine secretion, which drives mitochondrial fragmentation in distal tissues, and suppresses the effects of AMPK on systemic mitochondrial metabolism and longevity. These findings highlight the dominance of neuronal energy-sensing mechanisms and neuronal signals on systemic metabolic homeostasis and impact on aging process. This regulatory mechanism is consistent with the cell-non-autonomous role for neuronal AMPK in modulating peripheral lipid storage in worms (Cunningham et al. 2014).

18.6 Concluding Remarks

The AMP-activated protein kinase (AMPK) system was first discovered 35 years ago. Since that time, knowledge of the diverse physiological functions of AMPK has grown rapidly and continues to evolve. Most certainly, genetically modified mice have become instrumental to the study of AMPK. However, the use of fly and worm model organisms has also played a key role in delineating the physiological role of AMPK signaling pathway and will continue to contribute to the expanding field of AMPK.

Identifying the best animal model to study AMPK function or characterize AMPK agonists/antagonists is an important consideration and requires a thorough understanding of the advantages and disadvantages of each animal model, as there are important factors to consider. Animal models with comparable AMPK heterotrimeric composition to human cells and tissues are typically relevant models to examine AMPK pathophysiological role and develop tissue-specific therapeutic interventions. However, although mouse models possess many advantages for biomedical research, it has been reported that the composition of AMPK heterotrimers differs between rodent and human hepatocytes (Stephenne et al. 2011; Wu et al. 2013), addressing a major challenge for future preclinical translational studies.

References

Ahmad F et al (2005) Increased alpha2 subunit-associated AMPK activity and PRKAG2 cardiomyopathy. Circulation 112(20):3140–3148

Akman HO et al (2007) Fatal infantile cardiac glycogenosis with phosphorylase kinase deficiency and a mutation in the gamma2-subunit of AMP-activated protein kinase. Pediatr Res 62(4): 499–504

Andreelli F et al (2006) Liver adenosine monophosphate-activated kinase-alpha2 catalytic subunit is a key target for the control of hepatic glucose production by adiponectin and leptin but not insulin. Endocrinology 147(5):2432–2441

Apfeld J et al (2004) The AMP-activated protein kinase AAK-2 links energy levels and insulin-like signals to lifespan in C. elegans. Genes Dev 18(24):3004–3009

Arad M et al (2002) Constitutively active AMP kinase mutations cause glycogen storage disease mimicking hypertrophic cardiomyopathy. J Clin Invest 109(3):357–362

Arad M et al (2003) Transgenic mice overexpressing mutant PRKAG2 define the cause of Wolff-Parkinson-White syndrome in glycogen storage cardiomyopathy. Circulation 107(22): 2850–2856

Arad M et al (2005) Glycogen storage diseases presenting as hypertrophic cardiomyopathy. N Engl J Med 352(4):362–372

Banerjee SK et al (2007) A PRKAG2 mutation causes biphasic changes in myocardial AMPK activity and does not protect against ischemia. Biochem Biophys Res Commun 360(2): 381–387

Banerjee SK et al (2010a) Activation of cardiac hypertrophic signaling pathways in a transgenic mouse with the human PRKAG2 Thr400Asn mutation. Biochim Biophys Acta 1802(2): 284–291

Banerjee SK et al (2010b) SGLT1, a novel cardiac glucose transporter, mediates increased glucose uptake in PRKAG2 cardiomyopathy. J Mol Cell Cardiol 49(4):683–692

Barnes BR et al (2004) The 5′-AMP-activated protein kinase gamma3 isoform has a key role in carbohydrate and lipid metabolism in glycolytic skeletal muscle. J Biol Chem 279(37): 38441–38447

Barnes BR et al (2005) Changes in exercise-induced gene expression in 5′-AMP-activated protein kinase gamma3-null and gamma3 R225Q transgenic mice. Diabetes 54(12):3484–3489

Barre L et al (2007) Genetic model for the chronic activation of skeletal muscle AMP-activated protein kinase leads to glycogen accumulation. Am J Physiol 292(3):E802–E811

Baylis HA, Vazquez-Manrique RP (2011) Reverse genetic strategies in Caenorhabditis elegans: towards controlled manipulation of the genome. ScientificWorldJournal 11:1394–1410

Bayrak F et al (2006) Ventricular pre-excitation and cardiac hypertrophy mimicking hypertrophic cardiomyopathy in a Turkish family with a novel PRKAG2 mutation. Eur J Heart Fail 8(7): 712–715

Blagih J et al (2015) The energy sensor AMPK regulates T cell metabolic adaptation and effector responses in vivo. Immunity 42(1):41–54

Blair E et al (2001) Mutations in the gamma(2) subunit of AMP-activated protein kinase cause familial hypertrophic cardiomyopathy: evidence for the central role of energy compromise in disease pathogenesis. Hum Mol Genet 10(11):1215–1220

Bland ML et al (2010) AMPK supports growth in Drosophila by regulating muscle activity and nutrient uptake in the gut. Dev Biol 344(1):293–303

Boulin T, Hobert O (2012) From genes to function: the C. elegans genetic toolbox. Wiley Interdiscip Rev Dev Biol 1(1):114–137
Buhl ES et al (2002) Long-term AICAR administration reduces metabolic disturbances and lowers blood pressure in rats displaying features of the insulin resistance syndrome. Diabetes 51(7): 2199–2206
Burkewitz K et al (2014) AMPK at the nexus of energetics and aging. Cell Metab 20(1):10–25
Burkewitz K et al (2015) Neuronal CRTC-1 governs systemic mitochondrial metabolism and lifespan via a catecholamine signal. Cell 160(5):842–855
Burwinkel B et al (2005) Fatal congenital heart glycogenosis caused by a recurrent activating R531Q mutation in the gamma 2-subunit of AMP-activated protein kinase (PRKAG2), not by phosphorylase kinase deficiency. Am J Hum Genet 76(6):1034–1049
Cabreiro F et al (2013) Metformin retards aging in C. elegans by altering microbial folate and methionine metabolism. Cell 153(1):228–239
Chen ZP et al (2000) AMPK signaling in contracting human skeletal muscle: acetyl-CoA carboxylase and NO synthase phosphorylation. Am J Physiol Endocrinol Metab 279(5): E1202–E1206
Cheung PC et al (2000) Characterization of AMP-activated protein kinase gamma-subunit isoforms and their role in AMP binding. Biochem J 346(Pt 3):659–669
Ciobanu D et al (2001) Evidence for new alleles in the protein kinase adenosine monophosphate-activated gamma(3)-subunit gene associated with low glycogen content in pig skeletal muscle and improved meat quality. Genetics 159(3):1151–1162
Claret M et al (2007) AMPK is essential for energy homeostasis regulation and glucose-sensing by POMC and AgRP neurons. J Clin Invest 117(8):2325–2336
Cook M et al (2012) Increased RhoA prenylation in the loechrig (loe) mutant leads to progressive neurodegeneration. PLoS One 7(9), e44440
Cook M et al (2014) Increased actin polymerization and stabilization interferes with neuronal function and survival in the AMPKgamma mutant Loechrig. PLoS One 9(2), e89847
Cool B et al (2006) Identification and characterization of a small molecule AMPK activator that treats key components of type 2 diabetes and the metabolic syndrome. Cell Metab 3(6): 403–416
Costford SR et al (2007) Gain-of-function R225W mutation in human AMPKgamma(3) causing increased glycogen and decreased triglyceride in skeletal muscle. PLoS One 2(9), e903
Cunningham KA et al (2014) Loss of a neural AMP-activated kinase mimics the effects of elevated serotonin on fat, movement, and hormonal secretions. PLoS Genet 10(6), e1004394
Daniel T, Carling D (2002) Functional analysis of mutations in the gamma 2 subunit of AMP-activated protein kinase associated with cardiac hypertrophy and Wolff-Parkinson-White syndrome. J Biol Chem 277(52):51017–51024
Dasgupta B, Milbrandt J (2009) AMP-activated protein kinase phosphorylates retinoblastoma protein to control mammalian brain development. Dev Cell 16(2):256–270
Dasgupta B et al (2012) The AMPK beta2 subunit is required for energy homeostasis during metabolic stress. Mol Cell Biol 32(14):2837–2848
Davies JK et al (2006) Characterization of the role of gamma2 R531G mutation in AMP-activated protein kinase in cardiac hypertrophy and Wolff-Parkinson-White syndrome. Am J Physiol 290 (5):H1942–H1951
Deak P et al (1997) P-element insertion alleles of essential genes on the third chromosome of Drosophila melanogaster: correlation of physical and cytogenetic maps in chromosomal region 86E-87F. Genetics 147(4):1697–1722
Dietzl G et al (2007) A genome-wide transgenic RNAi library for conditional gene inactivation in Drosophila. Nature 448(7150):151–156
Dockendorff TC et al (2000) Genetic characterization of the 44D-45B region of the Drosophila melanogaster genome based on an F2 lethal screen. Mol Gen Genet 263(1):137–143
Dzamko N et al (2010) AMPK beta1 deletion reduces appetite, preventing obesity and hepatic insulin resistance. J Biol Chem 285(1):115–122

Enfalt AC et al (1997) Estimated frequency of the RN- allele in Swedish Hampshire pigs and comparison of glycolytic potential, carcass composition, and technological meat quality among Swedish Hampshire, Landrace, and Yorkshire pigs. J Anim Sci 75(11):2924–2935
Essen-Gustavsson B et al (2011) Muscle glycogen resynthesis, signalling and metabolic responses following acute exercise in exercise-trained pigs carrying the PRKAG3 mutation. Exp Physiol 96(9):927–937
Estrade M et al (1994) Enzyme activities of glycogen metabolism and mitochondrial characteristics in muscles of RN- carrier pigs (Sus scrofa domesticus). Comp Biochem Physiol Biochem Mol Biol 108(3):295–301
Folmes KD et al (2009) Distinct early signaling events resulting from the expression of the PRKAG2 R302Q mutant of AMPK contribute to increased myocardial glycogen. Circ Cardiovasc Genet 2(5):457–466
Fontana L et al (2010) Extending healthy life span—from yeast to humans. Science 328(5976): 321–326
Foretz M et al (2010) Maintenance of red blood cell integrity by AMP-activated protein kinase alpha1 catalytic subunit. FEBS Lett 584(16):3667–3671
Foretz M et al (2011) The AMPKgamma1 subunit plays an essential role in erythrocyte membrane elasticity, and its genetic inactivation induces splenomegaly and anemia. FASEB J 25(1): 337–347
Fu X et al (2013) AMP-activated protein kinase alpha1 but not alpha2 catalytic subunit potentiates myogenin expression and myogenesis. Mol Cell Biol 33(22):4517–4525
Fu X et al (2015) AMP-activated protein kinase stimulates Warburg-like glycolysis and activation of satellite cells during muscle regeneration. J Biol Chem 290(44):26445–26456
Fu X et al (2016) Obesity impairs skeletal muscle regeneration through inhibition of AMPK. Diabetes 65(1):188–200
Fujii N et al (2000) Exercise induces isoform-specific increase in 5′AMP-activated protein kinase activity in human skeletal muscle. Biochem Biophys Res Commun 273(3):1150–1155
Fujii N et al (2005) AMP-activated protein kinase alpha2 activity is not essential for contraction- and hyperosmolarity-induced glucose transport in skeletal muscle. J Biol Chem 280(47): 39033–39041
Fujii N et al (2007) Role of AMP-activated protein kinase in exercise capacity, whole body glucose homeostasis, and glucose transport in skeletal muscle -Insight from analysis of a transgenic mouse model. Diabetes Res Clin Pract 77(Suppl 1):S92–98
Fukuyama M et al (2012) C. elegans AMPKs promote survival and arrest germline development during nutrient stress. Biol Open 1(10):929–936
Fullerton MD et al (2013) Single phosphorylation sites in Acc1 and Acc2 regulate lipid homeostasis and the insulin-sensitizing effects of metformin. Nat Med 19(12):1649–1654
Galic S et al (2011) Hematopoietic AMPK beta1 reduces mouse adipose tissue macrophage inflammation and insulin resistance in obesity. J Clin Invest 121(12):4903–4915
Garcia-Roves PM et al (2008) Gain-of-function R225Q mutation in AMP-activated protein kinase gamma3 subunit increases mitochondrial biogenesis in glycolytic skeletal muscle. J Biol Chem 283(51):35724–35734
Gollob MH et al (2001a) Identification of a gene responsible for familial Wolff-Parkinson-White syndrome. N Engl J Med 344(24):1823–1831
Gollob MH et al (2001b) Novel PRKAG2 mutation responsible for the genetic syndrome of ventricular preexcitation and conduction system disease with childhood onset and absence of cardiac hypertrophy. Circulation 104(25):3030–3033
Gong H et al (2011) MEF2A binding to the Glut4 promoter occurs via an AMPKalpha2-dependent mechanism. Med Sci Sports Exerc 43(8):1441–1450
Granlund A et al (2010) Effects of exercise on muscle glycogen synthesis signalling and enzyme activities in pigs carrying the PRKAG3 mutation. Exp Physiol 95(4):541–549
Granlund A et al (2011) The influence of the PRKAG3 mutation on glycogen, enzyme activities and fibre types in different skeletal muscles of exercise trained pigs. Acta Vet Scand 53:20

Greer EL et al (2007) An AMPK-FOXO pathway mediates longevity induced by a novel method of dietary restriction in C. elegans. Curr Biol 17(19):1646–1656
Guigas B et al (2006) 5-aminoimidazole-4-carboxamide-1-{beta}-D-ribofuranoside and metformin inhibit hepatic glucose phosphorylation by an AMP-activated protein kinase-independent effect on glucokinase translocation. Diabetes 55(4):865–874
Gundewar S et al (2009) Activation of AMP-activated protein kinase by metformin improves left ventricular function and survival in heart failure. Circ Res 104(3):403–411
Halseth AE et al (2002) Acute and chronic treatment of ob/ob and db/db mice with AICAR decreases blood glucose concentrations. Biochem Biophys Res Commun 294(4):798–805
Hamilton SR et al (2001) An activating mutation in the gamma1 subunit of the AMP-activated protein kinase. FEBS Lett 500(3):163–168
Hardie DG (2014) AMP-activated protein kinase: maintaining energy homeostasis at the cellular and whole-body levels. Annu Rev Nutr 34:31–55
Hedegaard J et al (2004) UDP-glucose pyrophosphorylase is upregulated in carriers of the porcine RN- mutation in the AMP-activated protein kinase. Proteomics 4(8):2448–2454
Huang LS et al (2004) Genetic variations of the porcine PRKAG3 gene in Chinese indigenous pig breeds. Genet Sel Evol 36(4):481–486
Hue L, Taegtmeyer H (2009) The Randle cycle revisited: a new head for an old hat. Am J Physiol Endocrinol Metab 297(3):E578–E591
Jansen G et al (1997) Reverse genetics by chemical mutagenesis in Caenorhabditis elegans. Nat Genet 17(1):119–121
Johnson EC et al (2010) Altered metabolism and persistent starvation behaviors caused by reduced AMPK function in Drosophila. PLoS One 5(9), e12799
Jorgensen SB et al (2004a) The alpha2-5′AMP-activated protein kinase is a site 2 glycogen synthase kinase in skeletal muscle and is responsive to glucose loading. Diabetes 53(12): 3074–3081
Jorgensen SB et al (2004b) Knockout of the alpha2 but not alpha1 5′-AMP-activated protein kinase isoform abolishes 5-aminoimidazole-4-carboxamide-1-beta-4-ribofuranosidebut not contraction-induced glucose uptake in skeletal muscle. J Biol Chem 279(2):1070–1079
Kazgan N et al (2010) Identification of a nuclear export signal in the catalytic subunit of AMP-activated protein kinase. Mol Biol Cell 21(19):3433–3442
Kelly BP et al (2009) Severe hypertrophic cardiomyopathy in an infant with a novel PRKAG2 gene mutation: potential differences between infantile and adult onset presentation. Pediatr Cardiol 30(8):1176–1179
Kim M et al (2014) Mutation in the gamma2-subunit of AMP-activated protein kinase stimulates cardiomyocyte proliferation and hypertrophy independent of glycogen storage. Circ Res 114(6):966–975
Kohlstedt K et al (2013) AMP-activated protein kinase regulates endothelial cell angiotensin-converting enzyme expression via p53 and the post-transcriptional regulation of microRNA-143/145. Circ Res 112(8):1150–1158
Kone M et al (2014) LKB1 and AMPK differentially regulate pancreatic beta-cell identity. FASEB J 28(11):4972–4985
Laforet P et al (2006) A new mutation in PRKAG2 gene causing hypertrophic cardiomyopathy with conduction system disease and muscular glycogenosis. Neuromuscul Disord 16(3): 178–182
Lai YC et al (2014) A small-molecule benzimidazole derivative that potently activates AMPK to increase glucose transport in skeletal muscle: comparison with effects of contraction and other AMPK activators. Biochem J 460(3):363–375
Lantier L et al (2014) AMPK controls exercise endurance, mitochondrial oxidative capacity, and skeletal muscle integrity. FASEB J 28(7):3211–3224
Lee JH et al (2007) Energy-dependent regulation of cell structure by AMP-activated protein kinase. Nature 447(7147):1017–1020

Lee H et al (2008) The Caenorhabditis elegans AMP-activated protein kinase AAK-2 is phosphorylated by LKB1 and is required for resistance to oxidative stress and for normal motility and foraging behavior. J Biol Chem 283(22):14988–14993
Lee-Young RS et al (2009) Skeletal muscle AMP-activated protein kinase is essential for the metabolic response to exercise in vivo. J Biol Chem 284(36):23925–23934
Li FY et al (2012) Endothelium-selective activation of AMP-activated protein kinase prevents diabetes mellitus-induced impairment in vascular function and reendothelialization via induction of heme oxygenase-1 in mice. Circulation 126(10):1267–1277
Lieberthal W et al (2013) Susceptibility to ATP depletion of primary proximal tubular cell cultures derived from mice lacking either the alpha1 or the alpha2 isoform of the catalytic domain of AMPK. BMC Nephrol 14:251
Lindgren K et al (2007) Regulation of the muscle-specific AMP-activated protein kinase alpha2beta2gamma3 complexes by AMP and implications of the mutations in the gamma3-subunit for the AMP dependence of the enzyme. FEBS J 274(11):2887–2896
Lippai M et al (2008) SNF4Agamma, the Drosophila AMPK gamma subunit is required for regulation of developmental and stress-induced autophagy. Autophagy 4(4):476–486
Liu Y et al (2013) Identification of a novel de novo mutation associated with PRKAG2 cardiac syndrome and early onset of heart failure. PLoS One 8(5), e64603
Loffler AS et al (2011) Ulk1-mediated phosphorylation of AMPK constitutes a negative regulatory feedback loop. Autophagy 7(7):696–706
Luptak I et al (2007) Aberrant activation of AMP-activated protein kinase remodels metabolic network in favor of cardiac glycogen storage. J Clin Invest 117(5):1432–1439
Maarbjerg SJ et al (2009) Genetic impairment of AMPKalpha2 signaling does not reduce muscle glucose uptake during treadmill exercise in mice. Am J Physiol Endocrinol Metab 297(4): E924–E934
Mahmoud AD et al (2015) AMPK deficiency blocks the hypoxic ventilatory response and thus precipitates hypoventilation and apnea. Am J Respir Crit Care Med 193(9):1032–1043
Mair W et al (2011) Lifespan extension induced by AMPK and calcineurin is mediated by CRTC-1 and CREB. Nature 470(7334):404–408
Maixner DW et al (2015) Adenosine monophosphate-activated protein kinase regulates interleukin-1beta expression and glial glutamate transporter function in rodents with neuropathic pain. Anesthesiology 122(6):1401–1413
Medina PM et al (2006) A novel forward genetic screen for identifying mutations affecting larval neuronal dendrite development in Drosophila melanogaster. Genetics 172(4):2325–2335
Merrill GF et al (1997) AICA riboside increases AMP-activated protein kinase, fatty acid oxidation, and glucose uptake in rat muscle. Am J Physiol 273(6 Pt 1):E1107–E1112
Milan D et al (2000) A mutation in PRKAG3 associated with excess glycogen content in pig skeletal muscle. Science 288(5469):1248–1251
Mirouse V et al (2007) LKB1 and AMPK maintain epithelial cell polarity under energetic stress. J Cell Biol 177(3):387–392
Miura S et al (2009) Alpha2-AMPK activity is not essential for an increase in fatty acid oxidation during low-intensity exercise. Am J Physiol Endocrinol Metab 296(1):E47–E55
Morita H et al (2008) Shared genetic causes of cardiac hypertrophy in children and adults. N Engl J Med 358(18):1899–1908
Mounier R et al (2013) AMPKalpha1 regulates macrophage skewing at the time of resolution of inflammation during skeletal muscle regeneration. Cell Metab 18(2):251–264
Mu J et al (2001) A role for AMP-activated protein kinase in contraction- and hypoxia-regulated glucose transport in skeletal muscle. Mol Cell 7(5):1085–1094
Nakada D et al (2010) Lkb1 regulates cell cycle and energy metabolism in haematopoietic stem cells. Nature 468(7324):653–658
Narbonne P, Roy R (2006) Inhibition of germline proliferation during C. elegans dauer development requires PTEN, LKB1 and AMPK signalling. Development 133(4):611–619

Narbonne P, Roy R (2009) Caenorhabditis elegans dauers need LKB1/AMPK to ration lipid reserves and ensure long-term survival. Nature 457(7226):210–214

Ng CH et al (2012) AMP kinase activation mitigates dopaminergic dysfunction and mitochondrial abnormalities in Drosophila models of Parkinson's disease. J Neurosci 32(41):14311–14317

Nilsson EC et al (2006) Opposite transcriptional regulation in skeletal muscle of AMP-activated protein kinase gamma3 R225Q transgenic versus knock-out mice. J Biol Chem 281(11): 7244–7252

O'Neill HM et al (2011) AMP-activated protein kinase (AMPK) beta1beta2 muscle null mice reveal an essential role for AMPK in maintaining mitochondrial content and glucose uptake during exercise. Proc Natl Acad Sci U S A 108(38):16092–16097

Obba S et al (2015) The PRKAA1/AMPKalpha1 pathway triggers autophagy during CSF1-induced human monocyte differentiation and is a potential target in CMML. Autophagy 11(7):1114–1129

Onken B, Driscoll M (2010) Metformin induces a dietary restriction-like state and the oxidative stress response to extend C. elegans Healthspan via AMPK, LKB1, and SKN-1. PLoS One 5 (1), e8758

Pan DA, Hardie DG (2002) A homologue of AMP-activated protein kinase in Drosophila melanogaster is sensitive to AMP and is activated by ATP depletion. Biochem J 367(Pt 1): 179–186

Pehmoller C et al (2009) Genetic disruption of AMPK signaling abolishes both contraction- and insulin-stimulated TBC1D1 phosphorylation and 14-3-3 binding in mouse skeletal muscle. Am J Physiol Endocrinol Metab 297(3):E665–E675

Poels J et al (2012) Autophagy and phagocytosis-like cell cannibalism exert opposing effects on cellular survival during metabolic stress. Cell Death Differ 19(10):1590–1601

Poirier L et al (2008) Characterization of the Drosophila gene-switch system in aging studies: a cautionary tale. Aging cell 7(5):758–770

Porto AG et al (2016) Clinical spectrum of PRKAG2 syndrome. Circ Arrhythm Electrophysiol 9(1), e003121

Ramratnam M et al (2014) Transgenic knockdown of cardiac sodium/glucose cotransporter 1 (SGLT1) attenuates PRKAG2 cardiomyopathy, whereas transgenic overexpression of cardiac SGLT1 causes pathologic hypertrophy and dysfunction in mice. J Am Heart Assoc 3(4), e000899

Rockl KS et al (2007) Skeletal muscle adaptation to exercise training: AMP-activated protein kinase mediates muscle fiber type shift. Diabetes 56(8):2062–2069

Rolf J et al (2013) AMPKalpha1: a glucose sensor that controls CD8 T-cell memory. Eur J Immunol 43(4):889–896

Rubin GM, Spradling AC (1982) Genetic transformation of Drosophila with transposable element vectors. Science 218(4570):348–353

Russell RR 3rd et al (2004) AMP-activated protein kinase mediates ischemic glucose uptake and prevents postischemic cardiac dysfunction, apoptosis, and injury. J Clin Invest 114(4):495–503

Sag D et al (2008) Adenosine 5′-monophosphate-activated protein kinase promotes macrophage polarization to an anti-inflammatory functional phenotype. J Immunol 181(12):8633–8641

Sayers S et al (2016) Proglucagon promoter Cre-mediated AMPK deletion in mice increases circulating GLP-1 levels and oral glucose tolerance. PLoS One 11(3), e0149549

Schaffer BE et al (2015) Identification of AMPK phosphorylation sites reveals a network of proteins involved in cell invasion and facilitates large-scale substrate prediction. Cell Metab 22(5):907–921

Schertel C et al (2013) Systematic screening of a Drosophila ORF library in vivo uncovers Wnt/Wg pathway components. Dev Cell 25(2):207–219

Schonke M et al (2015) Skeletal muscle AMP-activated protein kinase gamma1(H151R) overexpression enhances whole body energy homeostasis and insulin sensitivity. Am J Physiol Endocrinol Metab 309(7):E679–E690

Schulz TJ et al (2007) Glucose restriction extends Caenorhabditis elegans life span by inducing mitochondrial respiration and increasing oxidative stress. Cell Metab 6(4):280–293
Scott JW et al (2004) CBS domains form energy-sensing modules whose binding of adenosine ligands is disrupted by disease mutations. J Clin Invest 113(2):274–284
Sidhu JS et al (2005) Transgenic mouse model of ventricular preexcitation and atrioventricular reentrant tachycardia induced by an AMP-activated protein kinase loss-of-function mutation responsible for Wolff-Parkinson-White syndrome. Circulation 111(1):21–29
Song XM et al (2002) 5-Aminoimidazole-4-carboxamide ribonucleoside treatment improves glucose homeostasis in insulin-resistant diabetic (ob/ob) mice. Diabetologia 45(1):56–65
Spasic MR et al (2008) Drosophila alicorn is a neuronal maintenance factor protecting against activity-induced retinal degeneration. J Neurosci 28(25):6419–6429
Stapleton D et al (1996) Mammalian AMP-activated protein kinase subfamily. J Biol Chem 271(2):611–614
Steinberg GR et al (2010) Whole body deletion of AMP-activated protein kinase {beta}2 reduces muscle AMPK activity and exercise capacity. J Biol Chem 285(48):37198–37209
Stenesen D et al (2013) Adenosine nucleotide biosynthesis and AMPK regulate adult life span and mediate the longevity benefit of caloric restriction in flies. Cell Metab 17(1):101–112
Stephenne X et al (2011) Metformin activates AMP-activated protein kinase in primary human hepatocytes by decreasing cellular energy status. Diabetologia 54(12):3101–3110
Sun G et al (2010) Ablation of AMP-activated protein kinase alpha1 and alpha2 from mouse pancreatic beta cells and RIP2.Cre neurons suppresses insulin release in vivo. Diabetologia 53(5):924–936
Sun G et al (2015) LKB1 and AMPKalpha1 are required in pancreatic alpha cells for the normal regulation of glucagon secretion and responses to hypoglycemia. Mol Metab 4(4):277–286
Sung MM et al (2015) AMPK deficiency in cardiac muscle results in dilated cardiomyopathy in the absence of changes in energy metabolism. Cardiovasc Res 107(2):235–245
Swick LL et al (2013) Isolation of AMP-activated protein kinase (AMPK) alleles required for neuronal maintenance in Drosophila melanogaster. Biol Open 2(12):1321–1323
Tamas P et al (2006) Regulation of the energy sensor AMP-activated protein kinase by antigen receptor and Ca2+ in T lymphocytes. J Exp Med 203(7):1665–1670
Thornton C et al (1998) Identification of a novel AMP-activated protein kinase beta subunit isoform that is highly expressed in skeletal muscle. J Biol Chem 273(20):12443–12450
Tohyama D, Yamaguchi A (2010) A critical role of SNF1A/dAMPKalpha (Drosophila AMP-activated protein kinase alpha) in muscle on longevity and stress resistance in Drosophila melanogaster. Biochem Biophys Res Commun 394(1):112–118
Tower J (2000) Transgenic methods for increasing Drosophila life span. Mech Ageing Dev 118(1–2):1–14
Treebak JT et al (2006) AMPK-mediated AS160 phosphorylation in skeletal muscle is dependent on AMPK catalytic and regulatory subunits. Diabetes 55(7):2051–2058
Tschape JA et al (2002) The neurodegeneration mutant lochrig interferes with cholesterol homeostasis and Appl processing. EMBO J 21(23):6367–6376
Tullet JM et al (2014) DAF-16/FoxO directly regulates an atypical AMP-activated protein kinase gamma isoform to mediate the effects of insulin/IGF-1 signaling on aging in Caenorhabditis elegans. PLoS Genet 10(2), e1004109
Ulgherait M et al (2014) AMPK modulates tissue and organismal aging in a non-cell-autonomous manner. Cell Rep 8(6):1767–1780
Vichaiwong K et al (2010) Contraction regulates site-specific phosphorylation of TBC1D1 in skeletal muscle. Biochem J 431(2):311–320
Viollet B et al (2003) The AMP-activated protein kinase alpha2 catalytic subunit controls whole-body insulin sensitivity. J Clin Invest 111(1):91–98
Viollet B et al (2009) AMPK: lessons from transgenic and knockout animals. Front Biosci (Landmark Ed) 14:19–44

Wang S et al (2010) AMPKalpha1 deletion shortens erythrocyte life span in mice: role of oxidative stress. J Biol Chem 285(26):19976–19985

Wang S et al (2015) Resveratrol induces brown-like adipocyte formation in white fat through activation of AMP-activated protein kinase (AMPK) alpha1. Int J Obes (Lond) 39(6):967–976

Winder WW, Hardie DG (1996) Inactivation of acetyl-CoA carboxylase and activation of AMP-activated protein kinase in muscle during exercise. Am J Physiol 270(2 Pt 1):E299–E304

Winder WW, Hardie DG (1999) AMP-activated protein kinase, a metabolic master switch: possible roles in type 2 diabetes. Am J Physiol 277(1 Pt 1):E1–E10

Wojtaszewski JF et al (2000) Isoform-specific and exercise intensity-dependent activation of 5′-AMP-activated protein kinase in human skeletal muscle. J Physiol 528(Pt 1):221–226

Wolf CM et al (2008) Reversibility of PRKAG2 glycogen-storage cardiomyopathy and electrophysiological manifestations. Circulation 117(2):144–154

Wu J et al (2013) Chemoproteomic analysis of intertissue and interspecies isoform diversity of AMP-activated protein kinase (AMPK). J Biol Chem 288(50):35904–35912

Wu Y et al (2015) Activation of AMPKalpha2 in adipocytes is essential for nicotine-induced insulin resistance in vivo. Nat Med 21(4):373–382

Xing Y et al (2003) Glucose metabolism and energy homeostasis in mouse hearts overexpressing dominant negative alpha2 subunit of AMP-activated protein kinase. J Biol Chem 278(31): 28372–28377

Yang J et al (2008) Chronic activation of AMP-activated protein kinase-alpha1 in liver leads to decreased adiposity in mice. Biochem Biophys Res Commun 370(2):248–253

Yang S et al (2015) Beta-Guanidinopropionic acid extends the lifespan of Drosophila melanogaster via an AMP-activated protein kinase-dependent increase in autophagy. Aging Cell 14(6):1024–1033

Yang C et al (2016) Chondrocyte-specific ablation of AMPKalpha1 does not affect bone development or pathogenesis of osteoarthritis in mice. DNA Cell Biol 35(3):156–162

Yavari A et al (2016) Chronic activation of gamma2 AMPK induces obesity and reduces beta cell function. Cell Metab 23:821–836

Yu H et al (2006) Muscle-specific overexpression of wild type and R225Q mutant AMP-activated protein kinase gamma3-subunit differentially regulates glycogen accumulation. Am J Physiol 291(3):E557–E565

Zhang BL et al (2013) Identification and functional analysis of a novel PRKAG2 mutation responsible for Chinese PRKAG2 cardiac syndrome reveal an important role of non-CBS domains in regulating the AMPK pathway. J Cardiol 62(4):241–248

Zhang BL et al (2014) Overexpression of G100S mutation in PRKAG2 causes Wolff-Parkinson-White syndrome in zebrafish. Clin Genet 86(3):287–291

Zhu YP et al (2015) Adenosine 5′-monophosphate-activated protein kinase regulates IL-10-mediated anti-inflammatory signaling pathways in macrophages. J Immunol 194(2):584–594

Zong H et al (2002) AMP kinase is required for mitochondrial biogenesis in skeletal muscle in response to chronic energy deprivation. Proc Natl Acad Sci U S A 99(25):15983–15987

Zou L et al (2005) N488I mutation of the gamma2-subunit results in bidirectional changes in AMP-activated protein kinase activity. Circ Res 97(4):323–328

Chapter 19
In Vitro Methods to Study AMPK

Xiaoqing Zhu, J. Willem Voncken, and Dietbert Neumann

Contents

Abstract AMPK studies in cell-free and cellular systems have significantly contributed to recent progress in the AMPK field. Biochemical characterization, structure determination and elucidation of AMPK-dependent signalling events benefit from application of state-of-the-art tools and methodology. This chapter provides a synopsis of recombinant protein expression systems and biochemical and cell-based study methods. We summarize three different expression systems for AMPK production: bacteria, insect cells and mammalian cells. In addition, kinase activity measurement, kinase substrate identification and determination of physical interaction with AMPK are discussed. The last part of this chapter focuses on the

X. Zhu • J.W. Voncken • D. Neumann (✉)
Department of Molecular Genetics, CARIM School for Cardiovascular Diseases, Maastricht University Medical Centre, P.O. Box 616, 6200 MD, Maastricht, The Netherlands
e-mail: d.neumann@maastrichtuniversity.nl

M.D. Cordero, B. Viollet (eds.), *AMP-activated Protein Kinase*, Experientia Supplementum 107, DOI 10.1007/978-3-319-43589-3_19

use of pharmacological activation, inhibition, and molecular genetic tools to study AMPK involvement in cellular signalling pathways.

Keywords Recombinant protein expression • Kinase activity determination • Substrate identification • Genetic tools • AMPK activity modulation

19.1 Expression/Production of AMPK Protein

To biochemically characterize AMPK and to study its involvement in cellular processes, AMPK protein expression is an essential step. For any expressed protein, quality and quantity are two crucial factors for further successful application in vitro or in vivo. Since cellular AMPK functions as a heterotrimeric protein complex comprising α, β and γ subunits, the formation of AMPK requires co-expression of all three subunits. To date, a number of expression systems are used in the AMPK field, including prokaryotic, insect virus-based and mammalian systems. These systems and their applications are discussed in the sections below.

19.1.1 Prokaryotic Expression Systems

As the easiest and most widely used method for recombinant protein production, the bacterial expression system combines several potential advantages, such as a rapid result, high yield, including the possibility of upscaling, and absence of posttranslational modifications (PTMs).

Production of AMPK heterotrimeric complexes has been a challenging task, due to low protein solubility and failure of co-renaturation from separately purified denatured single subunits. A polycistronic bacterial expression system has been developed to overcome these problems (Neumann et al. 2003). Expression of all three subunits from a single tricistronic vector in bacteria conveniently produces the functional heterotrimer. AMPK complexes can thus be obtained in the milligram range from *Escherichia coli*. This method was also shown to be valuable for production of uniform subunit isoform composition and was applied in the majority of biochemical and structural studies. However, the tricistronic expression strategy is not equally powerful for all subunit isoform combinations. For example, the human muscle-specific AMPK isoform combination α2β2γ3 was successfully expressed and purified from a single tricistronic transcript in *E. coli,* which offered insight into the function of this AMPK (Rajamohan et al. 2010), but expression levels of this isoform complex were too low to enable structural studies using, e.g., X-ray crystallography.

Notably, an activating phosphorylation event is often required in the case of recombinant protein kinases. Indeed, AMPK heterotrimers purified from the above-described tricistronic vectors are almost entirely inactive due to the absence of upstream activating kinases in bacteria. In order to activate AMPK, upstream kinases (e.g. LKB1 complex or CaMKK2) have thus been added to purified bacterially expressed AMPK complexes in the presence of ATP. CaMKK2 can be easily produced in bacteria, which makes this kinase suitable for applications where an activating AMPKα threonine 172 (T172) phosphorylation event is part of the assay (Kukimoto-Niino et al. 2011). It has also been reported that activated AMPK could be re-purified after in vitro activation by either recombinant CaMKK2 or LKB1 complex (Suter et al. 2006). Functional LKB1 complex (consisting of LKB1, MO25α and STRADα) can also be produced in bacteria, although purification yield was low (Neumann et al. 2007). In cases where bacterial production of active AMPK is desirable, LKB1 complex can be co-expressed with AMPK. For this purpose, a single hexacistronic expression vector encoding three AMPK and three LKB1 subunits was employed (Fig. 19.1a, b). This strategy directly yields active AMPK, as reported recently (Bullen et al. 2014; Oligschlaeger et al. 2015).

Tricistronic constructs for expression of AMPK usually encode a hexahistidine (His)-tagged α-subunit and non-tagged β and γ subunits (although this might not be exclusive). Thus, commonly, His-tagged AMPK protein obtained from a prokaryotic expression system can be purified by Nickel-affinity chromatography in a single-step using standard procedures (Fig. 19.2). To exclude possible problems from the presence of His tag for biochemical characterization of AMPK, especially crystallization, non-tagged mammalian AMPK isoform combinations were purified successfully from bacteria using a multistep chromatography approach that included anion exchange chromatography and size-exclusion chromatography (Riek et al. 2008). Automation of such multidimensional purification protocol was established even allowing for full Internet remote control (Riek et al. 2009).

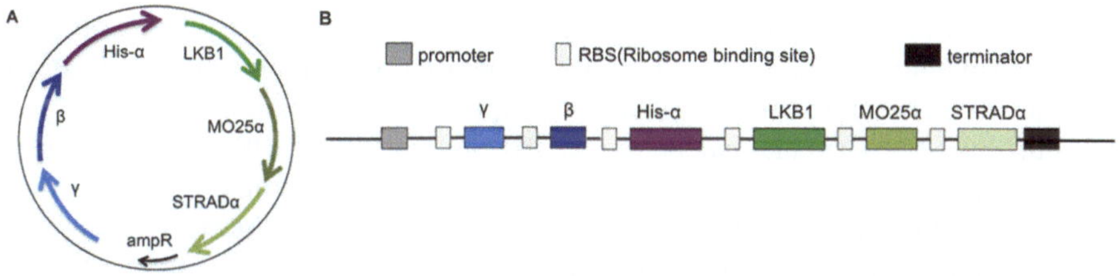

Fig. 19.1 Bacterial expression construct for production of active AMPK complexes. (**a**) Schematic representation of the hexacistronic expression vector encoding for the AMPK heterotrimer and LKB1 complex for expression of active AMPK in bacteria. (**b**) The hexacistronic gene construct consists of a single promoter, open reading frames of AMPK γ subunit, β subunit, hexahistidine-tagged α subunit, LKB1, MO25a and STRADa, respectively, interspersed with ribosome binding sites (RBS), and a transcriptional terminator sequence. Plasmid maps are not drawn to scale. This vector and its application were described recently (Bullen et al. 2014; Oligschlaeger et al. 2015)

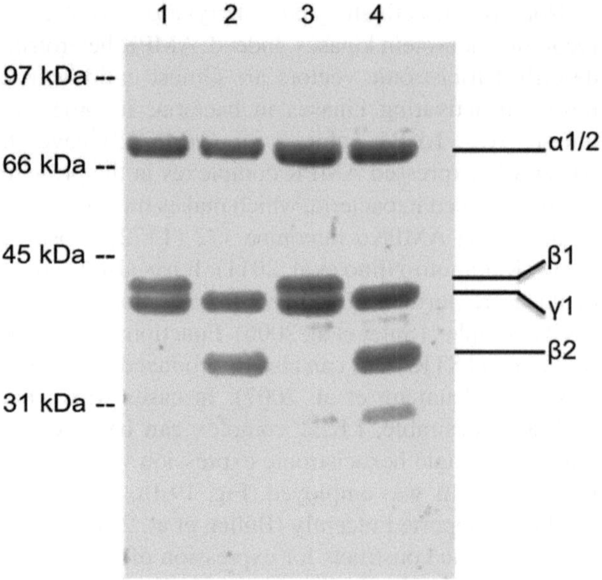

Fig. 19.2 Bacterially expressed purified AMPK isoforms. Tricistronic plasmids encoding various AMPK isoforms were transformed to competent *E. coli* and harvested after overnight expression in auto-inducing media. Proteins were purified by Nickel-affinity chromatography and purity was analysed by SDS-PAGE followed by Coomassie Brilliant Blue staining. His-α1β1γ1 (*Lane 1*), His-α1β2γ1 (*Lane 2*), His-α2β1γ1 (*Lane 3*) and His-α2β2γ1 (*Lane 4*)

To date, the bacterial expression system has been validated as a useful source of various AMPK isoforms as either active or inactive enzyme. Moreover, the bacterial system has been utilized for expression of functional mutants, such as kinase-defective mutants (e.g. α1D157Aβ1γ1) or truncation mutants (e.g. lacking the kinase domain in the α subunit).

19.1.2 Baculovirus-Insect Cell Expression Systems

Baculovirus-mediated protein expression can provide both high protein yield and eukaryotic protein PTMs (Jarvis 2009). Infected insect cells co-expressing all three AMPK subunits from three individual baculoviral vectors provided stable heterotrimeric complex, whereas separate expression/purification resulted in protein insolubility (Ramanathan et al. 2010). Successful AMPKα1β1γ1 complex expression was achieved (Iseli et al. 2008). However, also co-expression of only the γ1 and the C-terminus of β1 subunit resulted in a stable complex in the absence of the α subunit in insect cells (Iseli et al. 2008).

The main advantage of insect cell expression is the presence of eukaryotic PTMs, which is for example highlighted by the identification of a gatekeeper function for N-terminal myristoylation of the β subunit (Oakhill et al. 2010). Furthermore, by using recombinant human AMPK produced in insect cell systems, several additional phosphorylation sites have been identified (Steinberg and Kemp 2009). Similar to the bacterially expressed protein insect cell-based systems produce relatively high protein yields and for protein purification purposes affinity chromatography and anion exchange chromatography were applied yielding high purity (Iseli et al. 2008). Compared to the bacterial expression system, AMPK expression in insect cells offers the potential advantage that relevant eukaryotic PTMs are present on the artificially expressed AMPK protein.

19.1.3 Mammalian Tissues and Expression Systems

Mammalian tissues and cells are sources for endogenous AMPK complex proteins. In addition, mammalian cell cultures have been used for recombinant protein expression. The expression of mammalian proteins in mammalian cells generally results in correct folding, assembly of the required protein complex components and post-translational processing, which are all important for faithful biological activity.

In the early days of research into AMPK function, native AMPK was extracted from tissues and enriched in a series of subsequent purification steps; e.g. a combination of DEAE-sepharose, Blue-sepharose and size-exclusion chromatography was applied to purify AMPK from male Wistar rat livers (Carling et al. 1989). Later, a number of bio-affinity-based purification methods have been developed to increase the efficiency of AMPK isolation, for example a specific ATP-γ-Sepharose resin (Davies et al. 1994) and an α1 isoform-specific peptide substrate affinity-chromatography column with ADR1 (222–234)P^{229} (LKKLTRRPSFSAQ) (Michell et al. 1996).

For expression of exogenous recombinant protein, different vectors are available to transfer the target genes into host cells. These vectors range from non-selectable vectors to vectors carrying selection markers (usually eukaryotic antibiotic resistance markers) and include non-viral as well as viral (adeno-, retro-, lentiviral) vectors. Widely used and relatively easily transfectable mammalian cell lines include Human embryonic kidney (HEK 293) and Chinese hamster ovary (CHO) cells.

Although mammalian expression systems produce proteins harbouring native features (PTMs, folding, interaction), their main application has been in the exploration of cellular signalling pathways and interaction networks. Hence, overexpression of exogenous AMPK in mammalian cells is a widely applied method. To this end, mammalian expression constructs are transfected or virally transduced to a cell type of choice. As an example, both the constitutively active and dominant-negative forms of AMPKα have been used to define the role of AMPK in glucose uptake regulation by metformin in skeletal muscle cells (Turban et al. 2012). Notably, although single subunit overexpression is achievable and produces interpretable results, triple transfection (α, β, γ subunits co-expression) was shown to increase AMPK kinase activity up to 50–110-fold compared with single subunit expression in mammalian cells (Dyck et al. 1996). Therefore, co-expression of all three AMPK-complex subunits is highly recommended and widely applied in cellular signalling studies. For instance, co-expression of

AMPKα1, β1 and γ1 has contributed to the elucidation of the mechanism underlying AMPK activation by A-769662 and AMP (Scott et al. 2014) and the identification of an autophosphorylation event that regulates AMPK-glycogen binding affinity (Oligschlaeger et al. 2015).

For both endogenous and exogenous protein purification, highly specific and high-affinity reagents, such as (PTM-) specific antisera, have been useful to target and immune-capture proteins of interest from cell extracts (Kim and Eberwine 2010; Khan 2013). At present, most researchers choose to enrich endogenous AMPK protein by immunoprecipitation from mammalian cells using specific antibodies, e.g., by immunoprecipitating AMPK from cell lysate with highly selective anti-α1/2 antibodies raised in sheep and protein G sepharose (Stahmann et al. 2006).

19.2 Biochemical Characterization of AMPK

Structural and functional analysis of AMPK requires AMPK-specific knowledge and the availability of apt biochemical tools. Vice versa, structural data fertilized the biochemical characterization of AMPK. Thus, the interplay between structural biology and biochemistry has provided significant novel insights into the molecular functioning of AMPK.

19.2.1 AMPK Kinase Activity Measurement

In essence, protein kinases are phosphoryl transferases that transfer the γ-phosphate group of ATP to a threonine, serine or tyrosine residue on specific protein substrates. Therefore, protein kinase activity measurement is the determination of phosphate transfer, i.e. by quantifying the production of the phosphorylated products or by detecting the ADP production, which also should inversely mirror the decrease of ATP (Ma et al. 2008).

Among various assays for AMPK activity measurement, the radiometric assay is a common method, although the introduction of non-radioactive assay formats has reduced the frequency of its application. In any such assay, AMPK is incubated with a synthetic substrate peptide, in the presence of radiolabeled (^{32}P-γ-ATP or ^{33}P-γ-ATP) or non-labelled ATP. Incorporation of the radiolabel to the substrate peptide can be quantified by determining scintillation counts after spotting the reaction mix onto a phosphocellulose paper, i.e. the 'classical' kinase assay that has been successfully implemented for numerous kinases. A more rough estimation of kinase activity may alternatively be obtained by using a protein substrate and comparing autoradiography signals after separating the reaction mix by gel electrophoresis. If using nonradiolabeled ATP, various commercial kits are available for quantification, such as luminometer-based, fluorescent and colorimetric assays.

Different substrate peptides of AMPK have been reported. The synthetic SAMS peptide (HMRSAMSGLHLVKRR) was shown to be a specific target for AMPK from rat liver (Davies et al. 1989). Similarly, AMARA peptide (AMARAASAAALARRR) has been used to measure AMPK activity in cell lysates (Salt et al. 1998a). However, the vast majority of researchers utilized SAMS peptide in their AMPK studies; e.g. it was shown that LKB1 directly phosphorylates and activates AMPK kinase activity towards SAMS peptide (Shaw et al. 2004a). To determine AMPK activity, various additional technologies and platforms have been developed and are available from commercial suppliers that are based on the recognition of the phosphorylated substrate peptide by suitable phosphorylation-specific antibodies.

All of the above kinase activity measurements offer their specific advantages but focus on either substrate phosphorylation or on ATP consumption as a read-out. Analysis of the kinase activity reaction mixture by high-performance liquid chromatography (HPLC) has been developed as an alternative (Suter et al. 2006). HPLC-based analysis allows for quantification of all nucleotides (ATP, ADP and AMP), plus it provides an accurate determination of SAMS and phosphorylated SAMS (pSAMS) peptide; the latter requires a separate HPLC run using a different column chemistry, solutes and eluent. In contrast with other methods, the HPLC-based read-out enables monitoring of AMP levels, which may be present as contamination or produced during the assay in the reaction mixture. It is also possible to record dynamic changes of ATP/ADP and pSAMS/SAMS as a function of time (i.e. by taking samples of an activity assay at different time points). Moreover, this single assay-based technology provides important data cross-validation, since a perfect correlation between ADP formation and pSAMS detection within the duration of the AMPK assay was observed (Suter et al. 2006).

19.2.2 AMPK Structure Determination

The detailed atomic structure of AMPK has mainly been captured by X-ray crystallography. Thus, AMPK was purified in large quantities and subjected to crystallization trials. The mammalian AMPK structure and AMP-binding sites were elucidated recently (Xiao et al. 2011). Also, the binding of small molecule activators, such as A769662, was elucidated by X-ray crystallography (Xiao et al. 2013). These studies crucially rely on the ability to produce AMPK complex in sufficient amount, purity and homogeneity. Accordingly, the bacterial polycistronic expression strategy (Sect. 1.1, this chapter) has been used in virtually all AMPK structural studies published to date.

Nuclear magnetic resonance (NMR) spectroscopy is alternative method to obtain structural insight, which has been applied to isolated AMPK domains and even revealed dynamic changes or affinities. As an example, the carbohydrate-binding module of muscle-specific AMPK β2 subunit has been found to display higher affinity for carbohydrates than the corresponding domain of the AMPK β1 subunit (Koay et al. 2010). NMR-based analyses have also demonstrated that the isolated CBM domain alone shows weak interaction with A769662 (Calabrese et al. 2014).

19.2.3 Screening Methods for AMPK Downstream Target Identification

A multidimensional substrate screen for protein kinases (MudSeeK) in tissue homogenates has been recently developed to identify downstream targets for AMPK (Tuerk et al. 2007). This proteomic method utilises radiolabelling of substrate proteins by activated AMPK in the presence of radiolabelled ATP. Phosphorylated substrates are detected after separation of proteins by two-dimensional gel electrophoresis and subsequent autoradiography. Although the principle is simple, several steps are necessary and crucial for success (not further detailed herein). In this study and subsequent work (Thali et al. 2010), the MudSeeK approach has proven effective for AMPK substrate protein identification with a low number of false-positive hits.

A bioinformatics-based approach has been applied to search for the established AMPK substrate motif. Such method predicts potentially novel AMPK substrates and their AMPK-targeted phosphorylation sites. As an example, Serine 722 and 792 of the regulatory-associated protein of mechanistic target of rapamycin (Raptor) have been predicted as AMPK target sites by Scansite (http://mit.scansite.edu) and Prosite (http://ca.expasy.org/prosite/) (Gwinn et al. 2008). Moreover, a substrate-specific antibody was developed based on the known consensus of AMPK phosphorylation motif (LXRXXpS/pT). Using this antibody, 57 proteins recently were identified as potential AMPK targets (Ducommun et al. 2015).

Following up on the results of the screening methods above, each putative AMPK substrate protein requires rigorous validation as AMPK target. As a first step, this can be verified in in vitro kinase assay (IVK) with active AMPK using radiolabelled ATP and autoradiography. For phosphorylation site identification, a similar reaction mix may be subjected to mass spectrometry analysis, most often involving SDS electrophoresis and isolation of tryptic peptides from gel pieces. Site-directed mutagenesis in combination with IVK is frequently used to validate a newly identified phosphorylation site. To this end, the presumed serine or threonine residue of the AMPK target site is exchanged to alanine and the target protein is produced in a recombinant expression system. Upon incubation of wild type or the mutant with active AMPK, phosphorylation of the target protein can be detected by either autoradiography or antibodies (phospho-AMPK substrate antibody/custom-made phosphorylation-specific antibody). The workflow of AMPK downstream target identification is depicted in Fig. 19.3.

Recently, a chemical genetics approach was developed to identify AMPK targets (Banko et al. 2011). In this method, AMPK α2 subunit was genetically engineered to use N^6-modified ATPγS analogues for substrate thiophosphorylation. Next, the transferred thiophosphate on serine or threonine residues was chemically alkylated. This alkylation can be recognized by a thiophosphate ester-specific antibody that is used to enrich AMPK target proteins by immuno-affinity capturing for subsequent identification by mass spectrometry.

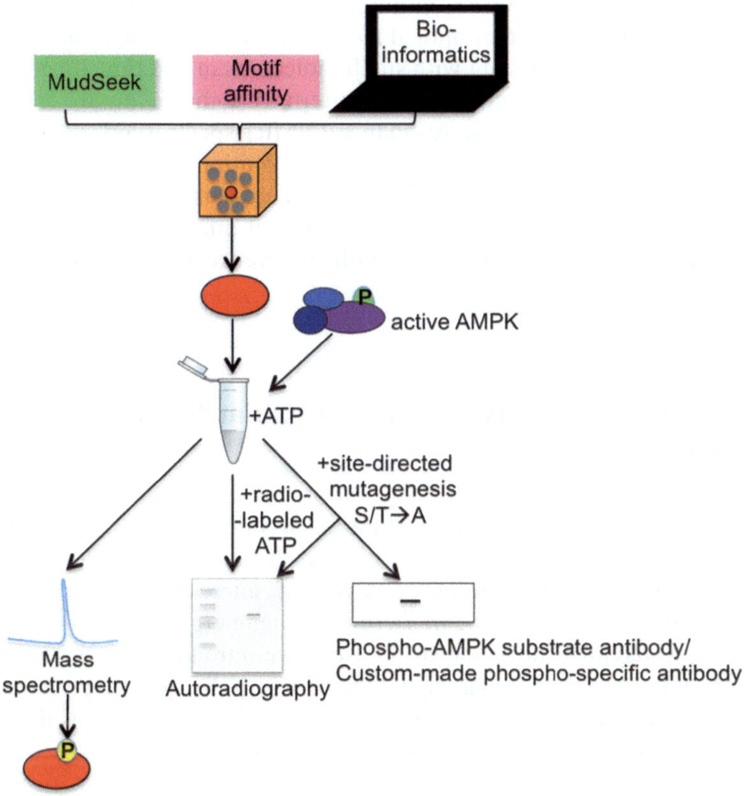

Fig. 19.3 Workflow of AMPK downstream target identification. Potential AMPK targets are identified via MudSeek or motif affinity screening. In silico prediction software yields potential AMPK substrates harbouring potential phosphorylation sites. Validation of a candidate target protein as AMPK substrate can be performed in kinase assays using active AMPK, radiolabelled ATP and autoradiography. Similar reaction mixtures can also be used for mass spectrometric analysis for phosphorylation site identification. Site-directed mutagenesis can then be employed by exchanging the putative serine/threonine target residues of candidate substrate proteins into alanine (non-phosphorylatable) to validate whether these amino acids are AMPK targets in vitro and in vivo (using phospho-AMPK substrate antibody/custom-made phosphorylation-specific antibody) and to study the biological relevance of phosphorylation at these sites

19.2.4 Physical Interaction Studies

Physical interaction studies are performed to explore binding affinity of nucleotides, drugs or interaction partners with AMPK. Among these studies, Surface Plasmon Resonance (SPR) is a widely used method, which is versatile and label free and has outstanding sensitivity. The technology uses purified AMPK protein that is captured on a sensor chip. Binding of drugs or nucleotides causes a detectable real-time angle shift of light reflected from the chip surface. Variation in binding response value makes the SPR technology useful to detect differences in binding affinity (Xiao et al. 2013). For example, glutathione S-transferases (GST) were identified as AMPK binding partners by using this method (Klaus et al. 2013). Similarly, another optical technique called Bio-layer interferometry (BLI) uncovers AMPK binding with small molecules in real time by measuring a wavelength shift in the interference pattern (Calabrese et al. 2014; Zhan et al. 2012). Nuclear magnetic resonance (NMR) spectroscopy, as previously discussed in the context of atomic structure resolution, can also be applied for this purpose. Extensive studies were performed using the yeast two-hybrid screening approach leading to the identification of multiple novel protein binding partners of AMPK subunits, i.e. with preference for the AMPKβ subunit (Sanz et al. 2013).

19.3 Methods to Study AMPK in Cellular Signalling Pathways

AMPK is a ubiquitous energy sensor of the eukaryotic cell. Thus, AMPK has been studied in many different cellular signalling pathways and specific methodology was applied to elucidate various processes of interest, such as cell growth, autophagy, metabolism and gene transcription (Steinberg and Kemp 2009; Hardie et al. 2012; Mihaylova and Shaw 2011). Here, we introduce the common starting points used in AMPK cellular signalling research by focusing on metabolic stressors and pharmacological compounds as input signals, AMPK as the integrating centre and downstream substrates as read-out (Fig. 19.4).

19.3.1 AMPK Isoforms and Cell Models

In principle, any eukaryotic cell is expected to express AMPK (or a close homologue) and thus can serve as a cell model. However, the composition of the heterotrimeric AMPK and the level of subunit expression and incorporation show differences in a cell type-specific context and also at the level of activation, interestingly. In mice, the distribution of different isoforms is well defined: α1

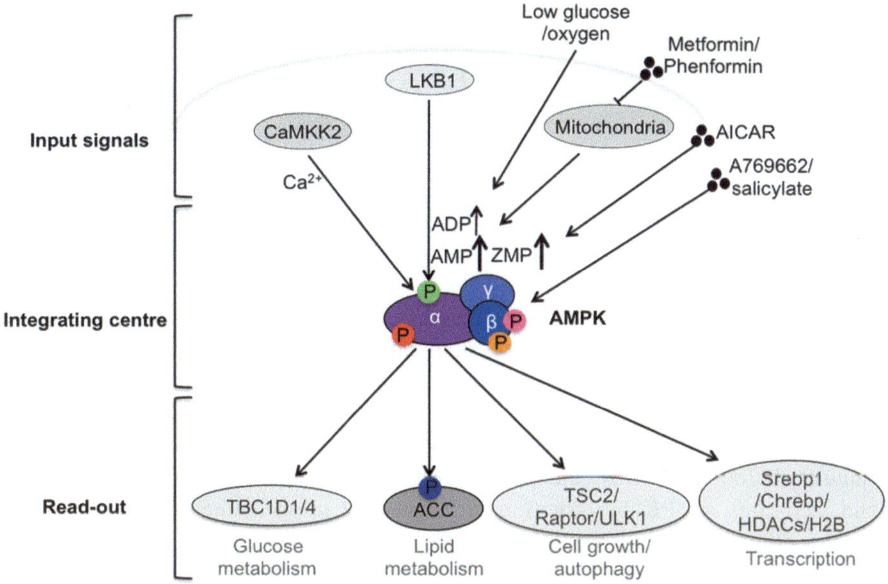

Fig. 19.4 AMPK in the context of cellular signalling studies. The depicted AMPK cellular signalling model comprises three conceptual levels: input signal/activation (by upstream regulatory molecules), AMPK as a signal sensor and convergence/integration point of signalling and downstream targets (biological/functional read-out). As input signals, metabolic stress, like low glucose or oxygen and small molecule activators metformin/phenformin, leads to increased cellular AMP/ADP, whereas AICAR forms AICAR-phosphate (ZMP), a structural analogue of AMP. Thus, LKB1 directly phosphorylates AMPK at T172 for activation, while CaMKK2 activates AMPK in response to calcium influx. Activity of AMPK is increased by activating phosphorylation, and/or pharmacologically, by binding of A769662/salicylate. The phosphorylation status of AMPK downstream targets changes in response to AMPK activation. As read-out for AMPK activity changes, phosphorylation status of ACC at S79 is most commonly detected, i.e. using commercially available antibodies. Phosphorylation of other substrates, involving in glucose metabolism, cell growth/autophagy and transcription, may also be used as biological endpoints

evenly presents across heart, liver, kidney, brain, spleen, lung and skeletal muscle, while α2 is most abundant in heart and skeletal muscle; similarly, the β2 subunit is expressed in the same muscular tissues whereas the β1 isoform is ubiquitous; γ1 and γ2 are widely expressed with γ2 levels highest in heart, while the γ3 isoform is restricted to skeletal muscle (Steinberg and Kemp 2009). These observations, therefore, emphasize the importance of choice of a relevant cell model with respect to specific signalling studies and organs. Also, AMPK α2 subunit has been shown with preferential nuclear localization and therefore is more indicated to be involved in transcriptional regulation (Salt et al. 1998b). Studies into processes that depend on AMPK activation often apply HeLa cells, due to their lack of LKB1 (Hawley et al. 2003). Likewise, neurons (Anderson et al. 2008) and T cells (Tamas et al. 2006) are explored for the role of CaMKK2 on AMPK activity. Therefore, the AMPK-activating mechanisms and controls may differ per cell type and so does the spectrum of AMPK downstream targets.

19.3.2 Physiological and Pharmacological Modulators of AMPK Activity

AMPK promotes ATP generation in response to increased AMP and ADP levels and thus various metabolic stressors, such as low oxygen or nutrient deprivation, induce AMPK activation (Mihaylova and Shaw 2011). In particular, glucose deprivation has been applied to cells in order to activate AMPK. In response to such energy stress, LKB1 directly phosphorylates AMPK at threonine 172 (T172) (Shaw et al. 2004a; Hawley et al. 2003; Woods et al. 2003a). CaMKK2, on the other hand, activates AMPK by targeting the same site upon increased calcium influx (Hawley et al. 2005; Hurley et al. 2005). For example, in endothelial cells thrombin has been shown to activate AMPK in a CaMKK2-dependent manner (Stahmann et al. 2006).

The compound 5-Aminoimidazole-4-carboxamide ribonucleotide (AICAR) is phosphorylated intracellularly to form AICAR phosphate (ZMP), a structural analogue of AMP, thus capable of activating AMPK and widely used as AMPK activator in cell studies (Guo et al. 2009). Metformin and phenformin are found to induce AMPK activation via inhibiting mitochondrial ATP synthesis (Andrzejewski et al. 2014) and therefore increasing the cellular AMP-to-ATP ratio. By directly binding to AMPK, A769662 (Cool et al. 2006) and salicylate (Hawley et al. 2012) induce allosteric activation of AMPK and thus are increasingly used for AMPK cellular function analysis. Importantly, A769662-mediated AMPK activation was also found to reduce the metastatic efficiency of breast tumour cells (Chakrabarti et al. 2015), illustrating the potential clinical importance of both AMPK and small molecule activating compounds.

To inhibit AMPK activity, compound C has been utilized as an ATP competitor. However, compound C and other kinase inhibitors lack specificity for AMPK (Liu et al. 2014). Although compound C undoubtedly inhibits AMPK and may still be considered useful, the experimental results obtained from using this inhibitor thus should be interpreted with caution. The CaMKK inhibitor STO-609 has been applied in studies of CaMKK2-mediated AMPK activation (Hurley et al. 2005). Inhibitors of LKB1 have remained elusive.

19.3.3 AMPK as the Integrating Centre for Activity Modulation

19.3.3.1 Antibodies for AMPK Activity Measurement in Cells

Phosphorylation sites in AMPK have been reported in α and β subunits mainly and site-specific phosphorylation-sensitive antibodies have been generated against several sites. Phospho-AMPKα threonine 172 (pT172) antiserum is commonly used as an indicator of AMPK activity status (Hawley et al. 1996). Phosphorylation at serine 485 (α1)/serine 491 (α2) can also be detected by phospho-specific antibodies and was linked to AMPK inhibition (Suzuki et al. 2013). In the CBM of β subunit, serine 108 phosphorylation has been monitored by a specific antibody to explore the activation mechanism of small molecule activators (Scott et al. 2014). Although the role of phosphorylation at serine 182 in β1 subunit is not well defined yet, phospho-specific antibody is also commercially available. Several research groups also developed their own antibodies, such as the phospho-AMPKβ threonine 148 (pT148) antibody (Oligschlaeger et al. 2015), which have remained non-commercial.

19.3.3.2 Genetic Intervention for AMPK Activity Modulation

In addition to pharmacological activation or inhibition, AMPK activity can be modulated by molecular genetic intervention to study and validate its biological relevance in a certain process. As discussed above (Sect. 1.3), AMPK can be expressed in cell models using transient transfection or viral transduction approaches. The expression of available and novel AMPK (subunit) mutants allows analysis of the involvement of specific protein domains or PTMs in regard to a cellular response. Mutation of aspartate 157 to alanine (D157A) within the α1 subunit renders AMPK catalytically inactive (Woods et al. 2003b). Constitutively active (CA-AMPK) and dominant-negative (DN-AMPK) forms were generated by mutating the threonine residue at 172 into an aspartic acid (T172D) and an alanine (T172A), respectively (Turban et al. 2012; Vingtdeux et al. 2010). Mutation of serine 485 within the α1 subunit into alanine (S485A) has been used to study possible suppressive signalling pathways of AMPK function (Ning et al. 2011). Deletion mutant analysis of the α1 subunit was designed to locate specific binding regions for its binding partners (Lee et al. 2011). Likewise, an autophosphorylation site mutant at serine 108 in β subunit (S108A) was used to study the mechanism by which A769662 activates AMPK (Sanders et al. 2007). An isolated carbohydrate-binding module (CBM) located in the β subunit and β threonine 148 mutant (T148A) was used in AMPK-glycogen binding study (Oligschlaeger et al. 2015). Interestingly, mutation of arginine 531 into glycine (R531G) in the γ2 subunit, which is also found in Wolff–Parkinson–White (WPW) syndrome, causes a lack of AMP/ADP binding (Scott et al. 2004) and thus has been used as an

AMP-insensitive mutant to distinguish upstream activating signals of AMPK that are not requiring elevated AMP levels (Hawley et al. 2010).

Loss of AMPK function can be obtained genetically (knock-out/knock-down) by use of gene-targeting technology or the use of RNA interference-based post-transcriptional silencing approaches, respectively. To date, various knock-out mice lacking AMPK-specific subunits have been created, which are valuable sources for genetically altered cell systems in in vitro studies. Notably, the animal-derived cell lines can be used in the primary constitution (Foretz et al. 2010) or immortalized to establish novel cellular models; e.g. AMPKα-deficient immortalized mouse embryonic fibroblasts (MEFs) were derived from AMPK α1 and α2 double-knockout mice (Bungard et al. 2010). To achieve AMPK knock-down, small interfering RNA (siRNA) and short hairpin RNA (shRNA) have been developed and verified mainly for α subunit: α1 and α2 subunit knock-down was effectively achieved in both human and mouse cells by siRNA or shRNA, by targeting a completely conserved sequence among human, mouse and rat AMPK (Tangeman et al. 2012). In muscle cells, AMPK knock-down has been achieved by both siRNA (Aguilar et al. 2007) and lentiviral shRNA (Turban et al. 2012). Commercial siRNA and shRNA are also available and have been verified in some studies (Harhaji-Trajkovic et al. 2009; Zhou et al. 2009).

19.3.4 Downstream Targets as Study Read-Out

With increasing knowledge of AMPK signalling network, multiple downstream targets in different pathways have been well studied and established as indicators of AMPK activity. Among all the substrates, acetyl-CoA carboxylase (ACC), a regulator in fatty acid synthesis, is most commonly used. Serine 79 in ACC was identified as the AMPK phosphorylation site (Ha et al. 1994), and phospho-ACC serine 79 (pS79) has since then been used as a general indicator of the effect of various treatments on AMPK activity (Sujobert et al. 2015; Hoogendijk et al. 2013). In cell growth signalling, phosphorylation of tumour suppressor tuberous sclerosis complex 2 (TSC2) (Inoki et al. 2003; Shaw et al. 2004b) and regulatory-associated protein of mechanistic target of rapamycin (Raptor) (Gwinn et al. 2008) has been associated with the inhibitory effect of AMPK on mechanistic target of rapamycin complex 1 (mTORC1). As downstream targets of mTORC1, phosphorylation of p70S6K1 (Thr389) and 4E-BP1 (Ser65) has also been utilized to check the net effect of AMPK signalling changes (Gwinn et al. 2008). In autophagy, a direct connection between AMPK and phosphorylation of the ULK1 complex was reported (Hardie 2011). In the context of signalling pathways in glucose uptake, phosphorylation of TBC1 domain family member 1 and 4 (TBC1D1/4) has been used as read-out of AMPK involvement (Treebak et al. 2006; Taylor et al. 2008). Further, the role of AMPK in metabolism has been revealed in transcriptional control as well, via directly phosphorylating the lipogenic transcription factor sterol regulatory element-binding transcription factor 1 (Srebp1) (Li et al. 2011), the glucose-sensitive transcription factor carbohydrate-responsive element-binding protein (Chrebp) (Kawaguchi et al. 2002), histone deacetylases (HDACs) 4/5/7/9 (McGee et al. 2008) and histone 2B (H2B) (Bungard et al. 2010).

19.4 Perspectives

Application of up-to-date biochemical and molecular biology techniques has been a major driver of research progress in the AMPK field. Implementation of novel technologies will facilitate advanced insight into AMPK function and functioning in the near future. With the CRISPR/Cas methodology on the rise, it is expected that the nuclear encoded expression of AMPK variants in various cell systems will become available at a much-accelerated rate, i.e. independent of the availability of a corresponding genetic mouse model. Thus, functional knockouts, domain and single-site mutants of the endogenous gene can be created by CRISPR/Cas-mediated genome editing to study their cellular effect. Super-resolution microscopy could facilitate the study of AMPK at subcellular and organelle level, thereby providing insight into AMPK function in relation to localization. Further, the application of imaging mass spectrometry will allow for analysis of AMPK in biological specimen to provide positional information in combination with identification of neighbouring molecules and/or posttranslational modifications of AMPK. Advancement in chemical genetics may further enhance the study of regulatory pathways and elucidation of causal relationships to AMPK. In conclusion, a new wave of in vitro and in vivo methods is awaiting implementation in the context of AMPK research. Such new data is expected to significantly contribute to the scientific advancement of the AMPK field and can be crucial for clinical application of novel AMPK modulators.

References

Aguilar V et al (2007) S6 kinase deletion suppresses muscle growth adaptations to nutrient availability by activating AMP kinase. Cell Metab 5(6):476–87

Anderson KA et al (2008) Hypothalamic CaMKK2 contributes to the regulation of energy balance. Cell Metab 7(5):377–88

Andrzejewski S et al (2014) Metformin directly acts on mitochondria to alter cellular bioenergetics. Cancer Metab 2:12

Banko MR et al (2011) Chemical genetic screen for AMPKalpha2 substrates uncovers a network of proteins involved in mitosis. Mol Cell 44(6):878–92

Bullen JW et al (2014) Cross-talk between two essential nutrient-sensitive enzymes: O-GlcNAc transferase (OGT) and AMP-activated protein kinase (AMPK). J Biol Chem 289 (15):10592–606

Bungard D et al (2010) Signaling kinase AMPK activates stress-promoted transcription via histone H2B phosphorylation. Science 329(5996):1201–5

Calabrese MF et al (2014) Structural basis for AMPK activation: natural and synthetic ligands regulate kinase activity from opposite poles by different molecular mechanisms. Structure 22 (8):1161–72

Carling D et al (1989) Purification and characterization of the AMP-activated protein kinase. Copurification of acetyl-CoA carboxylase kinase and 3-hydroxy-3-methylglutaryl-CoA reductase kinase activities. Eur J Biochem 186(1-2):129–36

Chakrabarti KR et al (2015) Pharmacologic regulation of AMPK in breast cancer affects cytoskeletal properties involved with microtentacle formation and re-attachment. Oncotarget 6 (34):36292–36307

Cool B et al (2006) Identification and characterization of a small molecule AMPK activator that treats key components of type 2 diabetes and the metabolic syndrome. Cell Metab 3(6):403–16

Davies SP, Carling D, Hardie DG (1989) Tissue distribution of the AMP-activated protein kinase, and lack of activation by cyclic-AMP-dependent protein kinase, studied using a specific and sensitive peptide assay. Eur J Biochem 186(1-2):123–8

Davies SP et al (1994) Purification of the AMP-activated protein kinase on ATP-gamma-sepharose and analysis of its subunit structure. Eur J Biochem 223(2):351–7

Ducommun S et al (2015) Motif affinity and mass spectrometry proteomic approach for the discovery of cellular AMPK targets: identification of mitochondrial fission factor as a new AMPK substrate. Cell Signal 27(5):978–88

Dyck JR et al (1996) Regulation of 5'-AMP-activated protein kinase activity by the noncatalytic beta and gamma subunits. J Biol Chem 271(30):17798–803

Foretz M et al (2010) Metformin inhibits hepatic gluconeogenesis in mice independently of the LKB1/AMPK pathway via a decrease in hepatic energy state. J Clin Invest 120(7):2355–69

Guo D et al (2009) The AMPK agonist AICAR inhibits the growth of EGFRvIII-expressing glioblastomas by inhibiting lipogenesis. Proc Natl Acad Sci U S A 106(31):12932–7

Gwinn DM et al (2008) AMPK phosphorylation of raptor mediates a metabolic checkpoint. Mol Cell 30(2):214–26

Ha J et al (1994) Critical phosphorylation sites for acetyl-CoA carboxylase activity. J Biol Chem 269(35):22162–8

Hardie DG (2011) AMPK and autophagy get connected. EMBO J 30(4):634–5

Hardie DG, Ross FA, Hawley SA (2012) *AMPK: a nutrient and energy sensor that maintains energy homeostasis.* Nature reviews. Mol Cell Biol 13(4):251–62

Harhaji-Trajkovic L et al (2009) AMPK-mediated autophagy inhibits apoptosis in cisplatin-treated tumour cells. J Cell Mol Med 13(9B):3644–54

Hawley SA et al (1996) Characterization of the AMP-activated protein kinase kinase from rat liver and identification of threonine 172 as the major site at which it phosphorylates AMP-activated protein kinase. J Biol Chem 271(44):27879–87

Hawley SA et al (2003) Complexes between the LKB1 tumor suppressor, STRAD alpha/beta and MO25 alpha/beta are upstream kinases in the AMP-activated protein kinase cascade. J Biol 2 (4):28

Hawley SA et al (2005) Calmodulin-dependent protein kinase kinase-beta is an alternative upstream kinase for AMP-activated protein kinase. Cell Metab 2(1):9–19

Hawley SA et al (2010) Use of cells expressing gamma subunit variants to identify diverse mechanisms of AMPK activation. Cell Metab 11(6):554–65

Hawley SA et al (2012) The ancient drug salicylate directly activates AMP-activated protein kinase. Science 336(6083):918–22

Hoogendijk AJ et al (2013) AMP-activated protein kinase activation by 5-aminoimidazole-4-carbox-amide-1-beta-D-ribofuranoside (AICAR) reduces lipoteichoic acid-induced lung inflammation. J Biol Chem 288(10):7047–52

Hurley RL et al (2005) The Ca2+/calmodulin-dependent protein kinase kinases are AMP-activated protein kinase kinases. J Biol Chem 280(32):29060–6

Inoki K, Zhu T, Guan KL (2003) TSC2 mediates cellular energy response to control cell growth and survival. Cell 115(5):577–90

Iseli TJ et al (2008) AMP-activated protein kinase subunit interactions: beta1:gamma1 association requires beta1 Thr-263 and Tyr-267. J Biol Chem 283(8):4799–807

Jarvis DL (2009) Baculovirus-insect cell expression systems. Methods Enzymol 463:191–222

Kawaguchi T et al (2002) Mechanism for fatty acid "sparing" effect on glucose-induced transcription: regulation of carbohydrate-responsive element-binding protein by AMP-activated protein kinase. J Biol Chem 277(6):3829–35

Khan KH (2013) Gene expression in mammalian cells and its applications. Adv Pharm Bull 3 (2):257–63

Kim TK, Eberwine JH (2010) Mammalian cell transfection: the present and the future. Anal Bioanal Chem 397(8):3173–8

Klaus A et al (2013) Glutathione S-transferases interact with AMP-activated protein kinase: evidence for S-glutathionylation and activation in vitro. PLoS One 8(5), e62497

Koay A et al (2010) AMPK beta subunits display isoform specific affinities for carbohydrates. FEBS Lett 584(15):3499–503

Kukimoto-Niino M et al (2011) Crystal structure of the Ca(2)(+)/calmodulin-dependent protein kinase kinase in complex with the inhibitor STO-609. J Biol Chem 286(25):22570–9

Lee KM et al (2011) Functional modulation of AMP-activated protein kinase by cereblon. Biochim Biophys Acta 1813(3):448–55

Li Y et al (2011) AMPK phosphorylates and inhibits SREBP activity to attenuate hepatic steatosis and atherosclerosis in diet-induced insulin-resistant mice. Cell Metab 13(4):376–88

Liu X et al (2014) The AMPK inhibitor compound C is a potent AMPK-independent antiglioma agent. Mol Cancer Ther 13(3):596–605

Ma H, Deacon S, Horiuchi K (2008) The challenge of selecting protein kinase assays for lead discovery optimization. Expert Opin Drug Discov 3(6):607–621

McGee SL et al (2008) AMP-activated protein kinase regulates GLUT4 transcription by phosphorylating histone deacetylase 5. Diabetes 57(4):860–7

Michell BJ et al (1996) Isoform-specific purification and substrate specificity of the 5'-AMP-activated protein kinase. J Biol Chem 271(45):28445–50

Mihaylova MM, Shaw RJ (2011) The AMPK signalling pathway coordinates cell growth, autophagy and metabolism. Nat Cell Biol 13(9):1016–23

Neumann D et al (2003) Mammalian AMP-activated protein kinase: functional, heterotrimeric complexes by co-expression of subunits in Escherichia coli. Protein Expr Purif 30(2):230–7

Neumann D et al (2007) Co-expression of LKB1, MO25alpha and STRADalpha in bacteria yield the functional and active heterotrimeric complex. Mol Biotechnol 36(3):220–31

Ning J, Xi G, Clemmons DR (2011) Suppression of AMPK activation via S485 phosphorylation by IGF-I during hyperglycemia is mediated by AKT activation in vascular smooth muscle cells. Endocrinology 152(8):3143–54

Oakhill JS et al (2010) beta-Subunit myristoylation is the gatekeeper for initiating metabolic stress sensing by AMP-activated protein kinase (AMPK). Proc Natl Acad Sci U S A 107 (45):19237–41

Oligschlaeger Y et al (2015) The recruitment of AMP-activated protein kinase to glycogen is regulated by autophosphorylation. J Biol Chem 290(18):11715–28

Rajamohan F et al (2010) Escherichia coli expression, purification and characterization of functional full-length recombinant alpha2beta2gamma3 heterotrimeric complex of human AMP-activated protein kinase. Protein Expr Purif 73(2):189–97

Ramanathan L et al (2010) Purification and characterization of truncated human AMPK alpha 2 beta 2 gamma 3 heterotrimer from baculovirus-infected insect cells. Protein Expr Purif 70 (1):13–22

Riek U et al (2008) Structural properties of AMP-activated protein kinase: dimerization, molecular shape, and changes upon ligand binding. J Biol Chem 283(26):18331–43

Riek U et al (2009) A versatile multidimensional protein purification system with full internet remote control based on a standard HPLC system. Biotechniques 46(6):ix–xii

Salt IP et al (1998a) AMP-activated protein kinase is activated by low glucose in cell lines derived from pancreatic beta cells, and may regulate insulin release. Biochem J 335(Pt 3):533–9
Salt I et al (1998b) AMP-activated protein kinase: greater AMP dependence, and preferential nuclear localization, of complexes containing the alpha2 isoform. Biochem J 334(Pt 1):177–87
Sanders MJ et al (2007) Defining the mechanism of activation of AMP-activated protein kinase by the small molecule A-769662, a member of the thienopyridone family. J Biol Chem 282 (45):32539–48
Sanz P, Rubio T, Garcia-Gimeno MA (2013) AMPKbeta subunits: more than just a scaffold in the formation of AMPK complex. FEBS J 280(16):3723–33
Scott JW et al (2004) CBS domains form energy-sensing modules whose binding of adenosine ligands is disrupted by disease mutations. J Clin Invest 113(2):274–84
Scott JW et al (2014) Small molecule drug A-769662 and AMP synergistically activate naive AMPK independent of upstream kinase signaling. Chem Biol 21(5):619–27
Shaw RJ et al (2004a) The tumor suppressor LKB1 kinase directly activates AMP-activated kinase and regulates apoptosis in response to energy stress. Proc Natl Acad Sci U S A 101 (10):3329–35
Shaw RJ et al (2004b) The LKB1 tumor suppressor negatively regulates mTOR signaling. Cancer Cell 6(1):91–9
Stahmann N et al (2006) Thrombin activates AMP-activated protein kinase in endothelial cells via a pathway involving Ca2+/calmodulin-dependent protein kinase kinase beta. Mol Cell Biol 26 (16):5933–45
Steinberg GR, Kemp BE (2009) AMPK in health and disease. Physiol Rev 89(3):1025–78
Sujobert P et al (2015) Co-activation of AMPK and mTORC1 Induces Cytotoxicity in Acute Myeloid Leukemia. Cell Rep 11(9):1446–57
Suter M et al (2006) Dissecting the role of 5'-AMP for allosteric stimulation, activation, and deactivation of AMP-activated protein kinase. J Biol Chem 281(43):32207–16
Suzuki T et al (2013) Inhibition of AMPK catabolic action by GSK3. Mol Cell 50(3):407–19
Tamas P et al (2006) Regulation of the energy sensor AMP-activated protein kinase by antigen receptor and Ca2+ in T lymphocytes. J Exp Med 203(7):1665–70
Tangeman L, Wyatt CN, Brown TL (2012) Knockdown of AMP-activated protein kinase alpha 1 and alpha 2 catalytic subunits. J RNAi Gene Silencing 8:470–8
Taylor EB et al (2008) Discovery of TBC1D1 as an insulin-, AICAR-, and contraction-stimulated signaling nexus in mouse skeletal muscle. J Biol Chem 283(15):9787–96
Thali RF et al (2010) Novel candidate substrates of AMP-activated protein kinase identified in red blood cell lysates. Biochem Biophys Res Commun 398(2):296–301
Treebak JT et al (2006) AMPK-mediated AS160 phosphorylation in skeletal muscle is dependent on AMPK catalytic and regulatory subunits. Diabetes 55(7):2051–8
Tuerk RD et al (2007) New candidate targets of AMP-activated protein kinase in murine brain revealed by a novel multidimensional substrate-screen for protein kinases. J Proteome Res 6 (8):3266–77
Turban S et al (2012) Defining the contribution of AMP-activated protein kinase (AMPK) and protein kinase C (PKC) in regulation of glucose uptake by metformin in skeletal muscle cells. J Biol Chem 287(24):20088–99
Vingtdeux V et al (2010) AMP-activated protein kinase signaling activation by resveratrol modulates amyloid-beta peptide metabolism. J Biol Chem 285(12):9100–13
Woods A et al (2003a) LKB1 is the upstream kinase in the AMP-activated protein kinase cascade. Curr Biol 13(22):2004–8
Woods A et al (2003b) Identification of phosphorylation sites in AMP-activated protein kinase (AMPK) for upstream AMPK kinases and study of their roles by site-directed mutagenesis. J Biol Chem 278(31):28434–42
Xiao B et al (2011) Structure of mammalian AMPK and its regulation by ADP. Nature 472 (7342):230–3

Xiao B et al (2013) Structural basis of AMPK regulation by small molecule activators. Nat Commun 4:3017

Zhan YY et al (2012) The orphan nuclear receptor Nur77 regulates LKB1 localization and activates AMPK. Nat Chem Biol 8(11):897–904

Zhou J et al (2009) Inactivation of AMPK alters gene expression and promotes growth of prostate cancer cells. Oncogene 28(18):1993–2002

Chapter 20
Genetically Encoded Fluorescent Biosensors to Explore AMPK Signaling and Energy Metabolism

Martin Pelosse, Cécile Cottet-Rousselle, Alexei Grichine, Imre Berger, and Uwe Schlattner

Contents

M. Pelosse • C. Cottet-Rousselle • U. Schlattner (✉)
Laboratory of Fundamental and Applied Bioenergetics (LBFA) and SFR Environmental and Systems Biology (BEeSy), University Grenoble Alpes, Grenoble, France

Inserm, U1055 and U1209, Grenoble, France
e-mail: uwe.schlattner@ujf-grenoble.fr

A. Grichine
Inserm, U1055 and U1209, Grenoble, France

Institute for Advanced Biosciences, University Grenoble Alpes, Grenoble, France

I. Berger
University of Bristol, Bristol, UK

M.D. Cordero, B. Viollet (eds.), *AMP-activated Protein Kinase*, Experientia Supplementum 107, DOI 10.1007/978-3-319-43589-3_20

Abstract Maintenance of energy homeostasis is a basic requirement for cell survival. Different mechanisms have evolved to cope with spatial and temporal mismatch between energy-providing and -consuming processes. Among these, signaling by AMP-activated protein kinase (AMPK) is one of the key players, regulated by and itself regulating cellular adenylate levels. Further understanding its complex cellular function requires deeper insight into its activation patterns in space and time at a single cell level. This may become possible with an increasing number of genetically encoded fluorescent biosensors, mostly based on fluorescence resonance energy transfer, which have been engineered to monitor metabolic parameters and kinase activities. Here, we review basic principles of biosensor design and function and the advantages and limitations of their use and provide an overview on existing FRET biosensors to monitor AMPK activation, ATP concentration, and ATP/ADP ratios, together with other key metabolites and parameters of energy metabolism.

20.1 Introduction

Cells and tissues are heterogeneous assemblies governed by localized and transient events that characterize metabolic and signaling processes (Tsou et al. 2011). Classical analytical methods are unable to combine the spatial and temporal resolution necessary to study these processes, since they either use extracts (e.g., biochemical and immunological methods) or average larger tissue volumes (e.g., NMR). Novel approaches are required that are noninvasive and quantitative and provide at the same time the necessary spatial and temporal resolution, i.e., allow tracing of metabolic and signaling events at a sub-micrometer scale and in real time. This type of data still represents a bottleneck for mechanistic insight and for the development of predictive mathematical models required for systems biology (Huang 2011).

Imaging techniques are more and more closing this gap. Ratiometric and environment-sensitive fluorescent dyes have proven their versatility in cell biology since decades, whether they monitor concentration changes like e.g. of calcium (Miyawaki et al. 1999) or biophysical parameters like membrane potential (Kuznetsov et al. 2005). During recent years, the engineering of an increasing number of genetically encoded fluorescent biosensors has shown that a specific sensor may be obtained for almost any putative target. Such protein sensors apply classical protein engineering to generate the required sensitivity and specificity and mostly exploit Fluorescence (or Förster) Energy Transfer (FRET) between two appropriate fluorophores. However, also biosensors based on a single fluorophore have emerged. This chapter summarizes the basic principles of fluorescent biosensors and provides an overview on available biosensors that can be used as readout of cellular energy homeostasis in general and activation of

AMP-activated protein kinase in particular. We emphasize their advantages and limitations, as well as their increasing application for monitoring processes at the single cell level.

20.2 Genetically Encoded Fluorescent Biosensors

Biosensors are analytical tools based on a biological element such as a macromolecule (protein or RNA), or even a cell, which is able to detect a specific ligand by interacting with it and generating a measurable signal. Such sensing elements are of course already present in nature, and it is obvious that development of biosensors should be bio-inspired. A genetically encoded biosensor is a molecular system which uses a macromolecular domain expressed by the cellular machinery (protein or RNA) as the biological recognition element for detection of a biological event within a cell. The genetically encoded biosensors that use fluorescence as a readout are based on protein constructs containing at least one fluorescent protein (FP) tag, capable of converting ligand detection into a fluorescence signal.

Genetically encoded biosensors represent a convenient, noninvasive way for monitoring signals within living cells, and an increasing number of such sensors have been developed and applied during recent years as a readout of various kinds of molecular events (Palmer et al. 2011; Lindenburg and Merkx 2014; Hochreiter et al. 2015). Protein-based biosensors can be easily engineered by modifying their DNA coding sequence and transfected to be constitutively expressed by the cellular protein synthesis machinery. This confers numerous advantages, such as the possibility to optimize specificity and affinity for the given biomolecule and to target them to specific subcellular compartments, protein complexes, or membranes. Genetically encoded biosensors can be used for sensing applications in cell culture, tissues, or even animals (Kamioka et al. 2013). In principle, they can offer high spatial resolution and versatility. As a noninvasive methodology, genetically encoded biosensors stably transfected into cells may also avoid the stress and metabolic alterations potentially occurring when non-encoded biosensors are loaded into cells. Depending on the required resolution, fluorescent biosensors in living cells can be analyzed by a large variety of common instruments, including fluorescence microplate readers, flow cytometers, wide-field fluorescence microscopes, or confocal and multiphoton laser scanning microscopes. However, the biosensor field has been mainly driven by the successful introduction of a large variety of fluorescent proteins derived from Green Fluorescent Protein (GFP) and their combination with FRET.

20.2.1 Two-Fluorophore Biosensors for FRET

Most fluorescence-based biosensors contain a couple of FPs capable of generating FRET (Fig. 20.1a, b). The theory of resonance energy transfer was originally developed by Theodor Förster. Preconditions for FRET are the presence of two different fluorophores, donor and acceptor, with an emission spectrum of the donor overlapping with the excitation spectrum of the acceptor, favorable orientation, and a distance between both fluorophores close enough to allow non-radiative dipole–dipole coupling (Förster distance, Fig. 20.2). Under these conditions, FRET leads to quenching of donor fluorescence intensity and reduction of its fluorescence lifetime, accompanied by an increase in acceptor fluorescence emission. All these parameters can be exploited for detection of the FRET phenomenon.

In biosensors based on FRET between two fluorophores, an element recognizing the desired signal (recognition element) is fused to a couple of FPs (Fig. 20.1a, b). The sensing mechanism consists in an alteration of the recognition element, e.g., by binding or dissociation of a biomolecule or a covalent modification such as phosphorylation, which has to trigger changes in overall conformation, intramolecular domain interactions, or oligomeric structure, finally resulting in an altered distance

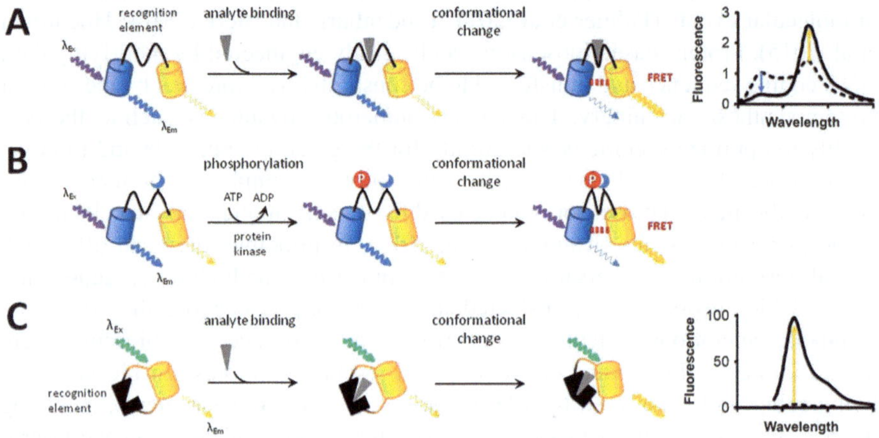

Fig. 20.1 Design, function, and typical readout of the most widespread types of genetically encoded fluorescent biosensors. *Left*: Design and principle, *right*: changes in fluorescence emission observed. (**a, b**) Dual fluorophore biosensors with recognition elements flanked by two fluorescent proteins. Events at the recognition element lead to changes in protein conformation and FRET efficiency. The recognition element either contains a binding site for a ligand (**a**) or the target for a posttranslational modification (e.g., phosphorylation (**b**)). (**c**) Single fluorophore biosensor based on a circular permutated fluorescent protein (cpFP) containing a recognition element. An event such as ligand binding to the recognition element leads to conformational changes that alter fluorescence emission. *Blue* and *yellow* cylinders: fluorescent proteins (e.g., CFP and YFP); *waved arrows*: excitation and emission light; *red dotted line*: non-radiative FRET; *gray triangle*: ligand; *red and blue dots*: posttranslational modification and its recognition sequence, respectively

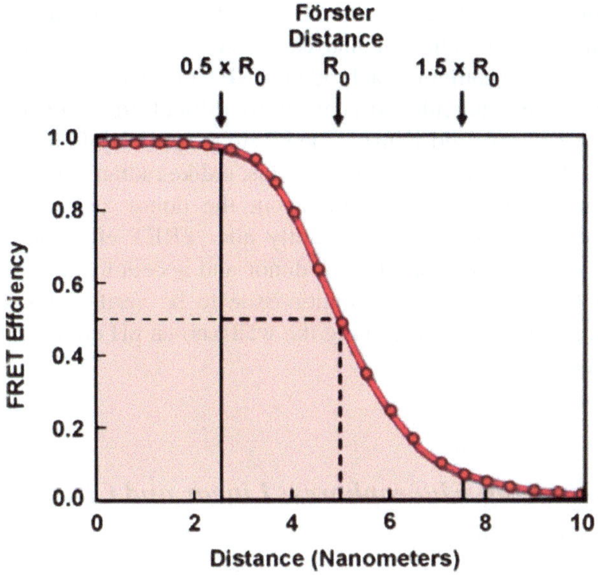

Fig. 20.2 Distance dependence of FRET efficiency. The Förster theory shows that efficiency E varies as the inverse sixth power of the distance between the two molecules (denoted by r). For any pair of fluorescent molecules, the Förster distance R_0 (also termed Förster radius) can be calculated as the value where E is at 50 % of its maximum. The useful range for measuring FRET is within 0.5 and 1.5 × R(0). For the fluorophore couple *enhanced cyan fluorescent protein* (ECFP) and *yellow fluorescent protein* (EYFP) shown here, R_0 is 4.9 nm

between the two fluorophores and thus translation of the primary signal into a change in FRET.

FRET efficiency is inversely proportional to the sixth power of the distance between donor and acceptor. This makes FRET extremely sensitive to small changes in a distance range between about 2 nm and 8 nm (Fig. 20.2), a distance range relevant to intramolecular conformational changes. FRET can thus be used as a sort of molecular ruler, as shown, e.g., by using precision spectroscopic approaches in structural biology (Mekler et al. 2002). However, for most applications in cell biology, the given signal-to-noise ratios limit FRET experiments to a more binary readout, distinguishing high-FRET and low-FRET situations. Experimentally, FRET is mostly quantified by the ratio between acceptor and donor emission peaks, but fluorescence lifetime is increasingly exploited as a more reliable readout (see below).

In addition to distance, there are other parameters affecting FRET that can be used to optimize a FRET biosensor. These include the degree of spectral overlap between donor and acceptor, as well as the quantum yield of the donor and the extinction coefficient of the acceptor. Thus, FRET can be maximized by choosing a fluorophore couple that has a large overlap in their spectral profiles and includes a donor of highest quantum yield and an acceptor with highest absorbance. FRET coupling also depends on the angle between the two fluorophores, i.e., donor emission dipole moment and acceptor absorption dipole moment, in a similar way as the orientation of a radio antenna can affect its reception. If the donor and acceptor are aligned parallel to each other, FRET efficiency will be higher than if they are oriented perpendicular. It is generally difficult to optimize this parameter experimentally, e.g., by rigidly fusing the FPs to the protein of interest, but this may have dramatic effects (Piljić et al. 2011). Thus, unlike radiative mechanisms, FRET can also yield structural information about the donor and acceptor pair. The surrounding solvent does not significantly alter FRET efficiency but can have major impact on spectral properties of donor and acceptor fluorophores and thus indirectly affect FRET. This is a major issue to be verified for a given FRET measurement, in particular concerning the well-known pH effects on GFP fluorescence (Willemse et al. 2007).

20.2.2 Optimizing Fluorophores: Linear and Circular Permutated

During recent years, many variants of GFP were engineered and optimized for FRET applications, in particular by altering spectral properties and increasing quantum yield, brightness, and photostability (Rizzo et al. 2004; Shaner et al. 2004). Some undesirable properties have been reduced, such as their sensitivity to environmental variations, in particular pH, and their tendency to oligomerize (Griesbeck et al. 2001; Nagai et al. 2002). While traditional mutational approaches preserve the linear sequence of GFPs, more recently also circular permutation was applied to GFP engineering. Here, the coding sequence is altered in a way fusing the natural N- and C-termini, while cleaving at another position (Baird et al. 1999). Although this creates novel N- and C-termini within the GFP sequence, the fluorescent protein generally refolds correctly, creating a circular protein with novel fluorescence properties, e.g., useful to fine-tune FRET efficiency (Nagai et al. 2004). These issues are not further discussed here, and the reader is referred to existing specialized reviews (Shaner et al. 2005; Palmer et al. 2011).

20.2.3 Optimizing Recognition Elements

An important issue in biosensor design is the recognition element. There are different possibilities for a recognition element to react to a biological parameter like binding of a metabolite or modification, e.g., in the form of a phosphorylation event. Such molecular recognition can trigger an association/dissociation event involving the biosensor or the induction of conformational changes within the biosensor. In the latter case, the recognition element either directly undergoes some conformational change, or the conformational change is indirectly induced by the occupied (or modified) recognition element and rather due to its interaction with a second binding domain. These mechanisms can be a natural property of the recognition element or be engineered according to rational design.

All FRET-based biosensors contain such conformationally responsive receptors which have to translate fluctuations in a biological parameter into changes of the fluorescence signal. Thus, the successful development of fluorescence-based biosensors largely depends on these signal-induced conformational changes or association/dissociation events. In short, success relies on the selection of the recognition element. Because recognition elements for most parameters, notably in energy metabolism, are already present in the nature, development of biosensors should be bio-inspired. Another key concern when engineering and using a genetically encoded probe is that it should be metabolically neutral, i.e., not perturb the biological system that it intends to measure, either by catalysis or buffering properties.

Once the recognition element is selected, further optimizations can be achieved to modify affinity/sensitivity and selectivity, notably through molecular engineering and mutagenesis. It is difficult to predict conformational changes and their amplitude as a result of ligand binding; thus, so far most successful examples of biosensors are using proteins that were already known to undergo conformational changes upon ligand binding (Berg et al. 2009; Imamura et al. 2009; Zhao et al. 2011).

20.2.4 Optimizing Detection: FRET Ratio Versus FLIM

A typical FRET experiment consists in the excitation of the donor at a given wavelength and the monitoring of the emission fluorescence spectrum, a combination of both donor and acceptor emission. If FRET occurs, the donor emission peak is reduced, while an acceptor emission peak (at higher λ) occurs. Changes in FRET efficiency due to variations in distance and/or orientation between donor and acceptor lead to variations in the relative importance of donor and acceptor emission peaks. In the simplest approach, these changes in FRET efficiencies can be quantified by the ratio between the acceptor and the donor maximum emission peak intensities. Alternatively, absolute values for FRET can be obtained by calculating the sensitized emission signal. Details on these approaches can be found elsewhere (Kedziora and Jalink 2015).

However, FRET can also be quantified by observing only the donor by fluorescence lifetime imaging (FLIM) which monitors lifetime of the fluorophore signal rather than its intensity. Fluorescence lifetime defines the average time span that a molecule spends in the excited state upon absorption of an exciting photon. After a pulse of excitation, fluorescence of the ensemble of molecules shows an exponential decay, and this decay is sensitive to the environment, in particular its quenching effects. Under conditions of FRET, the acceptor quenches donor fluorescence and thus shortens fluorescence decay time of the donor. FLIM is therefore a direct and more unambiguous readout of FRET efficiency. The advantage of measuring fluorescence lifetime of chromophores is that this parameter is directly dependent on excited-state reactions but independent of chromophore concentration and light-path length, parameters that are difficult to control inside a cell. Furthermore, FLIM is less prone to the spectral cross talk between fluorophores than the ratiometric approach, since only the donor emission is measured. FLIM data acquisition is now rapid enough to make measurements in live cells feasible. Early FLIM studies successfully analyzed the excited state of different fluorophores reporting pH (Sanders et al. 1995), Ca^{2+} (Lakowicz et al. 1994), or proteolytic processing (Bastiaens and Jovin 1996) in a time- or frequency-dependent manner and using confocal microscopy (Bastiaens and Squire 1999). Further details on this topic can be found in a recent review (Yellen and Mongeon 2015).

20.2.5 Single-Fluorophore Biosensors

Alternative fluorescent biosensors carry a single fluorescent protein that reports changes in its close environment without the need for FRET. Such sensors have been developed in particular to monitor parameters such as pH or redox potential. Although wild-type GFP fluorescence is not particularly sensitive to pH changes in the physiological range, pH-sensitive variants can be generated that have different pK_a, making them suitable for measurements in compartments that are more acidic such as Golgi (EGFP: pKa~6.0) or the more basic cytoplasm (pHluorin: $pK_a = 7.2$; YFP: pKa ~7.1) (Llopis et al. 1998). In fact, mutation of amino acids that are protonated or form hydrogen bonds can result in altered sensitivity to changes in pH, since H^+ transfer within the fluorophore is likely involved in fluorescence (Kneen et al. 1998). Through the same mutational approaches, fluorescent protein variants sensitive to reduction–oxidation were generated (Hanson et al. 2004) and successfully applied in cells (Liu et al. 2014). These reduction–oxidation-sensitive green fluorescent proteins (roGFPs) exploit the fact that chromophore fluorescence is also dependent upon the extent of disulfide bond formation between thiol groups,

Another strategy to obtain single-fluorophore biosensors is engineering of circular permutated fluorescent proteins (cpFP; Baird et al. 1999; Hung et al. 2011; Zhao et al. 2011) which can be more sensitive to the environment (Fig. 20.1c). Like in case of two-fluorophore biosensors, addition of a specific recognition element into the biosensor sequence allows to couple molecular events at the recognition element, such as ligand binding or phosphorylation, to conformational changes or association/dissociation of protein complexes. This alters the environment of the single fluorophore and thus leads to variations in its absorption or emission spectrum. These biosensors also have a ratiometric property, in case the recognition event triggers a change in their absorbance spectrum. The excitation-ratiometric detection then uses two different excitation wavelengths at a single emission wavelength.

20.2.6 Fluorescent Biosensors: Advantages and Disadvantages

As compared to other noninvasive methods such as NMR or expression of luminescent biosensors (e.g., luciferase; Kennedy et al. 1999), the use of genetically encoded fluorescent biosensors unites several conceptual advantages: (i) they obtain spatial resolution in a cell or a tissue, and they can be even addressed to a specific cellular subcompartment, (ii) they do not chemically modify the bound metabolite, (iii) in principle, any metabolite can be measured for which a recognition element with affinities in the physiological range exists and that can be linked to conformational changes or protein–protein interactions to obtain a FRET signal. Indeed, FRET is extremely sensitive to small changes in a distance range relevant for small conformational changes.

For all practical purposes, the coding sequence of biosensors (fluorophores and recognition elements) can be tailored to fine-tune the biosensor's biological and spectral properties, its subcellular targeting, or its expression level. Monitoring fluorescence of FRET and single cpFP signals is non-detrimental and nondestructive and thus suitable to observe spatial and temporal events such as fluctuations of ions, metabolites, or enzyme activities in living cells over longer periods of time. FRET also offers an inherent ratiometric property, which has a clear advantage over measuring mere intensity changes that are affected by unequal sensor expression, cellular distribution, or inhomogeneous cell morphology. Ratiometric FRET measurements allow normalization and reliable quantification of the signal, correcting for most of the cellular disturbances. Further, genetically encoded biosensors can be multiplexed, since in principle a set of biosensors with sufficiently different spectral properties can be used together to monitor different parameters in parallel at the same time under a given condition.

An important property of biosensors expressed in a cell as compared to analysis of cell extracts is temporal and spatial information. It should be kept in mind that in case of concentration measurements, the sensor will generally "see" the free local concentration of a given metabolite or ion and thus report the accessible pools, not the inaccessible or otherwise structurally bound or compartmentalized molecules.

Finally, for any given biosensor, there may be limitations despite all optimization attempts. The spectral properties of fluorescent proteins and the function of the inherent sensing mechanism (association/dissociation events, conformational changes) may not only respond to the specific signal but also to other cellular parameters such as pH, ion concentrations, membrane potentials, etc. Thus, efforts have to be made for each fluorescent biosensor, in particular for cellular in vivo application, to ensure that the readout truly depends only on the parameter of interest. For example, a very good negative control is a biosensor variant that is mutated only in its recognition element and thus no longer responds to the specific signal. Last but not least, the use of any genetically encoded sensor in primary cells may be limited, since their transfection is not always very efficient.

20.2.7 Fluorescent Biosensors for Energy Metabolism

Given the complexity of cellular energy metabolism, it is convenient to use suitable macroparameters that integrate information about a large number of cellular processes and metabolic pathways. Already a few of these macroparameters may be sufficient for a global description of cellular energy conversion (Fig. 20.3). Many genetically encoded fluorescent biosensors have been developed during recent years that can be harnessed to explore these key elements of energy metabolism. An overview is given in Table 20.1, and a more complete list can be found at http://biosensor.dpb.carnegiescience.edu/. There are several recent general reviews available on fluorescent biosensors (Alford et al. 2013; Lindenburg and Merkx 2014; Hochreiter et al. 2015), also addressing more in particular energy metabolism (San Martín et al. 2014b).

In this context, a cell signaling hub like AMPK is very informative, since it already integrates multiple cellular inputs. We will discuss fluorescent sensors for AMPK in detail in Chap. 3. Further sensors for kinase signaling have been reviewed

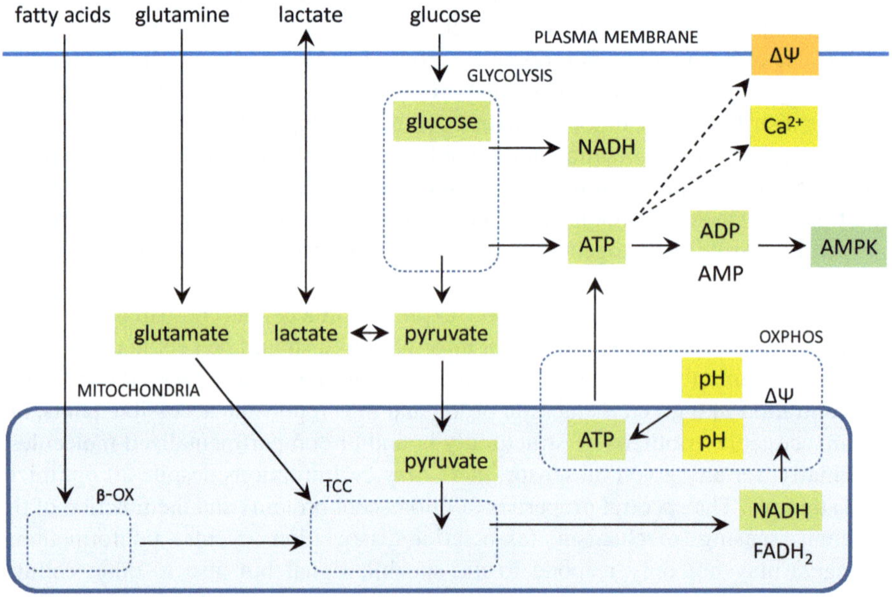

Fig. 20.3 Selected parameters in cellular energy metabolism detectable by genetically encoded fluorescent biosensors. Schematic representation of major components in the transformation of energy within a cell, starting from nutrients (glucose, fatty acids) through glycolysis and mitochondrial pathways to finally yield ATP and membrane ion gradients. Parameters for which biosensors are discussed in this chapter are highlighted: *yellow*, ion concentrations; *light green*, metabolite concentrations; *green*, enzyme activities; *orange*, electrical parameters. Other symbols: *blue lines*, membranes; *dotted blue lines*, metabolic pathways; *arrows*, metabolic flow; *dotted arrows*, regulatory effects. ΔΨ, membrane potential; *AMPK* AMP-activated protein kinase, *β-OC* β-oxidation, *TCC* tricarboxylic acid cycle, *OXPHOS* oxidative phosphorylation

Table 20.1 Selected genetically encoded fluorescent biosensors for cellular energy metabolism

Parameter	Biosensor name	Type	Fluorescent protein(s)	Reference(s)
Metabolite concentrations				
ATP	Ateam	FRET	mseCFP/cpVenus	Imamura et al. (2009)
	GO-Ateam	FRET	mEGFP/mKOk	Nakano et al. (2011)
	QUEEN	sFP	cpEGFP	Yaginuma et al. (2014)
ATP/ADP	Perceval	sFP	cpmVenus	Berg et al. (2009)
	PercevalHR	sFP	cpmVenus	Tantama et al. (2013)
NADH/ NAD^+	Peredox	sFP	T-Sapphire	Hung et al. (2011)
	Frex	sFP	cpYFP	Zhao et al. (2011)
Glucose	FlipGlu	FRET	CFP/YFP	Fehr et al. (2003)
	GBPcys	FRET	AcGFP1/mCherry	Veetil et al. (2010)
	FLII12Pglu-700μδ6	FRET	CFP/Venus	Bermejo et al. (2010)
Lactate	Laconic	FRET	mTFP-Venus	San Martín et al. (2013)
Pyruvate	Pyronic	FRET	mTFP-Venus	San Martín et al. (2014b)
Glutamate	iGluSnFR	sFP	cpGFP	Marvin et al. (2013)
Ion concentrations				
Calcium	Cameleons	FRET	eCFP/eYFP; eBFP/eGFP; eCFP/cpVenus	Miyawaki et al. (1997), Nagai et al. (2004)
	Pericams	sFP	cpeYFP	Nagai et al. (2001)
	TN-XL	FRET	eCFP/cpCitrine174	Mank et al. (2006)
	TN-XXL	FRET	Cerulean/Citrine	Mank et al. (2008)
	GCaMP	sFP	cpeGFP	Tian et al. (2009)
pH	ECFP, EYFP	sFP	ECFP, EYFP	Llopis et al. (1998)
	pHluorin	sFP	GFP	Miesenböck et al. (1998)
	mtAlpHi	sFP	GFP	Abad et al. (2004)
	phRed	sFP	mKeima	Tantama et al. (2011)
Electrical parameters				
Membrane potential	FlaSh	sFP	GFP	Siegel and Isacoff (1997)
	hVoS2.0	FRET	GFP/dipicrylamine	Wang et al. (2010)
	PROPS	sFP	Proteorhodopsin	Kralj et al. (2011)
	ArcLight	sFP	sepHluorin	Jin et al. (2012)
	ASAP1	sFP	cpGFP	St-Pierre et al. (2014)
	QuasAr2	sFP	Archaerhodopsin 3	Hochbaum et al. (2014)
	eFRET sensors	FRET	Diff. GFPs/QuasAr2	Zou et al. (2014)
	FlicR1	sFP	cpRFP (cpmApple)	Abdelfattah et al. (2016)
Kinase activities				
AMPK	AMPKAR	FRET	eCFP/cpVenus	Tsou et al. (2011)
	ABKAR	FRET	Cerulean3/cpVenus	Sample et al. (2015), Miyamoto et al. (2015b)
	bimABKAR	FRET	Cerulean3 + YPet	Depry et al. (2015)

Type of signal detection:- FRET, two FPs with FRET detection; *sFP*, single FP fluorescence recently (González-Vera and Morris 2015). The most classical macroparameters for energy metabolism are the concentrations of adenine nucleotides (ATP, ADP, and AMP) and nicotinamide adenine dinucleotides (NAD, NADP) which determine *adenylate energy charge* (Atkinson 1968; Hardie and Hawley 2001) and *redox state* (Williamson et al. 1967), respectively, both involved in the control of many metabolic reactions. An overview on available sensors for ATP and ATP/ADP ratios is given in Chap. 4. Fluorescent sensors for detection of NADH and NADH/NAD^+ include Peredox (Hung et al. 2011) and Frex (Zhao et al. 2011). More indirectly, the cellular calcium handling and membrane potential are sensitive to perturbations of cellular energy state. Research on calcium detection pioneered the field of fluorescent biosensors. Today, many different Ca^{2+} sensors are available, including two FP sensors such as Cameleons (Miyawaki et al. 1997; Nagai et al. 2004) or TN-XL (Mank et al. 2006), and also several single FP sensors such as Pericam (Nagai et al. 2001) and GCamP (Tian et al. 2009). A similar variety of sensors is available for probing the membrane potential, applied mainly for signal transduction in neuronal cells (Siegel and Isacoff 1997), and the pH, a factor to be controlled as it very often affects GFP fluorescence (for a review, see Benčina 2013).

Finally, cellular concentrations of some key metabolites can yield information on nutrient uptake or mobilization, as well as the intracellular carbon flow through different metabolic pathways. Several fluorescent biosensors have been developed for such metabolites (for a recent review, see (Mohsin et al. 2015)), including glucose (Fehr et al. 2003; Bermejo et al. 2010; Veetil et al. 2010; Deuschle et al. 2006; Behjousiar et al. 2012), lactate (San Martín et al. 2013), pyruvate (San Martín et al. 2014a), 2-oxoglutarate (Zhang and Ye 2014), citrate (Ewald et al. 2011), and other C4 carboxylates (Bourdès et al. 2012), as well as amino acids like glutamine (Gruenwald et al. 2012), glutamate (Hires et al. 2008; Marvin et al. 2013), tryptophan (Kaper et al. 2007), arginine (Bogner and Ludewig 2007), leucine, and histidine (Okada et al. 2009). Multiplexed application of such biosensors may provide sufficient detail for modeling and understanding cellular energy metabolism from a system's point of view.

20.3 Biosensors for AMP-Activated Protein Kinase

AMPK is at the heart of a complex signaling network, but primarily serves as a master sensor and regulator of energy homeostasis in eukaryotic cells (Hardie et al. 2011; Hardie et al. 2016). Since AMPK integrates a large number of intra- and extracellular signals that report the energy and nutritional state of the cell and the organism, respectively, its activation state is a suitable macroparameter for energy availability. Reporting the activity of the AMPK signaling pathway has become a basic requirement of an ever increasing number of studies. So far, this is done by endpoint assays using radiolabeling methods, fluorescence-based assays or immunodetection with phosphosite-specific antibodies. Applied to cells, the latter has limited specificity and spatiotemporal resolution. Given that protein kinases can be activated within seconds to minutes and in a potentially transient way, as well as in a nonhomogeneous manner within a given cell or between cells, these techniques clearly limit the possible insight into AMPK signaling. Genetically encoded fluorescent sensors appear here as the method of choice to overcome such limitations. This will be necessary to advance our understanding of the complex molecular mechanisms governing the role of AMPK in metabolic diseases (Weikel et al. 2016).

20.3.1 AMPKAR

Tsou and colleagues developed a genetically encoded FRET biosensor for monitoring AMPK activity called AMPKAR (Tsou et al. 2011). Its design principle is similar to many other FRET sensors for protein kinase activity that have been developed during recent years (for reviews, see (Oldach and Zhang 2014; González-Vera and Morris 2015). These kinase activity reporters (KARs) express a FRET pair of autofluorescent proteins flanking a construct consisting of a surrogate kinase substrate sequence and a phosphoamino acid-binding domain. In the case of AMPKAR, the FRET pair eCFP and cpVenus flanks the AMPK substrate sequence MRRVATLVDL (phosphorylated Thr underlined) fused to the phosphothreonine-binding forkhead-associated domain 1 (FHA1). When AMPK becomes activated, the substrate motif is phosphorylated and binds to the FHA1 module, thus bringing donor and acceptor fluorophores in close proximity to allow FRET. The initial report also excluded cross-phosphorylation of AMPKAR by other kinases like PKA, PKC, or PKD due to certain similarities in their recognized substrate motifs with the one used in AMPKAR.

The AMPKAR sensor was validated by comparing the FRET data with classical immunoblotting using anti-phospho antibodies against the activating alphaT172 site (Tsou et al. 2011). The observed variations in the AMPKAR FRET signal were directly correlating with AMPK activation and not affected by quenching properties of ATP as reported for an adenylate-sensitive probe based on the AMPK gamma subunit (Willemse et al. 2007). The AMPKAR FRET signal was also proportional to the level of applied activating stimuli. The observed cell-to-cell variability of the FRET signal was depending on the level of the applied stimuli: the more moderate it was, the more variable was its FRET signal and thus the determined AMPK response.

Since kinase signaling involves reversible processes, both activation and inactivation of AMPK should be detectable by a given FRET sensor. AMPKAR can indeed be dephosphorylated by endogenous phosphatases, thus allowing in principle also following AMPK inactivation. It should be kept in mind, however, that the AMPKAR FRET signal follows the fate of an AMPK substrate and not of the protein kinase itself. While this can be neglected for the AMPK activation step, which is rapidly translated into substrate phosphorylation, this may not be always the case for the inactivation step. AMPKAR dephosphorylation kinetics probably vary from those of AMPK inactivation (i.e., dephosphorylation) or even dephosphorylation of other AMPK targets.

Some follow-up studies using AMPKAR have allowed more insight into the advantages and disadvantages of this sensor. Two earlier studies used AMPKAR to visualize AMPK activation during the early phases of mitosis in U2OS cells, just before cell division (Banko et al. 2011), and as an endpoint in a novel Smo–Ca^{2+}–AMPK pathway that triggers reprogramming toward a Wartburg-like glycolytic metabolism in 3 T3-L1, HEK293, and other cell types (Teperino et al. 2012). All these studies used cell lines, given the fact that transfection was difficult in primary cells, but confirmed that AMPKAR is capable of detecting activation of AMPK as rapid as 20 s, e.g., after giving a Smo activator (Teperino et al. 2012).

A more recent study by Prehn, Huber, and colleagues (Connolly et al. 2016) went a step further, using AMPKAR together with FRET sensors for glucose (FLII12Pglu-700μδ6; Bermejo et al. 2010) and ATP (ATeam; Imamura et al. 2009) to analyze bioenergetics in different models of excitotoxicity or ischemia-like injury in single primary neurons. The FRET sensors were each multiplexed with a fluorescent dye detection of mitochondrial membrane potential (TMRM), and also intracellular Ca^{2+} was followed by an intracellular fluorescent probe (Fluo-4). The two types of examined insult, excitotoxicity and ischemia (oxygen and nutrient deprivation), resulted in transiently decreased mitochondrial membrane potential, but differed very much in the response of the different FRET biosensors. During the excitotoxic insult, ATP levels dropped over several minutes, while AMPKAR was activated very rapidly in the range of a minute and then in most cells started to decline slowly already before the end of the stimulus. Surprisingly, glucose accumulated until the end of the insult, mainly resulting from AMPK-triggered cell surface exposure of GLUT3, but also from using internal glycogen stores. After the short insult, AMPK activity state and ATP returned to previous levels. The latter was due to mitochondrial ATP generation, since it occurred also when using pyruvate as direct mitochondrial substrate, but not when mitochondrial ATP synthase was inhibited by oligomycin. Injury by oxygen and nutrient deprivation led to a similar decrease in ATP levels, but in contrast to excitotoxicity, without activating AMPKAR or leading to significant glucose accumulation. In addition, after ischemia, ATP levels did not entirely return to previous levels. These findings emphasize the plasticity of cellular energy metabolism, differing depending on the stimuli. Irrespective of clear cell-to-cell variability, the data confirm at the single cell level the amazing capacity for ATP homeostasis. The excitotoxicity data show that the AMPKAR signal occurs very rapidly and as a more transient-like process, which apparently is sufficient to trigger the neuronal responses like increased glucose uptake via GLUT3. Since AMPKAR dephosphorylation may lag behind AMPK inactivation, as stated above, the AMPK response could be even more transient. However, as also stated by the authors, one may not exclude a persistent low-level AMPK activation beyond the detection limit of AMPKAR. Indeed, AMPK activation often occurs as a minor shift in basic AMPK activity. Interestingly, the data obtained by the set of FRET sensors were suitable to develop a minimal mathematical model for basic parameters of glutamate excitotoxicity in neurons, including the dynamics of cytsolic concentrations of Ca^{2+}, ATP, and glucose, as well as AMPK activity (Depry et al. 2015). The model was able to correctly describe and predict the acute bioenergetic response to the excitotoxic insult and suggested AMPK-mediated intracellular glucose increase as a critical parameter for recovery of ATP levels.

20.3.2 ABKAR

More recent work is now putting a note of caution on the interpretation of AMPKAR data, at least in neuronal cells. AMPKAR phosphorylation is not as specific for AMPK as initially thought, but also occurs with the brain-specific kinases 1 and 2 (BRSK1, BRSK2) (Sample et al. 2015). The BRSKs belong to the family of 13 AMPK-related kinases that share high sequence homology and all lie directly downstream of LKB1. It cannot be excluded that also other close relatives of AMPK share some capacity to phosphorylate the AMPKAR substrate sequence; only microtubule affinity-regulated kinase (MARK) has been excluded so far. It should also be noted that the AMPKAR substrate motif does not entirely correspond to the emerging consensus phosphorylation motif of AMPK (Hardie et al. 2016).

An AMPKAR variant has been introduced more recently by Zhang, Inoue, and colleagues, containing a brighter fluorescence donor, cerulean3, and also available in a control version where the phosphosite threonine is mutated into an alanine (Miyamoto et al. 2015b; Sample et al. 2015). In keeping with the limitation of limited specificity, this variant was renamed ABKAR for AMPK/BRSK activity reporter. Glycolysis inhibition with 2-deoxyglucose (2DG) or CamKK stimulation with Ca^{2+} ionophores increased the ABKAR signal by about 40 % and 51 %, respectively, in Cos7 cells and by about 10 % and 30 %, respectively, in primary undifferentiated hippocampal neurons. Zhang and colleagues applied ABKAR for studying spatial heterogeneity of cell signaling in differentiating primary hippocampal neurons that already developed the axon–dendrite polarity (Sample et al. 2015). In contrast to other regions of the neuron, the distal region of the axon showed high basal BRSK activity, which already saturated ABKAR, such that Ca^{2+}-triggered AMPK activation had no additional effect here. However, BRSK knockdown recovered ABKAR signal increase in these axon tips upon Ca^{2+}-triggered AMPK activation. This suggests that the activity of both kinases, AMPK and BRSK1/2, is polarized, i.e., follows a clear gradient from distal to proximal parts of the axon of differentiated neurons. As determined by ABKAR, BRSK activity is not affected by the classical AMPK activators, i.e., a drop in ATP (by inhibition of glycolysis) and Ca^{2+} (by CamKK).

Another invention with ABKAR has been the generation of organelle-targeted variants, sticking to the cytosolic side of organellar membranes (plasma membrane, Golgi, ER, mitochondria, lysosomes), targeted into the nucleus, or retained in the cytosol (organelle-specific ABKAR, osABKAR). With these tools, Inoue and colleagues addressed the question of heterogeneity in AMPK activation within a single cell in more detail, using non-neuronal cells to avoid potential problems with

BRSK (Miyamoto et al. 2015b). Comparing control with AMPK double k.o. mouse embryonic fibroblasts (MEFs) first under nutrient-rich control conditions, ABKAR-reported basal AMPK activation appeared higher at membranes than in cytosol and nucleus. AMPK complementation of double-k.o. MEFS suggested that it is mainly alpha1 complexes that determine AMPK activity at the plasma membrane, Golgi, and lysosomes, while it is a mix of both alpha isoforms in other compartments. Nutrient limitation by glucose deprivation or 2DG increased ABKAR-reported AMPK activity in the cytosol but not in the nucleus and had heterogeneous effects in space and time on AMPK activity at the various examined intracellular membranes, depending on the way nutrient limitation was applied. However, the observed globally diminished response of membrane-anchored ABKAR, in particular at the plasma and ER membranes, could also reflect a diminished dynamic range due to membrane surface localization of the sensor (see below). Furthermore, this study introduced the use of subcellularly targeted peptides containing the classical AMPK substrate motif SAMS and acting as competitive AMPK inhibitors (Miyamoto et al. 2015b). Prolonged perturbation of mitochondria-associated AMPK led to an increase in cellular ATP levels, suggesting a role of this AMPK in metabolic reprogramming. Taken together, these data provide first evidence for a crucial role of spatial compartmentalization of AMPK signaling, which may provide specificity within complex downstream signal transduction (Miyamoto et al. 2015a).

20.3.3 BimABKAR

A most recent addition to AMPK-sensitive FRET probes is a bimolecular kinase activity reporter (bimKAR) for AMPK activity, bimABKAR (Depry et al. 2015). In this design, the kinase-specific substrate with one FP and the phosphoamino acid-binding domain with the second FP are expressed as two different polypeptides. This is thought to increase distance between the two FPs, thus diminishing basal FRET between the two FPs and enhancing dynamic range. The bimABKAR sensor consists of Cerulean fused to the phosphothreonine-binding domain and the AMPK substrate fused to YPet. Both polypeptides contain a C-terminal nuclear export sequence (NES) to retain them in the cytosol, which can be exchanged against a membrane targeting peptide in the AMPK substrate containing construct for subcellular targeting like e.g. to the plasma membrane (bimABKAR-PM). As compared to ABKAR, the FRET change reported by this bimolecular biosensor in 2DG-challenged Cos7 cells was preserved in the cytosol and increased several-fold when localized to the plasma membrane. This suggests that bimKARs are much more suitable to detect localized kinase activities close to membranes.

While bimABKAR seems to have a clear advantage in terms of dynamic range and responsiveness within surface layers, caution has to be taken for equal expression levels of the two polypeptides to avoid quantification problems. The presence of two individual modules may also reduce temporal resolution due to limitations in the diffusion necessary for interaction and favor interference with other cellular signaling mechanisms (e.g., via phosphosite binding).

Zhang and colleagues used bimABKARs for AMPK and PKA to examine their cross talk at the plasma membrane (Depry et al. 2015). The biosensor results suggest that at this localization, basal PKA activity reduces the activation response of AMPK to stimuli like glucose deprivation. However, stimulation of PKA, but only before glucose deprivation, was able to enhance the AMPK response, and this effect was linked to PKA-induced phosphorylation of the AMPK upstream kinase LKB1. These results confirm literature data on both activatory and inhibitory effects of PKA on AMPK and show that this can occur even together in a single cell.

20.3.4 Conclusions

Genetically encoded FRET sensors for AMPK activity have been applied so far in only a few studies. Based on this limited experience, AMPKAR and the derived sensors ABKAR and bimABKAR seem to be versatile tools that can provide new insight into AMPK signal transduction. They reveal temporal and spatial heterogeneity between and within cells by visualization of kinase signaling in real time at the single cell and the subcellular scale. In future, they may also represent promising tools for drug screening in cell culture. Similar principles as described for these AMPK sensors can be used to quantify AMPK activity in vitro by time-resolved FRET. This technique has been already applied to the in vitro screening of AMPK activators and inhibitors (Martin et al. 2015; Xu et al. 2015).

However, some limitations should be kept in mind, which concern their specificity for AMPK (at least in neuronal cells), their activation when bound to surfaces like membranes, and possibly also their rate of inactivation, if this is an issue. In any case, it is recommended to use appropriate controls. These include (i) cells incapable of AMPK signaling, i.e., knockdowns or double k.o. for both alpha subunits; (ii) mutant sensors that are insensitive to AMPK activation, i.e., lack the phosphosite; and (iii) a fluorescent pH sensor to control for the lack of major pH changes that may affect FP fluorescence. Further, the available AMPK sensors report the full activation state of AMPK and do not distinguish between the energy-sensing modulation of AMPK phosphorylation by AMP and ADP and covalent activation via phosphorylation/dephosphorylation at alpha-T172 by upstream kinases and phosphatases linked, e.g., to endocrine factors and Ca^{2+} (Hardie et al. 2016). Thus, although AMPK is sensitive to ATP/ADP and ATP/AMP ratios, AMPKAR is not an ideal reporter of changes in cellular energy state. Specific detection of allosteric AMPK activation would be useful to specifically monitor energy imbalance during metabolic stress.

Finally, from a practical point of view, expression levels of these sensors should be kept moderate. Any protein overexpression puts a certain workload on the protein synthesis machinery that can affect energy state. Minor energy stress can already increase ADP and AMP concentrations and lead to high basal activation levels of AMPK. Generally, this basic activation should be determined, by FRET or immunoblotting of cell extracts, since it diminishes the exploitable activity increase and change in sensor FRET signal, and may even misleadingly suggest a nonresponse.

20.4 Biosensors for Adenine Nucleotides

The cellular energy state is determined by adenylate ratios (ATP/ADP or ATP/AMP) or the adenylate energy charge as defined by Atkinson (Atkinson 1968). These macroparameters regulate metabolism and metabolic signaling, including AMPK (Hardie and Hawley 2001). Following cellular ATP concentration and ATP/ADP ratios is thus useful when studying AMPK signaling. The generation of genetically encoded fluorescent biosensors for ATP has been a milestone in the development of biosensors for energy metabolism, holding the promise to overcome the limitations of conventional techniques. Three major families of ATP sensors have been generated, one by the Yellen group reporting rather ATP/ADP ratios based on a single cpFP (Berg et al. 2009; Tantama et al. 2013) and two by Noji, Imamura, and colleagues reporting essentially ATP concentrations based either on FRET between two FPs (Imamura et al. 2009; Nakano et al. 2011) or on an excitation-ratiometric intensity signal with a single cpFP (Yaginuma et al. 2014).

In contrast to sensors for AMPK activity, which respond according to an on/off mechanism, i.e., a phosphorylation event, ATP sensors have to respond to gradual changes in a metabolite concentration, probably different for each cell and each stimulus. Thus, for any given cell type and experimental condition, the intrinsic properties of the ATP sensor have to be considered: the affinity of its recognition domain for ATP and the exploitable range for changes in ATP concentration that yield detectable FRET signals.

20.4.1 ATeam

ATeam sensors, developed by Imamura and colleagues, allow the dynamic monitoring of fluctuations in ATP concentration (Imamura et al. 2009). As a recognition element, these sensors exploit the ATP-binding domain of the ε subunit of bacterial F_0-F_1 ATP synthase from *Bacillus subtilis*, sandwiched in between an FP pair of CFP/YFP or GFP/OFP variants. This ATP-binding domain naturally evolved to sense ATP, thus enabling the ε subunit to regulate ATP synthase activity depending on ATP availability. Importantly, the ε subunit undergoes large conformational changes upon ATP binding, switching from an elongated shape when bound to Mg^{2+}-ADP to a contracted hairpin-like structure when bound to Mg^{2+}-ATP (Feniouk et al. 2006). These two features make this ATP-binding domain an ideal module for the engineering of an ATP-specific FRET sensor. Since the FRET dynamic range of an initial ATeam version was not sufficiently large, authors replaced the fluorescent acceptor mVenus by cpVenus, based on the studies on Cameleon Ca^{2+} sensors (Nagai et al. 2004). In addition, further ATeam versions were created differing in their FRET pair (mseCFP/cp173-mVenus or cp173-mEGFP/mKOk). This allows multiplexing ATeam for ATP monitoring with another fluorescent biosensor of choice, since excitation and/or emission fluorescence overlaps can be avoided. Another known issue of FRET sensors is the weak dimerization that may occur between its two GFP-based FPs. Since this may yield a FRET signal not related to recognition domain activity, the involved residues were identified and mutated to generate monomeric FPs. However, the ATeam FPs still contain Ala206 involved in such weak dimerization, and this had a rather positive effect on their FRET amplitude. Indeed, the weak dimerization property of FPs can yield high-performance biosensors with the greatest dynamic range (Kotera et al. 2010).

AT1.03 shows a K_d of 3.3 mM for ATP at 37 °C and can be used to measure ATP levels in the millimolar range. AT3.10 uses the ε subunit of *Bacillus* sp. PS3 and has a much higher affinity with a K_d for ATP of 7.4 μM at 37 °C; it may thus be used as a control always bound to ATP (Imamura et al. 2009). Another suitable control is the ATP-insensitive variant Ateam1.03$^{R122K/R126K}$ carrying two point mutations in the ATP-binding site ("dead variant"). Sequence comparison of the different ε subunits allowed further tuning of ATP affinity by swapping amino acids positioned nearby the ATP-binding pocket. This yielded AT1.03^{YEMK} and AT3.10^{MGK} which have affinities for ATP of 1.2 mM and 14 μM, respectively, suggesting that the ATP affinity of ATeam sensors can be adjusted for the needs of a specific experimental setting. Importantly, temperature plays an important role for ATP affinity, as observed with AT1.03 by a fivefold increase in K_d(ATP) over a range of 10 °C. Another important issue is nucleotide specificity which was examined in vitro: addition of as much as 10 mM GTP or ADP did not affect the FRET signal of AT1.03, but the one of AT3.10, showing a K_d for GTP and ADP of 2.6 mM and 230 μM, respectively. With both sensors, the FRET signal remained stable during pH fluctuations in the range of 7.1 to 8.5; thus, small pH variations in the physiological range should not affect ATeam measurements. However, if the cytoplasmic pH of about 7.3 (Llopis et al. 1998) drops below 7.1, which may at least occur in some cellular subcompartments (Llopis et al. 1998), the ATeam signal would be biased. Further, it became apparent during ATeam characterization that saturation of the FRET signal can be an issue and that the sensor response has to be calibrated to avoid situations of FRET signal saturation leading to non-interpretable data. The temporal resolution of this type of sensor depends on the association and dissociation constants of the metabolite, which in case of ATP/AT1.03 are rather fast with about 0.02 mM^{-1} s^{-1} and 0.10 s^{-1}, respectively.

In HeLa cells expressing ATeam1.03 (Imamura et al. 2009), an overall cytosolic Mg^{2+}-ATP concentration of about 5 mM can be estimated from FRET ratios determined in vitro, which may be a reasonable value. For more detailed spatial information, ATeam1.03 variants were targeted to different subcellular compartments of HeLa cells, revealing an ATP concentration that was significantly lower in mitochondria than in the cytosol or the nucleus. The physiological relevance of the ATeam readout was further tested by inhibition of glycolysis with 10 mM 2DG or of oxidative phosphorylation (OXPHOS) with 1 mM potassium cyanide (KCN). Addition 2DG reduced ATP concentration, whereas KCN alone had no effect. Interestingly, 2DG and KCN together had a synergistic effect leading to a dramatic drop in the ATP concentration. However, such a non-physiological extreme situation may alter many cellular parameters, including pH, and thus bias the FRET signal. In addition, it is unclear how altered Mg^{2+} concentrations would affect the ATeam FRET signal that is specific for the physiologically relevant Mg-complexed ATP.

The versatility of ATeam sensors is supported by an increasing number of studies. Most of them use the sensor to analyze ATP fluctuations in single cells upon a specific stimulus, in particular in primary neurons. In hippocampal neurons, ATP generation reported by ATeam is mainly linked to maintenance of Na^+/K^+ transmembrane ion gradients that ensure neuronal activities (Toloe et al. 2014). In primary rat cortical neurons, disruption of kif5c-mediated mitochondrial trafficking resulted in more punctate-shaped mitochondria and increased cellular ATP levels as indicated by ATeam (Iworima et al. 2016). In a comparison of rat embryonic and postnatal primary hippocampal neurons, the application of ATeam together with dyes sensitive to mitochondrial membrane potential (TMRM, JC-1) and NAD(P)H autofluorescence revealed a striking difference in energy metabolism (Surin et al. 2012). In contrast to mature neurons, ATP synthesis in embryonic neurons is predominantly glycolytic, and mitochondria are rather consumers than producers of ATP. The same authors employed ATeam for monitoring ATP and pH (using only ATeam acceptor excitation) as well as Ca^{2+}-sensitive dyes (e.g., Fura-FF) to study glutamate excitotoxicity in primary hippocampal neurons (Surin et al. 2014). Such approaches may help to better understand mechanisms governing the activity of neuronal networks, especially under pathological conditions. Valdebenito et al. studied the energy metabolism of astrocytes (Valdebenito et al. 2015) by applying FRET sensors for ATP (ATeam), pyruvate (Pyronic; San Martín et al. 2014b), and glucose (an FLII variant, Takanaga et al. 2008). They used both transfected primary astrocytes and organotypic hippocampal slice cultures from animals whose hippocampus was adenovirus-transfected with FRET sensors. In response to a chronic exposure to ketone bodies, this study revealed inhibition of astrocyte glucose consumption, a blunted response of glycolysis to diverse stimuli, and increased mitochondrial pyruvate metabolization.

The ATeam sensor was also used in many other non-brain cell types. In single HeLa cells, ATP and Ca^{2+} concentrations were monitored upon histamine stimulation (Nakano et al. 2011). In oocytes, mitochondria are the primary source of ATP, necessary for successful maturation and early development, while the layer of surrounding cumulus cells is highly glycolytic and exchanges metabolites with the oocytes. Expression of ATeam1.03 in mice oocytes and cumulus cells allowed observation of ATP levels in single cells over extended time periods (Dalton et al. 2014). This revealed ATP fluctuations in oocytes indicative for different ATP consumption rates associated with specific maturation events. Cumulus cells had higher ATP levels, but only as long as they were connected to oocytes via gap junctions.

Several studies used the capacity of ATeam to provide more detailed spatial information, either using the natural distribution of ATeam in cytosol and nucleus or generating targeted versions of the sensor. Forkink et al. targeted ATeam1.03 and its "dead variant" to cytosol and mitochondria to follow fluctuations of ATP in HEK293 cells (Forkink et al. 2014). Combined with pH monitoring by the SypHer sensor (Poburko et al. 2011) and other techniques, the authors confirmed lower ATP concentration in mitochondria as compared to cytosol and deciphered how respiratory complexes and oxidative phosphorylation behave after complex I inhibition. Vevea et al. developed mitoGO-ATeam, an ATeam variant for mitochondria carrying a Green/Orange FRET pair. Their study on mitochondrial bioenergetics used mitoGO-ATeam together with a genetically encoded mitochondrial redox state sensor, roGFP (Jiang et al 2006), in living yeast cells (Vevea et al. 2013). Another ATeam sensor targeted to the mitochondrial matrix, mit-Ateam1.03, was established as a suitable readout for oxidative phoshorylation (Kioka et al. 2014). This sensor characterized a transient increase of oxidative phosphorylation during the early hypoxic response of mouse cardiomyoctes which could be linked to the hypoxia-induced GO/G1 switch gene 2 (G0s2) that interacts with and stimulates F_0F_1-ATP synthase. Very little is known so far about the bioenergetics of the ER compartment. By using an ATeam variant targeted to the ER lumen (ERAT4.01) of different cell lines (HeLa, HEK293, and Ins-1), Vishnu et al. could study ATP fluctuations in parallel to Ca^{2+} dynamics (Vishnu et al. 2014) recorded with the fluorescent sensor D1ER (Palmer et al. 2004). This study could evidence a specific regulation of ER ATP levels depending on the cell type, and a coupling of Ca^{2+} release to an ATP increase within the ER, itself controlled by AMPK. The spatial resolution of ATeam was also exploited to study HuH7 cells infected by Hepatitis C virus (HCV) and to follow ATP concentration during viral replication (Ando et al. 2012). The energy demand linked to active HCV replication was detectable by reduced cytosolic ATP levels. Most interestingly, the cytosol of these cells showed dot-like domains with locally high ATP concentration (about 5 mM), co-localizing with nonstructural viral protein indicating sites of HCV replication. Only 1 mM ATP was measured at peripheral sites not involved in HCV replication and about 2 mM in virus non-replicating cells. These ATP concentrations are below those observed with ATeam in HeLa cells before (Imamura et al. 2009), suggesting variability among clonal HeLa cells. Finally, a recent study used ATeam to follow the relationship between ATP and changes in the shape of HeLa cells (Suzuki et al. 2015). It could describe ATP effects on peripheral microtubule and actin dynamics that regulate cellular motility and morphology.

Taking ATeam even a step further, the variant 1.03 has already been used in animal studies in vivo. Classical model organisms like *Caenorhabditis elegans* and *Drosophila melanogaster* are increasingly used also for bioenergetic and metabolic studies (Baker and Thummel 2007; Harrison and Haddad 2011). In a first approach, transgenic *Caenorhabditis elegans* expressing ATeam was created, but fluorescence decay with age and intrinsic fluorescence hindered FRET analysis in body muscle cells (Kishikawa et al. 2012). This interference was not present in the vulva cells of the worm, so that heterogeneity in ATP concentration between cytosol and mitochondria could be observed as before in HeLa cells. However, no further fluctuations of ATP levels were detected as, e.g., with aging. This could be due to an increased K_d(ATP) of ATeam with the lower body temperature of the worm, saturating the sensor at physiological ATP concentrations. To further exploit ATeam in model species like *C. elegans* and *Drosophila melanogaster*, but also for single cell imaging studies, the ATP affinity of ATeam was optimized for a temperature of 25 °C, yielding the variant Ateam1.03NL (Tsuyama et al. 2013). This illustrates a general advantage of genetically encoded biosensors mentioned above, i.e., the possibility to tune and optimize sensor affinities and specificities. As an application to plant physiology, transgenic *Arabidopsis thaliana* expressing the ATeam1.04-nD/nA variant (Kotera et al. 2010) was used to study the bacteria-induced hypersensitive death response of plant cells (Hatsugai et al. 2012). The study revealed that inoculation with non-virulent bacteria leads to an early increase of intracellular ATP levels, required for the hypersensitive response, while at a later time point a drastic drop in ATP occurs, albeit well before rupture of the vacuole.

In summary, ATeam sensors have proven in many studies to be a very useful tool. They allow monitoring dynamics of free Mg^{2+}-ATP with unprecedented resolution in space and time in single cells and also appear to be applicable in small model organisms in vivo. While they are suitable for many experimental settings that show rather large changes in ATP concentration, they may be limited in situations with smaller fluctuations as generally occurring during physiological mild energy stress which is sufficient, e.g., to activate AMPK. Further, for cellular reactions, more important than the absolute ATP pool size is the ATP/ADP ratio, which is not detected by ATeam sensors.

20.4.2 Perceval

Another genetically encoded fluorescent biosensor for spatiotemporal analysis of adenine nucleotides uses a single cpFP design and rather senses the ATP/ADP ratio instead of absolute ATP concentrations (Berg et al. 2009). Here, a cpFP variant (cpmVenus) is inserted within the T loop of the ATP-binding protein GlnK1 from *Methanococcus jannaschii*. Although the nucleotide-binding site accommodates both Mg^{2+}-ATP and ADP, only Mg^{2+}-ATP binding triggers closure of the T loop and a maximal increase in fluorescence emission.

Affinities determined in vitro for the initial sensor (QV5) are 40 nM for Mg^{2+}-ATP and 200 nM for ADP, far below physiological concentrations (Berg et al. 2009). The extremely high affinity for ATP would seem to cause permanent binding of this nucleotide, thus rejecting its use for ATP sensing. However, since Mg^{2+}-ATP and ADP bind at the same site, occupation of the sensor by ATP is efficiently competed by increasing ADP concentrations in the physiological range. In other words, competition by ADP is efficiently lowering the sensor's apparent affinity for Mg^{2+}-ATP, which becomes, e.g., 1 μM in the presence of 5 μM ADP (Berg et al. 2009). Thus, the sensor is best described as a reporter of Mg^{2+}ATP/ADP ratios. However, there are other limitations of QV5, including slow kinetics of ATP/ADP exchange (time constant of about 1 min) and an affinity of about 0.3 mM for 2-ketoglutarate (2KG), leading to a reduced response to ATP in the presence of 2KG.

The Perceval sensor, which evolved from QV5 by further molecular engineering, avoids these limitations (Berg et al. 2009). In addition, the sensitivity for changes in ATP/ADP ratio is closer to physiological conditions, with a half-maximal response at an ATP/ADP ratio of 0.5 instead of 0.2 for QV5. However, without particular workload, the cellular ratio of total ATP/ADP is rather above 5–10 (Nilsson et al. 1996; Erecińska and Silver 1994). Since much more ADP is bound to cellular structures than ATP, the ratio of free ATP/ADP that is effectively available for cellular reactions is even much higher, probably far above 20 (Mörikofer-Zwez and Walter 1989; Koretsky et al. 1990). Thus, the range detectable by Perceval, which goes up to about 5, is at the far lower end of what can be expected under energy stress. Perceval will therefore sense only extreme energy deficits as expected under ischemia, anoxia, nutrient starvation, or chronically high workload. Another limitation of Perceval is the pH sensitivity of its fluorescence that also occurs at pH values in the physiological range.

Despite its design as a single cpFP sensor, Perceval is suitable for excitation-ratiometric measurements that allow correction for most potential bias in cellular applications (Berg et al. 2009). Binding of ATP triggers a change in its absorbance spectrum, with a decrease at 405 nm and an increase at 490 nm of absorbance and emission determined at 530 nm. As a proof of concept, Perceval detected a decrease in ATP/ADP ratio upon inhibition of glycolysis by 2DG treatment of HEK293 cells.

Optimization of Perceval by structure-guided engineering further improved the range of detectable ATP/ADP ratios (Tantama et al. 2013). In a range of 10–92 % sensor saturation, this variant named PercevalHR can sense ATP/ADP ratio from about 0.4 to 40 with preserved kinetics of ATP/ADP exchange. PercevalHR transfected into astrocytes reported acute fluctuations in ATP/ADP ratios triggered by glucose supply. However, the pH sensitivity persists also with this variant. Physiological pH variations between about pH 6.7 and 7.8 affect the dynamic range of fluorescence ($\Delta F_{max}/F_{initial}$, between 2 and 5) and the ATP/ADP ratio at which half-maximal response is achieved (between 2 and 10). It is therefore essential to simultaneously monitor intracellular pH, either by using a pH indicator dye (e.g., SNARF-5 F; Berg et al. 2009) or a genetically encoded pH-sensitive fluorescence sensor (e.g., pHRed; Tantama et al. 2013). The latter is particularly suitable for cellular applications due to its spectral properties.

Only few studies involving Perceval sensors have been published so far. Most of them have analyzed the role of bioenergetics in pancreatic β-cells, where ATP links the glucose signal and mitochondrial bioenergetics to the closure of ATP-sensitive K^+ channels, membrane depolarization, Ca^{2+} influx, and insulin secretion. Studies by the Rutter group combined patch-clamp electrophysiology and simultaneous real-time fluorescence imaging using targeted Pericam (Nagai et al. 2001) and

Perceval sensors together with different fluorescent Ca^{2+}-sensitive dyes for the cytosol (Fura-Red), mitochondria (2mt8RP), and ER (D4ER). Glucose-induced compartmentalized changes of Ca^{2+} and ATP/ADP ratios were analyzed in single primary mouse β-cells (Tarasov et al. 2012). The pH was controlled but had negligible effects on the Perceval signal; over a pH range from 6.6 to 7.8, the cytosolic ATP/ADP ratio varied between 0.9 and 1.35. The study revealed a major role of mitochondrial Ca^{2+} uniporter (MCU) and a modulator role of mitochondrial Na^{+}/Ca^{2+} exchanger (NCLX) for β-cell glucose sensing and insulin secretion (Wiederkehr et al. 2011), in particular by regulating the cytosolic ATP/ADP ratio. Another study on β-cells from the same group followed Ca^{2+} concentrations in different cellular compartments together with cytosolic ATP/ADP ratios reported by Perceval. It confirmed the important role of MCU for glucose-stimulated insulin secretion (Tarasov et al. 2013) and indicated that mitochondria can translate the frequency of the cytosolic Ca^{2+} oscillation frequency into stable increases of mitochondrial Ca^{2+} and cytosolic ATP/ADP ratio. Finally, TIRF microscopy was applied to Min6 and primary mouse β-cells to directly observe sub-membrane concentrations of ATP (ATP_{pm}) with Perceval and Ca^{2+} (${Ca^{2+}}_{pm}$) with Fura red (Li et al. 2013), together with a pH control (BCECF). In control experiments with permeabilized cells, the authors surprisingly found a better correlation of the Perceval signal with ATP concentration, not with ATP/ADP ratios, an issue to be clarified. The study further revealed during the early glucose response an ATP_{PM} increase that preceded and was transiently interrupted by influx of Ca^{2+}. Later and at higher glucose concentrations, oscillations of ATP_{pm} inverse to oscillations in ${Ca^{2+}}_{pm}$ occurred at a minute timescale and were synchronized between neighboring cells, possibly connected to pulsatile insulin secretion.

There are also few studies using Perceval in other cell types. For example, the sensor has been applied for controlling ATP availability along the axons in primary neurons, which occurred to be homogeneous and sustained by both mitochondria and glycolysis (Zala et al. 2013).

Taken together, last generation Perceval sensors are suitable readouts of free ATP/ADP ratios, as long as the signal is corrected for pH fluctuations. Detailed protocols on how to use Percveal and how to multiplex it with other probes can be found in recent reviews (Tantama and Yellen 2014; Tarasov and Rutter 2014). In principle, a sensor that follows ATP/ADP ratios rather than absolute ATP concentration can be a more reliable indicator of energy state, since it is independent of variations in the absolute pool size of ATP, ADP, and AMP that can vary over time or in between cells. It is the free ATP/ADP ratio that determines the thermodynamic efficiency and thus the driving force of energy-requiring reactions, while total ATP is rather a determinant of total potential turnover. Like with ATeam, smaller changes of the free ATP/ADP ratio, in particular in cells with a very high basal ratio, may also escape detection by Perceval sensors.

20.4.3 QUEEN

The most recent ATP-sensitive FRET probe, QUEEN, also uses a single FP design. A cpEGFP fluorophore is inserted into the ε subunit of bacterial F_0F_1-ATPase (Yaginuma et al. 2014). The search for such a sensor was motivated by problems occurring with classical FRET sensors, in particular for determination of absolute ATP concentrations and for applications in cells with fluctuating proliferation rate. The signal of FRET sensors having two FPs can be biased by a maturation time lag between the two FPs or the presence of degradation intermediates, leading to sensors with only one functional FP, donor or acceptor. Such dysfunctional sensors occur mainly at high metabolic turnover rates, e.g., during rapid cell proliferation. Variations in proliferation rate therefore affect the precise quantification of the sensed parameter. This is a main problem for monitoring bacterial cultures with growth rates spanning from very high to almost zero. A single FP design can avoid such problems related to maturation or degradation of the sensor.

Queen was engineered in two versions having affinities for Mg^{2+}-ATP of 7 μM (QUEEN-7 μ) or 2 mM (QUEEN-2 m), as well as a control version mutated in the ATP-binding cassette (QUEEN-NA) (Yaginuma et al. 2014). The output signal of QUEEN is excitation-ratiometric, since it is calculated as ratio of excitation at either 400 nm or 494 nm and emission detected in both cases at 513 nm. This ratio increases in vitro with increasing ATP concentration, between about 1 and 70 μM for QUEEN-7 μ and between about 0.5 and 8 mM for QUEEN-2 m (all at 25 °C). From these data, only QUEEN-2 m is suitable to quantify ATP concentrations in bacteria or eukaryotic cells. QUEEN-7 μ may be used as a control that should not vary under physiological conditions. The signal of QUEEN-2 m is only slightly affected by ADP at physiological concentrations and by pH changes above pH 7.3. In contrast to ATeam sensors, QUEEN is not affected by the bacterial growth phase and offers a twofold higher dynamic range. Since in vitro and in vivo QUEEN signals differed only slightly, the absolute ATP concentration of bacterial cells could be calculated from the QUEEN fluorescence signals with only minor corrections. These ATP concentrations differed between cells in a population and showed a non-Gaussian distribution, with ATP in some cells reaching up to 9 mM.

20.4.4 Conclusions

Biosensors for adenylates are offering important advantages as compared to invasive analysis methods, in particular for dynamic monitoring of variations of ATP at the submicron scale in single cells. Since ATeam and QUEEN report ATP pool size, and Perceval rather the ATP/ADP ratio, the parallel use of both types of sensor may be recommended to dissect distinct aspects of changes in cellular energy state. Particular caution is required when larger changes in pH or proliferation rate could occur in a given experiment, since they can affect the fluorescence readout.

Further, it should be kept in mind that the available ATP sensors are best suited to report large changes as occurring by inhibition of entire metabolic pathways. This can yield valuable information about the contribution and the fluxes of specific pathways under a given experimental condition. However, the sensors may not always adequately report smaller physiological changes in ATP concentration or ATP/ADP ratios that are already relevant, e.g., for activating the AMPK signaling pathway. In this respect, following fluctuations in ADP and even more AMP would represent a more promising way to detect such small fluctuations in cellular energy state. Given their much smaller cellular pool size, together with the action of adenylate kinase, their concentrations increase several-fold upon a only partial hydrolysis of the large ATP pool (Neumann et al. 2003). Unfortunately, no genetically encoded fluorescent biosensors for ADP and AMP are available so far. There are fluorescent biosensors for ADP (Kunzelmann and Webb 2009; Kunzelmann and Webb 2010) and GDP (Kunzelmann and Webb 2011), but these are not genetically encoded. They use a diethylaminocoumarin fluorophore coupled to a bacterial actin homologue (ParM) and require an invasive experimental procedure which has limited their general application.

20.5 Fluorescent Biosensors: Conclusions and Outlook

Unlike conventional techniques, which mostly generate endpoint data for populations of cells, genetically encoded fluorescent biosensors allow for online spatiotemporal analysis of energy metabolism in single, live cells and even at the subcellular level. An ever increasing number of such biosensors have been designed and validated during the last decade. FRET-based probes are intrinsically emission-ratiometric which corrects for most potential bias, and FRET can be measured with different techniques. This is less the case for cpFP-based single fluorophore biosensors, which however can be excitation-ratiometric and often have a greater dynamic range and are less influenced by changes in proliferation rate. The sensitivity for pH changes remains a problem with GFP-based fluorescent biosensors, and sometimes the parallel measurement with a fluorescent pH sensor is indicated.

The current development in the field aims at multiplexing several biosensors for simultaneous recordings of multiple parameters in a single cell. Also the application of such biosensors in animals in vivo remains a long-term goal. While this can be technically achieved in principle in small animal models (e.g., *Drosophila*), this remains a challenge for larger organisms unless there are no brilliant GFP variants that work at long wavelengths able to penetrate tissue.

Finally, genetically encoded fluorescent biosensors are just one element in the emerging optogenetic toolbox. Optogenetics refers to techniques in which a combination of light and genetic manipulation is used to perturb, control, or analyze cellular functions (Alford et al. 2013). It thus also comprises optogenetic actuators, i.e., proteins that have a light-controllable biological function. This opens unprecedented possibilities of both manipulation *and* monitoring of single cells in a noninvasive manner.

Acknowledgments This work was supported by a grant from the Region Rhône Alpes, France (to U.S. and I.B.). Initial experiments of the authors on FRET sensors were supported among others by the EU 6th framework programs (contract LSHM-CT-2004-005272 EXGENESIS), *Fondation ARC* R2012 (CA 25/09/2014), and the *Agence Nationale de Recherche* (France, *chaire d'excellence* to U.S.).

References

Abad MFC, Di Benedetto G, Magalhães PJ et al (2004) Mitochondrial pH monitored by a new engineered green fluorescent protein mutant. J Biol Chem 279:11521–11529. doi:10.1074/jbc.M306766200

Abdelfattah AS, Farhi SL, Zhao Y et al (2016) A bright and fast red fluorescent protein voltage indicator that reports neuronal activity in organotypic brain slices. J Neurosci 36:2458–2472. doi:10.1523/JNEUROSCI.3484-15.2016

Alford SC, Wu J, Zhao Y et al (2013) Optogenetic reporters. Biol Cell 105:14–29. doi:10.1111/boc.201200054

Ando T, Imamura H, Suzuki R et al (2012) Visualization and measurement of ATP levels in living cells replicating hepatitis C virus genome RNA. PLoS Pathog 8, e1002561. doi:10.1371/journal.ppat.1002561

Atkinson DE (1968) The energy charge of the adenylate pool as a regulatory parameter. Interaction with feedback modifiers. Biochemistry 7:4030–4034

Baird GS, Zacharias DA, Tsien RY (1999) Circular permutation and receptor insertion within green fluorescent proteins. Proc Natl Acad Sci U S A 96:11241–11246

Baker KD, Thummel CS (2007) Diabetic larvae and obese flies-emerging studies of metabolism in Drosophila. Cell Metab 6:257–266. doi:10.1016/j.cmet.2007.09.002

Banko MR, Allen JJ, Schaffer BE et al (2011) Chemical genetic screen for AMPKalpha2 substrates uncovers a network of proteins involved in mitosis. Mol Cell 44:878–892. doi:10.1016/j.molcel.2011.11.005

Bastiaens PI, Jovin TM (1996) Microspectroscopic imaging tracks the intracellular processing of a signal transduction protein: fluorescent-labeled protein kinase C beta I. Proc Natl Acad Sci U S A 93:8407–8412

Bastiaens PIH, Squire A (1999) Fluorescence lifetime imaging microscopy: spatial resolution of biochemical processes in the cell. Trends Cell Biol 9:48–52. doi:10.1016/S0962-8924(98)01410-X

Behjousiar A, Kontoravdi C, Polizzi KM (2012) In situ monitoring of intracellular glucose and glutamine in CHO cell culture. PloS One 7, e34512. doi:10.1371/journal.pone.0034512

Benčina M (2013) Illumination of the spatial order of intracellular pH by genetically encoded pH-sensitive sensors. Sensors 13:16736–16758. doi:10.3390/s131216736

Berg J, Hung YP, Yellen G (2009) A genetically encoded fluorescent reporter of ATP/ADP ratio. Nat Methods 6:161–166. doi:10.1038/nmeth.1288

Bermejo C, Haerizadeh F, Takanaga H et al (2010) Dynamic analysis of cytosolic glucose and ATP levels in yeast using optical sensors. Biochem J 432:399–406. doi:10.1042/BJ20100946

Bogner M, Ludewig U (2007) Visualization of arginine influx into plant cells using a specific FRET-sensor. J Fluoresc 17:350–360. doi:10.1007/s10895-007-0192-2

Bourdès A, Rudder S, East AK, Poole PS (2012) Mining the Sinorhizobium meliloti transportome to develop FRET biosensors for sugars, dicarboxylates and cyclic polyols. PloS One 7, e43578. doi:10.1371/journal.pone.0043578

Connolly NMC, D'Orsi B, Monsefi N et al (2016) Computational analysis of AMPK-mediated neuroprotection suggests acute excitotoxic bioenergetics and glucose dynamics are regulated by a minimal set of critical reactions. PloS One 11, e0148326. doi:10.1371/journal.pone.0148326

Dalton CM, Szabadkai G, Carroll J (2014) Measurement of ATP in single oocytes: impact of maturation and cumulus cells on levels and consumption. J Cell Physiol 229:353–361. doi:10.1002/jcp.24457

Depry C, Mehta S, Li R, Zhang J (2015) Visualization of compartmentalized kinase activity dynamics using adaptable BimKARs. Chem Biol 22:1470–1479. doi:10.1016/j.chembiol.2015.10.004

Deuschle K, Chaudhuri B, Okumoto S et al (2006) Rapid metabolism of glucose detected with FRET glucose nanosensors in epidermal cells and intact roots of Arabidopsis RNA-silencing mutants. Plant Cell 18:2314–2325. doi:10.1105/tpc.106.044073

Erecińska M, Silver IA (1994) Ions and energy in mammalian brain. Prog Neurobiol 43:37–71

Ewald JC, Reich S, Baumann S et al (2011) Engineering genetically encoded nanosensors for real-time in vivo measurements of citrate concentrations. PloS One 6, e28245. doi:10.1371/journal.pone.0028245

Fehr M, Lalonde S, Lager I et al (2003) In vivo imaging of the dynamics of glucose uptake in the cytosol of COS-7 cells by fluorescent nanosensors. J Biol Chem 278:19127–19133. doi:10.1074/jbc.M301333200

Feniouk BA, Suzuki T, Yoshida M (2006) The role of subunit epsilon in the catalysis and regulation of FOF1-ATP synthase. Biochim Biophys Acta 1757:326–338. doi:10.1016/j.bbabio.2006.03.022

Forkink M, Manjeri GR, Liemburg-Apers DC et al (2014) Mitochondrial hyperpolarization during chronic complex I inhibition is sustained by low activity of complex II, III, IV and V. Biochim Biophys Acta 1837:1247–1256. doi:10.1016/j.bbabio.2014.04.008

González-Vera JA, Morris MC (2015) Fluorescent reporters and biosensors for probing the dynamic behavior of protein kinases. Proteomes 3:369–410. doi:10.3390/proteomes3040369

Griesbeck O, Baird GS, Campbell RE et al (2001) Reducing the environmental sensitivity of yellow fluorescent protein. Mechanism and applications. J Biol Chem 276:29188–29194. doi:10.1074/jbc.M102815200

Gruenwald K, Holland JT, Stromberg V et al (2012) Visualization of glutamine transporter activities in living cells using genetically encoded glutamine sensors. PloS One 7, e38591. doi:10.1371/journal.pone.0038591

Hanson GT, Aggeler R, Oglesbee D et al (2004) Investigating mitochondrial redox potential with redox-sensitive green fluorescent protein indicators. J Biol Chem 279:13044–13053. doi:10.1074/jbc.M312846200

Hardie DG, Carling D, Gamblin SJ (2011) AMP-activated protein kinase: also regulated by ADP? Trends Biochem Sci 36:470–477. doi:10.1016/j.tibs.2011.06.004

Hardie DG, Hawley SA (2001) AMP-activated protein kinase: the energy charge hypothesis revisited. Bioessays 23:1112–1119. doi:10.1002/bies.10009

Hardie DG, Schaffer BE, Brunet A (2016) AMPK: an energy-sensing pathway with multiple inputs and outputs. Trends Cell Biol 26:190–201. doi:10.1016/j.tcb.2015.10.013

Harrison JF, Haddad GG (2011) Effects of oxygen on growth and size: synthesis of molecular, organismal, and evolutionary studies with Drosophila melanogaster. Annu Rev Physiol 73:95–113. doi:10.1146/annurev-physiol-012110-142155

Hatsugai N, Perez Koldenkova V, Imamura H et al (2012) Changes in cytosolic ATP levels and intracellular morphology during bacteria-induced hypersensitive cell death as revealed by real-time fluorescence microscopy imaging. Plant Cell Physiol 53:1768–1775. doi:10.1093/pcp/pcs119

Hires SA, Zhu Y, Tsien RY (2008) Optical measurement of synaptic glutamate spillover and reuptake by linker optimized glutamate-sensitive fluorescent reporters. Proc Natl Acad Sci U S A 105:4411–4416. doi:10.1073/pnas.0712008105

Hochbaum DR, Zhao Y, Farhi SL et al (2014) All-optical electrophysiology in mammalian neurons using engineered microbial rhodopsins. Nat Methods 11:825–833. doi:10.1038/nmeth.3000

Hochreiter B, Garcia AP, Schmid JA (2015) Fluorescent proteins as genetically encoded FRET biosensors in life sciences. Sensors 15:26281–26314. doi:10.3390/s151026281

Huang S (2011) Systems biology of stem cells: three useful perspectives to help overcome the paradigm of linear pathways. Philos Trans R Soc B Biol Sci 366:2247–2259. doi:10.1098/rstb.2011.0008

Hung YP, Albeck JG, Tantama M, Yellen G (2011) Imaging cytosolic NADH-NAD(+) redox state with a genetically encoded fluorescent biosensor. Cell Metab 14:545–554. doi:10.1016/j.cmet.2011.08.012

Imamura H, Nhat KPH, Togawa H et al (2009) Visualization of ATP levels inside single living cells with fluorescence resonance energy transfer-based genetically encoded indicators. Proc Natl Acad Sci U S A 106:15651–15656. doi:10.1073/pnas.0904764106

Iworima DG, Pasqualotto BA, Rintoul GL (2016) Kif5 regulates mitochondrial movement, morphology, function and neuronal survival. Mol Cell Neurosci 72:22–33. doi:10.1016/j.mcn.2015.12.014

Jiang K, Schwarzer C, Lally E, Zhang S, Ruzin S, Machen T, Remington SJ, Feldman L (2006) Expression and characterization of a redox-sensing green fluorescent protein (reduction-oxidation-sensitive green fluorescent protein) in Arabidopsis. Plant Physiol 141:397–403

Jin L, Han Z, Platisa J et al (2012) Single action potentials and subthreshold electrical events imaged in neurons with a fluorescent protein voltage probe. Neuron 75:779–785. doi:10.1016/j.neuron.2012.06.040

Kamioka Y, Sumiyama K, Mizuno R, Matsuda M (2013) Live imaging of transgenic mice expressing FRET biosensors. Conf IEEE Eng Med Biol Soc 2013:125–128. doi:10.1109/EMBC.2013.6609453

Kaper T, Looger LL, Takanaga H et al (2007) Nanosensor detection of an immunoregulatory tryptophan influx/kynurenine efflux cycle. PLoS Biol 5, e257. doi:10.1371/journal.pbio.0050257

Kedziora KM, Jalink K (2015) Fluorescence resonance energy transfer microscopy (FRET). Methods Mol Biol 1251:67–82. doi:10.1007/978-1-4939-2080-8_5

Kennedy HJ, Pouli AE, Ainscow EK et al (1999) Glucose generates sub-plasma membrane ATP microdomains in single islet beta-cells. Potential role for strategically located mitochondria. J Biol Chem 274:13281–13291

Kioka H, Kato H, Fujikawa M et al (2014) Evaluation of intramitochondrial ATP levels identifies G0/G1 switch gene 2 as a positive regulator of oxidative phosphorylation. Proc Natl Acad Sci U S A 111:273–278. doi:10.1073/pnas.1318547111

Kishikawa J, Fujikawa M, Imamura H et al (2012) MRT letter: expression of ATP sensor protein in Caenorhabditis elegans. Microsc Res Tech 75:15–19. doi:10.1002/jemt.21103

Kneen M, Farinas J, Li Y, Verkman AS (1998) Green fluorescent protein as a noninvasive intracellular pH indicator. Biophys J 74:1591–1599. doi:10.1016/S0006-3495(98)77870-1

Koretsky AP, Brosnan MJ, Chen LH et al (1990) NMR detection of creatine kinase expressed in liver of transgenic mice: determination of free ADP levels. Proc Natl Acad Sci U S A 87:3112–3116

Kotera I, Iwasaki T, Imamura H et al (2010) Reversible dimerization of Aequorea victoria fluorescent proteins increases the dynamic range of FRET-based indicators. ACS Chem Biol 5:215–222. doi:10.1021/cb900263z

Kralj JM, Hochbaum DR, Douglass AD, Cohen AE (2011) Electrical spiking in Escherichia coli probed with a fluorescent voltage-indicating protein. Science 333:345–348. doi:10.1126/science.1204763

Kunzelmann S, Webb MR (2009) A biosensor for fluorescent determination of ADP with high time resolution. J Biol Chem 284:33130–33138. doi:10.1074/jbc.M109.047118

Kunzelmann S, Webb MR (2010) A fluorescent, reagentless biosensor for ADP based on tetramethylrhodamine-labeled ParM. ACS Chem Biol 5:415–425. doi:10.1021/cb9003173

Kunzelmann S, Webb MR (2011) Fluorescence detection of GDP in real time with the reagentless biosensor rhodamine-ParM. Biochem J 440:43–49. doi:10.1042/BJ20110349

Kuznetsov A, Bindokas VP, Marks JD, Philipson LH (2005) FRET-based voltage probes for confocal imaging: membrane potential oscillations throughout pancreatic islets. Am J Physiol Cell Physiol 289:C224–C229. doi:10.1152/ajpcell.00004.2005

Lakowicz JR, Szmacinski H, Nowaczyk K et al (1994) Fluorescence lifetime imaging of intracellular calcium in COS cells using Quin-2. Cell Calcium 15:7–27

Li J, Shuai HY, Gylfe E, Tengholm A (2013) Oscillations of sub-membrane ATP in glucose-stimulated beta cells depend on negative feedback from Ca(2+). Diabetologia 56:1577–1586. doi:10.1007/s00125-013-2894-0

Lindenburg L, Merkx M (2014) Engineering genetically encoded FRET sensors. Sensors 14:11691–11713. doi:10.3390/s140711691

Liu X, Wu J, Liu H et al (2014) RoGFP1 is a quantitative biosensor in maize cells for cellular redox changes caused by environmental and endogenous stimuli. Biochem Biophys Res Commun 452:503–508. doi:10.1016/j.bbrc.2014.08.107

Llopis J, McCaffery JM, Miyawaki A et al (1998) Measurement of cytosolic, mitochondrial, and Golgi pH in single living cells with green fluorescent proteins. Proc Natl Acad Sci U S A 95:6803–6808

Mank M, Reiff DF, Heim N et al (2006) A FRET-based calcium biosensor with fast signal kinetics and high fluorescence change. Biophys J 90:1790–1796. doi:10.1529/biophysj.105.073536

Mank M, Santos AF, Direnberger S et al (2008) A genetically encoded calcium indicator for chronic in vivo two-photon imaging. Nat Methods 5:805–811. doi:10.1038/nmeth.1243

Martin H, Burgess EJ, Smith WA et al (2015) JAK2 and AMP-kinase inhibition in vitro by food extracts, fractions and purified phytochemicals. Food Funct 6:305–312. doi:10.1039/c4fo00626g

Marvin JS, Borghuis BG, Tian L et al (2013) An optimized fluorescent probe for visualizing glutamate neurotransmission. Nat Methods 10:162–170. doi:10.1038/nmeth.2333

Mekler V, Kortkhonjia E, Mukhopadhyay J et al (2002) Structural organization of bacterial RNA polymerase holoenzyme and the RNA polymerase-promoter open complex. Cell 108:599–614. doi:10.1016/S0092-8674(02)00667-0

Miesenböck G, De Angelis DA, Rothman JE (1998) Visualizing secretion and synaptic transmission with pH-sensitive green fluorescent proteins. Nature 394:192–195. doi:10.1038/28190

Miyamoto T, Rho E, Inoue T (2015a) Deconvoluting AMPK dynamics. Oncotarget 6:30431–30432. doi:10.18632/oncotarget.5447

Miyamoto T, Rho E, Sample V et al (2015b) Compartmentalized AMPK signaling illuminated by genetically encoded molecular sensors and actuators. Cell Rep. doi:10.1016/j.celrep.2015.03.057

Miyawaki A, Griesbeck O, Heim R, Tsien RY (1999) Dynamic and quantitative Ca2+ measurements using improved cameleons. Proc Natl Acad Sci U S A 96:2135–2140. doi:10.1073/pnas.96.5.2135

Miyawaki A, Llopis J, Heim R et al (1997) Fluorescent indicators for Ca2+based on green fluorescent proteins and calmodulin. Nature 388:882–887. doi:10.1038/42264

Mohsin M, Ahmad A, Iqbal M (2015) FRET-based genetically-encoded sensors for quantitative monitoring of metabolites. Biotechnol Lett 37:1919–1928. doi:10.1007/s10529-015-1873-6

Mörikofer-Zwez S, Walter P (1989) Binding of ADP to rat liver cytosolic proteins and its influence on the ratio of free ATP/free ADP. Biochem J 259:117–124

Nagai T, Ibata K, Park ES et al (2002) A variant of yellow fluorescent protein with fast and efficient maturation for cell-biological applications. Nat Biotechnol 20:87–90. doi:10.1038/nbt0102-87

Nagai T, Sawano A, Park ES, Miyawaki A (2001) Circularly permuted green fluorescent proteins engineered to sense Ca2+. Proc Natl Acad Sci U S A 98:3197–3202. doi:10.1073/pnas.051636098

Nagai T, Yamada S, Tominaga T et al (2004) Expanded dynamic range of fluorescent indicators for Ca2+ by circularly permuted yellow fluorescent proteins. Proc Natl Acad Sci U S A 101:10554–10559. doi:10.1073/pnas.0400417101

Nakano M, Imamura H, Nagai T, Noji H (2011) Ca^{2+} regulation of mitochondrial ATP synthesis visualized at the single cell level. ACS Chem Biol 6:709–715. doi:10.1021/cb100313n

Neumann D, Schlattner U, Wallimann T (2003) A molecular approach to the concerted action of kinases involved in energy homoeostasis. Biochem Soc Trans 31:169–174

Nilsson T, Schultz V, Berggren PO et al (1996) Temporal patterns of changes in ATP/ADP ratio, glucose 6-phosphate and cytoplasmic free Ca2+ in glucose-stimulated pancreatic beta-cells. Biochem J 314(Pt 1):91–94

Okada S, Ota K, Ito T (2009) Circular permutation of ligand-binding module improves dynamic range of genetically encoded FRET-based nanosensor. Protein Sci 18:2518–2527. doi:10.1002/pro.266

Oldach L, Zhang J (2014) Genetically encoded fluorescent biosensors for live-cell visualization of protein phosphorylation. Chem Biol 21:186–197. doi:10.1016/j.chembiol.2013.12.012

Palmer AE, Jin C, Reed JC, Tsien RY (2004) Bcl-2-mediated alterations in endoplasmic reticulum Ca2+ analyzed with an improved genetically encoded fluorescent sensor. Proc Natl Acad Sci U S A 101:17404–17409. doi:10.1073/pnas.0408030101

Palmer AE, Qin Y, Park JG, McCombs JE (2011) Design and application of genetically encoded biosensors. Trends Biotechnol 29:144–152. doi:10.1016/j.tibtech.2010.12.004

Piljić A, de Diego I, Wilmanns M, Schultz C (2011) Rapid development of genetically encoded FRET reporters. ACS Chem Biol 6:685–691. doi:10.1021/cb100402n

Poburko D, Santo-Domingo J, Demaurex N (2011) Dynamic regulation of the mitochondrial proton gradient during cytosolic calcium elevations. J Biol Chem 286:11672–11684. doi:10.1074/jbc.M110.159962

Rizzo MA, Springer GH, Granada B, Piston DW (2004) An improved cyan fluorescent protein variant useful for FRET. Nat Biotechnol 22:445–449. doi:10.1038/nbt945

Sample V, Ramamurthy S, Gorshkov K et al (2015) Polarized activities of AMPK and BRSK in primary hippocampal neurons. Mol Biol Cell 26:1935–1946. doi:10.1091/mbc.E14-02-0764

San Martín A, Ceballo S, Baeza-Lehnert F et al (2014a) Imaging mitochondrial flux in single cells with a FRET sensor for pyruvate. PloS One 9, e85780. doi:10.1371/journal.pone.0085780

San Martín A, Ceballo S, Ruminot I et al (2013) A genetically encoded FRET lactate sensor and its use to detect the Warburg effect in single cancer cells. PloS One 8, e57712. doi:10.1371/journal.pone.0057712

San Martín A, Sotelo-Hitschfeld T, Lerchundi R et al (2014b) Single-cell imaging tools for brain energy metabolism: a review. Neurophotonics 1:11004. doi:10.1117/1.NPh.1.1.011004

Sanders R, Draaijer A, Gerritsen HC et al (1995) Quantitative pH imaging in cells using confocal fluorescence lifetime imaging microscopy. Anal Biochem 227:302–308. doi:10.1006/abio.1995.1285

Shaner NC, Campbell RE, Steinbach PA et al (2004) Improved monomeric red, orange and yellow fluorescent proteins derived from Discosoma sp. red fluorescent protein. Nat Biotechnol 22:1567–1572. doi:10.1038/nbt1037

Shaner NC, Steinbach PA, Tsien RY (2005) A guide to choosing fluorescent proteins. Nat Methods 2:905–909. doi:10.1038/nmeth819

Siegel MS, Isacoff EY (1997) A genetically encoded optical probe of membrane voltage. Neuron 19:735–741

St-Pierre F, Marshall JD, Yang Y et al (2014) High-fidelity optical reporting of neuronal electrical activity with an ultrafast fluorescent voltage sensor. Nat Neurosci 17:884–889. doi:10.1038/nn.3709

Surin AM, Gorbacheva LR, Savinkova IG et al (2014) Study on ATP concentration changes in cytosol of individual cultured neurons during glutamate-induced deregulation of calcium homeostasis. Biochemistry 79:146–157. doi:10.1134/S0006297914020084

Surin AM, Khiroug S, Gorbacheva LR et al (2012) Comparative analysis of cytosolic and mitochondrial ATP synthesis in embryonic and postnatal hippocampal neuronal cultures. Front Mol Neurosci 5:102. doi:10.3389/fnmol.2012.00102

Suzuki R, Hotta K, Oka K (2015) Spatiotemporal quantification of subcellular ATP levels in a single HeLa cell during changes in morphology. Sci Rep 17:16874. doi:10.1038/srep16874

Takanaga H, Chaudhuri B, Frommer WB (2008) GLUT1 and GLUT9 as major contributors to glucose influx in HepG2 cells identified by a high sensitivity intramolecular FRET glucose sensor. Biochim Biophys Acta 1778:1091–1099. doi:10.1016/j.bbamem.2007.11.015

Tantama M, Hung YP, Yellen G (2011) Imaging intracellular pH in live cells with a genetically encoded red fluorescent protein sensor. J Am Chem Soc 133:10034–10037. doi:10.1021/ja202902d

Tantama M, Martínez-François JR, Mongeon R, Yellen G (2013) Imaging energy status in live cells with a fluorescent biosensor of the intracellular ATP-to-ADP ratio. Nat Commun 4:2550. doi:10.1038/ncomms3550

Tantama M, Yellen G (2014) Imaging changes in the cytosolic ATP-to-ADP ratio. Methods Enzymol 547:355–371. doi:10.1016/B978-0-12-801415-8.00017-5

Tarasov AI, Rutter GA (2014) Use of genetically encoded sensors to monitor cytosolic ATP/ADP ratio in living cells. Methods Enzymol 542:289–311. doi:10.1016/B978-0-12-416618-9.00015-7

Tarasov AI, Semplici F, Li D et al (2013) Frequency-dependent mitochondrial Ca(2+) accumulation regulates ATP synthesis in pancreatic β cells. Pflüg Arch 465:543–554. doi:10.1007/s00424-012-1177-9

Tarasov AI, Semplici F, Ravier MA et al (2012) The mitochondrial Ca2+ uniporter MCU is essential for glucose-induced ATP increases in pancreatic β-cells. PloS One 7, e39722. doi:10.1371/journal.pone.0039722

Teperino R, Amann S, Bayer M et al (2012) Hedgehog partial agonism drives Warburg-like metabolism in muscle and brown fat. Cell 151:414–426. doi:10.1016/j.cell.2012.09.021

Tian L, Hires SA, Mao T et al (2009) Imaging neural activity in worms, flies and mice with improved GCaMP calcium indicators. Nat Methods 6:875–881. doi:10.1038/nmeth.1398

Toloe J, Mollajew R, Kügler S, Mironov SL (2014) Metabolic differences in hippocampal "Rett" neurons revealed by ATP imaging. Mol Cell Neurosci 59:47–56. doi:10.1016/j.mcn.2013.12.008

Tsou P, Zheng B, Hsu CH et al (2011) A fluorescent reporter of AMPK activity and cellular energy stress. Cell Metab 13:476–486. doi:10.1016/j.cmet.2011.03.006

Tsuyama T, Kishikawa J, Han Y-W et al (2013) In vivo fluorescent adenosine 5′-triphosphate (ATP) imaging of Drosophila melanogaster and Caenorhabditis elegans by using a genetically encoded fluorescent ATP biosensor optimized for low temperatures. Anal Chem 85:7889–7896. doi:10.1021/ac4015325

Valdebenito R, Ruminot I, Garrido-Gerter P et al (2015) Targeting of astrocytic glucose metabolism by beta-hydroxybutyrate. J Cereb Blood Flow Metab. doi:10.1177/0271678X15613955

Veetil JV, Jin S, Ye K (2010) A glucose sensor protein for continuous glucose monitoring. Biosens Bioelectron 26:1650–1655. doi:10.1016/j.bios.2010.08.052

Vevea JD, Wolken DMA, Swayne TC et al (2013) Ratiometric biosensors that measure mitochondrial redox state and ATP in living yeast cells. J Vis Exp. doi:10.3791/50633

Vishnu N, Jadoon Khan M, Karsten F et al (2014) ATP increases within the lumen of the endoplasmic reticulum upon intracellular Ca2+ release. Mol Biol Cell 25:368–379. doi:10.1091/mbc.E13-07-0433

Wang D, Zhang Z, Chanda B, Jackson MB (2010) Improved probes for hybrid voltage sensor imaging. Biophys J 99:2355–2365. doi:10.1016/j.bpj.2010.07.037

Weikel KA, Ruderman NB, Cacicedo JM (2016) Unraveling the actions of AMP-activated protein kinase in metabolic diseases: systemic to molecular insights. Metabolism 65:634–645. doi:10.1016/j.metabol.2016.01.005

Wiederkehr A, Szanda G, Akhmedov D et al (2011) Mitochondrial matrix calcium is an activating signal for hormone secretion. Cell Metab 13:601–611. doi:10.1016/j.cmet.2011.03.015

Willemse M, Janssen E, de Lange F et al (2007) ATP and FRET--a cautionary note. Nat Biotechnol 25:170–172. doi:10.1038/nbt0207-170

Williamson DH, Lund P, Krebs HA (1967) The redox state of free nicotinamide-adenine dinucleotide in the cytoplasm and mitochondria of rat liver. Biochem J 103:514–527

Xu Y, Wang Y, Xu Y et al (2015) Development of a novel phosphorylated AMPK protection assay for high-throughput screening using TR-FRET assay. J Biomol Screen 20:906–912. doi:10.1177/1087057115585471

Yaginuma H, Kawai S, Tabata KV et al (2014) Diversity in ATP concentrations in a single bacterial cell population revealed by quantitative single-cell imaging. Sci Rep 4:6522. doi:10.1038/srep06522

Yellen G, Mongeon R (2015) Quantitative two-photon imaging of fluorescent biosensors. Curr Opin Chem Biol 27:24–30. doi:10.1016/j.cbpa.2015.05.024

Zala D, Hinckelmann M-V, Yu H et al (2013) Vesicular glycolysis provides on-board energy for fast axonal transport. Cell 152:479–491. doi:10.1016/j.cell.2012.12.029

Zhang C, Ye B-C (2014) A single fluorescent protein-based sensor for in vivo 2-oxogluatarate detection in cell. Biosens Bioelectron 54:15–19. doi:10.1016/j.bios.2013.10.038

Zhao Y, Jin J, Hu Q et al (2011) Genetically encoded fluorescent sensors for intracellular NADH detection. Cell Metab 14:555–566. doi:10.1016/j.cmet.2011.09.004

Zou P, Zhao Y, Douglass AD et al (2014) Bright and fast multicoloured voltage reporters via electrochromic FRET. Nat Commun 5:4625. doi:10.1038/ncomms5625

Erratum to: AMPK Regulation of Cell Growth, Apoptosis, Autophagy, and Bioenergetics

Marina Villanueva Paz, David Cotán, Juan Garrido Maraver, Manuel Oropesa-Ávila, Mario de la Mata, Ana Delgado Pavón, Isabel de Lavera, Elizabet Alcocer Gómez, Mónica Álvarez Córdoba, and José A. Sánchez Alcázar

M.D. Cordero, B. Viollet (eds.), *AMP-activated Protein Kinase*, Experientia Supplementum 107, DOI 10.1007/978-3-319-43589-3_3

In the original version of this chapter, the surname and forename format of the authors was incorrect. The names should read

M. Villanueva-Paz,
D. Cotán,
J. Garrido-Maraver,
M. Oropesa-Ávila,
M. de la Mata,
A. Delgado-Pavón,
I. de Lavera,
E. Alcocer-Gómez,
M. Álvarez-Córdoba,
J.A. Sánchez-Alcázar.

The names are corrected through this erratum.

The updated original online version for this chapter can be found at http://dx.doi.org/10.1007/978-3-319-43589-3_3

M.D. Cordero, B. Viollet (eds.), *AMP-activated Protein Kinase*, Experientia Supplementum 107, DOI 10.1007/978-3-319-43589-3_21

MIX
Papier aus verantwortungsvollen Quellen
Paper from responsible sources
FSC® C105338

If you have any concerns about our products,
you can contact us on
ProductSafety@springernature.com

In case Publisher is established outside the EU,
the EU authorized representative is:
Springer Nature Customer Service Center GmbH
Europaplatz 3, 69115 Heidelberg, Germany

Printed by Libri Plureos GmbH
in Hamburg, Germany